D1130960

PHYSICAL SCIENCE

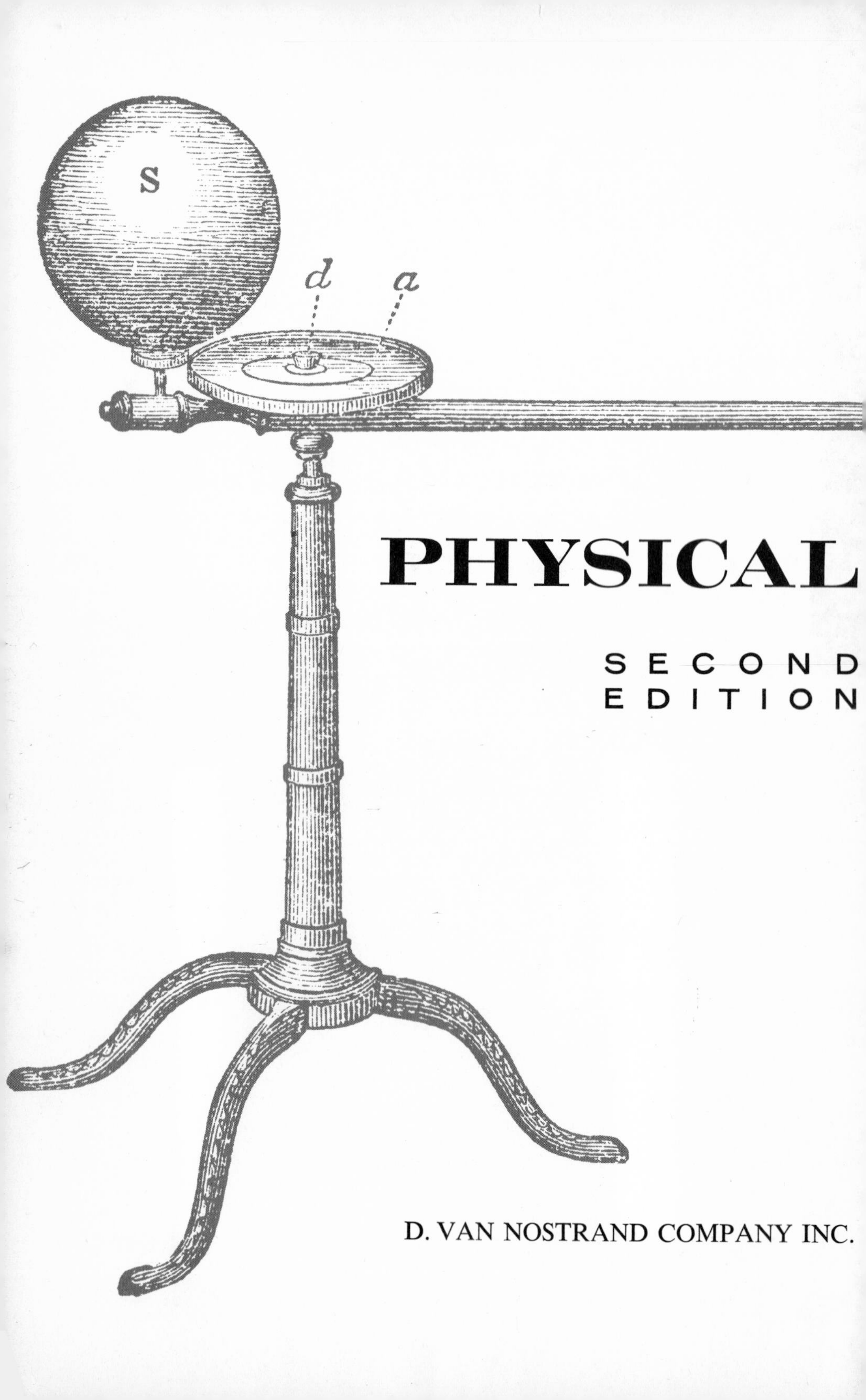

PHYSICAL

SECOND EDITION

D. VAN NOSTRAND COMPANY INC.

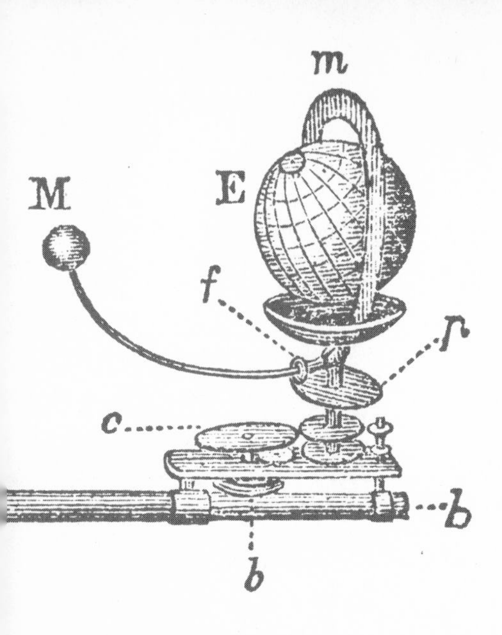

SCIENCE

DONALD S. ALLEN
Professor of Chemistry
State University of
New York at Albany

RICHARD J. ORDWAY
Professor of Geology
State University College,
New Paltz, New York

Princeton, New Jersey Toronto London Melbourne

VAN NOSTRAND REGIONAL OFFICES: *New York, Chicago, San Francisco*

D. VAN NOSTRAND COMPANY, LTD., *London*

D. VAN NOSTRAND COMPANY (Canada), LTD., *Toronto*

D. VAN NOSTRAND AUSTRALIA PTY. LTD., *Melbourne*

Published simultaneously in Canada by
D. VAN NOSTRAND COMPANY (Canada), LTD.

Library of Congress Catalog Card No. 68-20914

THE TITLE PAGE shows a mid-nineteenth-century Tellurian or Season Machine demonstrating the motions of sun, moon, and Earth. It is a system of spheres, gears, and pulleys, moved by a crank.

PRINTED IN THE UNITED STATES OF AMERICA

Preface

An introductory science text today, whether in a single area such as chemistry or in several areas, can introduce the student to only a relatively few concepts and principles. However, the second edition of *Physical Science* extends considerably (by about 15 percent) the scope of the earlier work. There is more extensive coverage of chemistry, including modern concepts of electronic structure and bonding. The original chapter on radioactivity has been extended and separated into two parts by introducing a new chapter on nuclear reactions. The new chapter on organic chemistry should appeal to students who feel they should have some basis for understanding the rapidly developing field of biochemistry.

The earth science section has been almost completely rewritten. Most illustrations and photographs have been replaced. Some areas have been shortened (astronomy), others expanded (meteorology). Two new chapters (oceanography and global geology) have been added.

The level of the text has been raised somewhat to take advantage of the improved science backgrounds that students bring to college today. The sequence of topics has been changed in a number of cases so as to give better unity.

Floyd Parker, Robert Schaffrath, and Eugene McLaren have made valuable suggestions for improving Chapters 1-16. Their assistance is gratefully acknowledged. Once again our wives, Kay Allen and Mary Jane Ordway, deserve a generous share of the credit for helping to get the second edition into print. Their assistance and encouragement are much appreciated.

Photographs and illustrations have come from many sources, and acknowledgments are made with each figure. However, the authors wish to acknowledge here their very great appreciation for the assistance of colleagues and institutions in obtaining these materials. In addition we wish to thank members of the editorial and technical staffs of D. Van Nostrand Company and the artists who have worked with us on both editions of this work.

January 1968

DONALD S. ALLEN
RICHARD J. ORDWAY

Contents

1 The length and breadth of measurement

TODAY the city of Syracuse, at the southeastern corner of the Mediterranean island of Sicily, has a population of some 65,000. Twenty-two centuries ago it was many times as large—one of the rich cities in the Mediterranean world, a rival of Athens and Alexandria in the realms of culture and the intellect. Syracuse was founded by Greeks, and in this town the physical sciences reached the peak of their development in Greek civilization, thanks largely to the great geometrician Archimedes (287-212 B.C.; Fig. 1-1), a man famous in legend for solving King Hieron's problem with the royal goldsmith.

The king gave the goldsmith a lump of pure gold and asked him to fashion a crown from it. When finished, the crown was a splendid work of art, but the king suspected that it did not contain all the gold of the lump that he had given. True, the weight of the crown was the same as that of the original lump; but had not some of the gold been replaced by a less expensive metal? Archimedes was consulted. Had a substitution been made? Could Archimedes find out without marring the crown in any way?

For a long time even this astute mathematician was baffled. Then one day as he was in a public bath, brooding as usual over the king's problem, he noticed the water level rising higher and higher on the wall as more and more of his body became submerged—a common enough observation. Anybody knows that the material substance of Archimedes' mortal frame could not occupy the same space as the water in the bath. An object which is immersed in water pushes out of the way—displaces—an equal volume of water (Fig. 1-2). Yet this common observation proved to be the clue to the goldsmith's guilt.

The problem of the gold crown hinged on a concept of *density*. Every

Fig. 1-1. *Archimedes (287-212* B.C.). *A modern artist pictures the moment when the aging geometrician, deep in thought in his study, is surprised by the Roman soldier who killed him. According to legend, his last words, as the intruder's footstep blurred the lines of geometric figures traced on the sanded floor, were, "Be careful! Don't spoil my circles!" The Roman victor had been delayed for months by defensive machines devised by this man of genius.*

youngster has been asked the catch question, "Which is heavier, a pound of lead or a pound of cork?" Answer: a pound is a pound. The difference is in the volume occupied. A pound of lead occupies a much smaller volume than a pound of cork. In a word, lead is *denser* than cork.

Fig. 1-2. *An object which is completely immersed in a liquid displaces a quantity of liquid equal to its own volume. An irregular object like a crown, whose volume could not be measured directly, is thus as easily measurable as a cube. In the kitchen the cook may use a similar method for making sure she has exactly half a cup of butter.*

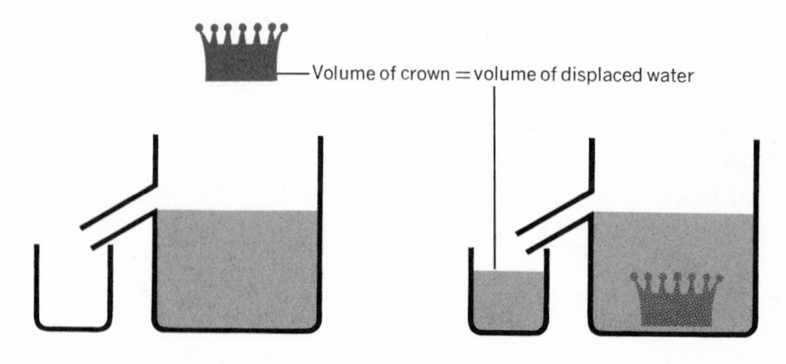

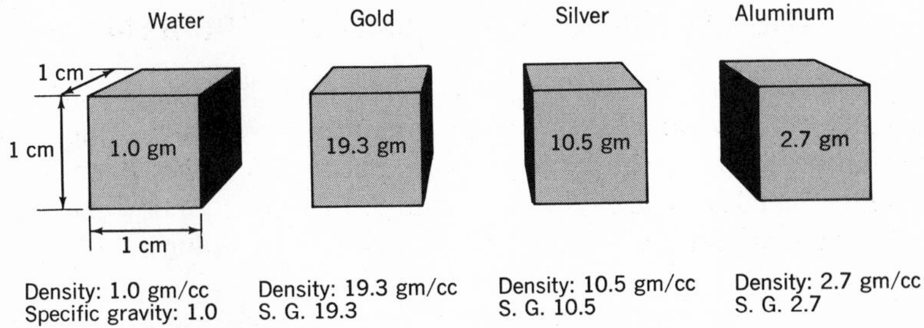

Fig. 1-3. *Centimeter cubes of different metals may have different masses. The mass of a unit cube is the density of the metal. It is numerically the same as weight per unit volume at sea level. Weight is distinguished from mass (see below, The Standard of Mass). The specific gravity of a substance is ascertained through dividing its density by the density of water. Since in the metric system the density of water at 4°C (Celsius) is 1 g/cm³, the specific gravities of other substances at this temperature are numerically the same as their densities; e.g., the density of gold is 19.3 g/cm³, and the density of water is 1 g/cm³, so the specific gravity of gold* $= 19.3$ *g/cm³* $\div 1$ *g/cm³* $= 19.3$. *(Notice that the units cancel out.)*

The Concept of Density If cubes of gold, silver, and aluminum, each 1 centimeter on a side (2.54 centimeters are equal to 1 inch) are weighed, their masses are as shown in Fig. 1-3. Density may be defined as *mass per unit volume.* The density of gold is 19.3 grams per cubic centimeter. (This may also be written 19.3 g/cm^3 or 19.3 g/cc.) To calculate the density of an object, divide its mass by its volume. Density may also be expressed in other units of mass and volume—for example, in pounds per cubic foot. Water has a density of 1 g/cm^3, equivalent to a density of 62.4 lb/ft^3; the numerical value depends on the units chosen.

ILLUSTRATIVE PROBLEM

Determine the density of gold if the volume occupied by 4825 g is 250.0 cm^3.*

Solution

$$\text{Density} = \frac{\text{mass}}{\text{volume}}$$

$$\text{Density} = \frac{4825 \text{ g}}{250.0 \text{ cm}^3} = 19.30 \text{ g/cm}^3$$

The mass given, 4825 g, is that occupying 250.0 units of volume (250.0 cm^3). Density is equal to the mass of a *single unit of volume.*

* The milliliter is often used for expressing liquid volumes. In 1964 the milliliter and the cubic centimeter were redefined and are now equal.

THE SOLUTION OF ARCHIMEDES' PROBLEM

Fig. 1-3 illustrates the meaning of density by emphasizing the difference in mass of equal volumes of material. Archimedes' report to King Hieron depended on the observation that when immersed in water, equal masses of different metals displace different volumes. In order to follow his reasoning, the equation, density = mass/volume, is solved for volume:

$$\text{Volume} = \frac{\text{mass}}{\text{density}}$$

The volume occupied by a gram of gold is

$$\text{Volume} = \frac{\text{mass}}{\text{density}} = \frac{1.00 \text{ g}}{19.3 \text{ g/cm}^3} = 0.0518 \text{ cm}^3$$

The volume occupied by a gram of silver is

$$\frac{1.00 \text{ g}}{10.5 \text{ g/cm}^3} = 0.0952 \text{ cm}^3$$

and by a gram of aluminum

$$\frac{1.00 \text{ g}}{2.70 \text{ g/cm}^3} = 0.370 \text{ cm}^3$$

Hence aluminum, when immersed in water, will displace a larger volume than will a similar mass of silver, which in turn will displace more than a similar mass of gold.

Since the finished crown weighed the same as the original gold, its volume must have been greater if the goldsmith substituted silver for some of the gold. What Archimedes suddenly realized in his bath was that the substitution of an equal mass of a metal having a lesser density would result in a crown of larger volume. This gave him the solution to the king's problem and this is why he raced home shouting "Eureka! Eureka!" (I have found it!). Once home, he promptly filled a vessel with water, plunged the king's crown into it, and measured the overflow. Sure enough, the amount of water displaced by the crown proved to be greater than the amount displaced by a lump of pure gold which balanced the crown in a pair of scales (Fig. 1-4). The king's goldsmith was dealt with accordingly.

Specific Gravity Specific gravity and density are closely related concepts; in fact, the two may have identical numerical values. Specific gravity may be defined as the ratio of the density of the object to that of water:*

$$\text{Specific gravity} = \frac{\text{density of object}}{\text{density of water}}$$

* For gases, air rather than water is usually the reference standard.

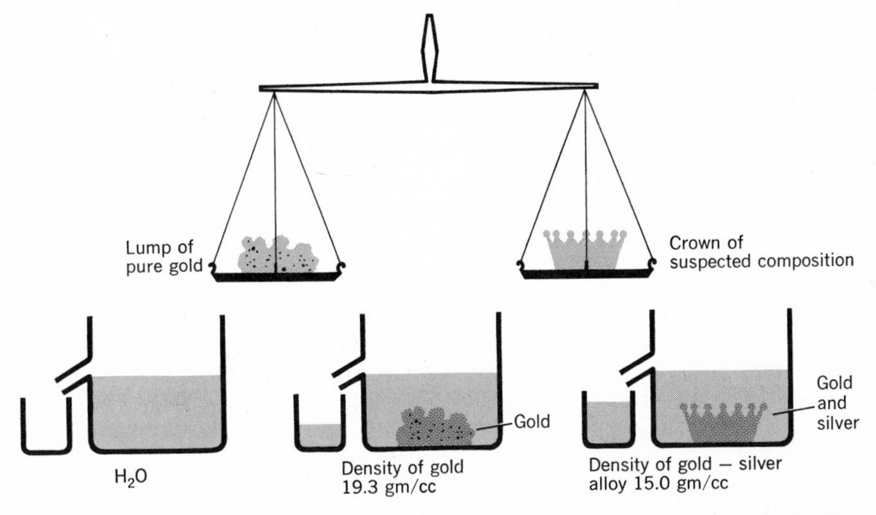

Fig. 1-4. *The solution of Archimedes' problem. Equal masses of metals having different densities will displace unequal volumes of water. The greater the density, the smaller the volume displaced. The lump of pure gold balances the crown in a pair of scales, but the crown proves to be made of an alloy, for it displaces a larger volume of water than the lump.*

At 4°C (Celsius) the density of water is 1 g/cm^3. Hence, at this temperature, specific gravity and density will have the same value. Since the density of water at room temperature (20°C) is very nearly 1 g/cm^3, normally the two values will be almost the same. Note, however, that specific gravity is the ratio of two densities, and is therefore unitless:

$$\text{Specific gravity} = \frac{\text{density of object (g/cm}^3)}{\text{density of water (g/cm}^3)}$$

Since the masses of the object and the equal volume of water are in the same ratio as their weights, specific gravity may also be expressed as

$$\text{Specific gravity} = \frac{\dfrac{\text{mass of object}}{\text{volume of object}}}{\dfrac{\text{mass of water}}{\text{same volume of water as object}}} = \frac{\dfrac{\text{weight of object}}{\text{volume of object}}}{\dfrac{\text{weight of water}}{\text{same volume of water as object}}}$$

Hence, specific gravity may also be expressed as the ratio of the *weight* of an object to the weight of an equal volume of water:

$$\text{Specific gravity} = \frac{\text{mass of object}}{\text{mass of equal volume of water}} = \frac{\text{weight of object}}{\text{weight of equal volume of water}}$$

ARCHIMEDES' PRINCIPLE

On the basis of his observations concerning bodies immersed in liquids, Archimedes eventually formulated two generalizations. The first of these noted that an object balanced by weights (standard masses) in air, would require fewer balancing weights if immersed in water. Again, the apparent loss of weight by an object suspended in water was equal to the weight of water which spilled over from the overflow can into the catch bucket (compare Fig. 1-5).

The generalization describing the above observations may be stated, "Objects which are *completely* immersed in a liquid weigh less (than in air) by an amount equal to the weight of the displaced liquid." This of course assumes that the density of the object is sufficiently large for it to become completely immersed.

ILLUSTRATIVE PROBLEM

What is the weight of 1 cubic centimeter of gold when immersed in water?

Solution

A cubic centimeter of gold weighs 19.3 g in air. When immersed in water, the cube displaces 1 g of water. Hence the immersed gold cube would weigh 1 g less, or 18.3 g.

Objects which are completely immersed in a liquid of lesser density than water, such as benzene (density = 0.879 g/cm^3) or ethyl alcohol (density = 0.789 g/cm^3) will show a smaller weight loss than when immersed in water. The liquid which spills into the overflow can weighs less than the same volume of water. Objects completely immersed in liquids denser than water, such as carbon tetrachloride (density = 1.595) will show a greater loss of weight.

To evaluate the specific gravity of an object, first weigh it in air; then weigh it again when completely immersed in water:

$$\text{Specific gravity} = \frac{\text{weight of object in air}}{\text{loss of weight in water}}$$

Because there is a sea of air around us (see p. 92), objects weighed in air will be slightly lighter than *in vacuo* [in a vacuum]. This slight loss of weight is due to the buoyant effect of the air displaced by the object.*

* This presumes that the weights will have much greater density than the object weighed and the buoyant effect on them is negligible. In an *in vacuo* correction, the air displaced by the weights would also have to be considered.

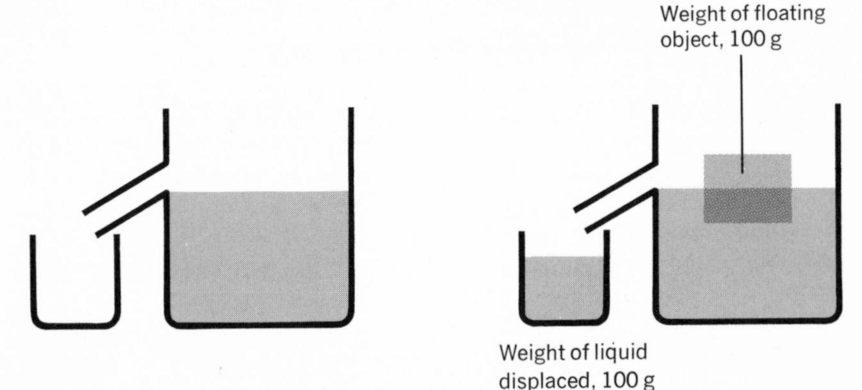

Fig. 1-5. *According to Archimedes' principle for* floating *bodies, the weight of liquid displaced equals the weight of the object.*

ILLUSTRATIVE PROBLEM

A lump of metal weighs 244.0 g in air. When completely immersed in water it weighs 224.0 g. Calculate the specific gravity of the metal.

Solution

The loss of weight in water is 20.0 g. Hence:

$$\text{Specific gravity} = \frac{\text{weight in air}}{\text{loss of weight in water}} = \frac{244.0\text{ g}}{20.0\text{ g}}$$

$$\text{Specific gravity} = 12.2$$

Archimedes' second generalization relates to floating objects, that is, objects which are not completely immersed in liquid. They sink in the liquid until they displace a volume of liquid equal in weight to their own weight. They are only partially immersed when the buoyant force of the liquid (upward) is equal to the downward force which is called the object's weight. If a floating object is lowered into an overflow can filled with water, the water which spills over into the catch bucket has the *same weight* as the object (Fig. 1-5).

If an additional force is exerted on the floating block until it is completely immersed in water, more water will spill into the catch bucket. This larger volume of water must have the same volume as the object. Its weight will be that of an equal volume of water.

$$\text{Specific gravity} = \frac{\text{weight of object}}{\text{weight of water same volume as object}} = \frac{\text{weight of water displaced by floating object}}{\text{weight of water displaced by completely immersed object}}$$

ILLUSTRATIVE PROBLEM

A wooden block is floated in a water-filled overflow can; 100 g of water spill over into the catch bucket. If the block is now completely immersed, an additional 80 g of water spill over into the catch bucket. Calculate the specific gravity of the block.

Solution

The block weighs 100 g. The weight of water having the same volume as the block is 180 g:

$$\text{Specific gravity} = \frac{100 \text{ g}}{180 \text{ g}} = 0.555$$

So much mention of grams, milliliters, and centimeters is bound to lead up to an important question: What are these units of measurement and how have they been established?

STANDARDS

The fact that Archimedes was able to solve the crown problem implies that he was able to make measurements of fair accuracy. The measuring devices available in his day allowed him to discriminate between the weights of equal volumes of material, or between the volumes of equal weights of material.

One of the marks of the advance of technical civilization has been an advance in the ability to make precise comparisons between measured quantities. As long as linear measures depended on the span of a human hand or the length of a foot, there was opportunity for a wide margin of error; the standard could vary with the individual hand or foot. Progress in the sciences has always been associated with the ability to define ever more precisely our standards of quantitative measure and to make more accurate comparisons with these standards. Modern analytical balances enable us to measure weights to the nearest millionth of a gram. Such accuracy was unattainable even a hundred years ago.

Standardization of weights and measures has also made an impact on our everyday life. In your city or perhaps in your county, there lives a man called the Sealer of Weights and Measures. It is his responsibility to check the accuracy of devices for weighing and measuring and so to assure you of full measure when you buy things, or even to protect merchants who may be overgenerous in dispensing their wares. You may have noticed the seals which the Sealer affixes to gasoline pumps in filling stations or to scales in grocery stores after he has made his inspection. He is the custodian of the local sets of standards. Once in about every five years he must take his sets of standard weights, standard measures, and standard tapes to the State Bureau of Weights and Measures for calibration. State standards are in turn checked at the National Bureau of Standards every ten years, and once in twenty years the national standards are checked against the international standards which are kept at the International Bureau of Weights and Measures at Sèvres in France. These standards are the ultimate standards for most of the countries of the civilized world today.

The Standard of Length The international unit of linear measure is the standard meter. It is simply the distance between two fine parallel lines ruled on a bar composed of an alloy of platinum and iridium. Since this alloy neither expands nor contracts to any appreciable extent with changing temperature, it is particularly valuable for maintaining constant dimensions. The standard meter is 39.37 inches long, i.e., some $3\frac{1}{3}$ inches longer than a yard. The yard is defined in this country as 3600/3937 meter.

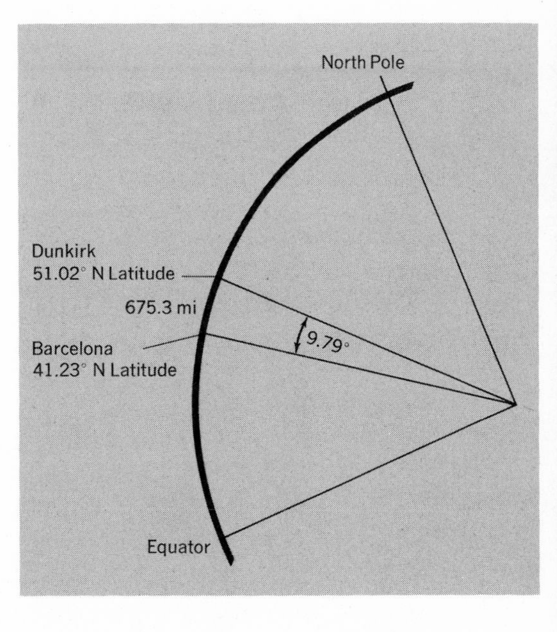

Fig. 1-6. *Calculating the distance from pole to equator:* $x/675.3\ mi = 90°/9.79°$; $x = 6208\ mi$. *A cross section of a quadrant of the earth, showing Barcelona and Dunkirk and the distance used in determining the length of the standard meter. The meter was intended to be one ten-millionth of the arc between the North Pole and the equator. Despite the inaccuracy of the original survey, the meter is very nearly the intended length.*

In the seventeenth century the French astronomer Mouton proposed that the length of a pendulum having a certain oscillation period should be taken as the linear standard. This length was remarkably constant for any given location. However, the suggestion had to be discarded when Richer, in 1671, discovered that the oscillation time for a pendulum of given length changed with geographical locations because of the difference in gravitational pull at different latitudes (see Chapter 4).

Late in the eighteenth century a movement was started in France to define the meter as one ten-millionth of the distance from the earth's equator to the pole, measured along a meridian (line of longitude). Since it was impractical to measure so long a distance, it was finally agreed to make a survey of the comparatively short distance between Dunkirk, on the northern coast of France, and Barcelona, Spain—cities which are located at sea level and on the same line of longitude. The difference between the latitudes of these two cities gave the angle subtended at the earth's center corresponding to the measured distance between them along the earth's surface. The distance from the equator to the pole could then be calculated, since this distance was in the same proportion to the measured distance between the two cities as 90° was to the difference in degrees between their latitudes (see Fig. 1-6). The meter was established as one ten-millionth of this calculated distance. In later years, sad to say, the survey was found to be in error. Thus the standard meter is actually just an arbitrary length rather than one ten-millionth of the distance from equator to pole.

Until recently the primary linear standard was a metal bar, as already mentioned. Then in 1960 an International Conference on Weights and Measures held in Paris redefined the standard meter as the length of 1,650,763 successive waves of the orange light emitted by krypton atoms of mass 86 (Kr-86).

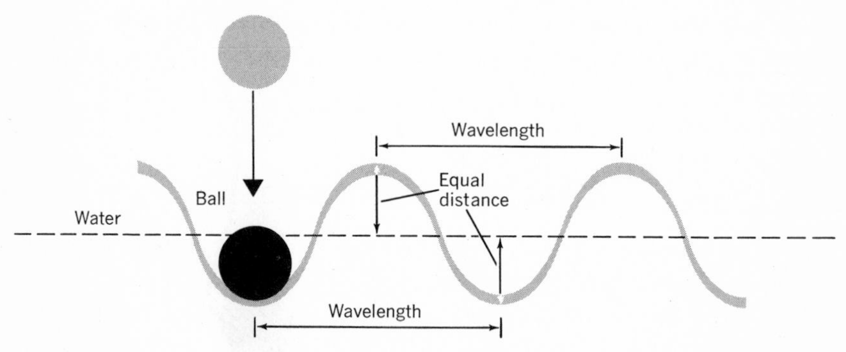

Fig. 1-7. *A wavelength is the distance between the corresponding parts of successive waves.*

To visualize the concept of wavelength, consider the waves produced when a ball drops into a placid pool of water. The ripples appear in cross section as shown in Fig. 1-7. A wavelength is simply the linear distance between corresponding points on any two successive waves—for example between two crests or between two troughs.

Light waves are many orders of magnitude smaller than water waves, but are measurable with far greater precision. Furthermore, this primary standard is not subject to variation under circumstances in which the meter bars would undergo minute linear changes.

The Standard of Mass Mass may be described as the amount of matter in a body. It is not synonymous with volume, which is the amount of space occupied by the body. It is not the same as weight, although masses may be compared by weighing and the weights in a weight set may be referred to as either masses or weights.*

The inertial effect of a kilogram mass (its resistance to the action of a force) is the same in different latitudes or at different elevations above sea level. However, the kilogram (1000-gram) mass suspended from a spring balance would *weigh* less on a mountain top than at sea level. Thus mass does not change with location on the surface of the earth. Weight on the other hand varies with location, decreasing as the object is moved farther from the earth's center.

The international standard of mass is a platinum-iridium cylinder whose diameter and height are both approximately $1\frac{1}{2}$ inches. It is called the standard kilogram and is one of the primary standards housed at the International Bureau of Weights and Measures near Paris. Our national standards of mass, housed in the National Bureau of Standards in Washington, D.C., are replicas of the international standard. The standard kilogram was originally intended to be the mass of a cube of water 10 centimeters on a side at 4°C, the temperature at which the density of water is a maximum. Subsequently refinements have shown the original measurement of this mass to be in error, and hence the standard kilogram is, like the meter, an arbitrary standard.

UNITS OF MEASUREMENT

When we measure a quantity, we single out a property such as length or mass for comparison with some standard. If we say the desk top is 2 meters long, we mean that two meter sticks laid end to end will equal the length of the desk. If an object has a mass of 3 kilograms, this means that three standard kilogram masses balance the object in the opposite pan of a pair of scales.

* The mass of a body can be measured by the resistance it offers to a change in motion—by its inertia (see Chapter 3) compared with the inertia of the standard kilogram.

Table 1–1 Metric Tables, with English Equivalents

LENGTH		VOLUME	
1 kilometer (km)	= 1000 m	1 kiloliter (kl)	= 1000 liter
1 hektometer (hm)	= 100 m	1 hektoliter (hl)	= 100 liter
1 dekameter (dkm)	= 10 m	1 dekaliter (dkl)	= 10 liter
1 meter		1 liter	
1 decimeter (dm)	= 0.1 m	1 deciliter (dl)	= 0.1 liter
1 centimeter (cm)	= 0.01 m	1 centiliter (cl)	= 0.01 liter
1 millimeter (mm)	= 0.001 m	1 milliliter (ml)	= 0.001 liter
1 m	= 39.37 in.	1 liter	= 0.264 gal (U.S.)
1 km	= 0.62 mi		

MASS			
1 kilogram (kg)	= 1000 g	1 decigram (dg)	= 0.1 g
1 hektogram (hg)	= 100 g	1 centigram (cg)	= 0.01 g
1 dekagram (dkg)	= 10 g	1 milligram (mg)	= 0.001 g
1 gram		1 kg	= 2.21 lb
		453.6 g	= 1 lb

The meter and kilogram are the common standards of measure, but these units are not convenient for all measurements. Many different linear units are in use, the unit chosen for a particular measurement depending on the length of the object or distance to be measured. If the distance between cities is to be measured, the mile is the linear unit used in the United States and Canada, the kilometer (about 0.6 mile) in Europe and many other parts of the world. The unit for estimating astronomical distances is the light-year (the distance traveled by light in 1 year), which is almost 10,000,000,000,000,000 times as long as the meter. On the other hand, light waves are extremely small and are measured in angstrom units (Å), which are only 1/10,000,000,000 as long as the meter.

Measurements made in terms of one linear unit, say the inch, can be converted into a corresponding number of a different size, say the centimeter, through multiplication or division by an appropriate constant factor. But linear units cannot be converted into units of mass, for example, by any such operation. Length and mass describe different aspects of matter; they are called *fundamental units.*

Units of area and volume are called *derived units,* since two or three units must be multiplied to obtain them. The area of a circle, for example, is $A = \pi r^2$. Here the radius (a linear unit) is squared (multiplied by itself).

Similarly, in the volume of a sphere, $V = \frac{4}{3}\pi r^3$, three linear units are multiplied—i.e., the radius is twice multiplied by itself.

Density, which is defined in terms of mass and volume, also involves more than one fundamental unit and so is a "derived unit."

Systems of Units Two systems of units are in general use today in this country, the metric and the English. There are also other systems, of course, such as the Troy weights of the jeweler and type sizes in printing, which belong to particular trades and industries. The metric system is almost universal in scientific research, but the English system still prevails in the United States and Canada in business and commerce. Some people feel that life would be simplified if we were to discard the English system altogether and use only the metric.

One of the advantages of the metric system is that the conversion from measurements in one metric unit to those in another (e.g., meters to centimeters) is performed entirely by multiplying or dividing by powers of 10—by moving a decimal point. This is far simpler than multiplying or dividing by such numbers as 12, 16, and 5280 as in the English system. The following problem illustrates this difference.

ILLUSTRATIVE PROBLEM

A distance of 2542 ft (775 m) is what part of 1 mile (or what part of 1 kilometer)?

Solution: English System

$$1 \text{ ft} = \frac{1}{5280} \text{ mi}$$

$$2542 \text{ ft} = \frac{2542}{5280} \text{ mi}$$

$$2542 \text{ ft} = 0.4814 \text{ mi}$$

$$\begin{array}{r} 0.4814 \\ 528\overline{)254.2000} \\ \underline{211\,2} \\ 43\,00 \\ \underline{42\,24} \\ 760 \\ \underline{528} \\ 2320 \\ \underline{2112} \end{array}$$

Solution: Metric System

$$1 \text{ m} = 0.001 \text{ km}$$

$$775 \text{ m} = 0.775 \text{ km}$$

Compare the ease of computing this problem in the metric system with the more cumbersome arithmetic in the English system.

Our adoption of the metric system as the sole system of weights and measures would also eliminate altogether the many conversions now necessary from one system of units to another—yards to meters, miles to kilometers, and so on. However, the issue is a controversial one. There are good arguments on both sides of the question. The English system has been adequate, the change would cause untold confusion, and converting or replacing all our measuring devices would be enormously costly.

Since there are in fact the two systems of units, conversions from one system to the other are common. At any time one may be confronted with such questions as, "How many kilometers are equivalent to ten miles?" or "What is the weight in kilograms of ten pounds of sugar?"

In determining whether to multiply or divide by the conversion factor, first ask, "Is the new unit larger or smaller than the given unit?" For example, in changing a given length in inches into an equivalent length in centimeters, the answer must contain more centimeters than the original number of inches. The centimeter is the smaller unit; there will be more in a given length (Fig. 1-8): 1 in. = 2.54 cm. Multiply inches by 2.54.

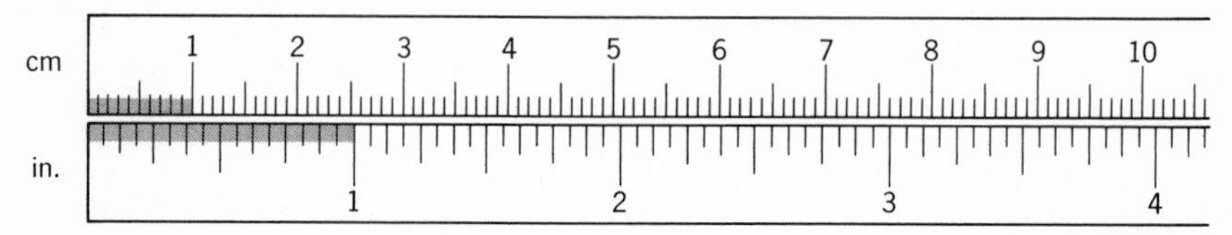

Fig. 1-8. *The centimeter is a smaller unit of linear measure than the inch. In a given length there will always be a larger number of centimeters than inches.*

Or one may ask, "What is the metric equivalent of 100 yards?" 1 meter = 3.28 feet and hence is longer than a yard. There will be fewer than 100 meters in 100 yards:

$$100 \text{ yards} = 300 \text{ ft} \qquad \frac{300 \text{ ft}}{3.28 \text{ ft/m}} = 91.5 \text{ m}$$

Remember, multiplying by a number larger than 1 increases the result; dividing by a number larger than 1 decreases the result. Multiplication by a number is equivalent to division by the reciprocal of the number. Dividing by $\frac{1}{2}$ is equivalent to multiplying by 2.

APPROXIMATION: SIGNIFICANT FIGURES

For the scientist, "12 grams" does not have the same meaning as "12.00 grams." He says the first measurement has two significant figures, the second has four. The former implies that the measurement is correct to the nearest

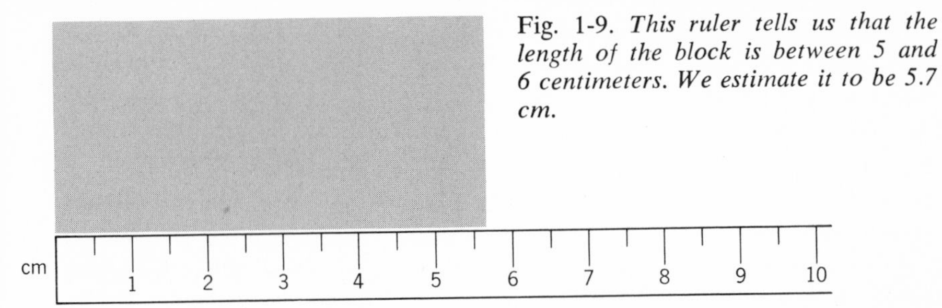

Fig. 1-9. *This ruler tells us that the length of the block is between 5 and 6 centimeters. We estimate it to be 5.7 cm.*

gram—i.e., lies between 11.5 and 12.5 grams; the latter, that it is correct to the nearest 0.01 gram—i.e., lies between 11.995 and 12.005 grams. The number of significant figures which may be obtained in a given measurement depends on the measuring instrument used. A micrometer caliper will ordinarily permit the recording of a larger number of significant figures in measuring the diameter of a steel cylinder than will a ruler.

Suppose that we measure the length of a block of wood by means of a meter stick having only centimeter divisions as shown in Fig. 1-9. The length of the block is certainly between 5 and 6 centimeters. We imagine the distance between the fifth and sixth centimeter marks to be divided off into ten equal subdivisions. We estimate that the edge of the block would coincide with the seventh subdivision; so we record 5.7 cm as the length of the block. Other observers might judge it to be 5.6 or 5.8 cm. Tenths of centimeters are then said to be "doubtful figures," but are nevertheless included as significant figures.

Suppose that the block is measured by a ruler actually having ten sub-

Fig. 1-10. *Measuring the block with a ruler marked in millimeters indicates a length between 5.6 and 5.7 cm. We estimate it to be 5.62 cm.*

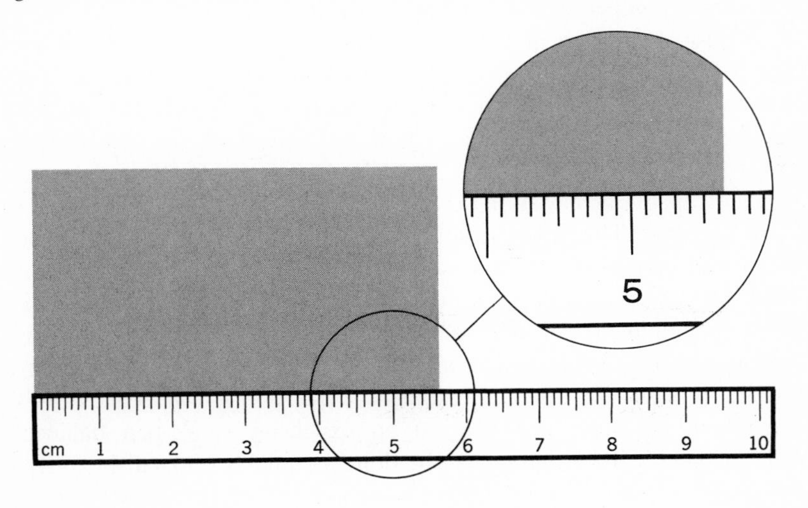

divisions between each pair of centimeter marks, so that we no longer have to imagine them. The length is between 5.6 and 5.7 cm, as shown in Fig. 1-10. The edge of the block is estimated to be $2/10$ of the distance from the 5.6 to the 5.7 mark, and the length is recorded as 5.62 cm. In this case we are certain of the 6, and the final 2 is the doubtful figure.

The most accurate possible measurement is sometimes of vital importance in scientific research. Near the end of the nineteenth century, the British scientist Sir William Ramsay noted that the density of the "nitrogen" produced by removing oxygen from air was 1.2572 grams per liter, whereas nitrogen prepared from ammonium compounds had a density of only 1.2505 g/l. The difference between the two was less than 0.007 g/l, but persisted even under the most careful possible measurement when the experiments were repeated.

In investigating possible causes of the discrepancy, Ramsay made use of the spectroscopic method of analysis developed several decades earlier by Bunsen and Kirchhoff. When a high-voltage electric current was passed through glass tubes containing the two different samples of "nitrogen," each glowed much like a neon sign. When each glowing tube was observed through a spectroscope, different spectral patterns were seen. The light emitted by the nitrogen from ammonia consisted of broad, colored bands; that from the air-filled tube showed as well some other very thin, colored lines. Ramsay was finally successful in demonstrating that the supposed nitrogen obtained from air actually contained a hitherto unknown gas, which he called argon. The discovery of the new element depended in the first place on the availability of measuring instruments capable of discriminating weights to the nearest 0.001 gram; a balance allowing only two-place accuracy would not have given Ramsay the hint he needed. The discovery depended also, of course, on the availability of the spectroscopic method, which will be discussed in Chapter 9.

A precise measurement was an indispensable factor in the discovery of the element argon. However, the impression that the scientist *always* makes the most precise measurements possible is not justified by the facts. Often he may be satisfied to ascertain the weight of an object to two decimal places, say 12.13 grams, when he has available in the laboratory a balance capable of measuring its weight to the fourth decimal place, say 12.1327 grams. The larger number of significant figures, while indispensable for certain purposes, is sometimes attained only with more time-consuming and painstaking operations and, in general, requires much more expensive equipment. An industrial company simply cannot afford to have a man spend fifteen minutes on an extremely accurate weighing when an adequate, though a less precise one, can be made in a few seconds. By the same token, the wear and tear on an expensive single-pan balance is hardly justified if the weighing can be done as quickly and satisfactorily on the far less expensive and less expensively maintained platform balance.

There is also a mathematical reason for simplifying some measurements. The product of two or more numbers cannot have more significant figures than the smallest number of significant figures in its factors, and the quotient of two numbers cannot have more significant figures than the smaller number in its dividend and its divisor. Therefore, if one factor of a product has only two significant figures, nothing is gained by making measurements to more than two significant figures for the other factor.

In adding or subtracting numerical quantities, one is justified in retaining only the smallest number of significant figures found in any *single* measurement.

ILLUSTRATIVE PROBLEM

Add	12.363 cm
	11.23 cm
	15.7826 cm
	39.37 cm

Solution

In the second measurement the final 3 is a doubtful figure, and hence the second decimal place in the sum will be a doubtful figure. There is then no justification for including more than two digits to the right of the decimal point in the answer.

Fig. 1-11. *The burette is used for measuring liquid volumes. A burette reading is shown which should be recorded with four significant figures.*

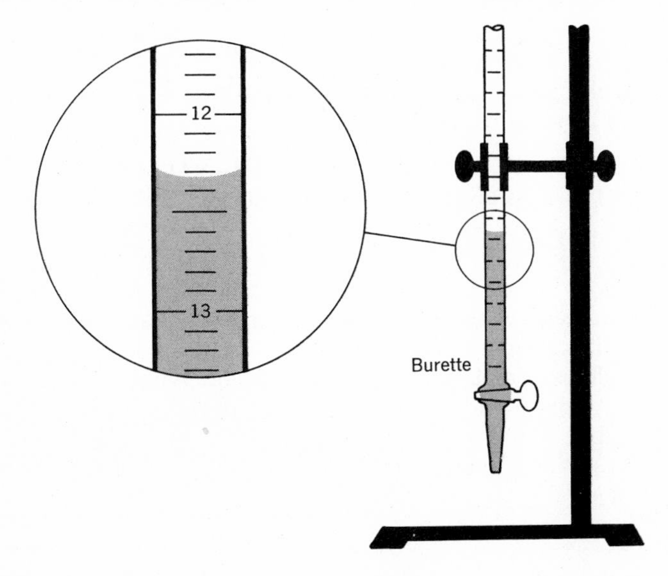

Final zeros are considered significant figures when located to the right of the decimal point as in 1.200. This implies that the third decimal place is a doubtful figure. However in .0024 (or 0.0024), the zeros preceding 24 serve to locate the decimal point and are not significant figures.

Consider the specific volume of water, defined as the volume occupied by a single unit of mass (usually the gram): specific volume = volume/mass. If the volume is measured in the burette shown in Fig. 1-11, only four significant figures, 12.32 cm^3, are recorded. Even though the mass could be determined to six significant figures, the specific volume would still have to be rounded to four significant figures.

Specific volume = volume/mass = 12.32 cm^3/12.31 g = 1.001 cm^3/g, when rounded to four significant figures. If the more accurate measure of 12.3089 g had been used for the divisor, more time would have been spent both in measurement and in arithmetic computation and yet nothing gained in the accuracy of the result. Unusable accuracy is waste motion, experimentally and mathematically.

Problems, Chapter 1

Be sure to label each answer with correct units and include an appropriate number of significant figures.

1. Calculate the height in inches of a 760 mm mercury column in a barometer tube.
2. The distance from New York City to Albany is 150 mi. Calculate the distance in kilometers.
3. What is the mass in kilograms of a cubic foot of water (62.4 lb)?
4. Calculate the number of grams equivalent to 1.00 oz avoirdupois.
5. Calculate the number of liters equivalent to a cubic foot of water; to a gallon of gasoline.
6. The density of ethyl alcohol (C_2H_5OH) is 0.790 g/cm^3. What is the mass of 75.0 cm^3 of ethyl alcohol?
7. Calculate the volume occupied by 100 g of mercury, density 13.6 g/cm^3.
8. A copper cube weighs 200 g in air when suspended from a spring balance. (a) The spring balance and cube are lowered until the cube is completely immersed in water. What is the weight of the cube in water? Why? (b) What is the specific gravity of the cube? What is its density? Assume the temperature is 4°C. (c) Would you expect the cube to weigh more in air or *in vacuo?* Why?

9. Metal cubes are lowered into and completely immersed in liquids contained in a filled overflow can, as described below. Calculate as directed the weight or volume of the liquid spilling into the catch bucket when the cube is *completely* immersed:

METAL CUBE	IMMERSING LIQUID	TO BE CALCULATED
(a) Aluminum (5 cm on an edge)	water	volume
(b) Aluminum (5 cm on an edge)	water	weight
(c) Silver (5 cm on an edge)	benzene	volume
(d) Silver (5 cm on an edge)	benzene	weight
(e) Aluminum weighing 200 g	water	volume
(f) Aluminum weighing 200 g	water	weight
(g) Silver weighing 200 g	water	volume
(h) Silver weighing 200 g	water	weight
(i) Silver weighing 200 g	benzene	weight
(j) Silver weighing 200 g	benzene	volume
(k) Aluminum weighing 200 g	benzene	volume
(l) Aluminum weighing 200 g	benzene	weight

10. When a floating block of wood is lowered into an overflow can filled with carbon tetrachloride, 175 cm^3, spill over into the catch bucket. What is the weight of the block?

11. Calculate the weight of water in a filled 2-gal can. (Assume a U.S. gallon.)

12. Calculate the mass of a spherical lead ball whose diameter is 2.50 cm.

13. An irregular block of metal has a weight of 160 g in air. When it is lowered into a graduated cylinder containing 70.00 cm^3 of water and completely immersed, the level rises to 130.0 cm^3. Calculate the specific gravity of the metal.

14. Calculate the mass of a cubic foot of aluminum metal.

15. In Canada and England gasoline is purchased in imperial gallons (1 imperial gallon = 277.4 $in.^3$) How many U.S. gallons are there in an imperial gallon?

16. How many gallons (U.S.) of water are in 1 ft^3?

Exercises

1. Define density. Describe how you could determine the density of an irregular object such as a stone.

2. How is specific gravity defined? Specific gravity and density are often equal. Is this always true? Explain!

3. State Archimedes' principle for (a) floating bodies; for (b) bodies whose densities are greater than the immersing liquid.

4. Explain in terms of Archimedes' principle the construction and use of a hydrometer. Why are the smaller specific-gravity values nearer the top of the scale? Explain! Name some of the specific-gravity scales in use today.

5. Lengths of light waves are now recognized as the primary standard of length. Why was the standard meter bar less reliable?

6. What is the difference between a fundamental and a derived unit? Give an illustration of each.

7. The standard meter was at one time defined as one ten-millionth of the distance from the Earth's equator to the pole. Describe how this length was determined in France and Spain during the late eighteenth century.

8. Would you favor the adoption of the metric system as the only system of weights and measures? State arguments pro and con.

9. Explain why a knowledge of the correct recording and correct use of significant figures in calculation is of great importance to the scientist.

10. In the numerical problems above, several liquids and metals were mentioned. If these were to be used in a laboratory exercise you might be wise to inquire whether any were particularly inflammable or toxic. Look up each in a copy of N. Sax, *Dangerous Properties of Industrial Materials,* 2nd ed., Reinhold Publishing Corp., New York, 1957.

2 Moments and machines

ABOUT 215 B.C. the Romans were storming the approaches to Sicilian Syracuse and met strong resistance in a form they had not expected. The engineering genius of Archimedes had created large-scale war machines—the sort of thing that is associated with much later ages—for the defense of his home city. Huge beams with heavy boulders attached swung out from the cliff tops over the advancing Roman ships. The boulders dropped and smashed the decks, making the ships toss as upon an angry sea. Enormous beaks were lowered over the prows and grasped the ships, lifting them up on end and dumping the crews into the harbor.

Despite the ingenuity of Archimedes' machines, they were limited to power supplied by the muscles of men and animals. They were adaptations of what are known today as the *simple machines*. The simple machines—the lever, screw, pulley, wheel and axle, and inclined plane—are devices which permit the moving of heavy objects with forces that are usually smaller than their weights. These machines were known thousands of years before Archimedes, and in one form or another their origin must have been in prehistoric times. It seems probable that man invented gadgets for making his work easier long before he had explanations of how they functioned.

In very ancient times implements were made of stone, later of bronze, and still later of iron and steel. Logs supplied the leverage necessary to move bulky things, or served as rollers on which things could be pushed. How did the Egyptians move into place the huge stone blocks with which they constructed the pyramids? The task called for ingenuity. Possibly the blocks were pushed on logs which served as wheels, up a gradually sloping road (inclined plane).

The great philosopher Aristotle (384-322 B.C.) contributed far more to the biological sciences than to physics, but he did give this explanation of the principle of the lever: "The weight which is moved is to the weight which moves it in the inverse ratio of the lengths of the arms of the lever; always in fact a weight will move as much more easily as the weight which moves it is further from the fulcrum."

THE PRINCIPLE OF MOMENTS: THE SEESAW

Any youngster knows that when he climbs on a seesaw, he must sit at the same distance from the point of rotation (fulcrum) as a partner of the same weight if they are to balance, or at a greater distance from the fulcrum if his partner weighs more. He also knows that if one youngster moves farther away from the fulcrum, his end of the seesaw will fall to the ground.

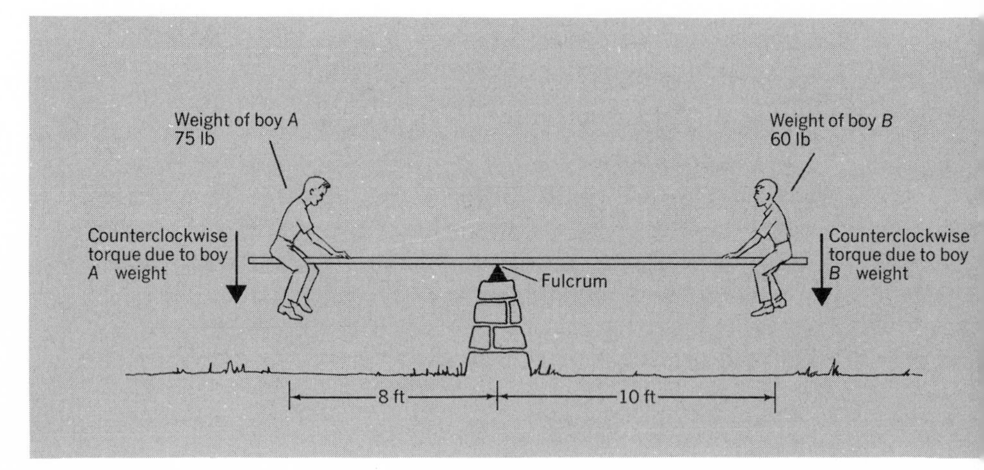

Fig. 2-1. *A seesaw balances when children of equal weights are seated at equal distances from the fulcrum; when the two weights are unequal, the lighter child must be seated farther from the fulcrum. The condition necessary for a seesaw to balance is that the rotational moments of force (the torques) shall be equal and in opposite directions.*

If boy A in Fig. 2-1 were to be so inconsiderate as to jump off, the seesaw would rotate in a clockwise direction. The rotational tendency would depend on two factors: the weight of the boy, and his distance from the fulcrum. The result obtained by multiplying these two quantities is called a *moment of force*. The *principle of moments* establishes the condition for a lever to be in a state of balance or equilibrium. It asserts that equilibrium will be maintained if the moments tending to clockwise rotation are equal to those tending to counterclockwise rotation.

A moment of force is also called a *torque*. The clockwise torque in Fig. 2-1 is the product of the weight of boy B (60 pounds) and his distance (10 feet) from the fulcrum; the counterclockwise torque is the weight of boy A (75 pounds) multiplied by his distance (8 feet) from the fulcrum. Since the torques are equal and opposite, the seesaw balances.

When a condition of equilibrium exists, there is an equality of two moments of force, and an equation may logically be written. The expression which represents the equality of the two torques (moments of force) is

$$\text{Counterclockwise moment} = \text{clockwise moment}$$

This equation may in turn be translated into the equality:

$$(\text{Weight of A}) \cdot \begin{pmatrix}\text{distance of A}\\ \text{from fulcrum}\end{pmatrix} = (\text{weight of B}) \cdot \begin{pmatrix}\text{distance of B}\\ \text{from fulcrum}\end{pmatrix}$$

The values for weight and distance from Fig. 2-1 are then substituted in the equation above:

$$(75\ \text{lb})\cdot(8\ \text{ft}) = (60\ \text{lb})\cdot(10\ \text{ft})$$

In a typical problem, three of the four quantities in the equation would be specified and the fourth would be unknown. Solving an equation amounts to finding a value for the unknown quantity which will make the two sides of the equation equal. Occasionally this unknown quantity can be obtained by inspection; more often it is obtained by algebraic methods.

ILLUSTRATIVE PROBLEM

A 100-lb boy sits 3 ft from the fulcrum of a seesaw. A 75-lb boy must sit how far from the fulcrum if the seesaw is to balance?

Solution

Let $x =$ the required distance.

$$\text{Counterclockwise moment} = \text{clockwise moment}$$

$$100\ \text{lb}\cdot 3\ \text{ft} = 75\ \text{lb}\cdot x\ \text{ft}$$

$$300\ \text{lb-ft} = 75\ \text{lb}\cdot x\ \text{ft}$$

$$x = 4\ \text{ft}$$

Solving this equation means finding the value of x which will make the right-hand member equal 300 lb-ft. This value of x is 4: the seesaw will balance if the 75-lb boy sits 4 ft from the fulcrum.

THE PLATFORM BALANCE AND THE ALGEBRAIC EQUATION

Every science laboratory makes use of a simple form of lever, the platform balance. Many platform balances have arms of equal length. An

Fig. 2-2. *How the platform balance illustrates the principles of solving an algebraic equation (Figs. 2-2 to 2-5).*

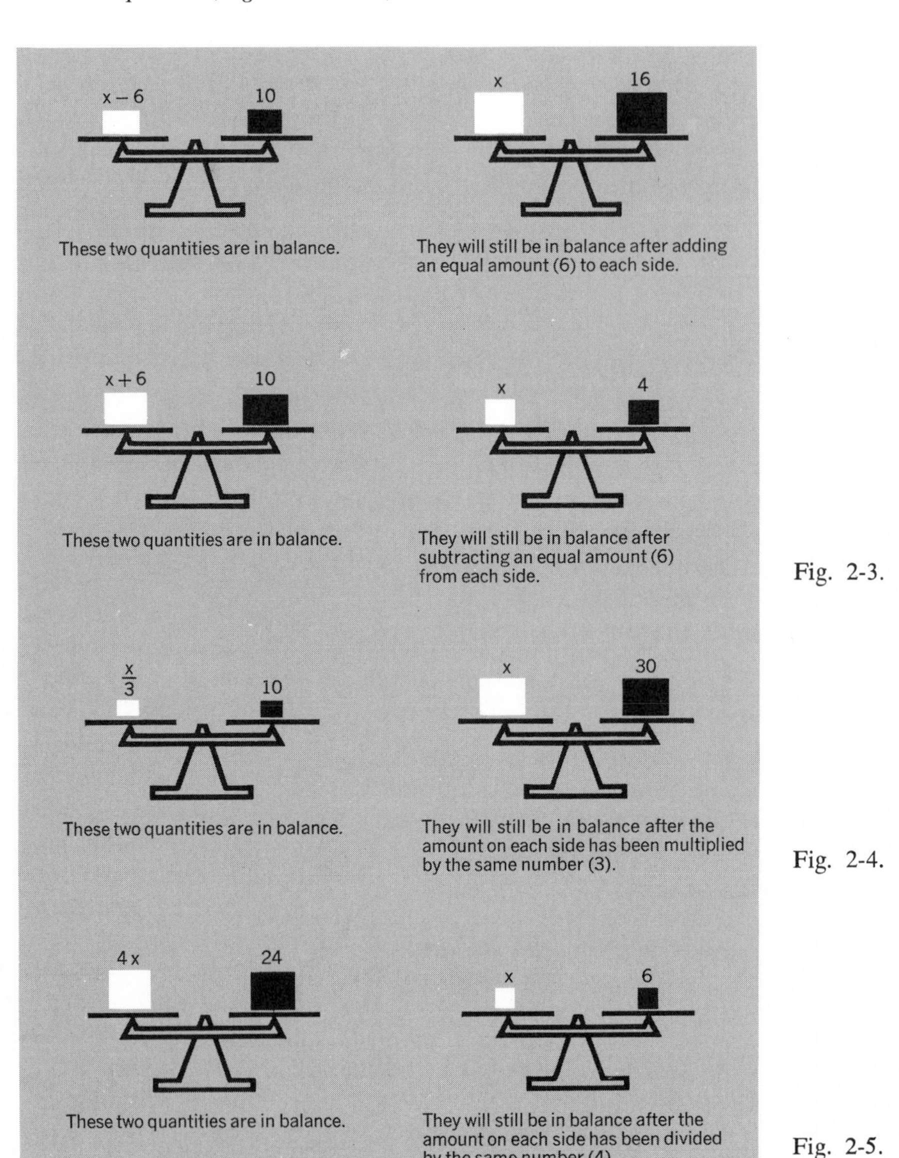

Fig. 2-3.

Fig. 2-4.

Fig. 2-5.

object to be weighed is placed on one pan and weights on the other. Thus there are equal torques which act in opposite rotational directions. The fundamental relation governing such a balance is the principle of moments.

The platform balance may be used to illustrate the four axioms of algebra:

I *If equals are added to equals, the results are equal.*
II *If equals are subtracted from equals, the results are equal.*
III *If equals are multiplied by equals, the results are equal.*
IV *If equals are divided by equals (other than zero), the results are equal.*

These axioms indicate the four ways in which one may operate on an algebraic equation. Think of any equation as a platform balance with equal weights on each pan. The equality sign implies that the quantities to the right and left are the same. No operation on the equation may upset this state of equilibrium of the platform balance.

THE AXIOM OF ADDITION

Illustrative Problem

Solve the equation $x - 6 = 10$

Solving the equation means finding the value of x which will make the left side of the equation equal 10. If $x = 16$ the two sides will balance, and hence 16 is the correct answer. In solving equations the aim is to obtain x all by itself on one side of the equality sign. If 6 is added to the left side of the equation it will cancel -6 and x will be the only term on the left. The axiom says that if 6 is added to the left side it must also be added to the right side. If there is a balance with $x - 6$ on one pan and 10 on the other, this balance will be preserved if 6 is added to both sides. The secret to correct operation on algebraic equations is to treat the two sides of the equation in the same way. Never let one side down (Fig. 2-2).

THE AXIOM OF SUBTRACTION

Illustrative Problem

Solve the equation $x + 6 = 10$

It is clear that if 4 is put in place of x, the condition of equality will be satisfied. If 6 is subtracted from the left side of the equation, x will be alone on that side. According to the axiom of subtraction, 6 must also be subtracted from the right side of the equation. The two sides must be treated alike or the equilibrium of the pair of scales will be upset (Fig. 2-3).

THE AXIOM OF MULTIPLICATION

Illustrative Problem

$$\text{Solve the equation } \frac{x}{3} = 10$$

If 30 is substituted for x, the two sides of the equation will be the same. If a pair of scales is in balance, it will remain in balance if the weights on each side are tripled (Fig. 2-4).

THE AXIOM OF DIVISION

Illustrative Problem

$$\text{Solve the equation } 4x = 24$$

If x has a value of 6, the above expression is a true equality. If a pair of scales is in balance, it will remain in balance if the weight on each side is reduced to $\frac{1}{4}$ its value (Fig. 2-5).

Fig. 2-6. *This crowbar has a mechanical advantage of 8 because a 50-lb force (effort) balances a 400-lb object (resistance).*

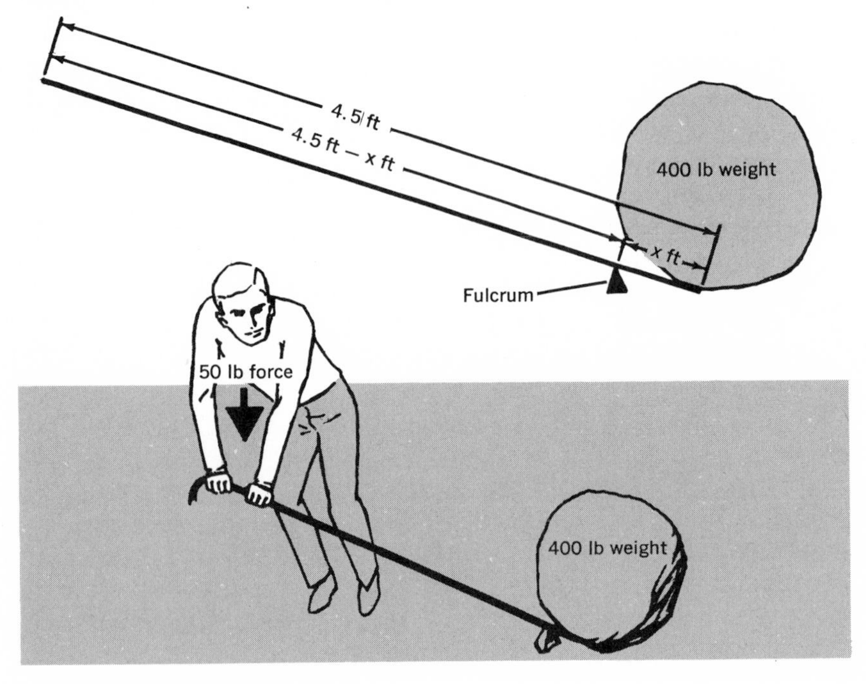

If in doubt concerning the correctness of an operation on an algebraic equation, visualize the platform balance; it will serve as an unerring guide. Treat both members of the equation in the same way; never fail to accord one member the same treatment as the other.

THE MECHANICAL ADVANTAGE OF A LEVER

Why do people make use of the so-called "simple machines" like the lever? Because the machines permit the moving of objects with *forces* smaller than would otherwise be required. In everyday terminology, they give us "leverage," i.e., force-multiplying ability. The physicist refers to this as the "mechanical advantage" of the machine. It is defined as the ratio of the resisting force to the applied force:

$$\text{Mechanical advantage} = \frac{\text{resisting force}}{\text{applied force}}$$

If an object like a 400-pound rock can be moved by exerting a 50-pound force on the end of a crowbar (Fig. 2-6), the crowbar is said to have a mechanical advantage of 8.

ILLUSTRATIVE PROBLEM

A 50-lb force applied at the end of a 4.5-ft crowbar just moves a 400-lb rock located under the opposite end of the bar. Where must the fulcrum be located?

Solution

The principle of moments is assumed to apply:

$$\text{Counterclockwise moment} = \text{clockwise moment}$$

$$\text{Weight of rock} \times \begin{pmatrix}\text{distance} \\ \text{to fulcrum}\end{pmatrix} = \text{applied force} \times \begin{pmatrix}\text{distance} \\ \text{from fulcrum}\end{pmatrix}$$

Let x = distance of 400-lb rock from fulcrum.

$$400\ \text{lb} \cdot x\ \text{ft} = 50\ \text{lb} \cdot (4.5 - x)\ \text{ft}$$

$$400x = 225 - 50x$$

$$450x = 225$$

$$x = 0.5\ \text{ft} = 6\ \text{in.}$$

The fulcrum must be located 6 in. from the rock or 48 in. from the point of application of the force. The mechanical advantage may also be expressed as

$$\text{Mechanical advantage} = \frac{\text{effort distance}}{\text{resistance distance}} = \frac{48\ \text{in.}}{6\ \text{in.}} = 8$$

Hence the mechanical advantage, which was first described as the ratio of two forces, may equally well be expressed as the inverse ratio of the lever arms:

$$\text{Mechanical advantage} = \frac{\text{effort distance}}{\text{resistance distance}} = \frac{\text{resistance (weight)}}{\text{effort (applied force)}}$$

The Wheel and Axle Two grooved wheels of different diameters which are bolted together (Fig. 2-7) are called a wheel and axle and may be con-

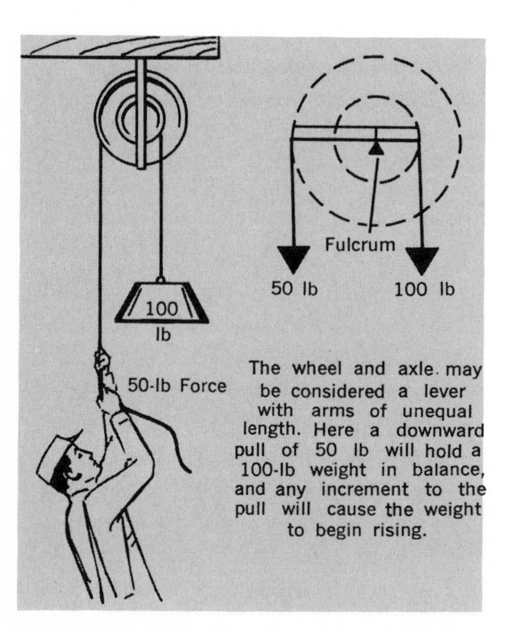

Fig. 2-7. *The wheel and axle may be considered a lever with arms of unequal length. Here a downward pull of 50 lb will hold a 100-lb weight in balance, and any increment to the pull will cause the weight to begin rising.*

sidered a lever with arms of unequal length and the fulcrum at the center of the axle. Suppose one wheel has a radius of 4 inches and the other 8 inches. The relation between force and weight may be determined in the manner described above. A 50-pound force applied to the radius of the larger wheel will balance a 100-pound weight suspended from the smaller. The mechanical advantage of this machine is 2. The weight of the object is twice as large as the force required to balance it.

The length of the pedal arm of a bicycle is larger than the radius of the geared wheel sprocket, to which it is attached. Hence the force exerted on the bicycle chain by the sprocket wheel will be larger than that exerted by the foot on the crank. The mechanical advantage will depend on the relation of the length of the crank to the radius of the sprocket. Once again the principle of moments applies.

SIMPLE MACHINES AND THE PRINCIPLE OF WORK

The problems considered above have all been based upon the principle of moments which defines the conditions for a balance of rotational tendencies.

Consider now another method which might have been used to solve these problems. It is based upon what may be called the work principle. When you climb a mountain, you do the same amount of work in reaching the top whether you take a path which is short and steep or one which is long and gradual. Work, as defined by the physicist, is the product of a force and a distance, both measured in the same direction. In the case of climbing a mountain, the force is the weight of the person, and the distance is the vertical height he rises above his starting point.

This principle will be employed in the solution of problems dealing with simple machines of several other types. A method depending on equilibrium conditions could also be applied.

The Inclined Plane A 300-pound box rests on the ground under the tailboard of a truck. If this box is lifted vertically against the pull of gravity, a force of 300 pounds will be required. If a 300-pound force acts vertically through a distance of 4 feet, 1200 foot pounds of work are done.

Fig. 2-8. *An inclined plane allows us to raise a heavy object by applying a force smaller than the weight of the object. In any actual case the total work done in moving the heavy object up the plane may be far greater than the work done in direct lifting, owing to the friction that must be overcome. Even so, it pays to use the inclined plane. With its help a much smaller force can raise the weight, even though it must also overcome friction. For calculation, the inclined plane is assumed to be frictionless.*

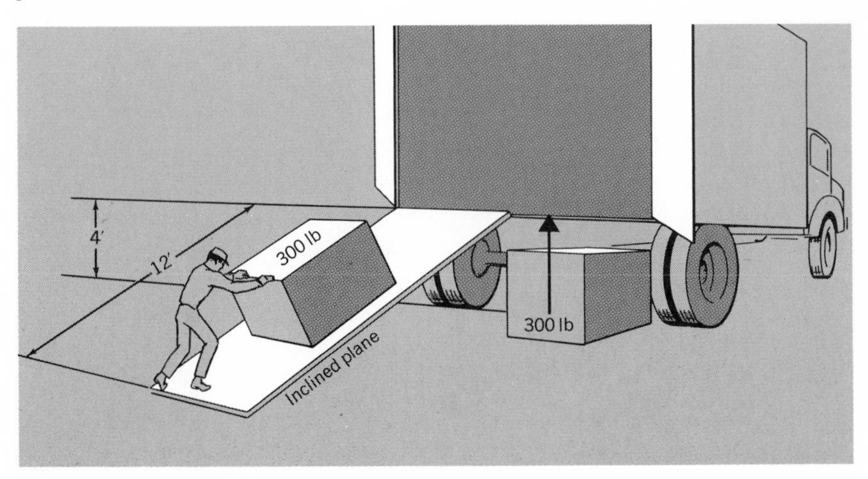

If a plank 12 feet long has one end resting on the ground and the other on the tailboard as shown in Fig. 2-8, the box may be pushed up the plank by exerting a force much smaller than 300 pounds:

$$\begin{pmatrix}\text{Work done in lifting}\\ \text{box vertically}\end{pmatrix} = \begin{pmatrix}\text{work done in pushing}\\ \text{box up plank}\end{pmatrix}$$

The equality is based on the assumption that the same amount of work will be required whether the box is lifted vertically or pushed up the plank. Keeping in mind that work is the product of a force and a distance, the above equality may be expressed

$$\begin{pmatrix}\text{Weight}\\ \text{of box}\end{pmatrix} \cdot \begin{pmatrix}\text{height}\\ \text{to be lifted}\end{pmatrix} = \begin{pmatrix}\text{number of lb force } F \text{ needed}\\ \text{to push box up plank}\end{pmatrix} \cdot \begin{pmatrix}\text{length}\\ \text{of plank}\end{pmatrix}$$

$$(300 \text{ lb})\cdot(4 \text{ ft}) = (F)\cdot(12 \text{ ft})$$

$$1200 \text{ lb} = 12F$$

$$F = 100 \text{ lb}$$

Thus anything over a 100-pound force will be sufficient to start a 300-pound object up the plank. This inclined plane has a mechanical advantage of 300 lb/100 lb = 3. The weight moved is three times as large as the applied force.

This calculation assumes that the plank is a frictionless surface along which the box may slide; in any actual case, additional force would have to be applied to overcome friction.

Fig. 2-9. *The jackscrew is a simple machine which has a very large mechanical advantage—the ratio of the circumference to the pitch in drawing* (a). *The screw can be considered an inclined plane wrapped round a cylinder.*

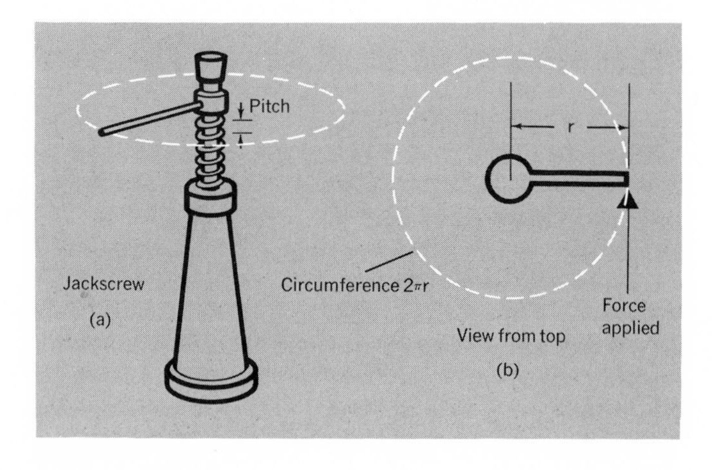

The Jackscrew In order to lift a very heavy object such as a building, a machine with very large mechanical advantage as, for example, the jackscrew (Fig. 2-9), must be used if a single workman is to move it. The work done on the heavy object is the product of the weight of the object and the vertical distance through which it is moved (note the similarity to the inclined plane). The workman applies the force at right angles to the handle of the jackscrew; the point of application of the force thus moves along the circumference of a circle. When the handle of the jackscrew moves through one revolution, the screw moves vertically a distance equal to the spacing between two successive threads (called the pitch of the screw). The basic pattern used in the solution of this problem is the same as for the inclined plane:

$$\begin{pmatrix}\text{Work done in lifting}\\ \text{house vertically}\end{pmatrix} = \begin{pmatrix}\text{work done in pushing}\\ \text{the handle}\end{pmatrix}$$

Keeping in mind again that work is force times distance, we may translate the equation above into the following:

$$\begin{pmatrix}\text{Weight}\\ \text{of}\\ \text{house}\end{pmatrix}\cdot\begin{pmatrix}\text{pitch}\\ \text{of}\\ \text{screw}\end{pmatrix} = \begin{pmatrix}\text{number of lb}\\ \text{force } F \text{ exerted}\\ \text{on handle}\end{pmatrix}\cdot\begin{pmatrix}\text{circumference of}\\ \text{circle through}\\ \text{which force moves}\end{pmatrix}$$

If the house weighs 4000 pounds, the pitch of the jackscrew is 0.250 inch and the length of its handle is 12.0 inches, the following substitutions may be made in the above equation:

$$\begin{aligned}(4000\ \text{lb})\cdot(0.250\ \text{in.}) &= (F)\cdot(2\pi\cdot 12.0\ \text{in.})\\ 1000\ \text{lb} &= 24\pi\cdot F\\ F &= \frac{1000\ \text{lb}}{24\pi} = 13.3\ \text{lb}\end{aligned}$$

Thus a force just over 13.3 pounds suffices to lift the 2-ton house. The mechanical advantage of this machine is 4000 lb/13.3 lb = 302.

Again it should be emphasized that idealizations have been assumed, as with the inclined plane. Frictional effects in the jackscrew have been neglected in the calculation, whereas in actuality they are of great importance. Furthermore, the calculation assumes that the applied force is constant as the workman turns the handle and that it always acts at right angles to the handle. In any actual case, we know that these conditions would be very difficult of realization. It is certain that in an actual case, a force substantially larger than 13.3 lb would be required to lift the house, using this jackscrew.

REDUCTION OF FORCE IS COMPENSATED BY INCREASED DISTANCE

Machines never save work; they merely make it possible to move an object by applying a force smaller than the object's weight. Such reduction of the applied force does make it possible to lift otherwise impossibly heavy objects, but is accompanied by a corresponding increase in the distance through which the force acts. In the inclined-plane problem the force pushing the box up the plane moved three times as far as if the box had been raised vertically. In the jackscrew problem the effort moved 302 times as far as the house moved vertically. In all cases, then, a reduction in the applied force is attained at a cost. Any fractional reduction in this force will be compensated by a corresponding increase in the distance through which it must act.

IDEALIZATION IN SCIENCE

Galileo is frequently called a scientific genius, possibly because of his uncanny ability to discover in a complex interplay of forces a simple principle. In the modern age a very important aspect of scientific discovery is that of finding patterns of orderliness in a complex and apparently chaotic world.

In discussing the principle of work, frictional effects were neglected; in a later chapter the effects of air resistance on the acceleration of falling bodies will be disregarded. Simplifying assumptions often lurk in the background in the generalizations of science.

The question may be asked, "If the generalizations of science are often inexact, why bother with them?" One important justification would be the economy of expression which results from the use of generalization. A single mathematical equation enables us to predict how a large number of similar devices (such as the inclined plane), each with its own distinctive characteristics, will operate. A simple equation relates forces, weights, and distances in all inclined planes even though the frictional effects are different in each. The generalization makes it unnecessary to devise a separate equation for every inclined plane.

A generalization may also be regarded as a limiting case which is approached as conditions are refined. Reduction of frictional effects in an inclined plane causes the actually measured forces to approximate more closely the predicted ones. Such idealizations form a part of the very fabric of science. The student should expect them; he should be on the alert to recognize them.

Problems, Chapter 2

1. A 500-lb barrel is rolled up a 15-ft plank to a platform whose vertical height above the ground is 3 ft. Neglecting friction, what effort is required?

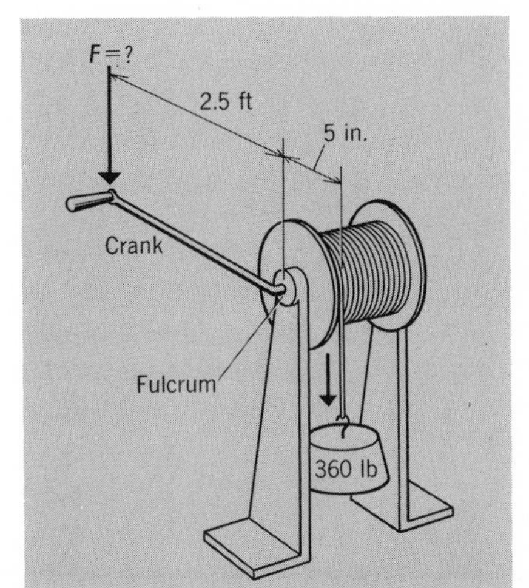

Fig. 2-10. *A windlass.*

2. The crank of a windlass is 2.5 ft long and the drum on which the rope winds is 5 in. in radius (Fig. 2-10). Neglecting friction, what effort is needed to raise an object weighing 360 lb?

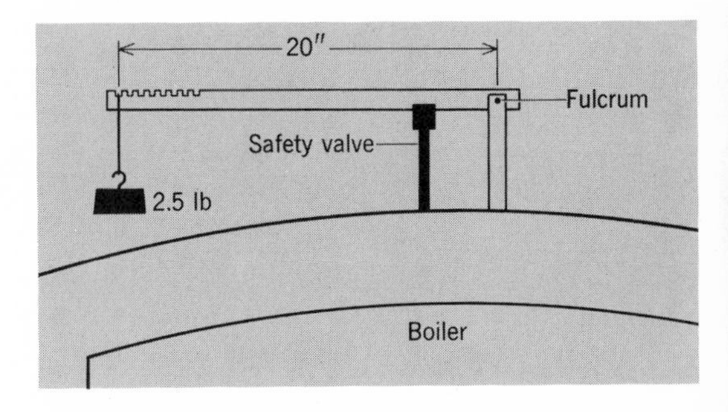

Fig. 2-11. *The safety valve on a boiler. Hint: Clockwise and counterclockwise moments must balance even though the fulcrum is at the end of the lever.*

3. The safety valve on a boiler is a horizontal metal bar 20 in. long which is hinged at one end (Fig. 2-11). What force exerted by steam acting on the valve 3 in. from the hinged end will balance a weight of 2.5 lb suspended from the other end of the bar?

4. What force must be exerted (vertically) on the handles of a wheelbarrow 3.5 ft from the axle of the wheel (Fig. 2-12) in order to lift a weight of 200 lb which is concentrated at a point 1.2 ft from the axle?

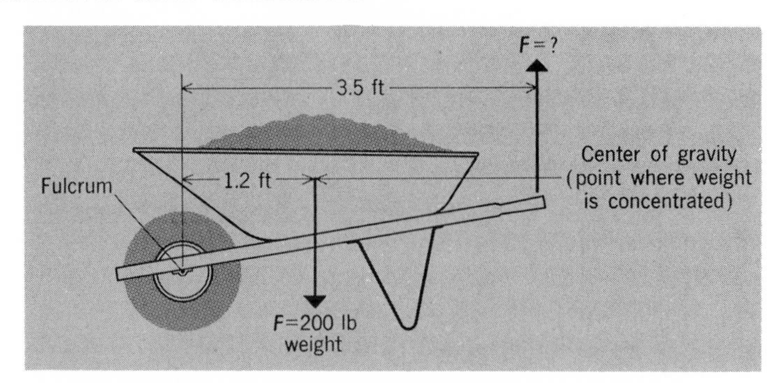

Fig. 2-12. *A wheelbarrow.*

5. What force exerted on a hammer handle 12.0 in. from the fulcrum (Fig. 2-13) will pull out a nail whose resisting force is 500 lb, if the distance from the nail to the fulcrum is 1.60 in.?

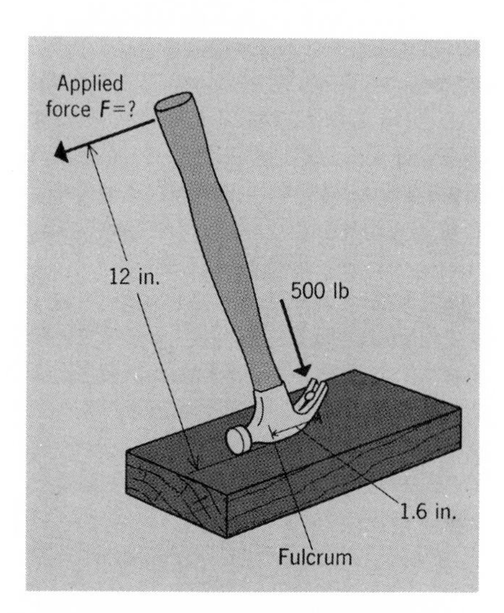

Fig. 2-13. *The hammer.*

6. A plank 18 ft long weighs 124 lb. This weight may be considered to be cencentrated at the mid-point of the plank. What is the minimum force which must be applied at one end of the plank in order to raise it (i.e., to cause it to rotate about the opposite end)?

7. What force must be exerted on the handle of a jackscrew 16.0 in. in radius if the pitch is 0.375 in. and the weight of the object to be moved is 6000 lb?

8. For every 24 ft measured along a road the vertical rise is 3 ft. If a 60-lb force applied parallel to the road just barely moves a wagon, what is the weight of the wagon? (Neglect friction.)

9. In the construction of the pyramids of Egypt, blocks of stone are said to have been pushed up an inclined road, possibly on tree trunks as rollers. How many laborers, each exerting a force of 50 lb parallel to the road, would have been required to push a 10-ton block of stone up a road rising 3 ft vertically for every 100 ft along the road?

10. A crowbar is used to pry loose a heavy rock. The resisting force of the rock is 750 lb, the perpendicular distance from the line of action of this force to the fulcrum is 8.0 in., and the force applied at the other end of the crowbar is 4.5 ft from the fulcrum. Calculate the force required to just move the rock. What is the mechanical advantage of the lever?

11. Calculate the mechanical advantage of a jackscrew whose pitch is $\frac{1}{8}$ in. and the radius of whose handle is 10 in. (Neglect friction.)

Exercises

1. Are the so-called simple machines of very ancient or of comparatively recent origin? Discuss briefly.

2. State the principle of moments. Illustrate its meaning with reference to the seesaw.

3. Make a list of commonly used levers.

4. Explain what is meant by solving an equation. Algebraic methods are usually preferred to inspection methods of solving algebraic equations. Why?

5. Illustrate how to calculate the mechanical advantage of a machine.

6. Machines are often referred to as labor-saving devices. Do simple machines really save us work? Explain carefully.

7. If scientific generalizations such as the principle of work or principle of moments do not accurately predict force and distance relationships, why do we continue to use them? Give additional illustrations of physical principles in which idealizations are assumed.

3 Forces, changers of motion

THE famous Greek scientist Aristotle made no really important contributions in physics, but some of his pronouncements were recovered in Europe some fifteen centuries after his death and revered as the gospel truth. The statement attributed to him that the speed of fall of an object was proportional to its weight became almost universally accepted. If true, this would have meant that a 10-pound ball would fall ten times as fast as a 1-pound ball. One would think that over a period of many centuries somebody would notice that very heavy and very light objects fell side by side. However there was the confusing fact that in air a light object such as a feather fell much slower than a stone; so the erroneous idea persisted.

Greek men of science were essentially philosophers and teachers; most of their efforts were devoted to speculation. The idea of experimentation was many centuries away, although individuals undoubtedly had curiosity about natural phenomena and even in Aristotle's time probed the secrets of nature in a simple fashion. But in general they were more concerned with the question of *why* rather than *how*.

In pondering the question of *why* objects fall, they were considering a matter which is at best only superficially understood even today; little is known about the cause of gravitation. Rather, in early modern times the question of *how* bodies fall—the relations between distance, rate, and time of fall—proved more fruitful for investigation. In the seventeenth-century world that accepted the view that the speed of fall of an object was directly proportional to its weight, there appeared on the scene a man who may justly be called the founder of the science of physics.

Galileo Upsets Authority The manner in which Galileo deduced generalizations concerning the fall of objects by "diluting their motion" with an inclined plane is discussed in his *Dialogues Concerning Two New Sciences,* a volume which is readily available today. Many consider his work the cornerstone of the modern science of mechanics.

Galileo was concerned with the question of *how* objects fell, not *why* they fell. Possibly through his observation of falling objects, he came to doubt the correctness of the Aristotelian statement. Aristotle had depended largely on chance observation. It occurred to Galileo that he might speed up the tempo of observation by making objects fall when he wished to observe them, and he was keen enough to consider a simpler aspect of free fall than the Greeks.

Accurate measurements of free fall were difficult to obtain directly, so he improved the technique of measurement by rolling objects down an inclined plane. This made it possible to measure, with fair reproducibility, distances traversed in a given time interval because of the "dilution" of the motion. The steeper the slope of the inclined plane the less the motion was diluted, the greater the distance covered in unit time, and the more closely the speed of the object approached that of free fall. Conversely, as the slope decreased, the time interval for rolling a given distance became longer, and hence the more accurate the measurement. In any case, at a given slope of the inclined plane, the acceleration imparted to a rolling marble was constant. As the slope was increased, the limiting value of the acceleration approached that value which is now called the acceleration of gravity.

The Method of Experimentation In the experiments which Galileo devised to learn something about the free fall of bodies he was making use of the experimental method. Undoubtedly others had done so before him, for by the end of the sixteenth century it was becoming evident to some men at least that they need not accept without reservations all of the wisdom which had been passed down to them; they should subject ideas to the test of experimentation. Aristotle said that heavy bodies fell faster than light ones, yet Galileo's experiments indicated that gravitational acceleration was the same for large as for small objects; it was independent of the weight of the falling object. Galileo felt sure he was right. The important point is that at last the authorities of the past were being challenged.

One man in particular dramatized this new spirit of inquiry and experimentation. Francis Bacon (1561-1626) deplored the fact that men had so long depended on the authorities of the past; they should rely more on their own resources and ingenuity in discovering information about the universe. He suggested a systematic method of collection, classification, and generalization of information. Through the intervening years his proposals have been incorporated into a pattern of investigations which is often called the "scientific method."

Bacon proposed the planning of experiments. Why sit around waiting for the chance happening of phenomena? Speed up the process by creating the situation you wish to investigate! In this way nature can be coaxed into divulging her secrets at a more rapid rate.

By the opening years of the seventeenth century there was in the air a new attitude concerning the investigation of natural phenomena; modern science was "aborning."

INERTIA: NEWTON'S FIRST LAW OF MOTION

When Newton "discovered" the first law (of motion), he really stood on the shoulders, to use his own expression, of those who preceded him. The growth of scientific concepts is gradual, and like trees, they do not mature until the environment is favorable. Every scientist is indebted to investigators who have preceded him and prepared the "conceptual climate"; scientists do not discover completely on their own.

Aristotle anticipated an important aspect of inertia when he said, "Everything resists being forced." All about us is evidence of this universal principle. If the motion of a speeding bicycle is suddenly halted, the rider is hurled over the handlebars. Bumps and bruises testify to the fact that a mass once set in motion tends to keep moving; it resists any change which would alter its motion. A locomotive with a train of loaded freight cars tugs and pulls with little effect as the train starts up. But once in motion, it is difficult to stop; the brakes must be applied far in advance of the desired stopping point.

Inertia is resistance to change of motion. If an object is at rest, inertia will oppose its being set in motion. If it is in motion, inertia will oppose its stopping. It might be described as a principle of contrariness. Try to set an object in motion and it prefers to stay at rest. Try to check it while it is moving, and it wants to go on moving at the same speed. There is a comparable principle in chemistry, the Le Châtelier principle, that refers to changes of pressure, temperature, or concentrations in equilibrium chemical systems. Another in electricity, Lenz's law, concerns opposition to change in the magnetic field surrounding a coil of wire. An alert mind will recognize other manifestations of this principle of resistance to change which was alluded to by Aristotle.

One further development in the understanding of inertia came as the result of Galileo's correcting an erroneous notion suggested by the famous early astronomer Johannes Kepler (1571-1630). Kepler asserted that the planets required some force to push them, to keep them moving in their orbits. Galileo appears to have been much more incisive in his thinking; he suggested that every body, even one as huge as a planet, *tends* to move in a straight line (because of inertia). Since the path of a planet in the heavens was actually an ellipse, one had to account also for the modification of this natural tendency (as will be discussed later) by a force.

While Galileo clearly understood the principle of inertia, Isaac Newton (1642-1727) is usually given credit for being the first to state it explicitly. His first law of motion says, in effect, "Every body which is at rest tends to remain at rest; every body which is in motion tends to remain in uniform motion in a straight line."

What is the meaning of *uniform motion?* It implies that an object traverses equal distances in equal time intervals. The usual motion of an automobile, for example, is not uniform. The statement that a car travels 40 mi/hr refers to an average speed, even though in solving algebraic problems it is often treated as if it were a uniform speed. The concept of uniform motion is frequently an idealization.

The first part of Newton's principle—bodies at rest tend to remain at rest—is consistent with everyday experience; inanimate objects such as stones and books do not start moving of their own volition. But since it is the usual circumstance for moving objects to come to rest, the second part of the principle seems less in accord with experience, and other forces (chiefly *friction*) must be invoked to account for the apparent inconsistency (see the section on Newton's second law below).

EQUATION FOR UNIFORM MOTION: CHOICE OF CORRECT UNITS

The equation for uniform motion, $d = r \cdot t$ (distance equals rate times time), is a familiar one. For many decades it has served the problem makers in algebra who have presented students with people rowing boats up and down rivers and with airplanes bucking headwinds. Sometimes the stated problems of algebra have appeared difficult enough in themselves, needing no added complications. But it is possible that reemphasizing a few points will make them immeasurably easier. It is most important to realize that an equation represents an equality of *units* as well as of *numbers*. Therefore, when substituting into an equation, always ask, "Are the units the same on both sides of the equation?"

ILLUSTRATIVE PROBLEM

How far does a car averaging 40 mi/hr travel in 30 min?

Solution

$$\text{Distance (miles)} = \text{rate}\left(\frac{\text{miles}}{\text{hours}}\right)\cdot\text{time (hours)}$$

$$\text{Miles} = \frac{\text{miles}}{\text{hours}}\cdot\text{hours}$$

$$\text{Time} = 30\text{ min} = \tfrac{1}{2}\text{ hr}$$

$$d = 40\text{ mi/hr}\cdot\tfrac{1}{2}\text{ hr} = 20\text{ mi}$$

Note that hours will cancel from numerator and denominator, and then the units on both sides of the equality sign will be the same.

Suppose the 30 min had been substituted directly into the equation:

$$\text{Distance (miles)} = \text{rate}\left(\frac{\text{miles}}{\text{hours}}\right)\cdot\text{time (minutes)}$$

$$\text{Miles} = \frac{\text{miles}}{\text{hours}}\cdot\text{minutes}$$

$$d = 40\text{ mi/hr}\cdot 30\text{ min} = 1200\,\frac{\text{mi min}}{\text{hr}}$$

While theoretically we might designate the distance traversed as 1200 mi min/hr, this is a scrambled collection of units which conveys little meaning. Take care to look over the units given in a problem and make such conversions as lead to recognizable equalities, as illustrated in the first solution. It would obviously be incorrect if written:

$$d = 40\text{ mi/hr}\cdot 30\text{ min} = 1200\text{ mi}$$

In the above problem the rate or speed of the car is expressed in miles/hour. The term *speed* might have been replaced by *velocity* if a direction had also been specified. Speed is a scalar, velocity a vector quantity. A scalar quantity involves magnitude only (e.g., 40 mi/hr); a vector quantity implies both direction and magnitude (e.g., 40 mi/hr in a north direction). Speeds or velocities are always expressed as linear units divided by a time unit—miles/hour, cm/sec, ft/sec, etc.

Be sure to check the units being substituted in every equation. Indiscriminate substitution of just any units given in a problem will surely lead to an incorrect result sooner or later. Failure to make sure that an equation represents an equality of units as well as of numbers has been the shoal upon which many a potentially fine student has been shipwrecked in his study of physics.

BUILDING ALGEBRAIC EQUALITIES

Somebody has said, "Mathematics is a language." The art of devising equations to represent scientific situations is primarily the art of expressing in mathematical language the fact that two quantities *are equal* under the conditions of a problem, or that they can be *made equal.* In lever problems, rotational tendencies are said to be *equal* if the lever is balanced. In the inclined plane, the amount of work done is assumed to be the *same* whether the object is lifted vertically against gravity or pushed up an inclined plank. Before starting to solve any such problem, first ask, "Under the conditions of the problem, what quantities are equal or can be made equal? What can we equate?"

Once again it must be emphasized that idealizations are involved. In a problem dealing with a rowboat on a river, to select a common example, it is often assumed that a boat starts at a certain point, rows up stream, and then returns to the original point *by an identical path.* This is obviously impossible as a practical matter. However, it is the assumed equality of the distance up the river and the distance back that justifies the writing of an equation. Usually in such problems, constant rowboat speeds and constant river-flow speeds are assumed—further idealizations.

Once an equality is discovered or constructed, it must be translated into mathematical symbols as an equation, and then solved.

ILLUSTRATIVE PROBLEM

A man drove to New York at an average speed of 50 mi/hr, and returned by the same route at an average speed of 40 mi/hr. If the round trip required 9 hr, how far did he drive?

Solution

What can be equated? Let $t =$ the number of hours driving to the city, let $9 - t =$ the number of hours returning, and equate the two distances ($d = r \cdot t$):

$$\text{Rate} \cdot \text{time (going)} = \text{rate} \cdot \text{time (returning)}$$

$$50 \frac{\text{mi}}{\text{hr}} \cdot t \text{ hr} = 40 \frac{\text{mi}}{\text{hr}} \cdot (9 - t) \text{ hr}$$

$$50t = 360 - 40t$$

$$90t = 360$$

$$t = 4 \text{ hr}$$

$$\text{So distance going:} \quad 50\,\frac{\text{mi}}{\text{hr}} \cdot 4\ \text{hr} = 200\ \text{mi}$$

$$\text{distance returning:} \quad 40\,\frac{\text{mi}}{\text{hr}} \cdot 5\ \text{hr} = 200\ \text{mi}$$

$$\text{and round trip:} \qquad = 400\ \text{mi}$$

Or let d = the distance each way and equate the total time (9 hr) with the sum of the times for the two parts of the trip, $t = d/r$:

$$\text{Time (going)} + \text{time (returning)} = 9\ \text{hr}$$

$$\frac{d\ \text{mi}}{50\,\frac{\text{mi}}{\text{hr}}} + \frac{d\ \text{mi}}{40\,\frac{\text{mi}}{\text{hr}}} = 9\ \text{hr}$$

$$\frac{d}{50} \cdot 200 + \frac{d}{40} \cdot 200 = 1800$$

$$4d + 5d = 1800$$

$$9d = 1800$$

$$d = 200\ \text{mi}$$

$$\text{so round trip} = 400\ \text{mi}$$

The dilution of aqueous solutions is another common type of algebraic problem. If such a problem deals with the addition of water to a salt solution, keep in mind that the *weight* of salt particles is just the same after the dilution as it was before. In other words the following statement is a logical one:

$$\text{Salt in original solution} = \text{salt in diluted solution}$$

It is also true that:

$$\text{Water in original solution} + \text{water added} = \text{water in diluted solution}$$

ILLUSTRATIVE PROBLEM

How much water must be added to 50 lb of a solution which is 60% alcohol (by weight) so that the resulting solution will be 25% alcohol (by weight)? See Fig. 3-1.

Solution in Outline

Adding water to the 60% solution does not change the *weight* of the alcohol molecules. Hence the following will be a logical equality:

$$\begin{pmatrix}\text{Weight of alcohol}\\ \text{in the 60\% solution}\end{pmatrix} = \begin{pmatrix}\text{weight of alcohol}\\ \text{in the 25\% solution}\end{pmatrix}$$

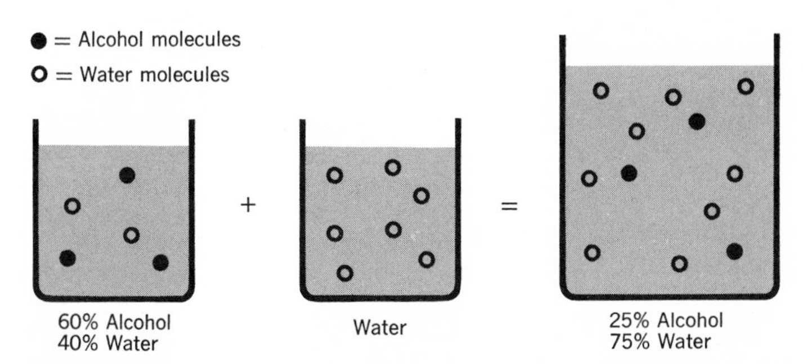

Fig. 3-1. *Changing the proportions of a solution. The addition of water to an alcohol-water mixture does not change the weight of the alcohol. The weight of the alcohol in the original solution is equal to that in the more dilute solution.*

Let $x =$ pounds of water added

$$0.60 \cdot 50 \text{ lb} = 0.25(50 + x) \text{ lb}$$

$$x = ?$$

Now complete the solution of this problem.

Note: It is assumed in this calculation that just two constituents, alcohol and water, are present. Furthermore, in this problem the volumes are not directly additive. When ethyl alcohol and water are mixed, the total volume is less than the sum of the original volumes. Hence an equality based upon the quarts of alcohol present before and after dilution would not be a correct one.

FORCE: NEWTON'S SECOND LAW OF MOTION

Newton's first law of motion described a natural tendency of objects, if at rest, to stay at rest, or if moving, to continue uniform motion in a straight line. But the resting body may be set in motion, or the uniformly moving object may be brought to rest. In either case forces are involved. The resting book on the table is set in motion if a force acts on it. The railroad train moving along a level track (idealization) gradually slows down and eventually comes to a stop, if the power is turned off. Air resistance and friction act as counterforces to oppose its inertial tendency to continue indefinitely its straight-line uniform motion. Forces then, are *changers* of motion.

If a car is traveling 30 mi/hr and its speed is increased to 40 mi/hr, a force must be acting upon it. If a train is traveling 60 mi/hr and slows down to 25 mi/hr, a force must be opposing its motion.

Just as lengths of object are measured in feet or meters, and masses in pounds or kilograms, forces are expressed in units called dynes, newtons, or poundals. Newton's second law of motion tells how to relate these units of force to other properties. The mathematical expression for this law is

$$F = ma$$

where F is the force, m is the mass, and a is the acceleration—a new concept.

Acceleration A body whose velocity *changes* 10 ft/sec—for example, from 40 ft/sec to 50 ft/sec—during a time interval of 1 second, is said to accelerate 10 feet per second per second, an expression usually written 10 ft/sec/sec or 10 ft/sec^2. The first part, 10 ft/sec, is the velocity change; the last "per second" is the time interval during which the velocity change takes place.

Acceleration, as the term is usually employed, is more correctly identified with *uniform* acceleration. Uniform velocity meant traversing equal distances in equal time intervals; uniform acceleration means equal velocity *changes* in equal time intervals. Uniform acceleration results from the action of a constant force. If the applied force increases or decreases, so will the acceleration. An accelerating body will also traverse increasingly greater distances in successive time intervals (see Fig. 3-2). As a matter of fact its

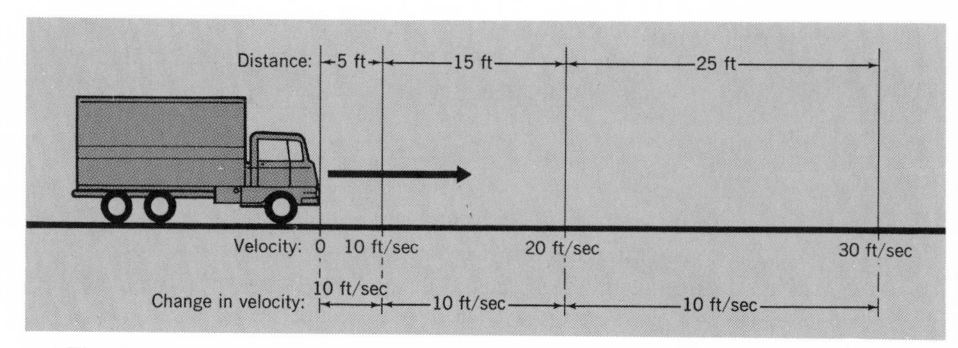

Fig. 3-2. *The velocity of a truck whose motion is accelerating uniformly 10 ft/sec^2 changes equal amounts in equal time intervals. The distance it travels in each time interval increases correspondingly.*

velocity is never the same from one second to the next, or from one instant to the next; it is changing continuously. (Think of an instant as a very tiny part of a second, say a billionth.)

Note that an acceleration of 10 ft/sec/sec is equivalent to an acceleration of 600 ft/sec/min; the *rate* of velocity increase is the same in the two cases, though the time interval during which it takes place is 60 times as large in the latter case.

The equation $F = ma$ reveals that to impart equal accelerations to each of two different masses a proportionately larger force must be exerted upon the larger mass. Ten times as much force is required to accelerate a 30,000-pound truck 20 ft/sec^2 as is required to increase the speed of a 3000-pound automobile by the same amount in 1 second. It is a matter of everyday experience that it is easier to change the speed of a child's cart than that of an automobile. The larger the mass, the greater the force necessary to impart a given acceleration.

Similarly, if different forces act upon bodies having equal masses, the accelerations imparted will be proportional to the forces. Twice as much force will be required to accelerate an object 12 ft/sec^2 as an object of equal mass by 6 ft/sec^2. Freely falling bodies accelerate 32 ft/sec^2, which is equivalent to 980 cm/sec^2 or 9.8 m/sec^2. This acceleration which is caused by the force called gravity is designated g and is called the gravitational acceleration. The gravitational force acting on a freely falling body is called its *weight*. Hence Newton's second law: $F = ma$ may also be written in the form:

$$\underset{\text{weight}}{W} = \underset{\text{mass}}{m} \cdot \underset{\text{gravitational acceleration}}{g}$$

Units: Newton's Second Law of Motion The units in the equation for uniform motion ($d = rt$)—those of distance, rate, and time—are quite familiar; units of force, mass, and acceleration are very likely less so. The dyne is defined as the force imparting to a 1-gram mass an acceleration of

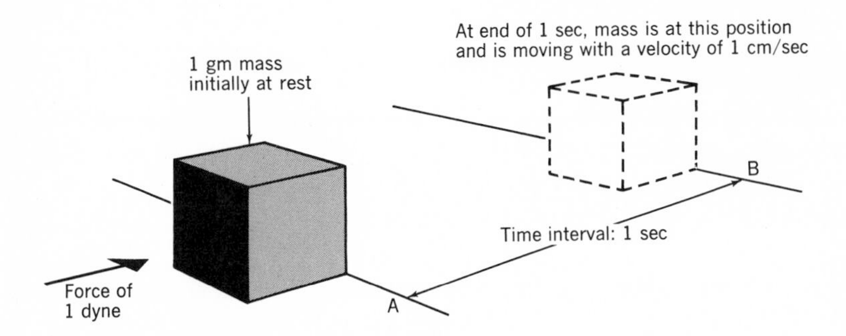

Fig. 3-3. *In the figure a constant 1-dyne force causes the velocity of the 1-gram mass to increase from rest at A to 1 cm/sec at B in 1 second.*

1 cm/sec^2 (see Fig. 3-3). The newton is a larger force unit, one imparting to a 1-kilogram mass an acceleration of 1 m/sec^2. The poundal is a force imparting to a 1-pound mass an acceleration of 1 ft/sec^2.

The table which follows should be helpful as a guide to the use of consistent units when solving problems by means of the equation $F = ma$. These force units are called "absolute" units.

ABSOLUTE UNITS

Force	=	*mass · acceleration*
Dynes	=	grams · cm/sec^2
Newtons	=	kilograms · m/sec^2
Poundals	=	pounds · ft/sec^2

ILLUSTRATIVE PROBLEM

What force imparts to a 10-lb *mass* an acceleration of 4.0 ft/sec^2?

Solution

$$F = ma$$

$$F = 10 \text{ lb} \cdot 4.0 \text{ ft/sec}^2$$

$$F = 40 \text{ poundals}$$

From the table it will be seen that when mass is expressed in pounds and acceleration in ft/sec^2, the force will be in poundals.

Newton's second law would be immeasurably easier if the concept of force were dismissed with consideration of absolute units only. But sooner or later the ghost of a 10-pound weight (rather than a 10-pound mass) would raise its ugly head. The difficulty arises because the pound, gram, and kilogram are sometimes expressed as masses and sometimes as weights. In dealing with problems, notice particularly whether the reference is to a *mass* or a *weight*. If the problem states a mass in pounds, grams, or kilograms, use the table of absolute units to determine the force and acceleration units which are consistent.

In the gravitational system the object's mass may be expressed as the ratio of its weight to g, since $W/g = m$. The table which follows show the units which are consistent in the gravitational system:

GRAVITATIONAL UNITS

$$\textit{Force} = \textit{mass} \cdot \textit{acceleration}$$

$$\text{Pounds} = \frac{\text{pounds (weight)}}{g \text{ (ft/sec}^2)} \cdot \text{ft/sec}^2$$

$$\text{Grams} = \frac{\text{grams (weight)}}{g \text{ (cm/sec}^2)} \cdot \text{cm/sec}^2$$

$$\text{Kilograms} = \frac{\text{kilograms (weight)}}{g \text{ (m/sec}^2)} \cdot \text{m/sec}^2$$

ILLUSTRATIVE PROBLEM

What force imparts to a 10-lb *weight* an acceleration of 4 ft/sec²?

Solution

$$F = ma = \frac{W}{g}a$$

$$F = \frac{10 \text{ lb}}{32 \text{ ft/sec}^2} \cdot 4 \text{ ft/sec}^2 = \frac{5}{4} \text{ lb}$$

Both illustrative problems in this section dealt with the same 10-pound object. The first problem referred to a 10-pound mass, the second problem to a 10-pound weight. Since an acceleration of 4 ft/sec² was imparted in both cases, the two forces calculated—40 poundals and $\frac{5}{4}$ pounds—must have been equivalent.

NEWTON'S THIRD LAW OF MOTION

Newton's third law of motion states that "for every action there is an equal and opposite reaction." It says in effect that forces always occur in opposing pairs. A pail of water on the table exerts a downward force (called its weight). The table resists the action of the downward force with an equal upward force; otherwise the table would collapse. The wheel of an automobile pushes back on the pavement. If the vehicle moves, the pavement also pushes forward on the wheel.

Thus far in the chapter forces have been considered as changers of motion. The equation $F = ma$ relates force to mass and acceleration. Actually this equation was first formulated as $F = mv/t$.* If both members of the equation are multiplied by t it becomes

$$Ft = mv$$

This equation indicates that the longer a force acts on a body of given mass, the greater its velocity and the larger the value of mv. The property depending on the mass of a body and the first power of its velocity is called the momentum of the body.

Two magnets with opposite poles facing, each floating on a cork in water, are mutually attracted. The pull that the first exerts on the second is **equal** to the pull that the second exerts on the first. If the mass of one is larger, it will move more slowly than the other. A principle of "conservation of momentum" requires the equality of mass times velocity for each magnet:

* In Chapter 4 there is discussion concerning the justification of the substitution $v/t = a$.

$$\underset{\text{large magnet}}{m_1v_1} = \underset{\text{small magnet}}{-m_2v_2}$$

The negative sign in this equation is a consequence of the velocity v being a vector quantity. Since the motions of the two magnets are in opposite directions, so also are their velocities v_1 and v_2. Hence their signs must be opposite.

If a bullet is fired from a rifle suspended by cords attached to the ceiling of a room, the recoil of the rifle can be plainly seen. The rapid release of gas molecules resulting from the explosion pushes the bullet in one direction, the rifle in the other. The rifle mass is very much larger than the bullet mass, so the rifle-recoil velocity is very much smaller than the bullet velocity.

The earth attracts a falling body. In turn the falling body pulls on the earth. The two forces are equal; the masses are so vastly different that the effect on the earth appears negligible.

Problems, Chapter 3

1. In the equation $d = rt$ state the units in which the *unknown* will be expressed if:

   (a) d = ft; t = sec
   (b) r = mi/hr; t = hr
   (c) r = m/sec; d = meters

2. A car travels at an average speed of 45 mi/hr. (a) How far will it travel in 5.5 hr? (b) What is its speed in ft/sec?

3. A car has a *weight* of 4800 lb. What (constant) force will give it a horizontal acceleration of 16 ft/sec^2?

Fig. 3-4.

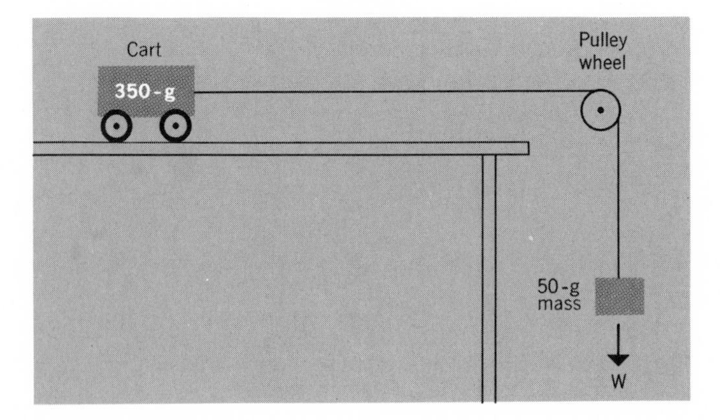

4. Calculate the (constant) force needed to accelerate 20 ft/sec^2 a 100-lb *mass*.

5. In a laboratory experiment illustrating Newton's second law of motion a toy car of 350-g mass moves along a horizontal track pulled by a falling 50-g mass. Calculate the acceleration. (See Fig. 3-4.)

6. Masses of 750 g and 850 g are suspended from a pulley by means of a cord as shown in Fig. 3-5. Calculate the acceleration.

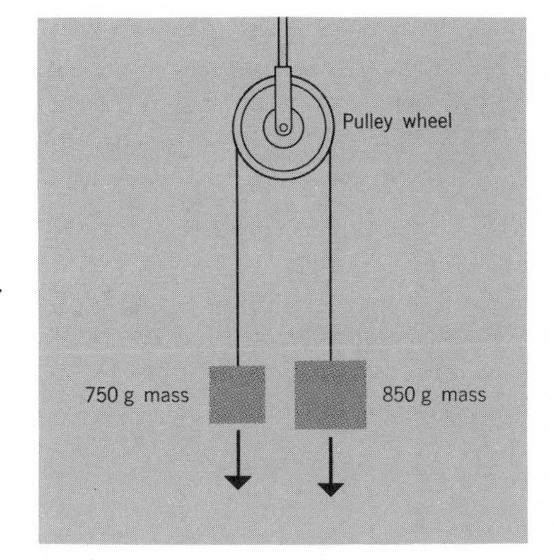

Fig. 3-5.

7. Calculate the ratio of the momenta of a 150-g bullet and a 1500-kg auto if their respective velocities are 900 m/sec and 15 m/sec.

8. What force is required to accelerate a 4.0-kg mass 10 m/sec^2?

9. What speed in ft/sec is equivalent to 60 mi/hr?

10. Calculate the acceleration imparted to an 800-g mass by a 2-newton force.

11. Assuming that Newtonian mechanics apply to a nuclear particle called the deuteron, mass 3.34×10^{-24} g, subject to a force of 2.25×10^{-9} dyne, calculate the acceleration.

Exercises

1. Explain what is meant by *inertia*. Illustrate.

2. According to Newton, how is inertia overcome? Illustrate.

3. Explain the difference between uniform motion and uniformly accelerating motion.
4. According to Newton's second law of motion how may the magnitudes of two forces be compared? Illustrate.
5. The same 10-lb object may under appropriate circumstances be described as either a 10-lb mass or a 10-lb weight. If it is described as a 10-lb mass, its weight cannot be 10 lb in the same system of units. Explain.
6. Give several illustrations of forces occurring in equal and opposite pairs.
7. How is momentum calculated? How is it related to force?

4 The force called gravity

ISAAC NEWTON, whose laws of motion were discussed in Chapter 3, is often regarded as the greatest scientific genius the world has ever known. He was born near Grantham, England, in 1642, the year in which Galileo died. Newton's ability must have been recognized early, since upon his graduation from Cambridge University (1665) he was invited to return as a fellow. But a plague visited England in that year and because of it the university was temporarily closed. So Newton withdrew to the family estate at Woolsthorpe, and here, during a two-year period, he carried on work of world-shaking importance. During this time he developed his method of fluxions, which laid the groundwork for the important branch of mathematics known as the calculus. He also carried out the well-known prism experiment by which he demonstrated that a beam of white light could be separated into components of different wavelength (see Chapter 9). Here too, according to the traditional story, he was reclining in the family orchard one day when a falling apple him hit on the head and launched him on his study of universal gravitation.

This story, like many others which have been told and retold for centuries, has accumulated details of dubious authenticity. Suffice it to say that Newton's thoughts appear to have gravitated from falling apples to the "fall of the moon." It occurred to him that if a mysterious force could tug at the apple on the twig, it might even extend as far as the moon.

So it was during this sojourn at the family estate that Newton made his first calculations on gravitation. According to one account, he used an incorrect figure for relating the distance along the earth's circumference to the corresponding angle at the center. Consequently he did not discover the hoped-for concordance between the known distance of fall in one second at the earth's surface and the corresponding "fall of the moon." Several different values for the number of miles in one degree of arc, including the correct value, 69 miles, were apparently available to Newton.* Had he made calculations with each one of these values he would have obtained the necessary concordance when using the correct value. In any case, this early work on gravitation was put aside for about twenty years.

Edmund Halley, discoverer of Halley's comet, was a close friend of Newton. Halley and several other colleagues were also interested in questions of gravitation. On one occasion Halley sought out Newton to ask him about certain details of a gravitational problem he was working on; he learned that Newton had worked out the same problem years before, and he later received a written solution of the problem from Newton. Only through Halley's urging did the report of Newton's work on gravitation ever reach the Royal Society. At a later date Halley even assumed much of the expense of publishing what is perhaps Newton's greatest scientific work, the *Principia*.

Newton seems to have been a very sensitive man, one who felt criticism keenly. He is said to have had a bitter exchange on one occasion with Robert Hooke concerning a problem in optics. He may have delayed publishing the third volume of the *Principia* out of a desire to avoid another such controversy.†

WHY IS GRAVITY A FORCE?

In Chapter 3 a force was described as a changer of motion. The speed of a falling object increases 32 ft/sec each second it falls; this acceleration indicates that a force must be acting. The force which causes all freely falling bodies at the Earth's surface to accelerate 32 ft/sec^2 is called gravity.

The Motion of Falling Bodies Galileo's experiment in which he rolled marbles down an inclined plane (see Chapter 3) had indicated a direct proportionality between velocity and time. For a given slope of the plane, the longer the marble rolled, the faster it rolled. This information is summarized mathematically by the expression $v \propto t$, read "v is proportional to t."

In order to convert any proportionality into an equality a constant factor must be introduced. In this case the proportionality factor is called the acceleration and is designated a. The equation may be written $v = at$.

As the slope of Galileo's inclined plane became steeper and steeper, the

* Florian Cajori, *A History of Physics,* Macmillan, 1924, p. 58.
† J. W. N. Sullivan, *Isaac Newton, Macmillan,* 1938, p. 90.

marble rolled faster and faster. Thus the value of a depended on the steepness of the slope of the plane. When the slope of the plane finally became vertical, the plane no longer had any influence on the velocity of fall; the object fell freely. The value of a had now reached its maximum, i.e., g, which is called the acceleration of gravity.

The expression $v = at$ is a general expression relating velocity, acceleration, and time. The expression $v = gt$ refers specifically to freely falling objects which are accelerated by gravity.

ILLUSTRATIVE PROBLEM

What is the instantaneous velocity of a body exactly 3.0 sec after it starts to fall? (Assume it starts from rest.)

Solution

$$v = gt$$

$$v = 32 \frac{\text{ft}}{\text{sec}^2} \cdot 3 \text{ sec}$$

$$v = 96 \frac{\text{ft}}{\text{sec}}$$

At the very instant that 3.0 sec have elapsed, the velocity is 96 ft/sec. Note the consistency of the units in this equation:

$$v = g \cdot t$$

$$\frac{\text{ft}}{\text{sec}} = \frac{\text{ft}}{\text{sec}^2} \cdot \text{sec}$$

The above discussion indicates that a graph of the equation $v = gt$ will be linear (Fig. 4-1). Suppose an object falls for 10 seconds. Its initial velocity is 0, its final velocity 320 ft/sec. During the 10-second interval the velocity will increase through all possible values between 0 and 320—i.e., 0, 1, 2, . . . 318, 319, and finally 320 ft/sec. It will not fall 3200 feet, for only at the very last instant will its velocity reach 320 ft/sec. At the 5-second point on the graph there will be as many velocities in excess of 160 ft/sec as there are less than this value. The velocity at 6 sec (192 ft/sec) and that at 4 sec (128 ft/sec) will average 160 ft/sec; similarly 7 seconds (224 ft/sec) and 3 seconds (96 ft/sec) will average 160 ft/sec. For every value of the velocity greater than 160 ft/sec there will be another value which is less, and with which it will average out to 160 ft/sec. Hence the accelerating falling body can be treated *as if* it had a constant average

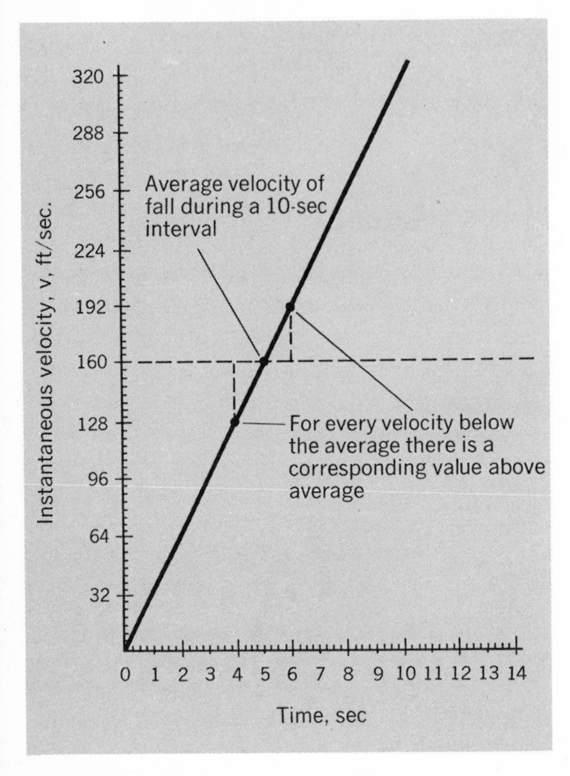

Fig. 4-1. *A falling body whose velocity is changing every instant may be treated as if it were falling with a constant velocity of $gt/2$.*

velocity of $gt/2$. The distance fallen is then calculated by means of the relation, distance = rate × time:

$$d = rt$$

$$d = \frac{gt}{2} \cdot t = \frac{1}{2} gt^2$$

This equation permits the calculation of the distance fallen in time t under the influence of gravitation.

ILLUSTRATIVE PROBLEM

An object hit the pavement exactly 8.00 sec after it was dropped from a skyscraper window. How far (feet) did it fall?

Solution

$$d = \tfrac{1}{2} \cdot g \cdot t^2$$

$$d\,(\text{ft}) = \frac{1}{2} \cdot 32.0 \frac{\text{ft}}{\text{sec}^2} \cdot 64.0 \text{ sec}^2$$

$$d = 1024 \text{ ft}$$

If the equations $d = \frac{1}{2}gt^2$ and $v = gt$ are equated through t^2:

$$\frac{v^2}{g^2} = t^2 = \frac{2d}{g}$$

$$\frac{v^2}{g^2} = \frac{2d}{g}$$

$$v^2 = 2gd$$

This equation permits the calculation of the instantaneous velocity of a falling object when it has fallen a distance d from rest.

ILLUSTRATIVE PROBLEM

How fast is an object falling at the instant it has fallen a distance of 49 ft?

Solution

$$v^2 = 2gd$$

$$v^2 = 2 \cdot 32 \frac{\text{ft}}{\text{sec}^2} \cdot 49 \text{ ft}$$

$$v^2 = 64 \cdot 49 \frac{\text{ft}^2}{\text{sec}^2} \quad \text{(What is the square root of 64? of 49?)}$$

$$v = 56 \frac{\text{ft}}{\text{sec}}$$

When the object has fallen a distance of 49 ft its instantaneous velocity is 56 ft/sec. Both d and g employ the same linear unit ($d = 49$ ft, $g = 32$ ft/sec²). The velocity squared has units of ft²/sec², and hence the velocity is in ft/sec.

Here in summary form are the equations which describe the motion of objects which start from rest and fall freely under the action of the constant* accelerating force of gravity:

$$v = gt \tag{4-a}$$

$$v^2 = 2gd \tag{4-b}$$

$$d = \tfrac{1}{2}gt^2 \tag{4-c}$$

where v = velocity of fall at a particular instant
t = time of fall (usually expressed in seconds)
d = distance of fall
g = acceleration imparted by the force called gravity. The numerical value assigned to g will depend on the unit of distance employed.

* Strictly speaking the accelerating force increases slightly as a body comes nearer the center of the earth. But over short distances of fall g is regarded as constant.

SUMMARY OF ASSUMPTIONS: EQUATIONS DESCRIBING FREE FALL

A great deal of stress has been placed on the fact that a falling object's velocity changes *every instant*. The letter v distinguishes the instantaneous value of the continuously changing velocity of a falling body, from the constant rate r in the equation $d = rt$. While both have a linear dimension divided by time, it is wise to distinguish them.

Each of the three equations just discussed (4-a to 4-c) assumes that the falling body starts from rest. If the body is already in motion, different equations have to be used.

Free fall is assumed; each equation would be strictly valid only for fall in a vacuum. But falling objects must push aside millions and millions of tiny air molecules. Even though these are individually very minute, they are so numerous that a measurable force is required to move them. Hence their effect is to retard the rate of fall. Over short distances the effects of air resistance are negligible, but the continuing acceleration of a body is accompanied by an increasing number of collisions with air molecules. The resisting force which results from these collisions gradually reduces the acceleration and eventually prevents further velocity increase; a constant (terminal) velocity is finally reached. The falling object is then traveling at a very great velocity, but there is no further *increase* in this velocity. Problems in this text will assume that air resistance is negligible.

Table 4–1 Freely Falling Bodies Starting from Rest and Undergoing Acceleration of 32 ft/sec²

TIME OF FALL (SEC)	INSTANTANEOUS VELOCITY AT END OF SECOND (FT/SEC)	AVERAGE VELOCITY DURING THE LAST SECOND (TREATED AS UNIFORM) (FT/SEC)	DISTANCE FALLEN DURING THE LAST SECOND (FT)	TOTAL DISTANCE FALLEN (FT)
0	0	0	0	0
1	32	$\frac{0+32}{2} = 16$	16	16
2	64	$\frac{32+64}{2} = \frac{96}{2} = 48$	48	$16 + 48 = 64$
3	96	$\frac{64+96}{2} = \frac{160}{2} = 80$	80	$64 + 80 = 144$
4	128	$\frac{96+128}{2} = \frac{224}{2} = 112$	112	$144 + 112 = 256$

The value of g is assumed to be constant. Newton's law of universal gravitation (discussed later in this chapter) suggests that it should change slightly with changes in latitude or altitude; these too will be neglected.

Table 4-1 summarizes information concerning free fall. The first column gives the time of fall, the second the instantaneous velocity after time t has elapsed. The third column, "Average Velocity During the Last Second," is the sum of the velocities at the beginning and end of the time interval, divided by 2. The average velocity, so calculated, is treated as a uniform speed r (in $d = rt$) during 1 second of fall (either the first, second, third, etc.). The distances in the fourth column are obtained through multiplication of the average velocity by a time of 1 sec. If the average velocity during the third second is 80 ft/sec, the distance fallen is 80 feet. The values of the distances in column four, in feet, are then numerically equal to the average velocities of column three, in ft/sec.

GRAVITATIONAL ACCELERATION AND GEOGRAPHICAL LOCATION

During the latter half of the seventeenth century the Dutch scientist Christian Huygens (1629-1695) made a study of the pendulum (see Chapter 5) and discovered that a pendulum just over 39 inches long was a useful device for keeping a clock ticking off the seconds uniformly. In the vicinity of Paris at least, a complete swing required 1 second; so it was called a seconds pendulum.

The French astronomer Richer was planning a trip to French Guiana in 1671 to observe an eclipse. Huygens urged Richer to take along a seconds

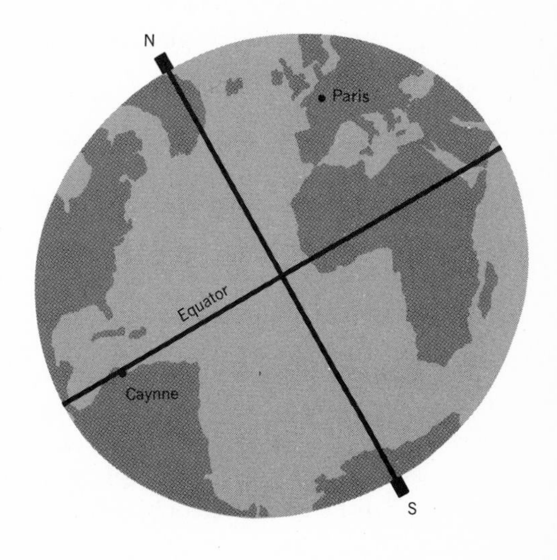

Fig. 4-2. *The earth is a slightly flattened sphere (greatly exaggerated as shown). As we move from the equator along the surface of the earth, our distance from the center of the earth decreases slightly. A given object will weigh slightly more at Paris than at Cayenne.*

pendulum. Apparently Huygens suspected that the period might be altered near the equator. Sure enough, Richer discovered that in the city of Cayenne a complete swing took longer than 1 second. How was this curious change in period to be explained?

The earth is a slightly flattened sphere (or according to recent evidence, very, very slightly pear-shaped); it bulges just a bit at the equator (Fig. 4-2). The North Pole is nearer the center of the earth than the equator. The pendulum was slightly farther from the earth's center in French Guiana than in Paris; when the pendulum was pulled aside, the gravitational restoring force was smaller in French Guiana. It fell more slowly; more time was required for a complete swing. The weight of the pendulum was less, but its mass was unchanged. This was one of the earliest hints of a distinction between mass and weight.

NEWTON'S LAW OF UNIVERSAL GRAVITATION

Every body in the universe attracts every other with a force which is directly proportional to the product of their masses and inversely proportional to the square of the distance between them. This is a statement of Newton's law of universal gravitation. It may be expressed mathematically by the equation

$$F = G\frac{m_1 m_2}{d^2} \qquad \text{(4-d)}$$

where F = force of attraction
G = gravitational constant (not the same as g)
m_1 and m_2 = masses of the two bodies
d = distance between (centers of gravity of) the bodies.

Equation 4-d has some interesting mathematical implications. First of all, if the distance between two bodies is constant, the force of attraction is directly proportional to the product of the masses. Masses of 100 kg and 200 kg, 10 meters apart, will attract four times as strongly as 50-kg and 100-kg masses the same distance apart (see Fig. 4-3). There is a direct proportionality between the force of attraction and the product of the masses.

Likewise, the force of attraction between any two given masses is inversely proportional to the square of the distance between centers. Suppose 50-kg and 100-kg masses 10 meters apart are moved so as to be 20 meters apart (see Fig. 4-4). The distance has been doubled, so the force of attraction is one fourth as large. Note that d is squared in Equation 4-d. If the distance between objects is increased three times, the force of attraction

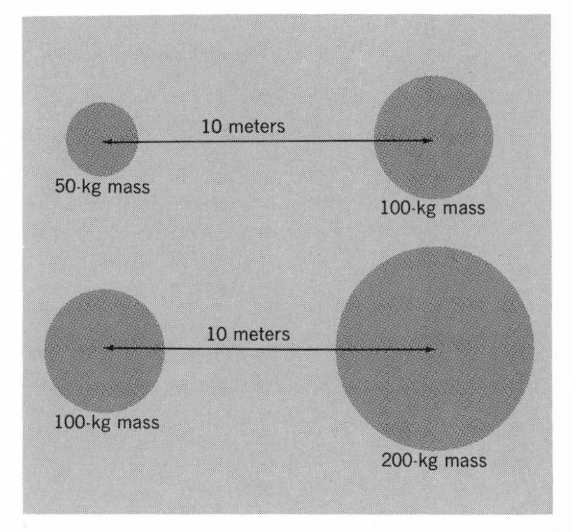

Fig. 4-3. *The forces of attraction between spheres the same distance apart are proportional to the product of their masses. The force of attraction of the lower pair of spheres is four times that of the upper pair.*

is one ninth as large. There is an inverse square relation between force of attraction and the distance between the masses.

G, the gravitational constant is what is called in mathematics a proportionality factor; it is equal to 6.67×10^{-8} dyne cm^2/g^2. It should not be confused with g, the acceleration imparted by the force of gravity on the earth's surface. G is not confined to the earth but is universal. It is also called the Newtonian constant, but its value was first actually measured in 1798 by Henry Cavendish, who was able to obtain a surprisingly accurate value of this constant by measuring in the laboratory the force of attraction between two metal spheres.

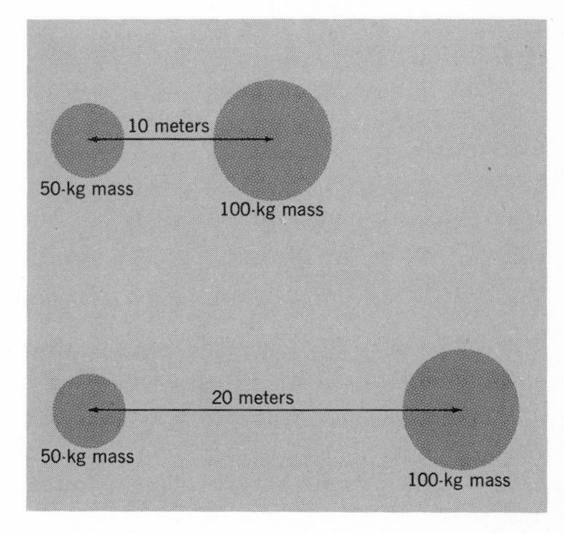

Fig. 4-4. *The forces of attraction between masses are inversely proportional to the squares of the distance separating them. The force of attraction between the upper pair of masses is four times that between the lower pair.*

ILLUSTRATIVE PROBLEM

Calculate the force of attraction between two 1-gram masses whose distance apart is 1 centimeter.

Solution

If the numerical value of G were omitted, it would appear that the force of attraction was 1 dyne.

$$F\,(\text{dynes}) = \frac{\text{dyne cm}^2}{\text{g}^2} \cdot \frac{1\text{ g} \cdot 1\text{ g}}{(1\text{ cm})^2} = 1\text{ dyne}$$

Experimental measurement of the force between these two masses actually gives a result of 6.67×10^{-8} dyne rather than 1 dyne. Introduction of this factor then brings the mathematical equation into accord with experimental results.

$$F\,(\text{dynes}) = 6.67 \times 10^{-8}\,\frac{\text{dyne cm}^2}{\text{g}^2} \cdot \frac{1\text{ g} \cdot 1\text{ g}}{(1\text{ cm})^2}$$

$$= 6.67 \times 10^{-8}\text{ dyne}$$

ILLUSTRATIVE PROBLEM

Two spheres of mass 250 grams and 500 grams, are 25.0 centimeters apart. Find the force of attraction in dynes.

Solution

$$F = G\,\frac{m_1 m_2}{d^2}$$

$$F = 6.67 \times 10^{-8}\left(\frac{\text{dyne cm}^2}{\text{g}^2}\right) \cdot \frac{250\text{ g} \cdot 500\text{ g}}{(25.0)^2\text{ cm}^2}$$

$$= 6.67 \times 10^{-8} \cdot 200\text{ dyne} = 1334 \times 10^{-8}\text{ dyne, or } 1.334 \times 10^{-5}\text{ dyne}$$

Gravitational forces are obviously largest when very massive bodies are involved. Hence some of the most important applications of this principle will be in astronomy. Tidal phenomena are explainable in terms of gravitational effects.

SCIENTIFIC NOTATION FOR WRITING NUMBERS

In the above illustrative problems an exponential method of expressing very small numbers has been used. Similar exponential numbers occur frequently in scientific writing. For instance, the numerical value of G is written 6.67×10^{-8} rather than 667/10,000,000,000, saving time and economizing space on the printed page. Expressions such as the former are

Table 4–2 Powers of 10

$10^3 = 1000$	$10^{-1} = 0.1$
$10^2 = 100$	$10^{-2} = 0.01$
$10^1 = 10$	$10^{-3} = 0.001$
$10^0 = 1$	$10^{-4} = 0.0001$

said to be written in *denary notation.* Table 4-2 shows the powers of 10 from 10^3 to 10^{-4}. By analogy it is possible to write the equivalent number for any power of 10. Remember that, just as

10^3 has 3 zeros after the 1, so
10^{20} has 20 zeros after the 1; but
10^{-3} has just 2 zeros after the decimal point before the 1, and
10^{-20} has just 19 zeros after the decimal point before the 1.

Remember, too,

to multiply by 10^3, move the decimal point 3 places to the right;
to multiply by 10^{-3}, move the decimal point 3 places to the left;

dividing by 10^3 is the same as multiplying by 10^{-3} $\left(\frac{1}{10^3} = 10^{-3}\right)$;

and dividing by 10^{-3} is the same as multiplying by 10^3 $\left(\frac{1}{10^{-3}} = 10^3\right)$.

ILLUSTRATIVE PROBLEM

Express 0.0000035 in denary notation.

Solution

$$0.0000035 = 0.0000035 \times \frac{10^6}{10^6} = \frac{3.5}{10^6} = 3.5 \times 10^{-6}$$

To change 0.0000035 to 3.5 multiply by 10^6. In order to retain the same value of the fraction, also divide by 10^6. Since $10^6/10^6 = 1$ the new fraction is equivalent to the original.

ILLUSTRATIVE PROBLEM

Express 602,000,000,000,000,000,000,000 in denary notation.

Solution

The number is to be expressed as 6.02 times some power of 10. The question is, By what power of 10 must 6.02 be multiplied to obtain

602,000,000,000,000,000,000,000 (to move the decimal point 23 places to the right)? Since multiplying by 10^{23} means moving the decimal point 23 places to the right, 602,000,000,000,000,000,000,000 = 6.02×10^{23}.

ILLUSTRATIVE PROBLEM

Multiply 6.25×10^{-27} by 8.40×10^{8}.

Solution

$$6.25 \times 10^{-27} \times 8.40 \times 10^{8} = 6.25 \times 8.40 \times 10^{-27} \times 10^{8}$$

$$6.25 \times 8.40 = 52.5$$

When 52.5 is multiplied by 10^{-27} the decimal point is moved 27 places to the left; when this is multiplied by 10^{8} the decimal point is moved back 8 places to the right; so the decimal point will end up 19 places to the left, and the answer will be 52.5×10^{-19} or 5.25×10^{-18}.

ILLUSTRATIVE PROBLEM

Divide 8.1×10^{15} by 2.7×10^{-4}

Solution

$$\frac{8.1 \times 10^{15}}{2.7 \times 10^{-4}} = \frac{8.1}{2.7} \times \frac{10^{15}}{10^{-4}} = 3.0 \times 10^{15} \times 10^{4} = 3.0 \times 10^{19}.$$

Problems, Chapter 4

1. (a) In the equation $v = gt$, in what units should v be expressed if g is 32 ft/sec^2 and t is expressed in seconds?
   (b) In the equation $v^2 = 2gd$, if v is expressed in m/sec, in what units should g and d be expressed?
   (c) In the equation $d = \frac{1}{2}gt^2$, if g is expressed as 9.8 m/sec^2 and t is expressed in seconds, in what units should d be expressed?

2. An object is dropped from a balloon which is at a height of 1600 ft. Calculate its velocity (a) at the end of 5 sec; (b) at the instant it strikes the ground.

3. Express the following in denary notation.
   (a) 2,500,000
   (b) 0.0000000137

4. Multiply 1.3×10^{7} by 3.8×10^{12}.

5. Divide 3.9×10^{23} by 2.6×10^{15}.

6. How many molecules are there in a pound of hydrogen gas if 2.02 g contain 6.02×10^{23} molecules?

7. The average diameter of a red blood corpuscle is 8.00×10^{-5} cm. Calculate its area of cross section. ($A = \pi r^2$; π is approximately 3.14.)

8. The charge on 6.02×10^{23} electrons is 96,500 coulombs. How many electrons constitute 1 coulomb of charge?

9. Two spheres, each of mass 8.00 kg, have their centers 40.0 cm apart. Calculate the attractive force between them in dynes.

10. A stone is thrown vertically upward with an initial velocity of 120 ft/sec. To what height does it rise? (*Hint:* Gravity decreases the upward velocity 32 ft/sec each second the stone rises.)

11. What is the height of the Tokyo tower if a stone dropped from its top reaches the ground in 8.27 sec? How fast is the stone falling at the instant it strikes the ground?

12. Calculate the value of gravitational acceleration in inches/sec^2.

13. An object falls a distance of 200 meters. In how many seconds does it strike the ground?

14. How far has an object fallen if its instantaneous velocity is 192 ft/sec when it strikes the ground?

Exercises

1. Why do we refer to gravity as a force?

2. The equation $v = gt$ was deduced from Galileo's experiment of rolling marbles down an inclined plane. Explain.

3. In this chapter the equation for linear motion: $d = rt$ was used in deriving the equation $d = gt/2 \cdot t = \frac{1}{2}gt^2$ for the distance fallen by an object under the accelerating influence of the force called gravity. Explain why the average rate of fall may be expressed by $gt/2$. Assume the object starts from rest.

4. State Newton's law of universal gravitation.

5. Gravitational acceleration varies with geographical location. Explain this in terms of the law of gravitation.

6. Explain the difference between g in the equation for the motion of falling bodies and G in the mathematical expression of Newton's law of universal gravitation.

7. Two objects, one of 4-lb and the other of 16-lb mass, are allowed to fall freely. Each is accelerated 32 ft/sec/sec. Compare the gravitational forces on these objects.

5 Force and work: The study of energy

WORK AND ENERGY

The physicist and the layman often have quite different conceptions of what constitutes work. The layman usually considers any exertion involving a push or pull to be work; the physicist demands that the push or pull shall cause an object to move. You will doubtless believe you are doing work if you have to wait at the door of your home with a heavy package of groceries in your arms. The physicist denies it unless you raise the package or otherwise exert a force through a distance.

Energy is formally defined as the ability to do work. In order to recognize forms of energy, ask yourself, "What force is acting and through what distance?"

If work is done, a force must act through a linear distance, *the direction of the two being the same.* This proviso is important, for in the discussion of the principle of moments (Chapter 2) the line of action of the force was at right angles to the measured distance. A state of balance existed; the force did not cause motion. If the force was increased slightly so as to cause rotation (unbalance the forces), the lever would move. The resulting work would then be the product of the force by the distance moved. Distinguish between a force holding another in balance and one causing the motion of an object. One is a static situation; the other, dynamic.

In discussing the inclined plane (Chapter 2) a *force* was exerted upon some object, causing it to move through a *distance.* Work was done. In the case of the jackscrew a *force* exerted at the end of a handle was multiplied by the *distance* it moved (the circumference of a circle). In the next chapter there will be discussion of a gas at high pressure confined in a cylinder by a movable piston. Work is done if the piston moves out against a smaller external pressure.

Potential Energy Some objects have the ability to do work by virtue of their relative position or state of strain. The book on the table has potential energy because of its elevated position; it can fall to the floor, exert a force on an object, and make it move. A wound-up watchspring possesses potential energy; as the spring unwinds, it can drive the mechanism of a watch.

Fig. 5-1. *A pile driver. When the elevated weight falls and strikes an object such as a wooden post (pile), a* force *is exerted on it and it is driven into the ground (distance).*

In engineering construction a device called a pile driver (Fig. 5-1) is often used. It consists of a huge weight and an engine for raising it. When the elevated weight is allowed to fall, a *force* is exerted on a vertical post (pile); this force drives the post a certain *distance* into the ground. Because of its elevated position the weight has potential energy; if released it can cause a force to act through a distance, i.e., to do work. The amount of work it can do, depends on its weight and the distance it may fall:

$$\text{Potential energy} = \text{weight} \times \text{height}$$

The heavier the weight and the greater its possible distance of fall, the larger its potential energy.

ILLUSTRATIVE PROBLEM

The 500-lb weight of a pile driver is 20 ft above the ground. Calculate its potential energy.

Solution

Potential energy = weight × height = 500 lb × 20 ft = 10,000 ft lb

A car parked on a hill has potential energy. The farther up the hill it is and the greater its weight, the greater is its potential energy.

Table 5–1 Potential Energy: P.E. $= w \times h$

POTENTIAL ENERGY WILL BE EXPRESSED IN:	IF UNITS OF WEIGHT ARE:	AND UNITS OF HEIGHT ARE:
	Absolute Units	
foot poundals	poundals	feet
dyne centimeters (ergs)	dynes	centimeters
newton meters (joules)	newtons	meters
	Gravitational Units	
foot pounds	pounds	feet
gram centimeters	grams	centimeters
kilogram meters	kilograms	meters

Kinetic Energy Just as some bodies at rest have power to do work—potential energy—so moving objects also have the ability to do work, i.e., have energy. They may strike other objects, exert a *force* on them, and cause them to move through a *distance*. The energy an object possesses by virtue of its motion is called *kinetic energy*.

What factors need be taken into account in determining the amount of kinetic energy? Potential energy was measured by the amount of work done in raising an object to its elevated position, the product of its weight (a force) and height (a distance). To calculate the kinetic energy, find the work needed to impart to a mass m a velocity v. A constant force acting on a body initially at rest sets it in motion and increases its velocity.

Gravity causes a body starting from rest and falling freely (Chapter 4) to attain an instantaneous velocity (squared) of $v^2 = 2gd$ after falling a distance d. A more general equation for *any* acceleration is $v^2 = 2ad$. This equation differs from the preceding in not being restricted solely to gravitational acceleration.

Suppose both members of the equation for Newton's second law, $F = ma$, are multiplied by d; then $F \times d = m \times ad$. On substituting $v^2/2 = ad$ in the latter equation, the work done in accelerating mass m to velocity v is found to be

$$F \times d = m\frac{v^2}{2} = \frac{1}{2}mv^2$$

This is the equation for kinetic energy:

$$\text{Kinetic energy} = \tfrac{1}{2}mv^2$$

The amount of kinetic energy which a moving object possesses, then, depends on two factors, its mass and its velocity. A moving heavy truck has more kinetic energy than a lighter automobile traveling at the same velocity. An automobile moving rapidly may exert a larger force on an object in its path than if it is moving slowly. Doubling the velocity of an object will change its kinetic energy more than will doubling its mass; velocity is squared in the expression for calculating kinetic energy; mass is not.

ILLUSTRATIVE PROBLEM

A 100-lb *mass* moves with a velocity of 12 ft/sec. Calculate its kinetic energy.

Solution

$$\text{K.E.} = \frac{1}{2}mv^2 = \frac{1}{2} \cdot 100\ \text{lb} \cdot 12^2\ \frac{\text{ft}^2}{\text{sec}^2}$$

$$= \frac{1}{2} \cdot 100\ \text{lb} \cdot 144\ \frac{\text{ft}^2}{\text{sec}^2}$$

$$= 7200\ \text{foot poundals (see Table 5-2)}$$

Table 5-2 gives the units of kinetic energy which are consistent with given units of mass and velocity.

The Pendulum The pendulum clock has been a familiar piece of furniture in New England homesteads for generations. The pendulum of the physics laboratory lacks the charm and grandeur of one of these old masterpieces—it is often merely a weight suspended from a string—but it is most useful for illustrating the relationship between potential and kinetic energy.

A pendulum is set in motion by pulling it aside from the vertical (Fig. 5-2). The sidewise pull also lifts it; work is done on the bob. The amount of work is the weight of the bob times the vertical distance of elevation

Table 5–2 Kinetic Energy: K.E. = $\frac{1}{2}mv^2$

KINETIC ENERGY WILL BE EXPRESSED IN:	IF UNITS OF MASS ARE:	AND UNITS OF VELOCITY ARE:
foot poundals	pounds	feet/sec
dyne centimeters (ergs)	grams	centimeters/sec
newton meters (joules)	kilograms	meters/sec
foot pounds	$\frac{\text{pounds weight}}{g\ (\text{ft/sec}^2)}$	feet/sec
gram centimeters	$\frac{\text{grams weight}}{g\ (\text{cm/sec}^2)}$	centimeters/sec
kilogram meters	$\frac{\text{kilograms weight}}{g\ (\text{m/sec}^2)}$	meters/sec

(height). Its potential energy then continually changes with its height above the low point of the swing.

Releasing the elevated bob subjects it to the accelerating force of gravity. Its velocity increases as it continues to fall. At the lowest point of the swing it is moving with maximum velocity; inertia keeps it moving

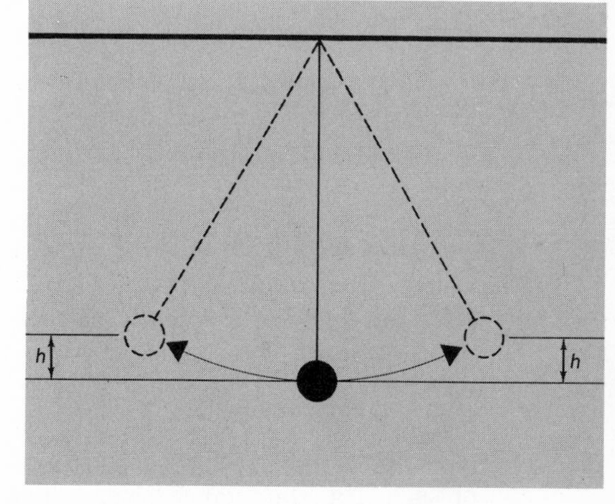

Fig. 5-2. *If a pendulum bob is pulled aside to a height* h *and then released, it cannot reach a height greater than* h *on the opposite side.*

and it rises with decreasing velocity as gravity opposes its motion on the upswing. Finally it comes to rest again on the far side and then the cycle is repeated.

At the ends of the swing the bob has no kinetic energy because it is momentarily at rest; its velocity is zero. At the bottom of the swing the bob has its maximum kinetic energy because there its velocity is greatest.

The potential energy is greatest at the extreme ends of the swing where the bob reaches its greatest height. The potential energy decreases as the bob falls and becomes zero at the low point of the swing where the height of the bob is zero.

The potential energy of the bob is at a maximum, then, when its kinetic energy is zero, and vice versa. As the bob falls and loses potential energy, there is a corresponding increase in the amount of its kinetic energy. The potential energy at the top of the swing is equal to the kinetic energy at the bottom, and at any intermediate position the sum of the two kinds of energy is equal to either the maximum potential or maximum kinetic energy.

ILLUSTRATIVE PROBLEM

Calculate the maximum kinetic energy and maximum potential energy of a pendulum bob having a mass of 3 lb if at the extreme of its swing its height is 2 ft above the lowest point.

Solution

Max P.E. $= wh = mgh =$ 3.00 lb · 32.0 ft/sec^2 · 2.00 ft = 192 ft poundals
Max K.E. $= \frac{1}{2}mv^2$
For a falling body: $v^2 = 2gd =$ 64.0 ft/sec^2 · 2.00 ft = 128 ft^2/sec^2
K.E. $= \frac{1}{2} \cdot$ 3.00 lb · 128 ft^2/sec^2 = 192 ft poundals

The maximum potential energy of the bob is attained at the highest point, the maximum kinetic energy at the lowest point of the swing. Effects of friction and air resistance are assumed to be negligible.

Other Forms of Energy Potential and kinetic energy together constitute what is called mechanical energy. But there are many other forms of energy. As additional forms—heat, light, electrical energy—are considered, the question frequently arises, What force is caused to act through what distance? Why is heat, for example, a form of energy? The inflammability of gasoline is well known. It is jolly well advisable to keep sparks or flames away from mixtures of gasoline vapor and air because of the tremendous force which may be released. However, when combustion of such a mixture takes place within a confined space like the cylinder of a gasoline engine, millions and millions of submicroscopic gas molecules are hurled at the piston with high velocity, and useful work is done. In one way or another, modern transportation depends largely on heat energy. As a result of the motion of enormous numbers of tiny but very rapidly moving particles (heat), a force has acted through a distance, work has been done. Heat is, then, a form of energy.

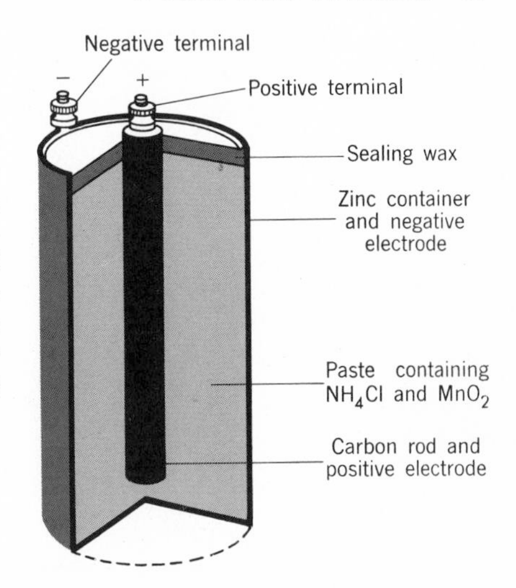

Fig. 5-3. *Cross section of a dry cell or flashlight battery. A chemical reaction in the cell creates a difference in electrical levels between the positive and the negative terminals. When the pole marked + is connected by a wire with the pole marked –, electrons flow from – to + and are capable of doing work.*

In the common dry cell (Fig. 5-3) a chemical reaction pushes along in a piece of wire tiny particles found in all matter, called electrons. If the wire through which they pass is the filament of a flashlight bulb, it becomes very hot (heat energy)—in fact, "white hot"—and light (another form of energy) is emitted. If the wire through which the electrons flow is placed between the poles of a magnet, the wire is subjected to a force which makes it move (Fig. 5-4). It is quite appropriate then, to refer to "chemical energy" and to "electrical energy."

A rapidly vibrating tuning fork (Fig. 5-5) alternately compresses and rarefies the millions and millions of air molecules which rebound from its surface. As the vibrating prong moves out and in, there is first a momentary increase in the density of air particles, then a decrease; a compressional wave is transmitted by the air. This sound wave may impinge on a human eardrum and cause it to move back and forth. Hence sound is a form of energy.

In 1903 Nichols and Hull carried out an experiment in which vanes

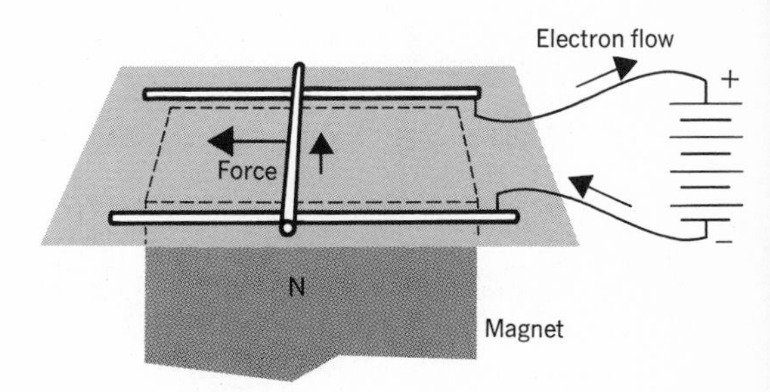

Fig. 5-4. *A metal bar through which an electric current passes experiences a sidewise thrust in a magnetic field. Work is done on it as a result of the passage of the electric current.*

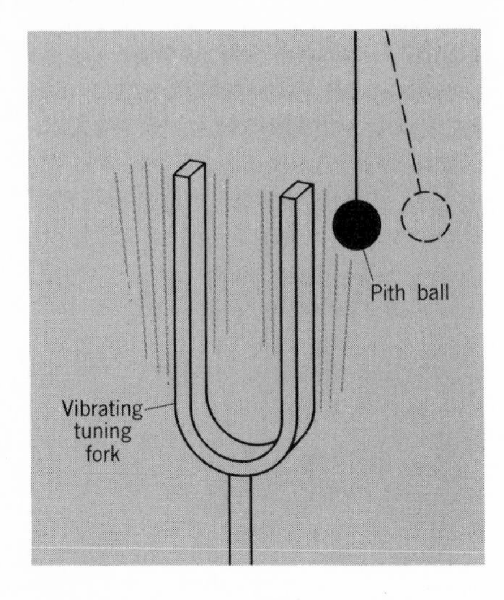

Fig. 5-5. *The pith ball shows how air molecules rebound from the prong of a tuning fork in setting up a compressional (sound) wave.*

suspended in a vacuum moved when they were struck by light waves. The experiment demonstrated that light could exert a force upon objects and move them. Light, then, is also a form of energy.

Perpetual Motion In all ages men have hoped to find something for nothing. One of the ever-alluring prospects is the possibility of building a machine that, once set running, will continue to operate without further energy being supplied to it—the perpetual-motion machine.

Efforts to devise such machines go back far in history. One would-be perpetual-motion machine was based on a kind of water pump attributed

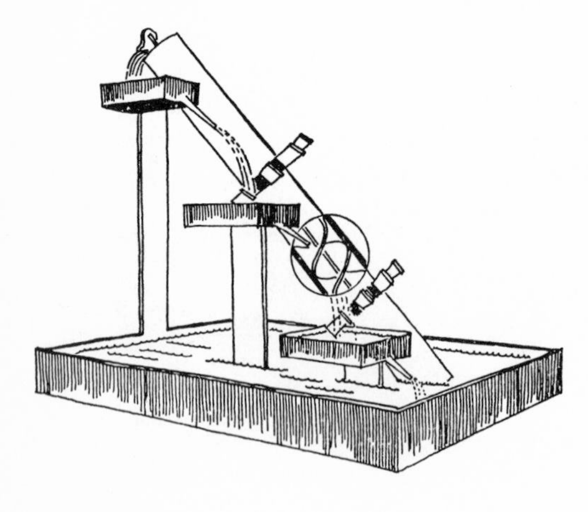

Fig. 5-6. *A would-be perpetual-motion machine based on a spiral tube used to raise water, the so-called Archimedes' screw. Water flowing out the top spills over the vanes causing the tube to turn. This pulls more water up the central spiral tube (an Archimedes' screw). Why does this model fail to operate?*

to that master designer in the ancient world, Archimedes—the so-called Archimedes' screw. This is a water-raising device still in use in the Near East. In the perpetual-motion version (Fig. 5-6) fins or vanes were rigidly attached to the exterior of the pipe. Water pumped up the tube issued from the top and thence poured over the fins. These were expected to rotate, and the pipe rotating with the fins would cause more water to issue from the top. Once started, the machine was supposed to operate by itself without any further supply of external energy. Unfortunately no model has ever worked.

In the Renaissance, Leonardo da Vinci chided those who speculated on motion for their will-o'-the-wisp perpetual-motion machines. Galileo also recognized the impossibility of constructing such machines, for he stated that a pendulum bob which was pulled aside and then released (Fig. 5-2) could not rise higher at the far end of the swing than its original height.*

Power vs. Energy *The Seven Follies of Science,* a delightful volume written many years ago by Phin† tells of a nineteenth-century inventor who thought he had at long last devised a machine which would turn out more work than the energy supplied to it. He knew that carbon disulfide (CS_2) was a highly volatile liquid. It readily changed into a vapor when heated. When the liquid was injected into the boiler of a steam engine, it vaporized quickly; many millions of tiny, rapidly moving gas molecules of CS_2 were introduced along with the steam already there. The gas pressure in the boiler increased; the engine took a new lease on life and its speed was miraculously increased. However, the ultimate energy source was the hot water (or steam) of the boiler; no new energy had been created. The inventor's exuberance turned to despair when he discovered he had only increased the rate of doing work—or the *power.*

Power is the *rate* of doing work; it depends on the time as well as the quantity of work. A common power unit is the horsepower. A 1-horsepower engine does 550 foot pounds of work in 1 second. Another power unit, the watt, is the joule/sec. Electric power is expressed in a larger unit, the kilowatt, which is 1000 times as large. A kilowatt hour is an energy unit (like an erg or joule). Electric light bills set charges on the basis of kilowatt hours used.

ILLUSTRATIVE PROBLEM

The 500-lb weight of a pile driver is lifted a vertical distance of 20 ft in 5 sec. What is the horsepower of the engine?

* Cf. Charles G. Fraser, *Half Hours with Great Scientists,* 1948.
† John Phin, *The Seven Follies of Science,* Van Nostrand, 1912, p. 66.

Solution

$$500 \text{ lb} \cdot 20 \text{ ft}/5 \text{ sec} = 10{,}000 \text{ ft lb}/5 \text{ sec} = 2000 \text{ ft lb/sec}$$
$$550 \text{ ft lb/sec} = 1 \text{ horsepower}$$
$$2000 \text{ ft lb/sec} = \frac{2000}{550} \text{ horsepower} = 3.64 \text{ horsepower}$$

TEMPERATURE AND THERMOMETERS

The English philosopher John Locke (1632-1704), a contemporary of Isaac Newton, suggested an experiment which demonstrates how unreliable the physical senses are in judging temperature, the intensity factor of heat. In the experiment a person plunges his hand into a basin of ice water and then into another basin of water at room temperature. He reports that the latter is "warm." A second person thrusts his hand first into a basin of hot water and then into the basin of water at room temperature. He reports that the latter is "cold." Each person makes a different judgment concerning the temperature of the same bucket of water. Apparently the human sensations of hot and cold are unreliable as temperature indicators.

Even before John Locke's time the need for temperature-recording devices had been recognized. Galileo's "thermoscope" (Fig. 5-7), invented near the close of the sixteenth century, was actually a combination of barometer and thermometer. When the bulb was warmed, the air inside expanded and pushed the liquid down the tube, thus indicating a temperature increase. However, since changes of air pressure also effected changes in the liquid level, the thermoscope was hardly a reliable temperature-measuring instrument.

Early in the eighteenth century the German physicist Fahrenheit (1686-1736) devised a thermometer consisting of a glass tube of small bore

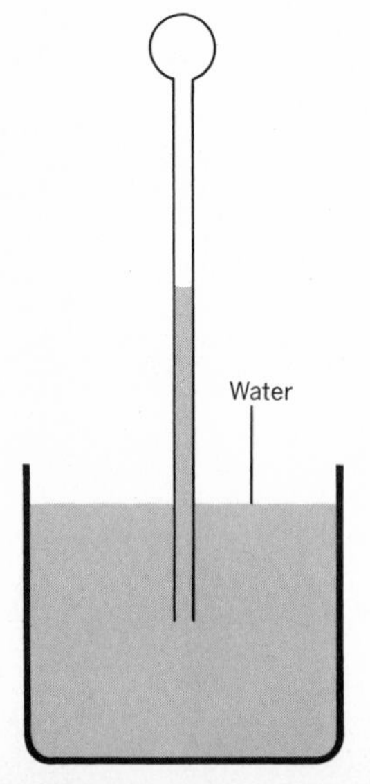

Fig. 5-7. *Galileo's thermoscope. One of the earliest temperature-measuring devices. Unfortunately the water level of this instrument responded to both temperature and pressure changes.*

(capillary) which contained a column of mercury as the working liquid. He selected as the 32-degree mark the freezing point of water and as the 212-degree mark its boiling point. This scale of 180 degrees between the freezing and boiling points was a bit less convenient than the scale selected somewhat later by Celsius, a professor at the University of Uppsala in Sweden. Celsius chose the freezing point of water as the 100-degree mark, and its boiling point at zero. The modern Celsius scale reverses these, putting the freezing point at zero degrees and the boiling point at 100 degrees (Fig. 5-8).

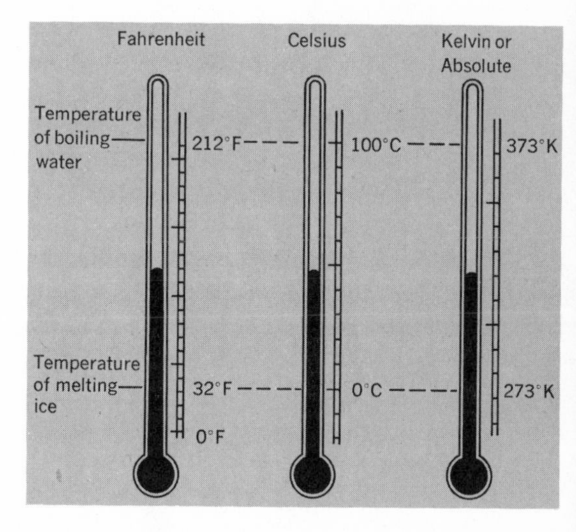

Fig. 5-8. *Common temperature scales. The Fahrenheit scale is familiar in the United States because of its everyday use. The Celsius and Kelvin scales are almost universally used in scientific work. Zero degrees Kelvin (absolute zero) is −273°C.*

The absolute or Kelvin scale selects as zero the temperature at which molecular motion is presumed to cease. It may be established by plotting the volumes of a gas against the corresponding temperatures (Fig. 5-9). The straight line so obtained is extended (extrapolated) beyond measurable temperatures, to the point at which the volume would be zero. The corresponding temperature is designated "absolute zero."

On the absolute (Kelvin) scale the freezing point of water is 273 degrees and the boiling point (at 1 atmosphere pressure) 373 degrees. "Absolute zero" is then 273 Celsius degrees* below the freezing point of water—i.e., −273°C.

Fixed Points of a Thermometer If a thermometer at room temperature is placed in a beaker containing both ice and water (Fig. 5-10), its mer-

* The exact value is 273.15°C below the freezing point of water or 273.16°C below the "triple point," where ice, liquid water, and water vapor are in equilibrium. Celsius degrees were formerly called centigrade degrees.

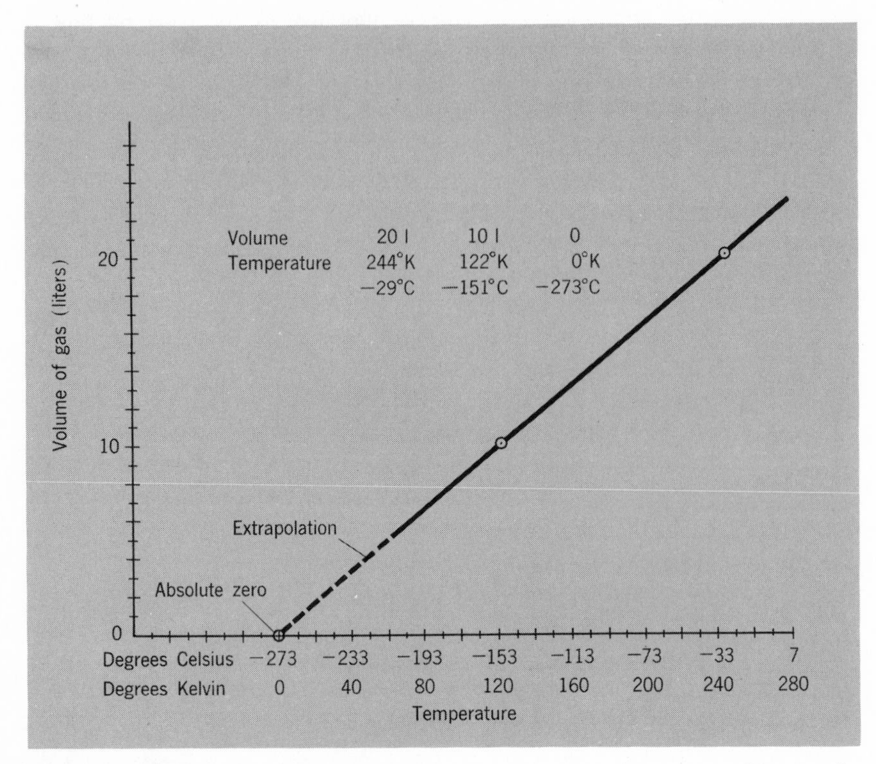

Fig. 5-9. *Absolute zero, theoretically the lowest temperature attainable, can be evaluated from a plot of gas volumes vs. Celsius temperatures. If this linear plot is extended (extrapolated) to zero volume, the corresponding temperature is absolute zero. It is 273.15 degrees below zero degrees Celsius. Temperatures very near this theoretical limit have been reached.*

cury column shortens as its temperature falls to that of the surroundings. Soon the length of the mercury column becomes constant. The position of the top of the mercury column on the scale represents the temperature called the freezing point of water. This temperature is designated 0° on a Celsius thermometer or 32° on a Fahrenheit thermometer.

If the thermometer is supported in a boiler (Fig. 5-11) so as to be bathed in steam (under a barometric pressure of 1 atmosphere), the mercury column lengthens and eventually reaches a fixed position farther up on the scale. This point, called the boiling point of water, is designated 100° on a Celsius thermometer or 212° on a Fahrenheit thermometer.

Two points have thus been established on each thermometric scale; they are called "fixed points." The distance between these fixed points is divided into 100 equal divisions on the Celsius scale and 180 on the

Fig. 5-10. *The freezing temperature of water provides a fixed point on a thermometer, zero degrees Celsius. It may be established by immersing a thermometer in ice and water.*

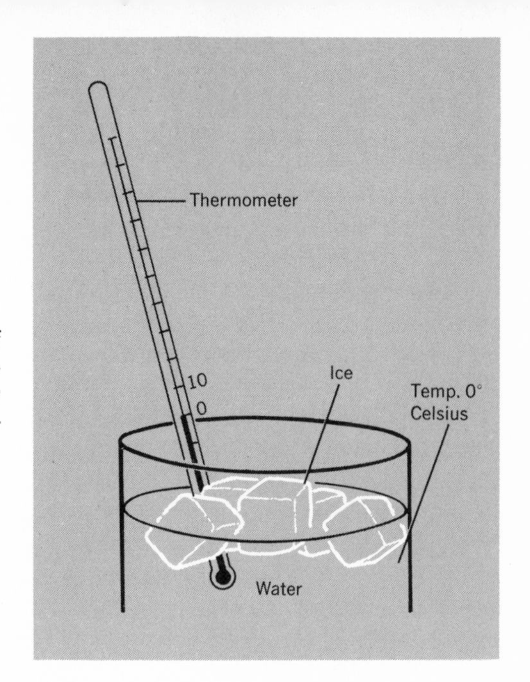

Fahrenheit scale. Thus a single division (degree) on the Celsius scale is equal to 180/100 or 1.8 Fahrenheit degrees.

The equation to convert a Celsius temperature into its corresponding value on the Fahrenheit scale is $F = \frac{9}{5}C + 32$, where F and C represent the Fahrenheit and Celsius temperatures. Note that F will be 32° if C is assigned a value of 0° (the freezing temperature of water).

Fig. 5-11. *The boiling temperature of water may be used to determine a fixed point on a thermometer scale. If the barometric pressure is 760 mm of mercury, the temperature in freely escaping steam will be 100°C.*

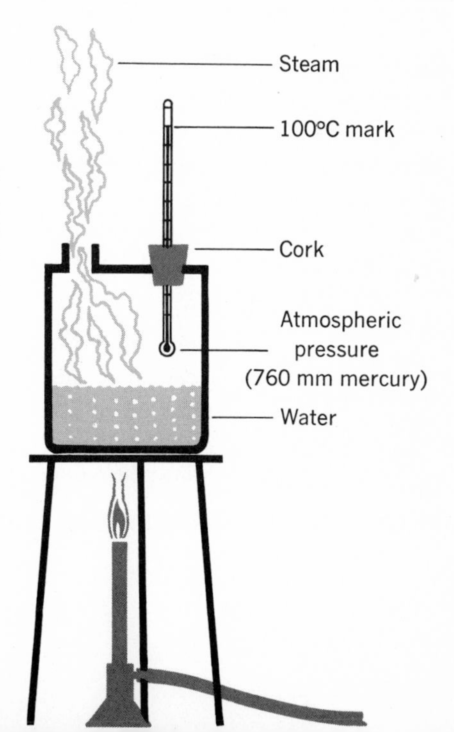

ILLUSTRATIVE PROBLEM

What Fahrenheit temperature corresponds to 20° Celsius?

Solution

$$F = \tfrac{9}{5}C + 32$$

$$F = \tfrac{9}{5}\cdot 20 + 32 = 68^\circ$$

In order to change Celsius to absolute temperatures (degrees Kelvin) simply add 273 to the Celsius temperature. Thus 25°C = 298°K and −50°C = 223°K.

THE MECHANICAL EQUIVALENT OF HEAT

Once a well-defined concept of energy had been established and different forms of energy had been recognized, it was to be expected that sooner or later some investigator would inquire into the equivalency between forms. The mechanical equivalent of heat is such an equivalency; it is the amount of work which must be expended to produce a given quantity of heat.

In 1840 the German physician Robert von Mayer (1814-1878) traveled to Java as a ship's surgeon. Apparently he had a good deal of free time to ponder this question for shortly after his return he predicted on the basis of purely theoretical considerations not checked by experiment a value of the mechanical equivalent. The accuracy of Mayer's prediction was amply confirmed a year later when the British physicist James Prescott Joule (1818-1889) carried out his famous *laboratory* measurement of the mechanical equivalent.

Mayer had predicted theoretically and Joule had proved experimentally that 778 * foot pounds of work must be expended in order to produce 1 Btu (British thermal unit) of heat.

QUANTITY OF HEAT

The Btu—British thermal unit—just mentioned is the quantity of heat needed to change the temperature of 1 pound of water 1 Fahrenheit degree. In order to calculate the number of Btu transferred to or from water, multiply the mass of the water in pounds by the temperature change in Fahrenheit degrees. Ten Btu are needed to raise the temperature of 2 pounds of water 5 Fahrenheit degrees.

* The original value proposed was actually smaller; this is a corrected modern value.

The *calorie* is a much smaller unit; it is the quantity of heat needed to change the temperature of 1 gram of water 1 Celsius degree. The calorie is more commonly used in scientific work than the Btu (Fig. 5-12).

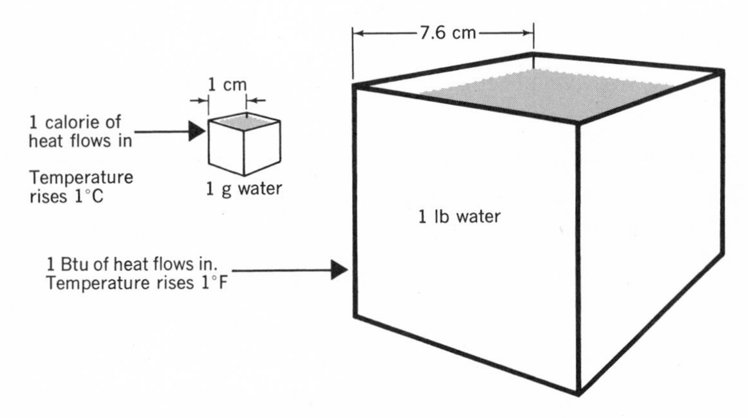

Fig. 5-12. *The relation between a calorie of heat and a British thermal unit. A pound cube of water weighs 454 g. But the Fahrenheit degree is 5/9 of the Celsius degree.*

There is still a third unit of heat also called the Calorie. The dietician uses it in measuring the energy values derived from foods. The Calorie is 1000 times as large as the calorie and is called the kilogram-calorie, or the *large* calorie. It is written with a capital letter (Calorie) in order to distinguish it from the smaller calorie.

One (small) calorie raises the temperature of 1 gram of water 1 Celsius degree. Ten calories change the temperature of 10 grams of water 1 Celsius degree. One hundred calories are needed to raise the temperature of 10 grams of water 10 Celsius degrees. One calorie is needed, then, for each gram of water for each Celsius degree of temperature rise.

Persons who have lived near the ocean or other large body of water know of the unusually high heat-absorbing capacity of water. In general larger quantities of heat are required to change the temperature of water than are required for equal temperature changes in the same masses of other materials—for example, aluminum or lead. Nearly five times as much heat is needed for a 1-degree temperature rise in a gram of water as for the same temperature increase in a gram of aluminum. Stated another way, aluminum requires 22/100 as much heat per gram per Celsius degree as does water. This fraction of the heat required per gram of material per degree, as compared with water, is called the *specific heat.* Aluminum has a specific heat of 0.22.

The quantity of heat to change the temperature of water is calculated by multiplying the mass of the water by the number of degrees temperature change, as illustrated above. For *any other material* such as aluminum, lead, or copper, this product must also be multiplied by the specific heat of the material. The general expression for calculating the quantity of heat is

$$(\text{Mass of substance}) \cdot (\text{temperature change}) \cdot (\text{specific heat})$$

This product is in calories if the mass is in grams and temperature change in Celsius degrees; it is in British thermal units if the mass is in pounds and temperature change in Fahrenheit degrees. The same value of the third factor—specific heat—is used with either calories or Btu. When dealing with temperature changes in water, the specific heat of water can be omitted, since it is 1.

ILLUSTRATIVE PROBLEM

Calculate the number of calories of heat absorbed by 10 g of lead if its temperature is increased from 20°C to 40°C. The specific heat of lead is 0.030.

Solution

$$10\ \text{g} \cdot (40°\text{C} - 20°\text{C}) \cdot 0.030 = \text{heat absorbed}$$

$$10\ \text{g} \cdot 20°\text{C} \cdot 0.030 = 6.0\ \text{calories}$$

Temperature is often described as a measure of *heat* intensity. The calorie is a measure of *heat quantity*. Heat quantity depends on the mass of material considered, temperature does not. Consider 10-g and 100-g masses of water, both at 90°C. If they cool to 80°C the former gives up 100 calories, the latter 1000 calories.

MEASURING HEAT TRANSFER: THE METHOD OF MIXTURES

Consider now the problem of how to measure experimentally a quantity of heat. The operation is usually carried out in a shiny metal cup called a calorimeter (Fig. 5-13). It contains water of known mass m_1 at temperature T_1. An object such as a brass cylinder, mass m_2, which has come to thermal equilibrium in a high-temperature bath (as for example, a steam bath at 100°C), is quickly transferred from the steam bath and immersed in the water of the calorimeter. The water is stirred and the highest temperature recorded. This temperature must be intermediate between T_1 and 100°.

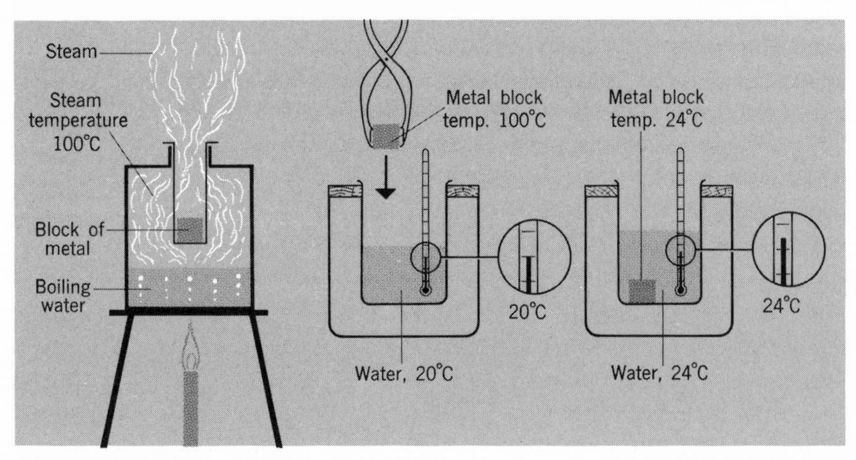

Fig. 5-13. *The method of mixtures. A block of metal of known mass is heated in freely escaping steam until its temperature is 100°C throughout. It is then quickly immersed in a known mass of water contained in a calorimeter, where the temperature increase is determined.*

When objects of different temperature are in thermal contact, heat flows from the warmer to the colder until both are at the same temperature, T_2. The basic assumption of the method of mixtures is the equality of the amounts of heat transferred: all of the heat lost by the warmer object in cooling is gained by the colder object in warming up to the final common temperature. The following equality expresses this basic assumption:

$$\begin{matrix}\text{Quantity of heat lost} \\ \text{by hot object (or objects)}\end{matrix} = \begin{matrix}\text{quantity of heat absorbed} \\ \text{by cold object (or objects)}\end{matrix}$$

The quantity of heat transferred (liberated or absorbed) depends on the three factors discussed earlier—the mass of the material, its specific heat, and the temperature change. The above equation may then be written

$$\begin{pmatrix}\text{Mass of} \\ \text{substance,} \\ \text{metal} \\ \text{cylinder}\end{pmatrix} \cdot \begin{pmatrix}\text{temperature} \\ \text{change}\end{pmatrix} \cdot \begin{pmatrix}\text{specific} \\ \text{heat}\end{pmatrix} = \begin{pmatrix}\text{mass of} \\ \text{substance,} \\ \text{water}\end{pmatrix} \cdot \begin{pmatrix}\text{temperature} \\ \text{change}\end{pmatrix}$$

$$m_2 \cdot (100 - T_2) \cdot \text{S.H.} = m_1 \cdot (T_2 - T_1)$$

ILLUSTRATIVE PROBLEM

Calculate the specific heat of a metal if 180 g at 100°C cool to 24.0°C when plunged into 250 g of water initially at 20.0°C.

Solution

Since 24.0°C is the final common temperature, the change in the temperature of the water is 24.0°C − 20.0°C = 4.0C°
By substituting the given values in the equation above:

$$(180\ \text{g})\cdot(76\ \text{C}^\circ)\cdot(\text{S.H.}) = (250\ \text{g})\cdot(4.0\ \text{C}^\circ)$$

$$\text{S.H.} = \frac{250\cdot 4.0}{180\cdot 76} = \frac{25}{342} = 0.073$$

The heat absorbed by the calorimeter cup and stirrer has been neglected in the above problem. This quantity of heat should be taken into account in laboratory practice. It is calculated by multiplying the three factors—mass, specific heat, and temperature change—as described above. The total heat absorbed or liberated is the sum of the individual amounts involved in warming or cooling.

JOULE'S EXPERIMENTAL EVALUATION OF THE MECHANICAL EQUIVALENT OF HEAT

The Joule experiment was designed to evaluate the quantitative relationship between work expended and heat produced. The apparatus Joule used is shown in Fig. 5-14. A weight is suspended from a string which passes over a pulley and attaches to a rotating drum. The falling weight causes the drum and attached paddlewheel to rotate. The frictional effect

Fig. 5-14. *Joule's experiment for determining the mechanical equivalent of heat.*

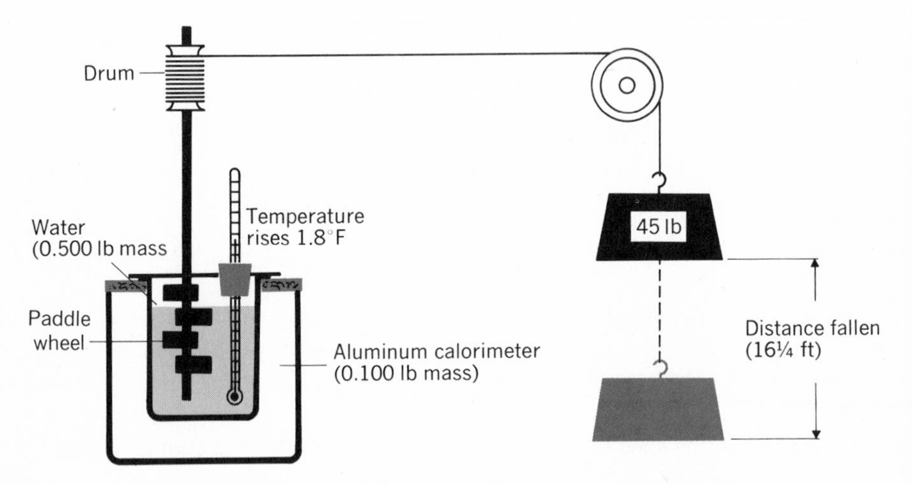

of the rotating paddlewheel produces heat, and the temperature of the water rises.

The weight has potential energy in its elevated position. As it falls, potential energy is lost and the work expended is equal to this loss (obtained through multiplication of its weight by its distance of fall). The quantity of heat produced is determined through multiplication of the mass of the water in the calorimeter by its temperature increase. In a more exact experiment, the heat absorbed by the calorimeter cup would be taken into account.

Joule spent many years in his efforts to perfect this experiment and to obtain an exact value of the equivalency which is called the mechanical equivalent of heat. He made use of the British thermal unit in his calculations.

ILLUSTRATIVE PROBLEM

A 45.0-lb weight falls 16.25 ft and causes a temperature rise of 1.80F° in an aluminum calorimeter containing 0.500 lb of water. The mass of the calorimeter is 0.100 lb, and its specific heat is 0.220. Calculate the mechanical equivalent of heat in ft lb/Btu (see Fig. 5-14).

Solution

Work done $= 16.25 \text{ ft} \cdot 45.0 \text{ lb} = 731 \text{ ft lb}$
Heat absorbed by water $= (0.500 \text{ lb}) \cdot (1.80\text{F}^\circ) = 0.900 \text{ Btu}$
Heat absorbed by calorimeter $= (0.100 \text{ lb}) \cdot (1.80\text{F}^\circ) \cdot (0.220) = 0.0396 \text{ Btu}$ (Note that this is small compared with 0.900 Btu)

$$\text{Total heat absorbed} = \text{work done}$$

$$0.940 \text{ Btu} = 731 \text{ ft lb}$$

$$1 \text{ Btu} = \frac{731}{0.940} \text{ ft lb} = 778 \text{ ft lb}$$

The mechanical equivalent of heat expressed in the metric system is 4.186 joules = 1 calorie.

CONSERVATION OF ENERGY

James Prescott Joule, whose name has figured so prominently in the story of the mechanical equivalent of heat, was a pupil of the Manchester (England) schoolmaster John Dalton, famous as a founder of the modern atomic theory of matter (Chapter 7). The flame of scientific interest appears to have been kindled early in Dalton's pupil, for Joule experimented with electric motors when he was a very young man.

In these experiments, Joule calculated the power of different electric motors by measuring the rates at which they were able to lift weights. He operated the motors by means of electric batteries. At first he seems to have regarded a battery as an inexhaustible source of energy,* but was not long in correcting this error. He soon recognized that the energy which could be derived from a given battery was limited and that it might appear in a number of different forms.

Joule seems to have grasped the idea of the conservation of energy—that energy may be neither created nor destroyed, but may be changed from one form to another. He thought that heat, electrical energy, and mechanical energy were all related; if one form appeared, less was available for conversion to another form.

The concept of the equivalence of different energy forms was not acceptable to Joule's contemporaries; he was a man ahead of his time. His paper published in 1842 entitled "On the Calorific Effects of Magneto Electricity and the Mechanical Value of Heat" was pretty much ignored. But again in 1847 Joule ventured with some trepidation to present at a scientific meeting his ideas on energy conservation. The program was crowded and running behind schedule. The chairman requested that Joule summarize his paper briefly rather than give it in its entirety.† At the close of his brief summary the chairman did not even invite discussion, since he regarded the contents as of trifling importance. This second paper too would likely have passed unnoticed except for William Thomson (Lord Kelvin), who rose and called to the attention of the meeting its far-reaching implications. Needless to say, Kelvin and Joule became lifelong friends.

The law of the conservation of energy denied the possibility of creating or destroying energy. Thus the energy of the universe was considered constant. Its form might be altered but not its amount. The law of the conservation of matter (see also p. 117 below) stated that the amount of matter in the universe was also fixed. Man could do nothing that would alter its amount. Atoms might hitch together or be detached, but no part of their masses could by any process of science or magic be obliterated. Such at least was the prevailing belief until the early years of the twentieth century, when Einstein informed the world that Lavoisier had been wrong about the conservation of matter. Einstein suggested that matter should be destructible, and if destroyed, a perfectly enormous energy release should result. At that time he proposed the now famous equation, $E = mc^2$, in which E is the new energy produced, m the amount of matter destroyed, and c the velocity of light.

* Alexander Wood, *Joule and the Study of Energy,* G. Bell & Sons, 1925, p. 46.
† Wood, *Joule,* p. 51.

In a sense there was nothing very startling about this equation. The equivalence between work and heat had been worked out by Joule, and the equivalence of other forms of energy was recognized in the law of the conservation of energy. The really exciting aspect of Einstein's new equa-

Fig. 5-15. *Albert Einstein, 1879-1955.*

tion was the extension of the concept of energy to include matter. If a method of annihilating matter could be discovered, an enormous new energy source, hitherto untapped, should be available. The enormous scale of the energy release accompanying the annihilation of matter becomes apparent when one considers the factor c^2. Its value is 9×10^{20} cm^2/sec^2 if the velocity of light is expressed in cm/sec. When this huge number is multiplied by even a very small mass change, say 10^{-6} (a millionth) gram, the resulting energy change is 9×10^{14} or nearly 10^{15} ergs, equivalent to approximately 21 million calories of heat. This is enough to change 500 pounds of water from the freezing point to the boiling point—a surprisingly large energy release, considering the tiny amount of matter destroyed.

ILLUSTRATIVE PROBLEM

Calculate the energy release accompanying the destruction of mass equal to that of a proton (hydrogen-1 nucleus), 1.67×10^{-24} g.

Solution

$$E = mc^2$$

$$E(\text{ergs}) = 1.67 \times 10^{-24}\ (\text{g}) \cdot \left(3.00 \times 10^{10}\ \frac{\text{cm}}{\text{sec}}\right)^2$$

$$E = 1.67 \times 10^{-24} \cdot (9.00 \times 10^{20})$$

$$E = 15.0 \times 10^{-4}\ \text{erg} = 1.50 \times 10^{-5}\ \text{erg}$$

For many years there was little or no direct experimental evidence to support the Einstein mass-energy equivalence principle. About 1940 the much-needed corroborating evidence was at hand. Under appropriate conditions (to be discussed in Chapter 14) atoms of uranium-235 could be made to fission (split into parts). The combined masses of the resulting fragments were slightly less than the mass of the original atom; the energy release was of enormous proportions. Here at last was experimental evidence which lent convincing support to the theoretical principle Einstein had announced over thirty years earlier. Suddenly Einstein's mass-energy equivalence principle emerged as a key factor in the development of nuclear energy. The utility of a theoretical principle is frequently not immediately apparent; often a new instrument or new technique is needed before supporting experimental evidence can be found.

Does the mass-energy equivalence principle invalidate the law of the conservation of energy? Insofar as the latter denies the possibility of creating new energy, it does; new energy is created when matter is destroyed. On the other hand it is still true that energy may be converted from one form to another; it is still true that in such a process, energy of one variety is produced only at the expense of an equivalent amount of another form.

Does the principle of conservation of mass still hold? The new principle of mass-energy equivalence allows for the destruction or creation of matter, which is denied by the law of the conservation of matter. On the other hand the law of the conservation of matter still remains the foundation of quantitative relations in the science of chemistry, as will be shown in Chapter 7.

The mass-energy equivalence principle is really an extension of the two laws of conservation. For a hundred years it has been known that 4.186 joules of work are equivalent to 1 calorie of heat. The Einstein principle simply establishes a similar equivalency between matter and energy. It explains that in the process of nuclear fission new energy is created at the expense of matter. It also suggests the possibility of the creation of particles (mass) from very high-energy radiation. It is really the sum of the energy and the matter in the universe then, which is conserved.

Problems, Chapter 5

1. Calculate the number of calories which are equivalent to 1 Btu.

2. A bomb having a weight of 4 tons is dropped from a height of 18,000 ft. What kinetic energy will the bomb have just before it strikes the ground? (Neglect air resistance.)

3. Calculate the kinetic energy of a speed boat having a mass of 900 lb and traveling at the rate of 20 mi/hr. Express the result in foot poundals.

4. A car is parked at the top of a 3000-ft incline which drops 1 ft vertically for every 12.0 ft along the road. If the car *weighs* 2900 lb, what is its potential energy?

5. A machine lifts an 80.0-kg weight 120 cm vertically in 3 sec. (a) How much work is done? (b) Calculate the horsepower. (c) Calculate the equivalent power in kilowatts.

6. Calculate the kinetic energy of a stone weighing 20 g and having a velocity of 500 cm/sec.

7. A brass cylinder (specific heat = 0.0890) whose temperature is 100°C has a mass of 200 g. It is dropped into an aluminum calorimeter (specific heat = 0.220) having a mass of 100 g and containing 150 g of water, both at 20.0°C. Calculate the final temperature.

8. Normal human body temperature is 98.6°F. What is the corresponding temperature on the Celsius scale?

9. Assuming that all of the kinetic energy of a falling body is converted into heat at the instant of impact with the ground, calculate the heat evolved (calories) when a 500-g mass falls a distance of 10.0 m.

10. Calculate the potential energy of a 500-g mass elevated 10.0 m above a floor.

11. In a nuclear process called "pair production," an electron-positron pair is created when a high-energy photon passes close to an atomic nucleus. What minimum energy in ergs is required if each of the two particles created from energy is assumed to have a (rest) mass of 9.11×10^{-28} g?

12. Carbon tetrachloride boils at 76.0°C (1 atm pressure). What is the corresponding Fahrenheit temperature? Absolute temperature?

13. Two grams (mass) of hydrogen gas contain 6.02×10^{23} molecules. Calculate the kinetic energy of a single hydrogen molecule which is moving at 1.66 km/sec. (Assume Newtonian mechanics.)

14. At what temperature are the readings of a Fahrenheit and Celsius thermometer scale equal?

Exercises

1. How are *work* and *energy* defined by the physicist?

2. Explain the meanings of *kinetic energy, potential energy, power,* and *mechanical equivalent of heat.*

3. Describe the changes in the potential and kinetic energies of a pendulum bob as it swings back and forth.

4. Draw a diagram representing each of the three commonly used temperature scales and showing the boiling and freezing points of water on each.

5. Describe how you might check the correctness of (calibrate) the scale on a laboratory thermometer.

6. Define (a) calorie, (b) Calorie, and (c) specific heat.

7. In calculating quantities of heat in Btu, do you use the same or a different value of the specific heat as in a calculation involving calories? Explain.

8. What information do you need to know in order to predict the temperature change which results when 200 calories of heat flow into an aluminum cylinder. (Assume uniform heat distribution throughout the mass.)

9. Describe the basic assumptions of the so-called method of mixtures.

10. Describe the Joule experiment for determining the mechanical equivalent of heat.

11. State the law of conservation of energy; the law of conservation of mass. Does the Einstein principle of mass-energy equivalence invalidate either or both of these? Explain.

12. Wound and unwound steel springs of equal mass are dissolved in equal volumes of an acid solution. Would you expect equal amounts of heat to be evolved? Explain.

13. What are the fixed points on the Réaumur temperature scale?

6 The Frenzy of molecular motion

THE idea that material substances might be composed of particles did not originate with modern science; it dates back at least to the time of the Greeks in the pre-Christian era. As early as 400 B.C. Democritus suggested that matter was composed of tiny rigid particles which moved through empty space. He believed these particles were indivisible and indestructible and had varying properties such as roundness or smoothness or roughness. Democritus' theory of matter was not accepted by many of the philosophers of his age. Aristotle, for example, rejected it. He favored the concept of a continuous structure of matter. The ideas of Greek atomists have reached us principally through the Roman poet Lucretius and his poem, *De Rerum Natura* (Concerning the Nature of Things).

In Galileo's time the argument over the structure of matter was revived and intensified. The "plenists" held that nature abhorred a vacuum. They argued that normally space was filled with matter and nature saw to it that matter came rushing in if the formation of a void was threatened. Descartes was a proponent of this school of thought.

Pierre Gassendi (1592-1655), on the other hand, was a "vacuist"; he held that matter was composed of particles which moved through a void. He believed that the distances between gas particles were very large compared with their diameters, that they were closer together in solids than in liquids, closer in liquids than in gases. In the seventeenth century, knowledge of the properties of gases had reached a stage of development in which a new theory might blossom forth at any time.

The honor of organizing this information and suggesting the first outline of the kinetic theory belongs to Daniel Bernoulli (1700-1782). He is called the father of the kinetic theory because of his famous volume, the *Hydrodynamica* (1739), which laid the foundations of the kinetic theory of gases. Bernoulli was one of the first to explain gas pressure as due to the force exerted by millions of tiny particles continually bombarding a surface. Certain passages in the *Hydrodynamica* indicate that he also understood there was a relationship between the temperature of a gas and the speeds of its particles.

THE KINETIC THEORY OF GASES

What is the nature of a gas as conceived by the modern kinetic theory? First of all, gases are assumed to be composed of submicroscopic particles—molecules—which move rapidly in all directions (Fig. 6-1). Imagine that the room in which you are sitting swarms with millions of minute Ping-Pong balls moving very rapidly in all directions. They collide with one another and rebound from the walls, floor, and ceiling of the room. Most of them move with approximately the same speed; occasionally one may be observed moving considerably faster or slower than the average.

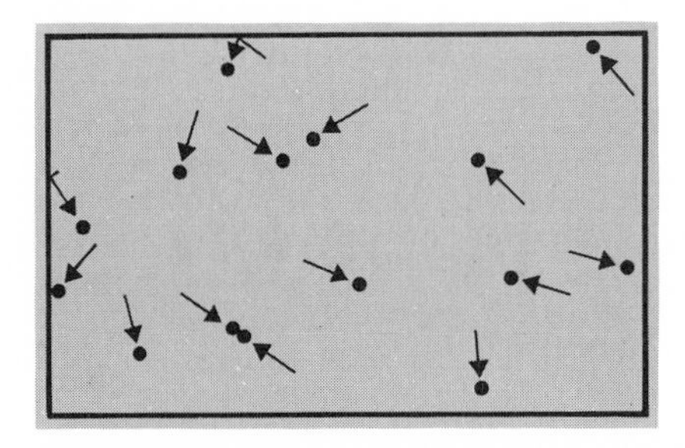

Fig. 6-1. *A gas is a collection of tiny particles, or molecules, moving rapidly in all directions. As they move freely through space, some rebound off one another or off the sides of the vessel. The distances between them are large in comparison with their diameters.*

Why are gas particles believed to be moving rapidly? If a tank of ammonia gas is opened, the odor of ammonia is soon detected in all parts of the room. The ammonia molecules must have moved in and out among the air particles. This movement of gas particles through space is called *diffusion.* If a jar containing air is placed over a jar of chlorine (whose molecules are much heavier), the yellow color of the chlorine will gradually appear in the covering jar. Thus even heavy gas particles such as chlorine will move upward against the force of gravity into all of the space available to them. These two experiments make it plausible to suppose that gas particles move rapidly and that they have diameters which are very small compared with their distances apart. A gas should then have a very porous structure, with the molecules themselves occupying negligible volume. This concept of gas structure is also supported by the known high compressibility of gases. Compressibility was the property of a gas referred to by Robert Boyle when he wrote about the "spring of the air." As gas molecules become confined in smaller and smaller spaces they do behave as if they were coiled springs, a property which will be evident if you push down on the handle of an old-fashioned tire pump whose outlet tube is closed.

Another attribute of gas molecules assumed by kinetic theory relates to their velocity of rebound from the walls of a container. A molecule rebounds from a stationary wall at the same velocity as that at which it approaches. Molecular impacts are described as *elastic* to distinguish them from impacts of objects like rubber balls which rebound with decreased velocity. Impacts such as the latter are called *inelastic.* If gas molecules were to lose energy at impact, their velocities would gradually decrease until eventually they would pile upon the bottom of the containing vessel. Gas molecules remain uniformly distributed throughout a confining space, and hence impacts are assumed to be elastic.

Newton's law of universal gravitation stated (Chapter 4) that "*every* body in the universe attracts every other with a force which is proportional to the product of the masses, and inversely proportional to the square of the distance between them." Gas molecules should then attract one another. However, since their masses are so very small and their distances apart comparatively very large, gravitational forces of attraction are always so minute as to be negligible. Gas molecules as conceived by kinetic theory, then, do not have attractive forces between them. (But real gas molecules often do attract one another with forces far stronger than gravitational.)

Consider again that the room in which you sit is filled with Ping-Pong ball molecules. If the room were heated, you would notice an increase in the average speed of molecular movement; the particles would move more briskly. Increased temperature is then to be associated with increased average speed of molecular movement.

In summary, the basic assumptions of the kinetic theory of gases are:

(1) Gases are composed of tiny particles called molecules.
(2) These molecules move rapidly in all directions at different speeds.
(3) The average distances between molecules are very large relative to their diameters.
(4) The average speeds of the molecules are increased when the temperature of the gas is raised.
(5) The impacts of gas molecules are elastic.
(6) The forces of attraction between gas molecules are negligible.

THE SEA OF AIR ABOUT US

The atmosphere, or "sea of air" around us, is today pretty much taken for granted. Such was not the case when the Pilgrims landed on the Massachusetts shore more than three centuries ago. Very little was known about the atmosphere or the constituents of the air, until later in the seventeenth century. Not until the 1640's did facts emerge which eventually led to the solution of a singular problem which had puzzled so many of the great scientific minds—the curious limits to nature's abhorrence of a vacuum.

If the air was pumped out of a pipe standing vertically with 40 feet of its length above the surface of a lake, water rushed in, as the Greeks said it should, to avert the occurrence of the abhorred vacuum. Curiously enough, if the water level in the pipe had risen 34 feet above the lake, no amount of pumping could lure more water into the remaining empty space. Nature's abhorrence of a vacuum mysteriously ceased at this level (Fig. 6-2).

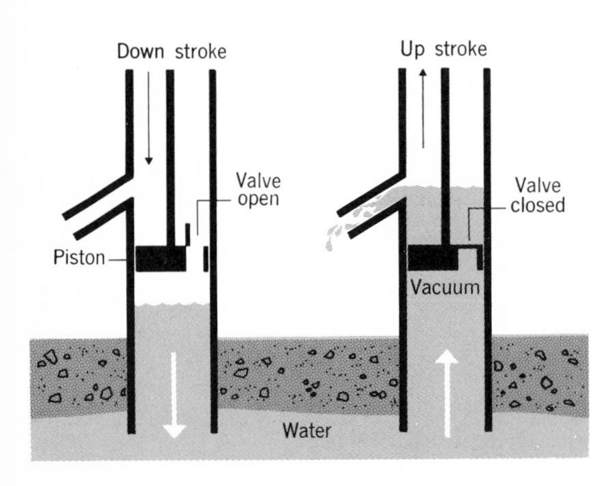

Fig. 6-2. *"Nature abhors a vacuum." The common pump, with its piston and valve, removes air from the pipe. Nature, abhorring the vacuum, causes water to rush in and fill the empty space. Galileo was perplexed by the fact that water fails to rush in after the column reaches a height of about 34 feet. The lift pump will not raise water to a greater height.*

Galileo pondered over this question. One of his co-workers, Torricelli (1608-1647), wondered if nature would also abhor a vacuum in a glass tube filled with the heavy liquid mercury. In 1643 he carried out the famous experiment by which he invented the barometer. A mercury-filled glass tube about a yard long was inverted in a dish of mercury (Fig. 6-3). In this device the mercury column was supported to a height of 30 inches above its level in the reservoir, not 34 feet as with water. During this same period there was a controversy concerning the reason for the liquid being supported at a level in the tube which was higher than that in the reservoir.

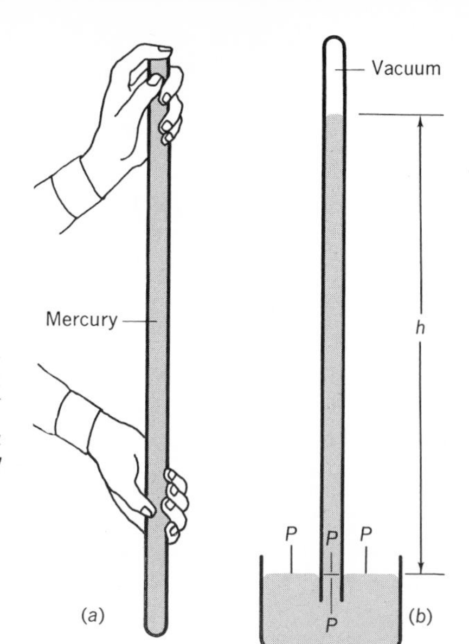

Fig. 6-3. *A Torricellian barometer. The height* h *of the mercury column varies around an average of about 76 cm or 30 in. at sea level. Is there a relationship between this height and the height that water will rise in a pipe exhausted by a pump? What is it?*

One of the interpretations was that air pressure held it up, and that changes in barometric levels were due to changes in air pressure.*

The famous French scientist Pascal (1623-1662) reasoned that if air pressure was responsible for the changes in the height of the mercury column in the barometer tube, it should be possible to reduce this height by merely climbing to a higher altitude above the bottom of the sea of air where a lesser weight of air would be pressing on the mercury in the bowl. One of the simplest ways to test this hypothesis was to carry a barometer to the top of a church steeple. However, the change in altitude proved to be so small that the experiment was not conclusive. But Pascal prevailed upon his brother-in-law, Périer, to take the instrument up the mountain known as the Puy de Dôme. This time the change in the mercury level was large enough to be observable, and as the ascent was made up the mountainside there was dramatic supporting evidence for those who believed that we live at the bottom of a vast sea of air.

BOYLE'S LAW

Robert Boyle (1627-1691), "the father of chemistry," was interested in natural phenomena of many types. About 1660 he published his *New Experiments Touching the Spring of the Air* in which he explained the "spring" by assuming that the particles of air behaved like minute watch-springs.† One of his experiments dealt with a problem concerning Tor-

* Pressure is formally defined as force per unit area.

† Louis T. More, *The Life and Works of the Honorable Robert Boyle,* Oxford University Press, 1944, p. 242.

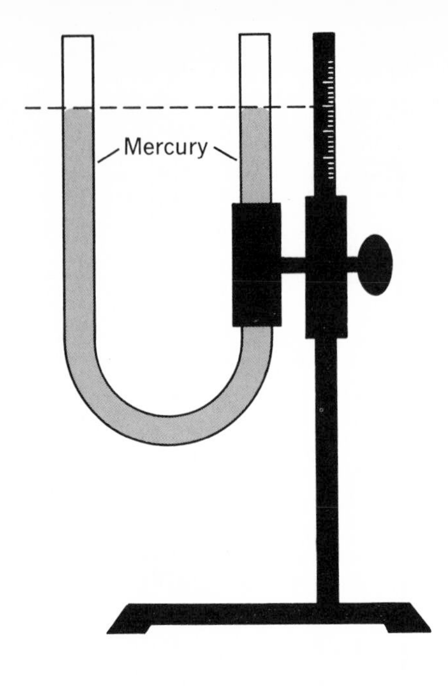

Fig. 6-4. *"Liquids seek their own level" in an open U tube: that is, in every one of many connected open vessels in which a liquid is standing, the height of the column is the same in the others.*

ricelli's barometer: It was well known that in an open U tube, a liquid would seek the same level in each arm of the tube (Fig. 6-4). Then what was holding the mercury in the barometer so high above the level of the mercury in the reservoir dish at its foot? One Franciscus Linus argued that an invisible cord—a *funiculus*—was responsible. Boyle, in order to disprove what he believed to be a crackpot idea, devised his J-tube experiment from which issued the data that suggested Boyle's law.

In this experiment, the short arm of the J is sealed and the long arm left open (Fig. 6-5). Before mercury is put in the J tube, the air pressure inside the tube is the same as atmospheric pressure.* At the instant that sufficient mercury is added to just close off the short (sealed) end of the J tube, the pressure is still about equal to atmospheric pressure. As soon as more mercury is added, the heavy liquid piles up higher in the open tube and its weight squeezes the air, confined in the sealed end, into a smaller space; the gas pressure in this closed tube is increased. The amount of increase is determined by the difference of level of the mercury in the two tubes. Suppose that atmospheric pressure is 760 millimeters of mercury and that the mercury in the open tube is 20 millimeters higher than in the sealed tube. The pressure of the gas in the sealed tube will be 780 millimeters.

In carrying out the experiment, Boyle had recorded data showing how the volume of gas trapped in the sealed-off portion of the J tube varied with the pressure applied by the weight of mercury in the open side. He put the notebook aside without noticing the mathematical relation which now

* Normal atmospheric pressure (also called "1 atmosphere") is 14.7 lb/in.^{2} or 760 mm of mercury.

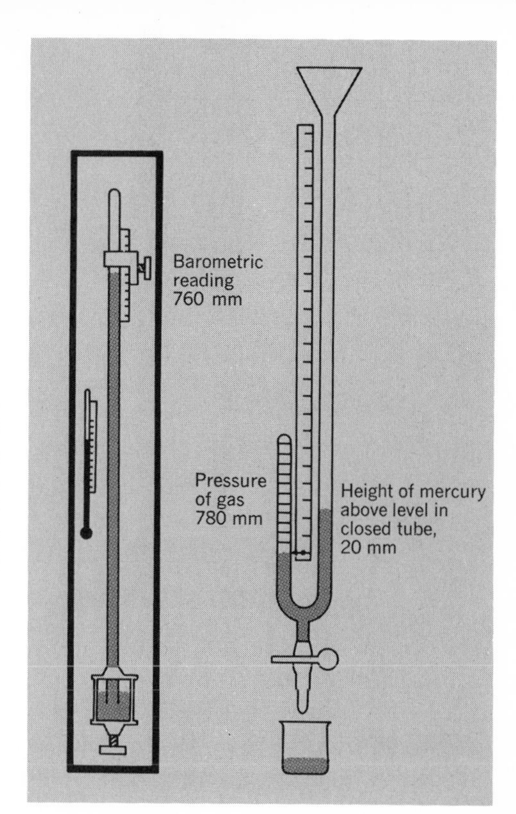

Fig. 6-5. *Boyle's J-tube experiment. The barometer is used to measure the air pressure in the open J tube. The amount by which the air pressure in the closed tube exceeds atmospheric pressure is measured by the difference in mercury level in the open and closed tubes.*

bears his name—that the product of any pressure and the corresponding volume was always the same for the same temperature and a given weight of air in the tube. Some friends observed this constancy of the PV product later when looking over the data. They insisted that Boyle deserved credit for the discovery of the principle* which now bears his name.

This scientific principle was discovered because Boyle wished to refute what he considered a very fanciful explanation of how the mercury column was supported in a barometer tube. Boyle's experiment was designed, not to prove a new principle, but to disprove what he regarded as a false conclusion. One of the by-products of the experiment was a set of data from which friends were able to discover a generalization which is called Boyle's law. This discovery illustrates the inductive method of reasoning. A generalization was drawn from apparently unrelated pressure-volume data concerning gases. The sequence of events hardly corresponds with the formal steps traditionally described as "the scientific method" of discovery.

Stated mathematically, Boyle's law is $PV = k$; in words, the product of the pressure of a gas and its volume is constant. This is true only as long as the temperature and mass of the gas under consideration do not change.

* James B. Conant, *Robert Boyle's Experiments in Pneumatics,* Harvard University Press, 1950, p. 62.

As an illustration, suppose that 1 gram of a gas is confined in a cylinder fitted with a piston as shown in Fig. 6-6a. The volume occupied by the gas is 10 liters and its pressure is 1 atmosphere. Substituting these values in the formula $PV = k$, we find that 1 (atm) $\cdot$ 10 (liters) $= k$, or that for these particular conditions k has a value of 10 liter atmospheres. If the piston is

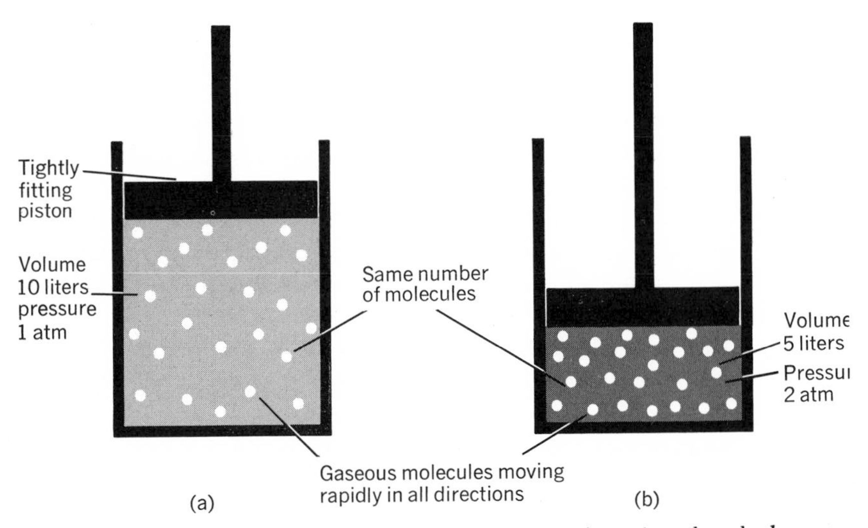

Fig. 6-6. *Boyle's law: If the volume of a given mass of gas is reduced, the pressure is proportionally increased. Temperature must remain constant during the change. Since a gas is warmed by compression and cooled by expansion, the apparatus shown would have to be immersed in a constant-temperature bath.*

moved down so that the new volume is 5 liters, as in Fig. 6-6b, the pressure must increase to 2 atmospheres, since the product, 10, must remain constant. If the volume is reduced to 2 liters, the pressure must increase to 5 atmospheres. As the amount of space available to the molecules—the volume—is altered, the product of P and V will always be 10 for 1 gram of gas at the particular temperature (presumed to remain constant).

Table 6–1 Sample Data, Boyle's Law

RESSURES IN ATMOSPHERES	VOLUME IN LITERS	PV PRODUCT
1	10	10
2	5	10
4	2.5	10
5	2	10

KINETIC THEORY INTERPRETATION OF BOYLE'S LAW

Why would Boyle's law be expected to hold in view of the assumptions of the kinetic theory of gases? If the kinetic theory is to be useful and inspire confidence, it must logically interpret known facts concerning gases. How does the theory explain this law?

If the volume of a gas is reduced to half, as in Fig. 6-6, twice as many gas molecules will be crowded into a given space. Since their average velocity does not change (temperature is assumed to be constant), the force exerted at each impact is the same. So the *number* of impacts must increase if the pressure increase is to be explained. In this case the number of impacts must double if the pressure doubles.

If the volume of the gas is increased, molecules make fewer impacts, and the force exerted on a unit area of surface is reduced. Boyle's law then seems to make sense when explained in terms of the assumptions of the kinetic theory of gases.

HEAT: MATTER OR MOTION?

The concept of heat as movement of the structural particles of matter was far from monopolizing the field in ancient times or in modern. The Greeks included fire among their four elements—earth, air, fire, and water. Most people thought of fire as being like earth, air, and water—a material substance. With the revival of the particle concept of matter in the seventeenth century, heat was so regarded, and in the eighteenth the famous Scottish chemist Joseph Black (1728-1799) referred to heat as a "subtle elastic fluid," which he called *caloric*.

With the passage of the years the conviction grew that heat was not a material substance after all, but rather the motion or vibration of the constituent particles of matter. An American-born scientist who spent most of his life in Europe made some observations on the boring of cannon which were crucial in establishing the vibration theory of heat.

Count Rumford: Cannon-Boring Experiments In the history of science there is scarcely a figure more colorful or more contradictory than Benjamin Thompson, a Massachusetts boy who became a count in the Holy Roman Empire (Fig. 6-7). Thompson was born in 1753 in North Woburn, Massachusetts. His willingness to walk from his home to Cambridge (about eight miles) to observe demonstrations in natural philosophy at Harvard is evidence of his early interest in science. As a lad he was seriously burned while making fireworks for the celebration attending the repeal of the Stamp Act. He probably witnessed the Boston Massacre and the Battle of Bunker

Hill. During the Revolutionary War he was questioned by local committees and was several times suspected of supporting the British cause. There is little doubt but he passed military intelligence to General Gage, the British commander in Boston, though no such charges were proved against him at the time.*

When the British evacuated Boston in 1776, Thompson went to England to serve in the British Colonial Office. After a brief return to America to lead a regiment of soldiers against his fellow countrymen, his path led to Munich, Bavaria, and it was here, while working with the Elector of Bavaria, Karl Theodor, that he became Count Rumford.

Fig. 6-7. *Benjamin Thompson, Count Rumford (1753-1814), in the portrait by Rembrandt Peale.*

As superintendent of the arsenal in Munich, Thompson was fascinated with the enormous amount of heat evolved in the operation of cannon-boring. Just as long as the horse supplying the motive power for the boring operation continued to keep the cannon rotating, the heat came streaming forth in apparently inexhaustible supply. The count could not understand how a material substance could continue to be emitted in such large quantities without causing at least some change in the mass or the specific

* Allen French, *General Gage's Informers,* University of Michigan Press, 1932, pp. 119-146.

heat of either cannon or shavings. His conclusion was inescapable; heat must be motion of the particles composing matter rather than the substance which had been called caloric. The cannon-boring experiments were thus a decisive factor in establishing the concept of heat as a mode of motion.

The Royal Institution: Humphry Davy Back once more in London, Thompson (now Rumford) in 1800 established the famous Royal Institution for the promotion of science and scientific research.

In 1801 a lad by the name of Humphry Davy came to London to be interviewed for a position at the Royal Institution (Fig. 6-8). Davy had been apprenticed to an apothecary and had become proficient as a chemical experimenter. His discovery of laughing gas (nitrous oxide) may have brought his name into prominence. Rumford needed a lecturer, and Davy

Fig. 6-8. *Humphry Davy (1778-1829).* [*Courtesy, N.Y. Public Library.*]

was invited to an interview. Rumford was not at all impressed with the appearance of his prospective employee and required him to give a private lecture. This task the young man carried out most capably and convincingly. So it was that the future discoverer of the elements sodium, potassium, and iodine took his position at the Royal Institution. It is quite possible too that Rumford knew of Davy's article, published two years earlier, entitled "An Essay on Heat, Light and the Combinations of Light." In this Davy described an experiment in which two pieces of ice were rubbed together. The frictional heat resulting converted ice at 0°C into liquid water at the same temperature. Joseph Black called this heat "latent heat" because no temperature change was produced as long as both ice and water were present together. The quantity of heat required to convert a gram of ice into a gram of water at 0°C is called the *latent heat of fusion* and is 80 calories per gram. The corresponding quantity of heat required to convert a gram of water into a gram of steam at 100°C, without change of temperature, the *latent heat of vaporization,* is 540 calories per gram.

THE SPEEDS OF GAS MOLECULES

Near the middle of the nineteenth century, James Prescott Joule asked the questions, "Just how fast do gas molecules move anyway? What speeds must they have in order to exert their observed pressures?" He made calculations of these speeds. The results astounded him. They showed that hydrogen molecules at standard temperature and pressure (0°C and 760 mm Hg) are moving about 1 mi/sec. A velocity of 1 mi/sec is equivalent to 3600 mi/hr, a very considerable speed even compared with that of jet planes. The first reaction to Joule's suggestion concerning the magnitude of molecular speeds was skepticism: the calculated speed sounded fantastic. Everybody knew that when a bottle of ammonia was opened, its characteristic odor was not detectable instantly in all parts of the room. Time had to elapse. But what should have been taken into consideration was that ammonia molecules were not traveling through evacuated space, but rather colliding with millions of air molecules en route. Individual gas molecules do travel through space at great speeds, but collide frequently with other molecules. The average distance between impacts, the *mean free path,* must be very short. If a gas molecule travels only about a hundred thousandth of a centimeter before colliding with another, it is understandable why gas diffusion is slow in spite of enormous molecular speeds.

MODERN KINETIC THEORY AND THE IDEAL GAS EQUATION

In 1857 the great German mathematical physicist Clausius (1822-1888) published the first comprehensive summary of the kinetic theory of gases. It included the assumptions of the theory outlined earlier in the chapter and an ideal gas equation, sometimes known as the Joule-Bernoulli equation:

$$PV = \tfrac{1}{3}nmu^2 \tag{6-a}$$

This equation predicted the pressure exerted by n molecules of gas, each of mass m, moving with speed u in a space of volume V. The earlier derivations of this equation assumed that all gas molecules had the same speed. Later, Maxwell and Boltzmann applied statistical methods to the treatment

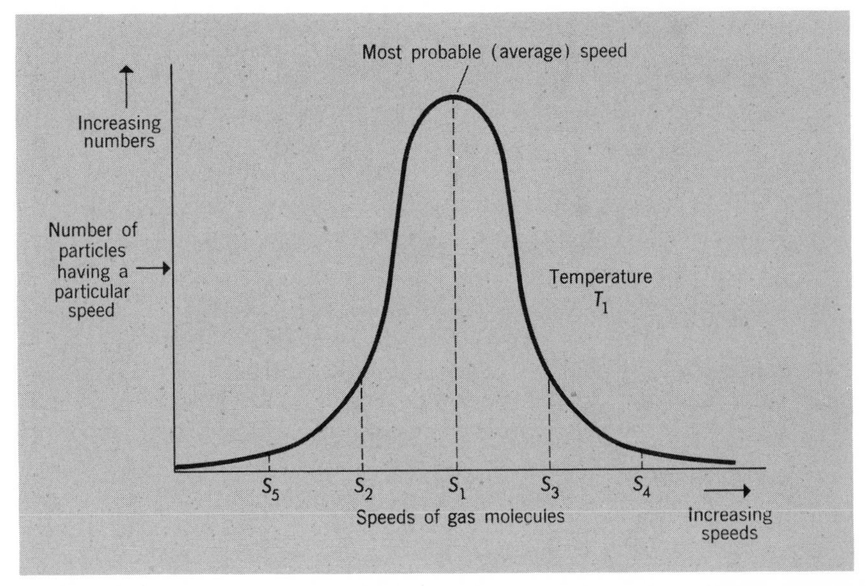

Fig. 6-9. *Speeds of gas molecules show a "normal" distribution. At temperature T_1 more molecules will have speed S_1 than any other; hence S_1 is the "most probable" speed. Smaller numbers of molecules will have speeds S_2 and S_3. Still fewer molecules will have speeds S_4 and S_5. (There is a very slight difference between the "most probable" and the "average" speeds.)*

of molecular speeds and showed that at a particular temperature the distribution of the values of these speeds was similar in form to the normal curve. Most gas molecules have speeds which cluster about a most probable value, as shown in Fig. 6-9. It will be assumed that the most probable speed is also the average speed of the molecules; it is very nearly so. The distribution curve of molecular speeds (Fig. 6-10) shows that comparatively few molecules have speeds *much* slower or *much* faster than the most probable value. In fact the farther removed a given speed is from the most probable value, the fewer the molecules which will have this speed. Fig. 6-10 also shows how the molecular speed distribution changes with temperature. As temperature increases, the peak of the curve shifts to the right, and the curve flattens slightly. The average speed is, then, increased and the range of speeds extended somewhat, at higher temperatures.

IDEAL GAS LAWS AND THE IDEAL GAS EQUATION

Equation 6-a, which is called an ideal gas equation, summarizes a great deal of information concerning the behavior of gases. Boyle's law has already been mentioned. How can it be derived from this ideal gas equation?

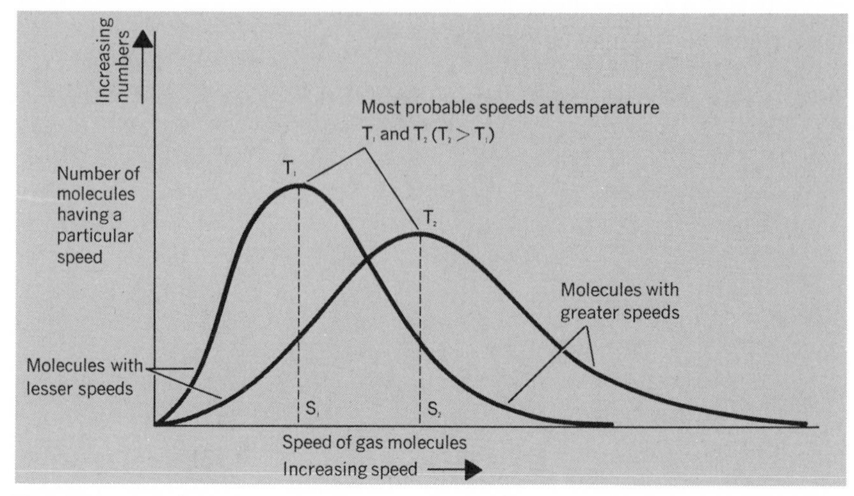

Fig. 6-10. *At a higher temperature T_2, the most probable speed is greater than at T_1. However, fewer gas molecules have the most probable speed S_2 than have S_1.*

Boyle's law states that for a given amount (mass) of gas at constant temperature, the product of pressure and volume must remain constant. The right-hand member of Equation 6-a contains nm, the total mass of the gas particles, and u^2, the square of their (average) speeds. As long as no gas molecules enter or leave the space, nm is constant. As long as the (absolute) temperature remains constant, the average molecular speed remains the same. Hence under these conditions $PV = k$,* this is the mathematical relation for Boyle's law.

ILLUSTRATIVE PROBLEM

50 liters of gas initially at 0.75 atm pressure are compressed until the pressure is 2.5 atm. Calculate the new volume. Assume that temperature and mass of gas are constant.

Solution

$$P_1V_1 = P_2V_2$$
$$2.5\ (\text{atm})\cdot V\ (\text{liters}) = 0.75\ (\text{atm})\cdot 50\ (\text{liters})$$
$$V = 15\ \text{liters}$$

Charles' law states that for a given mass of gas at constant pressure the

* k is a constant whose value depends on the mass of gas and the temperature. The larger each of these is, the larger k is.

quotient V/T (volume divided by absolute temperature) must remain constant. Equation 6-a may be rewritten in the form

$$\frac{V}{u^2} = \frac{1}{3}\frac{nm}{P}$$

For an ideal gas it is assumed that $T \propto u^2$ and $T = ku^2$ where k is a proportionality factor. Hence:

$$\frac{V}{T} = \frac{1}{3k} \cdot \frac{nm}{P}$$

As before, nm = the mass of gas and P = pressure, both of which remain constant. Hence V/T = constant. This is Charles' law.

ILLUSTRATIVE PROBLEM

Ten liters of an ideal gas at 0°C will occupy what volume if heated to 100°C?

Solution

$$\frac{V_1}{T_1} = \frac{V_2}{T_2}$$

T must be expressed in degrees Kelvin.

$$\frac{10.0 \text{ liters}}{273°\text{K}} = \frac{V_2}{373°\text{K}}$$

$$V_2 = \frac{3730}{273} \text{ liters} = 13.7 \text{ liters}$$

WHEN ARE IDEAL GAS LAWS APPLICABLE?

Boyle's law, Charles' law and other ideal gas laws do not apply under all temperature and pressure conditions. They apply more precisely under some conditions than others. What are the circumstances under which close agreement may be expected between theory and experiment?

Some gases—e.g., ammonia and sulfur dioxide—are easily liquefied at pressures near that of the atmosphere (14.7 lb/in.^{2}). Others—e.g., nitrogen and oxygen—must be cooled to very low temperatures (−147°C for nitrogen) before there is even a *possibility* of their liquefaction. Apparently ammonia molecules exert much stronger forces of attraction on one another than do nitrogen molecules. If nitrogen is compressed, the gas volume decreases and the product of P times V is substantially constant. On the other hand if ammonia is compressed, the volume decreases faster than Boyle's law predicts, and the values of the products PV do not remain constant as they should according to the theory. Why should the law apply in one case but not the other?

The kinetic theory of gases assumes there are no attractive forces between gas molecules. Molecules of easily liquefiable gases have large forces of attraction. These attractive forces cause the gas volume to decrease more than the law predicts as pressure is increased; the product PV decreases.

Even a gas like nitrogen, which is difficult to liquefy at room temperature, does not behave as predicted by Boyle's law at *extremely* high pressures, 100 or more atmospheres. The kinetic theory assumes that the distances between gas molecules are large compared with their diameters. At very high pressures gas molecules are crowded close together and do occupy an appreciable fraction of the volume. Further pressure increase does not bring about the expected volume decrease because some molecules are contiguous and cannot be forced any closer together. The product PV of Boyle's law does not remain constant as it should but increases with increasing pressure.

One way to tell whether the gas laws apply, is to consult a table of critical temperatures (the temperature *above* which the liquid cannot exist). If the critical temperature of the gas is low, it will be difficult to liquefy. At

Table 6–2 Critical Temperatures of Gases

GAS	CRITICAL TEMPERATURE (°K)
Sulfur dioxide	430
Hydrogen chloride	324
Carbon dioxide	304
Methane	191
Oxygen	154
Nitrogen	126
Hydrogen	33
Helium	5

room temperature the attractive forces between molecules will be negligibly small. If the critical temperature is high, attractive forces between molecules will be stronger; deviations from the gas laws will likely be observed.

Again, for the gas laws to apply, the pressure must not differ greatly from 1 atmosphere. The gas laws would not be expected to apply above 10 atmospheres. Certainly they would not apply at pressures as high as 100 atmospheres, even for gases with low critical temperatures.

Problems in this text dealing with gas volumes will employ conditions under which ideal gas laws are applicable. It should now be clear, however, that they should be used with understanding; they are not applicable under all conditions of temperature and pressure.

Problems, Chapter 6

1. What pressure in millibars (see Appendix) is equivalent to 770 mm of mercury?

2. At 770 mm (of mercury) pressure, what volume will be occupied by a mass of gas that occupies 10.0 liters at 740 mm pressure? (Assume mass of gas and temperature are constant.)

3. To what Celsius temperature must a mass of gas that occupies 20.0 liters at 10°C be heated if its volume is to increase to 30.0 liters, pressure remaining constant?

4. The mercury level in the open arm of a J tube is 35 cm higher than in the shorter, closed arm. If the barometer reading is 76.0 cm of mercury, calculate the pressure of the gas in the enclosed arm.

5. (a) Show by means of the ideal gas equation that for a given mass of gas whose volume is constant, the ratio of pressure to absolute temperature must remain constant, or that $P/T = k$.
   (b) If a gas initially at 273°K and 760 mm pressure is heated to 373°K, what is the final pressure (assume the volume is constant)?

6. (a) Show by means of the ideal gas equation that for a given mass of gas, the ratio $PV/T = k$ or that

$$\frac{P_1V_1}{T_1} = \frac{P_2V_2}{T_2}$$

   This is the so-called *general gas law*.
   (b) 22.4 liters of gas at 1 atm and 300°K are heated to 400°K. Calculate the volume if the final pressure is 1.25 atm.

7. In Hare's method for determining the specific gravity of an unknown liquid, vertical glass tubes project into each of two beakers containing the liquids as shown in Fig. 6-11. Suction applied to the central tube causes the liquids to rise to different heights. If liquid A (water) rises to a height of 24.3 cm while liquid B simultaneously rises 32.6 cm, calculate the specific gravity of liquid B.

8. It takes 1520 calories to heat 25.0 liters of air initially at 0°C (density 1.29 g/liter), to 273°C at constant volume. Calculate the

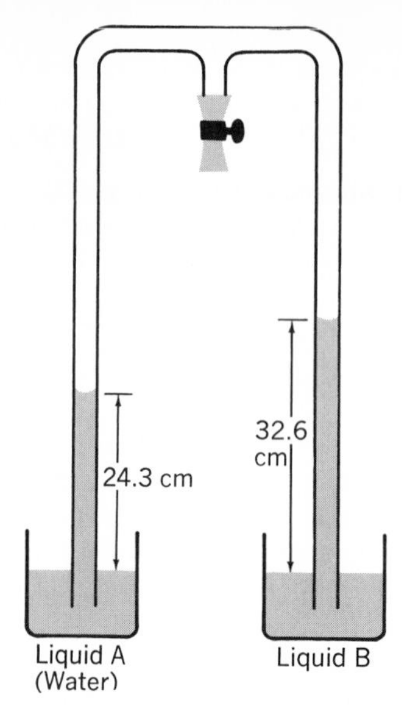

Fig. 6-11.

average heat capacity of air (cal/g), assuming it is constant over this temperature range.

9. According to Graham's law of diffusion the speed of gas diffusion is inversely proportional to the square root of the gas density. The densities of oxygen and hydrogen are respectively 1.43 g/liter and 0.0893 g/liter at 0°C. If the speed of diffusion of hydrogen gas at this temperature is 1 mi/sec, calculate the speed of diffusion of oxygen gas.

10. Pascal is said to have constructed and attached to the exterior of his house a water barometer. Calculate the length of the water column in this barometer if the barometric reading is 760 mm of mercury. If the mercury barometer were to decrease to 750 mm, how much would the water level fall in the water barometer? Explain.

11. Calculate the percent error which would result from failure to correct to *in vacuo* the weight of a water-filled Dumas bulb whose volume is 250.0 ml. It weighs 361.0 g when filled with water. (Neglect any buoyant effect of air on the brass weights used. Assume the density of air at the temperature of the experiment to be 1.293 g/liter.)

12. The boiling point of liquid benzene is 80°C. Calculate the corresponding Fahrenheit temperature; the corresponding absolute temperature.

13. If 12 liters of oxygen gas at 20°C are heated to 150°C, calculate the new volume. (Assume that pressure is constant and that no gas molecules enter or leave the space during the change.)

14. Calculate the final temperature attained if a 30.0-g cube of ice at 0°C is placed in a calorimeter containing 200 g of water at 30°C. (Assume no heat loss to surroundings or to the calorimeter.)

15. Calculate the amount of heat required to change 60 g of water from 20°C into steam at 100°C.

16. Steam at 100°C passes through 500 g of water in a calorimeter. If the water was originally at 25.0°C and its final temperature was 75.0°C (after the passage of the steam), calculate the mass of the steam used. (Assume all the heat available from the steam is absorbed by the water. Don't forget that all the steam which condenses to water also cools to the final temperature.)

Exercises

1. Describe the nature of a gas as conceived by the kinetic theory.

2. State the basic assumptions of the kinetic theory of gases.

3. Describe how to construct a mercury barometer. Does the diameter of the barometer tube affect the height of the mercury column? Explain.

4. A barometer tube has a cross-sectional area of 1 in^2. What is the weight of the mercury above the level in the reservoir when the pressure is 1 atm?

5. Describe how to carry out Boyle's J-tube experiment. What measurements must be made in order to verify Boyle's law by means of this experiment? Describe.

6. What is the approximate speed of gas molecules near room temperature? If they travel at such high speeds, why do they take such a long time to cross a room?

7. Show that Boyle's law and Charles' law may be derived from the ideal gas equation.

8. Boyle's law is an ideal gas law. It does not apply to all gases under all circumstances. Explain.

9. As its temperature rises, does a gas become more like or less like an ideal gas? Explain.

10. Look up the critical temperature of argon gas and ammonia gas. At room temperature which gas will behave more like an ideal gas? Why?

11. Gas molecules are contained in a cylinder 100 cm^2 in cross section; the length of the gas column enclosed by the piston is 10 cm. If the gas is compressed until the length of the gas column is 5 cm, the pressure is doubled. (Temperature and mass of gas are assumed constant.) The number of impacts per unit area of surface must therefore be twice as great as before. But each impact cannot be doubly effective because the surface area has not been reduced to half of its previous value. Calculate the surface area after the compression and compare the total forces exerted before and after the compression.

7 Atomic building blocks

ELEMENTS, COMPOUNDS, AND MIXTURES

Every material substance—solid, liquid, or gas—consists of tiny particles called atoms. A substance which is composed of only a single species of atom is an element. Elements cannot be reduced to simpler chemical substances by means of chemical reactions. The beautiful crystalline diamond is composed of atoms of carbon. The liquid metal mercury is composed of atoms of mercury. Carbon and mercury are elements. Helium gas is an element, and consists of single atoms of helium (Fig. 7-1). Oxygen gas consists of molecules containing two atoms each (Fig. 7-2), and such two-atom molecules are typical of many elements. (A molecule is a complex aggregate of atoms.) But since there is only a single atomic species in the molecules of oxygen, oxygen is also an element.

Molecules may consist of like atoms, as in oxygen, or of unlike atoms, as in carbon dioxide. If a substance is made up of more than one kind of atom, then it is not an element and must be either a compound or a mixture. One of the criteria of a compound is that each of the millions and millions of molecules composing it must be similarly constituted with respect to the number and kind of atoms. Every molecule of carbon dioxide gas is composed of a central carbon atom attached to two oxygen atoms (Fig. 7-3). Carbon dioxide is a compound. Water is also a compound: each of its molecules consists of two atoms of hydrogen and one atom of oxygen.

A sample of gas which consists of molecules having *different* numbers or kinds of atoms is called a mixture. Nitrogen and oxygen molecules in the same enclosing vessel (Fig. 7-4) constitute a mixture. Air consists of molecules of nitrogen, oxygen, argon, and carbon dioxide; it is a mixture.

Three of its constituents—nitrogen, oxygen, and argon—are elements, while the fourth, carbon dioxide, is a compound.

Under appropriate conditions, atoms of carbon and oxygen may combine, not in the ratio of one carbon to two oxygen, as in carbon dioxide,

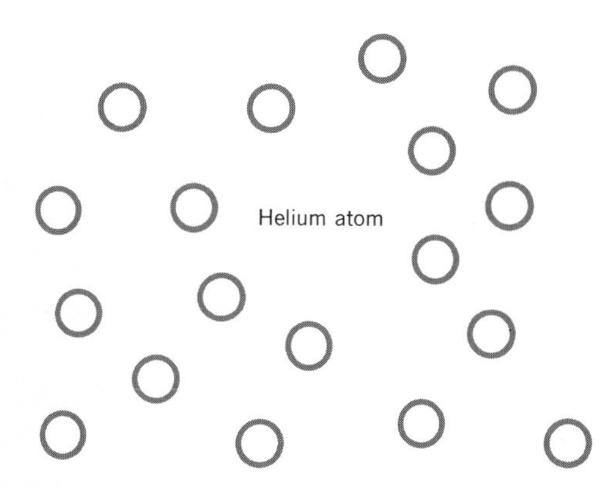

Fig. 7-1. *Helium gas is composed of single atoms. Helium is an element.*

but rather in the ratio of one to one. This chemical compound is called carbon monoxide (CO). If a vessel contains two kinds of molecules, some with one atom each of carbon and oxygen and some with one of carbon and two of oxygen (Fig. 7-5), it encloses a mixture.

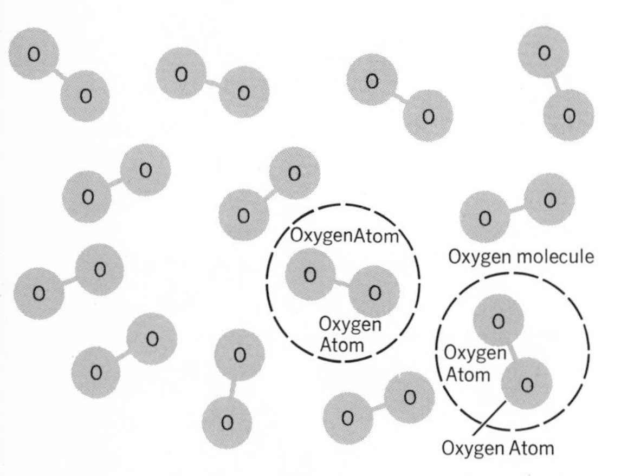

Fig. 7-2. *In oxygen gas, the atoms are combined in groups of two. Each group is called an oxygen molecule.*

In summary: in an element, all atoms are of the same kind, in a compound all molecules* are the same; and in a mixture, molecules differ in either the kind or the number of atoms constituting them.

* Some compounds are composed of structural units other than molecules.

The Concept of an Element: An Idealization Strictly speaking, the notion of an element is an idealization if by "element" is meant an isolated group of particles composed solely of a *single* atomic species. As a practical matter

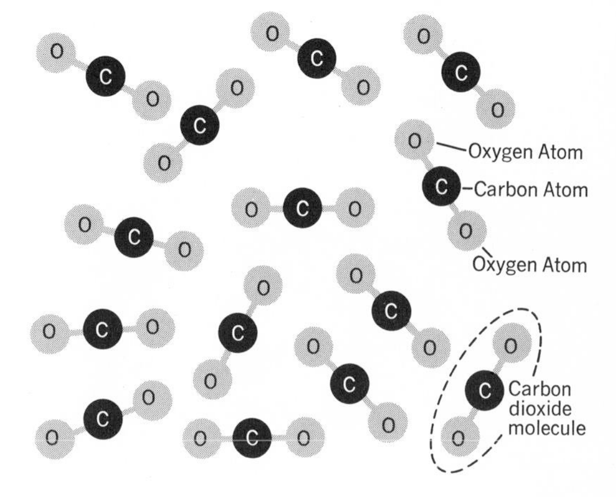

Fig. 7-3. *Carbon dioxide gas molecules have three atoms each, one of carbon and two of oxygen. Carbon dioxide is a compound.*

this is virtually impossible of attainment. There will inevitably be at least small amounts of other kinds of atoms—that is, impurities—present.

By way of illustration, consider one of the methods of preparing oxygen

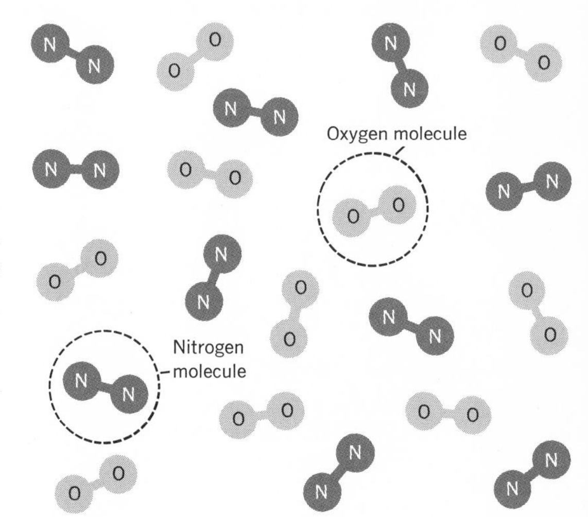

Fig. 7-4. *Nitrogen and oxygen molecules in the same vessel constitute a mixture.*

gas. A white solid compound called potassium chlorate ($KClO_3$) and a black powder, manganese dioxide (MnO_2), are mixed and put in a test tube. A delivery tube is attached as shown in Fig. 7-6. Air is contained in the test tube above the chemical substances and in the delivery tube. When the test tube is heated, the air in the tube expands and is forced out of the delivery tube and up through the heavier surrounding water. As the heating continues, the potassium chlorate ($KClO_3$) decomposes; a chemical change takes place. Oxygen atoms are split away from those of potassium and chlorine. The oxygen formed pushes air ahead of it into the delivery tube and up through the water. After the bubbling has continued long enough to sweep out all the air, a bottle filled with water is inverted over the end of the delivery tube. The oxygen now rises to the top, displacing the water downward, and soon the vessel is completely filled.

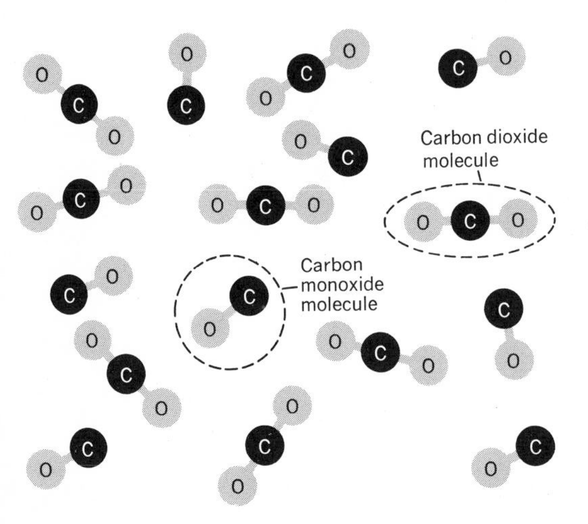

Fig. 7-5. *Carbon monoxide and carbon dioxide molecules in the same vessel constitute a mixture. The molecules of the two compounds contain the same kinds of atoms but different numbers of each in a given molecule.*

As a practical matter, are oxygen atoms the only atomic species present in the bottle? Very probably not. In the first place, the water doubtless contains some dissolved air, and as the oxygen passes through the water a small amount of this air will very likely be entrained (carried along). If the oxygen is collected over water, some of the most rapidly moving of the water (H_2O) molecules escape from the surface into the space above, so that water vapor molecules will also be present in the oxygen. Thus, when oxygen is prepared in the chemistry laboratory, although there may be an overwhelming majority of oxygen molecules in the enclosing container, there will undoubtedly be other kinds of atoms and molecules present in at least small amounts. Even if an attempt is made to remove the last traces of water, nitrogen, or carbon dioxide molecules by causing them to react and combine with other chemical substances, it is doubtless impossible to remove every last atom that is not oxygen. Similarly, the notion of a compound in the sense of a single *molecular* species is also an idealization. Even distilled water will contain small quantities of extraneous dissolved materials; when distilled water is poured into a glass vessel, there may be a

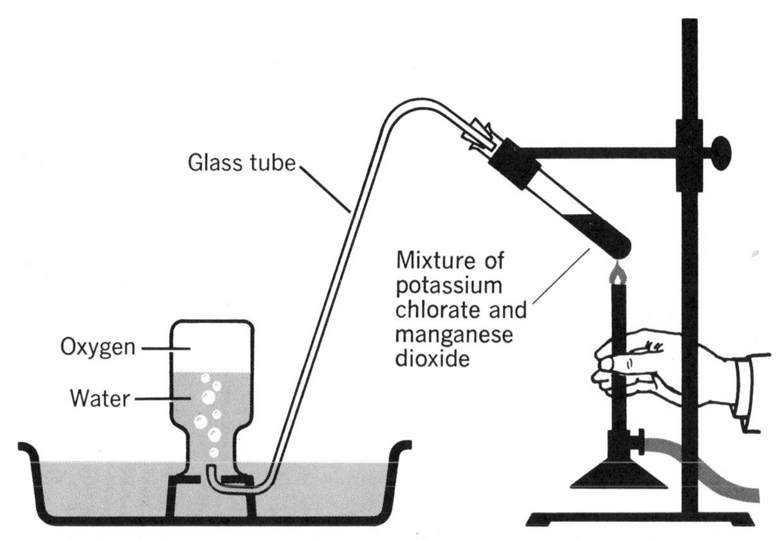

Fig. 7-6. *Oxygen gas may be prepared by heating potassium chlorate* ($KClO_3$) *in the presence of manganese dioxide* (MnO_2). *It is collected by the downward displacement of water.*

very slight dissolving of some of the sodium compounds of the glass. Gases from the air may also dissolve in the water. These limitations must be recognized when an element is described as consisting of a single species of atom. As a practical matter the concept of a pure single atomic or molecular species is as much of an idealization as the assumption of a frictionless inclined plane or of a vacuum.

CHEMISTRY THROUGH THE CENTURIES

The structural differences among elements, compounds, and mixtures described above are based on a concept of the atom not very different from that brought forward by John Dalton in the early years of the nineteenth century—that is, quite a recent concept in the long history of mankind.

The Beginnings of Chemistry Man was acquainted with materials now regarded as elements thousands of years before he made any distinction between elements and compounds or mixtures, just as he used gadgets and machines long before he understood the principles explaining their operation.

Gold was probably known several thousand years before the Christian era. Copper objects have been identified with some of the very ancient civilizations. Both of these metals occur in the uncombined or "native" state. Most metals, however, do not occur in the native state (as the element) but rather as compounds or mixtures, called "ores." The native nonmetallic element sulfur is mentioned in the Bible as "brimstone."

At an early date the art of converting ores into useful metals—metallurgy—was developed, no doubt as the result of some fortuitous event. Imagine the amazement of primitive man when a reddish earth (iron ore), accidentally mixed with charcoal and subjected to prolonged heating, produced a crude chunk of metal which could be shaped into useful objects of iron. As the art of the smith advanced, bronze (an alloy) and then iron became the favored metals, still thousands of years before Christ.

Baking and brewing, both chemical arts, have their origins in the very distant past. Even though the science of chemistry is only a few centuries old, many of the operations performed by today's chemist—for example, distillation, crystallization, and sublimation—were known long before the Christian era.

As early as the sixth century B.C. man groped for unifying principles and searched for simplifying relations in a bewildering world of apparently unorganized and unrelated materials. Was there any common denominator, any substance from which all others were derived? Thales (640-546 B.C.), one of the first Greeks who attained prominence in science, considered water such a material. He observed that when water evaporated it left behind a small amount of solid residue, and concluded that water had been converted into the solid. A modern explanation would be that the solid material was dissolved in the water originally and remained after the water had evaporated.

Early in the fifth century B.C. the Greek philosopher Empedocles developed a theory of matter based on four elements—earth, water, air, and fire. These were not considered elements in the current sense; they represented the three physical states of matter—solid, liquid, and gas—plus a fourth—glowing gas, "fire"—regarded as the basic principles which were responsible for conferring on materials their characteristic properties.

The name of Democritus (ca. 460-370 B.C.), another famous Greek philosopher, is usually associated with the origin of the concept of atoms. He opposed the view held by many of his contemporaries that matter was continuous in its structure, maintaining that it was composed of discrete particles or very tiny marblelike atoms. If one were to divide and subdivide matter into smaller and smaller bits, one would eventually obtain a particle incapable of further subdivision. This indivisible unit would be called an atom. Democritus' theory also assumed that atoms moved rapidly through empty space—a reminder that the kinetic theory of matter was anticipated many centuries before Daniel Bernoulli (1738).

Alchemy very likely had its origins in Alexandria, Egypt, shortly after the beginning of the Christian Era, and it flourished for nearly two millennia. At one time it was fashionable for European monarchs to have court alchemists. The alchemist's principal pursuit was the search for the philosopher's stone. He may have reasoned that since a reddish earth could be converted into a useful metal like iron, it should be possible to refine the commoner metals to the precious gold. And thus it was that hundreds of human beings were encouraged to spend their lives looking for the illusory "philosopher's stone," the mythical substance which was reputed to convert baser metals into gold (Fig. 7-7).

Fig. 7-7. *An alchemist of the late Middle Ages works at a furnace, helped by his apprentice. They are engaged in the "great work"—making gold.* [*Courtesy, Princeton University Library.*]

At about the time that Christopher Columbus made his celebrated trip to the New World, a Swiss child was born who was to exert a great influence on the future development of the science of chemistry. Paracelsus (1493-1541), as he was called, attempted to steer man away from the vain pursuit of the philosopher's stone to the search for medicines which would alleviate human suffering. His efforts gave rise to a new movement in chemistry, elevating its status to a higher plane than had been attained under the alchemists. The medical chemists—"iatrochemists"—in their search for the elixir of life brought forth much new factual information about the properties of chemical substances and provided grist for the mill from which some of the great generalizations of science were to evolve.

Seventeenth-Century Experimentation in Chemistry Though Paracelsus initiated a new trend in chemistry, alchemy was hardly banished in a single stroke. Even after Paracelsus' death, Jan Baptista van Helmont (1577-1644) of Brussels, who made important contributions to the understanding of chemistry, still clung to the belief that baser metals might be transmuted into gold. Van Helmont regarded water and air as the basic elements, since it appeared to him that they could not be converted into one another nor reduced into simpler elements.* In this respect he did not progress significantly beyond the thinking of Thales who had lived 2000 years earlier. In the experimental work of Van Helmont, however, a new trend in science makes its start, that of added attention to quantitative relations. This was the time of Francis Bacon ("father of scientific method"), a period which was ripe for discoveries of a quantitative nature. Van Helmont's famous willow-tree experiment shows his concern for such questions as "How much?" and "From whence?" (Fig. 7-8). In this experiment a tree weighing five pounds

Fig. 7-8. *J. van Helmont and his son (from his book on medical plants).* [*Courtesy, Princeton University Library.*]

was planted in a vessel containing 200 pounds of dried earth. During a five-year period he supplied only water to the growing tree. It continued to increase in size and at the end of this period weighed nearly 170 pounds. The earth in which the tree was planted was dried and weighed; its weight was found to be almost exactly the same as its original weight (within a few

* J. R. Partington, *A Short History of Chemistry,* Macmillan, 1939, p. 51.

ounces). The weight increase could not, then, have come from the soil. What was more natural than to suppose it was due to the water? Van Helmont's conclusion was wrong, as is now apparent, but his work represents an early attempt to gain knowledge through quantitative measurement, crude though it was.

Van Helmont also knew that a strip of silver metal dissolved in nitric acid (HNO_3) and was concealed in the liquid much as sugar or salt is concealed when dissolved in water (see Fig. 7-9). When all the water was

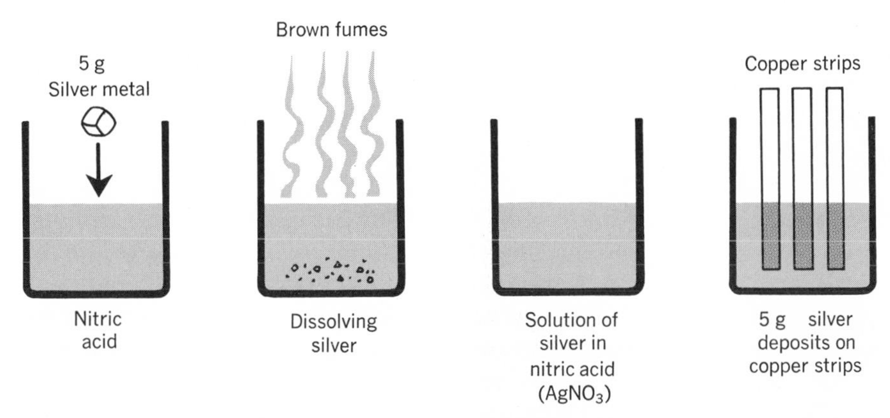

Fig. 7-9. *Van Helmont understood that a metal might be dissolved in a liquid and then be reclaimed from the solution in its entirety. Atoms are neither created nor destroyed as a result of chemical action.*

evaporated from a nitric acid solution of the silver, a white residue resembling salt was left in the evaporating dish. Van Helmont believed this white salt contained all the silver which had originally dissolved in the acid. This belief was confirmed by the fact that all the original silver could be regained from a water solution of the white salt by simply immersing strips of copper in it. When this was done, silver of the same weight as the original metal separated from the liquid and was deposited on the surface of the copper. Thus, early in the seventeenth century it was clear that metals could undergo chemical reactions without any matter being destroyed or created in the process. Although not explicitly stated, this could be regarded as recognition of the principle of the conservation of matter which states that in a chemical reaction matter may be neither created nor destroyed.

Emergence of the Concept of an Element As Van Helmont approached middle age, Robert Boyle, mentioned in Chapter 6, was born in Ireland. He not only carried out experiments dealing with the "spring of the air" but also made a name for himself as a chemist. His volume entitled *The*

Sceptical Chymist gives considerable insight into seventeenth-century chemistry, as well as dealing a death blow to the four elements of the Greeks. In this book Boyle stated clearly the concept of an element as now understood: an element was simply a substance which could not be resolved into a simpler form by means of chemical reactions. If a material defied all efforts to further resolve it into simpler substances it was to be considered an element. Thus Boyle is to be remembered not only for the gas law which bears his name, but also for his clarification of one of the most important concepts of chemistry, that of a chemical element. He did not, however, develop any experimental method of ascertaining which substances were elements, nor was a single gaseous element identified during his lifetime.

A surprisingly large number of the crucial early experiments in chemistry involved work with gases. Those relating to the process of combustion were especially important in establishing, on an experimental basis, the concept of an element. During the latter part of the seventeenth century the *phlogiston theory* of combustion, proposed by Becher and Stahl in Germany, gained wide acceptance. "Phlogiston" was the substance believed to be present in all combustible materials. When they burned, the phlogiston was thought to escape. The theory did give a plausible explanation of some features of burning. It accounted for the fact that combustible materials soon "went out" when only a limited supply of air was available. Air was needed to absorb the phlogiston from the burning substance. Once the air became saturated with it, the fire went out. Before it would again support combustion, the air had to be dephlogisticated. Logical enough! Other facts were more difficult to explain.

When wood was burned, the ash weighed less than the original material. This made sense; phlogiston was lost. On the other hand if metals such as tin were heated in air, the resulting calx (tin oxide) weighed *more* than the original tin. This was more difficult to explain. Why should there be a weight increase in one case and a decrease in another? The phlogistonists gave the explanation that phlogiston might have either positive or negative weight.

Such "ad hoc" hypotheses may prove difficult to refute at the time but often yield later to the onslaught of compelling logic. The days of the phlogiston theory were numbered.

In the early 1770's both the Englishman Priestley (1733-1804) and the French scientist Lavoisier (1743-1794) struggled with this and other troublesome questions concerning the nature of combustion and calcination (conversion of metals into their oxides). Lavoisier had proved that many combustible substances such as sulfur, phosphorus, and metals increased in weight when heated in air, the phlogiston theory notwithstanding. He suspected that this weight increase was derived from something in the air, but it took him several years to prove it. One of the important clues to the solution of the problem came from Priestley's discovery of "dephlogisticated air" (oxygen) by heating red mercuric oxide. Lavoisier was quick to realize that there might be a connection between Priestley's newly discovered element and the processes of calcination and combustion. He soon carried out the now famous experiment of heating mercury in contact with a confined volume of air (Fig. 7-10) and found that the mercury increased

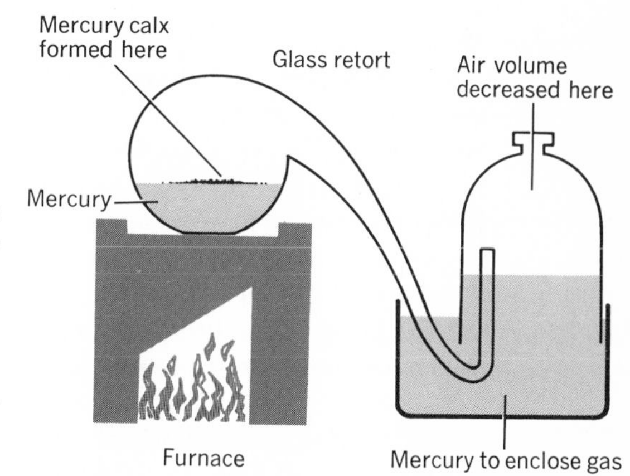

Fig. 7-10. *Lavoisier's demonstration that calcination depends on a constituent of the air.*

in weight while a fifth of the air disappeared. Oxygen was removed from air as it combined with mercury to form the heavier mercuric oxide, and the decrease in the weight of the air exactly matched the weight increase of the mercury.

Lavoisier had now put together the pieces of this complex puzzle of combustion and calcination; the element oxygen, a constituent of air, combined with a combustible material as it burned, or with a metal as it was calcined. When metals were heated in air, oxygen atoms attached, forming a solid, nonvolatile product. Its weight was greater than the original metal. But when combustible materials united with oxygen (burned), gaseous products formed. Their weight had to be taken into account. The sum of the weights of wood and oxygen taken from the air was the same as for ash and gaseous oxides.

One would think that in the face of experimental evidence convincingly supporting his logic, Lavoisier's explanation would at once become universally accepted. Unfortunately such was not the case. Even with all the facts available to him, Priestley remained a phlogistonist to his dying day.

Although Lavoisier had established himself as one of the immortals of science, he fell victim to the guillotine during the French Revolution. According to reports of his trial, a judge declared, "France has no need of scientists (*savants*)." The famous mathematician Lagrange commented that it took but a moment to cut off Lavoisier's head, though a hundred years might pass before there would be another like it.

THE LAW OF CONSTANT COMPOSITION

Just before the turn of the nineteenth century a French chemist, Joseph Proust (1754-1826), reported the results of an extensive study he had made concerning the composition of metallic compounds. In his report he announced the discovery of a *law of constant composition.* According to this law, the elements contained in a compound were always found in the same proportions by weight. In water the oxygen always weighed eight times as much as the hydrogen; the ratio of these weights never changed. To be sure, the elements carbon and oxygen combined to form carbon dioxide in one case, and carbon monoxide in the other, but these were two distinct compounds. In carbon dioxide there was *always* the same weight ratio of carbon to oxygen, and the same was true for the monoxide.

Proust's law of constant composition was hardly announced before it was challenged. Another French chemist, Berthollet (1748-1822), pointed out that when copper was heated in air, it might react with either a very small or a very large amount of oxygen. In fact a series of experiments convinced him that the amount of oxygen combining with copper could be varied almost continuously up to a certain maximum amount. It was absurd, then, to suggest that copper and oxygen combined in a definite ratio by weight.

Proust counterattacked; he argued that not all of the copper atoms had enough oxygen atoms with which to combine. But if the two elements united to form a given compound, then the numbers of atoms of each would always be the same. The argument was eventually resolved in favor of Proust. His law of constant composition is now regarded as one of the cornerstones of the quantitative basis of chemical reactions. The controversy between Proust and Berthollet regarding the nature of chemical composition illustrates very well how even famous scientists may interpret essentially the same set of data quite differently. This further suggests that the methods of science are far more complicated than merely extracting generalizations from masses of apparently unrelated data. The resolving of the Proust-Berthollet controversy immediately preceded and was undoubtedly related to the publication of Dalton's atomic theory, which eventually dominated the field.

DALTON'S ATOMIC THEORY

The concept of atoms, which had originated 2000 or more years earlier in the time of the Greeks, remained largely dormant until the seventeenth century. At that time the French philosopher and mathematician Pierre Gassendi (1592-1655) succeeded in reviving interest in the particle structure of matter. During that century and the next the words *corpuscle* and *particle* became increasingly prominent in scientific writing, but a comprehensive, unifying theory was still in the future.

John Dalton (1766-1844), the Manchester schoolmaster, published his atomic theory in the early years of the nineteenth century. According to this theory all substances were composed of tiny particles called atoms. Furthermore, all similar atoms—for example, all hydrogen atoms—were believed to have the same weight. But a hydrogen atom had a different weight than an atom of another element—for example, oxygen. The idea that all atoms of the same element had the same weight was later modified to conform to more modern experimental evidence, but for the present consider this to be true.

Dalton's theory also stated that atoms might combine to form compounds. If atoms of hydrogen, all of the same weight, combined with oxygen atoms all of the same weight, and the numbers of each in a cluster were the same, then Proust's law of constant composition followed. The significance of the term molecule was not clearly understood in Dalton's time, but Dalton did know that the oxygen in water weighed eight times as much as the hydrogen. Hence one ninth of the weight of water is due to its hydrogen atoms, the other eight ninths to its oxygen atoms.

To summarize the important points of Dalton's atomic theory:

(1) Atoms of the same element have the same weight.
(2) Atoms of different elements have different weights.
(3) Atoms may attach to form aggregates, combining in a fixed ratio in forming a given compound.

SYMBOLS AND FORMULAS

Dalton used the symbols shown in Fig. 7-11 to represent the atoms known in his day. More recently the symbol has become a capital letter (as O for oxygen) or a capital letter followed by a small letter (as Pb for lead). Each symbol represents one atom of an element; associated with the symbol is a number called the atomic weight. This number indicates the weight of the atom relative to a carbon atom, which has been assigned a

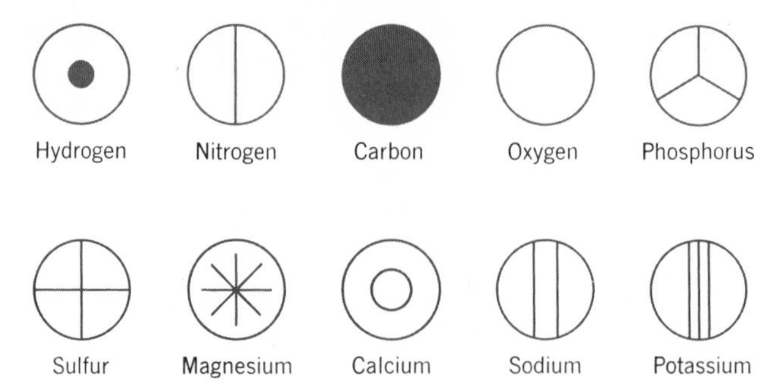

Fig. 7-11. *Dalton's symbols for chemical elements.*

value of 12. If magnesium has an atomic weight of 24, a single atom weighs twice as much as one of carbon. Table 7-1 lists the symbols of some of the most common elements. A more complete list, also including atomic weights, and atomic numbers, is given in the Appendix.

Table 7–1 Some Important Elements

Ag	Silver	Fe	Iron	N	Nitrogen
Al	Aluminum	H	Hydrogen	Na	Sodium
C	Carbon	Hg	Mercury	O	Oxygen
Ca	Calcium	I	Iodine	P	Phosphorus
Cl	Chlorine	K	Potassium	S	Sulfur
Cu	Copper	Mg	Magnesium	Si	Silicon
				Zn	Zinc

Atoms combine to form clusters called molecules. The molecule is represented in chemical shorthand by a formula: H_2O is a familiar example. The subscript 2 indicates that there are two hydrogen atoms for every

Table 7–2 Some Common Compounds

NH_3	Ammonia	MnO_2	Manganese dioxide
NH_4Cl	Ammonium chloride	HNO_3	Nitric acid
$BaCl_2$	Barium chloride	$KClO_3$	Potassium chlorate
$CaCl_2$	Calcium chloride	KNO_3	Potassium nitrate
$Ca(NO_3)_2$	Calcium nitrate	$AgNO_3$	Silver nitrate
CaO	Calcium oxide	NaCl	Sodium chloride
CO_2	Carbon dioxide	SO_2	Sulfur dioxide
CO	Carbon monoxide	H_2SO_4	Sulfuric acid
$Cu(NO_3)_2$	Copper nitrate	$ZnCO_3$	Zinc carbonate
		ZnS	Zinc sulfide

oxygen atom. Carbon dioxide (gas) molecules consist of clusters of three atoms—one of carbon and two of oxygen. Its formula is CO_2. The symbol for oxygen is O, but its formula is O_2, since each oxygen molecule consists of two atoms.

THE CHEMICAL EQUATION

The practical side of the science of chemistry is concerned with creating conditions under which atoms combine with one another or become separate from one another. Whenever either process occurs, a *chemical change* takes place. A chemical change alters the composition of the structural aggregates. An example is the union of carbon and oxygen to form carbon dioxide:

$$C + O_2 \rightarrow CO_2 \tag{7a}$$

Equation 7a indicates that atoms of the element carbon (represented by the symbol C) unite with atoms of the element oxygen (represented by the formula O_2) to form the compound carbon dioxide (represented by the formula CO_2).

The carbon may be a piece of charcoal, graphite, or even diamond. Each is a solid composed of carbon atoms, though charcoal is noncrystalline while graphite and diamond have different crystalline structures. The symbol tells nothing about the physical state of the substance, though this is often understood from a knowledge of properties. For example the oxygen used in burning the charcoal might come from the air, or it could be pure oxygen. As a result of heating the charcoal in oxygen or air, the atoms are activated so that they attach, a carbon atom uniting with two oxygen atoms to form clusters of three atoms each, called carbon dioxide molecules.

In other cases an equation may represent a single displacement reaction such as the following:

$$Pb + Cu(NO_3)_2 \rightarrow Cu + Pb(NO_3)_2 \tag{7b}$$

This reaction indicates that metallic lead (an element) is combining with (a water solution of) copper nitrate $Cu(NO_3)_2$ (a compound). The reaction takes place when strips of lead are immersed in a solution of copper nitrate (Fig. 7-12). (Nitrates are nearly all water soluble.) There is nothing in the equation to suggest that the copper nitrate is dissolved in water, but the chemist soon learns to recognize when a solution is implied. A reddish deposit of copper metal forms on the lead strips, and the original blue color of the solution gradually disappears. If the resulting clear, colorless solution is evaporated, a white salt—lead nitrate, $Pb(NO_3)_2$—is formed. Note that all of the atoms found on the left-hand side of the arrow in Equation 7b are also present on the right side. The net effect is to replace copper atoms by lead atoms; hence the type is a single replacement reaction.

Sometimes a chemical reaction is used to separate a compound into its elements, as in the equation

$$2\ NaCl \rightarrow 2\ Na + Cl_2 \quad (7c)$$

Equation 7c represents the chemical reaction taking place if an electric current (D.C., i.e., direct current) is passed through molten table salt (NaCl). The electric current separates sodium atoms from chlorine atoms. Thus two elements, sodium metal and chlorine gas, are produced from a compound containing them. No atoms have been created, none destroyed. The atoms (ions) of the salt crystal were separated.

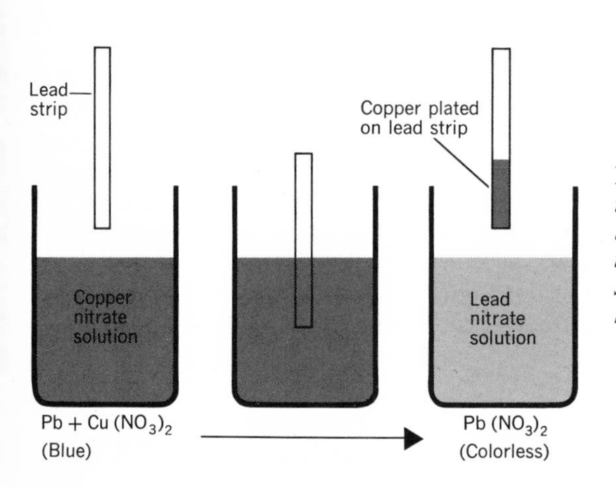

Fig. 7-12. *When a strip of lead metal is immersed in a copper nitrate solution, a coating of copper plates out on the lead. Lead nitrate is found dissolved in the liquid. A single replacement reaction has taken place.*

Sodium chloride, table salt, is a very abundant and hence inexpensive material. It occurs in huge underground deposits as shown in Fig. 7-13 and serves as a raw material from which to make chlorine gas for chlorinating water supplies and swimming pools or for use as a bleaching agent.

There are numerous other types of reactions, but those cited will suffice to illustrate the point that a chemical equation is a shorthand representation of the alteration in the manner of combination of atoms brought about by means of a chemical reaction. Our ability to cause atoms to combine in ways other than those in which they are found in nature makes possible the production of the great variety of chemical substances needed in everyday life.

Every equation expresses some sort of equality. In the chemical equations discussed above, note that there are the same numbers of each kind of atom at right and left of the arrow. The arrow is in effect an equality sign.

In the first two equations of this section there are no coefficients preceding the symbols and formulas. In the third equation there are coefficients.

Fig. 7-13. *Avery Island Salt Mine in Louisiana. The ceiling is about 100 ft high. The machinery in the background is a self-propelled platform rig used for drilling holes in the salt, for loading the holes, and for scaling the sidewalls and ceiling.*

Apparently there is an occasional need for a method of equalizing the numbers of each kind of atom. Such a method will be discussed later; it is called "balancing" an equation. From what has already been said it will be clear that there is not a true equation until balancing has been carried out.

ATOMIC WEIGHTS

Dalton said that each kind of atom had a definite weight which distinguished it from all other atoms. Two atoms which had different weights were simply atoms of different elements. He knew that the oxygen from water weighed eight times as much as the hydrogen, but the distinction between an atom and a molecule was not made during his lifetime. The values he used for the relative weights of atoms were then quite inexact compared with current values.

Whenever relative weights of atoms are involved a standard is implied. What is the chemist's standard atom? The atom of carbon which has been assigned a weight of 12.* The atomic weight of any other element is simply the weight of its atom relative to the weight of an atom of carbon. An atom of atomic weight 4 is one third as heavy as an atom of carbon; an atom with atomic weight 144 would be twelve times as heavy.

Until 1961 the standard atom was oxygen, with an assumed weight of 16, and many older books refer to the earlier standard, oxygen. The difference in any given atomic weight between the old figure and the new is rarely great enough to affect calculations at an elementary level.

How are the relative weights of single atoms determined? It is of course impossible to isolate and weigh single atoms. But it is possible to determine relative weights of atoms by means of a particle-counting principle proposed by the Italian scientist Avogadro. According to this principle, "equal volumes of gases under the same conditions of temperature and pressure contain the same number of molecules." A liter of hydrogen and one of oxygen would contain the same number of molecules if the temperatures and pressures were equal (Fig. 7-14). If the two gases were weighed, the oxygen would be 16 times as heavy as the hydrogen. There would be many billions of molecules in each liter, but as long as there was the same number of molecules in each, it might logically be concluded that a single oxygen molecule weighed 16 times as much as a single hydrogen molecule. Carrying this reasoning one step farther, since molecules of oxygen and hydrogen are known to contain 2 atoms each, it might also be concluded that a single oxygen atom weighed 16 times as much as a single hydrogen atom.

The foregoing is far from a complete story about atomic weights, and merely illustrates a method which has been used for determining them. Ultimately they are derived from a multitude of meticulously done quantitative experiments. One of the pioneers in the field of atomic weight determinations was the Swedish chemist Berzelius (1779-1848). His tables of atomic weights, published in the early decades of the nineteenth century, contained remarkably accurate values, considering the apparatus which was available to him.

Formula Weights and Molecular Weights Just as there is associated with every chemical element a numerical value called its atomic weight, there is also corresponding to every formula a number known as its formula weight. Often this quantity is called the molecular weight of the compound which the formula represents. (However, a formula does not necessarily imply that the aggregates making up the compound are molecules, and hence

* For a century or more chemists have been using atomic weights of the elements. This practice will be retained here. However, as the mass spectrograph and isotopes are described in subsequent sections, it will become evident that in view of recent revisions in the standard, it would be more appropriate to refer to the *atomic mass* of a C^{12} atom as 12.000 000 atomic mass units.

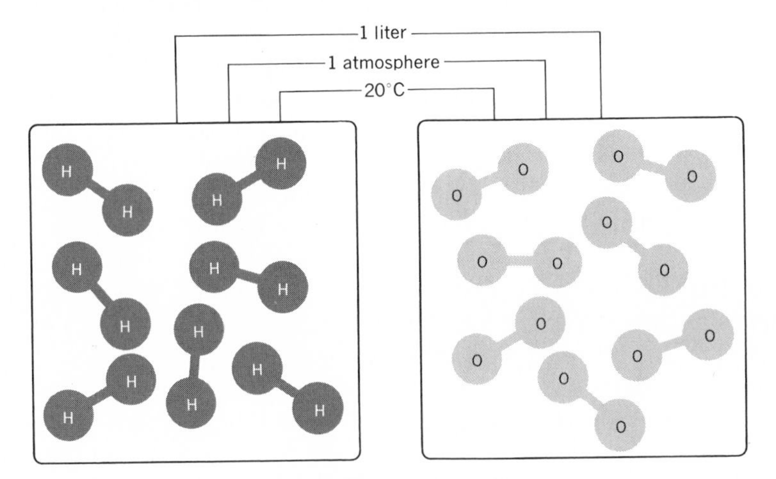

Fig. 7-14. *Equal volumes of gases under the same conditions of temperature and pressure contain the same number of molecules. A liter of oxygen weighs 16 times as much as a liter of hydrogen. Hence single oxygen molecules are 16 times as heavy as single hydrogen molecules. Since each oxygen molecule consists of two oxygen atoms and each hydrogen molecule consists of two hydrogen atoms, individual oxygen* atoms *are likewise 16 times as heavy as individual hydrogen* atoms.

it is perhaps better to speak of the formula weight, even though in many cases such as CO_2, SO_2, and CH_4, the term molecular weight would be equally appropriate.)

The formula weight of a compound is the sum of the atomic weights of the constituent atoms. The gram formula weight is the formula weight expressed in grams.

ILLUSTRATIVE PROBLEM

Calculate the gram formula weight of $Cu_3(AsO_4)_2$. See the Appendix for a table of atomic weights.

Solution

The subscript 3 after the Cu indicates three gram atoms* of copper. The subscript 2 following the parentheses has the effect of doubling everything within them. Therefore, in this formula there are

* A gram atom is a weight of the element equal to its atomic weight in grams. A gram atom of any element contains the same number of atoms as a gram atom of any other element.

$$
\begin{aligned}
\text{3 gram atoms of Cu} &= 3 \times 63.5 = 190.5\ \text{g} \\
\text{2 gram atoms of As} &= 2 \times 74.9 = 149.8\ \text{g} \\
\text{8 gram atoms of O} &= 8 \times 16 \quad = \underline{128.0\ \text{g}} \\
\text{Gram formula weight} &= 468.3\ \text{g}
\end{aligned}
$$

Earlier in the chapter the law of constant composition was discussed. Now it should be evident why it is true. If a compound always contains the same number of atoms, and if the atoms always have the same weight, each element must always constitute the same fraction of the total weight. In order to calculate this percentage, first form a fraction by dividing the weight of the element in the compound by the total weight (i.e., the formula weight). This fraction multiplied by 100 is the desired percentage.

ILLUSTRATIVE PROBLEM

Calculate the percentage composition of carbon dioxide (CO_2).

Solution

Gram formula weight of CO_2

$$
\begin{aligned}
\text{2 gram atoms of oxygen} &= 2 \times 16 = 32\ \text{g} \\
\text{1 gram atom of carbon} &= 1 \times 12 = \underline{12\ \text{g}} \\
\text{Gram formula weight} &= 44\ \text{g}
\end{aligned}
$$

$$\frac{12\ \text{gram}}{44\ \text{gram}} = 0.272 \times 100 = 27.2\%\ \text{carbon}$$

The percentage of oxygen is $100.0 - 27.2 = 72.8\%$ oxygen.

The Gram Molecular Weight and the Mole A gram molecular weight of a substance is called a *mole*. The molecular weight of methane, CH_4, for example, is 16; so 16 grams of methane is a mole. A mole of SO_2 (gas) is 64 grams. Furthermore, a mole of any "ideal" gas occupies a volume of 22.4 liters at STP (standard temperature and pressure: 0°C and 1 atm); 22.4 liters is called a gram molecular volume. It is a very important figure to remember, for it relates the weight of a gas and its volume: 22.4 liters of SO_2 weigh 64 grams at STP, because 64 is the molecular weight. The weight of a different volume of the gas may be determined from this relation.

ILLUSTRATIVE PROBLEM

Calculate the weight of 10.0 liters of ethane (C_2H_6) gas at STP.

Solution

A mole (molecular weight) of ethane is 30.0 g

2 gram atoms of carbon	$= 2 \times 12.0 =$	24.0 g
6 gram atoms of hydrogen	$= 6 \times 1.0 =$	6.0 g
Gram molecular weight or gram formula weight	$=$	30.0 g

$$\frac{30.0 \text{ g}}{22.4 \text{ l}} = \frac{x \text{ g}}{10.0 \text{ l}}$$

$$x = 13.4 \text{ g}$$

To further distinguish the terms molecular weight, formula weight, and mole, note the sum of the atomic weights of the elements in NaCl is 58.5; so 58.5 g is often called a mole. The structural units in a crystal of NaCl are electrically charged particles (ions, represented by Na^+ and Cl^-), not molecules. Hence the term *formula weight* is appropriate, but molecular weight is not.

Additional experience will be needed in distinguishing which substances actually exist as molecular aggregates. Indeed, the question is often not a simple one.

As already discussed, Avogadro's principle provides a means of determining the weights of single atoms or molecules without weighing them individually. The number of atoms in 1 gram atomic weight of an element, or the molecules in 1 gram molecular weight of a compound, has actually been measured. This number, called the Avogadro number, is of great importance in chemical calculation.

THE AVOGADRO NUMBER

The atomic weight of the element carbon is 12. Twelve grams of carbon contain 6.02×10^{23} atoms. Similarly, in four grams of helium gas (its atomic weight is 4) there are the same number of atoms. This number—6.02×10^{23}—is called the Avogadro number and is an important constant of chemistry and physics. The gram atomic weight of *every element* will contain this number of atoms. The gram molecular weight of many compounds* will also contain an Avogadro number of molecules. Thus there will be 6.02×10^{23} atoms in 40 grams of calcium metal, Ca, and the same number of molecules in 44 grams of CO_2 gas or in 342 grams of sucrose, $C_{12}H_{22}O_{11}$. With the aid of this number, then, it is possible to find the weight in grams of a single atom or molecule.

* In certain crystalline substances the structural particles are ions rather than molecules.

ILLUSTRATIVE PROBLEM

Calculate the weight of a helium atom.

Solution

$$4.003 \text{ g of helium contain } 6.02 \times 10^{23} \text{ atoms}$$

$$\frac{4.003 \text{ g}}{6.02 \times 10^{23} \text{ atoms}} = 6.65 \times 10^{-24} \text{ g/atom}$$

VALENCE AND OXIDATION NUMBER; FORMULA WRITING

Oxygen combines with many elements, forming compounds called oxides. The oxide of hydrogen is the familiar H_2O. Calcium oxide is CaO; aluminum oxide is Al_2O_3. Why are the subscript numbers used in some cases but not in others? Why are HCl, $CaCl_2$ and $AlCl_3$ all correct formulas? The questions concern the relative numbers of atoms making up the material. In liquid water there will be twice as many atoms of hydrogen as oxygen. In calcium oxide there will be one calcium atom for each oxygen. Apparently atoms of a given element such as oxygen may attach to other elements in different but definite integral ratios.

The phase of chemical science dealing with the manner in which atoms of the different elements are held together is called *bonding*. The nature of chemical bonding will be more clearly understood after the electronic structure of atoms has been studied in Chapter 15. In the meantime, however, use will be made of the concept of valence as an aid in the writing of correct formulas.

Valence may be described as the combining capacity of a single atom of an element in the particular substance being considered or the number of (single) bonds one atom of the element does form with the other atoms of the compound. Hydrogen is assigned a valence of 1 and it usually has an oxidation number of +1. (The oxidation number is simply the valence with a plus or minus sign attached.) Any other element such as chlorine, bromine, or iodine, which will combine with hydrogen atom for atom, will be assigned an oxidation number of −1. The writing of correct formulas is accomplished by balancing the positive and negative oxidation numbers of all elements in a formula. NaCl is a correct formula since the +1 of sodium balances the −1 of chlorine. Since the oxidation number of Ca is +2 and of oxygen −2, the correct formula for calcium oxide is CaO. The positive oxidation number of the calcium and the negative value of the oxygen are equal.

In writing formulas, subscript numbers are introduced to balance the positive and the negative charges. For example, $CaCl_2$ is the correct formula for calcium chloride, since a single calcium atom has an oxidation number of +2, and a single chlorine atom of −1. What is the correct formula for aluminum chloride if aluminum has a charge of +3? For tin(IV) chloride or stannic chloride, if tin has an oxidation number of +4? Why is Al_2O_3 the correct formula for aluminum oxide (aluminum, +3; oxygen, −2)? Cf. Formula Writing Sheets in the Appendix.

Occasionally a group of atoms may preserve its identity in a series of reactions. Such groups of atoms were for many years called radicals. Because in recent years the term radical has come to have other connotations, they will be called ions. (An ion is an electrically charged atom or group of atoms.) An example is the nitrate ion, a nitrogen atom and three oxygen atoms, which as a group has a charge of −1; the nitrate ion is written NO_3^-. This ion does not exist alone; it is always found in combination with an element of positive oxidation number. KNO_3 is potassium nitrate; $Ca(NO_3)_2$ is calcium nitrate; $Al(NO_3)_3$ is aluminum nitrate.

The sulfate ion is written $SO_4^=$. The two superscript minus signs following it indicate its ionic charge (since in this ion S has an oxidation number of +6 and 4 oxygens, −8). Like the nitrate, it exists only in combination with elements of positive oxidation number, for example, Na_2SO_4, $CaSO_4$, or $Al_2(SO_4)_3$. Other ions with negative ionic charges include the carbonate ($CO_3^=$), and the phosphate ($PO_4^{\equiv}$).

The ammonium ion, a nitrogen atom attached to four hydrogens and bearing a net charge of +1 is a positive ion. It exists only in combination with negatively charged ions. Ammonium chloride is NH_4Cl and ammonium sulfate $(NH_4)_2SO_4$.

Some elements have more than one oxidation number. For example, iron has oxidation numbers of +2 and +3. There are two chlorides of iron, $FeCl_2$ (iron II chloride or ferrous chloride) and $FeCl_3$ (iron III chloride or ferric chloride). Note that the *-ic* compound has a higher oxidation number than the *-ous*.

FROM CHEMICAL REACTION TO BALANCED EQUATION

A chemical equation represents an equality of atoms. The left member of an equation indicates the atoms present and their manner of combination (formulas) when the reaction starts. The right member shows how the same atoms are combined when the reaction is complete. Since atoms may be neither created nor destroyed, the numbers on one side must be made equal to those on the other side.

The reaction of sulfuric acid and aluminum may be written

$$Al + H_2SO_4 \rightarrow Al_2(SO_4)_3 + H_2\uparrow$$

This is not an equation, for there are two atoms of Al on the right side and only one on the left. The mass conservation law has been violated.

Why not drop the subscript after the Al in $Al_2(SO_4)_3$? Then the Al atoms would be balanced; but an incorrect formula for aluminum sulfate would result. In representing a reaction, correct formulas for reactants and products must be written. Thereafter subscript numbers must not be altered. Balancing must be achieved by introducing coefficients before the appropriate formulas.

To balance Al atoms, introduce a coefficient of 2 before Al on the left side of the equation. To balance S atoms, introduce a coefficient of 3 before H_2SO_4. To balance hydrogen atoms insert 3 before H_2. The equation is now balanced.

$$2\ Al + 3\ H_2SO_4 \rightarrow Al_2(SO_4)_3 + 3\ H_2\uparrow$$

Note: a coefficient preceding a formula affects each element within it. $3H_2SO_4$ means 6 hydrogens, 3 sulfurs, and 12 oxygens. The following "balance sheet" shows why the equation is now balanced.

LEFT MEMBER	RIGHT MEMBER
2 atoms of aluminum	2 atoms of aluminum
6 atoms of hydrogen	6 atoms of hydrogen
3 atoms of sulfur	3 atoms of sulfur
12 atoms of oxygen	12 atoms of oxygen

The vertical arrow after H_2 indicates that it is a gas. An arrow directed downward indicates an insoluble substance (precipitate). When clear solutions of calcium chloride and silver nitrate are mixed, the liquid becomes milky and a white solid (precipitate) settles. The reaction is

$$CaCl_2 + 2\ AgNO_3 \rightarrow Ca(NO_3)_2 + 2\ AgCl\downarrow$$

WEIGHT RELATIONS IN CHEMICAL REACTIONS

An amazing aspect of the preceding information concerning chemical formulas, formula weights, and equations is its power to predict the quantities of chemical substances which will react. Hydrogen burns in air (unites with the oxygen of the air); water is formed. The reaction is

$$H_2 + O_2 \rightarrow H_2O$$

Is the reaction balanced? No. It appears that as a result of the reaction, two atoms of oxygen have been reduced to one. The law of the conservation of matter denies this possibility. But if coefficients of 2 are introduced before hydrogen and water, then the equation is balanced.

$$\underset{4}{2\,H_2} + \underset{32}{O_2} \rightarrow \underset{36}{2\,H_2O}$$

In this balanced equation, hydrogen has a weight of 4 (two formula weights), oxygen of 32 (one formula weight), and water of 36 (two formula weights). The weight of the oxygen is eight times that of the hydrogen, or in other words, the weight ratio in which hydrogen and oxygen combine is 4:32. This ratio never changes. Once the equation has been worked out and there is assurance that this is the only reaction taking place, the formula weights correctly predict the reaction ratios. The weight of water formed in this reaction will always be 9 times the weight of the hydrogen used or $\frac{9}{8}$ the weight of oxygen used.

ILLUSTRATIVE PROBLEM

How much water is produced by burning 5.0 g of hydrogen?

Solution

Let x = the number of grams of water produced.
Since the weight of water is 9 times that of the hydrogen used,

$$\frac{H_2O}{H_2} = \frac{x}{5.0\text{ g}} = \frac{9}{1}, \text{ and } x = 45\text{ g}$$

45 g of water would be produced by burning 5.0 g of hydrogen.

Consider the following reaction. Is it balanced?

$$Mg + O_2 \rightarrow MgO$$

No. There are 2 oxygens on the left and 1 on the right. Introducing a 2 before MgO balances oxygen; a 2 before Mg balances magnesium.

$$2\,Mg + O_2 \rightarrow 2\,MgO$$

The formula weights may be calculated as follows

$$\begin{aligned} 2\,Mg &= 2 \times 24 = 48 \\ O_2 &= 2 \times 16 = 32 \\ 2\,MgO &= 2(24 + 16) = 80 \end{aligned}$$

The weight relation of the quantities is

$$\underset{48}{2\,Mg} + \underset{32}{O_2} \rightarrow \underset{80}{2\,MgO}$$

It is now possible to answer such questions as "What weight of magnesium is needed to react completely with 10 grams of oxygen?" Magnesium and oxygen always react in the ratio of 48:32, i.e., of 3:2. Stated another way, the weight of magnesium is always $1\frac{1}{2}$ times as large as that of the oxygen with which it unites: 15 grams of magnesium react with 10 grams of oxygen.

The fraction 48/32 is obtained from the formula weights of the two chemical substances under consideration. Even though there will often be more than two reactants in any chemical equation, in calculating weight problems, focus on only two at a time. Other necessary reactants are assumed to be present in appropriate quantities: other products than the one under consideration may be formed. The two quantities under consideration may both be reactants, or both products.

ILLUSTRATIVE PROBLEM

What weight of zinc sulfate is produced by the reaction of 10.0 g of zinc with an appropriate quantity of sulfuric acid?

Solution

The balanced equation for the reaction is

$$\underset{65}{Zn} + \underset{98}{H_2SO_4} \rightarrow \underset{161}{ZnSO_4} + \underset{2}{H_2}$$

The formula weights are given below each substance. The weight of zinc sulfate produced will be 161/65.0 the weight of the zinc used. Consequently, if $x =$ the number of grams of zinc sulfate produced by the reaction of 10 g of zinc,

$$\frac{ZnSO_4}{Zn} = \frac{161}{65.0} = \frac{x}{10.0\text{ g}}, \quad x = \frac{1610\text{ g}}{65.0} = 24.8\text{ g}$$

24.8 g of zinc sulfate will result if 10.0 g of zinc metal are placed in enough H_2SO_4 to react completely. The proportion relates only to the quantities of Zn and $ZnSO_4$. H_2SO_4 and H_2 also appear in the equation but are disregarded in this problem.

An entirely different problem might have been stated concerning this same equation:

ILLUSTRATIVE PROBLEM

How many grams of zinc must react with sulfuric acid in order to produce 5.00 g of hydrogen?

Solution

$$\begin{array}{llll} Zn + & H_2SO_4 \rightarrow & ZnSO_4 + & H_2\uparrow \\ 65 & 98 & 161 & 2 \end{array}$$

$$\frac{Zn}{H_2} = \frac{65.0}{2.00} = \frac{x}{5.00\text{ g}}$$

$$2x = 325\text{ g}$$
$$x = 162.5\text{ g}$$

162.5 g of zinc reacting with a sufficient amount of sulfuric acid (which is perfectly definite and predictable by means of the equation) produces 5.00 g of hydrogen gas. Note that there is no mention of the amount of zinc sulfate formed along with the hydrogen. The quantity of it formed from the given amount of zinc is definite and predictable even though of no consequence in this problem.

The principles just discussed form the cornerstone of stoichiometry, the study of quantitative chemical relations. The chemist employs them in predicting the amounts of reactants needed for a desired quantity of product. Thus the economical operation of chemical industry depends on their application.

THE METALLURGY OF ZINC

Chemistry is the science dealing with the conditions under which changes in the composition of matter are brought about. The metallurgy of zinc will illustrate how such composition changes are brought about in an industrial chemical process.

The most commonly occurring zinc ore is a sulfide called sphalerite. Carbonate, oxide, and silicate ores of zinc are also known. Since ores are often mixed with earthy materials, an early stage of refining is frequently one of concentrating the metal-containing portion. This is often accomplished by means of a flotation process which takes advantage of the difference in density between the ore and its impurities. Flotation leaves the ore richer in the desired metallic compound, but involves *physical change* only; there is no alteration in the structure of the constituent molecules, and hence no chemical change.

After concentration, sulfide ores are roasted, i.e., heated in the presence of an abundance of air. Substitution of oxygen for sulfur results. Zinc oxide is formed according to the equation

$$2\,ZnS + 3O_2 \rightarrow 2ZnO + 2SO_2\uparrow$$

What has this last step accomplished? Why is it better to have zinc oxide than zinc sulfide? Simply because oxygen is readily removed from the oxide by a comparatively inexpensive industrial material, coke. Coke is composed of carbon atoms, which react with oxygen, yielding the free metal. No equally inexpensive reactant is available for removing sulfur from zinc sulfide. Production of metals through reduction by carbon of metallic oxides is in wide use, although in some cases, metal atoms are so tightly bound to oxygen that more drastic reaction conditions are necessary.

The zinc oxide which results from the roasting process is molded with coke to form briquettes. These are fed into a *reduction furnace*. Inside the furnace, zinc oxide reacts with carbon according to the equation

$$ZnO + C \rightarrow Zn + CO \uparrow$$

The heat evolved in the reaction causes the zinc to vaporize. This vapor must be condensed at a temperature of about 420°C to obtain a liquid which can be cast. Otherwise, zinc powder will be formed, which is not easily melted to a solid mass. Fig. 7-15 shows modern zinc reduction furnaces.

Fig. 7-15. *Zinc reduction furnaces. Coked briquettes containing zinc are fed into the furnace. The heat resulting from the reaction of zinc oxide and carbon is sufficient to vaporize the zinc. The vertical retort furnace shown is equipped with a very efficient condenser capable of removing 96% of the zinc from the effluent products.* [*Courtesy, New Jersey Zinc Co.*]

After reheating, zinc may be alloyed with other metals, forming brass, bronze, and German silver. Much of the industrial output of zinc is used to protect the surface of sheet iron in the process known as "galvanizing." In this process sheets of iron are carefully cleaned and then dipped in the molten zinc, which adheres to the surface in a thin coat. This is one of the most effective methods of protecting an iron surface from corrosion. Another common use of zinc metal is for the casings of dry-cell batteries. In these it serves as the negative terminal as well as the container for the other cell materials.

In preparing zinc metal, then, the metallurgist starts with an ore containing only a small proportion of metal atoms. Many constituents are present in the ore; it is really a mixture. Concentration of the ore often involves physical separation of impurities, not chemical reactions. The chemical stages of the metallurgical process take place during the roasting of the ore concentrate and during the reaction of zinc oxide with coke. At the start, zinc atoms are mixed with a variety of other atoms; when the process is finished, the product consists almost wholly of zinc atoms. Thus a metallurgical process involves the elimination from the ore of unwanted kinds of atoms.

Problems. Chapter 7

1. Calculate the formula weights of the following compounds:

   $NaCl$ $NaOH$ $Al_2(SO_4)_3$ C_2H_5OH
   $NaHCO_3$ $CaCO_3$ $Ca_3(PO_4)_2$ HNO_3

2. Balance the following equations:

   (a) $H_2 + O_2 \rightarrow H_2O$
   (b) $N_2 + H_2 \rightarrow NH_3$
   (c) $CaCO_3 + HCl \rightarrow CaCl_2 + H_2O + CO_2$
   (d) $Al + H_2SO_4 \rightarrow Al_2(SO_4)_3 + H_2$
   (e) $MgCl_2 + AgNO_3 \rightarrow AgCl + Mg(NO_3)_2$
   (f) $BaCl_2 + K_2CrO_4 \rightarrow BaCrO_4 + KCl$
   (g) $Na_3AsO_4 + Ca(NO_3)_2 \rightarrow Ca_3(AsO_4)_2 + NaNO_3$
   (h) $NaBr + H_2SO_4 \rightarrow Na_2SO_4 + Br_2 + SO_2 + H_2O$

3. Write correct formulas for each of the following substances and balance the equation:

   (a) Silver nitrate + calcium chloride →
   silver chloride (precipitate) + calcium nitrate

   (b) Barium chloride + aluminum sulfate →
   barium sulfate (precipitate) + aluminum chloride

(c) Ammonium iodide + calcium hydroxide →
calcium iodide + ammonia + water
(gas)

(d) Hydrogen + chlorine → hydrogen chloride
(gas)

(e) Calcium carbonate → calcium oxide + carbon dioxide
(gas)

(f) Zinc + silver nitrate → silver + zinc nitrate
(precipitate)

(g) Copper + nitric acid →
dilute
copper(II) nitrate + water + nitric oxide
(gas)

4. (a) Calculate the weight of carbon dioxide gas (CO_2) produced when 25 lb of limestone ($CaCO_3$) react with hydrochloric acid (HCl) according to the equation of Problem 2(c) above.
   (b) Calculate the weight of barium chromate ($BaCrO_4$) precipitate formed when 10 g of barium chloride ($BaCl_2$) react with potassium chromate (K_2CrO_4) as indicated by the equation of Problem 2(f) above.
   (c) Calculate the volume of N_2 required to produce 10 liters of NH_3 according to the equation in Problem 2(b) above. (Assume STP)

5. Calculate the percentage composition by weight of:

   (a) $NaHCO_3$
   (b) HNO_3
   (c) NaOH
   (d) CCl_4

6. A sphalerite ore contains 8.0% ZnS. Calculate the weight of zinc metal theoretically obtainable from a ton of this ore.

7. (a) Calculate the weight of magnesium metal theoretically obtainable from a cubic meter of sea water, density 1.025 g/ml, containing 3.50% by weight of dissolved solids, of which 10.0% by weight is magnesium chloride ($MgCl_2$).
   (b) If the dissolved solids contain 35.0% sodium chloride by weight, calculate how much NaCl is available in a cubic meter of sea water.

8. $Pb(NO_3)_2 + K_2CrO_4 \rightarrow PbCrO_4 + KNO_3$. Calculate the weight of lead chromate ($PbCrO_4$) precipitated if a solution containing 50.0 g of potassium chromate is added to a sufficient amount of lead nitrate solution.

9. Ninety ml of a gas at STP weigh 0.176 g. Calculate the molecular weight of the gas. Is it heavier or lighter than air? How many times? A liter of air weighs 1.293 g.

10. A 750-lb load of limestone is 85.0% calcium carbonate. Calculate the weight of quicklime (CaO) obtainable by heating, if the following reaction takes place:

$$CaCO_3 \rightarrow CaO + CO_2$$

Assume that the reaction goes to completion.

11. Calculate the molecular weight of a gas 2.00 liters of which weigh 5.12 g at 20°C and 740 mm.

12. Copper (Cu) and nitric acid (HNO_3) may react as follows:

$$Cu + HNO_3 \rightarrow Cu(NO_3)_2 + NO_2 + H_2O$$

(a) Balance the equation.
(b) Calculate the weight of copper (II) nitrate, $Cu(NO_3)_2$, produced when 200 g of nitric acid react with an appropriate amount of copper metal.

13. (a) Calculate the weight of chlorine gas which reacts with 20.0 g of ethylene according to the equation:

$$C_2H_4 + Cl_2 \rightarrow C_2H_4Cl_2$$

(b) Calculate the weight at STP of 10.0 liters of ethylene gas.

Exercises

1. The concept of an element as a single atomic species is an idealization. Explain.

2. Did the Greeks of 2000 years ago have any knowledge concerning chemistry? Explain.

3. What great contribution did Paracelsus make to the development of chemistry?

4. Who first defined an element as now understood? State this definition.

5. One of the first great theories of chemistry was the phlogiston theory. It interpreted certain aspects of combustion adequately but failed in others. Explain.

6. What does a chemical symbol represent? What does a formula represent?

7. Every equation represents some equality. What is the nature of the equality represented by a chemical equation?

8. What is meant by the gram atomic weight of an element? By gram molecular weight of a compound?

9. How does the valence of an element differ from its oxidation number? Illustrate how oxidation numbers or ionic charges may be used in writing correct chemical formulas.

10. Why must chemical equations always be balanced before attempting to solve a problem dealing with weight relations in the equation?

11. A formula does not necessarily represent a *molecular* compound. What are the constituent particles in a crystal of salt, NaCl, believed to be?

12. Briefly describe the reduction-by-carbon metallurgical process for preparing zinc.

8 Atoms in families: The periodic table

THE early nineteenth century was marked by a flurry of discovery of new elements, soon followed up by careful studies of their properties and compounds. The most notable worker in this field was the man who discovered cerium, the Swedish chemist Berzelius (1779-1848) whose meticulous quantitative experiments produced a remarkably accurate series of atomic weights.

After 1860 the principle of Avogadro (who did not live to see it) won general acceptance and opened the way to more far-reaching classification of elements and their atomic weights (see Chapter 7). All the surmises and suggestions of half a century came to a head in the great work of the Russian chemist Mendeleev (1834-1907) who in 1869 brought forth his first periodic table of the elements, the basis of all classifications since. By that date the number of known elements had much increased and Dalton's short list of twenty had expanded to more than sixty.

THE PERIODIC TABLE OF MENDELEEV

Mendeleev spent many years studying the properties of the chemical elements. His "periodic table" of 1869 arranged the elements in horizontal rows (periods) in the order of increasing atomic weights. Other horizontal rows were started when elements closely resembling those in the preceding row fell in the same vertical column.

The table started with the element hydrogen, which stood alone. Two horizontal groups, of seven elements each, followed. (The noble gases—helium, argon, neon, and krypton—had not yet been discovered.) Each group of seven elements was called a "short period." In the third hori-

Fig. 8-1. *Dmitri I. Mendeleev, 1834-1907.*

zontal group (period), which started with the element potassium (K), the pattern of similarity was no longer consistent. The first two members of the period, K and Ca, each resembled the element 8 places behind it as expected. The space which followed Ca was left blank (for the element now known as scandium), for Mendeleev realized that the next then-known element, titanium, more closely resembled carbon and silicon than boron and aluminum.

Selenium (Se) was located 17 places after sulfur (S), which it resembled, and bromine was 17 places after chlorine (Cl), which it resembled. There were then spaces for 17 elements in this "long period." Additional periods were included in the Mendeleev table; more will be said about these later in the discussion of the periodic table based on modern atomic structure.

Mendeleev's success was in part due to his recognition of the necessity for a new pattern of arrangement after the first fifteen elements if family relationships were to be maintained. He was also imaginative enough to leave a blank space in the table whenever the properties of the next element did not resemble sufficiently closely the others in the same vertical column.

Mendeleev's Periodic Law Mendeleev had a very detailed knowledge of the properties of the elements, for he plotted a series of graphs each of which showed the fluctuation of some property—for example, atomic volume—as the atomic weight increased (see Fig. 8-2). Elements of the same

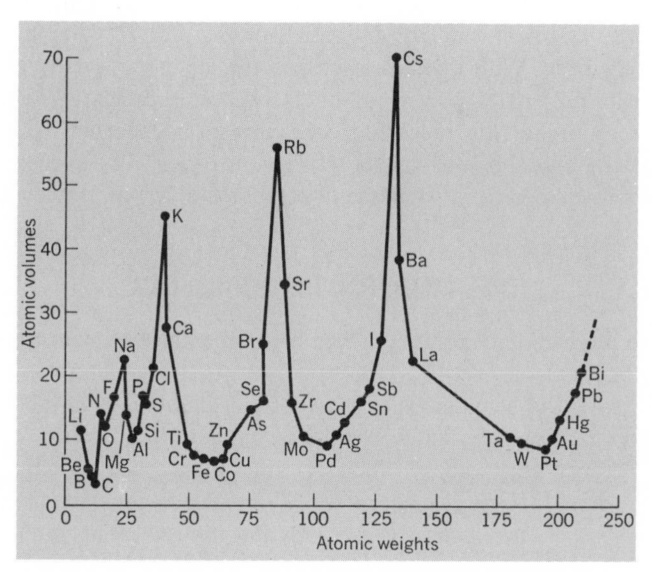

Fig. 8-2. *Graph showing how the atomic volumes of the elements vary with changes in atomic weight. Note the periodic (wavelike) character of the variations.*

family in his periodic table were in the same relative position in successive waves. If one member of a family was at the peak of one wave, the same would be true in the others; if an element was on the upward sloping portion, all other members would be similarly situated. In the graph, each of the alkali metals (Li, Na, K, Rb, and Cs) is at a peak; each of the alkaline earths (Ca, Ba, and Sr) is on the descending portion of a curve; each of the halogens (F, Cl, Br, and I) is on the ascending portion of a wave.

Mendeleev generalized his observations concerning the rise and fall in the numerical values of the different properties of elements, with the statement, "The properties of the elements are periodic functions of their atomic weights." This is known as the *periodic law*. As will be noted in the next section, this law has since been modified.

Mendeleev was uncanny in his power to predict the properties of several elements at that time still unidentified, for which he had left blank spaces in his chart. The element called scandium (Sc), unknown at the time, can illustrate his method. Calcium, on the left of scandium had an atomic weight of 40, and titanium on its right, 48.

It would be reasonable to estimate the atomic weight of the unknown element as 44. The atomic volume of the unknown element could then be read directly from the graph. Even if the atomic weight later proved to be slightly larger or smaller, the value of the atomic volume would be changed very little. With a series of graphs for other properties, Mendeleev predicted the properties of several undiscovered elements. When the elements were subsequently identified, the correspondence between the predicted and the experimental values was very striking. The success of these predictions was of great importance in establishing Mendeleev's reputation.

THE MODERN PERIODIC LAW

Early in the twentieth century, when experimental data concerning properties of the elements were far more abundant, it was discovered that an arrangement of the elements based on atomic numbers (see Chapter 11) eliminated some of the difficulties involved with that based on atomic weights. For example iodine, whose atomic weight was 126.9, preceded tellurium, whose atomic weight was 127.6. When thus arranged iodine did not appear in the same family with fluorine, chlorine, and bromine, all of which it closely resembled; nor was tellurium in the same family as oxygen, sulfur, and selenium, with which it was closely related. It was evident enough that the order of these elements based on their atomic weights should be reversed and certainly hinted that in placing elements with their correct families, there might be a property of the atom more fundamental than the atomic weight. The order of arrangement based on atomic numbers placed tellurium and iodine in their proper families (Table 8-1).

According to the modern statement of the periodic law the properties of the elements are periodic functions of their *atomic numbers,* rather than atomic weights as stated by Mendeleev. For the present, the atomic number will be considered an order number of the elements. Later (Chapter 13) it will be related to the number of protons in the atomic nucleus.

THE MODERN PERIODIC TABLE

Since the time of Mendeleev the periodic table has of course been modified many times. One modern arrangement is shown in Table 8-1. Its main features will be described here, leaving until Chapter 15 the question of its relation to atomic structure.

IA	IIA	IIIB	IVB	VB	VIB	VIIB	VIII			IB	IIB	IIIA	IVA	VA	VIA	VIIA	INERT GASES	
1 H 0.00797 ±0.00001																1 H 1.00797 ±0.00001	2 He 4.0026 ±0.00005	2
3 Li 6.939 ±0.0005	4 Be 9.0122 ±0.00005											5 B 10.811 ±0.003	6 C 12.01115 ±0.00005	7 N 14.0067 ±0.00005	8 O 15.9994 ±0.0001	9 F 18.9984 ±0.00005	10 Ne 20.183 ±0.0005	2 8
11 Na 22.9898 ±0.00005	12 Mg 24.312 ±0.0005											13 Al 26.9815 ±0.00005	14 Si 28.086 ±0.001	15 P 30.9738 ±0.00005	16 S 32.064 ±0.003	17 Cl 35.453 ±0.001	18 Ar 39.948 ±0.0005	2 8 8
19 K 39.102 ±0.0005	20 Ca 40.08 ±0.005	21 Sc 44.956 ±0.0005	22 Ti 47.90 ±0.005	23 V 50.942 ±0.0005	24 Cr 51.996 ±0.001	25 Mn 54.9380 ±0.00005	26 Fe 55.847 ±0.003	27 Co 58.9332 ±0.00005	28 Ni 58.71 ±0.005	29 Cu 63.54 ±0.005	30 Zn 65.37 ±0.005	31 Ga 69.72 ±0.005	32 Ge 72.59 ±0.005	33 As 74.9216 ±0.00005	34 Se 78.96 ±0.005	35 Br 79.909 ±0.002	36 Kr 83.80 ±0.005	2 8 18 8
37 Rb 85.47 ±0.005	38 Sr 87.62 ±0.005	39 Y 89.905 ±0.0005	40 Zr 91.22 ±0.005	41 Nb 92.906 ±0.0005	42 Mo 95.94 ±0.005	43 Tc (99)	44 Ru 101.07 ±0.005,	45 Ir 102.905 ±0.0005	46 Pd 106.4 ±0.05	47 Ag 107.870 ±0.003	48 Cd 112.40 ±0.005	49 In 114.82 ±0.005	50 Sn 118.69 ±0.005	51 Sb 121.75 ±0.005	52 Te 127.60 ±0.005	53 I 126.9044 ±0.00005	54 Xe 131.30 ±0.005	2 8 18 18 8
55 Cs 132.905 ±0.0005	56 Ba 137.34 ±0.005	57 La 138.91 ±0.005	72 Hf 178.49 ±0.005	73 Ta 180.948 ±0.0005	74 W 183.85 ±0.005	75 Re 186.2 ±0.05	76 Os 190.2 ±0.05	77 Ir 192.2 ±0.05	78 Pt 195.09 ±0.005	79 Au 196.967 ±0.0005	80 Hg 200.59 ±0.005	81 Tl 204.37 ±0.005	82 Pb 207.19 ±0.005	83 Bi 208.980 ±0.0005	84 Po (210)	85 At (210)	86 Rn (222)	2 8 18 32 18 8
87 Fr (223)	88 Ra (226)	89 †Ac (227)																

*Lanthanum Series

58 Ce 140.12 ±0.005	59 Pr 140.907 ±0.0005	60 Nd 144.24 ±0.005	61 Pm (147)	62 Sm 150.35 ±0.005	63 Eu 151.96 ±0.005	64 Gd 157.25 ±0.005	65 Tb 158.924 ±0.0005	66 Dy 162.50 ±0.005	67 Ho 164.930 ±0.0005	68 Er 167.26 ±0.005	69 Tm 168.934 ±0.0005	70 Yb 173.04 ±0.005	71 Lu 174.97 ±0.005	2 8 18 32 9 2

†Actinium Series

90 Th 232.038 ±0.0005	91 Pa (231)	92 U 238.03 ±0.005	93 Np (237)	94 Pu (242)	95 Am (243)	96 Cm (247)	97 Bk (247)	98 Cf (249)	99 Es (254)	100 Fm (253)	101 Md (256)	102 No (256)	103 Lw (257)	2 8 18 32 ? 9 2

() Numbers in parentheses are mass numbers of most stable or most common isotope.

Atomic weights corrected to conform to the 1963 values of the Commission on Atomic Weights.

The order of arrangement is according to atomic number. Horizontal groups are called *periods* and vertical columns, *families.* The zero vertical column at the right of Table 8.1 consists of the elements belonging to the family known as the noble gases. This column was not added until some thirty years after the publication of Mendeleev's periodic table. It accounts for the fact that the properties of the elements from lithium through fluorine, for example, repeat in every ninth element instead of in every eighth as suggested by Newland's law of octaves.

Hydrogen and helium are alone in the first horizontal row or period. Since hydrogen has a valence of 1, it is placed in vertical column I, although there is scarcely any further resemblance to the alkali metals which follow it. Helium (He) is the lightest of the noble gases.

The next (short) period consists of eight elements, beginning with lithium (Li) and ending with neon (Ne). The elements sodium (Na) through argon (Ar) likewise form a short period of eight elements. Each of the elements in this second period resembles that immediately above it.

The horizontal group starting with potassium (K) and ending with krypton (Kr) consists of 18 elements and is called a *long* period. The main-group elements K and Ca are followed by 10 "transition elements" (Sc to Zn). Their relation will be clearer after the discussion of electronic structure in Chapter 15. Gallium (Ga) is again a main-group element of this same period, as are the others as far as krypton (Kr).

Starting with rubidium (Rb) and strontium (Sr) in the second long period, there follow 10 transition elements (Y to Cd) and then 6 main-group elements, the last of which is the noble gas xenon (Xe).

A sixth and longer period of 32 elements starts with cesium (Cs), followed by barium (Ba), a group of 14 elements known as the *lanthanides,* or rare earths, then 10 transition elements and 6 main-group elements, the last of which is the noble (and radioactive) gas radon (Rn).

A seventh and final period, which is only partially complete at the present time, starts with the metal francium (Fr), followed by radium (Ra) and a group of radioactive elements known as the *actinides.*

Vertical columns of the periodic table marked I A through VII A, and also O (Table 8-1) are called groups or families. In general, the transition elements show similarities amongst themselves rather than in vertical columns, although the vertical columns including Cu, Ag, and Au (the copper subgroup) and Zn, Cd, and Hg (the zinc subgroup) show resemblances.

In summary then, two elements stand alone at the beginning of the periodic table, followed by two short periods of 8 elements each, two long periods of 18 elements each, one of 32 elements and one final group which is as yet incomplete.

The number of elements in each period (2, 8, 18, and 32) should be noted. These numbers will be significant later when the electronic structure of atoms is discussed (Chapter 15).

The discussion now turns to two chemical families, one of metals, the other of nonmetals, and to the transition from metallic to nonmetallic properties that is manifest in going from left to right in the periodic table.

THE ALKALI METALS: A CHEMICAL FAMILY

In column I A of the periodic table are the symbols H, Li, Na, K, Rb, Cs, and Fr. All except hydrogen are members of the alkali metal family. In its properties hydrogen differs markedly from the others. It is located here largely because like the alkali metals, it has an oxidation number of +1. There is, however, scarcely any further resemblance. Sodium and potassium are probably the most commonly known of the alkali metals, so attention will be focused on them. The properties of the other family members closely parallel these, differing only in degree.

Table 8–2 Densities of Metals (in g/cm^3 at 20° C)

ALKALI METALS		ALKALINE EARTH METALS	
Li	0.54	Be	1.86
Na	0.97	Mg	1.75
K	0.86	Ca	1.55
Rb	1.53	Sr	2.60
Cs	1.90	Ba	3.60

Table 8-2 shows the densities of the alkali metals and those of the adjacent Group II alkaline earth metals. This table shows that low density is a characteristic alkali metal family property. Note that all values are less than aluminum (Al), 2.70 g/cm^3, which is considered a "light" metal. Recall also that gold (Au) was 19.3 g/cm^3 and silver (Ag) 10.5 g/cm^3. Compared with the alkaline earths and still heavier metals such as silver, copper, lead, then, the alkali metals have low densities.

It is evident from Table 8-2 that Li, Na, and K metals will float on water. They move about rapidly on the liquid surface, decomposing some of the water molecules, setting free hydrogen gas and leaving the water soapy to the touch (due to the formation of an alkali, as for example, sodium hydroxide, NaOH). The equation for the reaction is

$$2\,Na + 2\,H_2O \rightarrow 2\,NaOH + H_2\uparrow$$

Red litmus paper changes to blue when immersed in the alkali-containing solution. This litmus test is characteristic of substances called bases.

Within the alkali metal family there is a transition in activity, from the slowly reacting lithium to the more active sodium to the still more reactive potassium. For potassium the heat evolved from the reaction with water is sufficient to set fire to the hydrogen formed.

Table 8–3 Melting Point of Metals (in °C)

ALKALI METAL		ALKALINE EARTH METAL	
Li	180	Be	1283
Na	98	Mg	650
K	63	Ca	850
Rb	39	Sr	770
Cs	29	Ba	704

Table 8-3 compares the melting points of the alkali metals with the alkaline earths. Once again it is evident that a second characteristic family property of the alkali metals is that of extremely low melting point.

Metals are often considered to be hard, like silver, iron, etc. But sodium and potassium are so soft they can be cut with a pocket knife. They have somewhat the consistency of hard cheese. The freshly cut surface has a characteristic metallic luster. This luster is quickly dulled by the formation of an oxide coating when the metal surface is exposed to air. Lithium is, to be sure, a bit harder than the others; but again, softness is a family characteristic. The alkali metals, like other metals, can be hammered into sheets (are malleable), can be drawn into wire (are ductile), and are good conductors of heat and electricity.

The metals react with water and dilute acids, producing hydrogen, and also with a wide variety of elements such as the halogens (F_2, Cl_2, Br_2, and I_2), forming salts. The table of alkali metal salts (Table 8-4) shows that all

Table 8–4 Formulas of Alkali Metal Salts

LiCl	LiBr	LiI	Li_2SO_4	Li_3PO_4
NaCl	NaBr	NaI	Na_2SO_4	Na_3PO_4
KCl	KBr	KI	K_2SO_4	K_3PO_4
RbCl	RbBr	RbI	Rb_2SO_4	Rb_3PO_4
CsCl	CsBr	CsI	Cs_2SO_4	Cs_3PO_4

have an oxidation number of +1. Almost without exception alkali metal salts are water soluble.

In summary, whereas there are transitions in properties from element to element within a family, there are nevertheless certain properties which characterize the members of a family as more closely related to each other than to other metals.

As a case in point, consider why silver, which has an oxidation number of +1, like the alkali metals, is not regarded as related to them. First of all, unlike the alkali metals, silver does not react with water or even with many acids—not at all with HCl, and with H_2SO_4 only if hot and concentrated. With H_2SO_4 the gaseous product is not hydrogen. In the second place, silver metal is unreactive enough to occur in the native state; the alkali metals are never so found in nature, but only in the combined state. The density of silver, 10.5 g/cm^3, is far greater than that of any alkali metal. Its melting point is 961°C, likewise far higher than that of any alkali metal. However, like all metals, silver has characteristic properties such as malleability, ductility, conductivity for heat and electricity, and luster. Silver is very clearly unrelated to the alkali metals.

Consider now the properties of a typical group of nonmetallic elements, the halogens.

THE HALOGENS: A FAMILY OF NONMETALS

In group VII-A of the periodic table occur the symbols F, Cl, Br, I, and At. Since At, astatine, exists only as a decay product of certain radioactive disintegrations, it will not be considered further. The remaining elements exist in the uncombined state as diatomic molecules; they are called the *halogens* or "salt formers." This designation is hardly a unique characterization of the elements, but the name has persisted through many years. The properties shared in common by the nonmetals are not as distinctly characteristic as those of the metals (luster, malleability, ductility, conduction of heat and electricity). However, certain family properties relate these elements more closely with one another than with other nonmetals.

Nonmetals are usually poor conductors of heat and electricity. Whereas metallic oxides often dissolve in acids and hence function as bases, the nonmetallic oxides often have acidic properties and "neutralize" bases. The water solutions of the binary (composed of two elements) halides, HF, HCl, HBr, and HI, are all acids. The halogens also form ternary acids (composed of three elements) such as HOCl, HOClO, $HOClO_2$, and $HOClO_3$. Table 8-5 shows that the formulas are of the same type, just as were members of the alkali metal family.

Table 8–5 Formulas of Halogen-Containing Compounds

NaF	CaF_2	AlF_3		
NaCl	$CaCl_2$	$AlCl_3$	HOCl	$HOClO_2$
NaBr	$CaBr_2$	$AlBr_3$	HOBr	$HOBrO_2$
NaI	CaI_2	AlI_3	HOI	$HOIO_2$

From the table it is clear that some halogens have oxidation numbers of −1, +1, and +5 (also +3 and +7).

ILLUSTRATIVE PROBLEM

What is the oxidation number of Br in $HOBrO_2$?

Solution

Almost without exception the oxidation number of hydrogen in an inorganic ternary compound is +1. Oxygen has almost without exception an oxidation number of −2. In order to balance the positive and negative charges in $HOBrO_2$,

$$\begin{matrix} H & O & Br & O_2 \\ +1 & -2 & ? & -4 \end{matrix}$$

Br must be assigned an oxidation number of +5.

The element fluorine, a gas, is unique in being the most reatcive of all the elements. Its late discovery (1886) was in part a consequence of this extreme reactivity; once set free, it would react with the materials of the reaction vessel.

Chlorine, also a gas under normal circumstances, is a heavy, yellow, choking gas which as long ago as 1918 was used as a poisonous gas of warfare. Today it is probably most commonly known because of its use in water purification systems.

Bromine, normally a deep-red liquid, is also a very caustic substance which *must not be spilled on the skin.*

Iodine is a shiny black solid which sublimes (changes directly from a solid to a beautiful purple vapor without melting to a liquid). The alcoholic solution of this solid is the familiar antiseptic solution. Iodine is also the element which is so essential to the normal functioning of the thyroid gland. In areas where goiter is endemic, iodized salt is often used to make up for possible iodine deficiency in the diet.

In the halogen family there is a transition from a gas of low density (fluorine) to one of greater density (chlorine) to a liquid (bromine) and to a solid (iodine). Enough has been said about these elements so that the parallels in properties, and the transitions in properties, should be evident. In Chapter 15 the discussion of family resemblances among the elements will be supplemented with evidences from atomic structure.

METALLIC AND NONMETALLIC ELEMENTS AND AMPHOTERISM

In the modern form of the periodic table, Table 8-1, there is a heavy gray zigzag line starting near the top of the page and continuing down to the lower right corner. The elements farthest from this line, at the lower left, are the most distinctly metallic elements (Cs, Ba, Rb, etc.). The elements farthest removed at the upper right (F, Cl, O, N) are the most distinctly nonmetallic.

One of the important characteristics of a metal oxide or its water solution, if soluble, is its ability to react with and neutralize the characteristic properties of acids such as HCl, H_2SO_4 or HNO_3. Metal oxides such as Na_2O, CaO or hydroxides such as KOH, and $Ba(OH)_2$ will, then, neutralize acids.

On the other hand the hydrogen compounds of nonmetallic elements usually have acidic properties—e.g., HCl, HF, H_2SO_4, HNO_3. Again the characteristic acid property will disappear if an appropriate amount of an hydroxide such as NaOH or $Ba(OH)_2$ is added.

Close to the heavy gray line of the periodic table are a number of in-between elements which exhibit *amphoteric* properties. If for example a sodium hydroxide solution is added to a clear solution of $Al(NO_3)_3$ a gelatinelike hydrous aluminum oxide precipitates. If an acid such as HNO_3 is added in sufficient quantity, the precipitate disappears. If a base such as KOH is added to the precipitate, it likewise dissolves. Because the oxide reacts with both acids and bases, it is called amphoteric. Similarly the oxide of Sb (Sb_2O_3) which is adjacent to the zigzag line, dissolves in HCl, giving chlorides (with antimony functioning as a metal) or in NaOH giving antimonites in which the oxide functions as if it were nonmetallic, or acidic. Antimony is thus said to be an amphoteric element. Arsenic likewise forms halides such as $AsCl_3$, in which it functions as a metal, and compounds such as Na_3AsO_4, in which the arsenic exhibits characteristic nonmetallic, or acidic, properties.

Observe now the transition from distinctly basic properties, to amphoteric properties, to distinctly acidic properties in a horizontal period of the periodic table. Sodium is characteristically a metal. Its hydroxide changes red litmus blue; it neutralizes the properties of acids. Magnesium hydroxide is insoluble. However, this hydroxide will also neutralize an acid such as HCl. Aluminum hydroxide (or more properly the hydrated hydrous oxide of aluminum) is likewise insoluble and will dissolve in either an acid or a base. It is amphoteric. The next element, silicon, has more distinctly acidic (nonmetallic) characteristics than aluminum. The element silicon dissolves in alkali hydroxides, giving hydrogen, a typically metallic reaction. On the other hand, in silicic acid and the silicates the nonmetallic character predominates. The compounds of phosphorus are distinctly acidic; sulfur and chlorine have progressively greater nonmetallic character. Thus at the left-hand side of the periodic table are the elements which are most distinctly metallic in character. The metallic characteristic diminishes in importance farther to the right. Next come the amphoteric elements, and finally the nonmetallic.

There is a similar but opposite transition in the properties of the members of the nitrogen family in vertical column V-A. The first two elements, nitrogen and phosphorus, are distinctly nonmetallic and hence related to acids. Arsenic and antimony, located on either side of the zigzag line, are amphoteric, exhibiting both acidic and basic character. The last element, bismuth, is the most distinctly metallic element in the family. Once again a transition from nonmetallic to amphoteric to distinctly metallic properties is observed. It should not be concluded, however, that all of the elements in the periodic table fit into this general pattern. Carbon, for example, can hardly be described as distinctly metallic or nonmetallic in character.

With the transition elements in the center of the periodic table such similarities as exist are likely to be among consecutive elements. Further description of trends in the periodic table will be deferred until the discussion of electronic structure in Chapter 15.

Exercises

1. What is meant by (a) a family of elements? (b) a period? Illustrate.

2. State (a) Mendeleev's periodic law; (b) the modern periodic law. (c) Why is the latter statement regarded as the correct one?

3. State some of the reasons for considering the elements Li, Na, K, Rb, and Cs to be members of the same chemical family.

4. Name the members of the halogen family. List some of the principal differences between the nonmetallic and the metallic elements.

5. What is an amphoteric element? Illustrate.

6. What changes in the character of the chemical properties of the elements take place going (1) from left to right in a period (2) from top to bottom in a family of the periodic table?

9 Rainbows from atoms and molecules

A RAINBOW spans the sky when rays of sunlight are dispersed by raindrops. The same effect may be observed in the spray of a garden hose in the bright sunlight. Both are reminders that white light is a complex of many colors, the colors of the rainbow. Though rainbows had been observed since the dawn of mankind, the scientific study of color began early in the seventeenth century when Galileo invented or developed the telescope. When objects were observed through one of these early telescopes, the images were blurred by color fringes, a consequence of the dispersion by the lens of the white light entering it. Two generations later, at Trinity College, Cambridge, the undergraduate Isaac Newton became interested in the properties of lenses and particularly in the annoying color fringes. His interest in correcting this defect of lenses led him to his study of color.* One sunny day he poked a hole in a curtain in a darkened room, thus allowing a narrow beam of sunlight to enter (Fig. 9-1). He placed a glass prism so that the beam met one of its surfaces obliquely, as shown in Fig. 9-2. To his amazement the spot of sunlight which fell on the wall was oblong in shape rather than circular like the hole in the curtain. Furthermore the light on the wall, instead of being white, was a rainbow of color, a "continuous spectrum"—violet at one end and a continuous blending into the other colors to red at the other end. The experiment indicated to Newton that white light was not homogeneous; certain components were bent (refracted) more than others in passing through the prism. If the beam of sunlight was taken as a line of reference, the beam of red light was deviated through a smaller angle than that of the violet light. He clinched this argument concerning the

* Abraham Wolf, *History of Science and Technology in the Sixteenth and Seventeenth Centuries,* Macmillan, 1950, p. 264.

Fig. 9-1. *A modern artist depicts Isaac Newton using a prism to disperse the components of white light. The sausage-shaped spectrum is a succession of overlapping images of the round hole in the curtain.* [*Courtesy Bausch & Lomb Optical Co.*]

complex nature of white light by showing that after the colors had been so separated, they could be brought back together again to produce white light like the original beam.

THE SPECTROSCOPE

If a triangular glass prism like Newton's is mounted as shown in Fig. 9-3 with a collimator tube on one side and telescope tube on the other, the device is called a spectroscope. An adjustable slit at one end of the *collimator* admits a very narrow vertical beam of light. A lens system within the tube renders the beam parallel as it passes toward the prism. The prism serves to spread out the components of the white light, bending the red least and the violet most. The lens system of the telescope tube gives the observer a magnified image of the slit for each component present.

The spectroscope (Fig. 9-4) is essentially a light analyzer. The light passing into the collimator depends on the physical state of the emitting material and on its temperature. If a metal ball is placed in front of the slit and heated, the observer sees nothing until a red color appears at one side

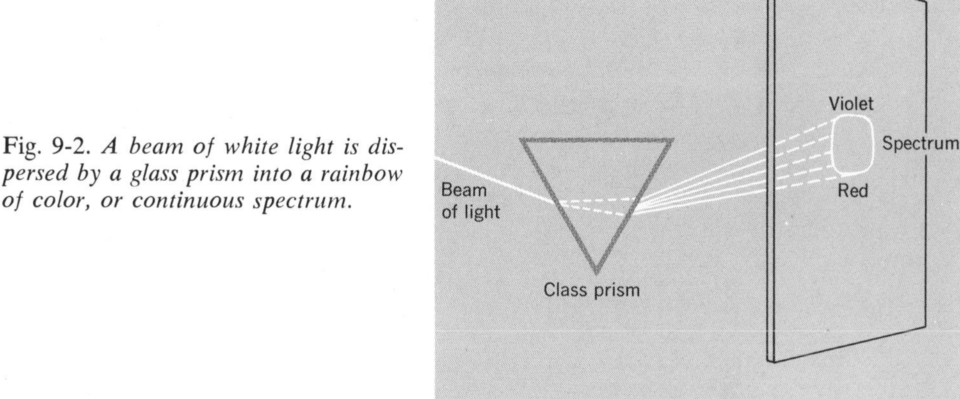

Fig. 9-2. *A beam of white light is dispersed by a glass prism into a rainbow of color, or continuous spectrum.*

of the field. The ball becomes red hot. As the temperature of the ball rises, orange appears, then yellow, green, and blue; and finally, when it is white hot, the rainbow is complete; the observer sees a continuous spectrum. Glowing *solids* give this type of spectrum. An incandescent light bulb gives a continuous spectrum since the source is the glowing tungsten filament.

Fig. 9-3. *Construction of a prism spectroscope or spectrometer. The collimator is a device for making divergent rays parallel. Its tube excludes all light from the source except a beam admitted through the slit in a diaphragm at one end. At the other end of the collimator tube a lens renders the rays of this beam parallel. The rays proceed through the prism and are dispersed into color bands that are magnified for an observer by the two lenses of the telescope.*

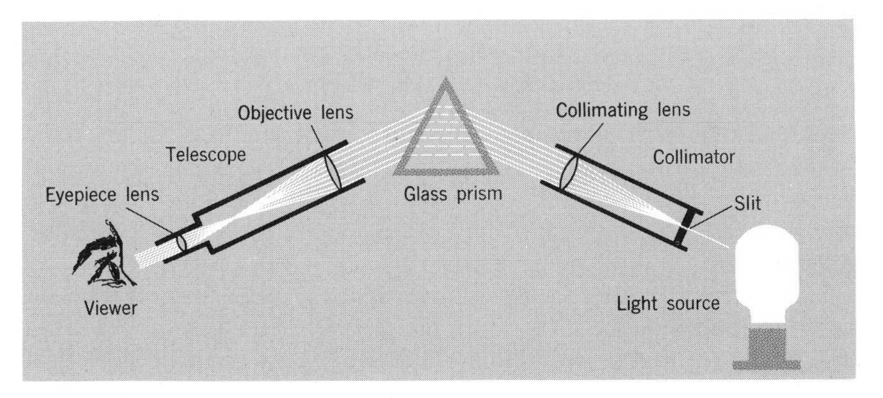

Fig. 9-4. *A spectroscope. The long narrow camera attachment at the left is used to photograph spectra. When the telescope tube is removed and replaced by the camera, the instrument is called a* spectrograph. *The black cover with side tube attached (shown at the right) fits over the table supporting the prism (center of the spectrometer). When the end of the side tube is illuminated, a graduated scale is superimposed upon the spectrum. This scale makes convenient the locating of spectral lines.* [*Courtesy, American Optical Co.*]

Unfortunately a continuous spectrum will not serve to identify the solid material which is emitting the light rays.

However, if a high-voltage electric current is passed through a glass tube containing a *gas* such as sodium vapor or neon (at low pressure) and the glowing gas is used as a light source, the observer sees in the spectroscope a pattern of thin vertical colored lines. This is called a *line spectrum* and in the hands of an experienced spectroscopist will serve to identify the emitting material. Much of the usefulness of the spectroscope depends upon the identification of elements by means of these *characteristic* patterns of lines (Fig. 9-5).

PROPERTIES OF WAVES

Many of the phenomena associated with light are best explained by assuming that it is propagated in the form of waves. The human eye cannot of course directly observe the wave form of light. But water waves are familiar to all, and observing them may assist in visualizing the behavior of light waves.

Consider for a moment the water waves which are set in motion when an object is dropped on the surface of a placid pool. The waves form a series of expanding concentric rings. Fig. 1-7 shows a cross section of a typical

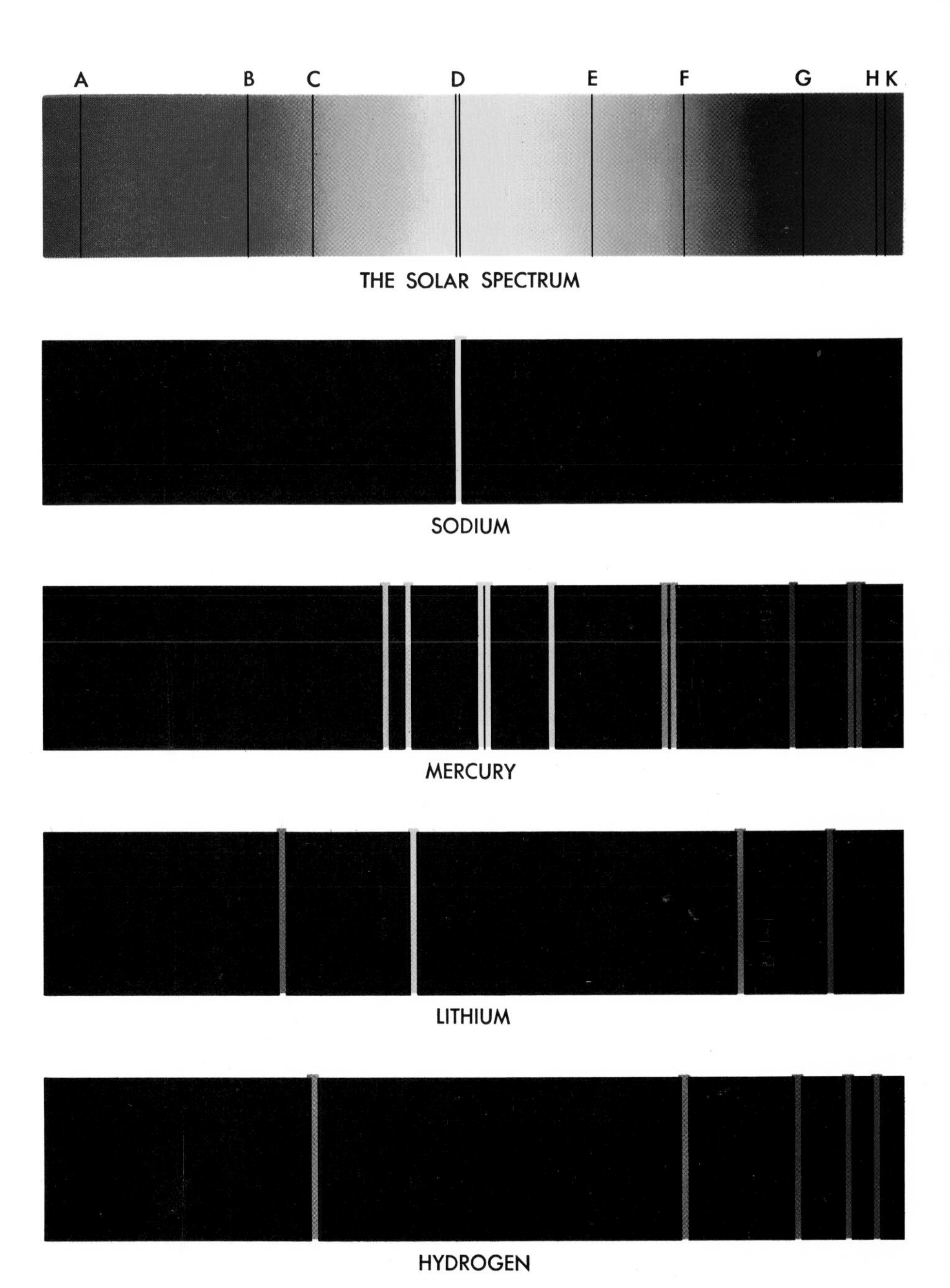

Fig. 9-5. Emission and absorption line spectra. The solar spectrum (top) shows the Fraunhofer (absorption) lines which enable us to identify elements on the sun. The remaining line (emission) spectra are characteristic bright-line spectra in the visible region. The patterns of lines shown would permit the identification of the excited gaseous atoms emitting them.

group of water waves. Though light waves differ in some respects from water waves, the latter permit readier visualization of the meaning of *wavelength.* Wavelength has been defined as the linear distance between the crests (or for that matter, between any corresponding parts) of two successive waves. Light waves also are characterized by their wavelengths. Red waves are the longest of the visible light waves, orange and yellow are shorter, and so on down to violet, which are the shortest. Differences in the color of light are associated with waves of different lengths.

The number of waves passing a given point (cf. Fig. 9-8) in 1 second is called the *frequency*. Since red light waves are longer than blue (see Fig. 9-6) and since all light waves travel with the same velocity, fewer red light

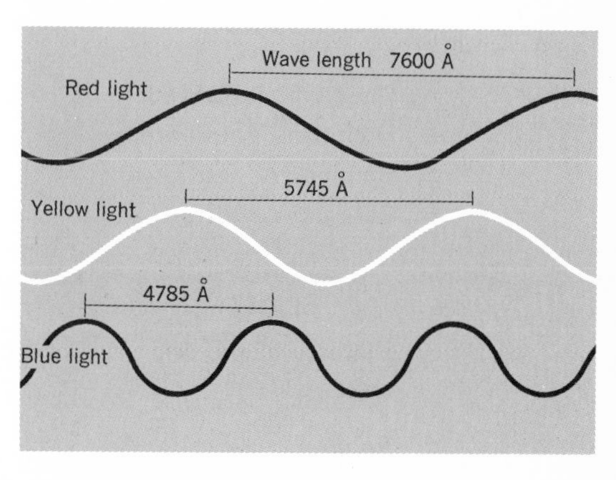

Fig. 9-6. *In the visible spetcrum, waves of red light are longer than those of yellow light, which in turn are longer than those of blue. The wavelengths are measured in angstrom units (Å), i.e., in hundred-millionths of a centimeter. The visible spectrum, between the infrared and the ultraviolet, is composed of light ranging from 4000 to 7800 Å in wavelength.*

waves than blue pass a given point in a second. Hence red light has a lower frequency than blue.

There are striking parallels between the behavior of light and sound. First of all, the wave equation $v = \nu\lambda$ applies to both wave forms: v is the velocity of the wave (186,000 mi/sec or 3×10^{10} cm/sec for light in vacuo, 1090 ft/sec at 0°C for sound in air); ν (nu) is the frequency or the number of waves passing a point in 1 second; λ (lambda) is the wavelength. Note that the linear unit in the velocity must be the same as that of the wavelength; also that the time unit of the frequency and the velocity must be the same. If the velocity of light is expressed in meters/sec, then the frequency must be in reciprocal seconds $\left(\frac{1}{\text{sec}}\right)$ and the wavelength in meters. There will then be an equality of units

$$v = \nu \cdot \lambda$$

$$\text{m/sec} = \frac{1}{\text{sec}} \times \text{m}$$

Table 9–1 Comparison of the Properties of Light and Sound Waves

SOUND WAVES	LIGHT WAVES
1. The wave equation $v = \nu\lambda$ applies to both sound waves and light waves	
2. Velocity in air at 0°C 1090 ft/sec	Velocity in vacuo 3×10^{10} cm/sec or 186,000 mi/sec
3. Longitudinal waves	Transverse waves
4. Velocity varies with temperature	No known variation of velocity with temperature
5. Requires air for transmission	Air not required for transmission
6. Cannot be polarized	Can be polarized
7. Both can be reflected	
8. Both can be refracted	
9. Both can be diffracted	

ILLUSTRATIVE PROBLEM

Find the length of a wave of violet light if the frequency is 7.5×10^{14} waves per sec.

Solution

$$v = \nu\lambda$$

$$3.0 \times 10^{10} \left(\frac{\text{cm}}{\text{sec}}\right) = 7.5 \times 10^{14} \left(\frac{1}{\text{sec}}\right) \cdot \lambda\ (\text{cm})$$

$$\frac{3.0 \times 10^{10}}{7.5 \times 10^{14}} = \frac{1}{2.5} \times 10^{-4} = 0.40 \times 10^{-4} = 4.0 \times 10^{-5} = \lambda$$

$$\lambda = 4.0 \times 10^{-5}\ \text{cm}$$

Wavelengths of spectral lines are usually expressed in linear units known as angstrom units. 1 Å (angstrom unit) = 10^{-8} cm. How many angstrom units there are in 4.5×10^{-5} cm?

$$\lambda = 4.0 \times 10^{-5}\ \text{cm} = 4.0 \cdot 10^{3} \times 10^{-5} \cdot 10^{-3}$$
$$\lambda = 4000 \times 10^{-8}\ \text{cm} = 4000\ \text{Å}$$

Sound waves are compressional, or longitudinal. They may result from the vibration of a piece of metal (for example, a bell) which alternately compresses and rarefies the air which strikes its surface. Individual air particles move very short distances before colliding with others; so they oscillate back and forth parallel to the direction of propagation of the wave (see Fig. 9-7). The series of concentric shells of compressed and rarefied air impinging on human eardrums produces the sensation of sound.

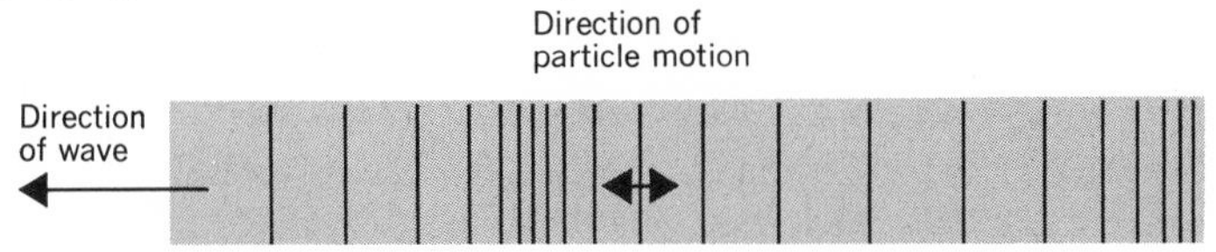

Fig. 9-7. *A longitudinal or compressional wave consists of alternate regions of compression and rarefaction in an elastic medium (such as air). In a sound wave the air particles move very short distances back and forth parallel to the direction of propagation of the wave. A vibrating bell sends out expanding spherical shells of compressed and rarefied air. The illustration shows a cross section through a few of these shells.*

The pitch of a musical note is determined by the frequency of the vibration (the number of waves per second). Thus the international standard of pitch "A 440" means a musical tone which emits 440 waves each second.

ILLUSTRATIVE PROBLEM

Calculate the length of the sound wave having a frequency of 440 waves/sec.

Solution

$$v = \nu\lambda$$

$$1090 \text{ ft/sec} = 440\left(\frac{1}{\text{sec}}\right) \cdot x \text{ (ft)}$$

$$440x = 1090$$

$$x = 2.48 \text{ ft}$$

The distance from one point in this wave to the next corresponding point, is 2.48 ft.

The high-pitch tones are those with the shorter, and the low-pitch tones those with the longer, wavelengths. The loudness of a musical tone depends on the amplitude of the waves, a concept perhaps best understood by examining Fig. 9-8.

Light waves differ from sound in being transverse rather than longitudinal. In Latin, *trans* means *across:* in light waves the vibrations are perpendicular to (across) the direction of propagation (see Fig. 9-8). Light waves can be plane-polarized (made to vibrate in a single plane), while sound cannot. Both can be reflected, refracted, and diffracted.

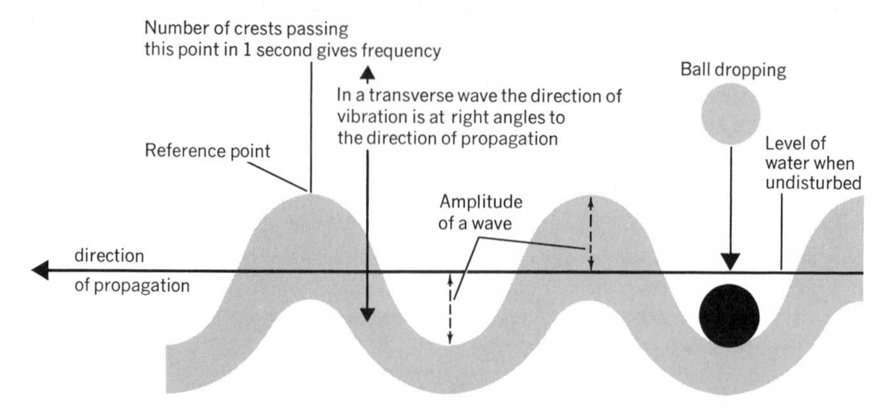

Fig. 9-8. *The propagation of a wave on water. In a transverse wave, the direction of vibration is at right angles to the direction of propagation. Electromagnetic waves, including light, are transverse, in contrast to the* longitudinal *waves of sound. A water wave is not a purely transverse wave, since there is a slight back-and-forth movement of water particles.*

What is set in motion as a light wave moves through space? A hundred or more years ago the answer would probably have been "the luminiferous ether" (not to be confused with the liquid ether, the anesthetic). For many years it had been known that a medium, air, was necessary for the transmission of sound waves. If the air was pumped out of a bell jar in which a ringing bell was suspended, the sound gradually diminished and eventually died out. But the vibrating bell hammer would still be visible, so light waves were able to travel in space which would not transmit sound. Since it was believed that a medium must be present for the propagation of any wave, it was suggested that the hypothetical medium, ether, was responsible for the transmission of light waves.

In 1887 Michelson and Morley in Cleveland, Ohio, devised an experiment by which they hoped to detect the drift of the ether past the earth. But their experiment was unsuccessful, and there is still no convincing experimental evidence for the ether, if it exists. Transverse waves require a rigid material for their transmission; but there certainly should be no problem in detecting the presence of a rigid material such as steel. The ether question must at present be relegated to the category of unsolved problems.

Less familiar than the wave properties of light are its particle properties. In 1887 Hertz discovered that high-voltage sparks would pass between two metal electrodes at lower than normal potentials (voltage) if ultraviolet light were incident upon the negative electrode (Fig. 9-9). Later, J. J. Thomson, and also P. Lenard, showed that when light fell on a

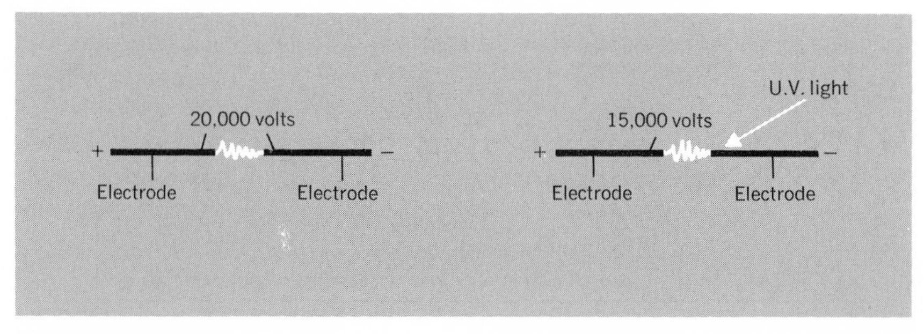

Fig. 9-9. *Hertz showed that sparks will jump a spark gap at a lower voltage if a beam of ultraviolet light is directed on the negative electrode.*

negatively charged zinc plate, tiny particles called electrons were released from it. This so-called photoelectric effect was readily explained if it was assumed that light had particle as well as wave properties. Hence if the observed behavior of light is to be adequately explained, both wave and particle properties must be invoked. In some cases light behaves as if composed of waves; in others as if composed of particles. With electrons there is a similar duality; they too behave like waves in some cases and like particles in others. Sometimes this curious state of affairs is referred to as the "wave-particle dilemma."

ELECTROMAGNETIC WAVES

Light waves are only one variety of a more general class known as electromagnetic waves. All electromagnetic waves travel with a velocity of 186,000 mi/sec, may be reflected or refracted, and behave similarly in many respects. But there are also minor differences. X rays, for example, will penetrate many materials which visible light will not. Ultraviolet rays will cause sunburn in people too long exposed to them, while longer waves will not. Electromagnetic waves vary in length all the way from a small fraction of an angstrom unit to several miles (Fig. 9-10). Waves of visible light have lengths ranging from about 4000 Å at the violet end of the spectrum to about 7800 Å at the red end; the spectral region between 4000 and 7800 Å is called the *visible* spectrum, since the human eye is able to detect these waves. Beyond the red end of the spectrum is a group of longer waves known as the infrared, ranging up to about 80,000 Å, or 0.008 mm. Beyond the infrared waves are radio waves, which extend to several miles in length.

On the short-wavelength side of the visible spectrum, next to violet rays, is the ultraviolet region. These rays have a wavelength range from about 150 to 4000 Å. X rays, which are so useful in medicine and dentistry, have

The electromagnetic spectrum consists of the types of radiation whose waves are transverse and have a velocity of 186,000 mi/sec (the so-called "speed of light"). X rays, ultraviolet, visible light, infrared, and radio waves are included in this classification .

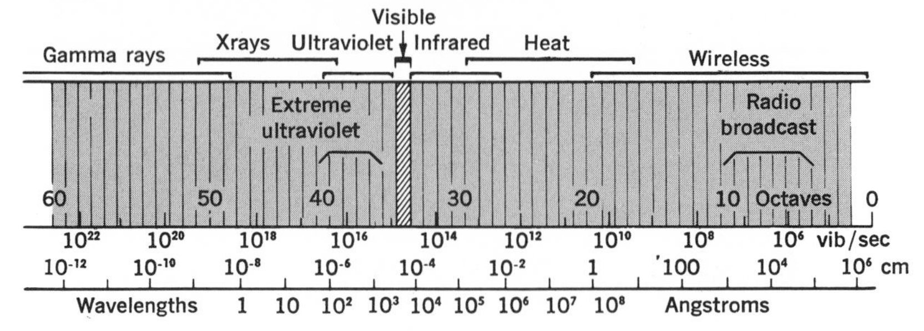

Fig. 9-10. *The electromagnetic spectrum consists of the types of radiation whose waves are transverse and have a velocity of 186,000 mi/sec (the so-called "speed of light"). X rays, ultraviolet, visible light, infrared, and radio waves are included in this classification.*

waves still shorter than the ultraviolet; gamma rays, which are emitted by the nuclei of radioactive atoms, are in turn shorter than X rays, although the distinction between the two has been unimportant since supervoltage X-ray machines have been available. The shortest of all electromagnetic waves and the most penetrating were supposed for a long time to be cosmic rays, but more recent work indicates that the effects of these rays may be explained on the basis of high-speed particles.

LINE SPECTRA

Though Newton's work with the prism-made rainbow was not forgotten, more than a century passed before a systematic study of the spectrum of sunlight was made—by a Bavarian, Fraunhofer (1787-1826). An expert optician, he was able to cover a glass plate with fine, even, parallel scratches less than a thousandth of a centimeter apart. Sunlight passing through such a plate—called a *diffraction grating*—yielded clearly defined spectra. Fraunhofer was fascinated by mysterious narrow dark lines he saw crossing the rainbow background (Fig. 9-5, top line). He counted hundreds of them and noticed that in every spectrum such lines occurred at the same points relative to the colored background. These lines in the so-called dark-line spectrum have ever since been known as *Fraunhofer lines* and everything about them has been carefully measured.

The mystery of the dark lines began to clear up when they were compared with bright lines in so-called *bright-line spectra* (Fig. 9-5, lower four lines). In the eighteenth century it was noticed that if common salt—sodium chloride—was sprinkled in a flame, a yellow flare resulted. When the yellow light passed through a prism, it produced a spectrum with a single narrow line (as in the second spectrum of Fig. 9-5). This became known as the sodium line. The glowing vapors of other elements produced other spectral patterns, e.g., mercury, lithium, and hydrogen, each typical of the element emitting light (Fig. 9-5). Foucault in 1849 found that the very line that glowed in the spectrum of sodium was the line that was darkened in the spectrum of sunlight if the sunlight passed through a tube of glowing sodium vapor. Evidently sodium vapor could act both as a source in producing a line spectrum and as an absorber, removing from white light the wavelengths which it produced as an emission source. Kirchhoff (1824-1887), a German physicist, soon discovered that the luminous vapors of many other elements would also absorb from a white-light source those wavelengths which the vapors would have emitted. White light in passing through a cooler luminous vapor will then have removed from it lines of those wavelengths which the vapor would give as an emitting source. The positions in the spectral field of the bright emission lines and the dark absorption lines would be exactly the same.

The two great German scientists Bunsen and Kirchhoff were the first to recognize the generality of the spectroscopic method. They discovered that the luminous vapor of a chemical element gives a characteristic pattern of spectral lines. The wavelength of each line is precisely determinable and characteristic of that element alone. The sodium pattern, for example, turns out to be two yellow lines, at exactly 5890 and 5896 angstrom units, and the other patterns of Fig. 9-5 each characterize an element uniquely. Thus an *emission spectrum* (bright-line spectrum) is produced by a gas in a light-emitting condition, and an *absorption spectrum* (dark-line spectrum) is produced when a glowing gas absorbs from white light those wavelengths which it would emit as a source.

Line spectra are produced by luminous vapors in the atomic (rather than the molecular) condition. The spectra produced by gases in the molecular condition, i.e., in which atoms are linked together as molecules, have broad emission color bands (or dark absorption bands against a rainbow background) instead of thin lines, and are called band spectra. Nitrogen gas, when excited to light emission, gives this type of spectrum. Glowing solids give continuous emission spectra (i.e., a complete rainbow).

THE DISCOVERY OF HELIUM

The invention of a new tool in science as the spectrometer has frequently opened up opportunities for a whole series of new discoveries. Shortly after devising the spectroscopic method, Bunsen made use of it in his discovery of the chemical elements rubidium and cesium. Even more spectacular was the prediction in 1868 by Lockyer of a new element on the sun. There was an absorption line in the yellow region of the solar spectrum for which there was no corresponding emission line. No known substance, when excited to a light-emitting condition, gave a bright yellow line in exactly the same position as this absorption line. Lockyer called the predicted element *helium,* deriving the name from the Greek word for sun, *helios*. Then in 1895 Ramsay examined the emission spectrum of a gas that came from the mineral cleveite. Lo, there was the unidentified yellow line of the sun's spectrum! Helium gas, still rare, has since become an article of commerce.

APPLICATIONS OF SPECTROGRAPHY

Suppose an industrial chemist needs to know whether a sample of steel contains traces of the metal vanadium. If he can make the sample into a glowing gas and can examine the spectrum of the light it emits, his problem is solved. To produce the glow he uses an arc lamp—a lamp that makes intense white light at the gap between two (carbon) electrodes as a high-voltage current crosses from one to the other. One electrode he tips with the

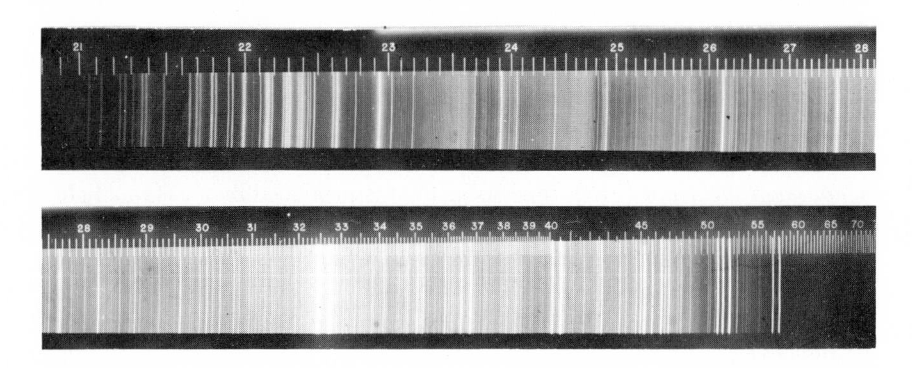

Fig. 9-11. *Metallic elements vaporized in an electric arc emit light which is dispersed by a spectrograph into a series of spectral lines. The experienced spectroscopist learns to recognize in a developed spectrogram the pattern of lines characteristic of each element present in the sample.*

sample metal, which is soon vaporized by the heat of the arc. Photographing the light with a spectrograph (Fig. 9-4), he has a record that he can study at leisure. In a moment the expert can identify many patterns in the picture of the bright-line spectrum and can declare whether the pattern of the vanadium spectrum occurs there (Fig. 9-11). In a few minutes he can even state the exact proportion of vanadium in the sample, by measuring the intensity of the bright lines that register in the photograph. About seventy elements (atoms) can be identified with such bright-line spectroscopic methods.

A still wider range of substances can be identified with absorption spectroscopy. White light passes through water or other liquid in which the unknown substance is dissolved. A spectrum is made and photographed. Or more usually, a device that uses a sensitive photoelectric cell to measure light intensity is applied to the spectrum. The tell-tale dark lines or their equivalent appear in patterns that identify the atoms present in the solution and their proportions (measured by the intensity of the dark lines). The human eye is limited to the range of visible light from red to violet. Not so the photographic film and the photoelectric cells. These make available spectral patterns in the infrared and ultraviolet ranges. Infrared patterns are especially valuable in identifying whole groups of combined atoms (Fig. 9-12).

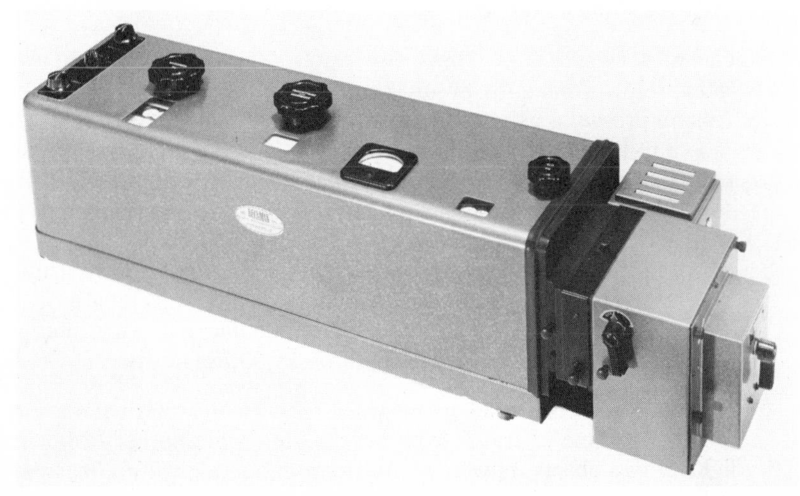

Fig. 9-12. *The spectrophotometer is an instrument which measures the percentage of the incident light which is transmitted by solutions placed in the light beam. Since different chemical substances in solution absorb light at particular wavelengths, this instrument may aid in the qualitative identification of substances. Since the amount of absorption at a particular wavelength and given thickness of solution is often a linear function of concentration, the instrument may also be used as a quantitative analytical device.* [*Courtesy, Beckman Instruments, Inc.*]

Electronic instruments like those mentioned have markedly increased the speed with which the chemist can carry out qualitative and quantitative analyses.

The science that depends altogether for the study of its objects upon light and other electromagnetic vibrations that reach the Earth from distant regions of space—astronomy—has drawn mostly upon the spectroscope for its astonishing discoveries about celestial bodies and their motions (see Chapter 23).

Problems, Chapter 9

1. The wavelength of a spectral line is 5880 Å. Calculate its frequency.

2. Calculate the wavelength in angstrom units of a spectral line whose frequency is 10^{16} waves/sec. In what region of the electromagnetic spectrum will this line be found?

3. Consult a newspaper and find the frequency of a local radio station. Calculate the length of the wave it sends out. (A kilocycle is the same as 1000 waves/sec.)

4. Chemists usually use millimicrons (mμ) rather than angstrom units to express wavelengths in absorption spectrometry.
   (a) Express the value of the micron in centimeters. The micron is a millionth of a meter.
   (b) Express the value of the millimicron (mμ) in centimeters.
   (c) What wavelength in mμ corresponds to a frequency of 5×10^{15} waves/sec?
   (d) In infrared absorption spectroscopy *wave numbers* are usually used. The wave number is the reciprocal of wavelength or the number of waves per centimeter. Calculate the wave number corresponding to a wavelength of 15,000 mμ.

5. Galileo reportedly tried to measure the velocity of light by estimating the time of travel from one mountain to another of lantern light. If two observers were stationed on mountains 5.0 miles apart, how long (microseconds) would light take to traverse this distance? (1 microsecond $= 10^{-6}$ sec.)

6. A radio wave (an electromagnetic wave) theoretically takes how long to travel 4950 mi from New York to Honolulu? How long would sound theoretically take to travel this distance at a velocity of 1090 ft/sec? (Assume that each travels in a straight line.)

7. The wavelength of a spectral line of helium is 3.885×10^{-5} cm. Calculate its frequency.

8. (a) The musical note middle C has a frequency of 264 vps (vibrations per second, or waves per second). What is the wavelength corresponding to this frequency? Assume that the speed of sound is 1090 ft/sec at 0°C.

(b) If the speed of sound increases 2 ft/sec for each 1°C temperature rise, calculate the velocity of sound at room temperature, 20°C.

9. The wavelength of the note given by a closed organ pipe is 4 times the length of the pipe. What is the length of a pipe giving the note high C? (This note is an octave above middle C, whose frequency is 264 vps. Doubling the frequency of a note gives the next higher octave.)

Exercises

1. How are the color fringes observed by Galileo in his telescope explained? What is the name of the lens defect producing these color fringes? What kind of a lens corrects for this defect?
2. Draw a diagram representing a prism spectroscope. Label the most important parts and describe the function of each.
3. Describe Isaac Newton's experiment which proved that white light was complex.
4. A red hot object looks different from a white hot object when viewed in a spectroscope. Describe.
5. What is meant by the wavelength of a wave? By its frequency?
6. Draw a diagram showing the regions of the electromagnetic spectrum.
7. What type of light source is used to produce the following: (a) line spectrum, (b) band spectrum, (c) continuous spectrum?
8. How are absorption spectra produced? Illustrate.
9. Describe some of the advantages of the spectroscopic method of analysis.
10. Helium was known on the sun before it was discovered on the earth. Explain.
11. Compare light waves and sound waves. In what respects do they behave similarly and in what respects do they differ?

10 Electricity, magnetism, and electrolysis

CHEMICAL effects of electric currents were known long before there was an ionic theory, and there was a body of evidence indicating the existence of a unit of electric charge decades before the existence of the electron was proved.

STATIC ELECTRICITY

The Greek scientist and philosopher Thales of Miletus (*ca.* 600 B.C.) knew that an amber rod which had been rubbed with fur would attract and pick up bits of paper and straw. Such a rod was later said to be electrically charged (*electron* is simply the Greek word for amber). The charge was called *static* or *electrostatic* because it seemed to remain at rest, in contrast with flowing charge or an electric current. Since the charged rod exerts a force on light objects in its vicinity, an electric field is thought to exist in the space surrounding it. A field, whether electric, magnetic or gravitational, is simply a region in which a force is exerted on a body placed there.

Benjamin Franklin suggested that the kind of electric charge from amber or hard rubber and fur (Fig. 10-1) be called "negative" and that from rubbing a glass rod with silk, "positive." These conventions are still in use today in defining positive and negative charge.

Two glass rods which are positively charged repel each other, or push apart. A hard-rubber rod (negative charge) and a glass rod (positive charge) attract each other and are drawn together. These observations may be generalized by stating that objects with like electric charges (either positive or negative) repel, those with unlike charges attract.

Before rubbing, the ebonite rod shows no evidence of being electrically

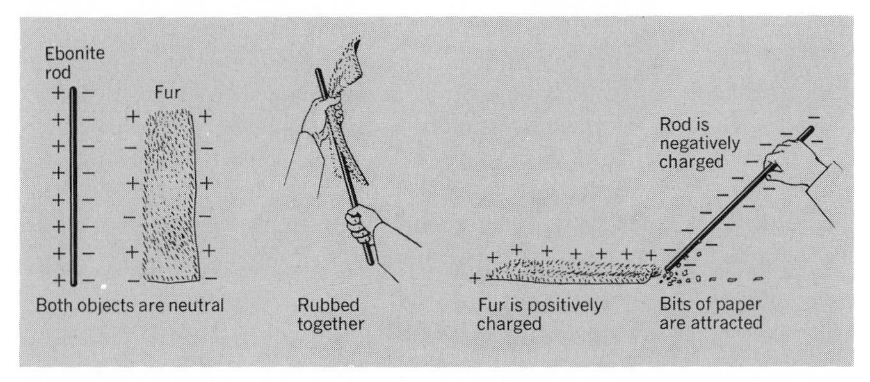

Fig. 10-1. *An ebonite (hard-rubber) rod may be electrified by rubbing with wool or fur.*

charged. It contains substantially equal numbers of positive and negative charges and hence is electrically neutral. When the rod is rubbed with fur, electrons (negative charges) are transferred from the fur to the ebonite, leaving the latter with more negative charges than positive. When the negative charges outnumber the positive, the rod is negatively charged. The fur, which was also neutral originally, has lost electrons and hence bears a net positive charge.

In 1780 a considerable body of knowledge concerning electrostatic phenomena was available. Dr. Luigi Galvani (1737-1798), professor of biology at the University of Bologna (Italy), was working in his laboratory with an electrostatic "influence machine" (a device which produced electric sparks). A metal scalpel came close to the machine at just about the time the blade touched one of the nerves in a frog's leg which was hung up close by. The frog's leg twitched. The follow-up of this simple observation culminated in the discovery of the electric battery.

ANIMAL ELECTRICITY

The incident excited Galvani's curiosity. He wondered if the same effect could be produced by lightning, so he hung the frog's leg by brass hooks from an outdoor iron railing. Although twitching did seem occasionally to accompany flashes of lightning, the same result could be realized by touching the frog's leg to the iron rail. It was soon clear that neither the lightning nor the influence machine was a necessary condition for the twitching. After repeating many such experiments Galvani erroneously concluded that the twitching was due to "animal electricity." While his conclusion was in error, his work nevertheless served to dramatize this new source of energy. The term "galvanic cell," which has persisted through the years, reminds us of the importance of his contribution to science. Even the statement of an erroneous principle has occasionally stimulated others to think about the implications of a problem and has led to great advances: Galvani's fellow countryman Alessandro Volta (1745-1827) challenged the idea of animal electricity. According to Volta the important aspect of repeating this effect was that two dissimilar metals, such as copper and iron, should come in contact. The contraction of the frog's leg was due to the conduction of the impulse; it was not the source of the impulse.

THE VOLTAIC PILE

Volta, the inventor of the battery, found that the electrical effect of a single pair of metals in contact was multiplied if several pairs of metal strips such as silver and zinc, separated by strips of leather or pasteboard soaked in a solution of table salt, were arranged in a "pile" (Fig. 10-2).

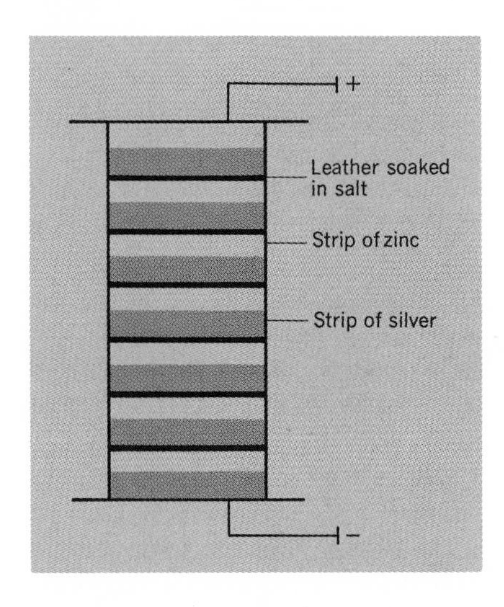

Fig. 10-2. *The voltaic pile, a forerunner of the modern electric cell or battery. The invention by Alessandro Volta opened the Age of Electricity in the year 1800. It was promptly applied to chemical and biological research.*

Anyone who completed the electrical circuit by touching the bottom silver strip and the top zinc strip would be reminded of this multiplying effect by receiving a jolting electric shock.

Shortly after Volta's discovery, the pile was used to separate water into its constituent elements, hydrogen and oxygen. Wires were attached to the pile and the electric current was led into the water by small metal pieces known as electrodes (Fig. 10-3). A few drops of sulfuric acid had to be added to the water in order to make it an electrical conductor. The process of bringing about a chemical change by means of an electric current was known as electrolysis.

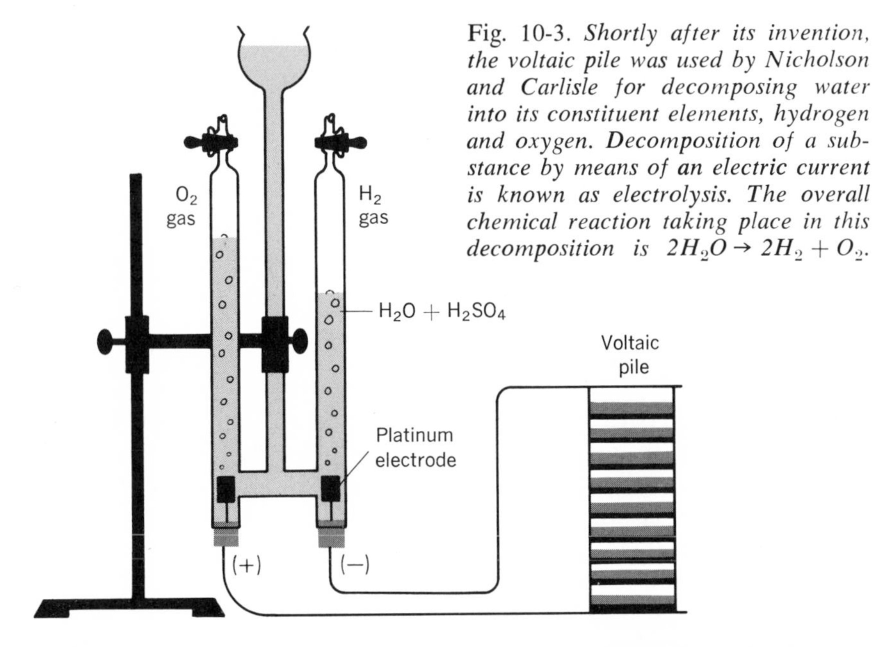

Fig. 10-3. *Shortly after its invention, the voltaic pile was used by Nicholson and Carlisle for decomposing water into its constituent elements, hydrogen and oxygen. Decomposition of a substance by means of an electric current is known as electrolysis. The overall chemical reaction taking place in this decomposition is* $2H_2O \rightarrow 2H_2 + O_2$.

Sir Humphry Davy used the voltaic pile to electrolyze water solutions of caustic soda (NaOH) and caustic potash (KOH) in attempts to prepare the metals, but obtained hydrogen and oxygen just as if he were electrolyzing water. He then tried electrolyzing the melted solid alkalis and eventually obtained at the cathode (negative terminal) tiny globules of a new metal which he called potassium. The discovery of barium (Ba), strontium (Sr), calcium (Ca), magnesium (Mg), and sodium (Na) followed in quick succession.

A SIMPLE ELECTRICAL CIRCUIT

An electric current is a flow of electrons, negatively charged particles.* They have very tiny masses compared with the other constituents of atoms. Protons and neutrons are nearly 2000 times as massive, and only electrons are set in motion when there is a difference of energy level between the terminals of the conductor.

If wires from a dry cell are connected to a light bulb and there is an unbroken path for electrons, the glow of the bulb serves as a detector of

* Actually movements of charged atoms in solution (ions) or an electron beam moving in a vacuum may also be considered "electric currents." Here the discussion will be limited to electron flow along a wire such as copper.

the electron movement or electric current. The cell, wire, and bulb constitute a simple electric circuit.

A voltaic pile or its modern counterparts, the dry cell and storage cell battery, are often referred to as electron pumps, being likened to water pumps. The latter must have an energy source to operate. The energy source for the windmill driven pump is air in motion, i.e., the wind. The engine-driven pump burns gasoline or kerosene. In the dry cell, chemical substances are the energy source. A chemical reaction at one electrode produces an abundant supply of electrons; at the other electrode a different reaction keeps removing electrons as they flow in. As a consequence, electrons move from the electrode where they are produced, through the wire (external circuit), to the electrode at which they are removed.

The flow of electrons (current) in an electrical circuit is often likened to the flow of water in pipes. The greater the horsepower of a water pump, the more gallons per second will be pumped. The greater the voltage of a cell, the more electrons (current) per second will pass through a wire (of given resistance).* If the interior of a pipe is rough, friction will be large, and there will be a corresponding large pressure loss per linear foot of flow. In like manner different kinds of metal wire offer different resistances to the electron flow. The "voltage drop" along a conductor is proportional to the resistance of the wire. Nichrome wire, for example, has about sixty times as much resistance as copper wire of the same length and diameter, and hence there will be about sixty times as much voltage drop in nichrome wire carrying an equal current.

An electrical circuit, then, requires a source of energy to set electrons in motion and a complete (closed) path to conduct them. The battery—the pumping device—has two electrodes, one at which a chemical reaction is pushing electrons into the external circuit (called the negative terminal), the other a reaction which removes electrons from the external circuit (the positive terminal).† The electric current is the electron flow which results from the difference in electrical pressure between the electrodes.

Georg Simon Ohm The relationship between the voltage, resistance and current in an electric circuit was worked out by Georg Simon Ohm (1784-

* Electric power depends on the product of voltage and currents (volts × amperes).

† There are two ways of describing the direction of flow of electric currents. Electric currents were known before the discovery of electrons. For many years it was assumed that currents (positive charges) left a battery at the positive terminal and passed through the external circuit and back into the cell at the negative terminal. Later, when the nature of the electron and its behavior were better understood, it was found that the *electron flow* was in just the opposite direction. By that time the convention had been established for more than a hundred years, and that is why we still find some authors describing the current as leaving the cell at the positive terminal, while others describe the electron flow as issuing from the negative terminal.

1854) and bears his name, Ohm's law. Ohm was the son of a Bavarian locksmith. In his earlier years he was a teacher in elementary schools and later taught in the secondary schools. His ambition was to become a university professor, and one way in which he was likely to attain this goal was to carry out an important research investigation. For years he spent much of his spare time studying the factors which affected the ability of metal wires to conduct an electric current. When the results of his research were published, a storm of criticism broke loose. Not only did he fail to receive the recognition needed for the prized university appointment; he even lost his secondary-school teaching position. Twenty years elapsed before the importance of his work was recognized by his fellow scientists, and not until late in life was he actually appointed to a professorship in physics at the University of Munich. The principle which he enunciated is one of the most important generalizations in the theory of current electricity.

OHM'S LAW

Ohm's law like many of the other generalizations of science, may be most concisely expressed by a mathematical equation

$$E = IR$$

where E stands for electromotive force, I for electric current, and R for resistance. E is expressed in volts, I in amperes, and R in ohms: E (volts) $= I$ (amperes)$\cdot R$ (ohms). As electrons pass along a conductor, a loss of electrical pressure, called the potential drop, takes place; the total of all the potential (voltage) drops in the circuit is equal to the electromotive force. Since there is always a small voltage drop due to the flow of current through the internal resistance of the cell itself, the voltage between the terminals will always be a bit less than the electromotive force of the battery.

Electromotive force, E, is measured with great accuracy by an instrument called the potentiometer. In this electrical instrument the voltage is measured by balancing it against that of a standard cell. Thus the driving force is measured without actually drawing any current from the cell. The potential drop across any part of a circuit is most conveniently measured by connecting a voltmeter in parallel with it.

In order to determine the voltage drop due to an electric light bulb, for example, the terminals of the voltmeter would be connected to those of the socket as shown in Fig. 10-4. The instrument so connected provides an alternative path for the current, which may go either through the light bulb or through the voltmeter: the larger the resistance of the bulb, the greater the fraction of the current diverted through the voltmeter, and hence the larger the voltage drop registered by the instrument.

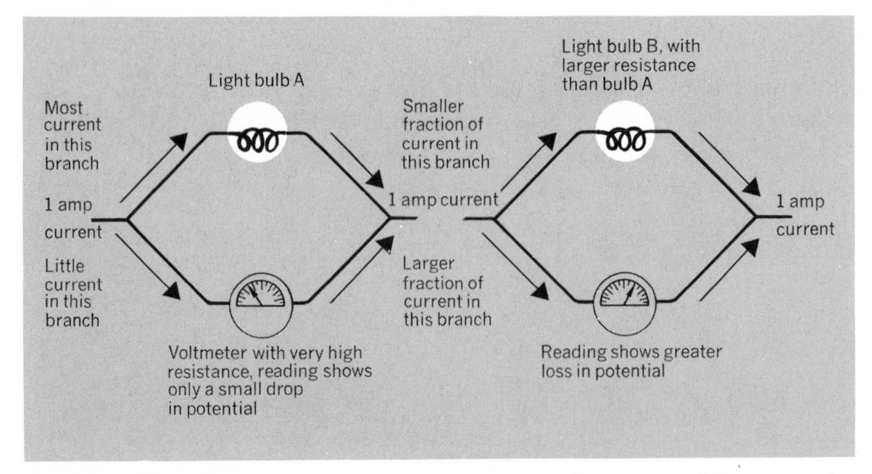

Fig. 10-4. *The voltage drop across a resistance may be measured by connecting a voltmeter in parallel with it. Since light bulb A has a lower resistance than light bulb B, the current after passing through A is still capable of more work than after passing through B. It retains a higher potential (available voltage).*

I, the electric current, is a measure of the number of electrons passing a given point in the circuit during 1 second. The greater the number of electrons that pass in a given time, the larger the current. The ammeter is the instrument used to measure electric currents. Its resistance is so

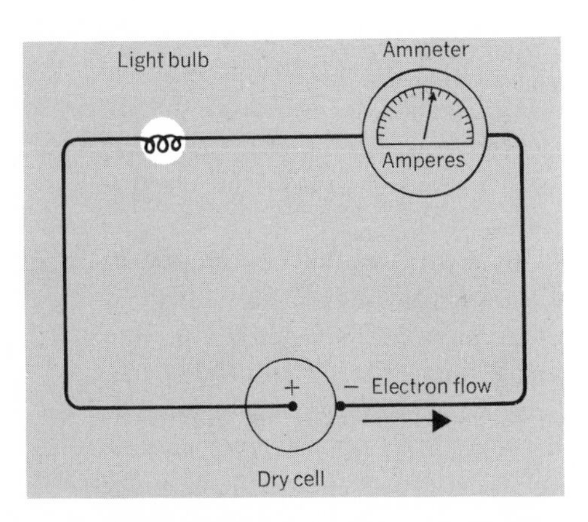

Fig. 10-5. *An ammeter is connected in an electric circuit in series with the resistance through which the current flow is to be measured.*

small that it measures a current without appreciably interrupting its flow. It is connected directly in the line in which the current flow is to be measured (i.e., in series rather than in parallel). Fig. 10-5 shows how an ammeter is connected in the circuit to measure the current flowing through the light bulb.

R, the resistance which a conducting wire offers to electron flow, depends on the material of the wire, its length, and its diameter. Iron wire has greater resistance than silver or copper wire of the same length and diameter. Wires of larger diameters have lesser resistances per linear foot than do smaller wires.

ILLUSTRATIVE PROBLEM 1

Calculate the current flowing if a 1.5-volt dry cell is connected in the circuit of Fig. 10-6A. The total resistance is 10 ohms.

Solution

$$E = IR$$
$$1.5\ (\text{volts}) = I\ (\text{amperes}) \cdot 10\ (\text{ohms})$$
$$1.5 = 10I$$
$$I = 0.15 \text{ ampere}$$

0.15 ampere will flow through the circuit.

ILLUSTRATIVE PROBLEM 2

Calculate the current flowing (Fig. 10-6B) if a 100-ohm resistance is attached to a 1.5-volt dry cell.

Fig. 10-6. *More current is driven through a 10-ohm resistance than a 100-ohm resistance by a given voltage. A 6-volt battery combination will push more current through a 100-ohm resistance than will a 1.5-volt cell.*

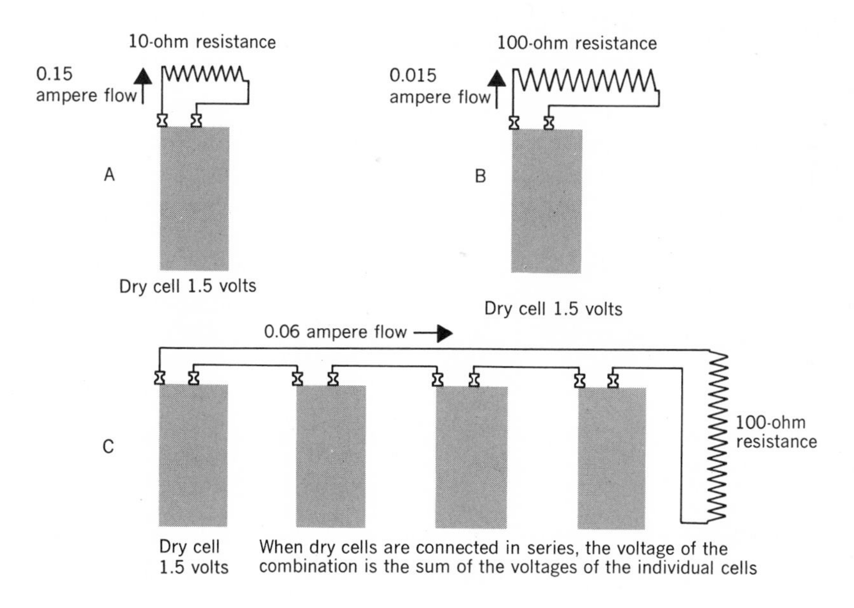

Solution

$$E = IR$$
$$1.5 = 100I$$
$$I = 0.015 \text{ ampere}$$

The current is one-tenth as large as in the preceding problem. Fewer electrons are pushed through the circuit in a given time by the same voltage, if the resistance is greater.

ILLUSTRATIVE PROBLEM 3

Calculate the current flowing if four dry cells are connected in series (voltages are additive: 4×1.5 volts $= 6.0$ volts) in a circuit whose resistance is 100 ohms.

Solution

$$E = IR$$
$$6.0 = 100I$$
$$I = 0.06 \text{ ampere}$$

A fourfold increase in the voltage has resulted in a fourfold increase in the current passing through the same resistance. Increased potential difference has resulted in setting more electrons in motion through the same resistance in unit time than in the second illustration.

THE STORAGE BATTERY

The storage battery (Fig. 10-7), a familiar part of the equipment of every motorcar, is a modern adaptation of the voltaic pile. Instead of alternate pairs of dissimilar metals, the storage battery has one set of lead (Pb) electrodes, and another set of electrodes of lead impregnated with lead dioxide (PbO_2). The electrolyte in the voltaic pile was a salt solution (soaked up by strips of leather or cardboard); the electrolyte used in the modern storage battery is dilute sulfuric acid. Both cells make use of two electrodes separated by an electrically conducting solution, or electrolyte.

The electrical energy produced by a storage battery is derived from chemical action: the reaction responsible for pushing electrons is

$$Pb + 2\,H^+ + 2\,HSO_4^- + PbO_2 \rightarrow 2\,PbSO_4 + 2\,H_2O$$

In order to show more clearly the electron-removing and electron-producing aspects of the above equation, the following half reactions are written:

$$Pb + HSO_4^- \rightarrow PbSO_4 + H^+ + 2e \quad \text{(Electron-producing half-reaction)}$$

$$2e + PbO_2 + 3\,H^+ + HSO_4^- \rightarrow PbSO_4 + 2\,H_2O \quad \text{(Electron-removing half-reaction)}$$

A more complete discussion of electrolytic cells is found in Chapter 15.

If a coil of wire or other suitable resistance is connected between the terminals of the battery, a path for electrons is completed. Electrons leave the cell at the negative terminal (lead plates), where an electron-releasing reaction takes place, creating an abundant supply. The electrons pass through the external circuit, and return to the cell at the lead dioxide plates, where an electron-accepting reaction takes place. The usefulness of the battery depends upon the fact that an electric motor, a horn, or a headlight may be operated if arranged appropriately in the external circuit. The energy supplied by the chemical reaction taking place in the battery may be converted into mechanical energy (by the electric motor), into sound (by the horn), or into light (by the headlight).

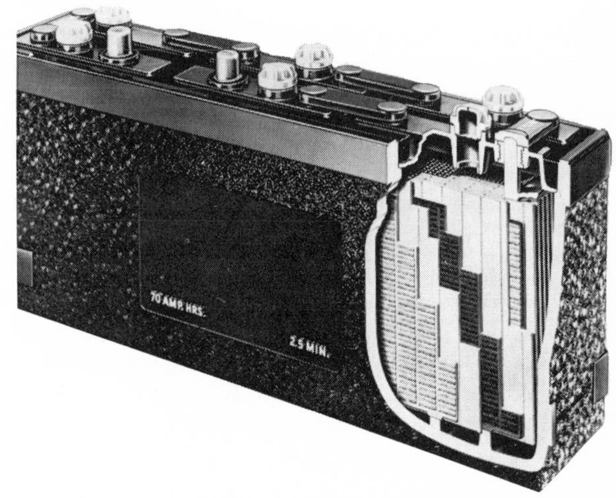

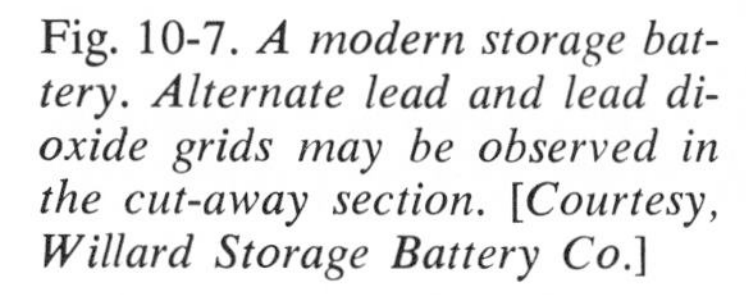

Fig. 10-7. *A modern storage battery. Alternate lead and lead dioxide grids may be observed in the cut-away section.* [*Courtesy, Willard Storage Battery Co.*]

Once again it should be emphasized that the law of the conservation of energy holds for each of these energy conversions. Energy cannot be conjured up from nothing; one form must be used up if a new form is to be produced. The potential energy of a stone resting on the top of a mountain is greater than its potential energy after it has rolled halfway down the mountainside: some of its potential energy has been converted into kinetic energy. Similarly, the chemical energy level of sulfuric acid, lead, and lead dioxide, the reactants, is higher than that of lead sulfate and water, the products formed: some of the energy of the former has been converted into electric energy.

Just as the stone on the mountaintop tends to roll down the mountainside, and just as heat tends to flow from a region of higher to one of lower temperature, so electrons tend to flow from a region where they have high energy to one where they have lower energy. The chemical reaction of the battery creates an abundant supply of electrons at the negative terminal and a scarcity of electrons at the positive terminal. The resulting difference in electrical levels causes a movement of electrons from the negative terminal of the cell, through the external circuit, to the positive terminal.

The lead sulfate ($PbSO_4$), a white solid, gradually collects on each electrode and covers more and more of its surface, thus reducing the amount of lead metal exposed to the action of sulfuric acid. The other product of this chemical reaction, water, tends to dilute the sulfuric acid as the cell discharges and thus to decrease the specific gravity of the acid. Recharging a battery reverses the chemical reaction, dissolving the lead sulfate from the plates and increasing the specific gravity of the electrolyte. The condition of battery charge is usually determined by measuring the electrolyte's specific gravity with an hydrometer.

HEATING EFFECTS OF AN ELECTRIC CURRENT

If a piece of copper wire is connected across the terminals of a new dry cell, the wire quickly becomes very hot. This is a brutal way to treat a cell, but it effectively demonstrates one of the effects accompanying the flow of electrons along an electrically conducting wire. No. 18 bell wire has a resistance of about 6.5 ohms for every 1000 feet. The resistance of a small piece of this wire, say 1 foot long, is negligibly small; and hence the current flowing, even with a small voltage, very large. Attaching a low-resistance wire across the terminals of a cell is called *short-circuiting* the cell and will ruin the cell in a short time.

The heating effect produced when a DC electric current passes through a wire is given by the equation

$$H = 0.24\, I^2Rt,$$

in which H represents the quantity of heat (calories), I the current flow (amperes), R the resistance of the wire (ohms), and t the time of flow (seconds). As this equation shows, doubling the amperage will have a greater effect on the heat produced than doubling the resistance or the time.

If a nichrome wire is connected to a 110-volt source of electric current by copper lead-in wires, as in an electric toaster, the nichrome will be heated red hot while the copper will not. Since there is only one path for the current, the number of amperes flowing in the copper wire must be the same as in the nichrome wire. Any difference in heating effects must be due to difference in resistance. Nichrome wire has a resistance approximately sixty times that of copper wire of the same diameter and length, so that the heating effect is about sixty times as great.

If two uninsulated wires in a 110-volt house circuit come in contact, there is a sputtering and shower of sparks; a short circuit is said to have occurred. A short circuit is a low-resistance path for the current; it allows an enormous current to flow and results in a dangerously excessive heating of the wires. To guard against these hazardous effects a fuse is put in the house circuit. The fuse contains a piece of easily melted wire. If a large amperage flows through it, the wire in the fuse heats and melts; thus the circuit is broken. Fuses, then, protect the house from the danger of overheated wires and hence from the possibility of fire within the partitions.

MAGNETISM

Since very ancient times the mineral lodestone, an oxide of iron, has been known to attract iron objects. Thales, the famous Greek wise man, knew of its attractive force (sixth century B.C.). He suggested that because of its attractive power for iron objects, it must have a soul.

Fig. 10-8. *The magnetic compass is simply a bar magnet suspended in such a way as to turn freely. The step from (a) to (c) was taken by the Chinese a thousand years ago or more.*

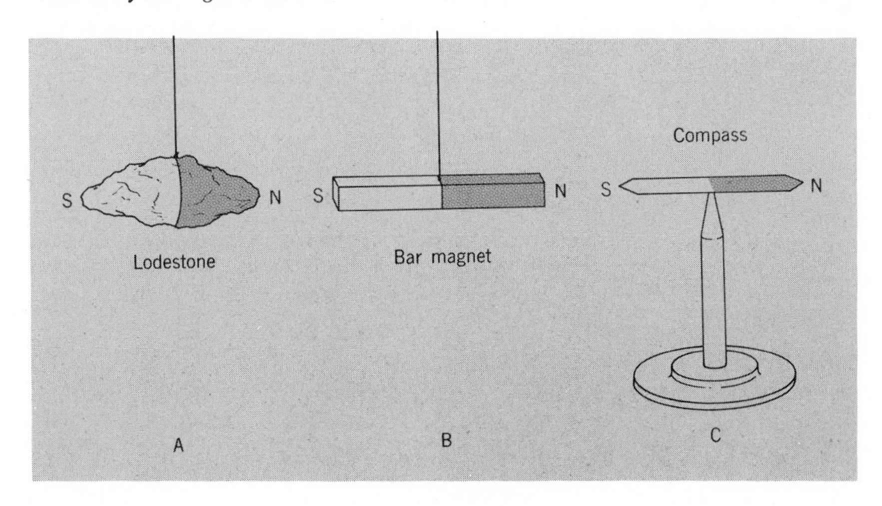

Fig. 10-9. *Sir William Gilbert demonstrates the effects of static electricity at the court of Queen Elizabeth. Painting by A. A. Hunt.* [*Courtesy, American Institute of Electrical Engineers.*]

There are many ancient stories concerning the power of the magnet. One recounts how sailors insisted that the ships on which they sailed should be built without iron nails lest too close an approach to one of the supposed "magnetic islands" might pull it apart at sea.

Very early in history there was curiosity concerning the strange behavior of a slender piece of lodestone suspended from a string and free to rotate in a horizontal plane. This lodestone pointer always oriented itself in a particular (north-south) direction. It was of course the forerunner of the

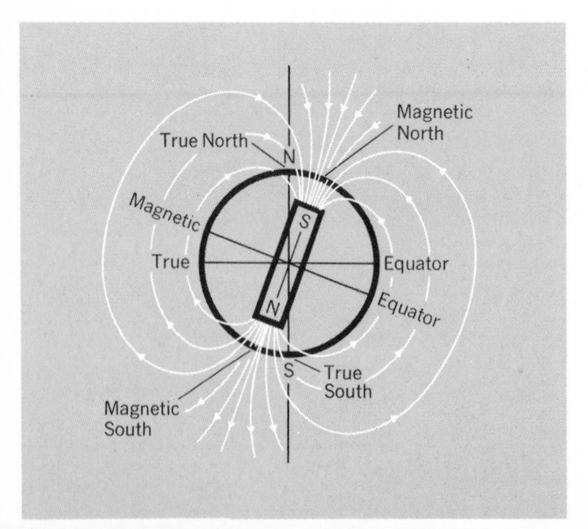

Fig. 10-10. *Sir William Gilbert believed the earth's magnetic field to be like that of a bar magnet imbedded in a sphere.*

DE MAGNETE, LIB. III.

CAP. XII.

Quomodo verticitas exiſtit in ferro quouis excocto magnete non excito.

HActenùs naturales & ingenitas cauſas, & acquiſitas per lapidem potentias declarauimus: Nunc verò & in excocto ferro lapide non excito, magneticarum virtutum cauſæ rimandæ ſunt. Admirabiles nobis magnes & ferrum promunt & oſtendunt ſubtilitates. Demonſtratum eſt anteà ſæpiùs, ferrum lapide non excitum in ſeptentiones ferri & meridiem; ſed & habere verticitatem, id eſt proprias & ſingulares polares diſtinctiones, quemadmodùm magnes, aut ferrum magnete attritum. Iſtud quidem nobis mirum & incredibile primùm videbatur: Ferri metallum ex vena in fornace excoquitur, effluit ex fornace, & in magnã maſſam indureſcit, maſſa illa diuiditur in magnis officinis, & in bacilla ferrea extenditur, ex quibus fabri rurſus plurima componunt inſtrumenta. & ferramenta neceſſaria. Ita variè elaboratur & in plurimas ſimilitudines eadem maſſa tranſformatur. Quid eſt igitur illud quod

conſeruat

Fig. 10-11. *Producing a magnet by pounding, from* De Magnete, *1600, by William Gilbert. Gilbert announced that iron bars showed polarity even though they had never been excited by a lodestone or another iron magnet.*

modern compass. The compass is simply a magnetized steel needle, appropriately pivoted and encased (Fig. 10-8).

The first comprehensive study of magnetism was made by an Englishman, William Gilbert (1540-1603), who was court physician to Queen Elizabeth (Fig. 10-9). In 1600 he published his famous work *De Magnete.* In this he noted that the earth behaved as a huge magnet (Fig. 10-10). He also observed that opposite magnetic poles attracted. Fig. 10-11, which is taken from his famous book, shows how a bar of iron or steel may be magnetized by pounding.

Magnetic Fields As mentioned earlier in the chapter, a field is simply a region in which forces act on a body. A magnetized bar attracts or repels other nearby iron or steel objects, and hence a magnetic field must exist in the region around it. To demonstrate the presence of a magnetic field, lay a bar magnet flat on a table, cover it with a sheet of paper, and scatter iron filings over the paper. The filings will arrange themselves in the pattern shown in Fig. 10-12. The string-like chains of iron filings

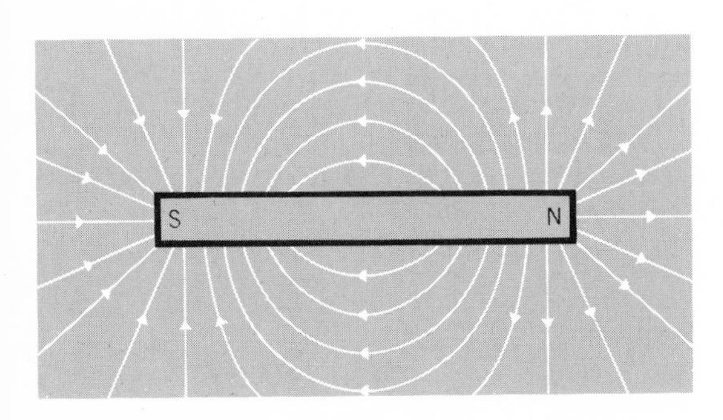

Fig. 10-12. *Lines of force about a bar magnet, from a photograph.*

are often referred to as *lines of force.* A tiny compass needle which is placed near the magnet will align itself parallel to them. In fact the pattern of lines may be mapped by using such a compass. The direction of a compass needle in a magnetic field will indicate the direction of the lines of force at that location.

Magnetic Poles The lines of force surrounding a bar magnet appear to crowd in at the ends of a magnet. The regions where the magnetic property seems to be concentrated are called magnetic poles. If a bar magnet is suspended from a string so as to move freely in a horizontal plane, one end will point in the direction of magnetic north and is called a north-seeking or a north pole. The opposite end will be a south pole.

If the north poles of two magnets are brought near together, they will push apart or repel. Two south poles will likewise repel. On the other hand two unlike poles will attract. The so-called laws of magnetism may be summed up by saying that like magnetic poles repel and unlike poles attract. The patterns of the lines of force between two like poles and two unlike poles are shown in Fig. 10-13.

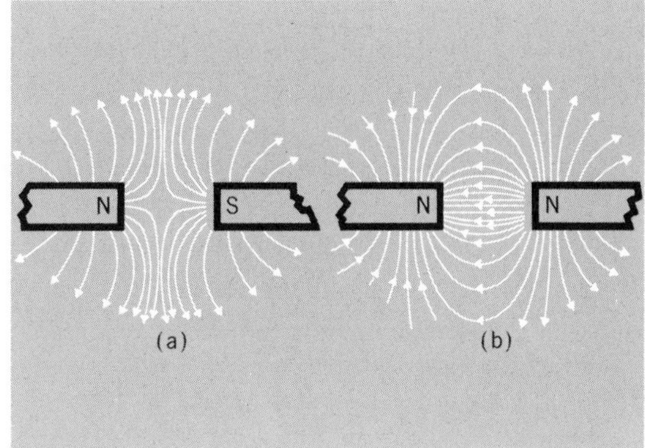

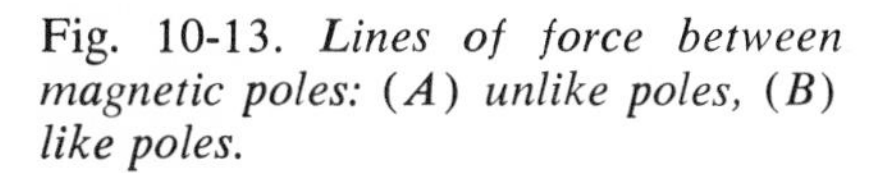

Fig. 10-13. *Lines of force between magnetic poles: (A) unlike poles, (B) like poles.*

Magnetic Effects of an Electric Current The discovery of magnetic effects due to electric currents was not made until more than 200 years after *De Magnete* was published, for the voltaic cell and current electricity were not available until near the close of the eighteenth century. In 1819, Hans Christian Oersted of Denmark discovered that a wire in which an

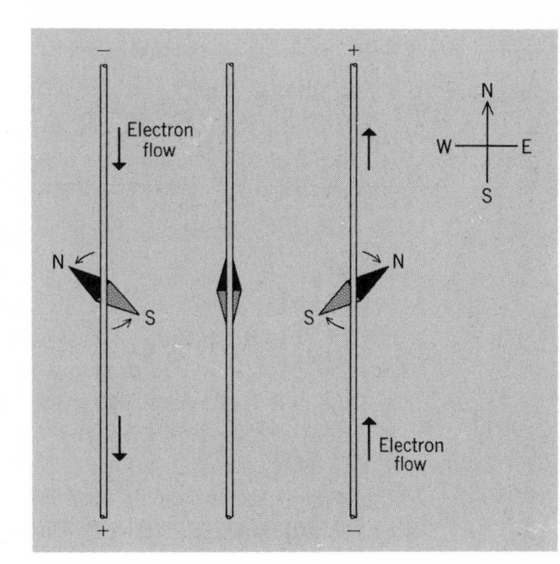

Fig. 10-14. *Oersted's experiment illustrating that a wire which is conducting an electric current is surrounded by a magnetic field. A compass needle which is placed beneath a horizontal wire and parallel to it turns at right angles when an electric current passes through the wire. Since a magnet (e.g., a compass needle) aligns itself with magnetic lines of force, these must be in a plane perpendicular to the wire.*

electric current was flowing was surrounded by a magnetic field. He noticed that a compass needle aligned parallel to a wire was rotated at right angles when an electric current flowed through the wire (Fig. 10-14). The presence of a magnetic field about a wire carrying a current may be shown by sprinkling iron filings on a piece of cardboard through which the wire is passed (Fig. 10-15). At first the iron filings are ran-

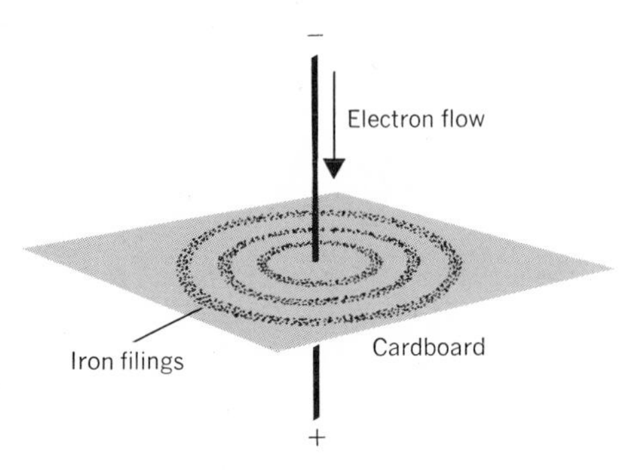

Fig. 10-15. *Iron filings line up in concentric circles around a wire carrying an electric current.*

domly oriented but when the electric current flows, they arrange themselves in concentric circles about the wire. A compass needle placed on the cardboard (Fig. 10-16) will be aligned tangent to these circular lines of force.

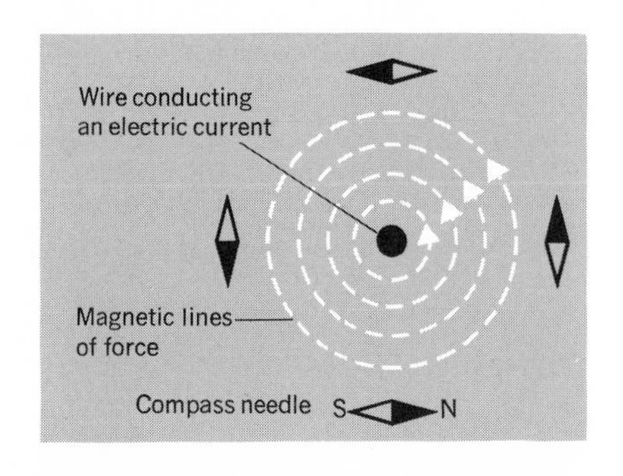

Fig. 10-16. *A compass needle aligns itself tangent to the concentric magnetic lines of force surrounding a wire carrying an electric current.*

An electric current flowing in a coil of wire (or air-cored solenoid) makes it a magnet (Fig. 10-17). Iron filings spread on a paper above such a solenoid are aligned in the characteristic pattern of a magnetic field. An exploring compass shows a north pole at one end of the coil and a south pole at the other—but only when a current flows through the coil. If the coil is wound about an iron nail, the magnetic property is intensified; the nail will pick up tacks (Fig. 10-18). Thus magnetism and electricity are related phenomena.

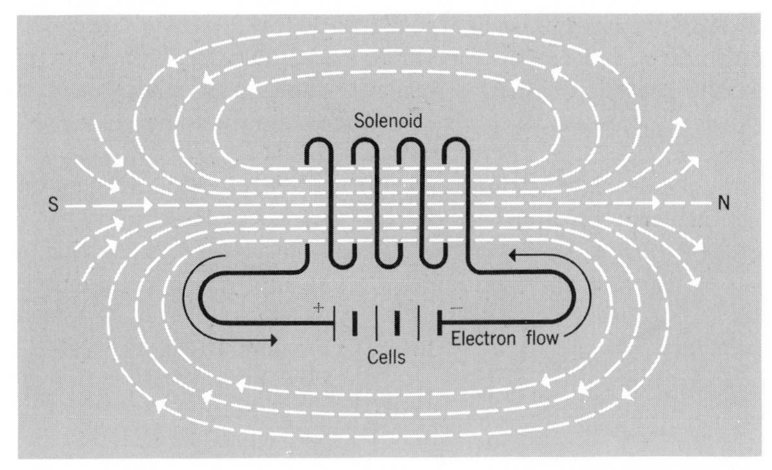

Fig. 10-17. *A coil of wire through which an electric current is passing, behaves like a magnet. Note the north pole at one end and the south pole at the other. Such a coil is called* a solenoid.

ELECTRIC AND MAGNETIC PHENOMENA COMPARED

Can any substance be given an electric charge? Can any substance be magnetized? In principle all materials may acquire an electric charge. However, as a practical matter, only the nonconductors, such as hard rubber, glass, and sulfur, are able to retain a charge for any considerable length of time. On the other hand, only a few metals—the most important is iron—and a few alloys can be magnetized.

Fig. 10-18. *An electromagnet may be constructed by winding wire around a nail or spike and connecting the end of the wire to a cell. The behavior of an electromagnet is very similar to that of a permanent bar magnet.*

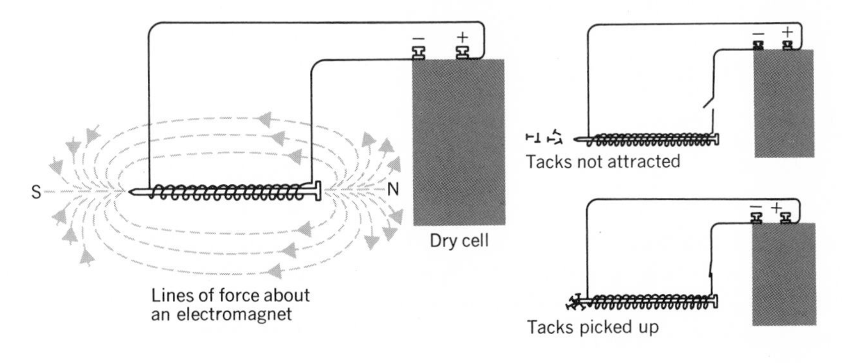

Magnets have regions, called poles, where the magnetic effect is concentrated. A bar magnet has a north pole at one end, a south at the other. Electrically charged objects, on the other hand, have an excess of either positive or negative charge, so the entire body appears to have a single kind of charge.

There is also an interesting parallel in the equations defining the forces between magnetic poles on the one hand and between electrically charged bodies on the other. According to Coulomb's law two bodies bearing quantities of electrostatic charge q_1 and q_2 (esu), and whose distance apart is d (cm) will attract or repel (depending on whether the charges are similar or dissimilar) with a force F (dynes). The equation relating these is

$$F = \frac{q_1 \cdot q_2}{Kd^2}$$

The corresponding relation for magnets is

$$F = \frac{m_1 \cdot m_2}{\mu d^2}$$

where m_1 and m_2 are pole strengths, d is the distance (cm) if the force F is in dynes. The constants K and μ depend on the medium between the charges or the poles. Each has a value of 1 for a vacuum, and is nearly 1 for air.

The parallels in the above equations are evident. In both cases the forces are directly proportional to the product of either electric charges or pole strengths, and inversely proportional to the square of the distance. Each contains a constant which depends on the medium.

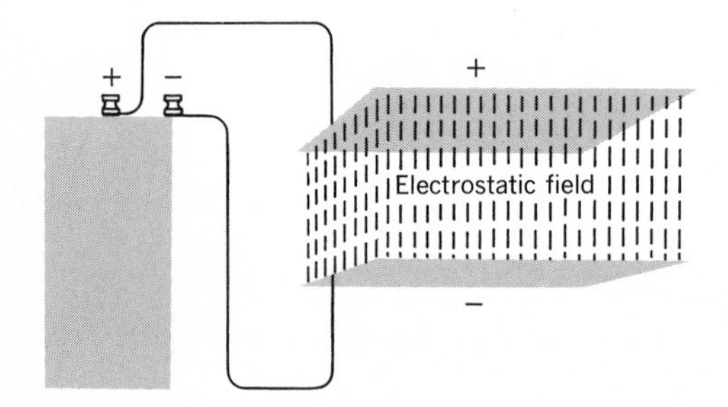

Fig. 10-19. *Electrically charged objects located between the metal plates are attracted to the plate of opposite charge. Since forces are exerted on the objects, an electric field is said to exist between the plates.*

Finally, the question may be raised, As a practical matter, how is an electrostatic or a magnetic field arranged experimentally? Between two parallel metal plates attached to a battery or other current source as shown in Fig. 10-19 there will be an electric field. The battery removes electrons from one plate, giving it a positive charge, and supplies electrons to the other, giving it a negative charge. An electrically charged body in the field will be attracted to the plate of opposite charge.

Recall that a coil of wire in which an electric current is passing behaves like a magnet, with a north pole at one end and south pole at the other. Thus between two coils arranged as shown in Fig. 10-20 there will be a

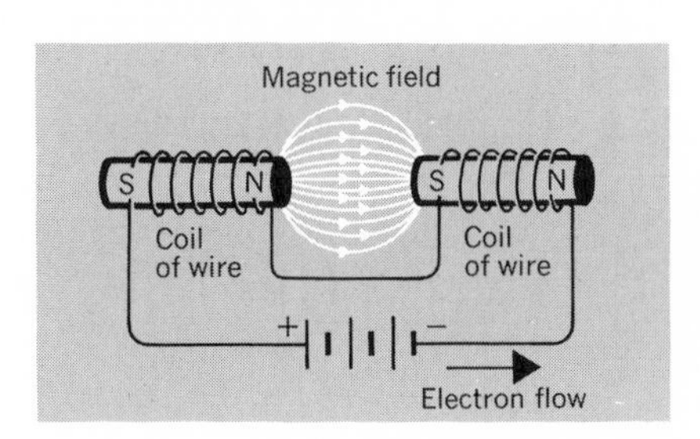

Fig. 10-20. *An electric current passing through two coils of wire will induce north and south poles as shown. There will be a magnetic field between the coils.*

magnetic field, much like the field between two bar magnets with unlike poles facing each other.

Though there are a number of similarities between electric and magnetic phenomena, it should never be overlooked that the two are distinct and different phenomena.

MICHAEL FARADAY: INDUCED CURRENTS

Michael Faraday (1791-1867) was born near London, of humble parentage. His father died when he was young, and it was necessary for him to assist his family. At an early age he was apprenticed to a London bookbinder. The bookbinding may have been dull at times but Michael discovered in the shop chemistry books which proved so fascinating he decided to attend some of the lectures given by Sir Humphry Davy at the Royal Institution. During one of these he took notes and sent Davy a copy. Davy was so impressed that he invited Faraday to come and talk with him. Soon thereafter he was appointed an assistant at the Royal Institution. Eventually Faraday became director of this world-famous scientific institution, which had been founded by Count Rumford.

Oersted had shown that a wire in which an electric current flowed was surrounded by a magnetic field. Faraday predicted the inverse effect, the inducing of electric current flow in the wire when magnetic lines of force swept over it. For a long time the predicted effect eluded him, but eventually both Faraday and Joseph Henry (who taught at one time in Albany Academy, Albany, N.Y.) almost simultaneously discovered the same principle of magnetic induction.

As long as a magnet and a nearby coil were stationary, no current flowed in the coil. But as soon as the magnet's lines of force *moved* over the coil, the deflection of the needle of a galvanometer connected in the circuit showed that an electric current was flowing (Fig. 10-21). Modern

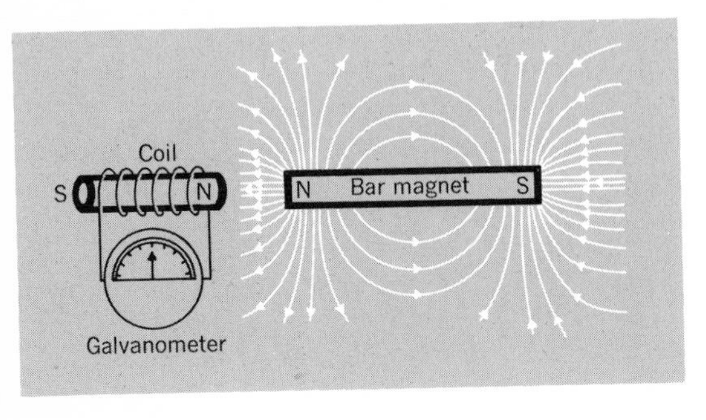

Fig. 10-21. *Magnetic lines of force surround a bar magnet. If there is no motion of the magnet with respect to the coil (or vice versa) there is no deflection of the galvanometer needle. But the needle does move when lines of force of the magnet pass over the wires of the coil; electrons in the wire are set in motion. An electric current is induced in the wire. The direction of the induced current is such as to make the polarity of the approached end of the coil the same as that of the approaching magnet (both north, as shown).*

commercial generators are based on this principle of magnetic induction.

Though Faraday and Henry reached essentially the same conclusions at about the same time, Faraday published his results earlier and is given credit for being the discoverer of magnetic induction.

Joseph Henry (Fig. 10-22) became professor of natural philosophy at

Fig. 10-22. *Joseph Henry, a great American scientist of the nineteenth century.* [*Courtesy, Smithsonian Institution.*]

Princeton and subsequently Director of the Smithsonian Institution in Washington, D.C. The unit of inductance* called the henry is named for him.

FARADAY'S LAWS OF ELECTROLYSIS

Faraday announced his laws of electrolysis some thirty-five years after Volta discovered the voltaic pile. According to his first law, the amount of a chemical substance formed at either electrode in an electrolysis depended on the quantity of charge passing in the circuit. The second law stated that equal quantities of charge produced the same number of chemical equivalents.

If two strips of silver (Ag) metal (electrodes) are immersed in a solution of silver nitrate ($AgNO_3$) and connected to a source of direct current—say, several dry cells—a chemical reaction takes place (Fig. 10-23).

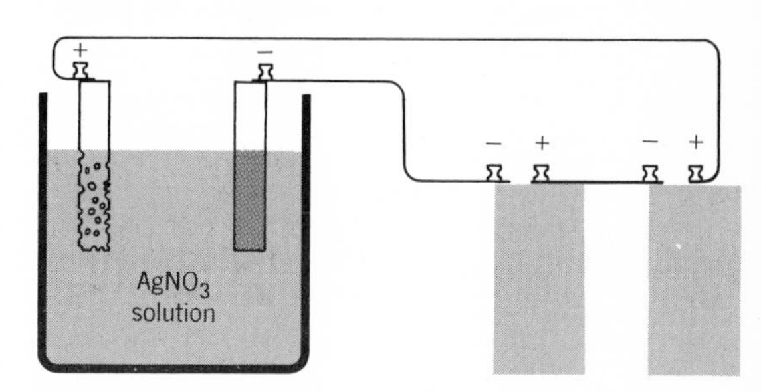

Fig. 10-23. *According to Faraday the amount of chemical action taking place during electrolysis is proportional to the number of coulombs of charge flowing through a solution.*

The electrode which is attached to the positive terminal gradually dissolves and loses weight as atoms are torn off and go into solution. As silver atoms at the electrode surface lose one electron each, they go into the solution as positive ions. The equation for the reaction is

$$Ag \rightarrow Ag^{+} + e$$

The other silver strip, which is attached to the negative terminal of the battery, increases in weight as silver ions from the solution deposit upon it. At this electrode, electrons are pumped in, and as silver ions reach

* Inductance is a property of a coil in an alternating current (AC) circuit and must be taken into consideration when calculating the "impedance" of such a circuit. Impedance in an AC circuit is roughly analogous to resistance in a direct current (DC) circuit.

the electrode, each accepts an electron, becomes a silver atom, and attaches itself to the electrode. The equation is

$$Ag^{+} + e \rightarrow Ag$$

The amount of silver reacting at either electrode depends on the quantity of electric charge passing. How is this quantity measured? Current flow is expressed in amperes, and amperage is proportional to the number of electrons passing a point in the circuit in one second. Amperage is read directly from an ammeter connected in the circuit and the time of flow measured in seconds. The quantity of charge is obtained through multiplication of these quantities. Coulombs = amperes × time (seconds). An electric current of 0.1 ampere flowing in a circuit for 10 minutes gives 60 coulombs of charge (0.1 ampere × 10 min × 60 sec/min). The amount of silver nitrate in the solution remains constant, since as much silver is added by the dissolving process at one electrode as is removed from solution by deposition on the other.

What does it mean to say that equal quantities of charge produce the same number of chemical equivalents?

Suppose three cups, each containing electrodes and electrolytes, are arranged in series with dry cells as shown in Fig. 10-24. Let the first cup contain two gold electrodes, separated from each other and immersed in a gold

Fig. 10-24. *A faraday (96,500 coulombs) of charge is equal to the charge on 6.02×10^{23} electrons. It is sufficient to deposit during electrolysis 107.88 g (the atomic weight) of silver, 31.8 g (half an atomic weight) of copper (II) or 65.7 g (a third of an atomic weight) of gold (III).*

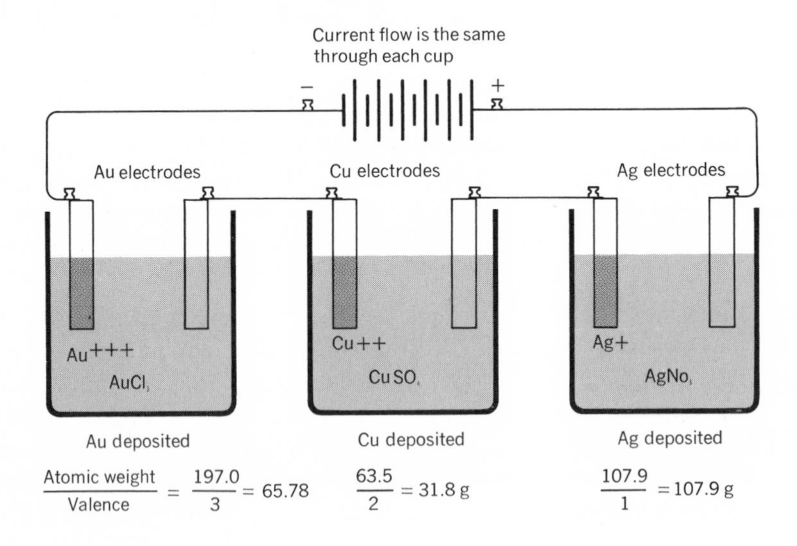

chloride ($AuCl_3$) solution; the second, copper electrodes in a copper II (cupric) sulfate ($CuSO_4$) solution; the third, silver electrodes in a silver nitrate ($AgNO_3$) solution. The dry cells push electrons into the external circuit at the negative terminal and thence to the first gold electrode, where Au^{+++} ions from the solution accept electrons, become atoms, and deposit on the electrode. The movement of ions through each solution constitutes the passage of electric current through that solution. At the right-hand gold electrode there is an electron-donating reaction. The number of electrons given up here exactly matches the number of those accepted from the electrode by Au^{+++} ions in depositing. Thus as many electrons flow out from the right-hand gold electrode to the left-hand copper electrode as enter the left-hand gold electrode from the dry cells. Similarly, as many electrons enter as leave the copper and silver electrodes. The same quantity of current flows in all parts of the circuit.

In each cup the metal electrode nearer the negative battery terminal gains weight. The quantity of current which causes 107.88 grams (1 gram atom) of silver to deposit on the silver electrode will cause 63.54/2 grams ($\frac{1}{2}$ gram atom) of copper to deposit and 197.0/3 grams ($\frac{1}{3}$ gram atom) of gold to accumulate on the gold electrode. The atomic weights of these three metals are: Ag, 107.88; Cu, 63.54; and Au, 197.0. Hence the quantity of electric current required for the deposition of one atomic weight of silver will deposit only half an atomic weight of copper and a third of an atomic weight of gold—an interesting fact when it is recalled that silver has an oxidation number of +1, copper of +2, and gold of +3. When a metallic ion deposits at an electrode, its oxidation number changes to zero. So the equivalent weight may be defined as the atomic weight divided by the change in oxidation number. (Here the term valance may be substituted for oxidation number.)

Since a *faraday* of charge is equivalent to 6.02×10^{23} electron charges, it will suffice to deposit in electrolysis all the atoms in a gram atomic weight of Ag, half those in a gram atomic weight of Cu, and $\frac{1}{3}$ of those in a gram atomic weight of Au. Hence one must ask, How many electrons are required per atom? For an element in which the change of oxidation number is 1, a faraday will provide sufficient electrons to deposit a whole atomic weight of a metal. If each atom requires two electrons, only half an atomic weight will deposit, and so on.

ILLUSTRATIVE PROBLEM

Calculate the weight of cadmium metal deposited on a negative electrode (cathode) from a solution of $CdSO_4$ if a current of 0.1 amp flows for 15 min. Cadmium has an atomic weight of 112.4 and an oxidation number of +2.

Solution

0.1 amp·900 sec = 90 coulombs of charge deposit cadmium metal.
96,500 coulombs deposit 1 g equivalent weight of cadmium.

90 coulombs deposit $\frac{90}{96{,}500}$ g eq. w.

$$\frac{90}{96{,}500} \text{ g eq. w.} = \frac{9}{9650} \cdot \frac{112.4}{2} \text{ g} = 0.0524 \text{ g}$$

The correspondence between the deposition in electrolysis of simple fractional parts of an atomic weight of metallic elements strongly suggested the existence of a unit of negative electrical charge. Hence it may be said that Faraday's laws of electrolysis offer an early hint of a unit charge of electricity, i.e., the electron.

Problems, Chapter 10

1. In a flashlight, two 1.5-volt batteries are connected in series. Find the current flowing through the filament of the light bulb if its resistance (assumed to be the entire resistance of the circuit) is 9.0 ohms.

2. What resistance in an electrical circuit will permit a current flow of 0.40 ampere, using a 6.0-volt battery?

3. A silver spoon is to be replated. It is made the cathode (negative electrode) in an electrolytic solution of a silver salt in which a silver anode (positive electrode) is also immersed. Calculate the weight of silver deposited on the spoon by a current of 0.500 ampere flowing for 30.0 min.

4. Calculate the resistance of a piece of No. 18 (copper) bell wire 6 in. long if the resistance of this wire is 6.5 ohms per 1000 ft. Calculate the amperes flowing if the wire is connected across the terminals of a 1.5-volt battery. (Neglect the internal resistance of the battery.)

5. The strength of an electric field is defined as the force (in dynes) exerted on a unit positive charge placed at a particular point in the field. What is the field strength at a point 10 cm from a tiny sphere carrying a charge of 25 esu? (Assume that there is air between the two charges and that its dielectric constant $K = 1$.)

6. The field strength at a point near a magnet is the force (in dynes) which would be exerted on a unit north pole at that location. Find the field strength at a point 15 cm from a magnet whose pole strength is 50 unit poles. (Assume that $\mu = 1$ for air.)

7. When direct current (DC) resistances are connected in parallel, the reciprocal of the total resistance is equal to the sum of the reciprocals of the individual resistances. (a) Calculate the total resistance of a 10-ohm and a 15-ohm resistance connected in parallel in a DC circuit. (b) If the remainder of the same circuit has a total resistance of 0.25 ohm calculate the current flowing (amperes) when the circuit is attached to a battery consisting of twenty 1.5-volt cells connected in series.

8. How much heat is liberated in one minute when 300 coulombs of charge pass through a heater whose resistance is 23.0 ohms? (Assume a DC power source.) What is the voltage of the power source?

9. How long must a current of 0.100 ampere flow in an electrolysis apparatus (see Fig. 10-3) in order to produce 10.0 ml of hydrogen gas at STP? (Assume that the cathode reaction is $2H^+ + 2e \rightarrow H_2$.)

10. How much metallic sodium is produced by Davy's electrolytic method of decomposing molten NaOH, if a current of 1.0 ampere flows for 10 minutes?

11. What weight of H_2SO_4 is removed from the electrolyte in a storage battery if 1.00 gram of $PbSO_4$ deposits on the electrodes? Consider the overall reaction to be

$$Pb + PbO_2 + 2H_2SO_4 \rightarrow 2PbSO_4 + 2H_2O$$

Exercises

1. How is positive electric charge defined? Negative?

2. State the laws of attraction and repulsion for electric charges; for magnetic poles.

3. To what did Galvani attribute the twitching when the frog's leg was touched with a metallic scalpel? How is the result interpreted today?

4. Draw a diagram of a voltaic pile and label the essential components.

5. Why was it essential for Davy to use molten sodium hydroxide for preparing the element sodium rather than electrolyzing an aqueous solution?

6. What is believed to be the nature of an electric current flowing in a conducting wire?

7. Show some of the analogies between electric current flow in wires and water flow in pipes.

8. Explain the meaning of each of the symbols in Ohm's law, $E = IR$.

9. Describe how to properly connect a voltmeter in an electric circuit. An ammeter.

10. Draw a diagram representing a storage battery. Label the parts. Write the overall chemical reaction taking place.

11. What is the energy source in a cell (or battery)? Explain.

12. What factors affect the amount of heat produced when an electric current flows along a conducting wire?

13. Explain the purpose of a fuse in an electrical circuit.

14. Describe two methods of demonstrating the presence of a magnetic field in the vicinity of a bar magnet.

15. How did Oersted show that there was a magnetic field about a wire carrying an electric current?

16. What is the "thumb rule" for predicting which end of a solenoid becomes a north pole when an electric current passes through it? (Look up in a physics text.)

17. State Faraday's laws of electrolysis.

18. What is meant by the equivalent weight of a chemical element?

11 Atomic structure: Evidence of discharge tubes

ATOMS are so tiny that they are invisible under the most powerful microscopes. Yet even the smallest atoms, those of hydrogen, are massive giants compared with electrons. It is natural to ask, What is the experimental evidence that attests to the existence of any such particles as electrons? The two most important sources of experimental evidence for them, for atomic structure, and for nuclear structure as well, are the subject of the present chapter and Chapter 12—discharge tubes and radioactive elements.

A discharge tube is often just a highly evacuated glass tube through whose wall metal electrodes are inserted and sealed. The neon sign is perhaps the best known adaptation of a discharge tube.

In 1853 the French scientist Masson noted that a high-voltage current passing through a partially evacuated glass vessel in which the electrodes were *far apart,* caused the vessel to glow instead of causing a sparking. The vacuum pump invented by Heinrich Geissler appeared in 1855. By the end of the next decade there was a great flurry of experimentation with "aurora tubes"—so called because the effects observed in them seemed to resemble the phenomenon known as northern lights (aurora borealis). Later they were called Geissler tubes in honor of the man who so skillfully fashioned them, and still later Plücker tubes (Fig. 11-1) as modified by Plücker, a professor of physics at the University of Bonn. Plücker also showed that the position of the glow in a tube was shifted by bringing near it a bar magnet. Thus he proved that the rays had a magnetic property. In 1869 Plücker and Hittorf devised a discharge tube in which one electrode was off to one side as shown in Fig. 11-2 rather than both being aligned as in Fig. 11-1. Passage of electric current through the

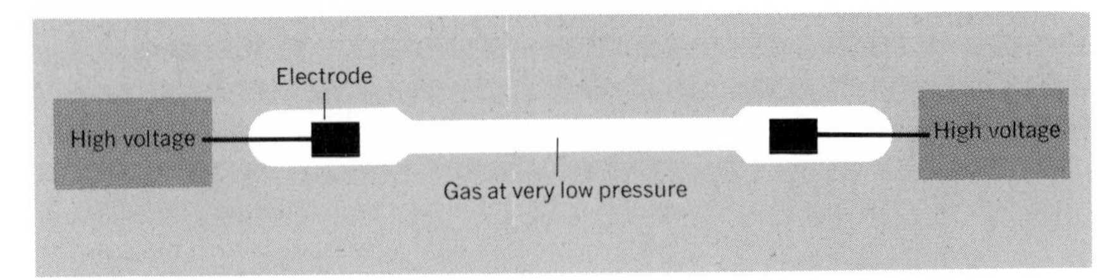

Fig. 11-1. *A Plücker tube. Metal pieces called electrodes which are sealed through the ends of the tube conduct electric current in and out of the tube. The tube is highly evacuated and a trace of gas such as helium, hydrogen or neon is introduced before it is finally sealed. When a high-voltage current is passed through it, light is emitted.*

tube caused a glow at the end of the tube, and objects placed inside, between the cathode and the end of the tube, cast a shadow. The sharpness of the shadow suggested that particles originating on the cathode and traveling in straight lines were responsible for the glow.

THE CROOKES EXPERIMENT

The English physicist Sir William Crookes (1832-1919) became much interested in the properties of cathode rays. With his famous Crookes tube (Fig. 11-2) he too observed that a metal object such as a cross placed in the path of the "cathode ray" cast a shadow in the luminescence produced at the far end of the tube. Crookes also constructed a tube with a pinwheel on a horizontal track, as shown in Fig. 11-3. When the high voltage from an induction coil was turned on, the pinwheel moved away from the negative terminal as though pushed. If the polarity of the electrodes was reversed, the pinwheel went in the opposite direction. It ap-

Fig. 11-2. *A Crookes tube. Streams of electrons from the cathode, called* cathode rays, *reach a luminescent material coating the interior of the end of the tube and cause it to light up. Where electrons are stopped by intervening objects a shadow is cast.*

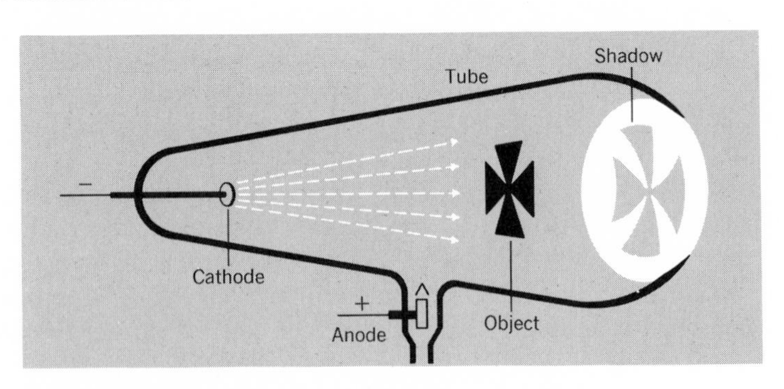

Fig. 11-3. *Cathode rays cause a pinwheel to move away from the cathode. They must consist of particles since motion is imparted to an object (the pinwheel).*

peared, then, that material particles must be leaving the cathode, since motion was imparted to the pinwheel.*

Near the close of the nineteenth century, Perrin (1870-1926), like Crookes a Nobel prize winner, repeated experiments to determine whether cathode rays were deflected in an electric field, found that they were attracted to the positive plate, and concluded that they actually were negatively charged. As the twentieth century approached, it was becoming well established that the rays emitted from the cathode of a Crookes tube under the influence of a high-voltage current were high-velocity particles of negative charge, and that their nature was independent of the metal electrode used or of the residual gas in the tube. The name given to such a negative particle by Johnstone Stoney remained with it—the *electron.*

THOMSON'S MEASUREMENT OF E/M

Sir J. J. Thomson (1856-1940), physicist at Cambridge University and one time professor at the Royal Institution in London, will long be remembered for his classic evaluation of e/m, the ratio of charge to mass for the electron. Until his time the atom had been regarded as the fundamental structural unit and hence was presumed to be indivisible. To be sure, a series of experiments had hinted that the atom might be complex, but as yet nobody had been able to make a quantitative comparison between the masses of atoms and of subatomic particles.

For his experiment Thomson used a modified Crookes tube as shown in Fig. 11-4. Electrons originating at the cathode (left) are formed into a narrow beam when passing through diaphragm D, travel across the tube at high velocity and produce a spot of light where they strike the fluorescent screen at A.

If now an electric current passes through the solenoids on either side of

* This point of view is contested. See *Journal of Chemical Education,* 38:480 (1961).

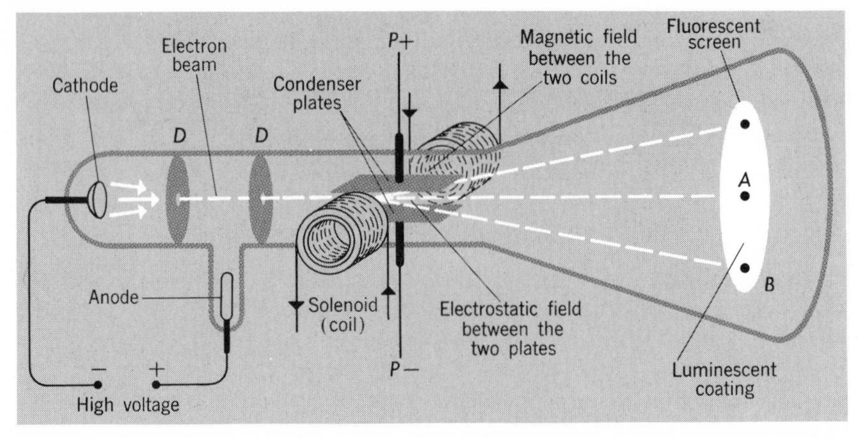

Fig. 11-4. *Apparatus used by Thomson in evaluating the ratio* e/m *for the electron. D, D, diaphragms with pinholes—P+, P−, plates that can be charged electrically—*A, *point at which the undeviated electron beam strikes when the solenoids carry no current—B, point struck by a ray bent in the magnetic field produced by current through the solenoids. The effect of electrostatic charges on the plates is to bend the electron beam upward, counteracting the effect of the magnetic field.*

the tube, the electron beam is bent downward to point B. The path of the beam inside the magnetic field is a portion of a circular arc whose radius is r.

The force exerted on the electron beam by the magnetic field is Hev, where H is the magnetic field strength, e the charge of the electron, and v the electron velocity. The magnetic force is equivalent to a centripetal force mv^2/r, where m is the electron mass, v the electron velocity, and r the radius of curvature of the electron beam. When the beam is stationary at B

$$Hev = \frac{mv^2}{r}$$

Dividing both members of the equation by Hmv gives

$$\frac{e}{m} = \frac{v}{Hr}$$

H and r are readily measurable. The electron velocity v is evaluated by turning on the electrostatic field and bending the electron beam back to exactly A. At A, the force exerted by the magnetic field is just balanced by the force exerted by the electric field (Xe). The electric field strength X is determined by the distance apart of the condenser plates and the voltage applied to them:

$$Hev = Xe \quad \text{or} \quad Hv = X \quad \text{or} \quad v = \frac{X}{H}$$

With v, H, and r known, e/m may be calculated. A modern value is

$$\frac{e}{m} = 1.76 \times 10^8 \text{ coulombs/g}$$

Note that e/m is a ratio and that knowledge of its value tells nothing about the value of either e or m.

From Faraday's laws of electrolysis, however, it was known that 96,500 coulombs of charge would deposit 1 gram of hydrogen gas in an electrolysis. If both electrons and hydrogen atoms (ions) in solution were assumed to have the same quantity of charge, then the mass of the hydrogen atom should be approximately 1800 times as large as that of the electron:

$$\frac{\dfrac{e}{m_{\text{electron}}} = 1.76 \times 10^8}{\dfrac{e}{m_{\text{hydrogen}}} = 9.65 \times 10^4} \quad \text{or} \quad \frac{m_H}{m_e} \cong 1800$$

Thomson's evaluation of e/m made possible an estimate of the ratio of the masses of the electron and the hydrogen atom.

Soon another great physicist performed the experiment which gave a measurement of the charge of the electron and thus made possible a precise evaluation of the electron mass.

THE MILLIKAN OIL-DROP EXPERIMENT

In 1911 Robert Andrews Millikan (1868-1953), a professor of physics at the University of Chicago and later at the California Institute of Technology, performed the famous oil-drop experiment by which the value of e, the charge of the electron, was measured. For this work he was awarded a Nobel prize in 1923.

The apparatus for this experiment is shown in Fig. 11-5. Two metallic plates suspended horizontally are attached to a high-voltage battery. In the top plate is a small hole through which an oil drop is admitted to the space between the plates by means of an atomizer. Since the oil drop is very tiny, the course of its movement can be followed by means of a short-focus telescope. An electric arc lamp at one side provides proper illumination of the field for observing the oil drop. An X-ray tube (or radioactive source) at the other side provides ionizing radiation to vary the amount of electric charge on the oil drop.

Oil drops are sprayed above the opening in the upper plate, and a few fall through the opening. The motion of a single drop is observed. Gravity

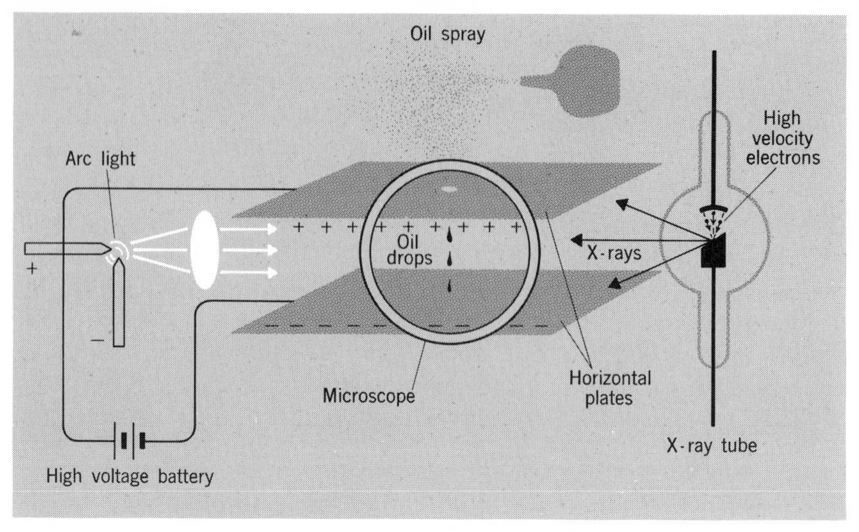

Fig. 11-5. *The Millikan oil-drop experiment established the unit which we call the electronic charge. Once the numerical value was known, it was possible to evaluate the electronic mass, 1/1840 that of the hydrogen atom. A convenient way of remembering the relation of these masses is to compare the mass of the electron to the pound and that of the hydrogen atom to the ton. The mass of the electron is very tiny compared with the mass of the smallest of all atoms, hydrogen.*

tends to accelerate the falling drop; frictional effects with rarefied gas tend to retard it. Experiments show that a tiny falling drop quickly reaches terminal velocity and thereafter falls with uniform velocity. Sir George G. Stokes described a method, applicable here, for calculating the radii of small particles falling through a viscous medium at constant velocity. The velocity is determined by noting the time required to fall a measured distance. This value is always the same for a given drop. Once the velocity of fall is known, the radius is obtained, and from it, the volume of the drop. From the volume of the drop and the density of the oil, the mass of the drop is determined.

So far the voltage source has not been connected to the plates and there has been no electric field (see p. 186) between them. Now, each time the oil drop approaches the bottom of the scale seen through the telescope, the high-voltage source is switched on in such a manner that the oil drop is driven upward in the field. When the drop reaches the top, the field current is disconnected and the drop falls again. Thus a given oil drop may be made to rise and fall many times. The downward velocity remains the same time after time. Irradiation of the drop by X rays causes the upward velocity to vary, becoming greater as the negative charge on the oil drop increases. By measuring the different upward velocities the different quantities of charge on the oil drop may be calculated. The striking feature of the values of these charges is that they are all multiples of a simplest number. If, for example, a series of values such as those in Table 11-1

Table 11–1 Discovering a Unit Value from a Series of Values

6	18	30	12
36	24	42	18
54	12	60	72

were obtained, it might be concluded that the smallest charge ever to be found—the unit charge—was 6, and that in the case of the drop containing a charge of 54 there would be 9 such units. Actually, the modern value of the smallest quantity of charge observed is 1.60×10^{-19} coulomb, the value of e, the quantity of charge on the electron.

If this value of e is substituted in

$$\frac{e}{m} = 1.76 \times 10^{8} \text{ coulombs/g}$$

the mass of the electron may be calculated, and is found to be 1/1840 as great as the mass of a hydrogen atom. Electrons, then, have masses which are very tiny compared with even the lightest of all atoms.

Here at last was the long-awaited link relating atomic and subatomic particles. Discharge tubes of still a different type made possible the precise measurement of atomic masses. An instrument called the mass spectrograph was soon central in experiments with *isotopes*.

ISOTOPES: THE MASS SPECTROGRAPH

In 1910, knowledge about radioactive atoms (to be discussed in Chapter 12) was accumulating rapidly. For a century Dalton's idea that all atoms of a particular element had the same mass had prevailed. Now it was becoming increasingly evident that certain radioactive atoms had different masses but identical chemical properties. The Daltonian concept was no longer tenable. The great English chemist and pioneer in the field of radioactivity, Frederick Soddy, suggested the name *isotopes* for these chemically similar atoms of different atomic masses.* Early in the twentieth century, Thomson carried out a series of experiments by which he determined the value of e/m for different positive rays. He used the apparatus shown in Fig. 11-6. The electric field between the metal plates and the magnetic

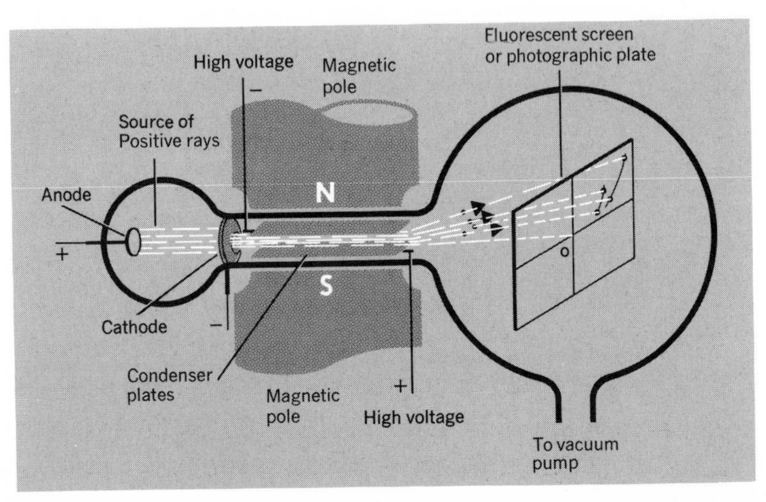

Fig. 11-6. *Thomson's mass spectrograph. Atoms with different values of* e/m *were brought to a focus on different parabolic curves. With this device, Thomson proved that neon was composed of atoms of two different masses, 20 and 22. Soddy called these atoms, with different masses but identical chemical properties,* isotopes.

field between the pole pieces were arranged so that the force exerted on a beam of positive rays by the one would be at right angles to that exerted by the other. The combination of the two forces would bring to a focus on a parabola all atoms of a given mass. In each experiment a trace of the gas to be investigated was introduced into the highly evacuated tube. If the gas was hydrogen, the beam of positive rays consisted of hydrogen ions, or

* Note that the mass spectrograph determines an atomic mass, not an atomic weight. One atomic weight unit on the older, chemists' scale (in which the oxygen isotopes are assigned a weight of 16) is equal to 0.999,957 atomic mass units on the new scale (which assigns to the carbon-12 atom 12 atomic mass units). The difference between the atomic *weight* unit and the atomic *mass* unit amounts thus to 43 parts in a million. For all but the most precise measurements, this difference is inconsequential.

protons. A photographic plate placed at the end of the tube was exposed to these positive ions. The developed photographic plate showed a characteristic parabolic curve. Atoms of different masses formed these parabolic curves at different locations on the plate.

About 1910 Thomson introduced neon gas into one of these tubes. Two parabolic curves were observed—a prominent one corresponding to an atomic mass of 20 and a less intense curve corresponding to an atomic mass of 22, rather than a single curve corresponding to the expected value of 20.2, the known atomic mass of neon. It appeared then that there were two kinds of neon atoms present, those of mass 20, predominating, and lesser numbers of mass 22.

Some years later, Thomson's student Aston redesigned the apparatus so that all particles having the same value of e/m would, by deflection in electric and magnetic fields, be brought to a focus, not on the same parabola as in Thomson's apparatus but on the same vertical line. A diagrammatic representation of Aston's apparatus is shown in Fig. 11-7. This *mass*

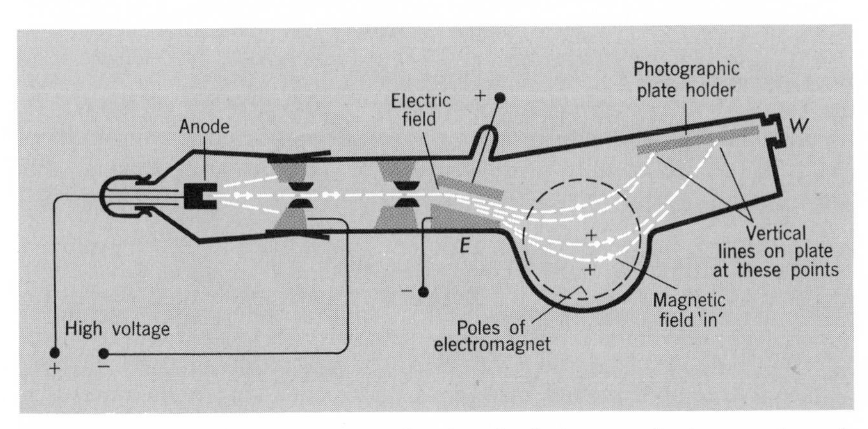

Fig. 11-7. *Aston's mass spectrograph. Aston's design greatly improved precision of measurement of atomic masses and ability to detect different isotopes and to measure their relative abundances.*

spectrograph provided the scientist with a precision instrument for identifying the isotopes predicted by Soddy.

Fig. 11-8 shows a mass spectrogram of the element hydrogen. The very intense line at A indicates that nearly all hydrogen atoms have a mass of 1; the faint line at B shows that a few atoms of mass 2 are also present. Hydrogen atoms of mass 2 are called deuterium atoms, or heavy hydrogen. Hydrogen and deuterium are isotopes.

The two isotopes of hydrogen undergo the same chemical reactions, but their masses are different and hence their densities are different too. Any physical properties depending on density, for example, their speeds of diffusion, are also different. Their density difference gives rise to different reaction rates and "isotope effects." Because of their heavier masses, deuterium atoms at a given temperature have somewhat smaller mean velocities than hydrogen atoms. Hence hydrogen atoms have a higher probability of undergoing collision reactions than do deuterium atoms.

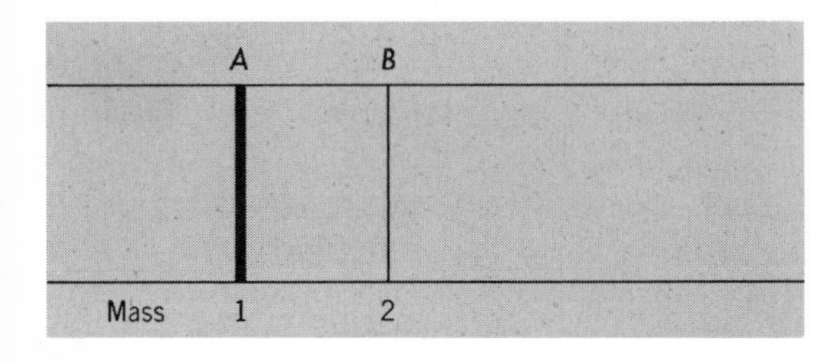

Fig. 11-8. *The mass spectrogram of hydrogen. The mass spectrogram is so called because it resembles a spectrogram of light. Just as the spectroscope separates the components of a beam of light and presents them as a series of bands along a scale of increasing wavelength, so the mass spectrograph separates the components of a stream of particles and presents them as a series of bands along a scale of increasing unit mass. Any such presentation may be called a spectrum.*

One final type of discharge tube, now to be considered, proved to be of great importance in probing the structure of the atom and in many other areas as well.

X RAYS

While experimenting with a Crookes tube in 1895 the German physicist Wilhelm Konrad Roentgen detected a very penetrating radiation issuing from the tube. This radiation affected a photographic plate even when wrapped in thick black paper. A screen coated with a fluorescent material such as zinc sulfide (ZnS) glowed when placed near the Crookes tube. An object placed between the radiation source and the screen cast a shadow, and the denser the object, the darker the shadow. Apparently heavier metals (such as lead) were more effective in attenuating this radiation than lighter metals such as aluminum. The rays also increased the electrical conductivity of gases through which they passed; i.e., they were ionizers of gases.

These penetrating rays were called Roentgen rays or X rays. They were shown to be electromagnetic waves whose wavelengths were a few angstrom units, compared with several thousand angstrom units for visible-light waves. Hence in many respects their characteristics (e.g., velocity, reflection, refraction, diffraction) resembled those of visible light. But they were far more penetrating and passed through objects which were opaque to visible light.

An X-ray tube is simply a Crookes tube modified by placing a piece of a heavy metal such as tungsten (called the target) in the path of the high-velocity electrons issuing from the cathode. Whenever streams of high-velocity electrons impinge on solids, X rays are produced. In general, the higher the voltage applied at the terminals of the tube, the higher the velocities of the electrons striking the target and the more penetrating the X rays.

The more modern Coolidge X-ray tube (Fig. 11-9) makes use of a

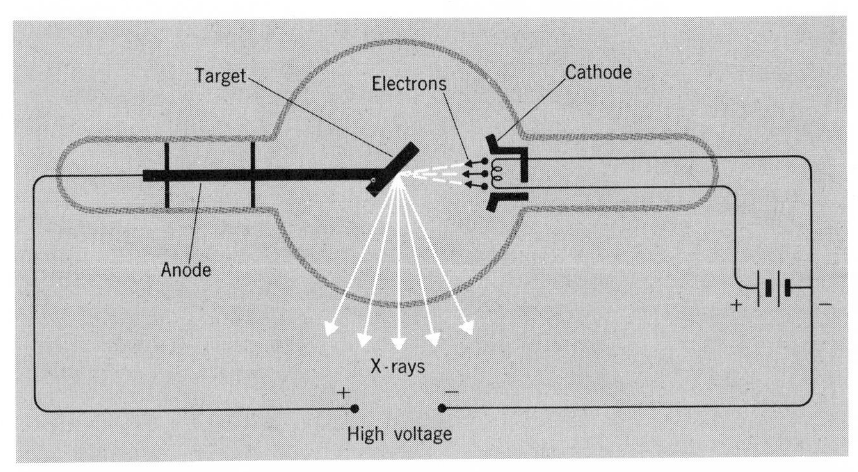

Fig. 11-9. *An X-ray tube is a modified Crookes tube in which the electron beam is focused on a metal target (of high atomic weight). The higher the voltage, the greater the speed with which the electrons strike the target atoms and the greater the penetrating power of the X rays emitted.*

coil of tungsten resistance wire recessed in the cathode. Electrons emitted from this hot wire are brought to a focus on the target. The intensity of the X rays may be altered by changing the current (amperes) flowing through the cathode coil.

As more and more penetrating X rays have been required, it has been necessary to apply higher and higher voltages to these tubes. At extremely high voltages, insulating against "arcing across" becomes a serious problem. In recent years a number of electron accelerators based on a principle other than high-voltage acceleration alone have been developed. One of these, the betatron (Figs. 11-10 and 11-11), whirls electrons in a doughnut-shaped evacuated tube and then hurls them at terrific speed into a heavy metal target, thus generating "super X rays." Another machine for ac-

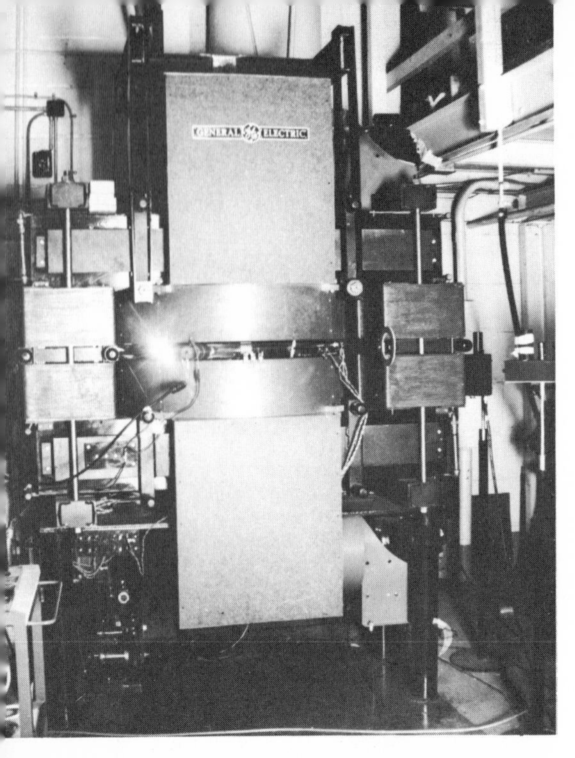

Fig. 11-10. *The betatron accelerates electrons to velocities which are much greater than those attainable in the conventional X-ray tube. High-velocity electrons from the betatron impinging on a target material give X rays of unusually high penetrating power.* [*Courtesy, General Electric Research and Development Center, Schenectady, N.Y.*]

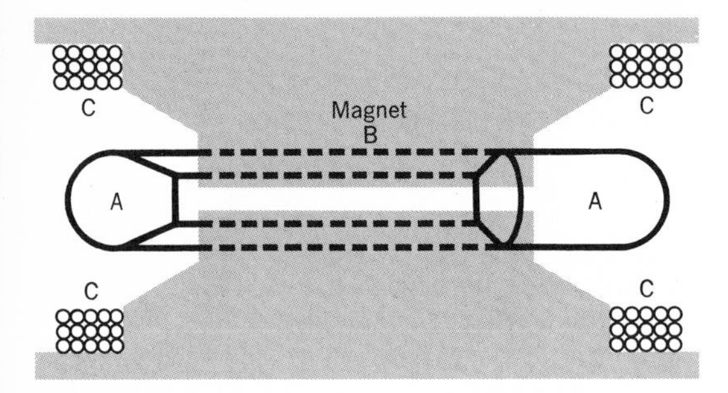

Fig. 11-11. *Cross-sectional diagram of a betatron showing the "doughnut"* (*A*), *a highly evacuated glass tube in which the electron beam is accelerated, and the electromagnet core* (*B*), *which is magnetized by an electric current* (*AC*) *passing through coils* (*C*). *By keeping the magnetic field strength increasing in step with the increase in electron velocity, it is possible to maintain constancy of the electron-beam radius. As soon as the peak velocity* (*of the AC current*) *is attained, the electron beam is deflected and directed at a target to produce X rays.* [*Courtesy, General Electric Research and Development Center, Schenectady, N.Y.*]

celerating electrons to even higher energies (and hence producing still higher-energy X rays) is the so-called linear accelerator.

The applications of X rays in dentistry for detection of cavities in teeth, and in medicine for viewing breaks in bones, making chest X rays for tuberculosis detection, and giving therapeutic treatments in cancer cases, are well known. X rays also play a very important role in engineering for detecting flaws in metal castings, in structural steel, and in welded joints.

X-RAY SPECTRA: ATOMIC NUMBERS

In physics X-ray spectra were of great value in determining the order numbers or atomic numbers of the chemical elements. The periodic law of Mendeleev was based on the order of arrangement of the elements' atomic weights; but (Chapter 8) strict adherence to this order led to certain pairs of elements such as tellurium and iodine falling in the wrong families of the periodic table. It was apparent that their order had to be reversed in spite of the fact that reversal would put them out of sequence with respect to their atomic weights. What basis was there for reversing the order of these elements?

One source of supporting evidence came from X-ray spectra. A luminous vapor, e.g., sodium or mercury, emits visible light waves which may be dispersed by a prism or diffraction grating into a series of characteristic lines, called a line spectrum (Chapter 9). Since visible light and X rays are both forms of electromagnetic radiation, it will not be surprising to learn of the existence of characteristic X-ray spectra, even though in wavelength X rays differ considerably from visible light.

In 1913 the brilliant young British physicist Henry G. Moseley photographed the X-ray spectra of a number of elements, using an X-ray tube with an interchangeable target: when one photograph had been made, before the next was taken, the metallic target was replaced with another target made of a different metal. Instead of a prism or ruled-line grating,

Fig. 11-12. *An X-ray spectrograph employs a source of electromagnetic waves (the X-ray tube), a crystal for dispersing the beam into different wavelengths and a photographic plate for recording the presence of the waves.*

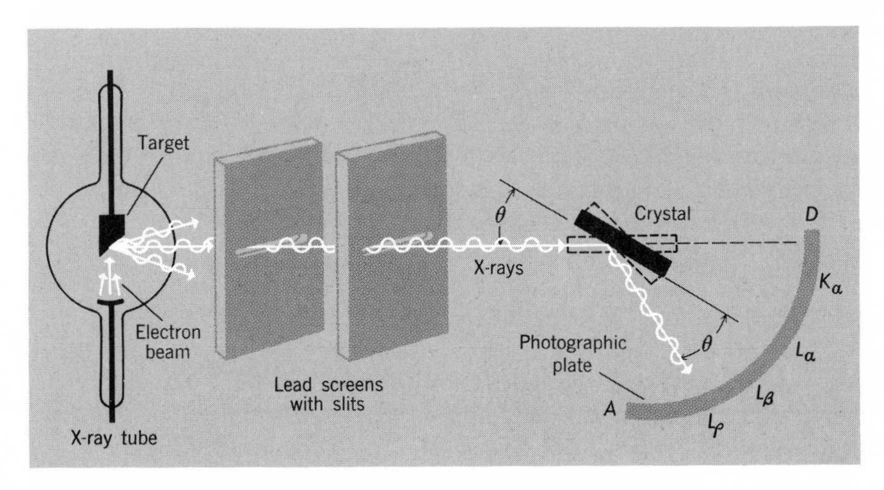

Moseley used a crystal of sodium chloride for diffracting the X rays, since the distances between the rows of atoms in crystals had about the values needed to let them behave like diffraction gratings.

A diagram of the X-ray spectrograph is shown in Fig. 11-12. The X-ray tube corresponds to the light source used in producing optical spectra. The X rays produced are characteristic of the target material placed in the path of the electron beam in the tube. The spreading out of the rays into a spectrum is accomplished by rotating the salt crystal on its axis so that the diffracted X-ray beam sweeps over the photographic plate. The spectrum so recorded is simple in structure compared with many optical spectra. When the photographic plate is developed, two sets of characteristic lines are obtained. Fig. 11-13 shows one set of these characteristic lines for each of several metallic elements.

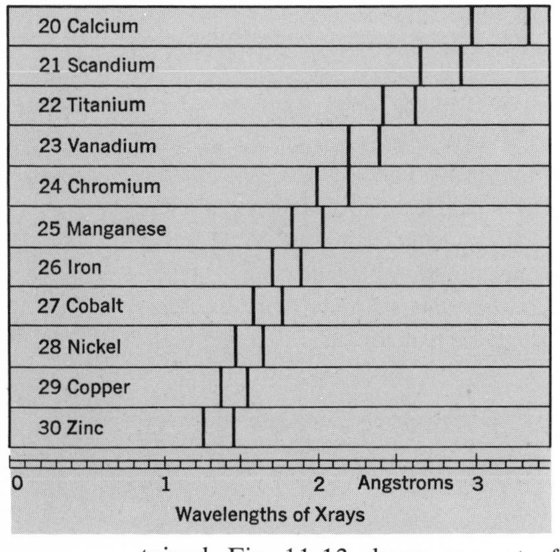

Fig. 11-13. *The order of the elements as determined by characteristic X-ray spectrum lines.*

Moseley arranged the elements in the order dictated by the shifting of the characteristic X-ray lines. He also discovered that by plotting the order number so obtained against the square root of the frequency of one of the characteristic lines, a straight line was obtained (Fig. 11-14).

If nickel and cobalt were placed in order of increasing atomic *weight* (Ni = 58.69, Co = 58.94), neither of them fell on the straight line just mentioned. However, when they were placed according to the order of their X-ray spectra, their order was reversed and they were on the line. Other pairs of similarly misplaced elements, resulting from the use of atomic weight as the order number, fell in the correct location on the line when the order numbers derived from X-ray spectra were used. The periodic law

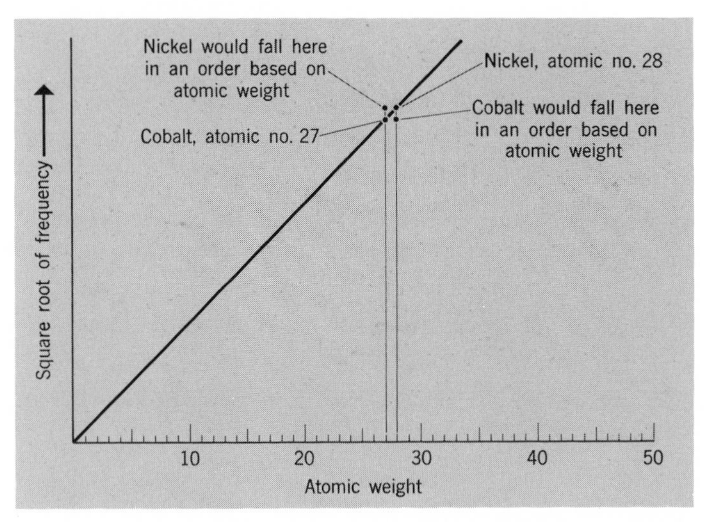

Fig. 11-14. *Plot of the square root of the frequency of a characteristic line in the X-ray spectrum of various elements vs. the atomic number of the element.*

(Chapter 8) was revised to state that the properties of the elements were periodic functions of their order numbers. These order numbers were called *atomic numbers.* Later experimental evidence obtained by Chadwick (see Chapter 13) linked the atomic number of the element with the number of positive charges (protons) on the nucleus of its atom.

Problems, Chapter 11

1. Calculate the number of molecules of hydrogen gas in a vacuum tube whose volume is 25 ml and whose pressure is 10^{-6} atm at 0°C. (2.0 g of hydrogen at 0°C and 760 mm contain 6.02×10^{23} molecules.)

2. (a) Calculate the density of deuterium gas at 0°C and 760 mm Hg. Assume that it is diatomic in the molecule and that its atomic weight is 2.01. (b) According to Graham's law of diffusion, the speeds of diffusion of gases are inversely proportional to the square roots of their densities. How does the speed of diffusion of hydrogen gas compare with that of deuterium?

Exercises

1. List several of the properties of cathode rays.

2. Describe Thomson's experiment for measuring e/m. Why is e/m such an important ratio in atomic physics?

3. Describe the Millikan oil-drop experiment. What is its great importance with relation to atomic structure?
4. What are isotopes? Illustrate.
5. What is a mass spectrograph? Describe the principle of its operation.
6. What are "isotope effects"? Why are they more pronounced with the lighter elements such as hydrogen and deuterium than with the heavier isotopes, as for example uranium?
7. Describe the construction of an X-ray tube. How are X rays produced?
8. Compare the characteristics of X rays with those of visible light rays.
9. Compare the construction of an X-ray spectrograph with one used for the detection of radiation in or near the visible region of the spectrum.
10. Explain how characteristic X-ray spectra were used in determining atomic numbers.

12 Exploding atoms: Radioactivity

SHORTLY after Roentgen made his world-shaking discovery of X rays, the French physicist Henri Becquerel (1852-1908) quite by chance on one occasion left some uranium ore on a photographic plate wrapped in heavy paper; when this plate was developed, he found, much to his surprise, that penetrating radiation had affected the plate. He thus discovered that penetrating radiation was continuously being emitted by these materials. These new radiations were soon named "Becquerel rays." If Becquerel rays were not identical with X rays, they were certainly very close relatives.

THE ELECTROSCOPE

An important property of X rays was their ability to discharge a charged electroscope. A leaf electroscope is shown in Fig. 12-1. It consists of a vertical fixed conductor, or leaf, to which is attached (as if hinged) a second thin metallic strip or leaf. Both are suspended inside a glass case to avoid interference of air currents.

If an electrically charged object is brought in contact with the knob, the two leaves acquire the same charge, repel each other, and separate. The electroscope is now said to be charged. Becquerel soon discovered that the penetrating radiation from uranium minerals discharged an electroscope (caused the movable leaf to fall) just as X rays did. He also found that the higher the percentage of uranium in an ore, the faster the rate of discharge of the electroscope. The ore pitchblende particularly excited his curiosity, for the rate at which it discharged an electroscope was far too rapid to be accounted for by the uranium content alone. He wondered if a more potent radioactive element was present. He was also successful in persuading one whose name has since become immortal in the field of radioactivity, Madame Curie (1867-1934), to work on the problem of searching for this more active element.

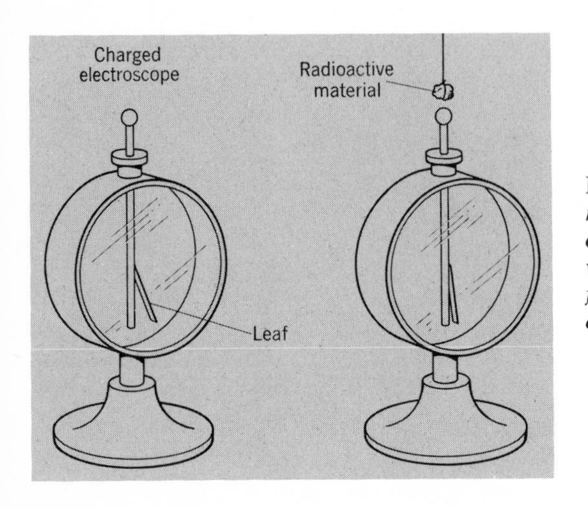

Fig. 12-1. *Radiations from radioactive materials produce ions (electrically charged atoms) in the air through which they pass. If these ions are formed in the vicinity of a charged electroscope, it discharges.*

MADAME CURIE'S DISCOVERY OF POLONIUM

Marja Slodowska Curie, better known as Madame Marie Curie, was born in Poland a few years after the close of the Civil War in the United States. In 1891 she had gone to Paris to study at the Sorbonne. Four years later, in the year that Roentgen discovered X rays, she married Pierre Curie, a physicist at the Ecole Municipale. They became so much fascinated with the prospect of finding in pitchblende a new element more powerfully radioactive than uranium that both decided to devote full time to this work.

The Curies dissolved the uranium ore in an acid and then bubbled hydrogen sulfide (H_2S, with its odor of rotten eggs) gas through the solution. A precipitate formed which was more highly radioactive than the original sample. The uranium remained in solution. The highly radioactive precipitate was again dissolved, and bismuth was precipitated from this solution; this time the activity remained in the solution. The new element responsible for the high degree of radioactivity was called polonium in honor of Madame Curie's native Poland.

THE DISCOVERY OF RADIUM

Shortly after the discovery of polonium, Madame Curie found that a material of still greater radioactivity could be separated from the dissolved uranium ore. It was precipitated from this solution along with the element barium, an impurity frequently associated with uranium. Barium chloride and the chloride of the element sought were both water-soluble, but when alcohol was added to the solution containing the two, the radioactive element precipitated as the metal chloride. By repeated dissolving and precipitation, a tiny quantity of a radioactive substance nearly a thousand times as radioactive as uranium was eventually isolated.

Once again the spectroscope proved to be a most valuable tool. The spectrum of the crystalline barium compound which first separated from the solution contained several spectral lines not found in a sample of pure barium. As the process of concentrating the new element progressed, the intensities of the characteristic lines of barium gradually disappeared, and finally a new spectral line became prominent. It became evident that the latter was a characteristic line of a new radioactive element, later called radium.

By 1902 the Curies had separated from two tons of pitchblende ore, by fractional crystallization, 100 milligrams of radium chloride ($RaCl_2$). Somewhat later they reported the atomic weight of the element as 225 and succeeded in preparing metallic radium by the electrolysis of the fused (molten) radium chloride.

Thereupon the characteristics of the radiation emitted by radioactive substances took on an absorbing interest for physicists all over the world.

RADIATION FROM RADIOACTIVE SUBSTANCES

There is nothing distinctive about the appearance of radioactive substances; they look just like any other substances. Their distinguishing characteristics are in the radiations they emit.

During the last few years of the nineteenth century Ernest Rutherford made some important discoveries, early in his distinguished career, regarding the radiation of radioactive materials. Sir Ernest Rutherford, one of the great men in the development of the physics of radioactive substances, was born in New Zealand in 1871. After graduating from college there, he traveled halfway around the world to study, first at McGill University in Montreal and later at the famous Cavendish Laboratory in Cambridge, England.

Rutherford's experiments showed that there were at least two components in the radiation emitted by radioactive materials. The first was stopped by an aluminum foil 1/50 mm thick, while the other was reduced to only half intensity by a foil 25 times as thick. The less penetrating component he called the alpha (α) ray; the more penetrating one, the beta (β) ray. Shortly thereafter, a third kind of radiation, the gamma (γ) ray, was discovered; it was shown to be a close relative of the X ray, though of slightly shorter wavelength than the X rays then in common use.

After Rutherford's discovery of the complex character of radioactive rays, developments came in rapid succession. A pencil of rays issuing from a radioactive source Fig. 12-2 was shown to be separable into the three

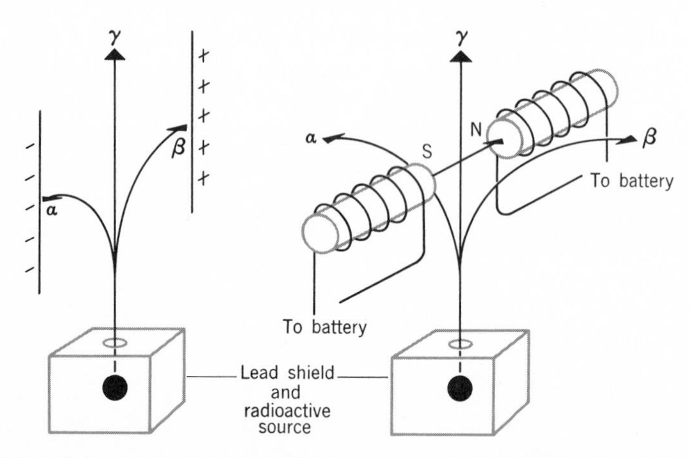

Fig. 12-2. *Alpha and beta rays, which consist of electrically charged particles, are deflected toward opposite poles in an electric field. Gamma rays have no charge and hence are unaffected. Since alpha and beta particles are electrically charged and travel at high velocities, they will behave like an electric current flowing in a wire, that is, produce a magnetic effect. They will be bent when they pass perpendicular to the lines of force in a magnetic field. Gamma rays are not deflected in a magnetic field.*

components by means of a magnetic or electric field. The radiation which was bent toward the negative pole in an electric field must carry a positive charge. It was shown to be the less penetrating of the radiations demonstrated by Rutherford. The value of e/m (ratio of charge to mass) was half the value of e/m for protons. Later, alpha particles were found to have four times the mass of the proton and double the quantity of charge:

$$\begin{pmatrix}\text{Value of } e/m \\ \text{for proton}\end{pmatrix} = \frac{e_p}{m_p} \qquad \begin{pmatrix}\text{Value of } e/m \text{ for} \\ \text{alpha particle}\end{pmatrix} = \frac{2e_p}{4m_p} = \frac{e_p}{2m_p}$$

Rutherford expected that alpha particles would prove to be helium ions, for he had noticed that helium gas was nearly always associated with uranium minerals. Several years passed, however, before he was able to verify his hunch that the two were the same.

Eventually in collaboration with Royds he designed a form of Geissler tube (Fig. 12-3) in the interior of which was suspended a long, narrow, thin-walled tube containing (radioactive) radon gas. Several days after the radon tube was introduced, high-voltage excitation of the Geissler tube gave a helium spectrum when it was used as a spectroscopic light source. Alpha particles hurled at high velocity from the radon atoms had penetrated the thin-walled tube and passed to the larger outer tube, where their excitation gave the helium spectrum. Thus the component of low penetrating power was proved to consist of helium ions (He^{++}).

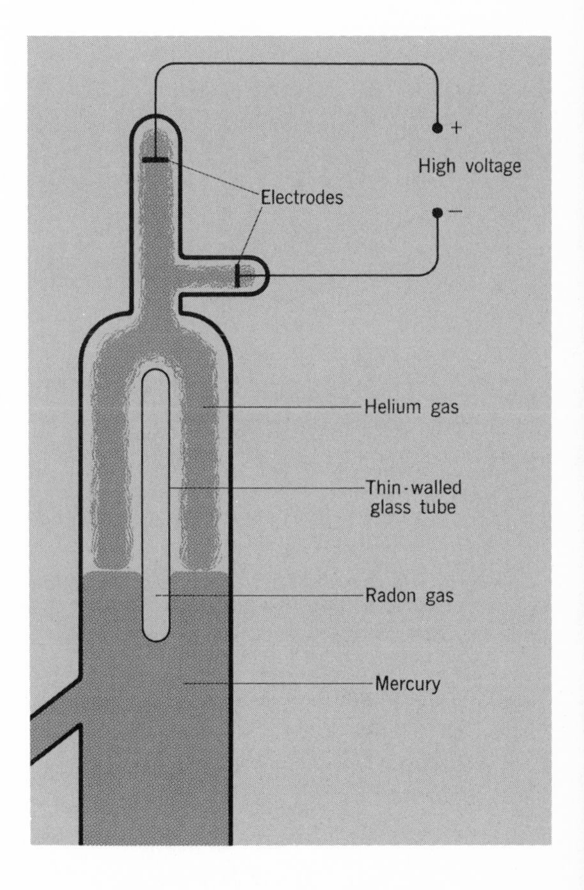

Fig. 12-3. *The Rutherford-Royds experiment. Radon atoms cannot escape from the inner tube, but alpha particles hurled at high velocities from these atoms readily pass through the thin-walled tube into the highly evacuated outer chamber. A high-voltage electric current passing between the electrodes causes a glow in the outer tube. If the light from this tube is examined in a spectroscope, the element is found to be helium.*

The component of the rays which was deflected toward the positive terminal in an electric field was found to have the same value of e/m as electrons. In fact the more highly penetrating beta particles proved to be none other than electrons which had been hurled from radioactive nuclei. They were very easily deviated in a magnetic field, whereas alpha particles were so little affected that their magnetic deflection was missed at first.

The mass of an alpha particle was more than 7000 times that of a beta particle, but it had the least penetrating power of any of the three types of radiation (Fig. 12-4). Alpha particles were stopped by a thick piece of paper, which was in turn very easily penetrated by beta particles and gamma rays. A piece of aluminum metal several millimeters thick stopped beta particles but readily transmitted gamma rays. Gamma rays could pass through a foot or more of concrete.

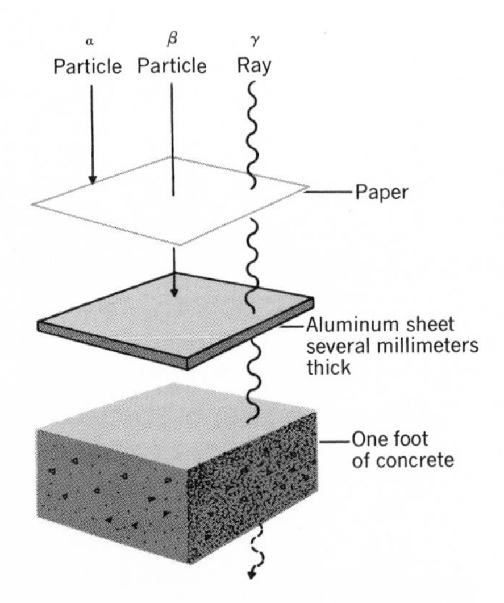

Fig. 12-4. *The relative penetrating powers of alpha, beta, and gamma radiation. Alpha rays are the least penetrating, being stopped by paper, clothing, and the like. Beta particles are more penetrating, requiring on the average about ⅛ inch of aluminum to stop them. Gamma rays are difficult to stop completely, even with considerable thicknesses of material such as lead or concrete.*

Here is a convenient way of remembering the relative ionizing and penetrating powers of the three types of radiation:

	ALPHA PARTICLES	BETA PARTICLES	GAMMA RAYS
Comparative penetrating ability	1	100	10,000
Comparative ionizing ability	10,000	100	1

On the average, beta particles have about 100 times the penetrating ability of alpha particles and 100 times the ionizing power of gamma rays. Likewise alpha particles are about 100 times more effective than beta particles as ionizing agents, and gamma rays 100 times as penetrating as beta particles.

While alpha particles had poor penetrating ability, they were excellent ionizers of any gas through which they passed. Gas molecules, normally neutral because there were equal numbers of positive charges on the nuclei and negative charges in their electron clouds, lost part of their negative charge when alpha particles passed through or near the electron clouds. The resulting gas molecules were left with a net positive charge, or as positive ions.

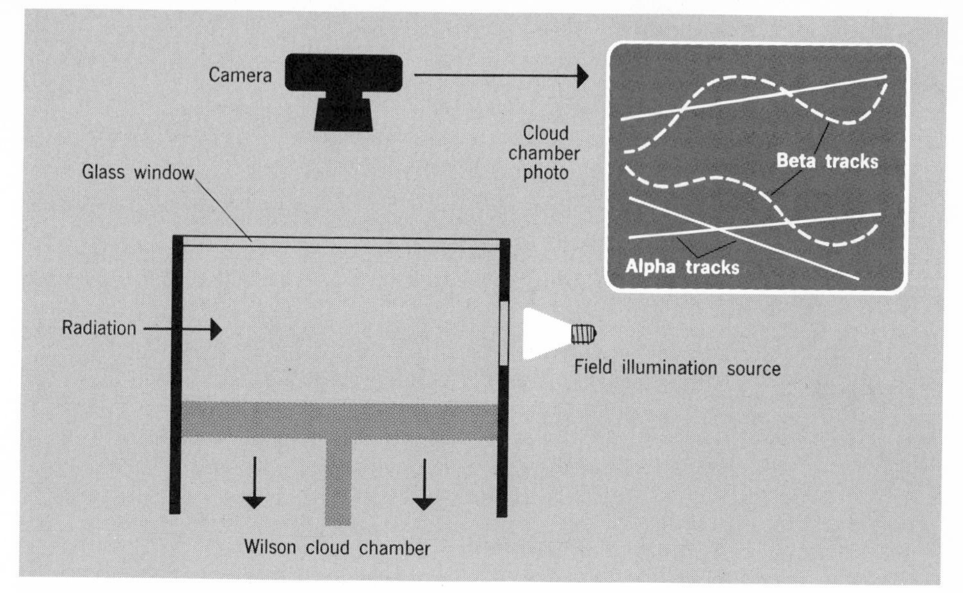

Fig. 12-5. *When the piston of a Wilson cloud chamber is lowered, the pressure decrease slightly cools the confined vapor. The saturated vapor momentarily becomes supersaturated. If a charged particle (ionizing radiation) now passes though the supersaturated vapor, the ions produced serve as centers of condensation of the vapor droplets and a cloud track is formed. Note the difference in appearance between an alpha and a beta cloud track, as here schematized.*

Fig. 12-6. *Cloud-chamber photograph of a neutron-proton collision produced by the Brookhaven Cosmotron. The long thin streaks originating at top center, and spreading out as they continue down, are tracks due mainly to protons and mesons. The three intersecting tracks, at center right, result when a neutron strikes a hydrogen nucleus (proton). The proton shoots downward toward the center and the two mesons downward and to the left respectively.* [*Courtesy, Brookhaven National Laboratory.*]

Alpha particles were better ionizers than beta particles because their velocities were much lower (alpha about $0.1c$,* beta as high as $0.9c$) and their charges twice as large. Gamma rays, the most penetrating of the three, were the least effective in ionizing gases.

Because of their ionizing ability, alpha and beta rays produce fog tracks in the supersaturated vapor of a Wilson cloud chamber (Fig. 12-5). In it, alpha tracks are thick, straight, and continuous; tracks due to beta particles are thinner, more curved, and interrupted. Gamma rays give no tracks because of their small ionizing power. The cloud chamber makes it possible to photograph or see the effects of alpha and beta particles as they pass through a supersaturated vapor.

RADIOACTIVE DECAY

The emission of alpha, beta, or gamma radiation by atomic nuclei is an evidence of an unstable energy condition of the atom. Loss of this surplus energy through the emission of particles or waves is called radioactive decay or radioactive disintegration.

The almost universal tendency for high-energy systems to fall to lower energy states was noted earlier. The stone tends to roll down the hillside, its potential energy decreasing as it descends. Likewise the radioactive atom, which is bursting at the seams with an overabundance of energy, may relieve this condition by emitting particles or waves (radiations).

The rate at which radioactive atoms decay depends on the number of radioactive atoms present. A million radioactive atoms of, say, radium, will emit twice as many alpha particles per second as a half-million, and a half-million twice as many per second as a quarter-million. Radioactive decay is sometimes described as "exponential," meaning that *equal fractions* of the radioactive atoms in samples will *disintegrate* in *equal time intervals*. If a million radioactive atoms decay in one second, then half of the 500,000 left, will decay in the next second.

Since the decay of a single radioactive atom is a matter of chance, radioactivity must be regarded as statistical in nature. Hence the decay rate will be proportional to the number of radioactive atoms present, only if the number of atoms is very large.

The proportionality between decay rate A (activity) and the number of radioactive atoms N may be expressed as $A \propto N$. By introducing a proportionality factor λ, the proportionality is changed into an equality, $A = \lambda N$. The decay constant λ is a characteristic property of the radioactive element and may be considered the fractional number of the atoms which decay in a second.

* c = the velocity of light, 3×10^{10} cm/sec.

ILLUSTRATIVE PROBLEM

If the decay constant for radium is $\lambda = 1.36 \times 10^{-11}$ sec^{-1}, calculate the number of radium atoms disintegrating per second when 1 g of radium is present.

Solution

The number of atoms of radium in 226 g is 6.02×10^{23}, since there is an Avogadro number of atoms in a gram atomic weight. Therefore in 1 g there will be $\frac{1}{226} \times 6.02 \times 10^{23}$ atoms.

$$A \text{ (disintegrations/sec)} = \lambda \left(\frac{1}{\text{sec}}\right) \times N \text{ (atoms)}$$

$$A = 1.36 \times 10^{-11} \left(\frac{1}{\text{sec}}\right) \times \frac{1}{226} \times 6.02 \times 10^{23} \text{ (atoms)}$$

$$A = 3.6 \times 10^{10} \text{ disintegrations/sec (dis/sec)}$$

Originally 1 gram of radium was identified with a *curie* and was thought to give 3.7×10^{10} disintegrations per second. Subsequently, more refined measurements showed that 1.02 grams of radium gave this number of disintegrations per second. The curie is now defined as that weight of *any* element which gives 3.7×10^{10} dis/sec. A millicurie is 1/1000 of a curie, or the weight giving 3.7×10^{7} dis/sec; the microcurie gives 3.7×10^{4} dis/sec. Table 12-1 shows the weights of 1 curie for several radioactive substances. It should be emphasized that a *gram* atomic weight contains an Avogadro number of atoms. Hence weights of radioactive material *must* be expressed in *grams* if the Avogadro number is to be used to calculate the number of atoms.

Table 12–1 Half-Lives of Some Common Radioactive Atoms

ELEMENT	RADIOACTIVE SPECIES	HALF-LIFE PERIOD	WEIGHT OF 1 CURIE
Carbon	C^{14}	5600 years	0.22 g
Cobalt	Co^{60}	5.3 years	8.8×10^{-4} g
Iodine	I^{131}	8.0 days	8.2×10^{-6} g
Phosphorus	P^{32}	14.3 days	3.5×10^{-6} g
Uranium	U^{238}	4.5 billion years	2980 kg

Attempts to change the rate of radioactivity decay by altering external conditions—temperature, pressure, and other factors—have been unsuccessful; the radioactive process continues as inevitably as time and tide. This suggests that the nucleus of the atom is involved in radioactive change, for it is well known that changes in temperature and pressure will alter the speeds of chemical reactions in which the extranuclear electrons are thought to be functional. The fact that the rate of decay is independent of the state of chemical combination seems to support this view. It seems to make no difference, as far as rate of decay is concerned, whether a radioactive material is in the form of an uncombined metal like uranium or in the form of one of its compounds like uranium hexafluoride.

Half-life Period The period of time during which one-half of the atoms of a radioactive sample decays is known as its half-life period. For example four and a half billion years must elapse before half the atoms in a sample of uranium disintegrate. Day by day decrease in activity will not be observable because of the extremely long half-life period. On the other hand, polonium has a half-life period of about four and a half months. So its day-to-day decrease in activity will be very evident.

I^{131}, which is widely used by the medical profession (see Chapter 14) in the diagnosis and therapeutic treatment of thyroid disorders, has a half-life of about eight days. After eight days, one-half of the original radioactive atoms have disintegrated. During the next eight days, one-half of those remaining will have disintegrated. After 24 days, only one-eighth of the original radioactive atoms will still be radioactive. The half-life period, then, tells whether the rate of decay is rapid or slow.

The half-life period of a radioactive substance is related to its decay constant by the equation

$$T_{1/2} \cdot \lambda = 0.693$$

According to this equation an isotope with a very long half-life has a correspondingly small decay constant. There is an inverse mathematical relation between $T_{1/2}$ and λ. The half-life and decay constant should therefore have reciprocal time units. Since decay constants are usually expressed in sec^{-1} the half-life must be expressed in seconds. Other time units could be used if the units of the decay constant were also changed.

ILLUSTRATIVE PROBLEM

If the decay constant for radium is 1.36×10^{-11} sec^{-1}, calculate its half-life.

Solution

$$T_{1/2} = \frac{0.693}{\lambda} = \frac{0.693}{1.36 \times 10^{-11}}$$

$$T_{1/2} = 5.10 \times 10^{10} \text{ sec}$$

$$1 \text{ year} = 365 \text{ days} \times 24 \text{ hrs} \times 60 \text{ min} \times 60 \text{ sec} = 3.15 \times 10^{7} \text{ sec}$$

$$T_{1/2} = \frac{5.10 \times 10^{10} \text{ sec}}{3.15 \times 10^{7} \text{ sec/year}} = 1619 \text{ years}$$

Half of any mass of radium will decay in 1619 years.

ILLUSTRATIVE PROBLEM

Calculate the number of curies in 0.001 g of H^3 (tritium) if its half-life ($T_{1/2}$) is 12.4 years.

Solution

$$A = \lambda N$$

$$\lambda = \frac{0.693}{T_{1/2} \text{ (sec)}} = \frac{0.693}{12.4 \times 365 \times 24 \times 3600}$$

$$N = \frac{0.001}{3.02} \times 6.02 \times 10^{23}$$

$$A \text{ (dis/sec)} = \frac{0.693 \times 0.001 \times 6.02 \times 10^{23}}{3.02 \times 12.4 \times 365 \times 24 \times 3600}$$

$$A = 3.5 \times 10^{11} \text{ dis/sec}$$

$$1 \text{ curie} = 3.7 \times 10^{10} \text{ dis/sec}$$

$$\frac{3.5 \times 10^{11} \text{ dis/sec}}{3.7 \times 10^{10} \text{ dis/sec/curie}} = 9.6 \text{ curies}$$

DETECTION OF RADIOACTIVITY—THE GEIGER COUNTER

It has already been shown that radiation may be detected by means of a photographic plate, by a cloud chamber, and by the discharge of a charged electroscope. There are, however, much more convenient radiation detection methods. Some years ago the Geiger counter had become the traditional detection device of the uranium hunters.

The Geiger tube, shown in Fig. 12-7 is a glass envelope containing a hollow metal cylindrical tube which is connected to the negative terminal of a high-voltage source or battery. A wire which is coaxial with the metal tube is attached to the positive terminal of the battery. A potential difference of 900-1000 volts is applied between the metal tube and the wire. Even though the glass envelope is highly evacuated, there are sufficient residual molecules to furnish enormous numbers of gas ions should an alpha, beta, or gamma ray penetrate the interior of the tube.

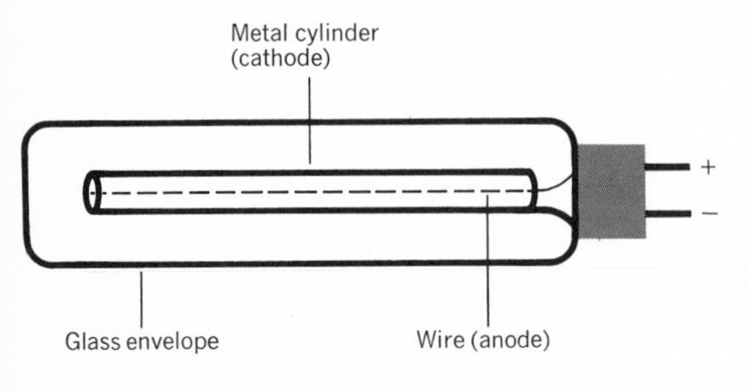

Fig. 12-7. *A Geiger tube consists of a thin central wire (anode) coaxial with a metal cylinder (cathode). Radiation passing through the low-pressure gas in the tube of a survey meter produces ionization, which reduces the potential difference between the electrodes. This actuates an electronic meter which is usually calibrated to read milliroentgens per hour.*

Radiation which passes through the atoms of low-pressure gas in the tube ionizes them by tearing loose electrons. These positive nuclei and electrons so formed are attracted to the electrodes of opposite charge, are accelerated by the electric field, gain kinetic energy, and create still more ions. An enormous avalanche of ions results almost instantaneously, and a pulse of current surges across between the electrodes. This surge of current lasts only for about 10^{-4} sec. If the pulse in the tube is fed into the electronic circuits of the survey meter, it may be heard as a click, seen as the flashing of a light bulb, or recorded by the movement of the needle on an electric meter. The Geiger tube may also be attached to a scaler which counts the pulses, if a more precise comparison of the strengths of radioactive sources is desired.

Some survey meters may be used to count all three types of radiation—(alpha, beta, and gamma) rays. The radiation actually detected depends to a large extent on the "window" thickness that admits the radiation to the tube. This must be thinner than a piece of paper if alpha particles are to enter. A window admitting alpha particles would permit the entry of the more penetrating betas and gammas. A thicker window would exclude alphas but still admit betas and gammas. A window of sufficient thickness to stop betas would permit only gamma rays to enter.

Even after all known radioactive substances are removed from the vicinity of an operating Geiger counter, there is still radiation reaching it; it continues to count in spite of everything. There are always radioactive materials present in tiny quantities in the walls of the room and the surroundings. This ever-present radiation is known as "background radiation." Cosmic rays coming from outer space also doubtless contribute to the background count.

Fig. 12-8. *A scintillation detection system. The energy of a gamma ray entering the NaI crystal is dissipated through excitation of electrons in the crystal's atoms. When these electrons (in excited states) fall to lower energy levels, light is emitted (principally in the ultraviolet region). Ultraviolet light falling on a photocathode causes electron emission, which is amplified in a series of dynodes until there is a measurable electric current. The quantity of the current accumulating in a very short time interval (roughly 10^{-6} sec) determines the "pulse height," which is then a measure of the energy of the incident gamma ray. Thus the scintillation counter serves not only to detect gamma rays but also to give information on its energy. The NaI crystal is coupled to the photomultiplier tube with an optically transparent compound such as a silicone grease. If an anthracene crystal replaces the NaI crystal, beta particles can be counted; or a ZnS crystal for alpha particles. Since the penetration of solids by alpha or beta particles is far less than for gamma rays, crystals for detecting the particles must be thinner.*

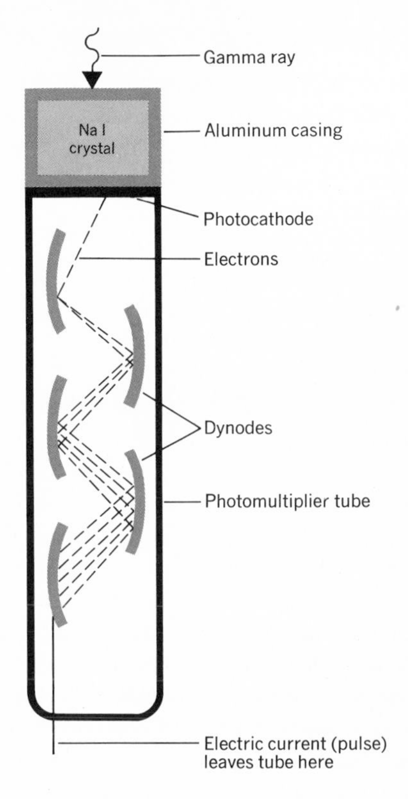

In counting a radioactive sample, then, counts which are due to the background must be deducted from the number of counts obtained from the sample. Thus a radioactive source which gives 1000 counts per minute (cpm), when the background count is 50 cpm, will have a net count of 950 cpm due to the sample.

While the Geiger counter was for a long time the most widely used and most commonly known radiation detector, it is not necessarily the most efficient detector. Another method was proved to be much more efficient in the detection of gamma rays.

The Scintillation Counter One of the earliest methods used to detect radiation depended on the tiny flashes of light (scintillations) produced when alpha particles struck a zinc sulfide screen. The procedure of counting the flashes was a very tedious one, however, and the method was rather limited in its usefulness. With the enormous advances in electronics taking place during and after World War II came the device known as the photomultiplier (PM) tube. In effect this tube converted flashes of light into measurable electric currents. When used in conjunction with certain crystals, called phosphors (substances which give flashes of light when radiation passes through them), a detector of gamma radiation much more sensitive than the Geiger counter was available. For the detection of gamma rays a crystal of sodium iodide (thallium activated) is used in conjunction with the PM tube (Fig. 12-8).

When gamma rays strike the sodium iodide phosphor, minute flashes of light result. These strike the photocathode of the PM tube where electrons are emitted, the tiny current pulses amplified, fed into a scaler, and counted in a manner similar to that described above in connection with the Geiger counter.

This type of detector, known as the scintillation counter, has largely replaced the Geiger tube as a means of counting gamma rays, because of its much higher efficiency. The cost of a scintillation counter tube is, however, considerably greater than that of the Geiger tube. The *well scintillation counter* employs a sodium iodide crystal with a test tube size hole (or well) drilled in the top. It is particularly useful in counting radioactive solutions because of the excellent geometry: a much larger fraction of the emitted radiation passes into the phosphor if the radioactive solution is surrounded by the crystal (Fig. 12-9).

There are of course many other methods of radiation detection besides Geiger and scintillation counters—for example, the ion chamber, pro-

Fig. 12-9. *On the left is a well scintillation counter, on the right a combination scaler and computer. The scaler counts the radioactive pulses from the source, the computer calculates the given count as a percentage of a standard sample and prints out the result on a tape.* [*Courtesy, Nuclear Chicago Corporation.*]

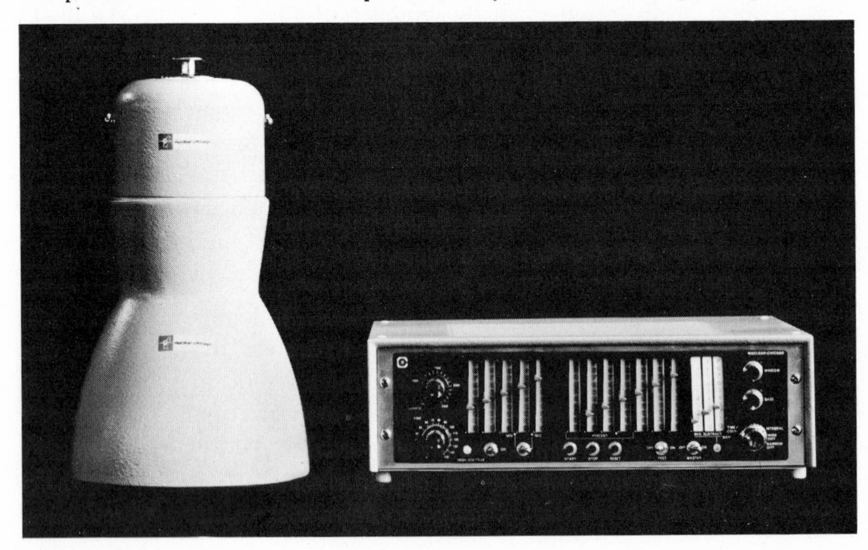

portional counter, and liquid scintillation counting methods. Each has its advantages for measuring the radiation from particular isotopes. Sometimes, for example, the energy of the radiation may be very weak and will not penetrate the wall of the Geiger tube or the solid crystal of the scintillation counter. Then the counting may best be done by dissolving the radioactive sample in a liquid phosphor and making use of the method of liquid scintillation counting. Alpha particles are often counted with a proportional counter.

Problems, Chapter 12

1. The half-life period of I^{131} is 8.0 days. How many of an original 10^8 atoms of I^{131} will still be radioactive after 40 days?

2. Calculate the decay constant (λ) of P^{32}.

3. (a) Calculate the kinetic energy in ergs of an alpha particle whose velocity is 2.00×10^9 cm/sec. (Remember that 6.02×10^{23} He atoms have a mass of 4.00 g. Neglect the small difference in mass between an alpha particle and a helium atom.)

   (b) Calculate the corresponding kinetic energy in electron volts (ev) if 1.60×10^{-12} erg = 1 electron volt.
   (c) Convert the answer in (b) to millions of electron volts (mev).

4. Calculate the weight of a radioactive sample which gives 1000 (beta) disintegrations per second if its half-life is 12.8 days and its atomic mass is 140.

5. Starting with the equation $A = \lambda N$, show that the weight (grams) of material in a curie of any radioactive isotope is given by the equation:

$$\text{wt} = \frac{3.7 \times 10^{10} \times \text{atomic mass}}{\text{decay constant } (\lambda) \times \text{Avogadro number}}$$

   and that this is in turn equivalent to

$$\text{wt} = \frac{3.7 \times 10^{10} \times \text{atomic mass} \times \text{half-life}}{0.693 \times \text{Avogadro number}}$$

6. The element potassium contains 0.012% of K^{40}, a beta-radioactive nuclide. Calculate the activity in microcuries of the K^{40} in a man weighing 80 kg if 0.35% of his body weight is potassium. The half-life of K^{40} is 1.3×10^9 years.

7. A 1.39×10^{-3} g tritium (H^3) sample, gives 4.80×10^{11} beta dis/sec. Calculate the half-life of tritium.

Exercises

1. Who was the discoverer of radioactivity? With what was his experimentation concerned when he made his discovery? Describe.
2. Madame Curie discovered in pitchblende ore an impurity which was more highly radioactive than uranium itself. Explain how she might have identified this highly radioactive element using an electroscope.
3. Describe how two chemical substances may be separated by means of a precipitation method. Explain how such a method could be applied in the separation of radioactive materials.
4. Compare the properties of alpha, beta, and gamma rays with respect to (a) charge, (b) mass, (c) ionizing power, (d) penetrating power.
5. What is meant by the half-life of a radioisotope? Illustrate.
6. Why is radioactivity believed to be a nuclear phenomenon? Explain.
7. Explain what is meant by the statement that "radioactive decay is exponential."
8. What is a curie of radioactivity? Is a curie always the same weight? Explain.
9. What is the physical significance of a decay constant? The decay constant of radium is 1.36×10^{-11} sec^{-1}. How many of an original 6.02×10^{23} atoms (on the average) would decay in 1 sec?
10. Distinguish between the rate of decay of a radioactive sample and its decay constant.
11. List the different methods by which radiation may be detected.
12. Describe briefly the method of detecting radiation by means of (a) a Geiger counter, (b) a scintillation counter. What are the advantages and disadvantages of each method?

13 Those unclear nuclear reactions

AS the twentieth century dawned, the notion of a complex atom was still in its infancy. Thomson had suggested that the atom was a neutral entity, containing negative corpuscles (electrons) imbedded in a sphere of positive charge, much like raisins in a plum pudding. The plum pudding model was not enthusiastically received by his contemporaries; there were too many irreconcilable facts. The time was ripe for a new theory of atomic structure.

EVIDENCE THAT ATOMS HAVE NUCLEI

Observations concerning the behavior of alpha particles hurled from the nuclei of radioactive atoms eventually resolved some of the inconsistencies of the plum pudding atomic model. Geiger, a co-worker of Rutherford, noted that the sharp image of an alpha particle beam due to the scintillations on a zinc sulfide screen became diffuse and smudgy when a foil was placed in the beam's path (Fig. 13-1). Apparently some of the alpha particles were scattered out of the path of the beam by the foil.

Geiger and Marsden made a detailed study of this alpha-particle scattering by thin metal foils. They discovered that there were a few scintillations on the screen even when it was removed from the direct path of the beam (Fig. 13-2). Some of the particles were being scattered through rather large angles, and occasionally one would even be turned back in a direction the opposite of that from which it came (i.e., turned through 180°). A plausible explanation was that the alpha particles were closely approaching other particles with similar charges and being repelled.

Rutherford calculations from wide-angle scattering indicated that the results could best be accounted for by assuming that the atom contained a central core having nearly all the mass of the atom, a diameter not greater than 10^{-12} centimeters, and a positive charge. Most of the alpha particles passed through the thin gold foil unaffected, as if the structure of the atom were very porous. But occasionally one which approached a nucleus was scattered through a wide angle or even had its direction reversed. Rutherford is reported to have expressed his amazement at this

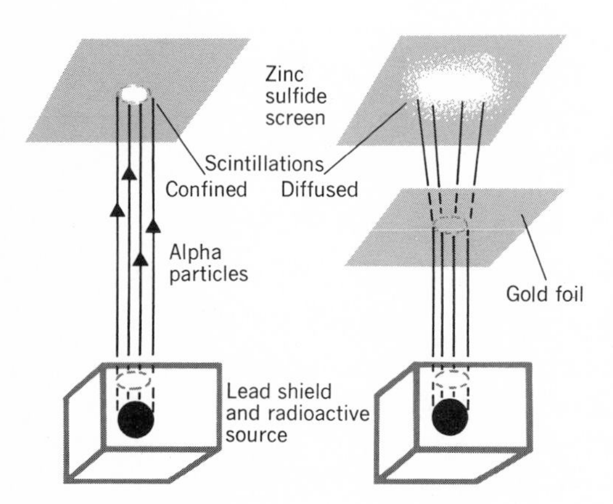

Fig. 13-1. *A beam of alpha particles from a radium source, producing a small, sharp image on a luminescent screen, becomes larger and more diffuse when a thin gold foil is interposed between the source and the screen. Many alpha particles are scattered through small angles in passing through the foil.*

phenomenon of a high-energy particle having its direction completely reversed. He is said to have commented that the result was about as incredible as if a cannon ball had been fired at a piece of tissue paper and been bounced back.

The classic Rutherford scattering experiment then led to the concept of an atom as a relatively porous structure with a tiny but very dense core or *nucleus* having a net positive charge. This was surrounded by electrons. According to this model, the atom resembled the solar system, the nucleus assuming the role of the sun and the electrons that of the planets.

But already events were leading up to the determination of the numbers of protons in the nucleus of an atom.

THE NUCLEAR CHARGE OF THE ATOM

As evidence accumulated that atoms were composed of electrons and protons, there was of course increasing interest in the question, "How many of each of these particles are there in a single atom of a particular element?" An approximate answer to this question was obtained from the alpha-ray-scattering experiments of Geiger and Marsden. The mathematical equation they used to calculate the number of scattered alpha particles contained a factor Z, the nuclear charge of the scattering element. For some elements the value of Z appeared to be roughly half the atomic weight. The atomic weight of carbon, for example, was 12, and the number of positive charges on its nucleus was 6. A series of experiments on

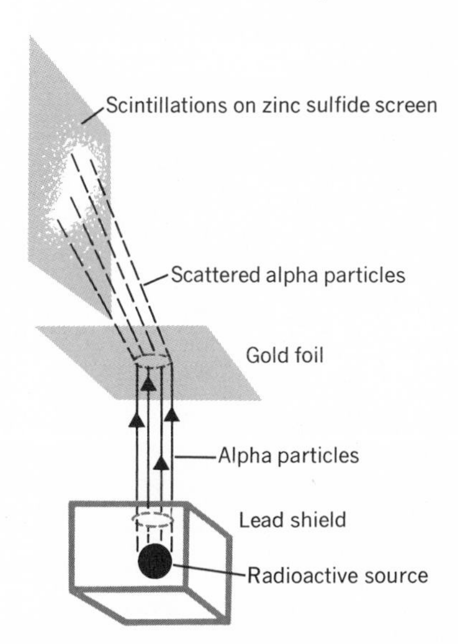

Fig. 13-2. *A small proportion of the alpha particles in the beam are scattered through wide angles as shown by the scintillations on the screen when it is completely out of the path of the main beam. From his scattering experiments, Rutherford concluded that an atom was a highly porous structure with most of its mass concentrated in a central positive core called the nucleus.*

the scattering of X rays carried out by Barkla at about the same time indicated that for some elements the number of electrons associated with an atom was also roughly half the atomic weight. Carbon would thus have 6 electrons. Unfortunately both of these relations held for only a few of the lighter elements in the periodic table. Since an uncombined atom would require equal numbers of positive and negative charges, these two results amounted to essentially the same thing.

In 1913 the Dutch physicist van den Broek made the suggestion that the number of positive charges in the nucleus of an atom was equal to the order number of the element in the periodic table, or its atomic number. In the same year Moseley (see Chapter 11) had demonstrated a method of obtaining atomic numbers based on an observed regularity in the shifting of characteristic X-ray lines from one element to the next. He concluded that this systematic shifting could only be due to changes in the nuclear charge.

After World War I, Chadwick in England made use of a precision method of measuring alpha-particle scattering to determine the number of positive charges in the nuclei of the elements copper, silver, and platinum. These results checked very closely with the accepted values of the atomic numbers. His data and those of other investigators give assurance that the order number of the element, the atomic number, is also the number of positive charges in the atomic nucleus.

THE FIRST CONTROLLED NUCLEAR REACTION

In 1919 another milestone in the field of nuclear physics was passed when Rutherford reported a nuclear reaction resulting from the collision of alpha particles with nitrogen nuclei. The two fused, momentarily forming a compound nucleus, which then erupted, shooting forth a high-energy proton. This was not, to be sure, the first nuclear reaction, for every time a naturally radioactive atom emitted an alpha or a beta particle, a nuclear reaction occurred. But there was nothing that could be done to start, stop, or alter natural radioactive processes. The nuclear reaction observed by Rutherford was controllable in the sense that the alpha particles could be directed at particular target atoms, in this case nitrogen.

When high-energy particles (alpha, beta, or protons) strike a screen containing zinc sulfide, tiny flashes of light (scintillations) are observed if the screen is examined through a microscope. The *range* of a high-energy particle is simply the maximum distance at which scintillations are observed. The apparatus for measuring alpha-particle ranges is shown in Fig. 13-3.

Rutherford observed that alpha-particle ranges had the expected values if oxygen or carbon dioxide were the only gases in a range-determining apparatus. But when dry air was used, the scintillations continued at a greater distance than expected. These unexpectedly long-range scintilla-

Fig. 13-3. *The distance from an alpha-particle source to a zinc sulfide screen is adjusted until scintillations are barely observed. This distance is called the range of the alpha particles.*

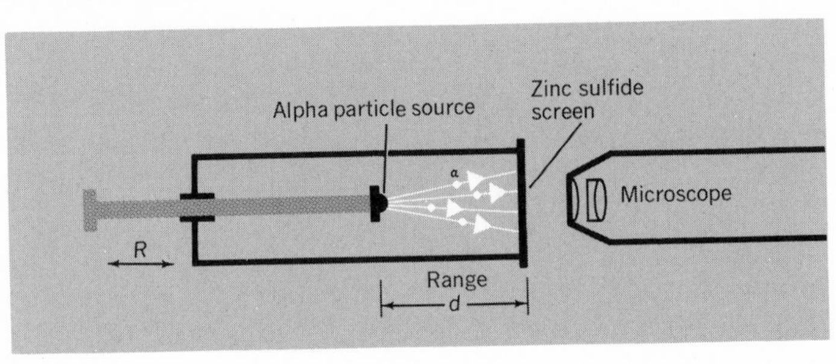

tions provided Rutherford with the clue which led him to suggest that a nuclear reaction was taking place. Additional experiments showed that the presence of nitrogen inside the range apparatus was necessary if long-range particles were to be observed. How could these results be interpreted?

Magnetic deflection of the long-range particles indicated that they were protons. Rutherford explained what had happened by assuming alpha particle had collided with and momentarily fused with a nitrogen nucleus. But the compound nucleus had far too much energy to be stable; almost immediately it erupted and a small portion of its nuclear matter exploded, sending off a high-energy proton. A nuclear reaction took place; a transmutation of elements occurred.

THE SYMBOLISM OF NUCLEAR REACTIONS

How can the physicist symbolically represent the above nuclear reaction? In the Rutherford reaction just considered, high-speed alpha particles crashed into nitrogen atoms and set protons free. The nuclear reaction may be represented by the equation

$$\underset{\text{alpha particle}}{{}_2\mathrm{He}^4} + \underset{\text{target atom of nitrogen}}{{}_7\mathrm{N}^{14}} \rightarrow \underset{\text{product atom of oxygen}}{{}_8\mathrm{O}^{17}} + \underset{\text{proton}}{{}_1\mathrm{H}^1}$$

The subscripts represent the atomic numbers (usually represented by Z), or the number of protons in the atomic nuclei. Occasionally subscripts are omitted because they are implied by the symbols of the elements. The superscript numbers are the mass numbers (usually represented by A), the sum of the numbers of protons and neutrons. There is nothing in the equation to suggest that the bombarding alpha particle is a He^{++} ion rather than a neutral atom. However, the number of commonly used, light bombarding particles is limited—protons, deuterons, neutrons, and alpha particles are the important ones—so it is usually easy to distinguish one of these from a target or product atom. It should also be pointed out that the oxygen atom produced contains one more neutron than the usual O^{16}, and hence is an isotope of that element. Note also that sums of the subscripts on each side of the equation are equal, as are the sums of the superscripts. Hence charge is conserved; there is the same total positive charge on the reacting atoms as on the product atoms. Likewise the sums of mass numbers of product atoms is the same as that for the reactant atoms.*

* The sum of the *atomic masses* of product atoms is usually not exactly equal to the sum of the atomic masses of reactant atoms. A nuclear reaction is exoergic (energy producing) or endoergic (energy absorbing) depending on whether the right-hand or left-hand member atoms have the same smaller net mass. In the exoergic reaction, also, the energy of ejected particles is attained at the expense of loss in mass.

This nuclear reaction is classified as an (α, p) reaction, meaning that the bombarding atomic projectiles are alpha particles and that protons are the light particles produced. The reaction might also have been written as N^{14} (α, p) O^{17} in which target and product atoms are also specifically designated.

ILLUSTRATIVE PROBLEM

Assuming that an (α, p) reaction is possible with ${}_{13}Al^{27}$ target atoms, determine the atomic number and mass number of the other product atom.

Solution

$$ {}_{2}He^{4} + {}_{13}Al^{27} \rightarrow ? + {}_{1}H^{1} $$

The sum of the atomic numbers (subscripts) on the left is 15. If a proton is formed (atomic number 1), the product atom must have an atomic number 14. The element must therefore be silicon. Since the sum of the mass numbers on the left is 31, the silicon atom must have a mass number of 30. It must be an isotope of silicon. The completed nuclear equation is

$$ {}_{2}He^{4} + {}_{13}Al^{27} \rightarrow {}_{14}Si^{30} + {}_{1}H^{1} $$

Transmutation has taken place in the above reaction. The alchemists' dream has, in a sense, been realized. While it is now possible to convert one kind of atom into another, relatively few of the target atoms are transmuted in any such reaction, and the problem of separating from the target atoms the minute quantities of new atoms produced is indeed a formidable one. The day when gram quantities of gold can be economically produced from baser metals still seems far distant, despite the fact that it is now possible to convert one element into another.

The above nuclear reaction, involving transmutation of elements, should be carefully distinguished from any chemical reaction. The chemical reaction of hydrogen and chlorine gases to form hydrogen chloride may be written

$$ H_2 + Cl_2 \rightarrow 2\ HCl $$

Originally hydrogen atoms are attached in groups of two, as are also chlorine. After the reaction is completed, exactly the same atoms are present, but they are attached differently. The new aggregates (molecules of HCl) consist of an atom of hydrogen attached to one of chlorine. Chemical changes result in alterations in the electronic structures of atoms (see Chapter 15) but the nuclei of the atoms remain the same.

The radioactive decay of a U^{238} atom by alpha-particle emission, on the other hand, is an example of a nuclear reaction, because when an exploding nucleus blasts part of the atom's mass off into space, the nuclear composition is changed. Hence, even though the product atom is usually designated as uranium X in the uranium disintegration series, it is really an isotope of the element thorium. The nuclear reaction may be written as follows:

$$_{92}U^{238} \rightarrow {}_{90}Th^{234} + {}_{2}He^{4}$$

While radioactive decay is a nuclear reaction, it is (unlike the Rutherford reaction of alpha particles with nitrogen nuclei) uncontrollable. Meanwhile, what is the mechanism of such reactions as these. Whose symbolism has just been explained?

THE NUCLEAR POTENTIAL BARRIER

Rutherford's pioneer atom-smashing experiment was followed by a series of alpha-particle bombardment reactions involving many different target nuclei. In these Rutherford and Chadwick found that nearly all of the elements up to potassium (K) in the periodic table emitted long-range protons just like nitrogen. But nuclei heavier than the K nucleus merely scattered (changed the direction of) the approaching alpha particles. The two particles did not fuse as in the case of the lighter nuclei.

Since an alpha particle had a positive charge, like all atomic nuclei, as it approached another nucleus closer and closer it experienced a larger and larger repulsive force, as predicted by Coulomb's law. It was as if the approaching alpha particle had to surmount a barrier before entering the nucleus of the target atom. When approaching smaller nuclei, the alpha particles had sufficient energy to get over the barrier; but as they approached nuclei with larger charge, their energy was insufficient to bring them into contact.

Picture an alpha particle as a marble rolling along a level tabletop (Fig. 13-4). As the particle encounters the ever larger repulsive force due to the nucleus approached, it is as if the marble had to roll up a hill with an ever steeper incline. Whether it goes over the crest and falls into the "potential well" of the nucleus or fails to do so depends on how fast the alpha particle is traveling (i.e., its energy) and on the charge of the target nucleus. If the latter is large, the "hill" will be very high, and unless the particle has a great deal of energy, it will not make it to the crest.

During the mid-1920's it was becoming apparent that alpha particles with energies greater than those emitted by radioactive nuclei were needed if the potential barriers of the atoms of larger nuclear charge were to be

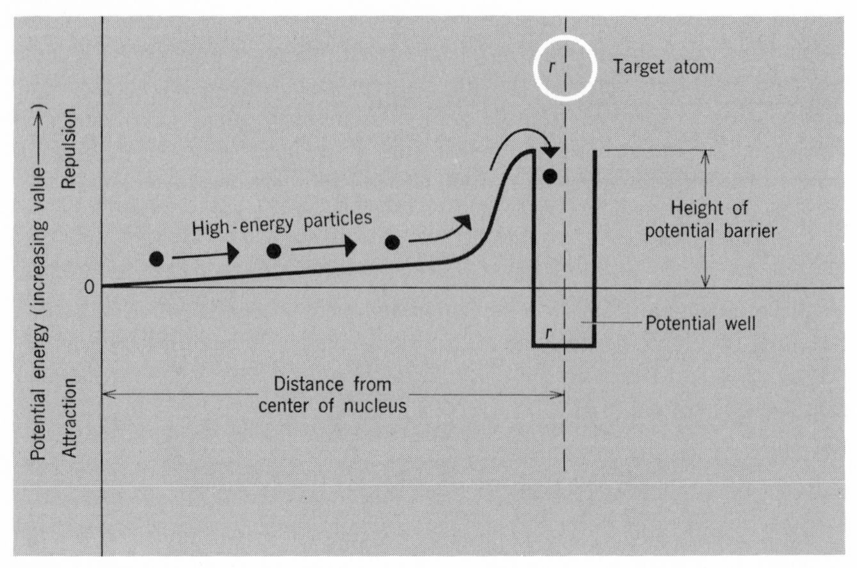

Fig. 13-4. *A proton (or other high-energy bombarding particle) approaching the nucleus of a target atom encounters a force of repulsion which makes its motion analogous to that of a marble rolling up an incline. If the bombarding particle has sufficient energy, it will enter the target nucleus, just as the marble with enough energy will roll over the crest of the incline. Think of the diagram as representing a set of coordinate axes, potential energy being plotted on the vertical axis and distance on the horizontal axis; consider the center of the target nucleus as the zero of distance. A nuclear particle which comes as close as* r *to the target nucleus will fuse with it. Up to this distance the force has been one of repulsion. Inside the nucleus the force becomes one of attraction.*

surmounted. Protons, because of their smaller nuclear charge, would encounter lower barriers than would alpha particles. But how could protons be accelerated to high energies?

ATOM SMASHERS

The acceleration of charged particles by an electric field had been known for many years. In a Crookes tube, for example, the higher the applied voltage, the faster the electrons were propelled from the cathode. Could voltage acceleration be used to speed up positive particles like protons?

Just before the close of the third decade of the twentieth century a number of scientific laboratories were experimenting with the construction of nuclear-particle accelerators. Before long these were called atom smashers. In 1932, Cockcroft and Walton, working in Rutherford's laboratory at

Cambridge University, devised a "voltage multiplier" which was the first *machine* to bring about a nuclear reaction. In this apparatus, protons were made to crash into lithium nuclei, producing alpha particles according to the equation

$$_3Li^7 + {_1H^1} \rightarrow {_2He^4} + {_2He^4}$$

With particle accelerators available, the physicist was no longer limited to alpha particles from naturally radioactive elements with a maximum energy of 10 mev for his nuclear bombardments. He could now use machines to accelerate charged particles to much higher energies. The era of atom smashing had been ushered in.

The Van de Graaff machine—another accelerator developed in the same period and still widely used (Fig. 13-5)—is a huge electrostatic generator operating on a principle somewhat similar to the Wimshurst electrostatic machine, so commonly used in the physics laboratory for demonstrating high-voltage sparking. The Van de Graaff generator is capable of producing accelerating potentials of over a million volts.

The energies obtainable are smaller than with some other accelerators, but the Van de Graaff machine has the virtue of giving an essentially monoenergetic beam (all particles having nearly the same energy). This

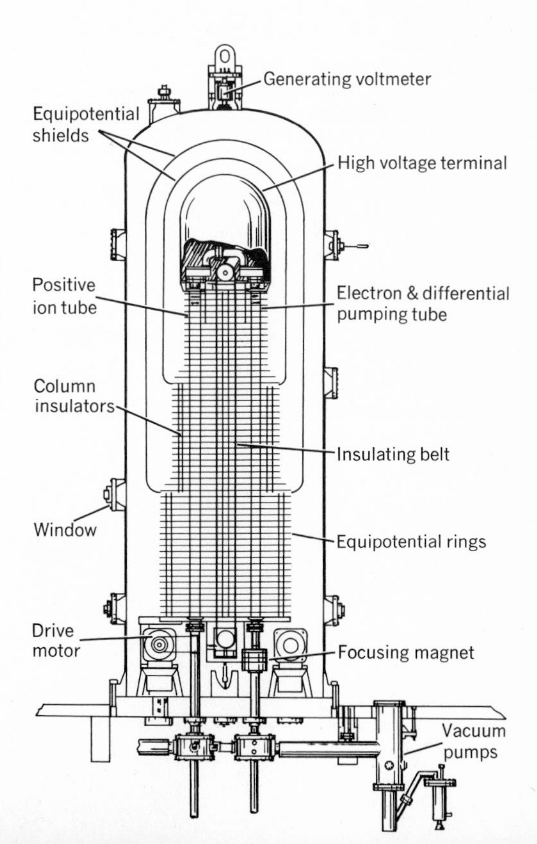

Fig. 13-5. *The Van de Graaff generator. The accelerating voltage in this atom smasher can be precisely controlled and hence the energies of bombarding particles exactly known. [Courtesy, General Electric Research and Development Center, Schenectady, N.Y.]*

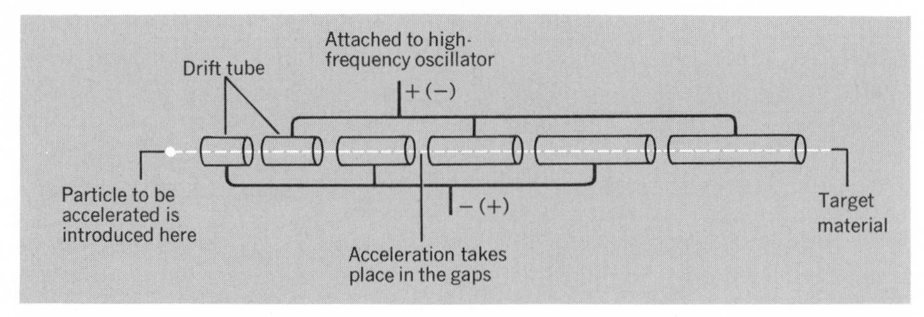

Fig. 13-6. *Alternate sets of drift tubes in a linear accelerator have opposite charges. A high-frequency oscillator changes the polarity on each drift tube millions of times per second. Positive ions are accelerated in the gaps and travel with essentially constant velocity in the field-free interior of each tube.*

is a valuable asset when it is essential to know precisely the energy of bombarding particles.

One form of linear accelerator (Fig. 13-6) consists of a series of hol-

Fig. 13-7. *The interior of the Hilac atom smasher at the University of California Radiation Laboratory. Doughnut-shaped drift tubes line the length of the barrel along which particles are hurled.* [*Courtesy, University of California Radiation Laboratory.*]

low cylindrical drift tubes of increasing length. Adjacent drift tubes always have opposite polarity. A high-frequency oscillator causes the polarity of a given drift tube to change from + to −, millions of times in a second. Positive ions are attracted toward the first drift tube when it is negative and are accelerated until they enter the field-free interior. By the time the ions have traveled through the tube and arrive at the next gap, the tube just beyond has become negatively charged. Once again the ion is accelerated in the gap. Thus, acceleration takes place in the gaps and ions move with constant velocity inside the drift tubes. After the ions have been accelerated to very high velocity they are directed at target materials and bring about nuclear reactions.

Different types of linear accelerators have been used for not only protons, deuterons and alpha particles, but also for ions with heavier masses—for example, argon. These are called heavy-ion linear accelerators (HILAC) (Fig. 13-7). The 2-mile long Stanford linear accelerator may also be used to impart to electrons energies of 20 bev or higher (Fig. 13-8).

The cyclotron was invented by Dr. Ernest O. Lawrence at the University of California about 1930. This particular accelerator captured the imagination of the nuclear scientists of the period, for many were constructed in different parts of the world. The cyclotron makes use of a magnetic field to bend positive ions into circular paths. The positive ions

Fig. 13-8. *The Stanford University linear accelerator makes use of radiofrequency waves to accelerate electrons to 20 bev or higher. Electrons are fed into one end of the 2-mile-long tube by means of an electron gun, accelerate very rapidly, and quickly reach essentially constant velocity, thereafter gaining energy by mass increase. The enormously energetic electrons at the far end of the tube impinge on target materials and may be used to determine size and structure of atomic nuclei, or perhaps in the discovery of new nuclear particles.* [*Courtesy, Stanford Linear Accelerator Center, Stanford University.*]

(e.g. protons) introduced at the center between the two hollow dees (see Fig. 13-9) are attracted to and enter the dee (D) of opposite charge. Here the ion beam is guided by the magnetic field into a circular path of small radius.

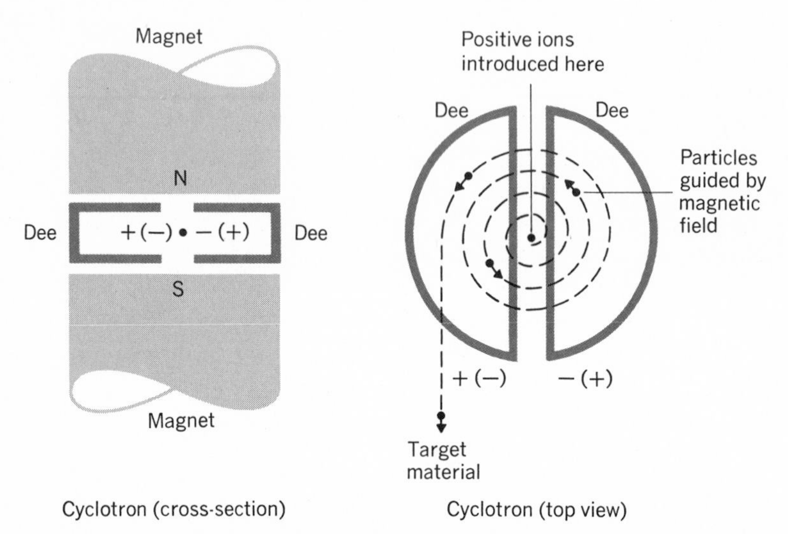

Fig. 13-9. *Dees of a cyclotron. If a pillbox were cut in half, each half would have a D shape. The dees of a cyclotron may have a diameter up to several feet. They are assigned opposite polarities, and a high-frequency oscillator changes these polarities very rapidly. Positive ions introduced at the center are attracted to the dee of opposite polarity, are accelerated in the gaps, and spiral out as they gather speed; they are guided in their paths by the magnetic field. The beam is finally deflected out at the periphery and directed at a target material.*

At any instant the charges on the two dees are always opposite—one positive, the other negative. However, the polarity on each dee is changed many times per second by a high-frequency oscillator. If the frequency of the oscillator is adjusted properly, ions emerging from one dee at the instant its polarity changes, are accelerated in the gap until they enter the next dee. Inside the next dee the ions travel with constant but higher velocity. Hence in the same time interval they travel farther in each successive passage in a dee. As a result, the particles spiral out toward the edge, where they are made to impinge on a target material (i.e., on the substance to be bombarded).

As the nuclear physicist wished to attain higher and higher energies, the size of the magnetic pole pieces had to be increased, and correspondingly the cost of construction (Fig. 13-10). Furthermore, as positive ions were

Fig. 13-10. *The cyclotron is a particle accelerator which whirls positive ions guided in circular paths by a magnetic field. The ions are accelerated as they pass between two dees which are rendered alternately positive and negative by a high-frequency oscillator. As ions travel faster, they spiral out and eventually go out through a "window" when they reach the periphery.*

accelerated to greater and greater velocities, relativity effects became more important; ion masses apparently increased. This in turn meant that the accelerating force did not speed up the particle as predicted by Newton's second law; apparently this law no longer applies at high particle velocities. At very high velocities, particles arrived late at the gaps, fell out of step with the frequency, and were no longer accelerated. To correct for these relativity effects the frequency-modulated (FM) cyclotron, or *synchro-cyclotron,* was invented.

Accelerators are now producing particles with energies well up in the bev (billion electron volt) range.* With the aid of these high-energy accelerators the depths of the atom are being probed and much information discovered concerning nuclear structure, nuclear forces, and nuclear reactions.

THE ENERGIES OF NUCLEAR PARTICLES

The masses of nuclear particles such as electrons, protons, and deuterons are so tiny that their kinetic energies are comparatively small even when their velocities are large. It is customary to express energies in multiples of a very tiny energy unit called the electron volt—the work done when

* Sometimes this is written gev (giga electron volt) so as to avoid the confusion which could arise from differing international usage of the term "billion." In both cases the symbol means 10^9 ev.

an electron is accelerated through a potential difference of 1 volt: 1 electron volt is equal to 1.6×10^{-12} erg, 1 mev = 1.6×10^{-6} erg, and 1 bev (gev) = 1.6×10^{-3} ergs.

ILLUSTRATIVE PROBLEM

Calculate the velocity (cm/sec) of a 5-mev alpha particle.

Solution

The mass in grams of a single alpha particle is

$$m = \frac{4.003 \text{ g}}{6.02 \times 10^{23}} = 6.65 \times 10^{-24} \text{ g}$$

$$\text{K.E.} = \tfrac{1}{2}mv^2$$

$$5 \text{ mev} = 5 \times 1.6 \times 10^{-6} \text{ erg}$$

$$\underset{\text{erg}}{8.0 \times 10^{-6}} = \tfrac{1}{2} \times \underset{\text{gram}}{6.65 \times 10^{-24}} \quad \underset{\text{cm/sec}}{v^2}$$

$$v^2 = \frac{8.0 \times 10^{-6}}{3.32 \times 10^{-24}} = 2.41 \times 10^{18} \text{ cm}^2/\text{sec}^2$$

$$v = 1.55 \times 10^{9} \text{ cm/sec}$$

A 5-mev alpha particle will have a velocity of 1.5×10^9 cm/sec or about $(1/20)c$ (c = velocity of light).

A similar calculation shows that the velocity of a 5-mev *electron* apparently has a velocity of 1.32×10^{11} cm/sec. Since this velocity exceeds the velocity of light in a vacuum (the limiting value of all velocities according to the theory of relativity) the result must be incorrect. According to this theory, as an object's velocity approaches that of light, its mass increases. Table 13-1 shows how the mass of an object increases with its velocity. The rest mass of the electron, 9.1×10^{-28} g, may accordingly be used in calculating the kinetic energies of only comparatively low-

Table 13–1 How Mass Increases with Velocity

PERCENT OF VELOCITY OF LIGHT	RATIO OF MASS TO REST MASS
1	1.000
50	1.15
78	1.59
90	2.35
99.6	12.8

speed electrons. It should now be apparent why an FM cyclotron or synchrotron is needed to accelerate very high-velocity positive ions.

In general, the higher the velocity of a particle, the higher its energy. On the other hand, particles of larger mass will, at the same velocity, have higher energy than particles of smaller mass. At particle velocities approaching the velocity of light, relativity effects increase the apparent mass of the particle.

THE NEUTRON

In the early 1930's, when particle accelerators were in their infancy, many important discoveries were made using as bombarding particles high-speed alpha particles from naturally radioactive nuclei. In 1930 Bothe and Becker in Germany discovered that high-speed alpha particles impinging on light metals such as beryllium produced radiation even more penetrating than gamma rays. For a while it appeared that this radiation might indeed be high-energy gamma rays.

In 1932 Irène Joliot-Curie (the daughter of Mme Curie) and her physicist husband, Frédéric Joliot, discovered that when a sheet of paraffin was put in the path of this very penetrating radiation, protons were produced in abundance. It seemed almost preposterous to think that this radiation could be gamma radiation because the energies required to kick protons out of paraffin would be so fantastically high. The idea did not make sense.

The *name* neutron had been proposed as far back as 1920. The atomic-structure theory of that day conceived of the atomic nucleus as containing both protons and electrons. The carbon nucleus, for example, atomic number 6, mass number 12, was presumed to contain 12 protons and 6 electrons. Six nuclear electrons neutralized 6 of the 12 protons, leaving a net nuclear charge of +6. Rutherford suggested there might be a close association of nuclear proton-electron pairs. Shortly thereafter the term *neutron* was proposed for such a pair. During the 1920's Rutherford's colleague Chadwick did experiments in an attempt to detect neutrons, but without success.

In 1932, then, following the paraffin-bombardment experiment of the Joliot-Curies, Chadwick made a study of the interaction of this new radiation with nitrogen nuclei in a cloud chamber. He concluded that in order to account more properly for the cloud-chamber nitrogen tracks, a *particle,* rather than high-energy gamma rays, had to be assumed. He came forward with the suggestion that the Joliot-Curie experiment might be more logically explained by assuming the formation of neutral particles (neutrons) having about the same mass as protons. Neutrons were produced by the reaction:

$$_2He^4 + {}_4Be^9 \rightarrow {}_0n^1 + {}_6C^{12}$$

Fast-moving neutrons underwent billiard-ball collisions with hydrogen atoms in paraffin, dislodging high-velocity protons. The absurdities of earlier explanations were cleared up by Chadwick's assumption of the neutron's existence.

Properties of the Neutron What was the nature of these elusive particles discovered by Chadwick? A highly capable scientist, he had searched in vain for a number of years for the neutron. Why was it so difficult to detect? In the first place, it was uncharged; therefore, unlike alpha and beta particles and protons, it was not deviated in either an electric or a magnetic field. Furthermore, because of its lack of charge, it did not produce appreciable ionization in passing through gases. Hence its presence could not be demonstrated by the usual detection instruments such as the Geiger counter, cloud chamber, or photographic emulsion. Only after it had interacted with nuclei, making them radioactive, could its presence be determined by the secondary ionizing radiation produced.

One method of detecting (slow) neutrons employs a boron-lined Geiger-counter tube. When a neutron enters the tube and is absorbed by a boron nucleus, the following reaction takes place:

$$_0n^1 + {}_5B^{10} \rightarrow {}_2He^4 + {}_3Li^7$$

Hence the detection of the neutron is actually due to the ionizing effect produced by an alpha particle.

The lack of electric charge which made the neutron so elusive also made it a most valuable nuclear bombarding particle. As the neutron approached a (positively charged) atomic nucleus it was not repelled as a positive particle (e.g., a proton or an alpha particle) would be. Often when it hit an atomic nucleus and fused with it, the product nucleus was radioactive and decayed by particle or wave emission. Perhaps even more important, as will be seen in Chapter 14, the neutron could also cause certain very massive atoms such as U^{235} to split or fission, releasing unbelievably large amounts of energy. The neutron became the key factor in sustaining a nuclear-fission chain reaction in a nuclear reactor.

ARTIFICIAL RADIOACTIVITY

Even though the neutron was known before 1934, its use as an agent in inducing radioactivity had not yet been discovered. In that year the Joliot-Curies carried out a series of experiments in which different light-metal foils were bombarded with high-speed alpha particle. As expected, neutrons were produced in a reaction such as the following:

$$_{13}Al^{27} + {}_{2}He^{4} \rightarrow {}_{15}P^{30} + {}_{0}n^{1}$$

Quite unexpected, though, was the continued emission of radiation even after the alpha-particle bombarding source was removed. Furthermore the rate of radiation emission decreased in a manner characteristic of radioactive substances. The inescapable conclusion was that radioactivity had been artificially induced.

P^{30} atoms produced in the above reaction were decaying by positron emission according to the equation

$$_{15}P^{30} \rightarrow {}_{+1}e^{0} + {}_{14}Si^{30}$$

The positron, designated $_{+1}e^{0}$, had the same mass as the electron but opposite charge. The positron had been predicted theoretically by Dirac in 1930 and discovered in 1932 by Anderson at California Institute of Technology while photographing fog tracks caused by cosmic rays. Since the positron was a charged particle, it was an ionizing particle, and the usual ion detection methods could be employed.

Stable phosphorus is P^{31}. The P^{30} nucleus has one less neutron and hence a higher ratio of protons to neutrons than P^{31}. It decays by positron emission, a new neutron being created at the expense of a proton:

$$_{1}H^{1} \rightarrow {}_{0}n^{1} + {}_{+1}e^{0}$$

Hence the decrease by one, in the atomic number of the product atom, Si^{30}.

Until the discovery of the Joliot-Curie reaction, radioactivity was a property of comparatively few elements, principally those with atomic numbers greater than 83. Now it was evident that the radioactive property could be artificially induced.

Shortly thereafter Enrico Fermi and his students were bombarding every conceivable element with the neutron, that new bombarding particle which encountered so little opposition in entering atomic nuclei.

A commonly occurring reaction was (n, γ). In this a target atom such as Cu^{65} interacting with low-energy (slow) neutrons became radioactive, emitting beta particles and gamma rays according to the equations:

$$\underset{\text{non-radioactive (stable) copper atoms}}{_{29}Cu^{65}} + \underset{\text{neutron}}{_{0}n^{1}} \rightarrow \underset{\text{radioactive copper atoms}}{_{29}Cu^{66}} + \text{gamma rays}$$

$$_{29}Cu^{66} \rightarrow \underset{\text{non-radioactive zinc atoms}}{_{30}Zn^{66}} + \underset{\text{electrons}}{_{-1}e^{0}}$$

A neutron entering the stable Cu^{65} nucleus increased the neutron-proton ratio. The resulting compound nucleus, Cu^{66}, decayed by beta (negatron) emission, a nuclear neutron being converted into a proton:

$$_0n^1 \rightarrow {}_1H^1 + {}_{-1}e^0$$

Conversion of a nuclear neutron into a proton results in the formation of an element whose atomic number is larger by 1.

The discovery of artificial radioactivity held many interesting possibilities for the nuclear scientist. If he could induce radioactivity in atoms, he might slip a few into different chemical substances and then follow the course of their reactions by means of the radiation emitted. This so-called "tracer technique" has indeed proved to be a most important research tool.

Today radioactive isotopes of all elements have been produced artificially, some in considerable quantities. For example, I^{131}, which is in common use today in the diagnosis and therapeutic treatment of thyroid dysfunction, is a by-product of a nuclear reactor or atomic pile (see Chapter 14). Artificially produced radioactive isotopes have made possible not only more effective diagnostic and therapeutic methods in medicine but also a multitude of new industrial uses.

Problems, Chapter 13

1. Element $_ZM^A$ is bombarded by alpha particles and a proton is set free, as in the first controlled nuclear reaction carried out by Rutherford. Write an expression for the new product nucleus X, putting in the appropriate values for Z and A.

2. Complete the following nuclear reactions and designate the type of reaction:

   (a) $_{20}Ca^{44} + {}_1H^1 \rightarrow {}_0n^1 + ?$
   (b) $? + {}_1H^2 \rightarrow {}_{15}P^{32} + {}_1H^1$
   (c) $_5B^{11} + ? \rightarrow {}_7N^{14} + {}_0n^1$
   (d) $_{13}Al^{27} + {}_0n^1 \rightarrow {}_{12}Mg^{27} + ?$

3. Nuclides having more neutrons than a stable nuclide tend to be negatron emitters, those with fewer neutrons, positron emitters. Predict which of the following radioactive nuclides will be positron emitters and which negatron emitters:

	STABLE NUCLIDE	RADIOACTIVE NUCLIDE
(a)	C^{12}	C^{14}
(b)	Na^{23}	Na^{22}
(c)	K^{39}	K^{40}
(d)	Co^{59}	Co^{60}
(e)	As^{75}	As^{74}

4. $_{80}Hg^{204}$ and $_{82}Pb^{204}$ are called *isobars*. They have the same atomic masses but different atomic numbers. Give other examples of isobaric pairs.

5. $_{15}P^{31}$ and $_{16}S^{32}$ are called *isotones*. They have the same number of neutrons but different mass numbers. Give other examples of pairs of isotones.

6. Nuclei having 2, 8, 20, 28, 50, or 82 nucleons (either protons or neutrons) are called *"magic number" nuclei* and are particularly stable. Give the nuclear symbol for one illustration of each of the above magic number nuclei.

7. Write the nuclear equation for:

   (a) α decay of $_{88}Ra^{226}$
   (b) β decay of $_{90}Th^{234}$
   (c) α decay of $_{86}Rn^{222}$
   (d) β decay of $_{82}Pb^{210}$

Exercises

1. Describe the alpha-particle scattering experiments of Geiger and Marsden and their interpretation by Rutherford.

2. Briefly outline the experiments which led to a knowledge of the number of protons in the nucleus of the atom.

3. Describe the circumstances which led to the discovery of the first controlled nuclear reaction.

4. Describe the significance of subscript and superscript numbers in the symbolic representation of nuclear reactions.

5. Interpret the meaning of the following:

$$Pu^{239}\ (\alpha,n)\ Cm^{242}$$

6. Describe some of the respects in which ordinary chemical reactions differ from nuclear reactions.

7. What is meant by the nuclear potential barrier? Explain briefly its significance with relation to nuclear reactions employing positively charged bombarding particles.

8. Explain the principle of particle acceleration in (a) the cyclotron, (b) the linear accelerator.

9. In what units are the energies of nuclear particles usually expressed. Illustrate.

10. What is the meaning of the "rest mass" and "relativity mass" of an atomic particle?

11. Describe the experiments which led to the discovery of the neutron.

12. Describe some of the characteristic properties of neutrons.

13. Describe a method by which slow neutrons may be detected.

14. Describe the experiment by which artificial radioactivity was discovered.

15. The kinetic energy imparted to a particle in a cyclotron may be calculated by the equation:

$$\text{K.E.} = \frac{1}{2}\frac{H^2e^2r^2}{m}$$

H = magnetic field strength
e = charge of the electron
r = radius of D
m = mass of particle

For a particular cyclotron what does the kinetic energy imparted depend upon? Explain.

14 Energy from splitting atoms

THE key to the release of energy in a nuclear chain reaction is the neutron. Neutron release resulting from the fission of U^{235} atoms makes possible nuclear explosions and the operation of nuclear reactors—and the nuclear reactor is the key to atomic power and large-scale production of radioactive isotopes. Thus neutrons play a vital role in applied nuclear science.

After the neutron was discovered in 1932, the Italian physicist Enrico Fermi was quick to realize its potential value. The neutron's lack of charge gave it a tremendous advantage over charged particles; in bombardment reactions neutrons did not encounter a force of repulsion as they approached target nuclei (also of positive charge). One of these neutron bombardment reactions led to the discovery of nuclear fission.

NUCLEAR FISSION

The neutron reaction which particularly fascinated Fermi was the one mentioned in Chapter 13, in which the product atomic number was one larger than that of the bombarded target metal. He was curious to know whether the same reaction would occur with uranium (the heaviest of the elements then known) as a target. If it did, then it seemed logical to suppose he might be able to prepare a new element of atomic number 93. However, nature sometimes divulges her secrets reluctantly. Element 93 was eventually discovered, but not until scientists had wrestled with many perplexing problems.

What was actually taking place in the reaction produced by neutron bombardment of uranium was difficult to explain for a number of reasons.

First of all, several different radioactive species resulted, and these in such small amounts that a special "carrier technique" was required to separate them. Quantities of material too tiny to remove from solution by ordinary precipitation methods were "carried" along when another substance was precipitated from the same solution. Since the carried material was radioactive, even though undetectable chemically, its presence in the precipitate was readily shown by the radiation it emitted.

In the neutron bombardment of uranium a barium isotope was produced, but was mistaken for radium, which is much closer to uranium in the periodic table. At the time, no nuclear reactions were known in which product atoms were as far removed from each other in the periodic table as U and Ba. But soon, Frisch and Meitner worked out the details of a theory of the splitting of the uranium nucleus. They predicted that fragments would be released with very high velocities. The ionization produced by such fragments should be very large and easily detectable. In 1939 several research laboratories in the United States verified this huge energy release. It had now been clearly established that neutron bombardment of uranium resulted in an altogether different type of nuclear reaction, in which a quivering atomic nucleus, set vibrating by a neutron, separated into two high-energy fragments. The new reaction type was called *nuclear fission*. The magnitude of the energy release accompanying nuclear fission set many minds to thinking.

The Energy Release Accompanying Fission The energy release occurring in a nuclear reaction can be predicted from a knowledge of the isotopic masses (determined by mass spectroscopy, Chapter 11) of reactants and products. If there is to be such an energy release, the total mass of all product particles must be smaller than the masses of the reactants, the mass difference being equivalent to the energy release as stated by the Einstein mass-energy equivalence principle.

ILLUSTRATIVE PROBLEM

Calculate the energy release when a slow neutron fissions a U^{235} nucleus. Assume the following reaction:

$$ {}_0n^1 + {}_{92}U^{235} \rightarrow {}_{40}Zr^{97} + {}_{52}Te^{137} + 2\,{}_0n^1 $$

Solution

(1) Add the isotopic masses on the left and right.
(2) Calculate the mass destroyed as a result of the reaction.
(3) Calculate the energy equivalent of the mass loss.

APPROXIMATE ISOTOPIC MASSES*

$1.009 + 235.044$	$96.911 + 136.909 + 2.018$
Sum of masses, left side: 235.044 + 1.009 = 236.053 amu†	Sum of masses, right side: 96.911 + 136.909 + 2.018 = 235.838 amu

Loss of mass as a result of fission: 236.053 − 235.838 = 0.215 amu

0.215 amu × 931 mev/amu = 200 mev, energy equivalent of mass lost

The enormous scale of the energy released by fission can be appreciated on comparing it with the energy released by chemical explosives. For years TNT (trinitrotoluene) was one of the most highly explosive materials available. But a single kilogram (2.2 lb) of U^{235} would yield, through fission, energy equivalent to 20,000 *tons* of TNT. This is in turn equivalent to 25,000,000 kilowatts of electrical energy, enough to keep ten 100-watt electric light bulbs burning continuously for 65 years. Even though as a practical matter only a small fraction of the energy theoretically available can actually be realized, a single pound of U^{235} still yields the energy equivalent of 1500 tons of coal. The discovery of nuclear fission thus marked the beginning of an era in which energy release was increased by another order of magnitude.

Furthermore, these developments led to a possibility that soon proved crucial—that of continued release of nuclear energy in a chain reaction.

THE FIRST NUCLEAR CHAIN REACTION

The enormous energy released by a fissioning uranium nucleus need not necessarily be useful energy. To be useful it must be controllable. This may mean control in the sense of determining the time of an energy release in an explosion or control of the process as it spreads from atom to atom in a continuous process.

In 1939 it was discovered that several neutrons, perhaps two or three, were released in each fission process. This was encouraging because it indicated at least the possibility that the fission reaction might be self-

* Isotopic masses usually include three more significant figures. Here the energy release is so large that the approximate values will suffice.

† 1 amu = 1 atomic mass unit = 1.66×10^{-24} g = 931 mev.

perpetuating (Fig. 14-1). Later it was shown that 2.5 *high-speed* neutrons per fissioning U^{235} nucleus was a more precise average figure when huge numbers of atoms were splitting. A nuclear chain reaction was certainly a possibility, though there was some apprehension lest a rapid neutron multiplication might result in a catastrophic explosion during the develop-

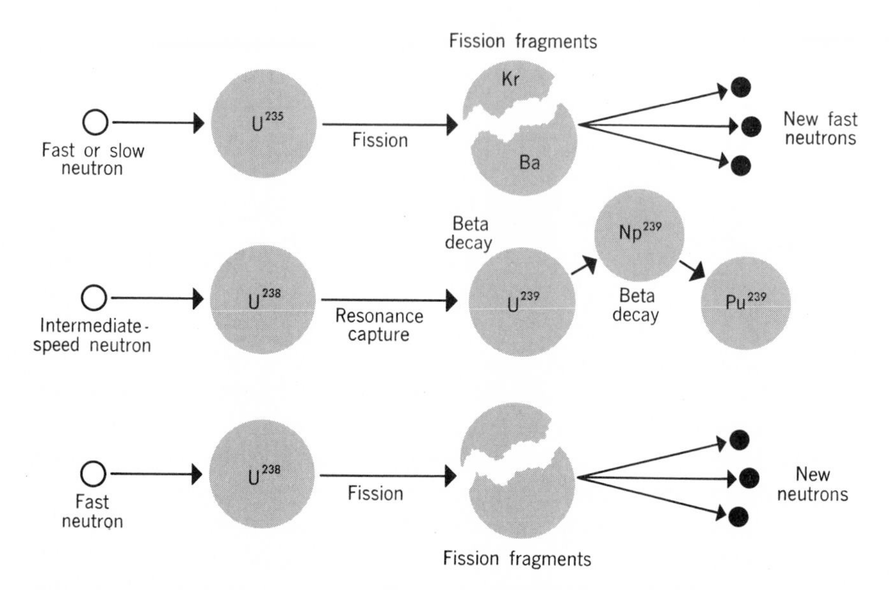

Fig. 14-1. *A nuclear chain reaction depends on each generation of neutrons reproducing itself. A fissioning atom of U^{235} provides on the average 2.5 neutrons. If the chain reaction is to be maintained, at least one of these must reach another U^{235} atom.*

ment stages. The fact that this did not happen was in part due to the fortuitous existence of "delayed neutrons" * but was also a tribute to the keen and courageous group of dedicated scientists who risked their lives to set into operation the first nuclear chain reaction.

If a neutron multiplication factor, k, could be kept at exactly 1—that is, if each fission process were to initiate just one more fission—a chain reaction would be possible. If the value were to exceed 1, and each fission initiated more than one additional fission, multiplication could proceed very rapidly and an explosion could result.

* Among the product atoms from the thermal neutron fission of uranium are several beta emitters of short half-lives. Through decay they give rise to daughter product nuclides which instantaneously emit neutrons. These "delayed neutrons" have the effect of increasing slightly the average time between neutron generations. Thus there is sufficient time for control of a nuclear chain reaction.

The earliest experiments with a uranium-graphite (subcritical) lattice* at Columbia University showed that neutrons were not reproducing themselves, i.e., *k* was somewhat below the hoped-for value of 1. Further purification of materials resulted in a *k* value very close to unity. Finally a quantity of uranium sufficient to sustain a nuclear chain reaction (critical quantity) was assembled under the bleachers at Stagg Field of the University of Chicago, and in December 1942 it was first demonstrated that a nuclear chain reaction was possible. Nerves were a bit on edge as the power level of the reactor rose on this momentous occasion, but delayed neutrons saved the day. It was now certain that uranium could sustain a chain reaction.

The most fissionable isotope of uranium, U^{235}, forms only 0.7 percent of natural uranium, most of which (over 99 percent) is U^{238}. U^{238} is also fissionable by very high energy neutrons, but the required energy is so high (2.0-3.0 mev) that such neutrons are not abundant and a nuclear chain reaction based on U^{238} fission is not possible. Yet the U^{238} will capture slower-moving neutrons and keep them from striking the rarer fissionable nuclei of U^{235}—that is, unless the neutrons are brought down to speeds comparable with the speeds of gas molecules near room temperatures—a little more than a mile a second (energies of about 0.025 ev). At these speeds, the so-called thermal neutrons are seldom captured by U^{238} but are far more effective in causing U^{235} to fission than are faster neutrons. The problem of slowing down the fast neutrons released whenever an atom split, was solved by the development of *moderators*.

MODERATORS

A moderator is a material whose atoms slow down the speeds of neutrons by colliding with them. The atoms, however, must not react appreciably with the neutrons. Moderator atoms whose masses are nearest those of neutrons will reduce neutron energies most in the fewest collisions. Hydrogen would appear to be the ideal "slowing down" agent for neutrons because its mass is so nearly the same. However, its comparatively high rate of neutron capture makes it impractical. Deuterium (H^2), because of its larger mass, requires more collisions than H^1 to slow neutrons to thermal speeds, but has a much lower rate of capture. So deuterium, despite its high cost, is among the moderators often employed. Graphite, because it is composed of much heavier carbon atoms, is a less effective thermalizer, but can nevertheless be obtained in sufficient purity, in sufficient quantity, and so much less expensively, that in the past it has been widely used (Fig.

* Though such an assembly could not support a chain reaction, it could give evidence whether neutron production was balancing utilization.

14-2). It also has a low rate of neutron capture. One drawback to the use of graphite has been the large size of installation needed when it is used with natural uranium. In recent years water and heavy water (D_2O) have been used as a combination moderator and coolant in reactors.

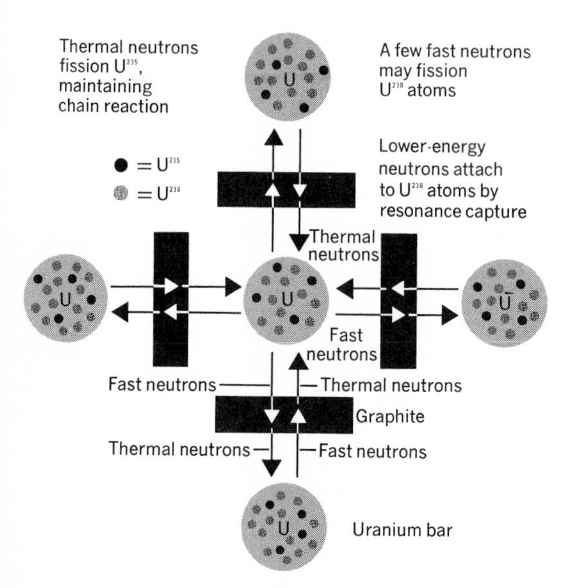

Fig. 14-2. *Neutrons resulting from fission of U^{235} atoms in a uranium bar are "fast neutrons." After they have passed through graphite, their energies are reduced so they will not be captured by U^{238} atoms (resonance capture), but are still capable of causing fission of U^{235} atoms in the next uranium bar.*

ENRICHING URANIUM WITH RESPECT TO U^{235}

Another method of improving the neutron economy for a chain reaction is to increase the U^{235} population relative to U^{238}. If the U^{235} atoms are not so hopelessly outnumbered as they are in naturally occurring uranium, there will be a higher probability of the all-important fission capture reaction. But solving the problem of separation of the uranium isotopes is not easy. First of all, chemical separation methods are ineffective because the chemistry of two isotopes is the same. On the other hand, a mass spectrograph is an instrument for sorting out isotopes of different masses and so a huge mass spectrograph might be effective. Another method of separation, the one which has proved most feasible, is based on gaseous diffusion. A light gas like hydrogen will diffuse through a porous barrier—say a clay cup—much more rapidly than a heavier gas like carbon dioxide.

In the separation of U^{235} and U^{238}, uranium is first converted into solid uranium hexafluoride, UF_6, which under appropriate conditions may be converted to a gas. Fortunately, fluorine has only one nuclide, so there is no complication in separation of uranium nuclides due to fluorine atoms of different masses. Because of its corrosive nature, however, UF_6 vapor is

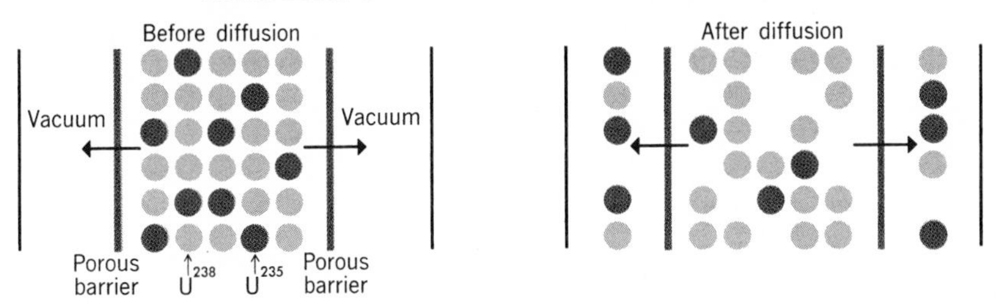

Fig. 14-3. *The gaseous fluoride of U^{235} (UF_6) diffuses slightly more rapidly than that of U^{238}. Even though the enrichment in a single diffusion unit is very small, if these gases are circulated through a sufficiently large number of units, say 1000 or more, eventually nearly pure U^{235} can be prepared.*

hardly the ideal substance to be pumped through a system of multiple diffusion units. The actual U^{235} enrichment taking place in a single diffusion unit is exceedingly small, but if the slightly enriched gas from one unit is fed into another similar unit and this process repeated 1000 or more times, a high degree of enrichment can be attained (Fig. 14-3). A minor miracle was accomplished during World War II in setting up, at the K-25 gaseous diffusion plant in Oak Ridge, Tennessee, thousands of such diffusion units, which made it possible to separate U^{235} from U^{238} (Fig. 14-4).

Fig. 14-4. *The gaseous diffusion plant at Oak Ridge, Tennessee* [*Courtesy, United States Atomic Energy Commission, Westcott, Oak Ridge.*]

The availability of enriched uranium permits construction of a nuclear bomb and smaller nuclear reactors than are possible with natural uranium alone.

THE ATOMIC BOMB

A nuclear explosion depends on the very rapid growth of the neutron population in a critical quantity of a fissionable material. As mentioned earlier, in fissioning U^{235}, an average of 2.5 fast neutrons are produced for every nucleus fissioned. A Pu^{239} nucleus (also fissionable with neutrons) produces a slightly larger number of fast neutrons on the average. Either U^{235} or Pu^{239} can be employed in a fission bomb. Fast neutrons striking other U^{235} atoms also cause them to fission, fewer to be sure than the same number of *thermal* neutrons would, but nevertheless U^{235} is capable of supporting a fast-neutron chain reaction. Assume that 3 new fast neutrons are available from each fission process; 3 will become 9, 27, 81, 243, etc. If one fission requires, say 10^{-6} sec, then in a twinkling, millions and millions of uranium atoms will be splitting—a nuclear explosion.

A critical quantity of U^{235}—the minimum quantity that will sustain a chain reaction—is needed for any nuclear chain reaction, but stray neutrons would set this quantity off if it were all in one piece. One way of initiating the reaction then, would be to fire into each other, at very high velocity, two subcritical masses of sufficient size to be critical when combined (Fig. 14-5).

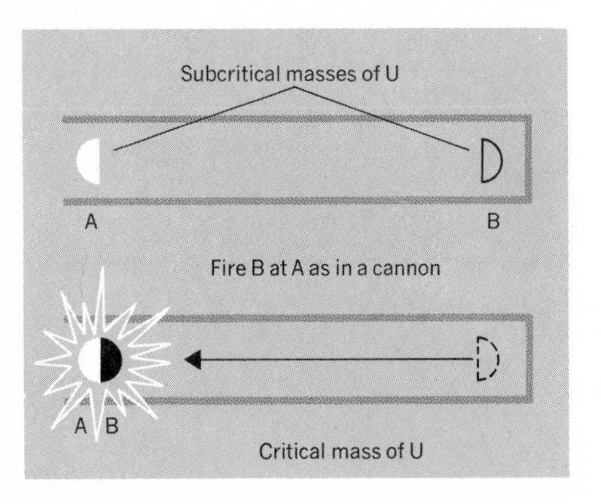

Fig. 14-5. *One subcritical quantity of uranium is fired at high velocity into another subcritical quantity. When the critical quantity is assembled, a nuclear explosion ensues.*

The terrifying effects accompanying nuclear explosions are now well known. Searing heat, intense gamma rays, and showers of neutrons extend out to considerable distances from "ground zero," the point directly under the explosion. A tremendous shock wave first pushes objects in its path and

Fig. 14-6. *A nuclear explosion.*

then reverses itself, pulling them back. The momentary void at the base of the explosion pulls air and dust up into a central shaft in the familiar mushroom (Fig. 14-6). Neutron irradiation of the dust renders it highly radioactive. Sooner or later it falls back to earth in turn irradiating whatever is in the path of the fallout.

NUCLEAR REACTORS

Some applications of nuclear energy appear to have greater prospect of being of constructive benefit to mankind than atomic bombs. The device in which uranium fission is brought about under controlled conditions is called a nuclear reactor. Both atomic bombs and nuclear reactors are dependent upon neutron fission of U^{235}, but there is an important difference. In bombs, neutron multiplication proceeds very swiftly; in nuclear reactors, the neutron levels are carefully controlled so that they will just reproduce themselves.

The balance is a delicate one. If too many neutrons are lost through escape or reaction with impurities, the chain reaction ceases. If not enough are used up, the neutron flux will increase; the device will run wild and destroy itself (though a violent *nuclear explosion* will not occur).

There are numerous types of nuclear reactors, and the development of others can be expected. Some operate with fast neutrons; others with thermal neutrons. Some make use of moderators such as graphite or heavy water (deuterium oxide, D_2O). One type uses fuel consisting of a uranium compound in water solution, with the water itself (especially D_2O) serving as the moderator. Another type makes use of metallic uranium bars, which may be either natural uranium or U^{235}-enriched. Some reactors produce more fissionable material than they use up and are known as *breeder reactors.* Some reactors use air as a coolant, others water, and some even molten sodium metal.

In a breeder reactor, more than two neutrons per fission process must be available. One is needed to keep the chain reaction going, the second to convert a U^{238} atom into a Pu^{239} (in a fast breeder). The availability of two neutrons creates only one replacement atom for each fissionable atom used up; there is no *gain* in fissionable material. If in a considerable number of fissioning Pu^{239} atoms, three fission neutrons are available, it would in principle be possible to produce more fissionable material than is used up. However, from what was said earlier concerning the problem of minimizing neutron losses, development of a successful breeder has its formidable aspects.

Graphite-moderated thermal reactors using natural uranium played a very important role in the development of nuclear energy. Fig. 14-2 shows a cross section of the core of such a reactor. The fuel sometimes consists of cylindrical slugs of natural uranium sealed in aluminum casings, inserted in cylindrical channels in a huge pile of "pure" graphite blocks. The uranium slugs are pushed in at the front face of the reactor. When fission products have accumulated in a fuel slug to such an extent that the neutron multiplication factor might be appreciably reduced, a new slug is pushed in at the front, displacing another at the rear face.

Control is effected by means of rods of boron steel, also located in slots in the graphite. Neutrons react readily with boron; it has a high cross section for neutron capture and may be regarded as a neutron remover. These boron rods may be pushed into the interior of the pile to reduce the neutron flux or withdrawn to allow it to build up.

If a reactor is to sustain a chain reaction there must be a critical quantity of uranium present. Smaller quantities suffice if the uranium fuel is enriched in U^{235}; larger quantities of natural uranium are needed. Thirty tons is about the requirement for criticality in a graphite-moderated reactor. The

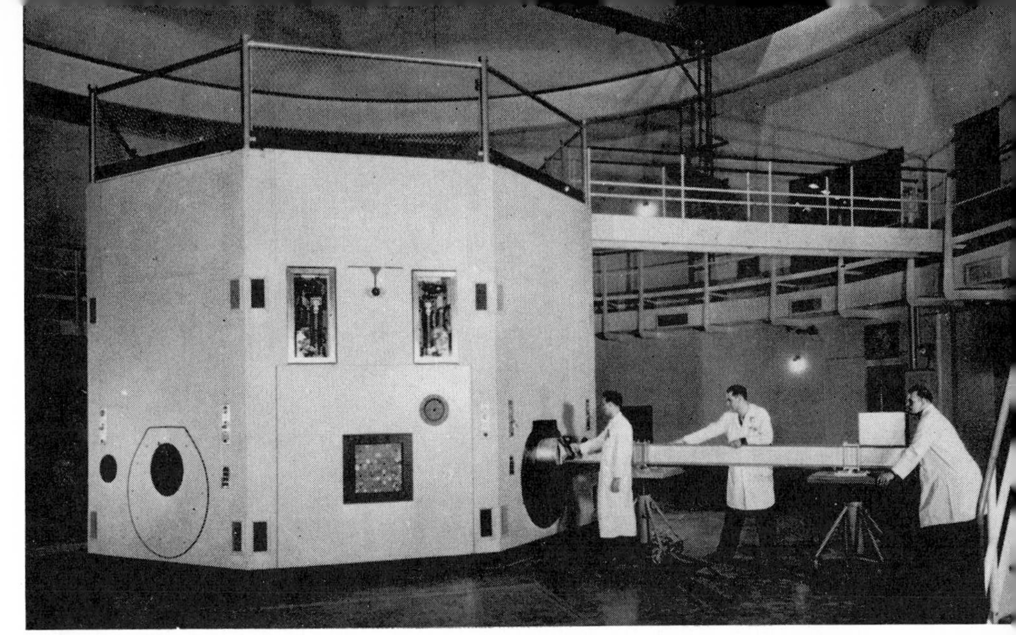

Fig. 14-7. *A research reactor. It uses U^{235} as a fuel and is cooled and moderated by heavy water. The octagon is 20 feet wide and 13½ feet high. Note the compactness attainable when an enriched fuel is used.* [*Courtesy, Argonne National Laboratory.*]

smaller, highly enriched reactors (Fig. 14-7) may use as little as a few pounds of uranium.

Uses of the Nuclear Reactor The nuclear reactor is the essential unit in the production of atomic power, of radioisotopes, and of fissionable materials. It is the most important neutron source available today, whether for research neutron bombardment or for production of commercial radioisotopes.

The nuclear reactor is the basic unit for the production of electric power from nuclear energy. Heat energy produced in a reactor as a result of the fission process may be absorbed by circulating a liquid through the pipes extending into the core of the reactor. Molten sodium is often the liquid, though water and hydrocarbons have also been employed. Hot liquid metallic sodium coming from the reactor is circulated through a heat interchanger where water circulating in a separate system is heated to a high temperature and the steam produced is used to operate conventional turbines.

As a result of exposure to the high neutron flux in the interior of the reactor, liquid sodium is rendered highly radioactive because of neutron bombardment. However, virtually no radioactivity is imparted to the water when it receives heat from the liquid sodium in the heat interchanger. Though the molten metal is highly radioactive, it is still not emitting neutrons, which are primarily responsible for inducing radioactivity in other substances.

The nuclear reactor is also of great importance as a source of radioactive isotopes. The spent uranium fuel slugs which are pushed out of the reactor may be processed for by-product isotopes such as the I^{131} which is so extensively employed in medicine. The actual weights of radioactive materials produced are very tiny. A curie of I^{131}, for example—an enormous amount of radioactive material—weighs only 8×10^{-6} g, or less than 1/100,000 g. Such a small amount of material would be barely measurable with the most sensitive microbalances. But this is enough radioactive iodine for 100,000 ten-microcurie samples such as are used in diagnostic thyroid tests in medicine. Stated another way, a single gram of I^{131} would be sufficient for about one hundred billion (10^{11}) such pills, or about 30 for every person on earth (1965).

In addition to by-product materials, the reactor may be used to induce radioactivity in a wide variety of materials, from chemicals to piston rings, by exposing them to the high neutron flux in the interior of a reactor. Radioisotopes are produced by introducing the materials to be irradiated into the interior of the reactor, sometimes through pneumatic tubes. Here they are subjected to the action of enormous neutron fluxes and are rendered highly radioactive. For example, if a chunk of cobalt metal is introduced into a reactor for a considerable length of time, it becomes a powerful emitter of gamma rays. Small pieces of it when appropriately irradiated may be used by the medical profession to perform nearly any function accomplished with our most powerful X-ray machines.

By proper choice of materials to be irradiated, a wide variety of radioisotopes with different half-lives, types of radiation, and energies can be made available. These may be purchased by the research scientist for a great variety of applications in research, medicine and industry. The radioactive atoms so produced are believed to undergo essentially the same chemical reactions as their nonradioactive counterparts. By introduction of the "hot" atoms into chemical molecules a method is available for tracing where they go as the parent molecule undergoes chemical reactions. A great deal has been learned about the mechanism of photosynthesis and many other chemical reactions by means of such tracer experiments. Finally, neutron beams from the reactor may be used for fundamental research investigations of the properties of neutrons and their interactions.

MEDICAL USES OF ISOTOPES

By their radiations radioactive isotopes make known their location in the body. Sometimes the count rate gives evidence of how well an organ or gland is picking up an element, as for example how I^{131} is picked up from the bloodstream by the thyroid gland. In other cases circulation of body

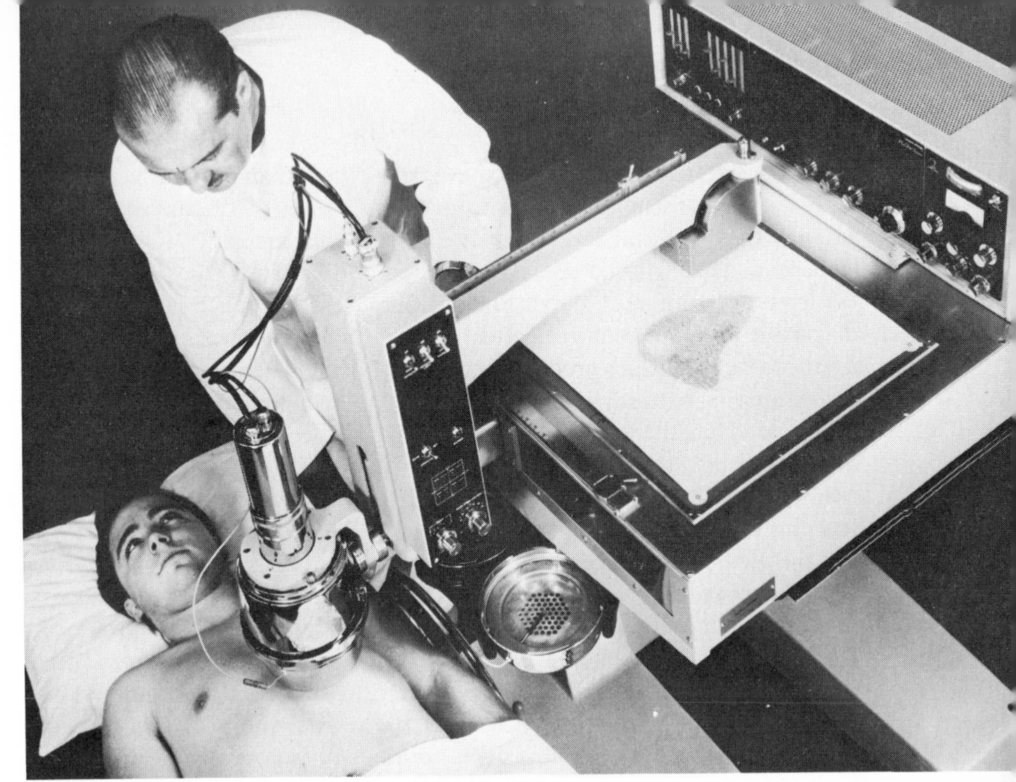

Fig. 14-8. *A medical radioisotope scanning system using a scintillation counter as detector (the vertical tube above the patient's chest). Note the profile of the isotope-containing organ at the right on the sheet of paper, the absence of activity at the upper left, and its greater concentration in the lower portion. By appropriate marking of the scan sheet, the location of isotope concentrations in the body may be determined by position on the profile sheet. This instrument can produce either photographic or dot scans.* [*Courtesy, Nuclear Chicago Corporation.*]

fluids may be followed by means of such labeled substances as Na^{24} in circulation studies. Sometimes the desired therapeutic effect is obtained by a selective inactivation of a particular kind of cell or tissue, as in the therapeutic treatment of hyperthyroidism by I^{131}.

Probably the most widely used of all isotopes is I^{131}, which emits both beta and gamma rays. In an iodine-uptake test (Fig. 14-8), the patient is given approximately 10 microcuries (or 10^{-10} g) of I^{131} in a water solution or a pill (the number of counts per minute being known at the time of ingestion). A day later the patient returns and a scintillation counter tube placed a measured distance from the thyroid gland, gives a smaller count than the original sample, similarly counted. If the corrected count over the thyroid is 1500 counts per minute (cpm) whereas the sample itself originally gave 6000 cpm, 25 percent of the radioactive atoms have been removed from the blood stream by the thyroid gland. This value indicates a normal uptake. A hyperactive gland would give a much greater-than-normal percentage of uptake, a hypoactive gland a less-than-normal value. The latter diagnosis is more difficult to make by means of the uptake test because of the low count and relatively high statistical error.

Besides its diagnostic use, I^{131} may be administered for slowing down the hyperactive thyroid gland. For this purpose doses much larger than for diagnostic tests are required, often about 5 millicuries (nearly 1000 times as much). Thyroxine-producing cells are inactivated, probably by the highly ionizing beta rays, which travel only a few millimeters in tissue inside the gland. Thus the action of the overactive gland is slowed. The patient is not subjected to the risks involved in surgical cutting of the gland, which is richly supplied with blood vessels. The I^{131} hyperthyroid treatment method has also been effective in over 90 percent of the cases. In most cases today it is the method of choice.

Radioactive tracers have proved to be of value in locating brain tumors, thus simplifying brain surgery. Such tumors preferentially concentrate certain elements—for example As^{74}, which is a positron emitter, or certain dyes incorporating I^{131}. The tumor may be located by determining with a detector such as a scintillation counter the region where the count is unusually high. Location of such concentrations of isotopes is often detected by means of a technique known as scanning. A radiation detector moves back and forth over an organ or region of the body registering an increased count by an increased concentration of dots on a piece of paper or increased exposure of a film. Thus it is possible to reproduce on paper or film a "profile" of any concentration of the radioactivity within the body or the organ.

Co^{60} is used both in the teletherapy unit, which performs essentially the same function as an X-ray machine, and in labeling cobalt atoms in vitamin B_{12} for testing pernicious anemia.

Other medical isotopes include P^{32} (a pure beta emitter), which is used in treatment of polycythemia vera, excessive proliferation of red blood cells, and Cr^{51}, which is used in determining the volume of blood in the body. I^{131} has also been used in connection with cardiac therapy and in the determination of blood volumes. Developments in the field of medical radioactive isotopes have been proceeding at a breath-taking pace in recent years. They have already enormously extended the scope and accuracy of medical diagnosis, to say nothing of the alleviation of human suffering.

INDUSTRIAL USES OF ISOTOPES

As in the medical field, there has been an enormous growth in the number of uses of radioactive isotope applications in industry. Thickness-gauging, for example, makes it possible to control the thickness of such diverse materials as rugs, paper, cigarettes, and sheet steel. A radioactive source is placed below the stock whose thickness is to be measured. A detector placed above registers an ionization current whose magnitude is inversely related to the materials thickness. If the material becomes too thick, a decrease in ionization current sets in operation a servomechanism for decreasing the thickness. If the material falls below proper thickness it allows an increased ionization current to pass, which activates a servomechanism increasing the thickness.

Beta-emitting isotopes can be used in gauging materials such as paper, but gamma emitters are needed for penetration of denser materials such as sheet iron. Gamma-emitting isotopes have also been especially valuable in radiography. Gamma rays from the isotopes perform much the same function as X rays. But an X-ray machine requires well-insulated high-voltage electric lines. Even though a radiography unit may require considerable heavy shielding, it is still a piece of equipment which can be more conveniently moved about than an X-ray unit. Also, it is often possible to put a radioactive source inside a metal casting. Location of flaws is thus often easier with isotopes than with an X-ray source.

Flow studies, location of leaks in pipelines, lubrication studies, gauging liquid level, irradiation of foods, and a multitude of new applications are being developed daily in industrial research laboratories across the nation.

FISSION OR FUSION?

In general, atoms of elements near the middle of the periodic table have greater stability than those of elements at the very beginning or end. Uranium, the heaviest of the naturally occurring elements fissions with the release of enormous quantities of energy. The fission fragments must be smaller in mass, and hence nearer the middle of the periodic table. They are then more stable and less likely to be fissionable. Similarly the lightest of all atoms, hydrogen, under appropriate conditions fuse, yielding heavier atoms of greater stability and in the process release enormous quantities of energy. Hence, elements near the middle of the periodic table are the most stable and the least promising as energy sources.

Fusion of the nuclei of light elements, just mentioned, offers a second way of liberating nuclear energy. If hydrogen or its isotopes were heated to a sufficiently high temperature and pressure, they could fuse into heavier atoms of helium having greater stability. Temperatures of a million or more degrees are momentarily realized in an atomic bomb explosion. These enormous temperatures might suffice to trigger a thermonuclear reaction of hydrogen. Energy releases of this type are of another order of magnitude or so above that of the atomic bomb. Theoretically there is no limit to the energy release attainable in this type of reaction.

Problems, Chapter 14

1. The average kinetic energy of a gas molecule is approximately

$$E = kT^*$$

k = the Boltzmann constant, or gas constant per molecule
T = absolute temperature
Show that the average kinetic energy of a slow neutron (0.025 ev) is about the same as that of a gas molecule at 20°C.

2. Calculate the mass of U^{235} available in a ton of natural uranium metal.

3. Calculate the weight of radioactive I^{131} in a 5-millicurie pill used in the therapeutic treatment of hyperthyroidism.

4. (a) How many fissioning U^{235} atoms would be required to supply the energy equivalent of sufficient electricity to keep a 100-watt light bulb burning for one hour? (Assume each fission releases 200 mev of energy). (b) What would be the mass (in grams) of this number of U^{235} atoms?

Exercises

1. Assuming the following reaction results in a beta-active compound nucleus, complete the nuclear reaction:

$${}_0n^1 + {}_{45}Rh^{103} \rightarrow$$

2. Why are neutrons particularly advantageous as bombarding particles in nuclear reactions?

3. What is the carrier technique of separating from a solution tiny quantities of radioactive tracers?

4. Why did scientists at first expect Ra rather than Ba to be one of the products resulting from the neutron bombardment of uranium.

5. What is meant by nuclear fission?

6. Compare the magnitude of the energy release accompanying uranium fission with that of commonly used (chemical) fuels.

* The exact equation for an ideal gas is $E = \frac{3}{2}kT$

7. Explain why a neutron multiplication factor k greater than 1 is necessary for a nuclear explosion and why it must be exactly equal to 1 for a nuclear chain reaction.

8. Explain why U^{235} is the key to nuclear energy.

9. Fission neutrons have an energy distribution approximating that of gas molecules. Using details given in Chapter 6 draw a possible neutron energy distribution curve and give a brief interpretation of its meaning.

10. Plutonium, which like U^{235} is fissionable with slow neutrons, forms as a result of resonance neutron capture by U^{238} atoms. Why then is it so important to make provision for rapidly thermalizing neutron energies?

11. Why is uranium enrichment so important? Explain.

12. Explain the principle involved in the separation of U^{235} from U^{238} by the gaseous-diffusion method.

13. What are the important characteristics of a moderator for a nuclear reactor. Compare the advantages of graphite and heavy water (D_2O) as moderators.

14. Describe what is meant by a critical quantity of uranium. The size of the critical quantity of uranium may range from a pound or so, to tons. Explain.

15. Describe a possible mechanism for exploding an atomic bomb.

16. Outline the different applications of a nuclear reactor.

17. I^{131} is probably the most widely used medical radioisotope. What is its principal source? What approximate levels (curies) are used in diagnostic and therapeutic thyroid applications?

18. What is meant by isotope scanning?

19. Briefly describe the use of radioactive isotopes in (a) thickness gauging, (b) radiography.

15 Chemical bonds and the electronic structure of atoms

THE story of chemical bonding is a story of the electron clouds surrounding atomic nuclei. Their importance in the study of chemistry is out of all proportion to the fraction of the atomic mass they occupy. Already by 1904 J. J. Thomson had predicted that the gradual changes observed in the properties of the elements in a horizontal row in the periodic table were related to their electronic structures. In his plum pudding model, an atom was pictured as a sphere of positive electricity with electrons imbedded in it in *concentric shells*. Then in 1911 Rutherford proposed the theory of the atom described in Chapter 13—the atom with a tiny but massive core (the nucleus) and negative charges at some distance.

Meanwhile, in 1900 Max Planck had announced his quantum theory, which was to be employed later by Bohr in his theory of the hydrogen atom. Planck's revolutionary idea was that radiation was emitted in chunks (called quanta) rather than in a continuous stream as conceived by the earlier classical theory. Classical theory considered radiation to be like a continuous stream of water issuing from a fire hose; Planck regarded it as more closely resembling bullets shot from a machine gun. Planck's new and revolutionary theory had been inspired by a predicament of classical theory. Nobody had ever devised a *single* mathematical relation applicable in all spectral regions on the assumption that radiation emission was continuous. Planck's genius lay in his break with the classical idea of continuous emission and his substitution of the concept of discrete bundles of energy.

THE BOHR THEORY OF THE HYDROGEN ATOM

Niels Bohr, a Nobel-prize-winning Danish physicist, studied with both J. J. Thomson and Lord Rutherford in England. His new theory of the hydrogen atom made use of the Rutherford atomic model according to which electrons revolved at some distance about a positive nucleus containing most of the atom's mass. He also accepted Planck's quantum theory, which assumed that atoms emitted energy only in chunks (quanta) whose size depended on the extent of the electron transition. The Planck equation, $E = h\nu$, related E, the energy of a quantum, ν, the frequency of the emitted radiation, and h, Planck's constant. Since E was proportional to ν, high-frequency waves should have large quanta.

ILLUSTRATIVE PROBLEM

Calculate the energy (ergs) in a photon (energy chunk) of ultraviolet light whose wavelength is 2000 Å.

Solution

$$E = h \cdot \nu$$

$$E\,(\text{ergs}) = 6.62 \times 10^{-27}\,(\text{erg sec}) \times \nu \frac{1}{\text{sec}}$$

$$\text{ergs} = \text{erg sec} \times \frac{1}{\text{sec}}$$

The frequency is calculated by $v = \nu\lambda$. Ultraviolet light is an electromagnetic wave which travels with the velocity of light, 3×10^{10} cm/sec. Hence

$$\nu = \frac{v}{\lambda} = \frac{3 \times 10^{10}\ \text{cm/sec}}{2000 \times 10^{-8}\ \text{cm}}$$

$$\nu = 1.5 \times 10^{5}\ \text{sec}^{-1}$$

Substituting in $E = h\nu$

$$E = 6.62 \times 10^{-27} \times 1.5 \times 10^{5}$$
$$E = 9.93 \times 10^{-22}\ \text{erg}$$

By using Coulomb's law expressing the force between an electron and a proton (both of charge e and r cm apart)

$$F = e^2/r^2 = mv^2/r$$

equating it to the centripetal force of the electron (mass $= m$ and velocity $= v$) and introducing certain quantum restrictions on electron transitions, Bohr was able to calculate the energies of electrons in their orbits.* In a given orbit the energy depended on the quantum number of the orbit. The wavelength of a spectral line emitted by the atom was related to the energy difference between two such orbits. It was assumed that an electron remaining in a particular energy shell neither absorbed nor emitted radiation.

* Bohr used the term "orbit." It is no longer considered appropriate.

Energy had to be supplied to the hydrogen atom to move its electron farther away from the attracting influence of the positive nucleus (proton). The atom whose electron was at greater-than-normal distance from the nucleus was said to be in an excited state. Since work had to be done to raise it to the excited state, it had greater potential energy. Just as a raised weight loses potential energy as it falls, so electrons in excited states fall to lower levels and emit the excess energy in the form of a spectral line. The greater the difference between energy levels (upper and lower), the larger the quanta and the higher the wave frequency. Spectral lines in the UV have larger quanta than those in the visible region, and these in turn larger than in the infrared.

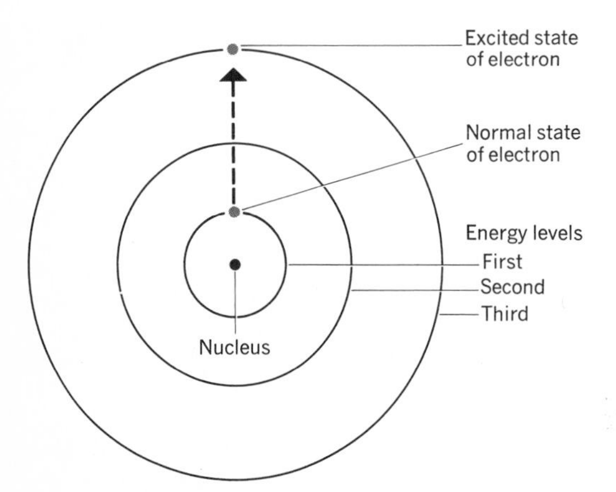

Fig. 15-1. *Energy supplied to atoms in the process of excitation drives electrons into a shell farther removed from the nucleus. Work is done as these negatively charged particles move against the attractive force of the positive nuclei.*

Fig. 15-1 shows a series of electron orbits and a single transition. An enormous number of atoms, each with its own electron undergoing a particular transition, are of course needed to account for all the observed spectral lines of hydrogen.

The energy level nearest the nucleus is called the K shell.* Electrons dropping back to it from excited states give spectral lines in the so-called Lyman series (in the UV). Those dropping back to the second shell, or L shell, emit lines in the so-called Balmer series (in the visible).

The Bohr theory was remarkably successful in predicting the properties of hydrogen and hydrogenlike atoms (e.g., He^+).† Its success raised hopes that spectra of other elements could be as adequately interpreted, but these hopes were unfulfilled. However, the notion of electronic (and subse-

* In modern quantum mechanical designation the K shell includes those electrons whose principal quantum number, n, = 1 (see Table 15-1 below).

† A helium atom which has lost one electron (He^+) has, like hydrogen, only one orbital electron. The helium nucleus contains 2 protons and 2 neutrons and hence differs from the hydrogen nucleus.

quently, nuclear) energy levels has been a most useful one in atomic physics, even though today what Bohr regarded as an electron orbit is conceived as being a region in space where there is a high probability of finding an electron.

WAVE PROPERTIES OF ELECTRONS

The Bohr theory was strikingly successful in its day, but as time passed more and more inadequacies appeared. Gradually it became apparent that electrons (and nuclear particles) could not be treated by the methods of classical mechanics. Heisenberg showed that an electron's location and its velocity (more correctly its momentum) could not both be precisely known (the Heisenberg uncertainty principle). If one were measured more exactly, the other became less certain. There was a sort of inverse variation between the errors in the two measurements. The calculation of a very precise Bohr radius of an electron orbit made no sense, because its motion would then be very poorly defined. Once again a new method of attack was needed.

During the 1920's De Broglie demonstrated that electrons (heretofore regarded as *particles*), like light waves, could be diffracted (a characteristic *wave* property). He proposed an equation for matter waves and suggested that the concept of matter waves could lead to a much more satisfactory interpretation of the energy levels of the hydrogen atom. Shortly thereafter Heisenberg and Schrödinger, working independently, proposed essentially equivalent theories explaining the hydrogen atom in terms of the wave concept.

The Bohr theory required only the orbit numbers to define electron energies. The Schrödinger equation required four numbers to define the properties of matter waves. These were the *quantum numbers.*

QUANTUM NUMBERS: ELECTRONIC CONFIGURATIONS

According to modern quantum mechanical methods it is possible to describe electron matter waves by their quantum numbers. Since the method used involves the solution of differential equations, this discussion will be limited only to the description of the quantum numbers and how electronic configurations of atoms are derived from them.

The four quantum numbers needed are

1. The principal quantum number, n.
2. The orbital (angular momentum) quantum number, l.
3. The magnetic quantum number, m.
4. The spin quantum number, s.

The principal quantum number n is a measure of the average distance of a particular group of electrons from the nucleus; the larger n is, the greater is this average distance. As a general principle, too, the larger n is, the greater is the electron energy. Table 15-1 shows the correspondence between n and the shell designation.

Table 15–1 Principal Quantum Number and Shell Designation

n =	1	2	3	4	5	6
Shell	K	L	M	N	O	P

The maximum number of electrons in a given energy shell may be expressed in terms of the principal quantum number, n; it is $2n^2$. The maximum numbers of electrons so calculated are shown in Table 15-2.

Table 15–2 Maximum Numbers of Electrons in an Energy Shell

Shell	K	L	M	N	O
Maximum number of electrons	2	8	18	32	50

From Table 15-2 it will be evident why (Chapter 8) there are 8 elements between Li and Ne, 8 between Na and Ar, 18 between K and Kr, 18 between Rb and Xe, and 32 between Cs and Rn.

Table 15–3 Electronic Subshells

l =	0	1	2	3
Electron designation	S	P	D	F

The orbital quantum number l has values from 0 to $n - 1$. Thus for $n = 1$, only the value $l = 0$ is possible. For $n = 2$, l may be 0 or 1; for $n = 3$, l may be 0, 1, or 2, etc. Values of l show the subdivisions (sublevels) of each main quantum level. Table 15-3 indicates how electrons are described according to the sublevels occupied. An electron with $l = 0$ is called an "*s* electron," with $l = 1$ a "*p* electron," etc.

The magnetic quantum number m may have values from $-l \ldots 0 \ldots +l$. For $n = 2$, $l = 0$ or 1. For $l = 0$, m may be 0 only. But for $l = 1$, m may be either -1, 0, or $+1$, and hence there may be three different orientations of the orbital in space, when there is a magnetic field.

The spin quantum number s may have values of only $+\frac{1}{2}$ and $-\frac{1}{2}$, the axes being the same, but the direction of spin opposite.

According to the Pauli exclusion principle no electron can have all four quantum numbers the same as any other electron of the same atom.

Two electrons with the same m values but having opposed spins (spin up and spin down) are said to occupy the same orbital. The shape of *s* orbitals is spherical, as shown in Fig. 15-2 and *p* orbitals are dumbbell-shaped. The

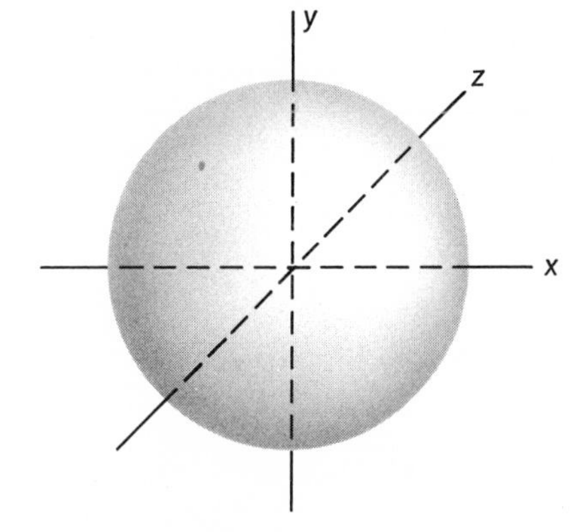

Fig. 15-2. *An electron in an atomic* s *orbital may be represented by a spherical electron cloud.*

planar representation of a *p* orbital looks more like a figure eight (Fig. 15-3). There are three *p* orbitals, each of whose axes is perpendicular to the others. Hence *p* orbitals have direction in space.

In the method of representing electronic configurations in terms of quantum numbers, the first atom considered is the hydrogen atom. The element hydrogen has an atomic number (Z) of 1 and hence in the neutral atom there is one electron. It is in the shell nearest the nucleus (the K shell), so $n = 1$. The maximum value of l is n $- 1$, so $l = 0$. Also, m must be zero (from $-l$ to $+l$) and s, the spin quantum number, + or $-$ ($+\frac{1}{2}$ or $-\frac{1}{2}$). The next element, He, has $Z = 2$, so there are two electrons. The electronic configuration is written $1s^2$, the superscript 2 indicating that there are two 1*s* electrons. The first electron of helium is assumed to be identical

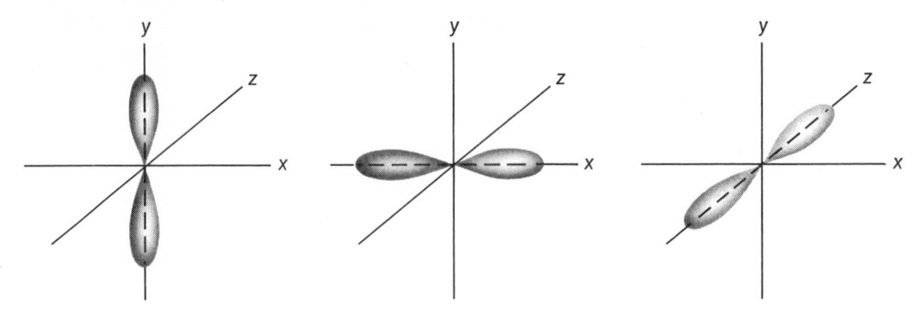

Fig. 15-3. *Electrons in atomic* p *orbitals have a twin-lobed shape directed along the three orthogonal axes in space. Overlap of two or three* p *orbitals in a given atom with orbitals of another atom or other atoms would be expected to give bond angles of 90°.*

with that of hydrogen. The exclusion principle requires that a second electron shall spin in the opposite direction from the first. Table 15-4 summarizes the discussion.

Table 15–4 Electronic Configurations of Hydrogen and Helium

ATOMIC NUMBER	SYMBOL	n	l	m	s	ELECTRONIC CONFIGURATION
1	H	1	0	0	+	1s
2	He	1	0	0	–	$1s^2$

The Li atom has 3 electrons, whose quantum states may be symbolically represented by $1s^22s$. Table 15-5 shows the state of each added electron in the first period of 8 elements. As with all subsequent atoms, each element must be presumed to contain an underneath $1s^2$ shell.

As in the first quantum level there are two s electrons. There are never more (why?). Starting with the element boron, B, p orbitals are beginning to fill. According to one of the Hund rules, one electron first enters each of the three p orbitals singly before two enter a given orbital. Since m may have a value of -1 or 0 or $+1$ and since the electrons in a given orbital may have either a $+$ or a $-$ spin, 6 p electrons are possible.

The third period of the periodic table closely parallels the second. The student is urged to prepare a table similar to Table 15-5 showing the appropriate electronic configurations. It is important to remember that underlying each new electron added is a $1s^22s^22p^6$ structure.

Table 15–5 Electronic Configurations of Period II Elements

ATOMIC NUMBER	SYMBOL	n	l	m	s	ELECTRON CONFIGURATION		
3	Li	2	0	0	+	$1s^2$	$2s$	
4	Be	2	0	0	−	$1s^2$	$2s^2$	
5	B	2	1	−1	+	$1s^2$	$2s^2$	$2p$
6	C	2	1	−1	−	$1s^2$	$2s^2$	$2p^2$
7	N	2	1	0	+	$1s^2$	$2s^2$	$2p^3$
8	O	2	1	0	−	$1s^2$	$2s^2$	$2p^4$
9	F	2	1	+1	+	$1s^2$	$2s^2$	$2p^5$
10	Ne	2	1	+1	−	$1s^2$	$2s^2$	$2p^6$

With potassium in the fourth quantum level, it may come as a surprise to find that the next added electron is found in a $4s$ rather than a $3d$ level. Electrons always occur in the lowest energy state available; hence $4s$ electrons must have slightly lower energy than $3d$. So far in the periodic table, electrons have been added sequentially much as would be expected from the directions given.* However, there are apparently some inversions of order. Various schemes have been proposed for remembering these. Table 15-6 shows one which has proved useful.

Table 15–6 The Order of Filling of Electronic Subshells

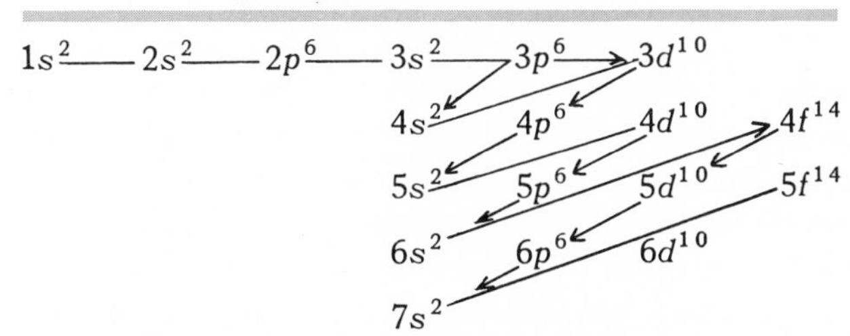

As an illustration of the use of this table, Francium (Fr), atomic number 87, would be expected to have the electronic configuration:

$$1s^2 2s^2 2p^6 3s^2 3p^6 4s^2 3d^{10} 4p^6 5s^2 4d^{10} 5p^6 6s^2 4f^{14} 5d^{10} 6p^6 7s$$

* For the sake of a vivid explanation it is usual to speak of "adding" electrons one by one to the atomic structure and forming elements of higher and higher weight. But there should be no suggestion that the elements have in cosmic history been formed by successive additions of electrons. The story of the formation of the elements is still to be told and will doubtless prove to be very different.

The order of filling of the shells should be checked by means of Table 15-6.

The so-called $n + l$ rule may also be applied to predict the order of filling of sublevels.* According to this rule, for two electrons with the same $n + l$ values—as for example, $n = 4, l = 0$ and $n = 3, l = 1$, the electron with the lower n value will have the lower energy. Hence a $3p$ electron $(n + l = 4)$ will have lower energy than a $4s$. But a $3d$ electron $(n + l = 3 + 2 = 5)$ will have a higher energy than $4s$ $(n + l = 4 + 0 = 4)$.

The filling of the $3d$ subshell with electrons *already in* the $4s$ level gives rise to "transition elements" located in the middle of the periodic table (see page 146). Table 15-7 shows that the d orbital does not fill up in a strict

Table 15–7 Subshell Structures of Certain Transition Elements

ATOMIC NUMBER	SYMBOL	$n = 1$	$n = 2$		$n = 3$			$n = 4$
		s	s	p	s	p	d	s
21	Sc	2	2	6	2	6	1	2
22	Ti	2	2	6	2	6	2	2
23	V	2	2	6	2	6	3	2
24	Cr	2	2	6	2	6	5	1
25	Mn	2	2	6	2	6	5	2
26	Fe	2	2	6	2	6	6	2
27	Co	2	2	6	2	6	7	2
28	Ni	2	2	6	2	6	8	2
29	Cu	2	2	6	2	6	10	1
30	Zn	2	2	6	2	6	10	2

sequence. For these structures it is wise to consult a table of configurations† if accurate knowledge of the order of filling is of great importance. Structures for the representative elements may however be written as discussed above.

IONIZATION ENERGIES

Energy must be supplied to atoms to raise electrons to higher (excited) energy levels. If the energy input is sufficiently large, an electron may be completely separated from nuclear attraction, that is, the atom is ionized.

* There are a few inversions such as copper, $Z = 29$, but by and large this works out correctly.

† See for example the *Handbook of Physics and Chemistry,* Chemical Rubber Publishing Co.

The energy required to remove this first electron is called the (first) ionization energy. The unit is usually the electron volt (ev) per atom, or the kcal per gram atom. "Ionization potential" is an equally appropriate term as long as the energy is expressed in ev. Values of the first ionization energy vary periodically with increasing atomic number as stated in the periodic law and as shown graphically in Fig. 15-4.

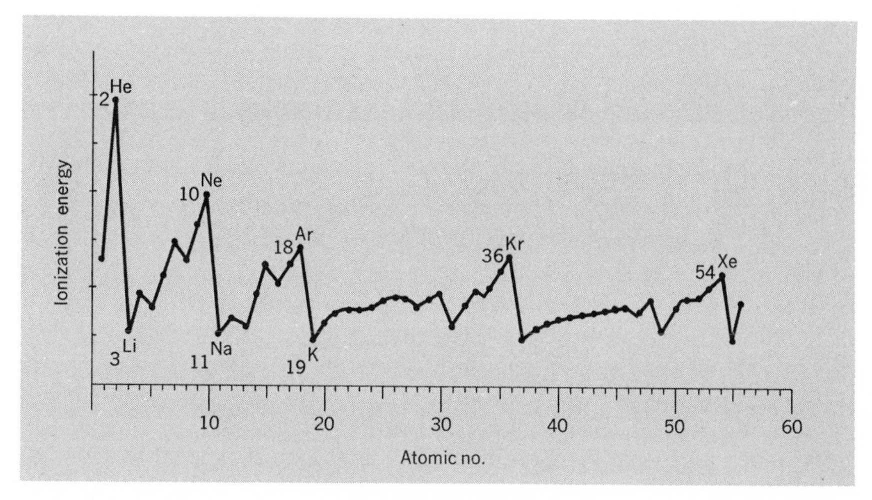

Fig. 15-4. *Variation of ionization energy of first electron with atomic number.*

Notice that the noble gas elements are located at peaks of the graph. Their large ionization energies suggest the stability and unreactivity of these atoms. Helium's first ionization energy has a value of 24.6 ev, the largest for any atom. Removal of the second electron would be expected to be more difficult, since the nuclear charge would be surrounded by an electron cloud with one less electron. The second ionization energy is always larger than the first.

The first ionization energy for Li is 5.4 ev. This is one of the lowest values on the graph. Note that all the alkali metal atoms have very low ionization energies; the electron is removed quite easily. The second ionization energy of Li is 75 ev, a very considerable increase. Li has an oxidation number of $+1$ in its compounds. It loses 1 electron before exposing an underneath shell of 8 electrons (a noble gas structure).

The element Be has ionization energies of 9.3 ev (first), 18.2 ev (second) and 153.9 (third). In this case the huge increase in ionization energy comes after the second electron has been removed. Be has an oxidation number of $+2$ in its compounds.

Elements in groups I and II of the periodic table readily lose 1 or 2 electrons, respectively, before an extremely large ionization-energy value is encountered. In each case an underneath shell of 8 electrons is bared. An outer shell of 8 electrons is called an octet. There are additional reasons for believing an octet a particularly stable configuration. More will be said later about trends of ionization energy values in the periodic table. In the meantime it is suggested that the graph of ionization energy values be examined to discover if other features of atomic structure already discussed can be observed.

ELECTRON AFFINITY AND ELECTRONEGATIVITY

The alkali and alkaline earth atoms have comparatively low ionization energies and tend to lose electrons when they combine, forming positive ions. Certain nonmetallic elements such as the halogens are powerful electron attracters, removing electrons from elements (such as the two families just mentioned) which readily lose them. As a result of electron gain by a nonmetallic atom, a noble gas structure of 8 electrons comes about. The energy measure of this electron-capture process is called the *electron affinity,* defined as the energy change accompanying the uniting of an electron (with a gaseous atom) to form a negative ion.

Of much more interest and importance is a quantity known as the electronegativity. It may be regarded as an element's attraction for a shared electron pair in a compound. In principle its value depends on the difference between the ionization energy and electron affinity. On the Pauling electronegativity scale, fluorine is assigned the value 4, the largest for any element. Elements adjacent to fluorine in the periodic table have comparatively large electronegativity values. The alkali metal atoms have comparatively low values (e.g., 0.7 for Cs).

On the basis of electronegativity *differences,* particular bonds are said to have "more covalent character" or "more ionic character." If the electronegativity difference is large as in the compound KF ($4.0 - 0.8 = 3.2$), a predominantly ionic bond is expected. In the case of smaller electronegativity differences such as in CCl_4 ($3.0 - 2.5 = 0.5$), covalence is expected to predominate. Both types of bonding will be discussed in the sections which follow.

ELECTRONIC STRUCTURE AND THE PERIODIC TABLE

Two elements, hydrogen and helium, stand alone in the first period of the table (see Table 8-1). They have *s* electrons only. The very large value of helium's ionization energy, and the sharp decrease in value following it, suggest the completion of the first energy shell.

The 2*s* orbital is being filled as a third and fourth electron are added, giving first Li and then Be. There is a slight decrease in ionization energy when an electron enters the first *p* orbital (the element boron). There is likewise a slight drop when each of the three *p* orbitals has a single electron (after nitrogen). With the element Ne a maximum value of the ionization energy is reached for the second period; the *p* orbitals are filled. The L shell now has its full complement of electrons ($2n^2 = 8$). Once again a sharp drop in the value of the ionization energy between the elements Ne and Na suggests the completion of a noble gas structure in the valence shell.

The radii of the atoms in this second period decrease from left to right. Since each successive atom adds one more electron to the cloud and one more proton to the nucleus, the added nuclear charge is apparently more influential in determining the size of the atom in a given period than the added electron. Examination of a table of electronegativities (Table 15-8) shows that they increase from left to right in a period.

Table 15–8 Electronegativity Values for Selected Elements

H 2.1						
Li 1.0	Be 1.5	B 2.0	C 2.5	N 3.0	O 3.5	F 4.0
Na 0.9	Mg 1.2	Al 1.5	Si 1.8	P 2.1	S 2.5	Cl 3.0
K 0.8	Ca 1.0	Ga 1.6	Ge 1.8	As 2.0	Se 2.4	Br 2.8
Rb 0.8	Sr 1.0	In 1.7	Sn 1.8	Sb 1.9	Te 2.1	I 2.5
Cs 0.7	Ba 0.9					

In the third period the same general pattern prevails. The differentiating electron for Na is added in a 3*s* orbital as is that for Mg. With the elements Al, Si, P, S, Cl, and Ar, electrons are being added one at a time to 3*p* orbitals. The sharp drop in the ionization energy value following the element

Ar suggests completion of another noble gas, 8 electron, structure. However, the third quantum level is not complete with 8 electrons. The maximum should be 18 (2×3^2). As noted earlier there is some overlap of third and fourth quantum level energies. Since 4*s* electrons have slightly lower energies than 3*d*, the differentiating electrons of K and Ca go into 4*s* levels. Then starting with the element Sc, the 3*d* levels begin to fill. There is very little change in the values of the ionization energies of these elements. The elements from Sc through Zn, in which an underneath shell (the 3*d*) is filling while there are electrons in a higher quantum level (the 4*s*), are called "transition" elements. With the element Ga, the first electron enters a 4*p* orbital and then the familiar pattern of a steeper increase in ionization energy follows, reaching a maximum value for the period in the element Kr. The fourth period consists, then, of 18 elements. With the first two, K and Ca, the electrons are added in the 4*s* level. They are followed by a group of 10 transition elements in which 3*d* orbitals are filling, and finally a group of 6 elements in which 4*p* levels are filling.

The fifth period, consisting of 18 elements, follows the same general pattern as the fourth. The differentiating electrons of the first two elements Rb and Sr are 5*s*. The 10 transition elements which follow, Y through Cd, are adding electrons in 4*d* orbitals. The last six elements, In (indium) through Xe are adding electrons in 5*p* orbitals.

Following Cs and Ba in the sixth period (6*s* differentiating electrons), the 4*f* levels start to fill, with electrons already in 5*s*, 5*p*, and 6*s* orbitals. The group of 14 elements in which 4*f* levels are filling, is known as the lanthanum or lanthanide series. These are also known as "rare earth" elements and are said to constitute an "inner transition" series. Starting with Lu, electrons enter 5*d* levels, with underneath shells now complete but electrons in 6*s* levels. These must therefore be a third group of transition elements. The period is completed with the filling of the 6*p* orbitals, the final element being the noble (and radioactive) gas, Rn.

The atomic radii of the elements in a *family* increase with greater atomic number. The K atom has a larger atomic radius than Li, and Br is larger than F. Positive ions have smaller radii than the corresponding atoms. The Mg^{++} ion is smaller than the Mg atom. After removal of 2 electrons the electron cloud is pulled in tighter. Negative ions are larger than their atoms. The addition of an electron to Cl gives Cl^-, with a slightly larger ionic radius.

Electronegativity values would be expected to parallel, approximately, ionization energy values. Both increase from left to right and from bottom to top in the periodic table. Heavier atoms have larger atomic radii, larger nuclear charges, and more shells of screening electrons. Since larger atoms have lower ionization energies and lower electronegativities, the effects of larger radius and larger screening apparently outweigh that of increased nuclear charge.

THE NONPOLAR COVALENT BOND: THE HYDROGEN MOLECULE

A hydrogen atom consists of a positive nucleus (proton) surrounded by a spherical *s* electron cloud of negative charge. Think of this atom as looking like a balloon at the center of which is located a tiny, dense nucleus whose radius is about 1/10,000 as large. If two such atoms approach, the nuclei repel and the charge clouds repel, but these two attract *each other*. At still closer approach the electron charge clouds overlap and the density of negative charge increases along the line between the nuclei; a covalent bond is formed. It is nonpolar because the charge cloud is symmetrically arranged with respect to the nuclei; each nucleus attracts it equally (Fig. 15-5).

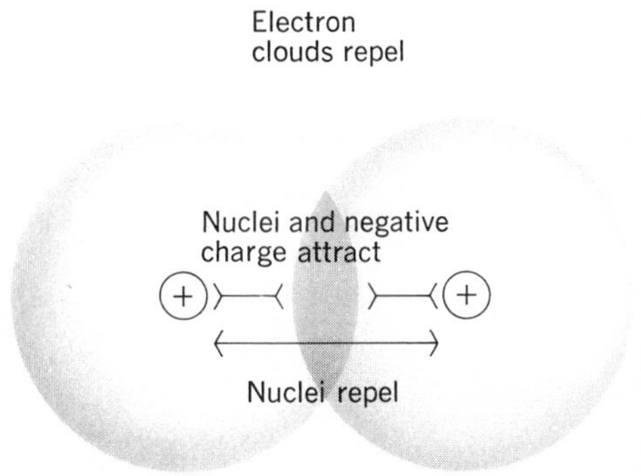

Fig. 15-5. *A hydrogen atom consists of a proton nucleus and a surrounding spherical electron cloud. An equilibrium between attractive and repulsive forces is attained when the electron clouds of two such atoms overlap and form a valence bond. Since the charge arrangement is symmetrical with respect to the nuclei, the electron concentration between the atoms is shared equally by both. The bond is nonpolar.*

The hydrogen molecule is represented by H:H, the pair of dots between the symbols representing an electron pair which is shared equally by the two atoms. The molecule may also be written H—H, the line representing the electron pair.

A POLAR COVALENT MOLECULE: HYDROGEN CHLORIDE

The molecule of hydrogen chloride (gas)* consists of an atom of hydrogen and one of chlorine. The formula H—Cl indicates there is an electron

* The water solution of this gas, hydrochloric acid, is also sometimes written HCl.

pair (covalent bond) between the atoms, but it is a different bond than that between two hydrogen atoms. The molecule may also be represented as H :Cl, indicating that the electron pair is more strongly attracted by the highly electronegative chlorine atom than by hydrogen. There is then an unsymmetrical distribution of charge, the chlorine end of the molecule

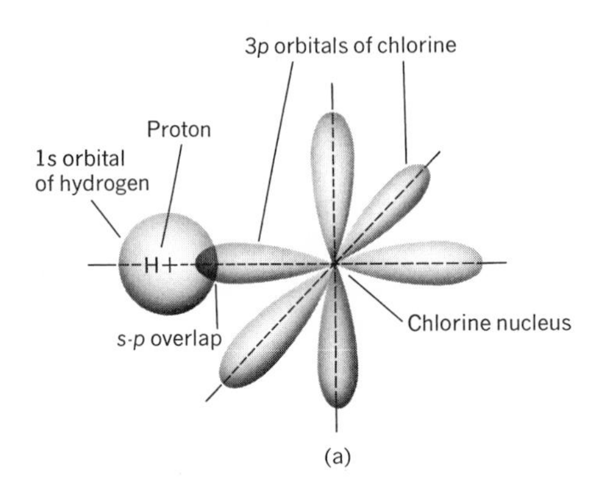

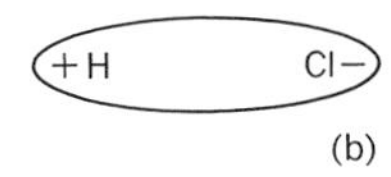

Fig. 15-6. *The 1*s *orbital of the hydrogen atom overlaps one of the 3*p *orbitals of chlorine forming a polar covalent bond. Owing to the unsymmetrical arrangement of charge, the molecule is a dipole with positive charge at the hydrogen end and negative charge at the chlorine end.*

being negative and the hydrogen end positive. The molecule is a dipole (two regions of charge) and the bond is called polar covalent. As a whole the molecule is electrically neutral, but the electron pair spends more time in the vicinity of the chlorine atom than near hydrogen; hence there is charge asymmetry (Fig. 15-6).

THE IONIC BOND: A CRYSTAL OF TABLE SALT

The active metallic elements at the extreme left of the periodic table, the alkali and alkaline earth elements, have low electronegativities; those in the second column from the right, the halogens, have the highest values. The electronegativity differences between the two are comparatively large; the bond type is ionic. In ionic compounds the element of high electronegativity has captured an electron, with the formation of a negative ion. The element of low electronegativity has lost an electron and formed a positive ion.

The crystal of sodium chloride is composed of positive sodium ions (Na^+) and negative chloride ions (Cl^-). These are arranged alternately in a crystal lattice. Any Na^+ ion is surrounded by six Cl^- ions and any

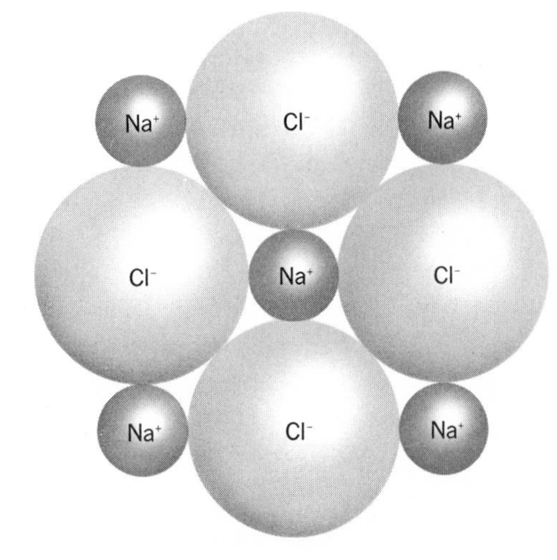

Fig. 15-7. *Sodium chloride, an ionic crystal. The crystal of table salt is composed of alternate sodium and chloride ions. Any given ion is surrounded by six ions of opposite charge.*

Cl^- ion by six Na^+ ions (Fig. 15-7). The forces holding the crystal together are coulombic (the attraction of opposite charges). Thus there is in solid NaCl no discrete molecule.

VALENCE, OXIDATION NUMBER, AND ELECTRONIC STRUCTURE

In Chapter 7 valence and oxidation numbers were presented as numbers, without discussing their relation to electronic structure. This relationship should now be understandable.

Earlier in the chapter, hydrogen chloride gas was described as a polar covalent molecule. The valence bond between the hydrogen and chlorine atoms may be pictured (Fig. 15-6) as resulting from the overlap of the spherical *s* orbital of a hydrogen atom, with the unfilled *p* orbital * of the chlorine atom. Hydrogen and chlorine atoms form one bond and hence have a covalence of 1. Chlorine is the atom with the higher electronega-

* The valence shell of chlorine has seven electrons. There are two *s* electrons, a filled subshell, and five *p* electrons. According to the Hund rule of maximum multiplicity mentioned earlier, one electron must go into each orbital before two can enter any orbital. With five electrons in three *p* orbitals, one must contain a single electron, and hence that orbital will be available for bonding.

tivity; the electron cloud is displaced toward it. Chlorine has an oxidation number of -1 * and hydrogen one of $+1$.

One valence bond structure of water (Fig. 15-8) shows the spherical *s* orbitals of each of two hydrogen atoms overlapping each of the two unfilled *p* orbitals of oxygen. Thus oxygen forms two bonds and has a covalence of 2. Each hydrogen atom forms a single bond and has a cova-

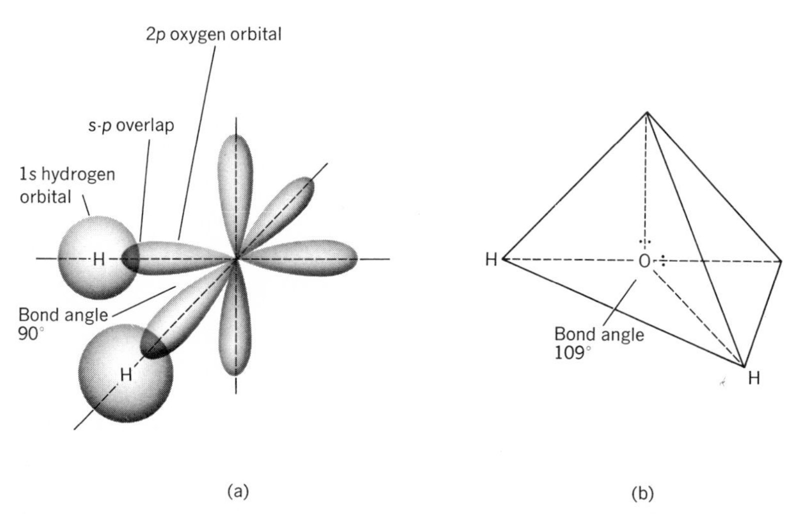

Fig. 15-8. *A water molecule. The 1*s *orbitals of each of two hydrogen atoms overlap two of the 2*p *orbitals of an oxygen atom, forming two polar covalent bonds. The concentration of negative charge is nearer the oxygen atom. The molecule is a dipole with negative charge at the oxygen end. One might expect the bond angle to be 90°. Actually it is closer to 105°.—Treatment of the structure of water as an* sp^3 *hybrid (see methane, p. 282) places the oxygen atom at the center of a tetrahedron, hydrogen atoms at two of the vertices, and electron pairs at the remaining two vertices. The H-O-H bond angle is 109° rather than the observed 105°.*

lence of 1. Because of the high electronegativity of oxygen, the molecule will be a dipole with a negative charge at the oxygen end. Oxygen has an oxidation number of -2 and hydrogen one of $+1$.

Probably a better representation of the structure of the water molecule is to employ hybrid sp^3 orbitals (Fig. 15-8) with the oxygen atom at the center of a tetrahedron, hydrogen atoms at two of the vertices and an electron pair at each of the others. The H—O—H hybrid bond angle of

* The shared electron pair is arbitrarily allotted to the more electronegative element for the purpose of assigning oxidation numbers to covalently bonded atoms.

109° is much closer to the experimental value than that predicted using pure orbitals. Neither representation is exact.

The nitrogen atom of ammonia (NH_3) has five valence electrons, two *s* electrons in a filled subshell and three *p* electrons. According to the Hund rule (see above) each of the three electrons must occupy a separate orbital. Hence three *p* orbitals of nitrogen will be available for bonding (Fig. 15-9). Each of three *s* orbitals of hydrogen overlaps one

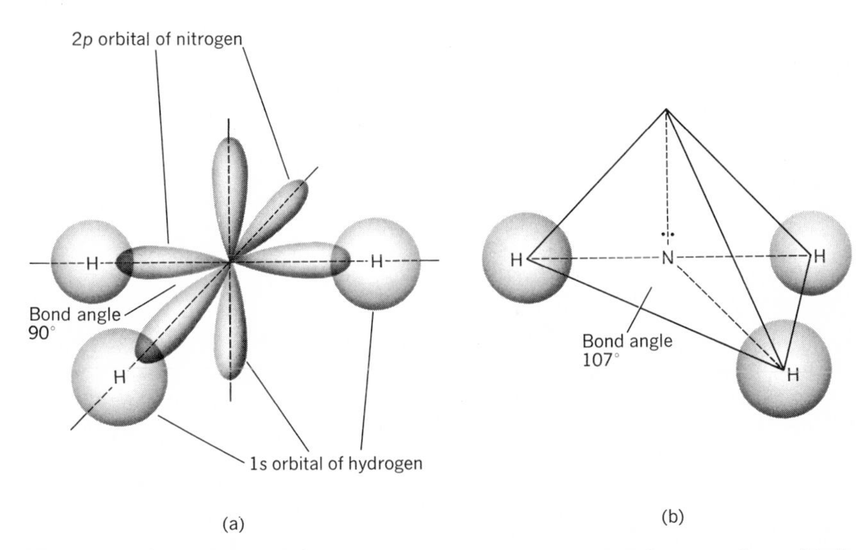

Fig. 15-9. *A nitrogen atom has five valence electrons, three in 2*p *orbitals. These* p *orbitals can form covalent bonds by overlap with the 1*s *orbitals of each of three hydrogen atoms. An alternative structure would be that using* sp^3 *hybrid orbitals with nitrogen at the center of a tetrahedron, three hydrogen atoms and an electron pair at the four vertices. The predicted H-N-H bond angle is 109° for a tetrahedron; the experimental value is 107°.*

of these three *p* orbitals, forming a bond. Hence nitrogen will have a covalence of 3, hydrogen a covalence of 1. Nitrogen will have an oxidation number of -3 and hydrogen one of $+1$.

Since the measured H—N—H bond angle in NH_3 is 107°, the use of tetrahedral sp^3 hybrid orbitals of nitrogen gives a much closer approximation (109°) to the observed value than do pure *s* and *p* orbitals (90°).

The element carbon, atomic number 6, has four electrons in its valence shell. Two electrons are in a filled *s* subshell and two in *p* orbitals. Since only the *p* orbitals would be singly occupied, it would suggest that carbon should form only two bonds. However, in the great majority of the very numerous organic compounds (see Chapter 16) carbon has a valence of 4, not 2. In the compound methane, CH_4, for example, there are four C—H bonds, and studies show they are all the same length. How account for four bonds with only two vacant *p* orbitals available for bonding?

A way of getting around the difficulty in this molecule is to "promote" one of the *s* electrons to a *p* level. This leaves only one electron in the *s* orbital, and thus permits it to form a bond. Furthermore there is now a single electron in each of three *p* orbitals so each of these may form a bond. Since with this arrangement one electron (the *s* electron) would appear to be different from the other three (*p* electrons), the orbitals are considered to be hybridized, all four having similar characteristics, and with intermediate properties such as bond length, bond energy, etc.

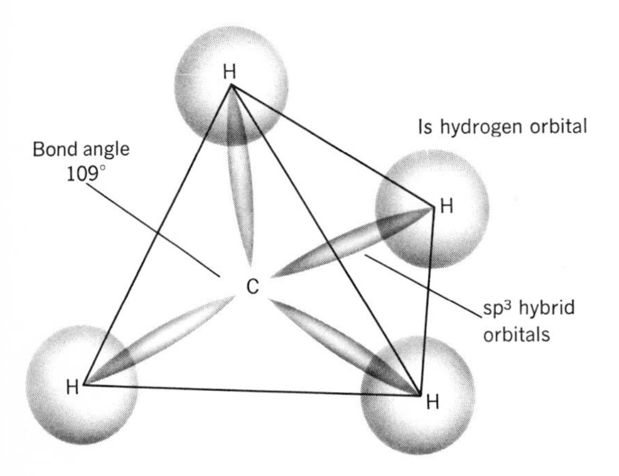

Fig. 15-10. *The methane molecule is tetrahedral. Each of four equivalent* sp³ *orbitals of the central carbon atom overlaps a 1*s *orbital of a hydrogen atom forming four covalent bonds. The H-C-H bond angles are 109°. The charge distribution in the methane molecule is symmetrical; it does not have a dipole moment.*

The methane molecule is represented as a tetrahedron (Fig. 15-10) with a carbon atom at the center and a hydrogen atom at each vertex. The *s* orbital of each hydrogen atom overlaps one of the four sp^3 hybrid orbitals of carbon. The H—C—H bond angles are each about 109 degrees. The molecule does not have a dipole moment, and hence there must be charge symmetry within the molecule. In methane, carbon has a covalence of 4, hydrogen a covalence of 1.

The Lewis structure for each of the molecules discussed in this section is written below. Notice that in each formula there is an octet composed of electron pairs and that the covalence is the number of bonding pairs.

Hydrogen Chloride	Water	Ammonia	Methane
H : C̤̈l :	: Ö : H ¨ H	H : N̈ : H ¨ H	H ¨ H : C : H ¨ H

In ionic compounds, most commonly formed by the representative elements of groups I, II, and VII, the metals have positive oxidation numbers and the nonmetals negative. Thus in the positive ion Na^+, one electron has been removed from a neutral atom, and in Ca^{++}, two. In the non-

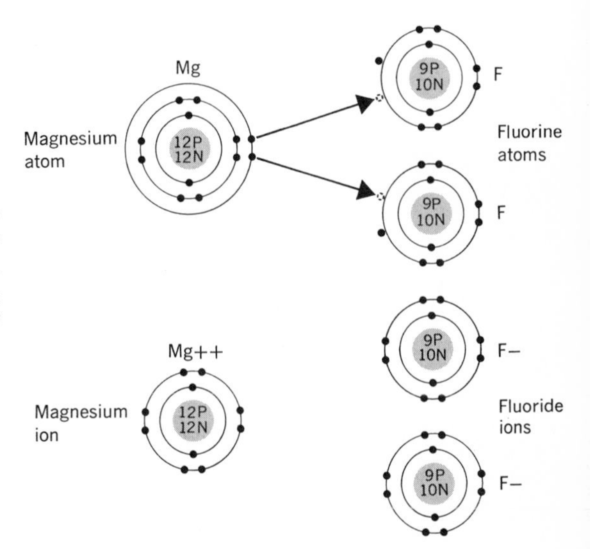

Fig. 15-11. *A magnesium atom has two electrons in excess of an inert gas structure; fluorine atoms need an additional electron. Magnesium has an oxidation number of +2, fluorine of −1.*

metal ions the charge of the negative ion shows the number of electrons gained. In the ion Cl^-, one electron has been gained by a neutral Cl atom. Fig. 15-11 represents the electron transfer in the production of the ionic compound MgF_2.

SOLUTIONS OF ELECTROLYTES AND NONELECTROLYTES

The properties of water solutions of ionic solids differ markedly from those of covalently bonded solids. The nonelectrolyte urea, $H_2N{-}\overset{\overset{\displaystyle O}{\|}}{C}{-}NH_2$, for example, is a water-soluble solid whose molecules contain 8 atoms each. The sodium chloride crystal is a lattice composed of alternate Na^+ and Cl^- ions. When NaCl is placed in water, the polar molecules of the water attract ions of opposite charge and pull apart the crystal. The ion-containing solution is an excellent conductor of an electric current. Because of its excellent conduction of electric current, NaCl is called an electrolyte. The urea solution conducts a current no better than pure water; hence the solute urea is a nonelectrolyte.

All dissolved solids reduce the freezing point of water below its usual value (0°C) and raise its boiling point above the normal value (100°C at 1 atm). Boiling-point elevation and freezing-point depression are called colligative properties, meaning they are dependent on the number of dissolved particles per gram of solvent, but not on the type of particle. Non-ionic solids such as urea give "normal" boiling-point elevations and freezing-point depressions. Ionic solids, however, give values which are close to integral multiples of the "normal" values—2,3,4, etc.—times as great, depending on the number of ions in the formula weight. NaCl with 2 ions (Na^+ and Cl^-) gives double the normal value; $CaCl_2$ with 3 ions (Ca^{++} and $2Cl^-$), three times the normal value, etc.

In principle a 1-molal solution (containing 1 mole of dissolved solute in 1000 g of solvent) depresses the freezing point of water 1.86C° and elevates the boiling point of water 0.52C°. As a practical matter this is a "dilute solution" principle, and hence illustrations given will be those for 0.1-molal solutions (Table 15-9). Since these contain only 0.1 as many particles per 1000 g of solvent, they will give only 0.1 of the "normal" boiling point elevation (0.052C°) and 0.1 of the "normal" freezing point depression (−0.186C°).

Table 15–9 Boiling-Point Elevations and Freezing-Point Depressions

SOLUTION	BOILING POINT ELEVATION, C°	FREEZING POINT DEPRESSION, C°
Urea, 0.1 molal	0.052	−0.186
NaCl, 0.1 molal	0.104	−0.372
$CaCl_2$, 0.1 molal	0.156	−0.558

Note: These are idealized values which neglect interionic attraction effects.

One formula weight of urea, 60 g, contains an Avogadro number of molecules (6.02×10^{23}). One formula weight of NaCl, 58 g, contains an Avogadro number of Na^+ ions and a similar number of Cl^- ions. An ion is as effective as a molecule in lowering freezing point or elevating boiling point. Therefore a 0.1-molal solution of NaCl will give twice the boiling-point elevation and twice the freezing-point depression of a 0.1-molal solution of urea (Fig. 15-12).

Nonelectrolytes, then, establish the "normal" values of freezing-point depressions and boiling-point elevations. Solids whose water solutions are good conductors of an electric current give elevations and depressions which are (very nearly) integral multiples of those of nonelectrolytes. When the formulas of ionic substances are examined, those giving twice the "normal" elevations or depressions, such as NaCl, KBr, and $MgSO_4$,

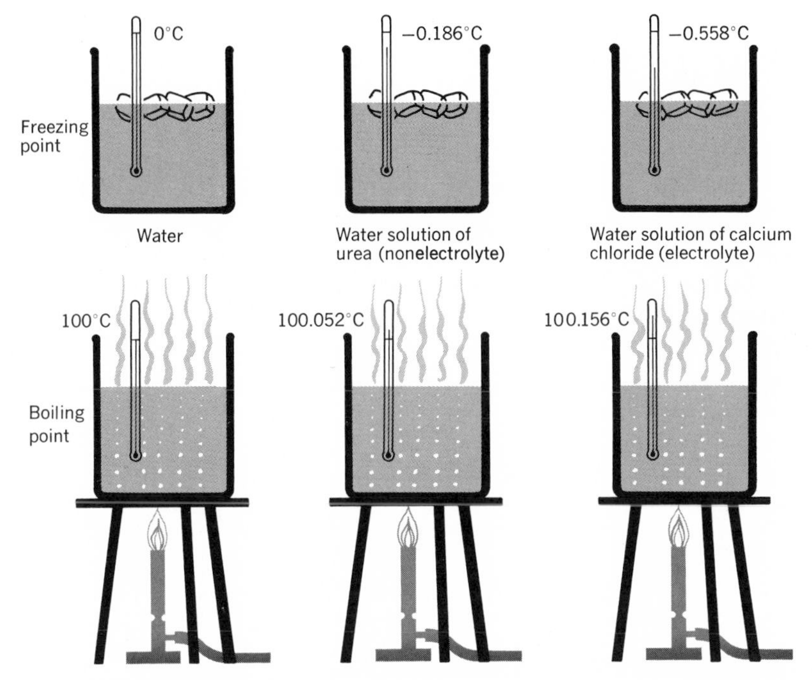

Fig. 15-12. *The freezing-point lowering of a dilute electrolytic solution is an integral multiple of that of a nonelectrolytic solution at the same concentration. Boiling-point elevations follow a similar pattern.—The freezing-point lowerings of 0.1-molal solutions of calcium chloride (an electrolyte) and urea (a nonelectrolyte) are in the ratio 3 : 1. Boiling-point elevations are in the same ratio. For every molecule in the urea solution there will be three ions from* $CaCl_2$ *(one Ca and two Cl).*

are the ones which have $2 \times 6.02 \times 10^{23}$ ions per formula weight. Other ionic solids, such as Na_2SO_4($2Na^+$ and $SO_4^{=}$) and $MgCl_2$ with $3 \times 6.02 \times 10^{23}$ ions per formula weight, give three times the "normal" elevation or depression values.

ILLUSTRATIVE PROBLEM

A solution containing 0.400 g of a solute (nonelectrolyte) in 100 g of water, freezes at −0.124°C. Calculate the molecular weight of the solute.

Solution

A gram molecular weight of a non-ionic solute *depresses* the freezing point −1.86C° (the freezing point is also −1.86°C). Since the observed depression is −0.124C°, the solution must be

$$\frac{0.124}{1.86} = 0.066 \text{ molal}$$

0.400 g of solute in 100 g H_2O = 4 g in 1000 g H_2O

$$\text{Hence } 4 \text{ g} = 0.066 \text{ mole and } 1 \text{ mole} = \frac{4.000 \text{ g}}{0.066 \text{ mole}} = 60.6 \text{ g/mole}$$

The molecular weight is 60.6.

The molal depression constants given so far are for water only. Other solvents have different constants. The molal freezing point constant for benzene (C_6H_6) is −5.12°C. Since the freezing point of benzene is 5.50°C, a 0.1 molal solution of a nonelectrolyte in benzene would freeze at

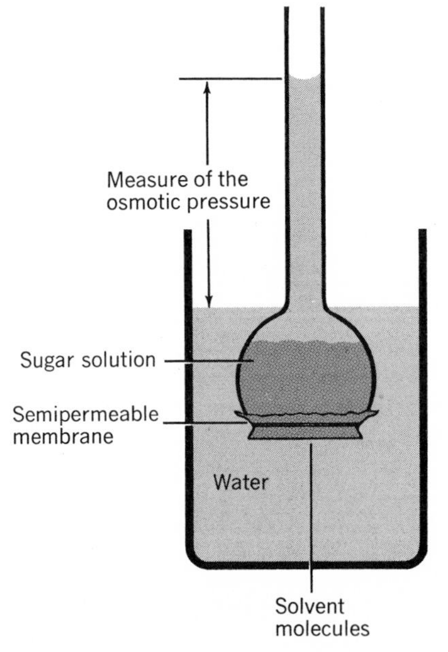

Fig. 15-13. *Immersion in water of a glass tube covered at one end by a semipermeable membrane and containing a sugar solution causes water to flow through the membrane, rising in the tube to a level higher than the water outside. The pressure due to the elevation of the liquid in the tube is called osmotic pressure. Ionic solutions give higher elevations than nonelectrolytes of the same concentration.*

5.50 − 0.51 = 4.99°C. The molal boiling point elevation constant for benzene is 2.60°C. Since the boiling point of benzene is 80.1°C, a 0.1 molal solution of a nonelectrolyte would boil at 80.10 + 0.26 = 80.36°C.

The freezing-point depressions and boiling-point elevations just discussed are also related to the tendency of dissolved substances to decrease the evaporation of a liquid. Water containing a dissolved substance evaporates less readily than does pure water. The quantitative measure of this evaporation tendency is called the "vapor pressure" of the liquid. Once again the vapor pressure lowerings caused by the dissolving of ionic substances are larger than for nonelectrolytes at the same concentrations.

If a sugar solution contained in a semipermeable* membrane sac, is immersed in water as shown in Fig. 15-13, water flows through the membrane into the sugar solution, giving rise to an "osmotic pressure." The osmotic pressure is measured by the difference in level between the sugar solution inside the tube and the water outside. Once more, ionic substances at the same concentration give rise to larger osmotic pressures than nonelectrolytes.

Nonelectrolytes or molecular substances, then, establish the norms of comparison for a number of solution properties. Electrolytes, or ionic solutes, give larger effects which are nearly integral multiples of these norms, the particular multiple depending on the number of ions per formula weight.

ELECTRON ATTRACTION OF IONS IN SOLUTION: ELECTROLYTIC CELLS

The electron flow in conducting wires, such as copper, that can be induced as the result of chemical reaction (Chapter 10) is the result of differences in the electron-attracting abilities of two chemical species in solution. First, consider a simple chemical reaction involving direct electron transfer. It will be demonstrated (by a very ingenious separation of the reactants) that the *same reaction* may be made to cause an electron flow through a conducting wire. The voltages of such cells are a measure of the relative electron-attracting abilities of the chemical constituents in the two half-reactions.

A copper(II) sulfate solution contains both Cu^{++} ions and $SO_4^{=}$ ions. If a thin strip of zinc metal is immersed in this solution (Fig. 15-14), a reddish brown deposit of copper metal quickly forms on the zinc strip; at the same time the deep blue color of the solution fades. Copper ions in the solution striking the zinc strip remove electrons from Zn atoms. Copper ions gain electrons according to the equation $Cu^{++} + 2e \rightarrow Cu$. An equal number of zinc atoms must each lose two electrons, forming a zinc ion according to the equation $Zn \rightarrow Zn^{++} + 2e$. By adding algebraically the above half-reactions, the overall reaction is obtained: $Cu^{++} + Zn \rightarrow Cu + Zn^{++}$.

* In principle only solvent molecules are able to penetrate such a membrane. In practice this is not strictly true.

Copper(II) *ions* in solution are able to remove electrons from Zn *atoms* in contact with the solution. When the two species are together the copper ions gain two electrons each, and zinc atoms lose two electrons each. Zinc atoms have a lesser hold on these electrons than do copper ions.

Now consider how this same reaction may be employed to cause an electron flow through a circuit—for example, through the filament of a flashlight bulb. If a zinc metal strip is immersed in a zinc sulfate ($ZnSO_4$) solution, and a copper strip in a copper sulfate ($CuSO_4$) solution, the two solutions being separated by a porous barrier as shown in Fig. 15-14,

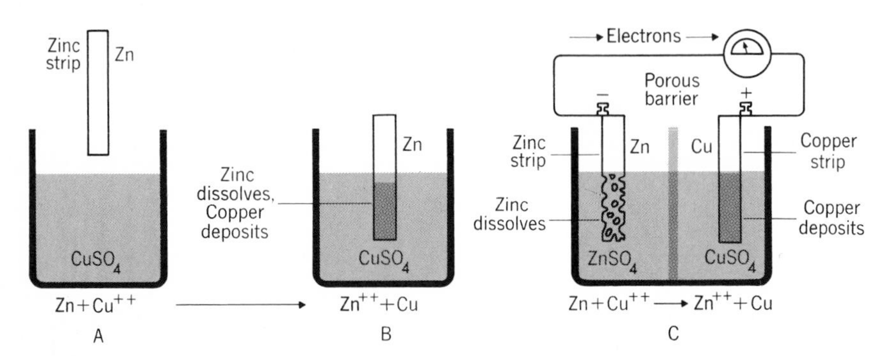

Fig. 15-14. *The same reaction takes place in B as in C. In one case electron transfer is direct, in the other via a piece of copper wire.*

a current-measuring instrument connected between the metal strips shows that electrons are flowing. Cu^{++} ions reaching the Cu electrode accept electrons from it, producing copper atoms which deposit on the electrode, increasing its weight. Deposition of Cu^{++} ions causes an electron deficiency at this electrode. Electrons flow in from the external circuit to make up this deficiency. Ultimately atoms on the zinc electrode are deprived of electrons and go into solution as Zn^{++} ions. The Zn electrode decreases in weight as a result. The chemical reaction taking place is just the same as in the simple displacement reaction discussed earlier, $Cu^{++} + Zn \rightarrow Cu + Zn^{++}$.

If the concentrations of the $CuSO_4$ and $ZnSO_4$ solutions are both 1-molal, a table of oxidation-reduction potentials may be used to predict the voltage of a cell. Table 15-10 shows that the half-reaction $Zn \rightarrow Zn^{++} + 2e$ has a potential of 0.76 volts. The half-reaction $Cu^{++} + 2e \rightarrow Cu$ is the reverse of the reaction shown in the table, and hence the sign of the electrode potential is changed to +0.34 volt. The potential of the cell is the sum of these two values, or 1.10 volts.

A cell using a zinc electrode in zinc sulfate solution and a silver electrode in a silver nitrate ($AgNO_3$) solution has a higher voltage.

$$\begin{array}{ll} Zn \rightarrow Zn^{++} + 2e & 0.76 \text{ volt} \\ \underline{2(1e + Ag^{+} \rightarrow Ag)} & \underline{0.80 \text{ volt}} \\ Zn + 2\,Ag^{+} \rightarrow 2\,Ag + Zn^{++} & 1.56 \text{ volts} \end{array}$$

Silver ion has an even greater electron-attracting power than Cu^{++}. Thus the farther apart two elements are in the oxidation potential series, the greater is the electron transfer tendency and the higher the potential of the cell.

Table 15–10 Oxidation Potentials at 25°C

HALF-REACTION	POTENTIAL, VOLTS
$Na \rightarrow Na^{+} + 1e$	2.71
$Al \rightarrow Al^{+++} + 3e$	1.66
$Zn \rightarrow Zn^{++} + 2e$	0.76
$H_2 \rightarrow 2H^{+} + 2e$	0.00
$Cu \rightarrow Cu^{++} + 2e$	–0.34
$Ag \rightarrow Ag^{+} + 1e$	–0.80

The ion in each of the above half-reactions is assumed to be in a 1-molal solution.

The neutral atoms at the top of the table are those which in solution part with their electrons most readily. Those at the bottom of the table are the atoms which most strongly hold their electrons. The inverse is true of the tendency of the ions to regain electrons. An ion of a metal which is very low in the table will readily remove electrons from a metal atom which is high in the table. A reaction between Cu metal and $Al_2(SO_4)_3$ (containing Al^{+++} ions) would not be expected. Al^{+++} ions have little tendency to capture electrons, particularly from an element such as copper which is lower in the table. The Cu atom holds its valence electrons much more tightly.

OXIDATION–REDUCTION

The half-reactions taking place in electrolytic cells serve as excellent illustrations of oxidation and reduction processes. The discussion of oxidation-reduction in this chapter will relate to ionic reactions in solution, but the principles are generally applicable to chemical reactions.

Consider a cell composed of a copper electrode in a $CuSO_4$ solution and a silver electrode in an $AgNO_3$ solution. Since Cu is higher than Ag in the oxidation potential series, the Cu electrode will dissolve, Ag will deposit. Electrons will flow out of the cell at the Cu electrode.

The cell reaction involving copper is $Cu \rightarrow Cu^{++} + 2e$. Copper metal is *oxidized* in this reaction. Electrons are given up by copper atoms; they become Cu^{++} ions and go into the $CuSO_4$ solution. In this process the oxidation number of Cu increases from 0 to +2. Oxidation may be defined, then, as either (1) loss of electrons or (2) algebraic increase in oxidation number.

In the half-reaction $Ag^{+} + 1e \rightarrow Ag$, a silver ion has gained an electron and deposited on the electrode as a silver atom. The oxidation number of silver has decreased by 1. The silver ion has been *reduced*. The process of reduction is defined as (1) gain of electrons or (2) algebraic decrease in oxidation number.

The overall cell reaction is

$$2\,Ag^{+} + Cu \rightarrow Cu^{++} + 2\,Ag$$

Note that a sort of "law of the conservation of electrons" applies. The reaction takes place because Ag^{+} attracts electrons more strongly than Cu atoms. The number of electrons gained by Ag^{+} must then be equal to the number lost by Cu. Cu loses 2 electrons per atom; Ag^{+} gains 1 electron per atom. For an electron balance there must be twice as many Ag^{+} ions accepting electrons as Cu atoms losing them. The substance oxidized functions as a *reducing agent* (copper metal in the above reaction). The substance reduced must therefore be the *oxidizing agent* (Ag^{+} ion in the above reaction).

SUMMARY: THE STRUCTURE OF TWO COMMON ATOMS

Compare, as a typical metal, magnesium with the elements adjacent to it in the periodic table (Table 15-11), and similarly with a typical nonmetal, bromine (Table 15-12). The Mg atom has an electronic configuration, $1s^22s^22p^63s^2$; it has therefore two valence electrons. Removal of these exposes an underneath noble gas structure, so the element is a metal capable of forming an ion with an oxidation number of +2. The oxidation number is further confirmed by the low ionization energies of the first two electrons, 7.6 and 15 ev respectively, but a very much higher value for the third electron, 80 ev.

Consider now the properties of Na, Mg, and Al, elements which are adjacent in the same period. Note that the sizes of the atoms (atomic radii) decrease from left to right. Each added electron would be expected to expand the atomic radius somewhat, because of the repulsion between electrons. But the added nuclear charge would tend to reduce its size. Apparently increase of nuclear charge is more important than adding another electron in the same quantum level. Increased ionization energies and increased electronegativities, from left to right, tend to support this point of view. Na metal has the highest oxidation potential of the three, loses its valence electron more readily than do the other two, and hence is the best reducing agent.

The atoms Be, Mg, and Ca, all adjacent in vertical column II A, have increasing atomic radii. Each successive element has 8 electrons and 8 nuclear protons more than its predecessor. This time, however, in going from one element to the next, an extra electron shell has been interposed between valence electrons and nucleus. Thus the nuclear charge is better

Table 15–11 Magnesium and Its Neighbors in the Periodic Table

Symbol	Be		
Ionization Energy (e)	9.3		
Atomic Radius (Å)	0.89		
Electronegativity	1.5		
Oxidation Potential (volts)	1.85		
Density (g/cm^3)	1.86		
Symbol	Na	Mg	Al
Ionization Energy (e)	5.1	7.6	6.0
Atomic Radius (Å)	1.57	1.36	1.25
Electronegativity	0.9	1.2	1.5
Oxidation Potential (volts)	2.71	2.37	1.66
Density (g/cm^3)	0.97	1.74	2.70
Symbol	Ca		
Ionization Energy	6.1		
Atomic Radius	1.74		
Electronegativity	1.0		
Oxidation Potential	2.87		
Density	1.55		

shielded and the larger atoms would be expected to attract their valence electrons less strongly. Thus ionization energies and electronegativities decrease in the larger atoms.

The increased oxidation potentials of the larger atoms are also suggestive of greater chemical activity, although oxidation potentials are in general more difficult to relate to periodic fluctuations than are certain other properties such as ionization energies. Finally, even though Be is a Group II element, a number of its compounds are quite different from those of the typical alkaline earths, for example, melted chlorides and fluorides of Be do not conduct an electric current, as do other molten alkaline earth halides. Also, Be forms many more complexes than do the other alkaline earths.

The bromine atom (Table 15-12) has an electronic configuration

Table 15–12 Bromine and Its Neighbors in the Periodic Table

Symbol		Cl
Ionization Energy		13.01
Atomic Radius (Å)		0.99
Electronegativity		3.0
Oxidation-Reduction Potential (volts)		−1.36
Symbol	Se	Br
Ionization Energy (e)	9.8	11.84
Atomic Radius (Å)	1.17	1.14
Electronegativity	2.4	2.8
Oxidation-Reduction Potential (volts)	—	−1.07
Symbol		I
Ionization Energy		10.44
Atomic Radius		1.44
Electronegativity		2.5
Oxidation-Reduction Potential (volts)		−0.54

$1s^2 2s^2 2p^6 3s^2 3p^6 4s^2 3d^{10} 4p^5$, indicating that there are 7 electrons in the valence shell. This is corroborated by the large increase in the ionization energy following the removal of the seventh electron. By removing an electron from an active metal (forming an ionic bond) or sharing one with another element (forming a covalent bond) a noble gas structure of 8 electrons will be completed. Bromine will have an oxidation number of -1 when ionically bonded, or covalently bonded to an element of smaller electronegativity. This oxidation number as well as its comparatively high ionization energy and electronegativity, indicate its nonmetallic character.

Comparison of Se and Br, adjacent elements in the same period, shows the same general trends as with the metals Na, Mg, and Al—decrease in atomic radius, increase in ionization energy and electronegativity.

The discussion of oxidation-reduction potentials thus far has been limited to metals. Lest the impression be left that they apply only to metals, Table 15-12 includes potentials for the oxidation of each of the three halide ions. If these are arranged in the order of decreasing potential, the series is

$$I^- \rightarrow \tfrac{1}{2} I_2 + e \qquad -0.54 \text{ v}$$
$$Br^- \rightarrow \tfrac{1}{2} Br_2 + e \qquad -1.07 \text{ v}$$
$$Cl^- \rightarrow \tfrac{1}{2} Cl_2 + e \qquad -1.36 \text{ v}$$

As in the earlier discussion of oxidation reduction potentials, I^-, highest of the reactant ions, will most readily lose electrons (just as the most active metals, such as Na, did). I^- is then the best reducing agent of the three. Cl_2, the lowest product (right-hand side of the equation) molecule will be the best oxidizing agent, just as metal ions low in the metal series were the best oxidizers.

With this all-too-brief discussion, the story of chemical bonding must close. It need hardly be added that in recent years enormous advances have been made in the interpretation of physical and chemical properties of elements and their compounds in terms of their electronic structures.

Problems, Chapter 15

1. The diameter of an average atom is about 1 Å and that of its nucleus approximately 10^{-12} Å. Calculate the fractional part of the atom's volume which is occupied by its nucleus. (Assume that the atom is spherical.)

2. Assuming that the (rest) mass of the electron is 9.1083×10^{-28} g and the atomic mass (which includes electrons and nucleons) of Mg is 23.98505 atomic mass units, calculate the fraction of the Mg atom's mass accounted for by its electrons. [1 atomic mass unit (amu) $= 1.660 \times 10^{-24}$ g].

3. Calculate the freezing point of an antifreeze solution containing 10 liters of ethylene glycol (density 1.43 g/ml) in 40 liters of water. (Density = 1.00 g/ml.)

```
    H   H
    |   |
H—C——C—H
    |   |
   OH  OH
```

Exercises

1. State the assumptions of the Bohr theory of the atom. Explain how it accounted for spectra.

2. What was the revolutionary assumption of Planck's quantum theory? Explain the meaning of each term in Planck's equation, $E = h\nu$.

3. Explain what is meant by the first ionization energy of an atom. In what units is it expressed? Show how its value is related to electronic binding.

4. Name the quantum numbers needed to define the electronic configurations of the electrons in an atom. What aspect of the electron's motion or location is determined by each? How many values may each quantum number have? Illustrate.

5. State the Pauli exclusion principle.

6. Write the electronic configurations for the following elements: Ca, I, Rn. Include numbers of electrons in subshells and the correct order of filling.

7. Describe the electronic characteristics of transition elements. Illustrate. Do the same for inner transition elements.

8. What is the difference between a nonpolar and a polar covalent bond? Give an illustration of each.

9. What is the difference between a covalent and an ionic bond? Give an illustration of each.

10. How are metals and nonmetals characterized in terms of their electronegativity values? Illustrate.

11. Write the oxidation numbers of each element in the following formulas: AgI, $CuCl_2$, K_2O, $MgSO_4$, $KMnO_4$, K_2CrO_4.

12. Compare the properties of solutions of electrolytes and nonelectrolytes. Give illustrations.

13. What is meant by osmotic pressure? Describe how it may be demonstrated experimentally.

14. Explain why you would expect fluorine and sodium to be very active elements in view of their atomic structures. Why you would expect the noble gases to be unreactive?

15. Na^{+} and Mg^{++} ions have the same electronic structures (they are called isoelectronic). How do you account for their different sizes?

16. How is the operation of an electrolytic cell or battery explained in terms of differing electron attractions of two chemical substances? Illustrate.

16 Organic chemistry

IN the early nineteenth century chemistry was divided into two principal branches, inorganic and organic. Substances derived from the mineral world, those which did not have characteristics of living things, were called inorganic compounds. Organic compounds were believed to be found only in living things, plants and animals. Berzelius thought they contained a "vital force." All the organic substances known in that day, such as alcohol and acetic acid (produced by the fermentation of grape juice or cider) were products of living organisms.

Then in 1828, according to the traditional story, the German chemist Friedrich Wöhler (Fig. 16-1) heated the inorganic compound ammonium cyanate and obtained the organic chemical, urea, a product found in the urine of mammals. This urea synthesis touched off a rash of experiments in which organic compounds were synthesized from inorganic substances. Needless to say the theory of vitalism had to be abandoned.

The element carbon was ultimately found to be common to all organic compounds. To be sure, other elements such as oxygen, sulfur, nitrogen and hydrogen were sometimes present—nearly always hydrogen—but *always* carbon. Thus it was that organic chemistry became the study of the compounds of carbon.

More than half a million carbon compounds are known and others are being synthesized daily. Petroleum, foods, coal, drugs, dyes, plastics, and biological materials all fall within the realm of organic chemistry. Compounds in which carbon is absent number about fifty thousand for all the other hundred-odd chemical elements known.

THE ELEMENT CARBON

In 1859 August Kekule proposed a brilliant yet simple explanation for the strikingly large number of compounds resulting from the union of carbon atoms with themselves and with other atoms. He suggested that carbon had a valence of four, whether in simple compounds such as CH_4

Fig. 16-1. *Friedrich Wöhler, the first man to synthesize an organic compound (urea)—1828, when he was twenty-eight.*

and CO_2, or in complex molecules such as $C_{12}H_{22}O_{11}$. Through his proposal concerning the tetravalency of carbon Kekule succeeded in establishing the structural theory of organic chemistry on a firm foundation but left unanswered such questions as to whether molecules such as methane were planar or three-dimensional.

In about 1875 Le Bel, and independently van't Hoff, proposed a three-dimensional structure for methane, with four hydrogen atoms symmetrically attached to a central carbon. The four hydrogen atoms were equidistant from each other. Thus the arrangement was tetrahedral with 109° 28′ bond angles (Fig. 16-2). It is noteworthy that Le Bel and van't Hoff came to this conclusion some twenty years before the discovery of the electron.

Atoms with four electrons in their outer shells tend to form covalent rather than ionic bonds. Thus the carbon atom has a tendency to share its four electrons with other atoms and forms stable covalent bonds with hydrogen and other elements. Carbon atoms also join with each other,

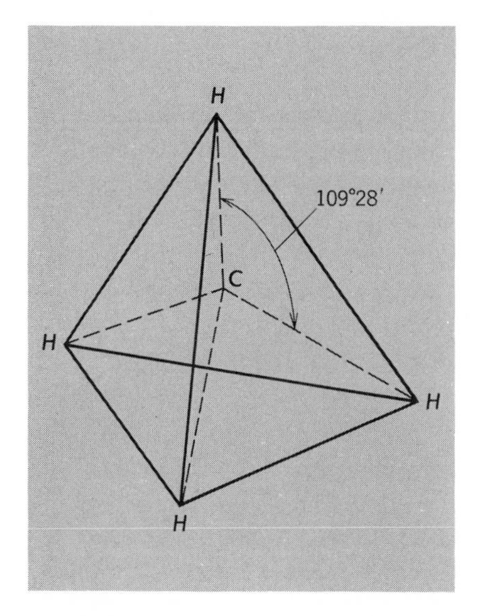

Fig. 16-2. *The tetrahedral configuration of the methane molecule. The solid straight lines represent the edges of the tetrahedron, while the dotted lines are the valence bonds within the tetrahedron connecting the carbon atom with the hydrogen atoms.*

a unique property called *catenation* which in large measure accounts for the vast number of organic compounds. Silicon also forms tetravalent covalent bonds, since it is in the same family as carbon. The Si—Si bond, however, is not as stable as the C—C bond and there are far fewer silicon compounds than carbon compounds.

HYDROCARBONS

Saturated Hydrocarbons Most carbon compounds contain hydrogen, and those containing carbon and hydrogen only are called hydrocarbons. The simplest hydrocarbon, methane, was discussed in Chapter 15. Ethane, C_2H_6, is the simplest hydrocarbon with a C—C bond:

```
   H  H            H  H  H
   |  |            |  |  |
H—C—C—H        H—C—C—C—H
   |  |            |  |  |
   H  H            H  H  H
  Ethane          Propane
```

Propane C_3H_8, has two C—C bonds and eight C—H bonds. In the propane molecule, carbon atoms are not in a straight line but have a partly folded carbon chain structure. Chains of three, or more carbon atoms have this zigzag arrangement (Fig. 16-3).

Compare the ethane and propane molecules and note that as the carbon chain length increases by one carbon atom, the formula of the compound increases by a CH_2 group. The succeeding members of this series will have the molecular formulas C_4H_{10}, C_5H_{12}, C_6H_{14}, and so on. These compounds are named respectively, butane, pentane, and hexane. Chemists have isolated or prepared similar hydrocarbons with over one hundred carbon atoms in a single molecule.

Fig. 16-3. *These models represent the methane molecule (left) and the ethane molecule (right). The dark spheres represent carbon atoms, the light spheres hydrogen. Note especially the three-dimensional folding of the ethane molecule.*

A series of compounds in which each member differs from its predecessor by a CH_2 group is called a *homologous* series. Each member of the particular series under consideration, known as the alkane or paraffin series, may be represented by the general formula C_nH_{2n+2}; n stands for the number of carbon atoms in a molecule. Octane, for example, has eight carbon atoms; its formula would be C_8H_{18}.

Another reason for the large number of organic compounds is the existence of branched-chain as well as straight-chain compounds. Butane, for example, may be represented by a structure in which no carbon atom is attached to more than two other carbons. This is called *normal* butane. Another butane has one carbon atom which is attached to three other carbons. The latter is called *isobutane*. Both compounds have the same formula, C_4H_{10}, but slightly different physical and chemical properties. They are called structural isomers. Increasing the number of carbon atoms in the molecule dramatically increases the number of possible isomers. There are five isomeric hexanes (C_6H_{14}), 18 octanes (C_8H_{18}), and seventy-five decanes ($C_{10}H_{22}$).

The paraffin hydrocarbons are comparatively unreactive, but they do burn in air or oxygen, yielding carbon dioxide and water vapor. For example:

$$2\,C_4H_{10} + 13\,O_2 \rightarrow 8\,CO_2 + 10\,H_2O + \text{heat}$$
$$CH_4 + 2\,O_2 \rightarrow CO_2 + 2\,H_2O + \text{heat}$$

Because they burn exothermically, paraffin hydrocarbons make excellent fuels. Methane is the principal constituent of city gas. Bottled gas is a mixture of propane and butane. The hydrocarbons in gasoline are principally pentanes, hexanes, heptanes, and octanes.

In the reaction of paraffin hydrocarbons with halogens, hydrogen atoms are replaced. Accordingly they are called *substitution reactions.* For example:

$$CH_4 + Cl_2 \rightarrow CH_3Cl + HCl$$
$$CH_3Cl + Cl_2 \rightarrow CH_2Cl_2 + HCl$$
$$CH_2Cl_2 + Cl_2 \rightarrow CHCl_3 + HCl$$
$$CHCl_3 + Cl_2 \rightarrow CCl_4 + HCl$$

Thus one or more atoms of hydrogen in methane may be replaced by a halogen atom.

Unsaturated Hydrocarbons The alkane hydrocarbons discussed thus far have single bonds between carbon atoms. There are, however, some hydrocarbons such as ethylene, C_2H_4, in which carbon atoms are joined by a double bond:

```
  H   H              H  H
  ..  ..             |  |
H:C: :C:H   or   H—C=C—H
```

In acetylene, C_2H_2, the carbon atoms are held together by a triple bond:

```
H:C⋮ ⋮C:H   or   H—C≡C—H
```

Ethylene is the first member of an homologous series whose members are called the *olefin* or alkene hydrocarbons. They have the general formula C_nH_{2n}. Acetylene is the first member of the alkyne series. Its members have the general formula C_nH_{2n-2}. The names and formulas of several members of each series are listed in Table 16-1.

Table 16–1 Names and Formulas of Some Alkenes and Alkynes

ETHYLENE SERIES (ALKENES)		ACETYLENE SERIES (ALKYNES)	
MOLECULAR FORMULA R COMPOUND	NAME	MOLECULAR FORMULA OR COMPOUND	NAME
C_2H_4	Ethylene	C_2H_2	Acetylene
C_3H_6	Propylene	C_3H_4	Propyne
C_4H_8	Butylene	C_4H_6	Butyne
C_5H_{10}	Pentene	C_5H_8	Pentyne
C_6H_{12}	Hexene	C_6H_{10}	Hexyne

The alkane hydrocarbons have single bonds between carbon atoms and are called saturated. Compounds such as ethylene and acetylene, containing double or triple bonds between carbon atoms, are called unsaturated. In general the C—C single bond length is greater than double bond length, C=C, which is in turn larger than the triple bond length C≡C. Furthermore the amount of energy required to separate carbon atoms in C≡C (the bond energy) is greater than for C=C, which in turn is larger than for C—C.

Unsaturated compounds would then be expected to be more stable as a result of the presence of multiple bonds. The opposite is actually true. Unsaturated compounds have a marked tendency to react and become saturated. The characteristic reaction of unsaturated compounds is *addition*. Two examples of addition reactions are shown in the formulas below.

$$\begin{array}{c} \text{H}\ \ \text{H} \\ |\quad | \\ \text{H–C=C–H} \end{array} + H_2 \xrightarrow{\text{catalyst}} \begin{array}{c} \text{H}\ \ \text{H} \\ |\quad | \\ \text{H–C–C–H} \\ |\quad | \\ \text{H}\ \ \text{H} \end{array}$$

Ethylene (unsaturated) — Ethane (saturated)

$$\begin{array}{c} \text{H}\ \ \text{H} \\ |\quad | \\ \text{H–C=C–H} \end{array} + Br_2 \rightarrow \begin{array}{c} \text{H}\ \ \text{H} \\ |\quad | \\ \text{H–C–C–H} \\ |\quad | \\ \text{Br}\ \text{Br} \end{array}$$

Ethylene (unsaturated) — Ethylene dibromide (saturated)

The addition reaction in which hydrogen attaches at the center of unsaturation is known as *hydrogenation*. An important commercial application is the hydrogenation of cottonseed or safflower oils to produce cooking fats or shortenings (Spry and Crisco). When a halogen adds itself at the center of unsaturation, the reaction is called *halogenation*. An example is the addition of bromine to ethylene, to form ethylene dibromide, an important constituent of high-test gasoline.

Cyclic Hydrocarbons So far all hydrocarbons discussed have had straight or branched chains. There are also hydrocarbons, called cycloalkanes, in which carbon atoms are arranged in a ring. Several members of this series are listed in Table 16-2. The general formula of the series is C_nH_{2n}. Compounds are named by prefixing *cyclo* before the name of the corresponding alkane. The simplest member of this series is cyclopropane—simplest because to form a ring at least three carbon atoms are needed. The chemical properties of the cycloalkanes are similar to those of the alkanes.

There is also an homologous series corresponding to the olefins. The first three members of the cycloalkene series are cyclopropene, cyclo-

Table 16–2 Formulas and Names of Some Cycloalkanes

NAME	MOLECULAR FORMULA	STRUCTURAL FORMULA
Cyclopropane	C_3H_6	H_2C——CH_2 CH_2
Cyclobutane	C_4H_8	CH_2 H_2C CH_2 CH_2
Cyclopentane	C_5H_{10}	CH_2 H_2C CH_2 H_2C——CH_2
Cyclohexane	C_6H_{12}	H_2 H_2 C—C H_2C CH_2 C—C H_2 H_2

butene, and cyclopentene. Once again the characteristic reaction is addition at the double bond.

Fig. 16-4. *Friedrich August Kekule, who originated the theory of the benzene molecule as a closed ring.*

Aromatic Hydrocarbons The most important of all cyclic hydrocarbons is benzene, C_6H_6. Determination of its structure puzzled and fascinated chemists at mid-nineteenth century. The low hydrogen-to-carbon ratio suggests multiple bonding, yet benzene does not undergo the characteristic addition reactions. In 1865 Kekule (Fig. 16-4) proposed that benzene was composed of one or both of the following structures diagramed below.

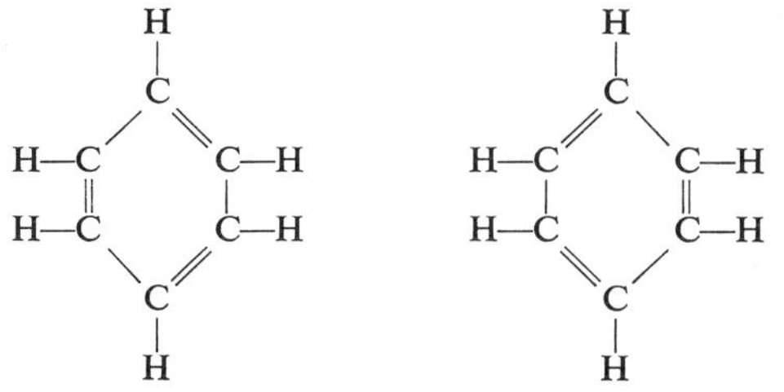

```
        H                    H
        |                    |
        C                    C
      /   \\              //   \
H—C         C—H      H—C         C—H
   ||       |           |        ||
H—C         C—H      H—C         C—H
      \   //              \\   /
        C                    C
        |                    |
        H                    H
```

Note that in both structures carbon is tetravalent. But since benzene does not undergo addition reactions, it should have no double bonds and hence Kekule's structures must be incorrect. Modern X-ray studies have shown the benzene molecule is planar and hexagonal. Experimental techniques have also shown that all six carbon-carbon bonds are equivalent. Thus there is no conventional way of representing the correct benzene structure. To resolve the problem one proposal has been that the benzene molecule exists as a structure intermediate between I and II. The carbon-carbon bonds in benzene are then of a special type and may be represented as resonance hybrids, shorter than single bonds, longer than double:

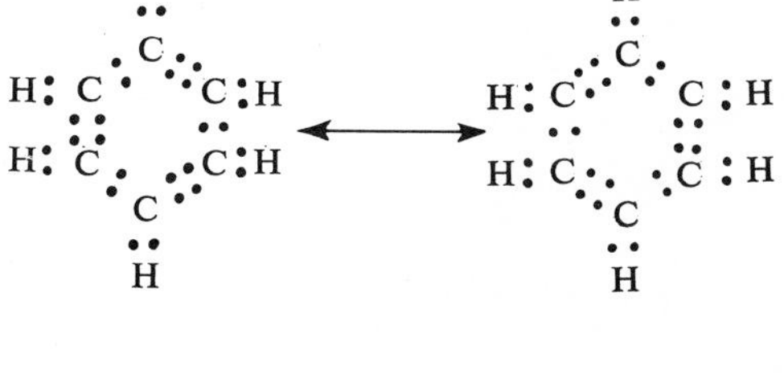

```
       H                          H
       ••                         ••
      .C..                       .C.
H: C•    ••C:H             H:C••    •C:H
   ••      ••      <——>      ••       ••
H: C.    .•C:H             H:C:.    .C:H
      •C••                      ••C
       ••                         ••
       H                          H
```

Note that the intermediate (resonance) phenomenon exists when the molecule may be correctly represented by two or more similar structural frameworks differing only in the location of electron pairs.

Compounds which are derivatives of benzene are described as aromatic. Benzene is usually represented as a regular hexagon, but if one wishes to indicate the presence of a delocalized pi-bond, a dotted circle is written inside the hexagon:

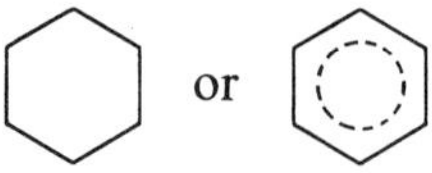

or

Hydrogen atoms of the benzene ring may be replaced by numerous substituents. Alkyl groups, for example, may replace hydrogens and form the compounds in Chart I.

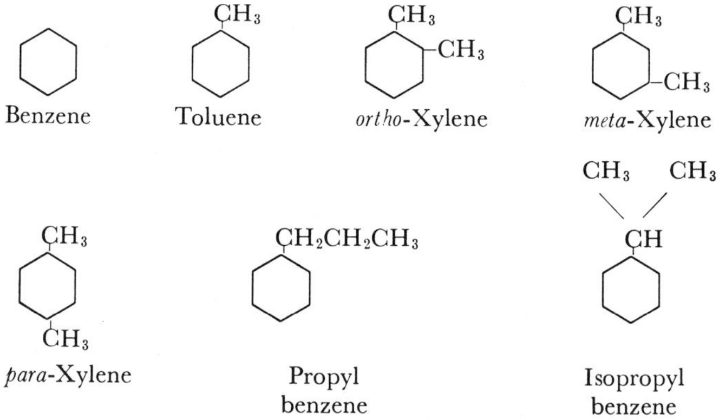

CH_3 CH_3 CH_3 CH_3 CH_3

Benzene Toluene *ortho*-Xylene *meta*-Xylene

CH_3 CH_3 CH_3 $CH_2CH_2CH_3$ CH CH_3

para-Xylene Propyl benzene Isopropyl benzene

CHART I

Note the isomerism of the three xylenes and that propyl and isopropyl benzene are isomeric.

Inorganic substituents may also replace one or more hydrogen atoms. Several examples are shown in Chart II. Since benzene is a comparatively stable molecule its reactions are likely to be sluggish, reactions usually taking place only at higher temperatures or in the presence of a catalyst.

HYDROCARBON DERIVATIVES

Many organic compounds contain atoms other than just carbon and hydrogen. In cases where such compounds are closely related to hydro-

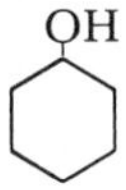

OH / OH, OH / Cl, Cl

Phenol (carbolic acid) | Hydroquinone 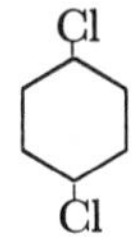 | *para*-Dichlorobenzene (moth-repellent)

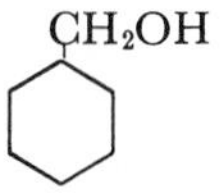

CH_2OH | NO_2 | CH_3, O_2N, NO_2, NO_2

Benzyl alcohol | Nitrobenzene 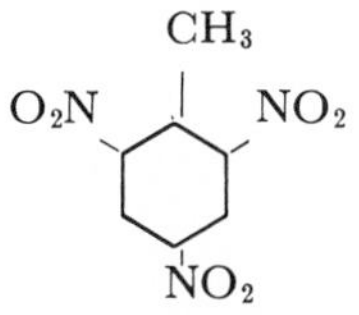 | Trinitrotoluene (TNT)

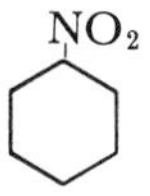

CHART II

carbons they are regarded as *derivatives*. The replacement of a hydrogen atom of C_2H_6, for example, by a hydroxyl (OH) group giving C_2H_5OH drastically changes the properties of the hydrocarbon. Alkanes are insoluble in water and comparatively unreactive. Ethyl alcohol on the other hand is not only water-soluble but quite reactive. A whole series of parallel reactions is characteristic of the alcohols, so the hydroxyl group is called a *functional group*. The carboxyl (COOH) and amino (NH_2) groups are other examples of functional groups.

HYDROXY COMPOUNDS

When a hydrogen atom of a non-benzenoid molecule is replaced by the —OH group, an alcohol results. When this group takes the place of a hydrogen atom in a benzene ring, the resulting compound is called a phenol and has distinctly different properties. For its structural formula see Chart II above. Phenol (carbolic acid) is a crystalline, colorless solid which has potent antiseptic properties. In contact with the skin it may cause severe burns.

Alcohols are named with reference to the hydrocarbons from which they are derived. CH_3OH (methyl alcohol) is the hydroxyl derivative of methane, CH_4. Methyl alcohol is also called wood alcohol because at one time it was produced by heating wood in the absence of air. Today the inefficient wood distillation process has been replaced by one in which a mixture of carbon monoxide and hydrogen is heated under pressure in the presence of a suitable catalyst.

Methyl alcohol or methanol is sometimes used as an antifreeze in automobiles, although its low boiling point makes it unsuitable for year-round use. It is extremely poisonous when taken internally, sometimes causing blindness and even death. Indeed, an intake of as little as 30 milliliters has been known to be fatal. In order to make tax-free industrial ethyl alcohol readily available and yet unfit for drinking, methyl alcohol is added as a *denaturing agent.*

Ethyl alcohol (grain alcohol) is commonly produced by the fermentation of the starches or sugars in grains and fruits. Starch, present in all grains, is mixed with malt from sprouting barley. Malt contains enzymes which convert starches into sugars. Yeast is added and its enzymes further change the sugars into ethyl alcohol:

$$\text{Starch} + H_2O \xrightarrow{\text{malt}} \text{sugar} + H_2O \xrightarrow{\text{yeast}} C_2H_5OH + CO_2$$

The first step is unnecessary if the starting material already contains sugars, as when grapes are fermented. Ethyl alcohol may also be produced by the catalytic hydration of ethylene:

$$H_2C = CH_2 + H_2O \xrightarrow{H_2SO_4} CH_3CH_2OH$$

Ethyl alcohol is present in such beverages as beer, wine, and whiskey. It is also used as a fuel, in drugs, toilet waters, shellacs, as rubbing alcohol, and for the manufacture of organic chemicals.

Of the alcohols with more than one hydroxyl group, among the most important are ethylene glycol, $C_2H_4(OH)_2$, used as a permanent antifreeze, and glycerol, $C_3H_5(OH)_3$, often called glycerin, which is used to manufacture nitroglycerin, drugs, and lotions.

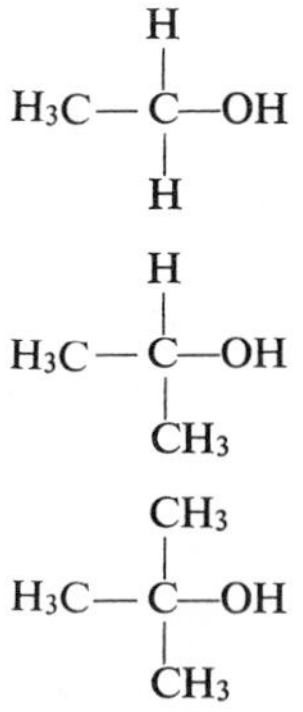

```
       H
       |
H3C—C—OH      A primary alcohol
       |
       H

       H
       |
H3C—C—OH      A secondary alcohol
       |
       CH3

       CH3
       |
H3C—C—OH      A tertiary alcohol
       |
       CH3
```

CHART III

When alcohols are treated with phosphorus trichloride, PCl_3, or other phosphorus halide, an organic halide is produced:

$$3\ CH_3OH + PCl_3 \rightarrow 3\ CH_3Cl + H_3PO_3$$

There are three classes of alcohols—primary, secondary, and tertiary. In primary alcohols the —OH is attached to a carbon atom to which are attached two other hydrogens. In secondary alcohols the functional group is attached to a carbon atom holding only one hydrogen atom. Tertiary alcohols have the —OH group attached to a carbon atom which has no attached hydrogen atoms (Chart III).

The oxidation of primary alcohols gives aldehydes:

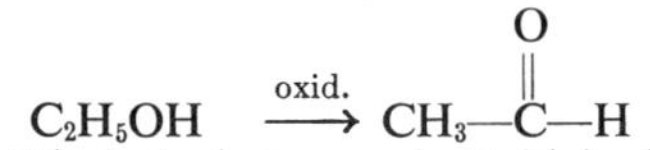

$$\underset{\text{Ethyl alcohol}}{C_2H_5OH} \xrightarrow{\text{oxid.}} \underset{\text{Acetaldehyde}}{CH_3{-}\overset{\overset{\displaystyle O}{\|}}{C}{-}H}$$

The oxidation of secondary alcohols gives ketones:

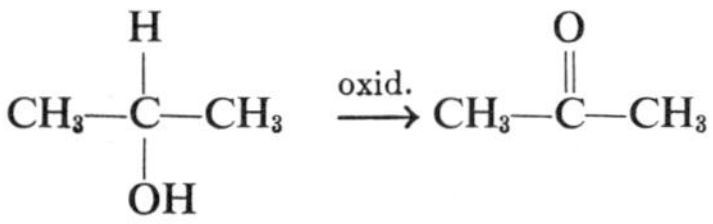

$$\underset{\text{Isopropyl alcohol}}{CH_3{-}\underset{\underset{\displaystyle OH}{|}}{\overset{\overset{\displaystyle H}{|}}{C}}{-}CH_3} \xrightarrow{\text{oxid.}} \underset{\text{Acetone}}{CH_3{-}\overset{\overset{\displaystyle O}{\|}}{C}{-}CH_3}$$

CARBONYL COMPOUNDS

Aldehydes and Ketones Aldehydes and ketones contain a carbonyl group:

$$-\overset{\overset{\displaystyle O}{\|}}{C}-$$

In aldehydes one of the bonds must be attached to a hydrogen atom. In ketones both free bonds are attached to hydrocarbon (alkyl) groups, i.e., CH_3— (methyl), C_2H_5— (ethyl), C_3H_7— (propyl), etc. Examples of other carbonyl compounds are:

Formaldehyde $\quad H{-}\overset{\overset{\displaystyle O}{\|}}{C}{-}H$

Methyl ethyl ketone $\quad CH_3{-}\overset{\overset{\displaystyle O}{\|}}{C}{-}C_2H_5$

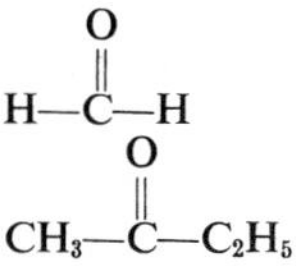

Formaldehyde is used principally as an embalming fluid, for preserving biological specimens, and as an ingredient in the synthesis of certain plastics. The ketone acetone is a solvent used in nail polishes, nail polish remover, and lacquers.

Carboxylic Acids Aldehydes, which are oxidation products of primary alcohols, may in turn be further oxidized to *carboxylic* acids:

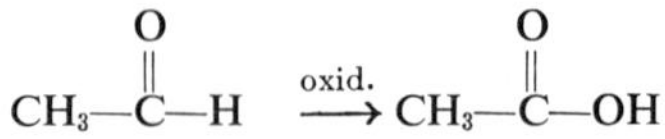

$$\underset{\text{Acetic aldehyde}}{CH_3-\overset{\overset{\large O}{\|}}{C}-H} \xrightarrow{\text{oxid.}} \underset{\text{Acetic acid}}{CH_3-\overset{\overset{\large O}{\|}}{C}-OH}$$

The —COOH (carboxyl) group is the functional group of the carboxylic acids. The most familiar of these is acetic acid. Vinegar, an impure dilute solution of this acid is prepared by fermentation of fruit juices. The action takes place in two stages: the first produces ethyl alcohol and the second, acetic acid. Formic acid, the material injected under the skin by bees and other stinging insects, and butyric acid, responsible for the odor of rancid butter, are also homologues of the same series as acetic acid.

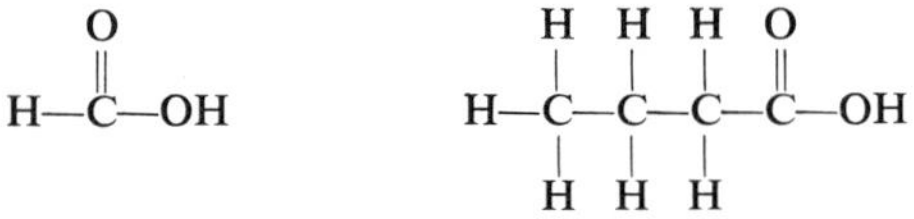

$$\underset{\text{Formic acid}}{H-\overset{\overset{\large O}{\|}}{C}-OH} \qquad \underset{\text{Butyric acid}}{H-\overset{H}{\underset{H}{C}}-\overset{H}{\underset{H}{C}}-\overset{H}{\underset{H}{C}}-\overset{\overset{\large O}{\|}}{C}-OH}$$

In addition to the saturated acids already mentioned, there are also unsaturated acids such as oleic,

$$CH_3(CH_2)_7CH{=}CH(CH_2)_7\overset{\overset{\large O}{\|}}{C}-OH$$

with one double bond, and linoleic acid,

$$C_{17}H_{31}\overset{\overset{\large O}{\|}}{C}-OH$$

with two double bonds. Long-chain carboxylic acids are frequently derived from fats or oils and are sometimes referred to as fatty acids.

Acids containing more than one carboxyl group, or both carboxyl and hydroxyl groups in the same molecule, are also known. Several of these are listed here.

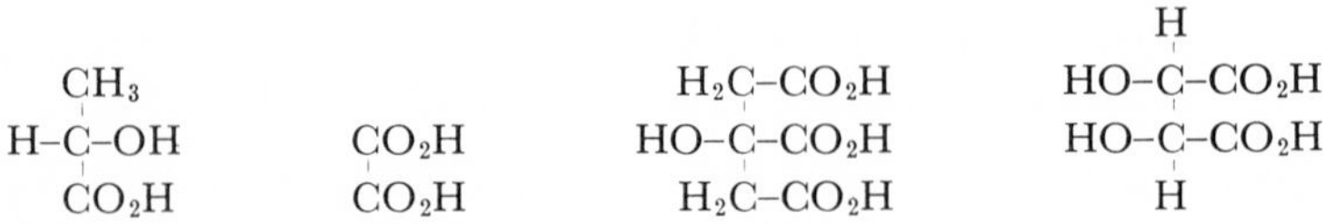

$\begin{array}{c} CH_3 \\ H-C-OH \\ CO_2H \end{array}$	$\begin{array}{c} CO_2H \\ CO_2H \end{array}$	$\begin{array}{c} H_2C-CO_2H \\ HO-C-CO_2H \\ H_2C-CO_2H \end{array}$	$\begin{array}{c} H \\ HO-C-CO_2H \\ HO-C-CO_2H \\ H \end{array}$
Lactic acid (in sour milk)	Oxalic acid (in rhubarb and other tart fruits)	Citric acid (in citrus fruits)	Tartaric acid (in grapes)

Esters Alcohols and carboxylic acids react in the presence of suitable dehydrating agents forming products called *esters*. For example:

$$\underset{\text{Ethanol}}{C_2H_5OH} + \underset{\text{Acetic acid}}{CH_3COOH} \rightarrow \underset{\text{Ethyl acetate}}{CH_3\overset{\overset{\displaystyle O}{\|}}{C}—O—C_2H_5} + H_2O$$

Esters are named as derivatives of the alcohol and acid from which they are prepared. Esters are responsible for the odors and tastes of many fruits and flowers. Table 16-3 lists several common esters. The reader should determine the formulas of the alcohols and acids from which each ester is derived.

Table 16–3 Some Esters

NAME OF ESTER	FORMULA	ODOR
Amyl acetate	$CH_3CO_2C_5H_{11}$	Pear
Isoamyl acetate	$CH_3CO_2C_5H_{11}$	Banana
Octyl acetate	$CH_3CO_2C_8H_{17}$	Orange
Ethyl butyrate	$C_3H_7CO_2C_2H_5$	Rum
Methyl salicylate	$C_6H_4OHCO_2CH_3$	Wintergreen
Methyl butyrate	$C_3H_7CO_2CH_3$	Pineapple
Butyl butyrate	$C_3H_7CO_2C_4H_9$	

Many perfumes contain synthetic esters. Lilac perfume, for example, is synthetic, for there is at present no known method of extracting the fragrance from lilac blossoms. Waxes are esters of high-molecular-weight acids and alcohols. Beeswax, for example, is principally myricyl palmitate:

$$C_{15}H_{31}\overset{\overset{\displaystyle O}{\|}}{C}—O—C_{30}H_{62}$$

Esters of glycerol and high-molecular-weight acids are called glycerides, or more commonly, fats and oils. Fats are glycerides of saturated acids such as palmitic, $C_{15}H_{31}\overset{\overset{\displaystyle O}{\|}}{C}—OH$. Oils are liquid glycerides of unsaturated acids, e.g., oleic acid (structure given above). Natural fats are mixtures of solid and liquid glycerides. Palm oil, for example, is a mixture containing largely glyceryl oleate, a liquid, and glyceryl palmitate, a solid.

When fats are saponified—that is, allowed to react with an alkali such as sodium hydroxide—they form glycerol and the sodium salt of a fatty acid. Saponification of glyceryl stearate would for example give:

$$\begin{array}{l} H_2\text{—C—O—}\overset{\displaystyle O}{\overset{\|}{C}}\text{—}C_{17}H_{35} \\ \quad\;\, | \\ H\text{—C—O—}\overset{\displaystyle O}{\overset{\|}{C}}\text{—}C_{17}H_{35} \\ \quad\;\, | \\ H_2\text{—C—O—}\overset{\displaystyle O}{\overset{\|}{C}}\text{—}C_{17}H_{35} \end{array} + 3\,NaOH \rightarrow \begin{array}{l} H_2\text{—C—OH} \\ \quad\;\, | \\ H\text{—C—OH} \\ \quad\;\, | \\ H_2\text{—C—OH} \end{array} + 3\,C_{17}H_{35}COONa$$

Sodium stearate, a salt of a metallic ion and a high-molecular-weight carboxylic acid, is a constituent of soap.

When fats are eaten they are digested slowly, being broken down by enzymes into their component acids and glycerol. These lower-molecular weight substances are absorbed into the blood stream through the wall of the intestine and eventually recombine to form fats. Once in the blood, fat may be transported to and stored around organs such as the liver and spleen as fatty tissue, serving as an insulator or as a reserve energy source.

CARBOHYDRATES

Sugars, starches and cellulose are members of a class of compounds known as *carbohydrates.* The word carbohydrate is a hold-over from earlier times. These compounds were observed to contain hydrogen and oxygen in the ratio of 2:1 as in water and were believed to be "hydrates" of carbon. The assumption was reasonable since carbohydrates such as glucose ($C_6H_{12}O_6$) and sucrose ($C_{12}H_{22}O_{11}$) may be represented by the general formula $C_x(H_2O)_y$. It is now recognized that not all carbohydrates have the 2:1 hydrogen-oxygen ratio. Carbohydrates are defined as polyhydroxy aldehydes or polyhydroxy ketones, or compounds which upon hydrolysis (reaction with water), yield either or both of these. Two sugars, glucose and fructose, will serve as examples:

$$\begin{array}{cc} \begin{array}{l} \\ H\text{-C=O} \\ \quad | \\ H\text{-C-OH} \\ \quad | \\ H\text{-C-OH} \\ \quad | \\ H\text{-C-OH} \\ \quad | \\ H\text{-C-OH} \\ \quad | \\ H\text{-C-OH} \\ \quad | \\ \quad H \end{array} & \begin{array}{l} \quad H \\ \quad | \\ H\text{-C-OH} \\ \quad | \\ \quad C\text{=O} \\ \quad | \\ H\text{-C-OH} \\ \quad | \\ H\text{-C-OH} \\ \quad | \\ H\text{-C-OH} \\ \quad | \\ H\text{-C-OH} \\ \quad | \\ \quad H \end{array} \\ \text{Glucose} & \text{Fructose} \end{array}$$

Glucose and fructose are similar in having five hydroxyl groups each, but differ in that fructose is a ketone, whereas glucose is an aldehyde.

Sugars may be classified as monosaccharides or disaccharides. Monosaccharides are considered the fundamental unit of sugars because they cannot be broken down into simpler carbohydrates. Most of the monosaccharides have the molecular formula $C_6H_{12}O_6$.

Glucose is a constituent of fruit juices and is found in the blood of animals. Less sweet than sucrose, it is produced in large amounts for the candy industry by the hydrolysis of starch. Medical patients who are unable to take nourishment orally are often given glucose intravenously. Fructose, which is many times sweeter than glucose, is responsible for the exceptional sweetness of honey.

Disaccharides, the best known of which is sucrose (table sugar), have the general formula $C_{12}H_{22}O_{11}$. They can be thought of as composed of two monosaccharide units joined together by elimination of a water molecule. In acid solution a sucrose molecule hydrolyzes giving one molecule each of glucose and fructose.

Only monosaccharides can be absorbed by the intestine and passed into the blood stream. Sucrose must therefore be converted by enzyme action of the digestive juices into fructose and glucose before utilization by the body. These monosaccharides may serve two functions when they have entered the blood stream, the production of heat and energy by instantaneous oxidation, or storage in the liver for future use.

Starch and cellulose are *polysaccharides* having the molecular formula $(C_6H_{10}O_5)_x$. In the cellulose molecule, x usually has a value of 1000 to 2000. Cellulose is the principal constituent of the heavy cell walls of plants. Wood is more than 50 percent cellulose, cotton nearly 100 percent. Starch molecules may have as few as 300 or as many as 6000 $C_6H_{10}O_5$ units. Starches may be converted into sugars in the human body and hence are digestible.

AMINO ACIDS AND PROTEINS

The amino acids constitute another important group of biologically important compounds. They are carboxylic acids which also contain an amino ($—NH_2$) group. Amino acids are the basic structural units of proteins. In the molecule of the protein-forming amino acid both the amino and carboxyl groups are attached to the same carbon atom. Three illustrations of amino acids are:

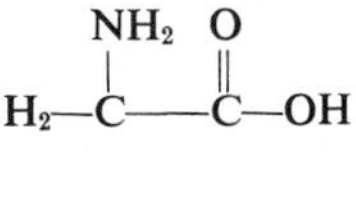

$$\begin{array}{ccc} & NH_2 & O \\ & | & \| \\ H_2— & C — & C—OH \end{array} \qquad \begin{array}{ccc} & NH_2 & O \\ & | & \| \\ CH_3— & C — & C—OH \\ & | & \\ & H & \end{array} \qquad \begin{array}{cccc} & H & H & O \\ & | & | & \| \\ CH_3— & C — & C — & C—OH \\ & | & | & \\ & CH_3 & NH_2 & \end{array}$$

Glycine 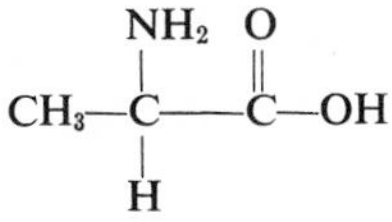 — Alanine 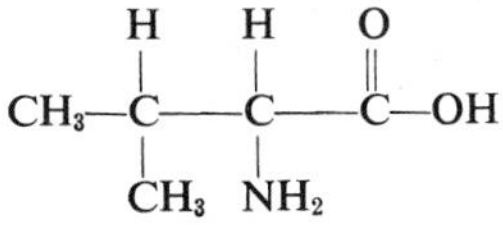 — Valine

Two amino acids may react with one another to form a dipeptide. For example:

$$H_2N-\underset{\displaystyle H}{\overset{\displaystyle H}{\underset{|}{\overset{|}{C}}}}-\overset{\displaystyle O}{\overset{\|}{C}}-\boxed{OH + H}-NH-\underset{\displaystyle H}{\overset{\displaystyle CH_3}{\underset{|}{\overset{|}{C}}}}-\overset{\displaystyle O}{\overset{\|}{C}}-OH$$

$$\Big\downarrow -H_2O$$

$$H_2N-\underset{\displaystyle H}{\overset{\displaystyle H}{\underset{|}{\overset{|}{C}}}}-\overset{\displaystyle O}{\overset{\|}{C}}-\overset{\displaystyle H}{\overset{|}{N}}-\underset{\displaystyle H}{\overset{\displaystyle CH_3}{\underset{|}{\overset{|}{C}}}}-\overset{\displaystyle O}{\overset{\|}{C}}-OH + H_2O$$

A dipeptide

The reaction may be represented as one in which a water molecule is eliminated between two molecules of the amino acid. The peptide linkage $-\overset{\displaystyle O}{\overset{\|}{C}}-\overset{\displaystyle H}{\overset{|}{N}}-$ is said to result. In a similar way dipeptide and tripeptide linkages may be formed. Compounds in which several peptide linkages are found are called *polypeptides.*

Proteins are present in all living cells. They are giant molecules with molecular weights ranging from about 10,000 to many millions. Muscles, nerves, nails, skin, and hair are rich in proteins. The casein of milk and the albumin of eggs are foods which have an abundant supply of proteins. About twenty-five amino acids have been isolated as products of the hydrolysis of different proteins. X-ray studies show that the polypeptide chain of proteins is a coiled helical or corkscrew structure.

Proteins are digested in the body by enzyme action, the peptide linkages are broken, and the lower-molecular-weight amino acids are formed. These are absorbed into the blood stream and circulated for use in particular tissues. Thus growth is sustained.

The nucleic acids are a fascinating group of compounds; they are found in the cells and perform a most important function in the building of proteins. The exact manner in which they perform this function is not yet known.

There are two types of nucleic acids, ribonucleic acid (RNA) and deoxyribonucleic acid (DNA). The latter is found in the cell nucleus and is the medium which passes on to progeny the genetic history of an individual. The function of the RNA is to transmit the hereditary message present in the DNA molecule to the site of the protein synthesis. RNA also serves as a pattern or template upon which new protein material is constructed with the correct amino acid sequence.

Nucleic acids break down by hydrolysis into smaller units called *nucleotides*. The different arrangements of these nucleotides account for the differences between nucleic acid molecules. Nucleotides are thus analogous to the amino acids of a polypeptide chain. Nucleotides may be represented by the word formula:

Nitrogenous base—sugar—phosphoric acid

The sugar contains five carbon atoms (a pentose). In RNA it is *ribose;* in DNA the sugar has one less oxygen atom and is called *deoxyribose.*

H O
C
H—C—OH
H—C—OH
H—C—OH
CH_2OH

Ribose

H O
C
H—C—H
H—C—OH
H—C—OH
CH_2OH

Deoxyribose

A nucleotide usually contains one of five nitrogen-containing bases of which there are two types, *purines* and *pyrimidines*. Purine bases are double-ring compounds such as adenine and guanine:

NH_2
C
N C—N
C—H
H—C C—N
N H

Adenine

O
C
H—N C—N
C—H
H_2N—C C—N
N H

Guanine

The pyrimidine bases are single-ring compounds and include uracil, thymine and cytosine:

OH
C
N CH
HOC CH
N

Uracil

OH
C
N CCH_3
HOC CH
N

Thymine

NH_2
C
N CH
HOC CH
N

Cytosine

A typical nucleotide is adenylic acid:

```
          NH2
           |
           C
         //  \
        N     C—N
        |     ||  \\
        |     ||    CH
        |     ||   /
     H—C      C—N
        \\   /     \        O                    O
          N          \   /     \                //
                      C   OH   OH  C—CH2OP—OH
                      | \  |    | / |       \
                      H   C——C    H        OH
                          |    |
                          H    H
```

Adenylic acid

Note that adenylic acid is composed of adenine, deoxyribose, and phosphoric acid.

The structures of the nucleotides have now been established but their manner of combination in the nucleic acid molecule is still not completely understood.

RNA and DNA differ not only in the sugar they contain but also in their pyrimidine bases. DNA contains adenine, guanine, *thymine,* and cytosine, whereas RNA contains adenine, guanine, *uracil* and cytosine. A portion of the DNA structure is shown in Fig. 16-5.

In 1954 the British biophysicist Crick and the American biologist Watson proposed that the DNA molecule consists of a double helix wound cylindrically. The helix consists of paired elongated nucleotide chains. Each chain may duplicate another but extends in the opposite direction so the two fit together much as two ropes might be intertwined. Hydrogen bonds between the nitrogenous bases lie across one another in the double helix and hold the two chains together.

The DNA molecule has the ability to reduplicate itself. This property permits daughter cells to contain the same genetic message as do parent cells.

CONCLUSION

The development has now been traced of a few of the important concepts of chemistry and physics. At the dawn of history man turned his thoughts to his everyday needs without paying much attention to the question of how the practical gadgets that he devised operated. The Greeks sought to explain many of the natural phenomena they observed, but made relatively little use of experimental methods. As science flowered in the sixteenth and seventeenth centuries, one of its important characteristics

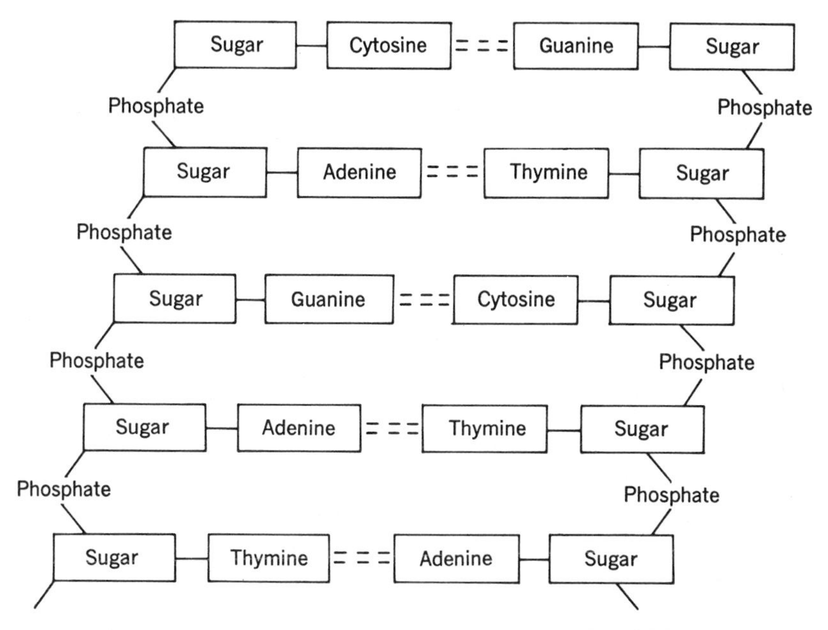

Fig. 16-5. *A representation of part of the DNA helix.*

was its increased attention to experimentation. Scientific societies, such as the Royal Society in England, undoubtedly played an important role in the communication of ideas and hence in the growth of science. The greater availability of printed materials and increased opportunity of publishing results of scientific investigations were important contributing factors. Many scientists and historians regard this seventeenth-century period as the beginning of the era of modern science.

The eighteenth and nineteenth centuries witnessed a wide variety of scientific developments. So much progress was made that an outstanding scientist is reputed to have commented that "all the important scientific discoveries have now been made. Henceforth, all that is left to do is to improve the precision of measurement of the important constants of nature"—in other words to add a few more significant figures to such quantities as the mechanical equivalent of heat. Nearly all the developments which have been discussed in the last five chapters have taken place since this statement was made.

Enormous scientific advances have been made in our twentieth century. Since World War II many multimillion-dollar industries have been made possible by scientific research. Research is no longer almost exclusively a pursuit in which people may dabble during their spare time; it is, for many men, a profession by which they earn their livelihood. There are now research institutes in all parts of the United States and in all the leading countries of the world. The future of scientific research and investigation seems almost limitless.

Exercises, Chapter 16

1. State several reasons why organic compounds are so numerous.
2. Three isomers of pentane all have the same molecular formula, C_5H_{12}. Write their structural formulas.
3. What was the significance of Friedrich Wöhler's experiment of 1828?
4. What is the difference between a substitution and an addition reaction? Illustrate each.
5. Write balanced equations for the complete oxidation of (a) nonane, C_9H_{20}; (b) hexane, C_6H_{14}.
6. Explain why benzene is regarded as having resonance structures.
7. The chemistry of silicon is less extensive than that of carbon. Explain.
8. Name four different alcohols and write the formula for each. Which if any are isomers? Which are primary, secondary, or tertiary? Why?
9. Describe how amino acids are believed to link together to form protein molecules.
10. What functional group do an aldehyde, a ketone, a carboxylic acid, and a carbohydrate molecule have in common? How do they differ? Illustrate.
11. Starting with acetaldehyde and any other reagents of your choice, how would you prepare (a) acetic acid, (b) methyl acetate?
12. How would you prepare a soap from a fat? Write an equation for a typical reaction.
13. What very unusual property does the DNA molecule have?
14. Discuss the digestion of fats, carbohydrates, and proteins in the human body.
15. To what class of organic compounds does each of the following belong? CH_3OH, CH_3COCH_3; $CH_2(OH)CH(OH)CH_2OH$; CH_3COOH; $CH_3COOC_2H_5$; $CH_3CH(NH_2)COOH$.
16. Why are starch and cellulose called polysaccharides?
17. Comment on the following statement: Nucleotides are to nucleic acids as amino acids are to polypeptides.

17 A journey through time and space

SUN, MOON, and stars, planets, meteors, and comets—these are the age-old objects of the science of astronomy. To them recent years have added satellites, space travel, and in the farthest reaches of space "radio stars," clouds of gas and dust, and distant galaxies. For thousands of years the serious study of celestial objects was linked at every point with religion and ritual. This linkage was severed a scant three centuries ago. Modern astronomy is a natural science; the gods and superstitions banished from it have retreated into a body of lore with a similar name—astrology—which remains among us as an entertaining curiosity.

The Solar Family As we embark on a quick journey through space to the galaxies, let us take a glance at the nearest celestial object—the planet we live on. The Earth, about 7900 miles in diameter, is almost a sphere, slightly flattened at its poles and bulging at the equator. With respect to the stars, the Earth rotates or spins on its axis through exactly 360 degrees in 23 hours and 56 minutes. This rotation causes the apparent daily east-to-west movement of all celestial objects, including the sun, and subdivides time into day and night. The Earth also revolves about the sun in a nearly circular orbit, making a complete swing of 360 degrees every $365\frac{1}{4}$ days, thereby subdividing time into years (Fig. 17-1). The Earth's average distance from the sun, nearly 93 million miles, is called an *astronomical unit*. (This is equal to the semimajor axis of Earth's elliptical orbit, p. 347.)

The sun's diameter is about 109 times the Earth's (864,000 miles vs. 7900 miles) and is less than a hundredth of the distance between the sun and the Earth. About 108 suns, arranged like beads on a string, would be needed to span 1 astronomical unit.

There are eight other major planets in the solar family (Fig. 17-2).

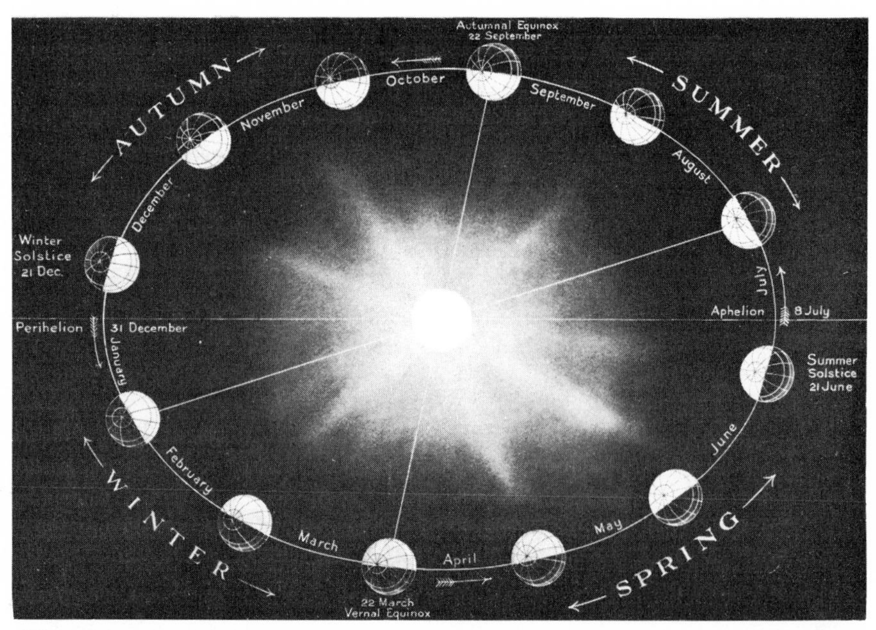

Fig. 17-1. *The Earth's position and motions in space. One end of the Earth's axis of rotation points almost at the North Star (Polaris, not shown). The axis makes an angle of 66½ degrees with the plane of the Earth's orbit about the sun; it is tilted 23½ degrees from a perpendicular to the plane. As the North Star is about 300 light-years away, the Earth's axis points toward it from all orbital positions.—To understand counterclockwise motion, imagine that you are in space somewhere beyond the North Pole looking "down" on the solar system and at a gigantic clock that is face up in the plane of the sun's equator. Look toward the clock at all times and consider only the motion on the part nearest you.—Again, imagine that you are in the Northern Hemisphere and facing north. West is to your left. Considered in this manner, the Earth always rotates from west to east, and this is the common direction of movement in the solar system. Planets revolve about the sun in this direction, the sun and the planets (except Uranus and perhaps Venus) spin on their axes in this direction, and most of the satellites revolve and rotate similarly. East and west are directions. The locations referred to by these terms are not fixed, but vary with the position of the observer.* [*Courtesy of the American Museum of Natural History.*]

Although exceptions occur, each planet is about twice as far from the sun as the next nearest. They all revolve around the sun in the same west-to-east direction, and their orbital planes are nearly parallel.

The sun is an enormous, gaseous, intensely hot body. Its mass is some 333,000 times that of the Earth, but its volume is about 1.3 million times greater. Thus its average density is about one fourth that of the Earth. The sun comprises about 699/700 of the matter in the entire solar system and is the ultimate source of most of its energy. With its retinue of planets, asteroids (planetoids), comets, and meteors, it speeds through space at about 12 mi/sec toward the bright star Vega. Despite its great size, the sun is not outstanding as a star. Many stars are larger, hotter, and brighter than the sun, but an even greater number seem to be smaller, cooler, and less luminous. Some stars are much denser than the sun, whereas others are less dense. On the one hand, the sun is so huge that if the Earth were placed at its center, the moon (about 240,000 miles from the Earth) would lie about halfway to the sun's surface. On the other hand, the sun is so tiny that if it were placed at the center of the giant star Antares, the orbits of Mercury, Venus, the Earth, and Mars would all fit inside Antares, with room to spare!

On a scale in which 1 yard represents 1 million miles, the sun could be a balloon about 30 inches in diameter. Mercury, Venus, the Earth, and Mars would then be about 36, 67, 93, and 141 yards away, respectively. On this greatly reduced scale, Mercury and Mars could be represented by peas, and Venus and the Earth by small marbles. Jupiter, Saturn, Uranus, Neptune, and Pluto would be located at distances of approximately $\frac{1}{4}$, $\frac{1}{2}$, 1, $1\frac{1}{2}$, and 2 miles, respectively. Oranges about 3 inches in diameter could represent Jupiter and Saturn; plums could be used for Uranus and Neptune; Pluto could be the size of a pea. The moon could be a sand grain 9 inches from the marble representing the Earth.

Fig. 17-2. *Approximate sizes of planets compared with the sun. There is a mnemonic sentence for the relative distances of the planets from the sun: Mary's Violet Eyes Make Anguished* (*asteroids*) *John Stay Up Nights Permanently.*

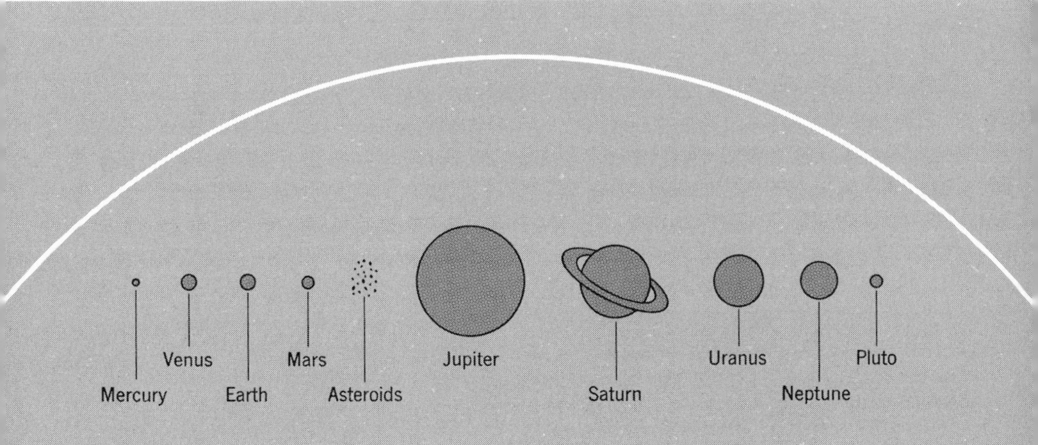

On this greatly reduced scale, on which 1 yard equals 1 million miles, the nearest star to the sun would be some 15,000 miles away. Perhaps an average star, less than 1 yard in diameter, is located some 10,000 to 20,000 miles from its nearest stellar neighbor (within a galaxy, average distances between neighboring stars may approximate 4 light-years). However, such averages encompass very wide ranges in sizes and distances. Some stars would be the size of corsage pinheads, whereas others would have diameters larger than a football field. Because many stars are members of clusters or of multiple-star systems, their distances apart are far from uniform.

The Light-Year Seeking a unit of distance adequate for the vast reaches of interstellar space, astronomers have had recourse to the *light-year*—the distance that light travels in 1 year while moving, as it does, at the astonishing rate of about 186,000 mi/sec: $186,000 \times 60 \times 60 \times 24 \times 365\frac{1}{4}$, or about 6 trillion miles. If 1 astronomical unit (93,000,000 miles) is reduced to an inch, then a light-year equals about a mile in length. At the unimaginable speed at which light travels, a beam of light sent from the Atlantic to the Pacific across the United States (nearly 3000 miles) and back again would make about 31 round trips in just 1 second.

The star nearest to the Earth (other than the sun) is approximately 4.3 light-years away. This star, Alpha Centauri, is the third brightest star in the Earth's sky and is located about 30 degrees from the south celestial pole. It is not visible from the latitude of New York. All stars are so very distant from the Earth that the hugest of them registers as a mere pinpoint of light in even the largest telescope. Astronomers have no direct means of determining whether any of the stars, even the nearest ones, have planetary systems of their own. (The very slightly sinuous path of Barnard's star is apparently caused by the gravitational attraction of a planet-sized object orbiting about it; see p. 461.)

Seeing the Past Light from the sun requires eight minutes to travel to the Earth. When we look at the sun, we see it as it actually appeared eight minutes earlier. The light we see when we look at the nearest star left that star more than four years ago and has been traveling at the tremendous speed of 186,000 mi/sec ever since. If a terrific explosion occurred on this star two years ago, astronomers will not find out about it for another two years! In 1972 we shall see the star as it was in 1968; in 1973, we shall see light that left the star in 1969, etc. Other stars are measured by astronomers to be tens, hundreds, millions, and even hundreds of millions of light-years away. Whenever we look at the stars, we are "seeing the past"; we may be seeing one star as it appeared a generation ago, another as it appeared twenty centuries ago, and still another

Fig. 17-3. *Clouds of gas and dust in space form dark and bright nebulae in the Horsehead Nebula in Orion. Bright nebulae are illuminated by the light of stars located within them. Dark nebulae obscure the light of the stars beyond them. Relatively few stars occur between us and the dark nebulae or within them.* [*Mount Wilson and Palomar Observatories.*]

as it appeared in the remote geologic past.* This is an astronomer's "time machine."

Thus our knowledge of the distances separating objects involves both space and time. If the distance to a certain galaxy is measured as 1 billion light-years, we know that the location of this galaxy 1 billion years ago was 1 billion light-years from our present location. Where we were in space when the light left this galaxy is uncertain, as is the location of this galaxy at the present time. In effect, we see remote celestial objects by "fossil starlight."

Stars and Nebulae Vast clouds of gas and dust, called *nebulae* (Fig. 17-3), occur within some galaxies. Though these clouds may be many

* The estimated approximate distances in light-years to some of the brightest stars follow (arranged in order of decreasing brightness as viewed from the Earth): Sirius, 9; Vega, 26; Capella, 46; Arcturus, 36; Rigel, 650; Procyon, 11; Altair, 16; Betelgeuse 650; Aldebaran, 68; Spica, 160; Antares, 170; and Deneb, 540.

Fig. 17-4. *An edgewise view of a spiral galaxy (NGC 4565 in Coma Berenices). Our galaxy would probably have a similar shape if seen edge on. Individual stars seen in the photograph belong to our own galaxy, and we look past them to see other galaxies far beyond. In some photographs stars show four spikes caused by the diffraction of starlight around supports for a mirror located inside the telescope. The dark strip represents interstellar clouds of gas and dust in the central plane of the galaxy—along its own equator.* [*Mount Wilson and Palomar Observatories.*]

light-years across, they are so thin and diffuse that an average sample is more nearly a vacuum than can be produced on Earth. Hydrogen may constitute three fourths of all the mass in the nebulae and in the entire universe, and helium may make up about one quarter. All of the other elements combined constitute only 1 to 2 percent or so of the total.

Stars apparently have formed, and are forming now, wherever gravitational attraction can cause part of a cloud of gas and dust to contract into a sphere. Temperatures and pressures increase in the central portion of the gaseous contracting sphere until thermonuclear reactions can occur—as if a number of so-called hydrogen bombs were being exploded each second. At this stage the sphere of gas can be called a star, and contraction ceases. The gravitational attraction that tends to cause contraction is in equilibrium with the thermonuclear reactions, high temperatures, and radiation that tend to cause expansion.

The amount of matter existing as clouds of interstellar gas and dust is fundamental in speculations concerning the future of the universe. New stars cannot form in regions where this interstellar matter is absent or too diffused. How much exists? Some estimates suggest that only 1 to 2 percent of all the matter in the universe is interstellar matter; other estimates are considerably higher. In some regions of space, e.g., the arms of a spiral galaxy, interstellar matter may make up about half of the total, and in such regions stars can continue to form and evolve.

Galaxies and the Universe Although galaxies differ in shape, many of them—the spiral varieties—have a thin, disklike form and spiral arms; in edge view, they look like two saucers placed rim to rim with the bottoms outward (Figs. 17-4 and 17-5). Other galaxies are elliptical,

Fig. 17-5. *A spiral galaxy viewed at a 90-degree angle to its equatorial plane (M101 in Ursa Major). Our galaxy may have a similar pinwheel shape, with the solar system in one of its spiral arms about 30,000 light-years from the center. Our galaxy may have a diameter of about 100,000 light-years and its central disk a maximum thickness of some 10,000 light-years. The shape of the galaxy indicates that it is rotating; i.e., the stars in it are revolving in nearly circular orbits about the central nucleus. If such a galaxy were not spinning, the mutual gravitational attraction of its stars and interstellar gas clouds would cause it to collapse.* [*Mount Wilson and Palomar Observatories.*]

globular, or irregular in shape, and gradations occur among the different types; the gradations suggest that galaxies may evolve, but very little is known about galactic evolution.

The Milky Way—the familiar faint band of light in our sky that is visible to the unaided eye—marks the equatorial plane of the galaxy which contains the solar system and which we commonly call the Milky Way galaxy or "our" galaxy (Fig. 17-6).

Fig. 17-6. *Panorama of the Milky Way, equatorial view. At the left are bright regions in Cassiopeia and Cygnus, with the great dark rift in Cygnus and Aquila left of center. In the lower right we see the Magellanic Clouds.* [*Prepared by Martin and Tatyana Keskühla; courtesy, Lund Observatory, Sweden.*]

The solar system forms a very tiny part of a single galaxy. All the stars we ever see at night with the unaided eye are in this galaxy and relatively near to us. Our sun is merely one of the billions of stars in our galaxy, not distinguished from many of its distant companions by size, composition, temperature, or other known properties. Astronomers assume that planetary systems are also commonplace (though they lack direct evidence of this). The magnitudes of the universe are thus on a scale that far outdistances the range of the imagination.

Galaxies appear to be distributed throughout the universe in clusters

(Fig. 24-11), and the number of galaxies in any one cluster ranges from a few to more than one thousand. Average distances separating neighboring galaxies within a single cluster may approximate a million light-years or so, but the average distance separating neighboring clusters may be a hundred times greater.

Galaxies differ greatly in size and probably range from about 7000 to 150,000 light-years in diameter. Thus our galaxy, about 100,000 light-years across, may be several times larger than average.

One of the really fascinating aspects of the universe is the possibility that it may be expanding. According to one view, the distances between clusters of galaxies are increasing, and at rates that are proportional to their distances from us. Nearer galaxies are moving away from us at slower rates than remoter galaxies. The evidence for this involves a phenomenon known as the red shift (p. 475). Three competing hypotheses—the Oscillating Universe, the Big Bang, and the Steady State—attempt to explain this relationship. We shall discuss these later (p. 476).

Although the enormous size of the universe fills us with wonder, perhaps even more amazing is the knowledge that, so far as man has been able to determine, the same physical laws that apply to the Earth hold true throughout the universe; matter appears to consist of the same familiar elements, although hydrogen and helium are far more abundant than on the Earth; light seems to be produced and transmitted in a similar manner; celestial bodies apparently obey the laws of motion and gravitation; and the energy-production processes in other stars appear to be similar to those in the sun. Since uniformity seems to be present throughout, we can extrapolate with some confidence to the rest of the universe the knowledge which has been obtained by studying the tiny portion accessible to us at relatively close range.

As astronomy developed, it upset prevailing ideas concerning man's place and role in the universe. Once the Earth was thought to be the center of a rather small universe. Then the sun became the center, and the status of the Earth was reduced to that of a rotating, revolving planet; but the universe was still considered to be relatively small. In the next revolutionary discovery, the sun's status was shifted to that of a peripheral star in a galaxy containing millions of other stars. Still later it was learned that other galaxies existed, in vast numbers and at great distances. Now we are in the midst of yet another great change in our ideas about the universe. Are we alone and unique in space, or do intelligent beings inhabit other planets orbiting about other stars?

The Constellations Groups of brighter stars within our own galaxy form patterns for the skywatcher—dippers, crosses, squares, and circles—and

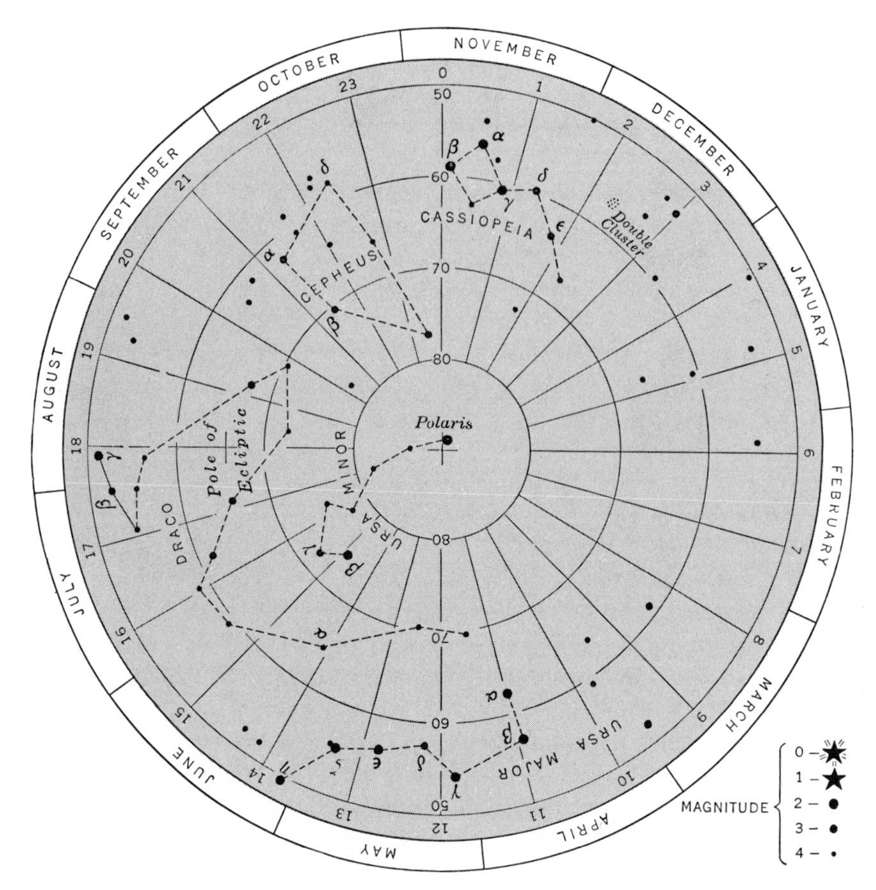

Fig. 17-7. *The northern constellations as viewed from 40 degrees north latitude. The stars in a constellation are designated by small letters of the Greek alphabet in order of brightness: the brightest star is alpha, the next brightest is beta, etc. Some of the brighter stars are known by their own special names.—Hold the map with the current month at the top to see the constellations as they appear at 9:00* P.M. *standard time. For any time other than 9:00* P.M., *rotate the map through the proper number of hours, counterclockwise for a later time and clockwise for an earlier time. For example, the Big Dipper will be high in the northern sky at 9:00* P.M. *in May or at 9:00* A.M. *in November.*

are called *constellations.* Most of the familiar constellations were named long ago by people with vivid imaginations, who traced in the sky the lines and symbols of earthly objects—bears, fish, heroes (Fig. 17-7). It need hardly be said that the resemblances are not as striking in all instances as they are in some. Yet constellations and bright stars are useful in designating areas in the sky, much as states and cities serve to locate places on the Earth's surface. In modern usage, the constellations represent specific areas bounded by straight lines, like some state boundaries.

To become familiar with the outstanding constellations is a pleasant and rewarding task. The joy of a clear, starlit evening will forever be enhanced by this familiarity, once gained. Inexpensive star charts are available which show where the constellations appear in the sky at any hour during the year. These will prove most helpful.

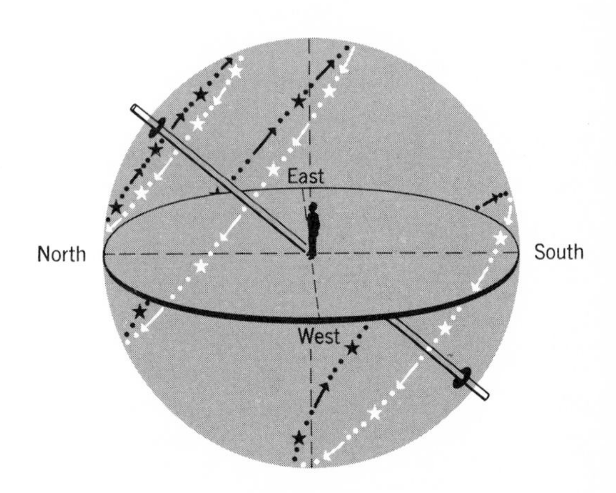

Fig. 17-8. *The Earth, sky, and stellar motions as they appear to an observer at 40 degrees north latitude. The Earth seems flat and the sky looks like the inside of half a sphere. The North Star remains about 40 degrees above the horizon at all seasons, and other stars appear to move from east to west (counterclockwise) in circular paths. Stars near Polaris never set, whereas stars farther away (i.e., making a larger angle with it) rise and set* [*After Rey.*]

The stars in a given constellation need not be closely associated in space as they are for the eye. Two stars shining side by side may look like intimate neighbors and yet be hundreds of light-years apart. When a given celestial object is said to lie in a particular constellation, this means that it will be visible in the general direction of the constellation. The object itself may be situated between the Earth and the greater number of the stars in the constellation, or it may be far out in space beyond them.

The Celestial Sphere Although some stars appear brighter than others, our eyes have no accurate depth perception at great distances and so one star does not seem closer to the Earth than another. All stars appear to be at the same distance from us, and together they seem to form a gigantic inverted bowl. Since we can make this same observation from any position on the Earth, the Earth seems to be at the center of a huge hollow sphere. The stars appear to line the inside of the sphere which seems to turn on a sloping axis daily from east to west (Fig. 17-8). The North Star is very close to the northern end of this axis (the *north celestial pole*) and follows a small circular orbit in the sky. In the Northern Hemisphere, the altitude of the North Star above the horizon is approximately equal to the latitude of the observer (Fig. 19-1).

The name *celestial sphere* has been given to this imaginary shell. Distances separating celestial objects are measured by the angles they subtend on the celestial sphere. Perhaps it is helpful in estimating the angle between two stars to imagine that a string extends in a straight line from each star to your eye. Obviously, it is meaningless to state that a certain star is "100 feet above the horizon," but the altitude of the star can be given in angular degrees. The following angular distances can be used as a scale in estimating other distances:

- The angular diameters of the sun and moon are about one-half degree.
- The pointers in the Big Dipper are about 5 degrees apart.
- The square of the constellation Pegasus is 15 degrees on a side.

Exercises and Questions, Chapter 17

1. Get a star chart and learn to use it (e.g., a Bennett-Rice Star Explorer, American Museum–Hayden Planetarium, New York, N.Y. 10024). After a little practice, you should be able to identify the more prominent constellations and stars in the sky at different times of the year and at different hours during any one night.

   Find a location for your observations where interfering lights can be blocked out. A red-bulbed flashlight will be useful in checking your star chart at night. (Coat the glass with red nail-polish or cover it with red cellophane.)

2. Observe at least one constellation in each of the four main compass directions—north, east, south, and west. Make a star-map sketch of each of these and label the date and hour of observation. Make similar observations during subsequent months at the same hour (and now at different hours of the night). Plot these different positions on your original star maps so that comparisons can be made.

   Perhaps you can contact students who live at different latitudes in other parts of the United States or in other countries. Persuade them to make star-map sketches of the same constellations and then exchange copies. Explain any differences.

3. With the aid of a star chart, list five winter constellations (prominent in the early evening sky in the winter), five summer constellations, and five nonseasonal constellations.

4. How do you "see the past" when you make observations in astronomy?

5. One evening a friend asks you to point a flashlight beam (a powerful five-cell flashlight makes an excellent pointer in studying stars) parallel to the Earth's axis and to swing it in the sky parallel to the Earth's equator. How can you do this?

6. Calculate the approximate dimensions of each of the following on a scale of 1 inch = 1000 miles. On this scale, the Earth is 7.9 inches, somewhat smaller than a basketball. This is an exercise to develop an awareness of relative sizes and distances of various celestial objects within our galaxy. Therefore, do not perform tedious additions or multiplications. Round off quantities wherever convenient (e.g., change a number such as 383,450 to 400,000). Use 5000 ft/mi instead of 5280 feet and use 60,000 in/mi instead of 63,360 inches:
   (a) Diameter of the moon (2160 miles).
   (b) Distance from the Earth to the moon (239,000 miles).
   (c) Diameter of the sun (864,000 miles).
   (d) Distance from the Earth to the sun (93,000,000 miles).
   (e) Distance from the sun to Pluto (3,670,000,000 miles).
   (f) Distance from the sun to Alpha Centauri (about 4.3 light-years; 1 light-year is about 6,000,000,000,000 miles).
   (g) Approximate diameter of our galaxy (i.e., the distance across the equatorial plane from one edge, through the nucleus, to the opposite edge); use 100,000 light-years as the diameter.

7. Calculate the approximate speed of light in mi/hr.

8. At a uniform rate of 5 mi/sec, how long would it take to move from the sun to Pluto? from the sun to Alpha Centauri?

9. What is the approximate ratio between the diameters of the sun and moon? What is the approximate ratio between the distances separating the sun and moon from the Earth? Do these ratios explain why the sun and moon appear to be about the same size?

10. How many astronomical units (mean distance between Earth and sun) does the Earth travel in making one orbit about the sun (assume a circular orbit)? Calculate the Earth's average orbital speed.

18 Does the earth move?

AS viewed from New York or California, the sun rises each morning above the eastern horizon, slants upward until it is highest in the southern sky at noon, and then slants gradually downward to disappear below the western horizon in the evening. This motion is known to all. Other celestial objects—the moon, stars, and planets—share this daily east-to-west movement. Although all these bodies appear to move through the heavens (we use words such as sunrise and sunset), we learn early in life that this is only an apparent motion caused by the spinning of the Earth in the opposite, west-to-east direction. Seeing is not believing in this instance. The explanation of celestial motion which is based on the Earth's rotation has been generally accepted for only about 300 years. The opposite explanation, that the Earth was motionless and that the sky moved, had been believed more or less without opposition for the preceding two millennia, or indeed as long as there have been men to ask for explanations.

DAILY WESTWARD MOVEMENT OF CELESTIAL BODIES

The circular paths made by star trails during the daily (*diurnal*) westward movement of the stars grow larger in diameter toward the equator and smaller in diameter toward the poles (Fig. 18-1). The North Star completes a tiny circular orbit each day (radius: 1 degree).

Careful observation shows that stars complete their daily westward swings through the sky in about 23 hours and 56 minutes—the *sidereal day*. On the other hand, the sun requires about 24 hours to complete its daily journey through the skies—*the solar day*. This difference between

Fig. 18-1. *Star trails in the northern sky. The photograph was made in two parts: (1) A time exposure of about 1⅔ hours was made through a motionless camera pointed at the northern sky. The daily east-to-west rotation of the heavens produced the concentric star trails shown. (2) A few minutes after Part 1 had been completed, the camera was rotated so that it moved at the same rate and in the same direction as the stars. Thus light from any star was constantly in focus on the same spot on the film. A photograph of the constellations resulted. The trail of a passing airplane can be seen as well. [Photograph, John Stofan, Teaneck, N.J.]*

the sidereal day and the solar day is caused by the revolution of the Earth around the sun (Fig. 18-2).

Since the Earth moves about 1 degree each day in its orbit, the sun appears to move eastward about 1 degree a day with respect to the fixed stars. This makes the sidereal day 4 minutes shorter than the mean solar day. Stated in another way, there is a daily westward shifting of the stars with respect to the sun.

Since our clocks keep time with the mean solar day, each evening any given star is observed to rise 4 minutes earlier than it rose on the preceding evening and to set 4 minutes earlier. If a certain star appears above the eastern horizon at 9:00 P.M. on, say, September 1, it will appear at 8:56 P.M. of the evening of September 2, and at about 7:00 P.M. on the evening of October 1 (30 days at 4 minutes per day). At 9:00 P.M. on October 1, the star will be located $\frac{1}{12}$ of a complete circle, or 30 degrees, west of the position it occupied a month earlier at 9:00 P.M. Since this daily westward shift of the stars is caused by the Earth's revolution around the sun, the stars will have shifted 360 degrees in a year, and the same star will again appear above the eastern horizon at 9:00 P.M. the following September 1.

Constellations which are prominent in the early evening hours during the winter months are known as winter constellations; those prominent in the early evening hours of the summer are known as summer constellations.

If we observe the positions of stars on a certain hour and date, we can

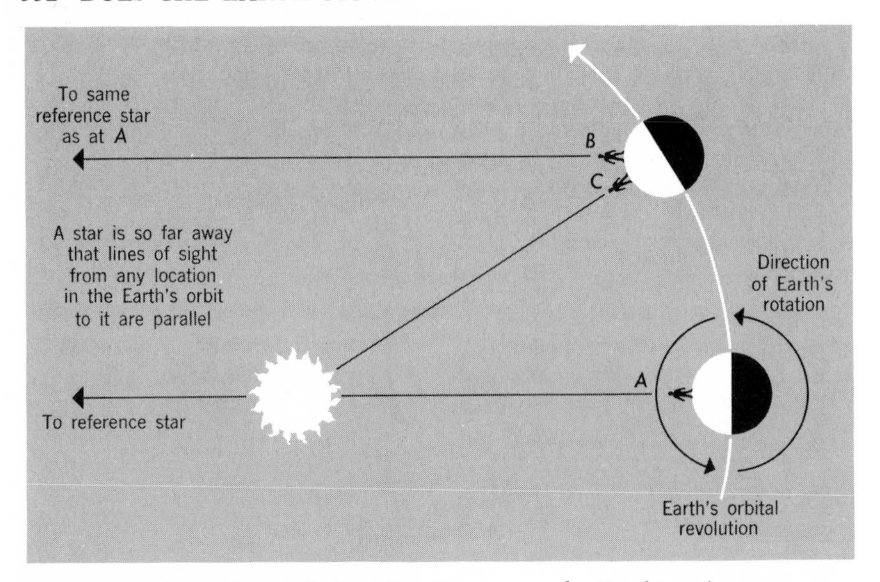

Fig. 18-2. *Solar vs. sidereal day. An observer on the Earth at A at noon sees the sun and a reference star lined up exactly, due south and high overhead. The observer is at B 23 hours and 56 minutes later because of the Earth's rotation and revolution. The Earth has rotated completely through 360 degrees. The same reference star is again due south and high overhead, but the sun will not reach a due-south position until the Earth has spun the observer from position B to position C, which takes about 4 minutes.—The Earth's orbital velocity varies. It moves fastest in January when closest to the sun; so in January the distance from A to B is greater than it is in June (the Earth covers more miles during the sidereal day of 23 hours and 56 minutes). The distance from B to C is also greater in January than in June; i.e., slightly more than 4 minutes of rotation is needed to bring the sun again to the due-south position. Thus variation in orbital velocity tends to make solar days longest during the winter months. Another factor: the sun shifts eastward more rapidly at a solstice than at an equinox (as projected to and measured along the celestial equator). Solar days average 24 hours.*

predict their positions at any other hour and date during the year because: (1) Stars complete their daily paths (diurnal circles) through the sky in about 24 hours; therefore, they shift $\frac{1}{12}$ of the distance in 2 hours, $\frac{1}{4}$ in 6 hours, and $\frac{1}{2}$ in 12 hours. This movement is in a counterclockwise direction for the observer facing north and clockwise for the observer facing south. (2) If observed at the same hour night after night, stars are also seen to shift their positions through 360 degrees once each year; they shift $\frac{1}{12}$ of the distance in 1 month, $\frac{1}{4}$ in 3 months, and $\frac{1}{2}$ in 6 months. The annual shift is in the same direction as the daily shift. To illustrate, we see Cassiopeia in a certain position at 8:00 P.M. tonight. Three months later and at 2:00 A.M. it will be shifted halfway around from its present position.

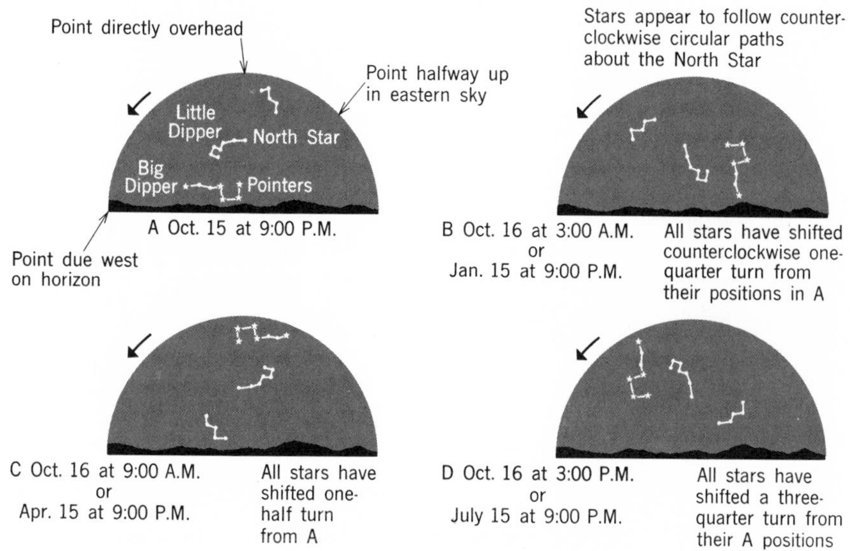

Fig. 18-3. *Star maps of the northern sky. Three familiar constellations are shown as observed from New York (40° latitude). The sky map shows one half of the sky visible to an observer, or one quarter of the celestial sphere. The star map shows two motions: (1) the daily rotation of the celestial sphere, which is counterclockwise as one faces the North Star (the three constellations shown here are too close to the North Star ever to set below the horizon at latitude 40 degrees) and (2) the annual shifting of the celestial sphere as the Earth revolves about the sun. This is also counterclockwise for an observer facing north.*

Fig. 18-3 illustrates both the nightly and the yearly westward movement of the stars.

In observing the motions of stars in the sky—motions that are actually caused by the rotation and revolution of the Earth—we note that stars move along concentric circles that are parallel to the Earth's equatorial plane. These circles center on the north and south celestial poles, and thus any one star keeps the same angular distance from Polaris. Stars near Polaris pass over it from right to left and beneath it from left to right (we face north to see this). Stars that have a greater angular separation from Polaris follow similar paths, but the horizon obscures parts of these. Thus they rise in the east and set in the west.

EASTWARD SHIFT OF THE MOON, SUN, AND PLANETS RELATIVE TO THE FIXED STARS

Stars actually are moving through space, some at rates of tens of miles each second, but they are located at such enormous distances from the

Earth that their movements are usually not discernible except over periods of a few hundred years (the most rapid shift is about ½ degree in 175 years). The Big Dipper did not always have its present shape, and since the stars in it are moving in different directions, its shape will continue to change over the centuries. Nevertheless, within a lifetime, the shapes of the Big Dipper and the other constellations do not change. During their diurnal and annual motions (caused by the rotation and revolution of the Earth), the stars maintain the same positions with respect to each other, and so they are termed "fixed." This easily distinguished them from planets, which look like stars in the sky to the unaided eye but are not "fixed" against the background of stars. (*Planetes* is the Greek word for "wanderer.")

Because we observe the sun from a moving Earth, the sun appears to move always to the east relative to the stars. If stars were visible when the sun is observed, this motion would be quite apparent. The moon, circling the Earth in a counterclockwise direction as seen from the north, also appears to move at a reasonably steady pace, and always to the east. Over a long period of time, the average motion of each of the planets is to the east (direct), but over shorter periods they sometimes move with retrograde (westward) motion. All of these motions are with respect to the fixed stars, and are, of course, superimposed on the much more rapid westward diurnal motion which results from the daily rotation of the Earth.

The Moon The eastward shift of the moon occurs at the rate of about ½ degree (the width of the full moon) per hour and is readily noticeable if the moon happens to be nearly in front of a bright star that can serve as a reference point. The moon rises and sets like the sun and for the same cause, but it rises at noticeably different times from one day to the next (Fig. 18-4). It may rise in the east during the daylight hours and not be a conspicuous object until it is viewed in the west after dark.

The Sun and the Stars Because of the Earth's orbital motion, the sun shifts eastward relative to the background of fixed stars and completes its swing in 365¼ days. This motion can be observed by noting the stars which appear in the eastern sky shortly before sunrise or in the western sky shortly after sunset. In addition, the sun makes an angle of 180 degrees with stars that are due south at midnight. Different stars will be seen in these parts of the sky with the advancing seasons. One full year is needed for a star to complete the cycle and return to its original position with respect to the sun. This is the basis for the subdivision of time into years. The circle was subdivided into 360 degrees by the ancient

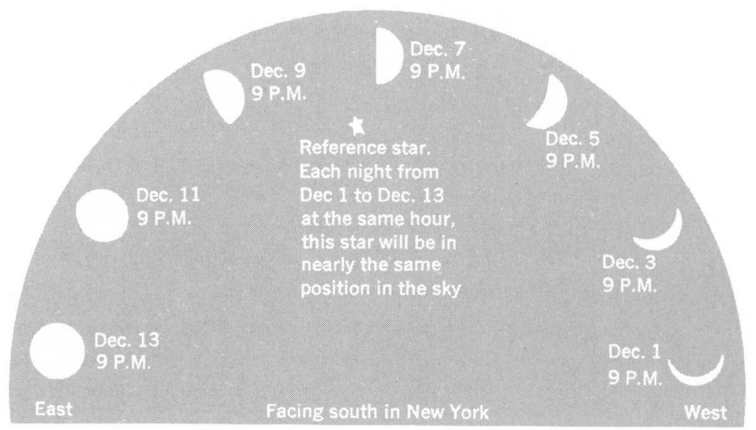

Fig. 18-4. *The moon's eastward shift relative to the background of fixed stars. Each evening at 9:00* P.M. *the moon is almost an hour east of its position on the preceding evening relative to the reference star. In about 1 month the moon completes an eastward swing among the stars and returns to its starting point. Of course the moon and the reference star participate each evening in the westward rotation of the heavens, but the relative positions of the moon and reference star change slowly and continuously. The same phases do not occur on the same days of a given month from one year to the next.*

sages of Mesopotamia because the eastward shift of the sun approximates 1 angular degree each day.

The ancients were puzzled for some ages before they could explain the disappearance of the stars during the day. Stars are, of course, still present in the sky during the day. A constellation that is high in the sky during a winter evening will be high in the sky during the day in the following summer. But in the same way that the stars become dimmer on successive nights as the moon waxes fuller, so also do they become completely invisible in the much brighter glare of the sun's light. Stars can be seen during the day during a total eclipse of the sun.

Planets Planets look like stars in the sky. Venus, Jupiter, Mercury, Mars, and even Saturn at certain times are brighter than any of the stars. Uranus is barely visible to the unaided eye. Neptune and Pluto can be seen only through a telescope. The brightness of a planet varies, depending in part upon its constantly changing distance from the Earth. Planets are not visible when their orbital positions place them in line with the sun (exception: a position 180 degrees from the sun). For the observer, planets tend to shine with a steady light, especially when high above the horizon, and stars tend to twinkle. However, small planets do twinkle vigorously when

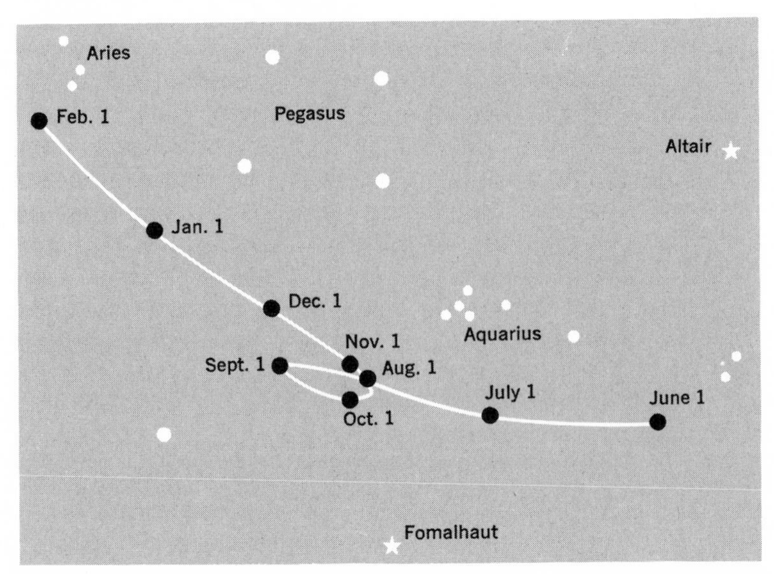

Fig. 18-5. *The path of Mars among the stars before and after opposition in 1939. See Fig. 18-9B for an explanation. The westward part of the loop (to the right in the diagram) is called retrograde motion.*

seen near the horizon. In a telescope, planets appear as disklike objects, whereas stars are tiny dots of light. Another distinction is the slow shifting of the planets against the background of fixed stars. The planet Mars, for example, rises and sets with the other celestial bodies, but each night the position of Mars is seen to have changed slightly relative to the fixed stars near it (Fig. 18-5). Usually the shift is eastward. Occasionally Mars shifts westward (retrogrades) for several weeks before resuming its eastward motion. Depending upon their distances from the sun, the planets take varying lengths of time to complete their eastward swings. The planets "wander" slowly, and the length of time a planet spends within the boundaries of a particular constellation varies with its distance from the Earth. For Mars, it averages about 2 months; Jupiter, 1 year; Saturn, 2 years; and Neptune, 13 years.

ORIGIN OF ASTRONOMY

Astronomy is one of the oldest of the sciences, partly because answers were desired a long time ago for such simple questions as: What time is it? When will you be back? Before watches, calendars, and the compass were invented, exact replies were difficult.

The following scene, with local variations, may have occurred regularly

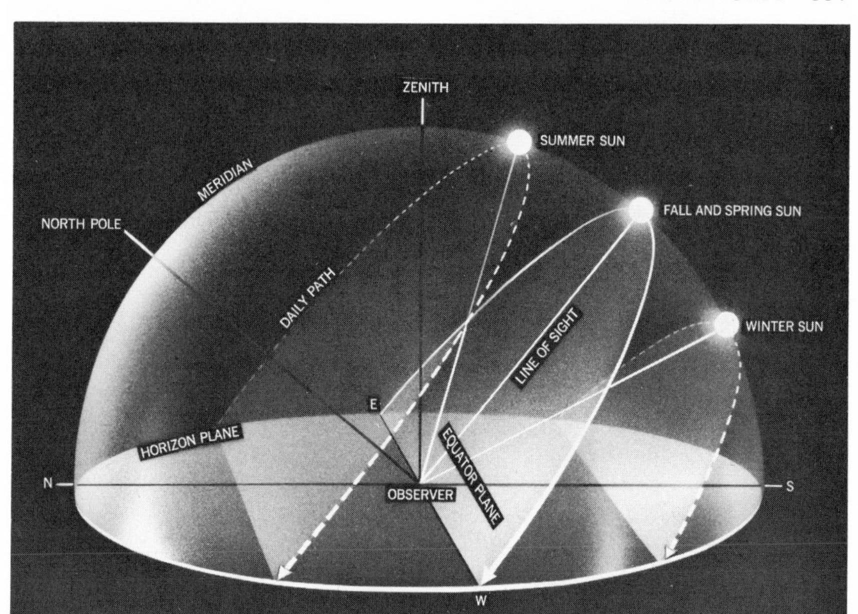

Fig. 18-6. *The sun's daily path through the sky during winter, spring, summer, and fall as seen from our northern latitudes. Such changes in the sun's path are caused by the Earth's tilt on its axis, its rotation, and its revolution (Fig. 17-1). During the winter the sun is visible for the shortest period of time and is not high in the sky at noon. It rises late in the southeast and sets at an early hour in the southwest. In the spring or fall, the sun rises in the east and sets in the west. In the summer the sun is visible for the longest time; it rises in the northeast, travels high overhead, and sets at a late hour in the northwest. The difference in altitude of the noon sun between winter and summer is 47 degrees.* [*Courtesy of the American Museum of Natural History.*]

in remote ages. A prehistoric man picks up a stout club and prepares to leave his cave. A woman calls to him, and the man points to a stone resting on the ground in front of the cave. This seems to satisfy the woman, and the man leaves. As the day wears along, the sun's shadow moves closer and closer to the stone (Figs. 18-6 and 18-7). When the shadow nears the stone, the woman builds a fire and begins to look expectantly into the distance. Soon her mate appears with a small deer which he has killed for their evening meal. His morning gesture told her that he would return when the sun's shadow reached the vicinity of the stone. Although he could not see the shadow while he was hunting, the man knew approximately when it would reach the stone by the sun's position in the sky.

Prehistoric man had a much greater awareness and knowledge of the existence and motions of celestial bodies than do most of today's civilized,

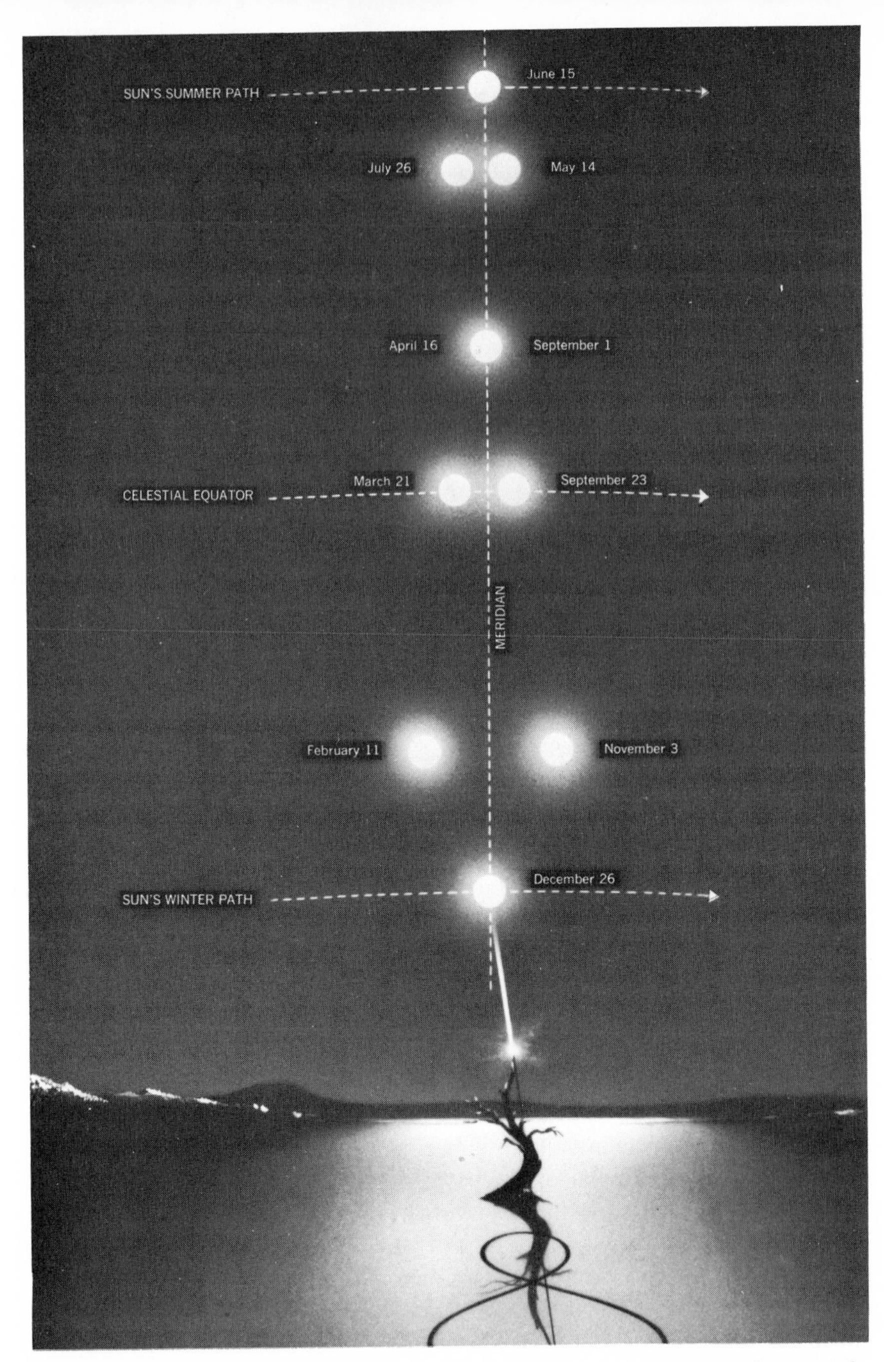

Fig. 18-7. *The sun's shadow at clock noon during a year traces a figure eight (analemma) on the ground. The sun is not due south at noon during most of the year because solar days vary slightly in length. A vertical stick (gnomon) can thus be used to subdivide time into years and lesser units. Its shadow always points northward at noon in northern latitudes. It is shortest in the summer when the sun attains its highest noon altitude. It thus takes six months for the shadow to shift from its longest to its shortest noontime position.* [*Courtesy of the American Museum of Natural History.*]

educated individuals. The splendor, beauty, and mystery of the heavens doubtless appealed to him; showers of meteors, the sudden appearance of a bright comet, a spectacular eclipse, stirred, frightened, and excited him. When men began to live in groups, knowledge of the heavens became necessary to them. Dates had to be arranged for future hunts and campaigns. The phases of the moon provided a ready calendar; it was soon learned that there were nearly 30 days between two successive full moons, and that approximately 12 full moons elapsed before a certain star rose again in the same place at the same time. On this basis the year was subdivided into twelve 30-day months. The light of the moon was once much more important in the conduct of life than it is today; e.g., evening journeys were best taken at times of full moon.

Shepherds may similarly have subdivided their night watches by the diurnal movements of familiar stars or constellations, each watch lasting the length of time necessary for certain stars to move through definite angular distances in the sky. The pole star was an important aid in determining directions. Specific seasonal events came to be associated with the positions of certain stars in the sky: the dates to pick elderberries or to plant corn, the mating seasons for cattle and deer, and the seasons of floods and frosts were all associated with the locations of certain bright stars in the sky that marked the time of the year. From this association of a star position with a seasonal event, it was only a step to ascribing the actual cause of the event to the star; e.g., the flooding of the Nile was attributed to the rising of Sirius in the east at dawn.

GEOCENTRIC AND HELIOCENTRIC EXPLANATIONS FOR OBSERVED CELESTIAL MOTIONS

The age-old view that the earth is at the center of the universe can still teach a lesson. Here is an erroneous concept that yet was believed almost without challenge for more than 2000 years. It is one of many examples in the history of science of abandoned theories that once went unchallenged. Doubtless many concepts which seem logical in the light of present knowledge will similarly have to be discarded or radically changed as additional data become available. If the long argument between the *geocentric* (Earth-centered) and the *heliocentric* (sun-centered) theories can teach us this, it is worth following.

In the sixth century B.C. Greek scholars developed the first scientific explanation of the motions described in the beginning of this chapter. Thales (624-545 B.C.) has been called the founder of Greek astronomy, but he was only one of a number of men who made valuable contributions to the early development of the geocentric theory. The earliest Greek

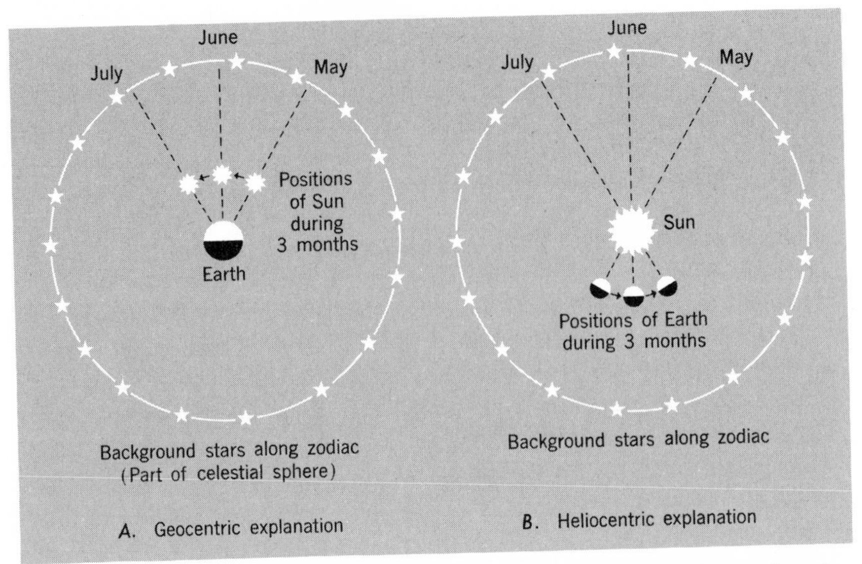

Fig. 18-8. *Explanation for the eastward shift of the sun. A similar explanation applies to the moon's eastward shift. Although stars are known to be situated at different distances from the Earth in the heliocentric theory, they seem to be equidistant.*

astronomy drew largely upon the lore of Egypt and Mesopotamia. The Greeks, however, being gifted with imagination, logic, and curiosity, attempted to explain the operation of the universe; other peoples placed more emphasis upon mapping the positions of celestial objects and assigned responsibility for the guiding mechanism to the separate wills of gods.

The Greeks generally held a geocentric theory. The Earth was round, motionless, and at the center of the universe. The sun, the moon, and the five bright planets known to them (the "seven stars" for which the days of the week are named), all revolved about the Earth (Fig. 18-8). The universe itself was a huge hollow sphere which rotated completely from east to west once each day. Some learned men believed all stars to be the same distance from the Earth and fastened firmly to the inside of the sphere—perhaps like the heads of great golden nails.

Celestial objects within the sphere accompanied it in this daily movement. That is why they were observed rising in the east and sinking in the west. However, special theories were needed to explain why the sun, moon, and planets all lagged behind the stars in this daily westward rotation, shifting eastward relative to the distant background of stars (Figs. 18-9A and 18-9B). As a matter of observation, the moon lags behind the stars the greatest distance in one day, the sun a lesser distance, and the planets shift eastward at rates which vary.

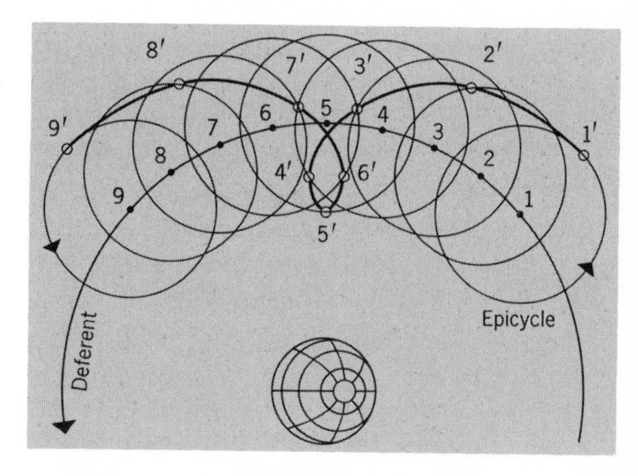

Fig. 18-9. *A. Geocentric explanation for planetary motions. Each 24 hours, it was supposed, the planet revolved almost completely around the Earth from east to west, but lagged slightly behind the stars and so shifted slowly eastward in front of them. To account for this and the occasional retrograde loop, the planet was conceived as revolving about an imaginary point which in turn revolved about the Earth (along the deferent in the sketch), like a point on the rim of a spinning wheel revolving about the Earth. Thus the planet moved through a looped path in space. When the center of the epicycle is at 1, the planet on the epicycle is at 1′. Other numbers show the locations of the center of the epicycle and the planet at regular intervals. Eastward is in the counterclockwise direction.*

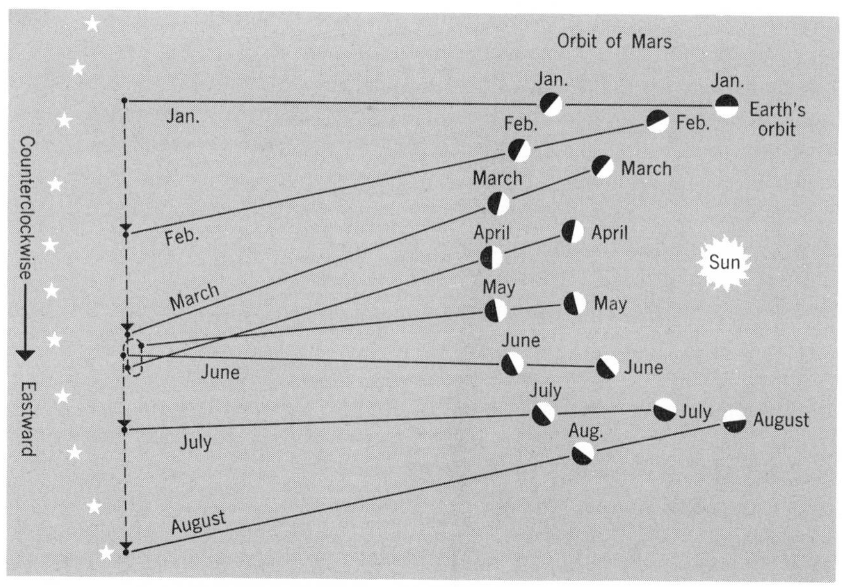

Fig. 18-9. *B. Heliocentric explanation for planetary motions (Mars). The apparent positions of Mars among the stars, as viewed from the Earth, are plotted for each month. The Earth revolves at a faster rate than Mars and covers a greater angular distance in any one month. Thus the Earth occasionally overtakes Mars and passes between it and the sun. At such times Mars appears to shift westward in front of the stars and to follow a looped path (Fig. 18-5). At all other times Mars shifts eastward against the stellar background at rates which vary. For Mars the retrograde motion may last about two months and occurs about every other year.*

A few of the early Greeks realized that celestial motions could be explained on the basis of a moving Earth, but their concepts of sizes and distances in the universe had far too small a scope, and the explanation involving a motionless Earth seemed both more understandable and more logical. When the idea of a moving Earth was brought forward in ancient times, as indeed it was, many difficulties kept it from being plausible. If the Earth is moving so rapidly, some argued, where are the powerful winds which should exist? Why are men not flung into space? And why do objects thrown into the air always return to the ground at the point where they were projected upward?

An apparently valid scientific test of the theory of the Earth's revolution around the sun was made by Aristotle in the fourth century B.C. (perhaps our old friend Archimedes in the third century B.C. should be credited with this). He reasoned that if the Earth actually revolved about the sun, then near stars should be displaced against the background of more distant stars. A similar displacement of Saturn would occur if the stars were equidistant and part of a celestial sphere located outside of the orbit of Saturn. (This displacement is called parallax; Fig. 18-10.) The brighter stars were assumed to be nearer ones (this is not necessarily true). Aristotle searched for evidence of parallactic displacement, found none, and concluded that the Earth was motionless. Aristotle's reasoning was correct; parallax does occur, but stars are so distant and show such small parallactic displacements that they could not be measured. The false assumptions about distances in the universe thus invalidated his check and led to an incorrect conclusion—another example of the eternal need for diligence in checking the accuracy of scientific hypotheses and theories.

Accordingly, the Greeks clung to a geocentric view and elaborated special explanations of why the moon, sun, and planets shifted in the sky. In one version, each of these celestial objects was assigned its own transparent "glassy" sphere, like an enormous bubble, rotating at a rate slower than the outermost sphere of the fixed stars.

In the second century after Christ, Ptolemy, the last of the great ancient astronomers, summarized, refined, and added to the geocentric theory, which afterward bore his name. He published a book, his *Almagest,* which remained the outstanding astronomical work for nearly 1500 years. We associate the epicycle (Fig. 18-9A) with Ptolemy's account of planetary motions. By now the geocentric theory had lost all vestige of simplicity, and several dozen epicycles were required to explain the observed motions of the planets.

Yet amazingly accurate results were achieved by the ancient astronomers in calculating the distances from the sun to various planets in terms of the distance from the Earth to the sun (the astronomical unit) and in

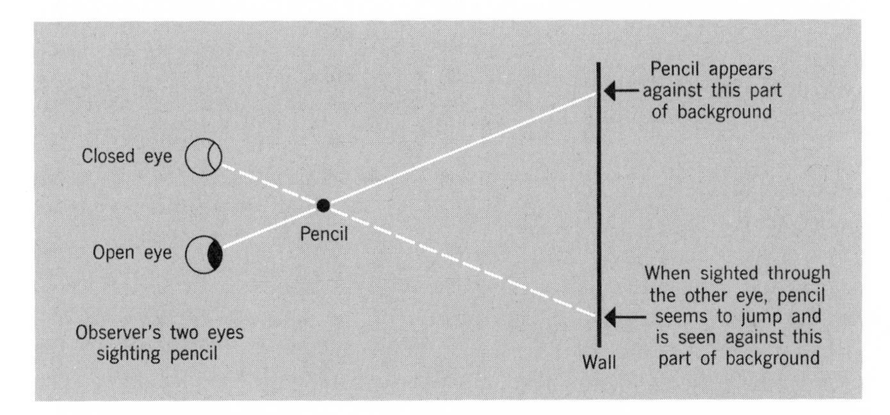

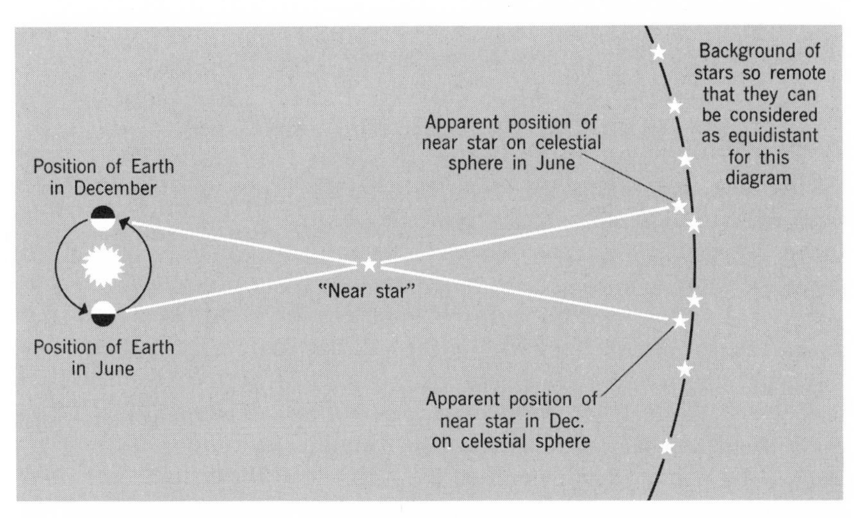

Fig. 18-10. *A. Parallax illustrated. Hold a pencil motionless in front of your face. Close one eye, and align the pencil with a point some distance away. Now open the closed eye and close the open eye. Line up the pencil again with a point in the distance. The pencil appears to shift against the background of distant objects because the base of the line of sight has been changed.—B. Parallactic displacement of a near star against a background of distant stars. Parallax proves that the Earth revolves about the sun, but parallactic displacements are so small that the first one was not measured until 1839. Measuring the largest parallactic displacement is approximately equivalent to measuring the diameter of a marble at a distance of a few miles. Parallax likewise provides a trigonometric method for obtaining the distances to the nearer stars. A right-angle triangle is formed by the Earth, sun, and near star. The base of the triangle is known (93 million miles) and the angle of parallax can be measured.*

measuring the sizes of the Earth and moon and the distance between them (Chapter 19). One is filled with admiration for such intellectual triumphs.

Aristarchus, in the third century B.C., seems to have been the first advocate of the heliocentric theory. He noted that the first-quarter moon was due south at sunset, evidence that sunlight strikes both the Earth and the moon at the same angle (Fig. 20-4). The sun must thus be very far away. Furthermore, to appear about the size of the nearby moon, the sun has to be a huge object. Therefore, to Aristarchus it seemed more logical that observed celestial motions resulted from the Earth's rotation and revolution than from the movement of the sun and celestial sphere.

The geocentric explanation persisted almost without challenge from the sixth century B.C. until the time of Copernicus. The theory lasted so long because it was based on straightforward observation of celestial phenomena, and because with its aid accurate predictions concerning future planetary positions and the times of eclipses could be made.

FOUNDERS OF MODERN ASTRONOMY

Five men had a very powerful impact on astronomy and science in general in the sixteenth and seventeenth centuries—Copernicus, Tycho Brahe, Kepler, Galileo, and Newton. During these centuries, new instruments were produced, e.g., the telescope, microscope, barometer, pendulum clock, vacuum pump, and thermometer. More precise quantitative measurements became possible, much factual knowledge accumulated, and mathematics was applied more widely and intensively. Mathematics and science were gradually accepted as subjects worthy of study by top scholars. Conclusions that had been accepted for centuries because they had been reached by some great authority, by Aristotle for example, were now questioned and discarded if found wanting. Experimentation and observation to test hypotheses became more common, and the study of astronomy was a great stimulus in this fundamental change in outlook. Cause-and-effect relationships were sought to explain natural phenomena; inquiring men searched for the fundamental natural laws that would explain the universe they observed. The regularities of many celestial motions suggested that such laws existed. A revolution gradually occurred in the manner in which men thought about and studied natural phenomena, and astronomy, with its attempt to account for celestial motions, played a central vital role in these changes.

Copernicus In the sixteenth century Nicolaus Copernicus (1473-1543), a Polish churchman-astronomer, stated the heliocentric (Copernican) theory. The idea had come from the writings of a few ancient natural philosophers, notably Aristarchus. The new and startling hypothesis—in direct opposition to a belief which had been firmly established for 2000 years—needed powerful supporting evidence in order to be accepted, but Copernicus could not offer convincing data in its favor. He could merely state that the heliocentric theory was simpler and that it seemed more logical for the tiny Earth to rotate and revolve than for the rest of the universe to do so.

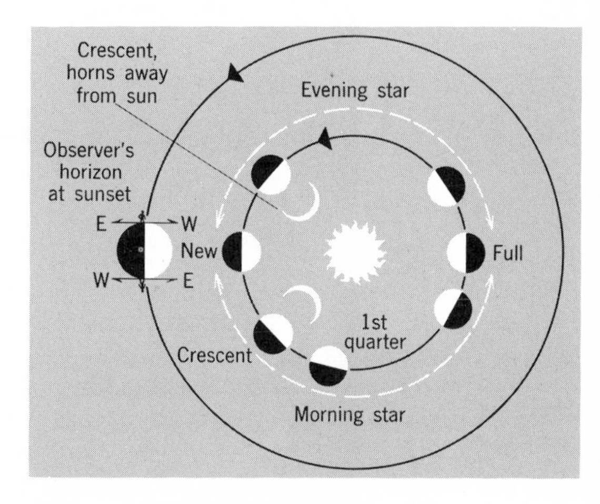

Fig. 18-11. *Phases of Venus. Positions of the Earth, Venus, and the sun are shown as viewed from "above." As Venus moves more rapidly in its orbit than the Earth, it eventually overtakes the Earth, passes between it and the sun, and emerges on the other side. Half of Venus is always illuminated by the sun's rays, and half is always in darkness. The phases of Venus and of the moon depend upon the proportion of the lighted half which can be seen from the Earth. The apparent diameter of Venus changes. Venus is not visible when it is directly in line with the sun at a "full-moon" position.*

As Copernicus believed that the paths which the planets followed about the sun were circular, predictions of future astronomical events were not significantly more accurate than those based on the Ptolemaic theory, and some epicycles were still necessary to explain planetary motions. Copernicus did predict that Mercury and Venus would show phases similar to the phases of the moon (Figs. 18-11 and 18-12), which could not be explained by the geocentric theory; but the accuracy of this prediction was not vindicated until Galileo trained his telescope upon the planets early in the seventeenth century.

Tycho Brahe This Danish nobleman (1546-1601), was sent to school as a youth to study subjects befitting his status. His tutor was responsible for selecting this curriculum, and astronomy and mathematics were excluded, for they were not respectable subjects at this time. However, when Brahe was only fourteen years of age, a partial eclipse of the sun occurred. Brahe was amazed that a few men could actually foretell such events by their studies of the heavens, and he developed a great desire to learn about astronomy. He obtained a copy of Ptolemy's *Almagest* and some books on mathematics. At first he had to read the books secretly, but he soon persuaded his tutor to permit the full-time study of astronomy.

Fig. 18-12. *Venus in different phases as seen through a telescope. The diameter of Venus appears smallest when the planet is in its full phase and appears about six times as great in the crescent phase. This apparent diameter change is explained readily by the heliocentric theory: Venus appears small on the far side of its orbit from the Earth and large when it is near the Earth.—Mercury goes through phases similar to those of Venus. Planets farther than the Earth from the sun show chiefly the full phase.*

At the age of sixteen, Brahe realized that the Ptolemaic-Copernican controversy could not be resolved without accurate information on the positions of the planets, sun, moon, and stars. Accurate data as a basis for later theorizing, now commonplace in scientific research, was an entirely new technique in the sixteenth century. Brahe was determined to make the necessary observations and in fact made them in the course of his life. He is remembered principally for these observations and for his insistence on the need for reliable data as a prerequisite for all theories.

Like Aristotle before him, Brahe realized that the stars on the celestial sphere should seem to shift their positions if the Earth revolved about the sun—that neighboring stars on the celestial sphere should have a greater angular separation when the Earth was nearest them than when viewed from the Earth on the opposite side of its orbit. As he could not find any displacement, he believed that the sun and moon revolved about the Earth. However, he thought that the other planets revolved about the sun.

Kepler Johannes Kepler (1571-1630), a German mathematician-astronomer, was a student of Tycho Brahe at Prague and had access to his observations after Brahe's death. These two scientists complemented each other's talents nicely—Brahe the observer and Kepler the theorizer. No hypothesis then known was completely satisfactory in accounting for planetary motions, and Kepler spent many years trying to devise a better explanation. He attempted to reconcile planetary motions with circular orbits, epicycles, and other geometric figures. Finally he hit upon elliptical paths as the solution. In the first part of the seventeenth century, he formulated the following laws of planetary motion, for which he earned the epithet, Legislator of the Heavens.

(1) *Every planet follows an elliptical orbit about the sun; the sun is located at one focus of the ellipse* (Fig. 18-13).

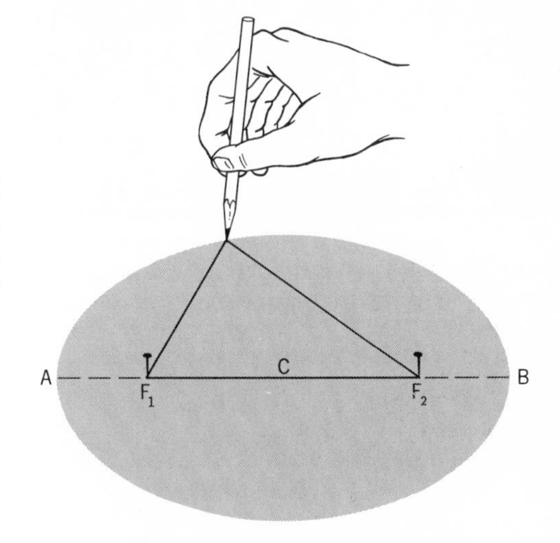

Fig. 18-13. *Drawing an ellipse. Loop a string around two fixed pins, the foci of the ellipse. An ellipse is a closed curve which is the path of a point moving in such a way that the sum of its distances from two fixed points, the foci, is a constant.*

A circle may be considered a special type of ellipse—both foci are at the center. *AB* in the figure represents the major axis of the ellipse, and either *AC* or *CB* is the semimajor axis or mean distance. If the distance between the foci is relatively great, the ellipse has a flattened shape and is said to have a large eccentricity. The eccentricity of an ellipse is numerically equal to the distance between the foci divided by the major axis. The eccentricity of an ellipse may range from zero (a circle) to 1 (a straight line).

In the solar system, the sun is at a focal point for each of the elliptical orbits followed by the celestial objects that revolve about it: planets, asteroids, comets, and meteoroids. *Aphelion* refers to that point in any one orbit that is most distant from the sun, whereas *perihelion* refers to the point that is closest to the sun. *Apogee* and *perigee* are similar terms used for objects orbiting the Earth (Greek *helio-* for "sun"; *geo-* for "Earth").

(2) *A line joining the center of each planet with the center of the sun sweeps over equal areas in equal periods of time* (Fig. 18-14).

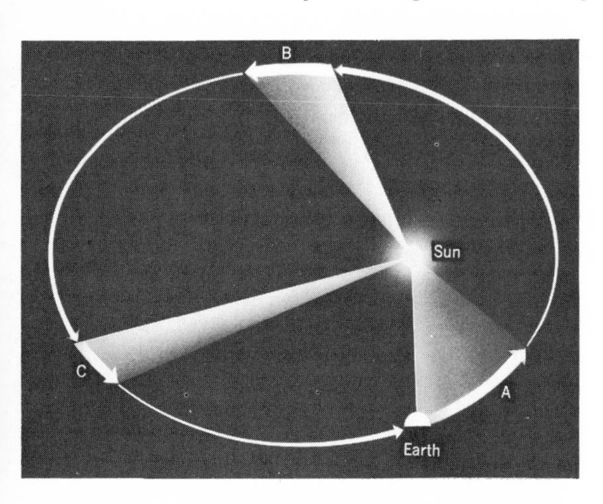

Fig. 18-14. *Kepler's equal areas law. The light-toned areas labeled A, B, and C are equal in areal extent, and the Earth traverses orbital segments A, B, and C in equal times. The Earth moves fastest when it is closest to the sun (at perihelion during winter in the Northern Hemisphere) and slowest at aphelion in the summer. Thus the length of an imaginary line between the centers of the Earth and sun varies inversely as the Earth's orbital velocity, and equal areas of space are swept over by this line in equal periods of time. [Courtesy of the American Museum of Natural History.]*

(3) *The squares of the periods of revolution of any two planets are in the same ratio as the cubes of their mean distances from the sun:*

$$\frac{(\text{Period of revolution of planet } A)^2}{(\text{Period of revolution of planet } B)^2} = \frac{(\text{distance of planet } A \text{ from sun})^3}{(\text{distance of planet } B \text{ from sun})^3}$$

A and *B* in law (3) may represent any two planets. If *B* represents the Earth, then both denominators will be equal to 1: the period of revolution of the Earth is 1 year, and its distance from the sun is 1 astronomical unit (93 million miles). Thus, if a planet occurs at a mean distance of 4 astronomical units from the sun, its period of revolution must be 8 years.

The law indicates that with increasing distances from the sun, planets not only require longer periods of time to make complete revolutions about the sun, but also move more slowly in their orbits; e.g., Mercury has an average orbital velocity of about 30 mi/sec, the Earth about $18\frac{1}{2}$ mi/sec, and Pluto approximately 3 mi/sec. As a similar relationship holds for any one planet in its own orbit—it moves fastest when closest to the sun—a fundamental relationship is evident between distance to the sun and orbital velocity. In fact, this relationship applies to all orbiting bodies; e.g., to the moons of Jupiter, to the particles forming Saturn's rings, and to artificial satellites orbiting the Earth.

Application of these laws simplified the explanations of planetary motions, permitted predictions of greater accuracy than was formerly possible, and thus gave strong support to the heliocentric theory. (The third law has been modified slightly; see p. 455.) Note that the laws do not explain why these motions occur; Newton later offered an explanation with his laws of motion and universal gravitation.

Galileo Galileo Galilei (1564-1642), the Italian astronomer-physicist, was converted to the Copernican system early in his career. A passage in his book, *Dialogues on the Two Chief Systems of the World, the Ptolemaic and the Copernican,* published in 1632, illustrates his reasoning:

> I was a very young man. . . . Being firmly persuaded that this opinion [Copernican] was a piece of solemn folly . . . I began to inquire. . . . Considering then that nobody followed the Copernican doctrine, who had not previously held the contrary opinion, and who was not well acquainted with the arguments of Aristotle and Ptolemy; while on the other hand nobody followed Ptolemy and Aristotle, who had before adhered to Copernicus, and had gone over from him into the camp of Aristotle; weighing, I say these things, I began to believe that, if anyone who rejects an opinion which he has imbibed with his milk, and which has been embraced by an infinite number, shall take up an opinion held only by a few, condemned by all the schools, and really regarded as a great paradox, it cannot be doubted that he must have been induced, not to say driven, to embrace it by the most cogent arguments. On this account I have become very curious to penetrate to the very bottom of the subject.

Galileo is perhaps best known for his telescopic studies in astronomy. Near the beginning of the seventeenth century, he learned that a Dutch spectacle-maker had constructed an instrument which magnified distant objects. The value of such an instrument in astronomy was immediately realized by Galileo, and he soon built a telescope of his own. Imagine the surge of intense interest and excitement which Galileo felt as he pointed this new instrument at various celestial objects, the first person ever to study astronomical phenomena with its aid! By this time he had become convinced of the validity of the Copernican explanation; but he had no proof, and he was even forced to teach the Ptolemaic system at the university where he was stationed. Now the telescope enabled him to gaze upon celestial phenomena which strongly supported the heliocentric theory. He observed that planets looked like disks or parts of disks, whereas the stars remained mere pinpoints of light, as they do today even in the largest of telescopes. The light-gathering power of the telescope also enabled him to see distant stars whose light had previously been too faint to affect the unaided eye, and he discovered that the Milky Way consisted of the combined light of thousands and thousands of stars, each invisible without optical assistance. He gives the following account of his discovery of four of Jupiter's moons:*

> On the 7th day of January in the present year, 1610, in the first hour of the following night, when I was viewing the constellations of the heavens through a telescope, the planet Jupiter presented itself to my view, and as I had prepared for myself a very excellent instrument, I noticed a circumstance which I had never been able to notice before, owing to want of power in my other telescope, namely, that three little stars, small but very bright, were near the planet; and although I believed them to belong to the number of the fixed stars, yet they made me somewhat wonder . . . but when on January 8th, led by some fatality, I turned again to look at the same part of the heavens, I found a very different state of things. . . . I, therefore, waited for the next night with the most intense longing, but I was disappointed of my hope, for the sky was covered with clouds. . . . But on January 10th the stars appeared. . . . These observations also established that there are not only three, but four, erratic sidereal bodies performing their revolutions round Jupiter. . . .

From observation of these motions, Galileo determined that four satellites revolved about Jupiter in the plane of its equator (eight other moons have been found since). Clearly the Earth was not the center of all celestial motion. Furthermore, Jupiter and its moons resembled a miniature solar system. Perhaps the Earth and the other planets moved about the sun in a similar manner.

Other startling discoveries followed at once. Through the telescope, the moon was seen to have an irregular surface marked by craters, mountains, and other topographic features; it definitely was not smooth. On the sun's surface he saw small black spots. During a series of observations these spots moved as a group from one side of the sun to the other (Fig. 24-2). Galileo interpreted this movement correctly as one caused by the sun's rotation. Most devastating to the geocentric theory, however, was Galileo's discovery that Venus and Mercury showed phases (Fig. 18-11 and 18-12) just as Copernicus, about sixty years earlier, had predicted that they would. Moreover, the observation that the diameters of the two planets seemed to

* *The Sidereal Messenger* (1610), translated by E. S. Carlos, 1880.

change was particularly embarrassing to the Ptolemaic theory which demanded that Venus and Mercury revolve about the Earth along epicycles at nearly unchanging distances from it.

Galileo's support of the Copernican theory eventually led him into conflict with the church, and he is reported to have remarked that the Bible was written to help men get to heaven, not to tell men how the heavens move. Galileo's many telescopic observations, especially those of sunspots, brought him to a state of nearly complete blindness before his death.

Newton It remained for Sir Isaac Newton, whose name is as central in astronomy as in physics and mathematics, to explain why planets follow paths about the sun. The force that keeps the planets moving continuously in their orbits had puzzled inquisitive men for hundreds of years.

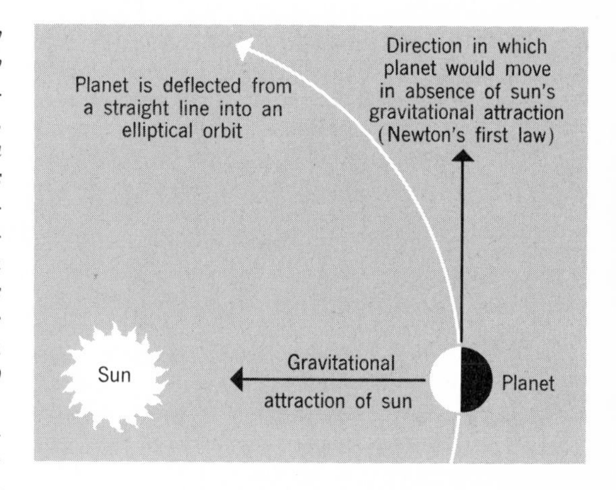

Fig. 18-15. *Forces governing planetary orbits. The planet "falls" continuously toward the sun. For the Earth this distance is less than ⅛ inch in 1 second, and during this second the Earth moves forward about 18½ miles in its orbit. The moon likewise falls continuously toward the Earth as it revolves, and Newton used this motion to prove the inverse square part of his law of universal gravitation. The moon's mean distance from the Earth (about 240,000 miles) is about 60 times longer than the Earth's radius. Thus the Earth's gravitational attraction at the moon's distance should be only* $(1/60)^2$ *of what it is for an object at the Earth's surface—it should be about 3600 times smaller. Thus the moon should fall toward the Earth much more slowly than does a falling object at the Earth's surface. Newton was able to show that this is the case.*

Experience on Earth indicates that objects move only when a force is applied to them. In part, the answer to this problem of planetary motion is given by Newton's first law of motion: *Every object remains at rest or in uniform motion in a straight line unless it is acted upon by some force* (see Chapter 3). In the absence of friction, motion is as natural a state as rest. The planets began to move a long time ago; presumably the initial impetus is related to the origin of the solar system. As interplanetary space is a nearly perfect vacuum, friction cannot slow celestial bodies as it does all objects on the Earth. Therefore, the planets should continue to move—but in straight lines at uniform rates of speed through space. Instead they follow elliptical orbits about the sun. Here Newton's great unifying principle, the *law of universal gravitation,* proved to be the key: *Every particle of matter in the universe attracts every other particle with a force which is proportional to the product of their masses, and inversely proportional to the square of the distance between them* (see Chapter 4).

In order to force a moving object such as a planet to curve out of a straight path, a force must be exerted at an angle to this path (Figs. 18-15 and 18-16). A planet's momentum or inertia makes it tend to move in a straight line at a uniform rate. On the other hand, the sun exercises a gravi-

Fig. 18-16. *Motion in an elliptical orbit discussed qualitatively and briefly. The lengths of the solid arrows tangential to the orbital positions A, B, C, and D are proportional to the speeds of the planet at these positions, and the lengths of the dashed arrows are proportional to the gravitational pull of the sun, but neither is to scale. At the start a planet is at position A in its orbit about the sun at S. Because the planet's speed here is too small to carry it in a circular orbit about the sun, the planet is pulled inward by the sun's gravitational attraction and moves to position B. The planet is now moving more rapidly because the acceleration toward the sun has a component along the orbit in the direction the planet is moving. The planet's speed increases as it falls closer to the sun and its orbit is curved sharply. The sun's gravitational attraction increases rapidly as the distance decreases. At position C the planet's speed exceeds that for a circular path about the sun at this distance, and the planet swings outward from the sun to position D. At D the sun's gravitational pull is slowing down the planet. Therefore, it is deflected toward the sun in its orbit and reaches position A with precisely the same speed that it had at the start. Thus it follows the same orbital path time after time (see also Fig. 22-3). [After McLaughlin, Introduction to Astronomy, 1961, p. 160.]*

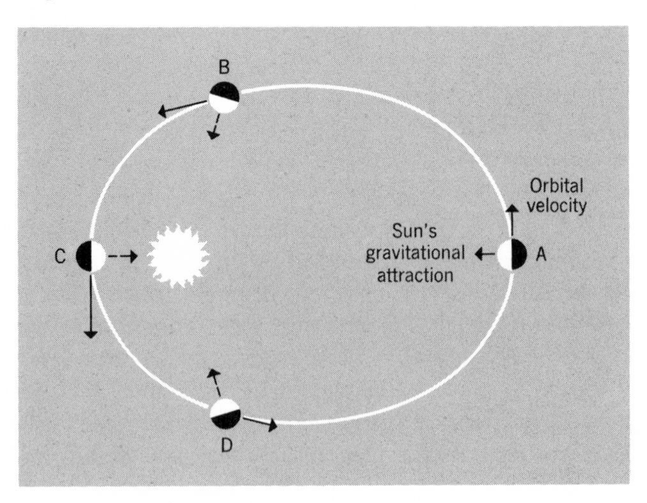

tational pull directly toward itself on a planet, and this is an unbalanced force, one component of which acts at right angles to the direction of the planet's momentum. Thus a planet is deflected from a straight line into an elliptical orbit. In other words, to keep a planet from moving off along a tangent, an unbalanced force toward the sun is required (a centripetal or "center-seeking" force) and this is furnished by the gravitational attraction of the sun.

If the sun's gravitational attraction were increased, the Earth should follow a path closer to the sun. If the Earth were stopped motionless and then released, it should move directly toward the sun.

One can think of the motion of a planet about the sun in terms of angular momentum (p. 478) and energy (p. 65). The angular momentum of the planet must be conserved. Therefore, its orbital speed must be least at aphelion, greatest at perihelion, and changing continuously between these points. At perihelion a planet's kinetic energy will be greatest and its potential energy will be least. At aphelion these will be reversed. Thus some of the kinetic energy of a planet is continuously converted into potential energy, or vice versa, as the planet's distance and orbital speed change continuously.

According to Newton's third law of motion, *To every action there is always an equal and contrary reaction*. Familiar examples are the recoil of a rifle and the movement of a small boat as one dives off. This has led to a common error: the assumption that there is an outward force on a planet (a centrifugal or "center-fleeing" force) that is equal in magnitude but opposite in direction to the inward gravitational pull of the sun. Two equal and opposite forces do exist, but the two forces act on two different objects. The gravitational attraction of the sun produces an inward unbalanced (centripetal force) on a planet. This is a very large force because of the enormous mass of the sun; if it were to be eliminated and a steel cable substituted to keep the Earth in its orbit, the cable would need a diameter of several thousand miles to prevent its breaking. Following Newton's third law of motion, the gravitational attraction of a planet produces an equal and opposite force on the sun, but the sun's motion is changed only a small amount because it is so very massive.

In the familiar and comparable example of a ball being whirled around a circular path at the end of a string, equal and opposite forces exist, but there is no outward force on the ball. There is, however, an unbalanced force on the ball that is directed inward toward the hand. In a sense, one end of the string pulls the ball inward (the centripetal force), the other end of the string pulls the hand outward (this is the actual centrifugal force). Thus if the string is cut, the ball flies off along a tangent to its former path; it does not fly directly outward.

According to Newton's second law of motion, *The product of the mass of an object times its acceleration varies directly as the resultant force, and the change in motion takes place in the direction of that force.* This is familiar to many as the equation $F = ma$, where F is the force (in dynes), m is the mass (in grams), and a is the acceleration (in centimeters per second squared) or change in the velocity per unit of time. Note that acceleration may consist of a change in direction without a change in speed; e.g., revolution in a circular orbit. This law indicates that a certain force will accelerate a large mass more slowly than a small one; thus the sun and Earth attract each other with equal and opposite forces, but the sun is accelerated very little because it is a third of a million times as massive as the Earth.

This law applies to all freely falling bodies on the Earth, because their masses are negligible relative to that of the Earth. If air resistance is disregarded, bodies of different masses located at the same distance from the Earth will fall toward it at the same rate; a larger mass is acted upon by a proportionally larger force. A feather and a lead ball fall at the same rate in a vacuum. However, a marble located at the moon's distance from the Earth would be accelerated toward the Earth more slowly than the moon, because the gravitational pull between the Earth and the moon is greater than that between the Earth and the marble.

Newton suggested that stars must be situated at tremendous distances from the sun not to be influenced by its gravitational attraction. The sun, which he considered a star, would have to be thousands of times as far from the Earth as it is, he said, in order to appear as dim as the stars. Newton was thus one of the first scientists to realize the enormous distances which lie between the Earth and the stars.

EVIDENCE SUPPORTING THE LAWS OF KEPLER AND NEWTON

About the middle of the nineteenth century there occurred an outstanding intellectual achievement which was also a striking proof of the law of universal gravitation. The planet Uranus had been discovered more or less by accident in 1781; the discovery was a by-product of astronomical research and not the result of a planned search. Subsequently, an orbit was calculated for it. For about forty years the location of Uranus in the sky checked accurately with the positions which had been calculated for it. Then Uranus slowly began to deviate from these calculated positions. To be sure, the deviations were slight, but they seemed too large to be mathematical errors. Two explanations appeared possible. Either the law of universal gravitation was not completely accurate or universal, or some unknown body was exerting a gravitational attraction upon Uranus.

Two young men, Adams of England and Leverrier of France, were convinced that the gravitational pull of an unknown planet, orbiting at a greater distance from the sun, was causing the deviation in the path of Uranus. They undertook independently the very difficult task of calculating the position of a planet of unknown size and distance whose gravitational pull might be diverting Uranus slightly from its normal path (Fig. 18-17). Still working independently, each was finally able to determine the position of the unknown planet Neptune in space. Both predictions proved amazingly accurate.

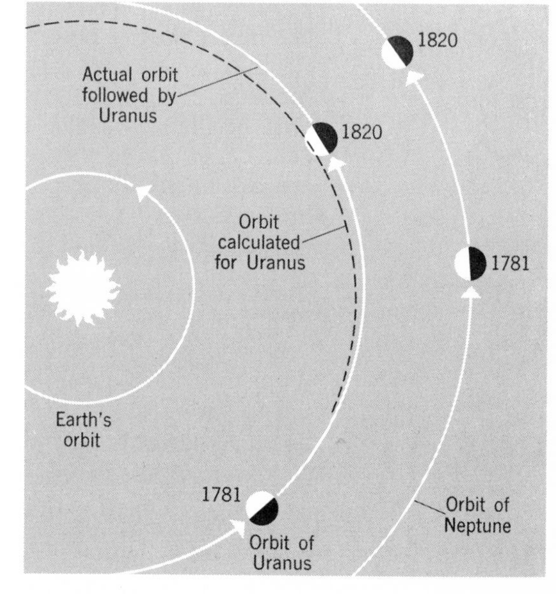

Fig. 18-17. *The discovery of Neptune. (Diagram is not to scale.) In 1781 Uranus was too far from Neptune to be influenced. But Uranus moves more rapidly than Neptune in its orbit, and as it approached Neptune it began to deviate noticeably from the orbit calculated. The orbit of Neptune is also affected by the gravitational attraction of Uranus.*

Pluto was the last planet to be discovered and was finally located in 1930 after a search lasting some twenty-five years. The discovery might have been made about ten years earlier, but a flaw on a photographic film, as well as the proximity of Pluto at that time to a bright star, prevented its detection. If perusal of a series of photographs had shown that one starlike object moved among its neighbors, then that object might have been identified as a planet. Predictions had been made concerning Pluto's existence also, but the discrepancies it caused in the orbit of Neptune were so small that its position could not be calculated accurately.

Slight unexplained perturbations may still occur in the orbits of Uranus and Neptune, and one must wonder if yet another planet, at present unknown to astronomers, orbits the sun. If so, it seems unlikely that it will be discovered. Possibly enough comets are moving beyond Pluto to have a slight gravitational influence on the outermost planets.

Perhaps the most convincing evidence supporting the validity and accuracy of Kepler's and Newton's laws is the prediction of the positions of celestial bodies for years into the future. These positions are of pinpoint accuracy in time and space.

Exercises and Questions, Chapter 18

1. According to one explanation of the Christmas Star, a few planets may have seemed so close together in the sky that they appeared to observers for a short time as one very bright starlike object. Kepler observed such a close approach in the seventeenth century and calculated that a similar close approach occurred about 7 B.C., which may actually be the date of the birth of Christ. Imagine that the planets are Mars, Jupiter, and Saturn and that they appear as one object near the eastern horizon at sunrise. Make a sketch (orientation: looking "down" on the solar system) which shows these three planets and the sun and the Earth in suitable locations.

2. Make a star map which shows the Big Dipper in four positions: (1) as it is now at 9:00 P.M., (2) at 3:00 A.M., (3) at 9:00 A.M., and (4) at 3:00 P.M. Mark these positions *A*, *B*, *C*, and *D* respectively. Next give the position of the Big Dipper at each of the following times: (1) 3 months later at 3:00 A.M., (2) 9 months later at 3:00 P.M. and (3) 3 months ago at 9:00 A.M.

3. About midnight one evening, you see a bright starlike object high in the sky. A friend states the object is one of the following: (1) Mercury, (2) Venus, (3) Jupiter, or (4) Uranus. Which is it?

4. What is the relationship between the velocity of a planet in its orbit and its distance from the sun? Does the same relationship hold for the moon about the Earth? For comets around the sun?

5. A certain star rises in the east at 6:00 P.M. now. When did it rise in the east 3 months ago? Another star sets in the west at midnight now? When will it set in the west 2 months from now?

6. Distinguish carefully between the actual and apparent motions of the sun, stars, and planets.

7. State the position taken by each of the following concerning the geocentric-heliocentric controversy and give at least one type of evidence or argument cited by each: Aristarchus, Aristotle, Copernicus, Brahe, Kepler, and Galileo.

8. What measurement can you make on the photograph in Fig. 18-1 to show that the time exposure was approximately $1\frac{2}{3}$ hours?

9. One of Jupiter's satellites is about 260,000 miles from Jupiter and completes a revolution in about 2 days. Compare these quantities to corresponding data for the Earth and moon. Explain why Jupiter's satellite moves so much more rapidly.

10. How would the mean distance of the Earth from the sun and its mean orbital speed be changed in each of the following fictitious cases?
    (a) If the mass of the sun were reduced.
    (b) If the mass of the sun were increased.
    (c) If the Earth's mean orbital velocity were increased but the mass of the sun remained unchanged.
    (d) If the Earth were to stop rotating.
    (e) If the Earth were to stop revolving.

11. An ellipse has a major axis of 100 cm and an eccentricity of 0.2. What is the distance from the center of the ellipse to a focal point?

12. Calculate the eccentricity of the Earth's elliptical orbit. The Earth's perihelion and aphelion distances from the sun are approximately 91.5 and 94.5 million miles respectively.

13. Assume that a meteoroid revolves about the sun in a period of 27 years. What is its mean distance from the sun in astronomical units? If the eccentricity of its orbit is 0.2, what is its aphelion distance?

14. If a comet requires 1000 years to complete one orbit about the sun, what is its mean distance? If the eccentricity of its elliptical orbit is 0.6, what is its aphelion distance? its perihelion distance?

19 The earth in space

THE simplicity stressed by Copernicus, the more accurate predictions of planetary positions obtained by using Kepler's laws, the many telescopic discoveries of Galileo, and the unifying explanations of Newton all strongly supported the heliocentric concept. However, actual proof that the Earth rotated on an axis and revolved about the sun was not obtained until the nineteenth century. The pendulum experiment of Foucault in Paris, the first determination of parallax, and the discovery of the aberration of starlight provided final proof.

Evidence for the Curvature of the Earth's Surface When Columbus set sail for the west, many people still believed that the Earth was flat, although learned men from the time of the early Greeks had realized that it was spherical. Ancient astronomers were aware of several lines of evidence which indicate the nearly spherical shape of the Earth.* The shadow of the Earth on the moon during a partial lunar eclipse is curved, and repeated observations show that the edge of the Earth's shadow is always curved in the same arc. Thus the Earth is spherical and its diameter is about four times that of the moon (Fig. 20-12).

Early scientists also observed changes in the positions of stars as an observer traveled north or south. The North Star rises in the sky as the traveler moves north, and some familiar stars in the southern sky no longer appear at all above the horizon.

* Measurements from satellites orbiting the Earth suggest that the North Pole is about 120 feet farther from the Earth's center and that the South Pole is about 120 feet closer to the Earth's center than was previously estimated. A cross section through the Earth at the equator is not a true circle, but an ellipse with one diameter about 1400 feet longer than a diameter oriented 90 degrees to it.

Fig. 19-1. *Stars appear at different altitudes in the sky at different latitudes because the Earth is spherical. Polaris (the North Star) makes an angle of about 90 degrees with the horizon at the north pole (Polaris is 1 degree from the north celestial pole), makes an angle of about 30 degrees at B, and is on the northern horizon as viewed from C. If the Earth were flat and in the plane of the equator, Polaris would be directly overhead from all parts of the Earth. To prove that the latitude of B (angle number 4) is approximately equal to the altitude of Polaris above the horizon (angle number 1), we know that angle 2 equals angle 3 (why?), and that the sum of angles 3 and 4 is 90 degrees. Also, the sum of angles 1 and 2 is 90 degrees. Thus angle 4 equals angle 1.*

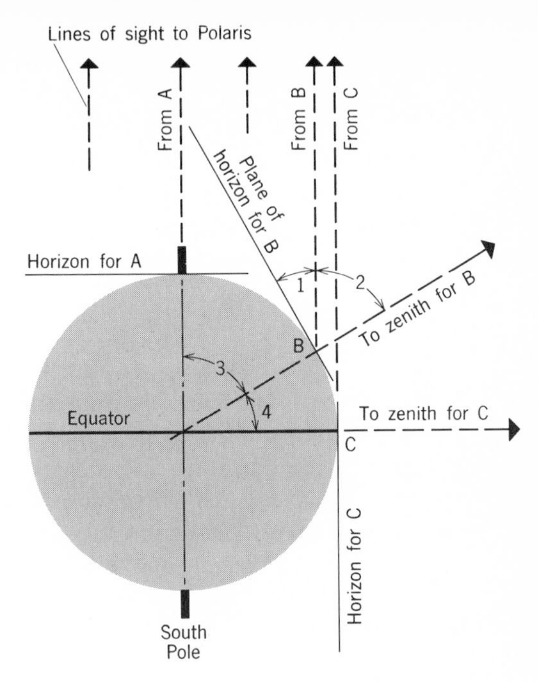

An observer's latitude is always exactly equal to the altitude of the celestial pole, a fact that is of basic importance in navigation and geodesy. In the Northern Hemisphere, the altitude of Polaris above the northern horizon is always approximately equal to the observer's latitude because Polaris is about 1 degree from the north celestial pole (Fig. 19-1).

All stars are so far away that light rays from any star are virtually parallel throughout the solar system. Thus if observers on different parts of the Earth at different times of the year point at a star such as Polaris, they all point in the same direction.

Observations of the altitude above the horizon of Polaris (actually of the north celestial pole) also show that the Earth is flattened at the poles and bulging at the equator. Measurements indicate that an observer has to move an average distance of about 69 miles in a north-south direction (along a meridian) to change the altitude of Polaris by 1 degree. However, this north-south distance is somewhat greater near the poles (about 69.4 miles) and less near the equator (about 68.7 miles). If the Earth were a true sphere, one could start at the equator where the altitude of Polaris is 0 degrees and travel due northward along a meridian for $\frac{1}{90}$ of the distance from the equator to the North Pole. At this new location, Polaris would be exactly 1 degree above the northern horizon. Traveling northward another $\frac{1}{90}$ would raise the altitude of Polaris to 2 degrees, another $\frac{1}{90}$ to 3 degrees, etc. This change in the altitude of Polaris is caused by the Earth's curvature. Therefore, where the Earth's surface is curved more than average, as it is at the equator, the north-south distance is less than average.

Astronomers also observed the way in which a ship disappears from sight: its deck disappears before its superstructure. If the Earth were flat, the ship would look smaller with increasing distance from an observer, but all of the ship would remain visible until it became a mere speck and then disappeared. In addition, a wider area of the Earth's surface is visible from a high altitude than from a low altitude because one can "see over the curvature."

Why Planets and Stars are Spherical There is a definite reason why the Earth is not shaped irregularly nor like a cube, cylinder, or tetrahedron. The planets, sun, stars, and many other objects in space are also spherical for the same reason—gravitational attraction. In an isolated body in space, the gravitational attraction of that body acts to concentrate the greatest amount of matter into the smallest possible volume, and the sphere is the most economical space saver of all three-dimensional figures. If some gigantic force could distort the Earth's shape, its self-gravity would level irregularities and again make it spherical.

On the other hand, few celestial objects are perfect spheres, because they rotate, and spinning on an axis causes polar regions to flatten and equatorial regions to bulge. The size of the equatorial bulge is approximately proportional to the rate of rotation, although other factors such as density and distribution of mass also enter in. Jupiter and Saturn spin very rapidly and have noticeably flattened polar regions (Figs. 21-3 and 21-5), whereas Venus spins more slowly and is more nearly a sphere. Thus self-gravity acts to make a sphere of a planet or star, whereas rotation keeps it from being perfectly spherical.

Weight refers to the gravitational pull of the Earth on an object, which varies inversely as the square of its distance. *Mass,* on the other hand, does not vary (p. 11). We might think of the mass of an object as equal to the total number of protons, neutrons, electrons, and other subatomic particles that are contained in this object.

An object weighs more at the poles than at the equator because sea level is about 13½ miles closer to the center of the Earth at the poles than it is at the equator. Thus the pull of gravity is greatest at the poles because the gravitational attraction of the Earth acts as if it were all concentrated at the Earth's center. The Earth's rotational speed is also a factor.

On the moon, a man would weigh about one-sixth of his weight on the Earth, but his mass would remain the same. Depending upon his skill and muscular development, he might be able to jump over a small house, put the shot the length of a football field, leap without harm from a 30-foot cliff, and send a rifle bullet 100 to 200 miles above the surface. Three factors influence one's weight on the moon and on other celestial bodies: one's

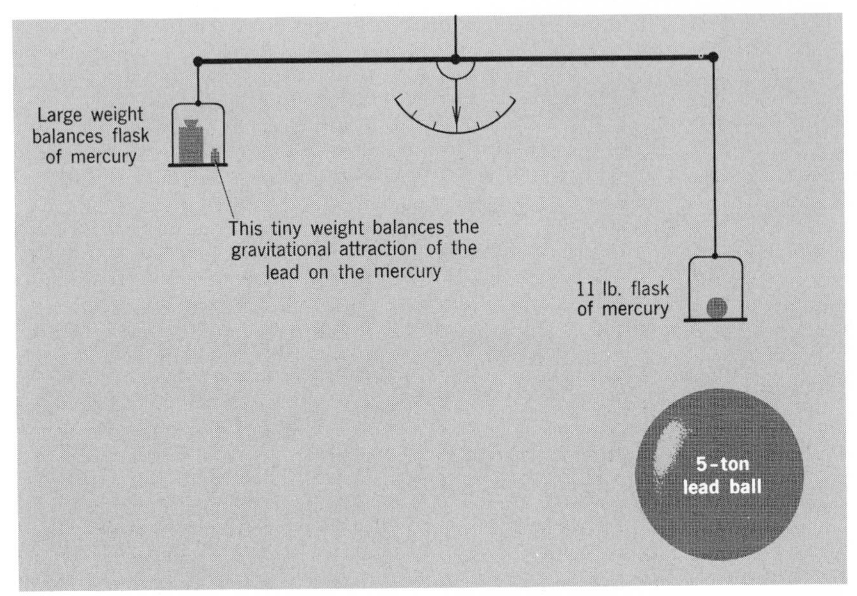

Fig. 19-2. *Jolly's method of "weighing" the Earth (1881). A large weight was put in exact balance with an 11-lb spherical flask of mercury. Then a 5-ton lead ball was moved under the mercury. Its gravitational attraction was sufficient to pull the mercury down a short distance, and a small weight was added on the left to restore the balance. Very accurate measurements were required.—The following equations can be set up. As the only unknown, x, is the mass of the Earth, it can be determined:*

$$\frac{(\text{Small weight}) \cdot x}{(\text{Earth's radius})^2} = \frac{(\text{mass of mercury}) \cdot (\text{mass of lead})}{(\text{distance between their centers})^2}$$

The lead and the mercury are put into spherical form because the gravitational attraction of a sphere can be considered to be concentrated at the center.

distance from its center, its total mass, and its rotation (this reduces the weight of an object more at the equator than near the poles).

The irregular manner in which many of the smaller asteroids reflect sunlight suggests that they are not spherical in shape. Apparently their irregular shapes are caused by their small sizes. The very weak self-gravity of a small asteroid cannot cause its strong rocky or metallic material to flow into a spherical shape (perhaps such asteroids are fragments of larger bodies that collided).

Fig. 19-2 illustrates one method of "weighing" the Earth.

Evidence of the Earth's Rotation In discussing diurnal motion, it is useful to refer to the *celestial meridian* of an observer, which is the great circle on the celestial sphere that passes through his zenith, the north and south celestial poles, and the north and south points on his horizon. From the latitude of New York the sun, moon, and planets are always highest in the sky and due south of the observer when they cross the meridian—in fact, the A.M. and P.M. of our clocks (*ante meridiem* and *post meridiem*) refer to the sun's passage across the meridian.

An experiment performed by Foucault in Paris in 1851 provided the first actual evidence that the Earth turns on an axis. Foucault suspended a heavy iron ball at the end of a 200-foot wire fastened to the roof of the Pantheon. A pointer was attached to the base of the weight, and a ring of sand was placed beneath the pendulum so that lines were drawn in the sand as the pendulum swung back and forth. The pendulum's inertia keeps it swinging back and forth in the direction in which it was originally started. The pendulum has no tendency to change its direction even though the ground beneath it is turning. However, as indicated by the lines in the sand, the plane of the pendulum's swing appears to turn in a clockwise direction. The apparent turning is caused by the actual rotation of the Earth in a counterclockwise direction.

This phenomenon is most easily understood if we imagine the pendulum to be suspended from a point in a roof over the North Pole. In this case the plane of the pendulum's swing will rotate with respect to the Earth's surface through 360 degrees in 23 hours and 56 minutes. At the South Pole, the plane of the pendulum's swing rotates in a counterclockwise direction with respect to the Earth's surface. At latitudes below the North Pole, it takes more than 24 hours for the pendulum's plane to rotate through 360 degrees; it takes about 31 hours in Paris. The apparent rotation becomes increasingly slower until finally at the equator there is no rotation at all; i.e., the pendulum always swings in the same, fixed plane with respect to the Earth's surface. South of the equator the rotation begins again in the opposite (counterclockwise) direction, and its rate increases at locations closer to the pole.

All motion is relative. An experience familiar to most of us is that of sitting in a railroad car at a station beside a second train on the next track. When one of the two trains begins to move, it is often impossible to tell, without using nearby buildings as reference points, which train is actually moving. The sphere of the Earth and the celestial sphere are like the two trains, except that there are no fixed reference points anywhere. Everything on the Earth moves together—air, water, buildings, people—so that even though the rate exceeds 1000 mi/hr at the equator, we have no way of detecting the movement, and therefore it seems to be the celestial sphere which is moving. No wonder the belief in a motionless Earth persisted for more than 2000 years!

Evidence of the Earth's Revolution Evidence of the Earth's revolution about the sun is given by the parallactic displacement of the nearer stars (p. 343), an annual Doppler effect (p. 448), and the aberration of starlight (not discussed).

Seasons of the Year The seasons of the year—fall, winter, spring, and summer—are caused by the Earth's revolution about the sun and the tilt of its axis to the plane of this orbit (Figs. 17-1 and 19-3). The varying

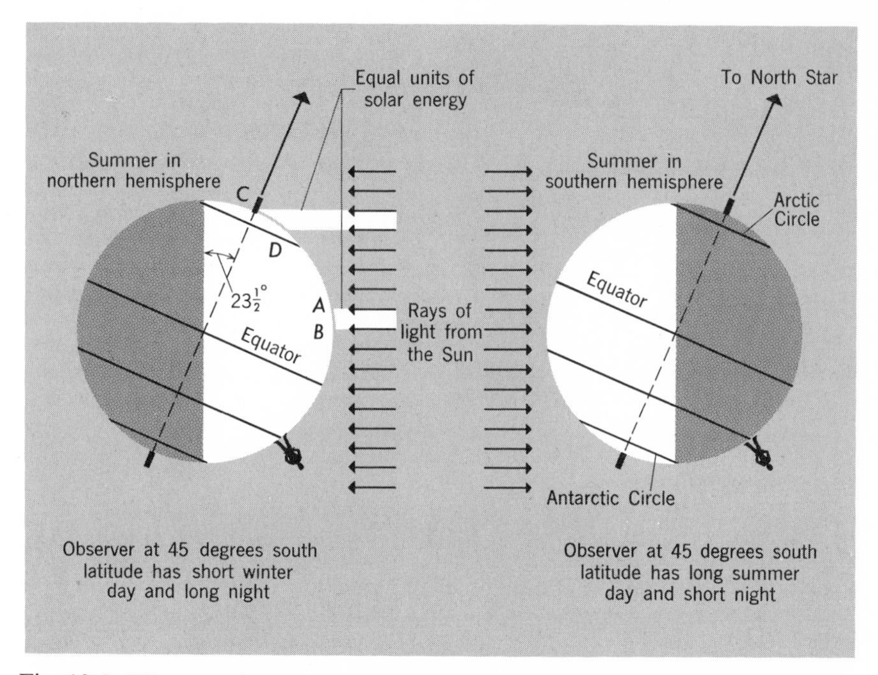

Fig. 19-3. *Winter and summer. Each day the Earth spins completely on its axis. As the direction of rotation for an observer on the Earth is parallel to the equator, summer days are longer than summer nights, and the sun cannot set at either pole during its summer.—The sun's rays are most concentrated when they strike the Earth's surface vertically (AB) and less concentrated when they approach at an angle (CD) because a larger area must be covered with the same amount of energy. The beam of a flashlight can be concentrated or spread out in much the same way.*

distance of the Earth from the sun, 94.5 to 91.5 million miles, is not an important factor in seasonal changes. As the angle of tilt is always the same, the Northern Hemisphere at one time is tilted toward the sun (its summer season). Six months later, when the Earth is on the opposite side

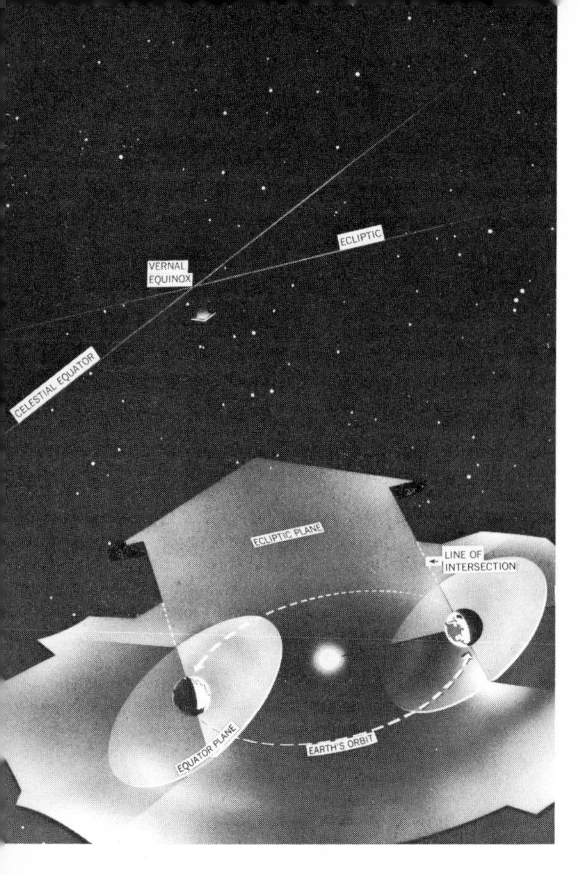

Fig. 19-4. *The angle between the ecliptic and the celestial equator remains constant at 23½°. The ecliptic is the apparent annual eastward path of the sun among the stars, but it may also be defined as the great circle formed by the intersection of the Earth's orbital plane (extended) with the celestial sphere. Planets are nearly always seen among the stars near the ecliptic. The Earth's equatorial plane stays parallel to itself as the Earth revolves about the sun. The equinoxes (equal nights) are the two opposite points on the imaginary celestial sphere where the ecliptic intersects the celestial equator. When the sun appears to be at either of these positions, day and night (including twilight) are equal in length—the line of intersection shown on the diagram passes through the sun. The solstices are the points midway between the equinoxes where the sun is located farthest above or below the celestial equator—hence the term solstice which means "the sun stands still."* [*Courtesy of the American Museum of Natural History.*]

of the sun, the Northern Hemisphere is tilted away from the sun (its winter season). At the equinoxes about the twenty-second of March and September, daylight and nighttime (which includes twilight) are equal in length everywhere on the Earth, and the Earth's axis is not slanted toward or away from the sun. Fall and winter are shorter seasons than spring and summer because the Earth is closer to the sun and travels faster in its orbit during these months.

If the Earth's axis were perpendicular to its orbital plane, there would be no seasons. Each degree of latitude would experience the same sort of weather phenomena all year long—an extended combination of fall and spring. Locations near the equator would be warmest, and those near the poles would be coldest.

Two factors combine to make summer days warmer than winter ones: there are more hours of daylight, and the sun's rays are more direct (Fig. 19-3). Opposite seasons occur simultaneously in the Northern and Southern Hemispheres. The sun is not visible in arctic regions during winter, and perpetual daylight occurs during arctic summers. Equinoxes and solstices are explained in Figs. 19-4 and 19-5.

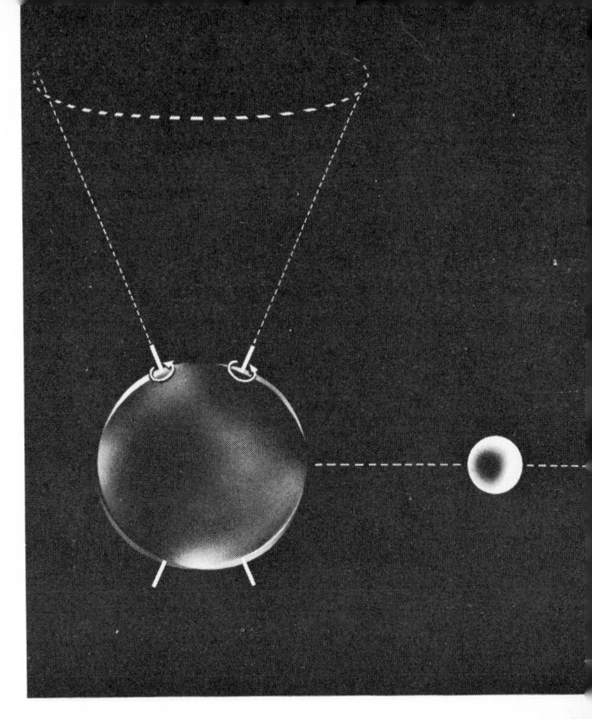

Fig. 19-5. *The Earth's axis shifts like that of a spinning top. If the Earth were not rotating, the gravitational pull of the moon (and sun) on the Earth's equatorial bulge would shift the Earth's axis so that it would make a 90-degree angle with the ecliptic (the plane of the equator would coincide with the ecliptic). Because the Earth is spinning, its axis describes a cone-shaped path in space but in the opposite direction to that of the rotation. This slowly shifts the positions of the celestial poles and celestial equator on the celestial sphere. The equinoxes are thus shifted along the ecliptic and make the year of the seasons—from equinox to equinox—about 20 minutes shorter than the period of the Earth's revolution relative to the stars. The equinoxes are shifted westward, whereas the Earth revolves eastward.* [*Courtesy of the American Museum of Natural History.*]

CALCULATIONS OF SIZES AND DISTANCES IN ASTRONOMY

Man has devised measuring rods mighty enough to reach into space for distances so vast that they are given in billions of light-years, and the first nearer measurements were made hundreds of years ago.

It was necessary to begin by measuring the size of the Earth (Fig. 19-6). Once this was achieved, it was possible to determine the distance to the moon and its size. The next step involved measurement of the mean distance from the Earth to the sun, the *astronomical unit,* which is an important yardstick in the scale of celestial distances. The distances between the Earth and some of the nearer stars were measured next. Several procedures were then discovered which made it possible to figure out the number of light-years between the Earth and remote stars and galaxies.

Distances and Diameters of the Moon and Sun About a hundred years after Eratosthenes measured the size of the Earth, an Alexandrian astronomer, Hipparchus, calculated the distance to the moon (Fig. 19-7) and the size of the moon (Fig. 19-8).

The distance to the sun cannot be measured accurately in a direct manner; its angle of parallax is too small, reference stars cannot be seen during the daytime, and the sun's center is difficult to locate precisely. However, relative distances in the solar system can all be calculated in terms of the

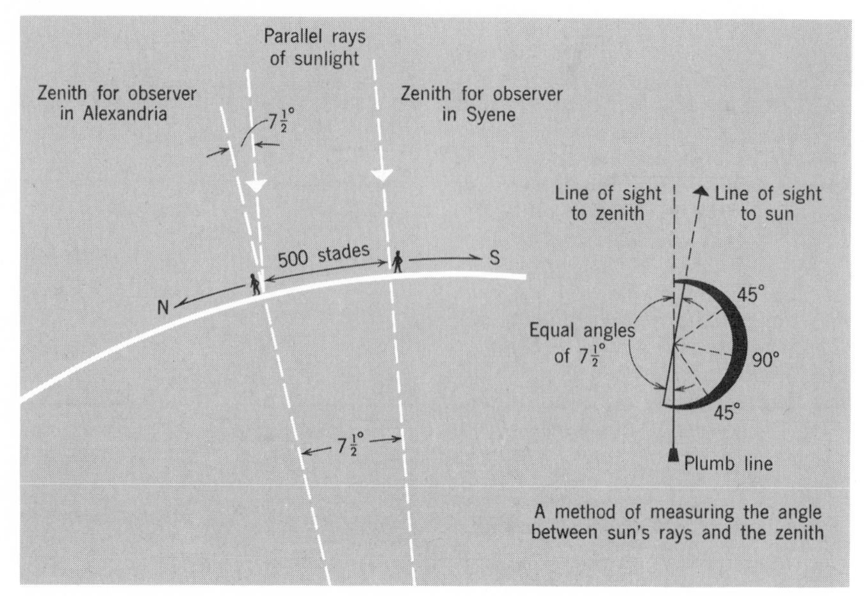

Fig. 19-6. *Method used by Eratosthenes, about 235* B.C., *to measure the size of the Earth. At noon in Alexandria, Eratosthenes measured the angle between the sun and his zenith at about 7¼ degrees. At that same moment (assuming that the sun's rays are parallel) the sun was at the zenith for an observer at Syene. Each zenith line can be projected downward along a radius to the center of the Earth. There they meet at an angle of 7¼ degrees (about 1/50 of a circle): if two parallel lines are cut by a third straight line, their corresponding angles are equal. Thus 5000 stadia, the measured distance between Alexandria and Syene, represents 1/50 of the Earth's circumference (360 ÷ 7¼). Therefore, the Earth's circumference equals 50 times 5000 or 250,000 stadia. Unfortunately we do not know the exact length of the stadium, but it may have been about 1/10 of a mile. The insert in the lower right shows an ancient method of measuring angles.*

astronomical unit by application of the laws of Kepler and Newton. Therefore, once the actual distance of one of the members has been obtained, the distances of the other members can be calculated. A few asteroids approach close enough to the Earth for their parallaxes to be measured accurately.

Distance to the Nearer Stars by Parallax Astronomers measure the parallactic displacement of a near star by taking lines of sight to it at time intervals of six months when the Earth is on opposite sides of its orbit. The angle at the star formed by the intersection of these two lines of sight can be determined (astronomers refer to half of this angle as parallax). Geometrically, the two points on the Earth's orbit and the star form an isosceles triangle in which the length of the base is twice the distance from the Earth to the sun: 186 million miles or 2 astronomical units. This triangle can be subdivided into two congruent right triangles in which a side

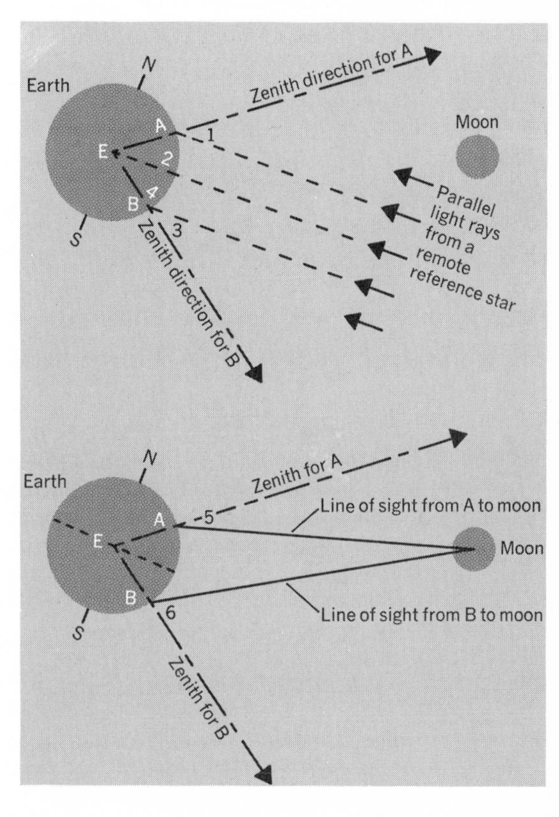

Fig. 19-7. *Measurement of the distance from the Earth to the moon. Diagram is a side view and not to scale. Hipparchus used a similar method in the second century* B.C. *In each diagram, two observers are located on the same north-south meridian at points A and B (they could be in the same hemisphere). The upper diagram indicates how the observers determine the angle AEB at the Earth's center between their respective radii. At the same moment, each measures the angle from his zenith to a certain reference star located on their celestial meridian (angles 1 and 3). The sum of these two angles equals the sum of angles 2 and 4. Why? Next, in the lower diagram, each observer simultaneously measures the angle between his zenith and the moon when it is on their celestial meridian (angles 5 and 6). Since the radius of the Earth was known at this time, it was possible to draw a diagram carefully to scale: lines of sight from A and B intersect at the center of the moon.* [*After Rogers.*]

(base) and an angle (the measured parallax) are known. The length of the hypotenuse, or the distance from the star to the Earth, can then be calculated. The distance from the Earth to the more distant stars, those 100 to 300 or more light-years away, cannot be determined by parallax because the angles are too small.

Distances to Remote Stars In modern times, a number of methods have been used in determining distances to remote stars and galaxies. Most of them depend directly or indirectly upon the *inverse-square law of light intensity: The apparent brightness of a source of light varies inversely as the square of its distance from an observer.* To illustrate the basic principle, imagine that on a dark night you are located on a long, straight, level stretch of road. In one direction, you see the faint headlights of a

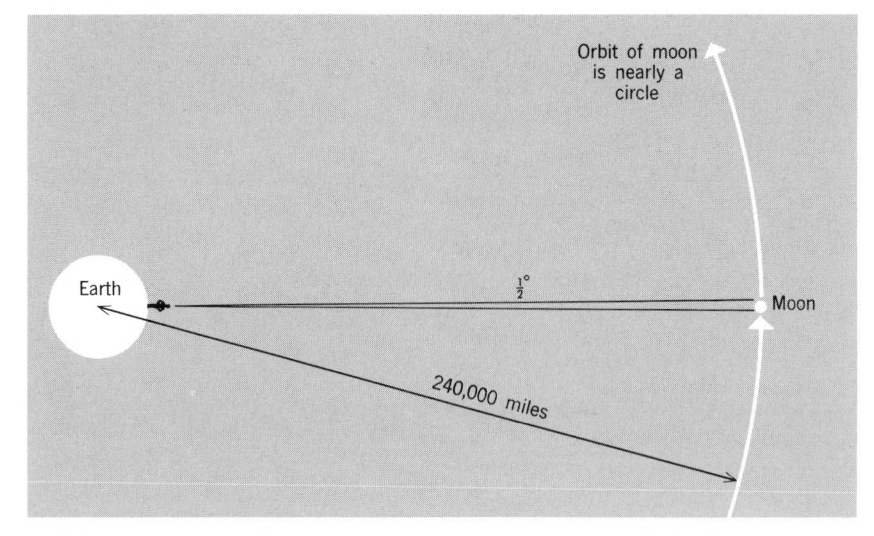

Fig. 19-8. *A method of ascertaining the size of the moon. (Diagram is not to scale.) The entire distance around the moon's orbit can be calculated because the orbit is nearly circular; it is approximately equivalent to the circumference of a circle whose radius is 240,000 miles. The angular diameter of the moon in the sky is measured. This is about ½ degree. Measuring the angular diameter of the moon from the Earth's surface rather than from its center does not make a significant difference in the size of the angle. As the moon's orbit is very large, an arc of ½ degree can be considered as a straight line and thus equal to the moon's diameter. Thus:*

$$\text{Moon's diameter} = \tfrac{1}{2} \cdot \frac{2\pi r}{360} = \tfrac{1}{2} \cdot \frac{2 \times 3.14 \times 240{,}000}{360} \text{ miles}$$

which is about 2000 miles. An accurate determination would have to consider the moon's orbit as elliptical, its angular diameter as slightly more than ½ degree, and certain other refinements.

car and conclude that it is far away; in the opposite direction, the headlights of a car appear moderately bright, and you conclude that this car is closer to you. To reach these conclusions, you assumed that the two cars had headlights of equal luminosity, and that a source of light appears dimmer as its distance from you increases (Fig. 19-9).

In the absence of intervening gas and dust, the observed brightness of a star depends upon two factors: its actual or intrinsic brightness (this can be called its luminosity) and its distance from the observer. Remote stars appear faint even if they are actually very luminous. The inverse-square relationship makes it possible to determine the distance of a remote object by comparing the apparent brightness of the object, which astronomers can measure readily, with its actual luminosity which is known or assumed. Conversely, if the apparent brightness and the distance are known, the actual luminosity can be calculated. Thus in determining the distance to a star by this method, astronomers use equations in which three factors are involved; if any two of these are known, the third can be calculated.

Astronomers have various methods of measuring the actual luminosities of stars, and we shall discuss briefly the one involving the Cepheid variable stars (another: the spectral class of a star is related to its luminosity and mass, p. 465).* During a study of our nearest galactic neighbors (the Magellanic Clouds, about 150,000 to 200,000 light-years away), it was noticed (by Leavitt and Shapley) that certain stars, subsequently named Cepheid variables, changed in brightness systematically over a period of time which in some instances lasted more than one month. Careful study showed that the time necessary for the light from such a star to vary from minimum brightness to maximum, and then back to minimum again—

* "Laboratory Exercises in Astronomy—Variable Stars in M15," by Owen Gingerich, *Sky and Telescope,* October 1967. Reprints are available from Sky Publishing Corporation, 49-50-51 Bay State Road, Cambridge, Massachusetts 02138.

Fig. 19-9. *The apparent brightness of a source of light varies inversely as the square of its distance from an observer. Suppose we place a 1-foot square a distance of 3 feet from a paint sprayer and spray it with black paint. If straight lines are drawn from the sprayer opening through the corners of this square and on into space, we find that the lines are 2 feet apart at a distance of 6 feet from the sprayer, 4 feet apart at a distance of 12 feet, etc. This follows from the principle of similar triangles. Doubling the distance doubles the length of a side of the square and thus increases its area by four times. If we double the distance a few additional times we obtain successive areas of 1, 4, 16, 64, 256, etc.: each preceding area is increased four times. Now we take another square that is 2 feet on a side (4 square feet of area), place it 6 feet from the paint sprayer and spray it with black paint (same quantity as before). It follows that all of the paint that would fall on a 1-foot square at a distance of 3 feet must now be spread across an area of 4 square feet, and only ¼ as much paint falls on each square foot of the larger square. At 12 feet, only ¼ as much paint falls on each part as at the 6-foot distance (or 1/16 as much as at the 3-foot distance). Thus, doubling the distance decreases the intensity to ¼ of the original; tripling it decreases the intensity to 1/9 of the original; and quadrupling it decreases the intensity to 1/16 of the original, etc.*

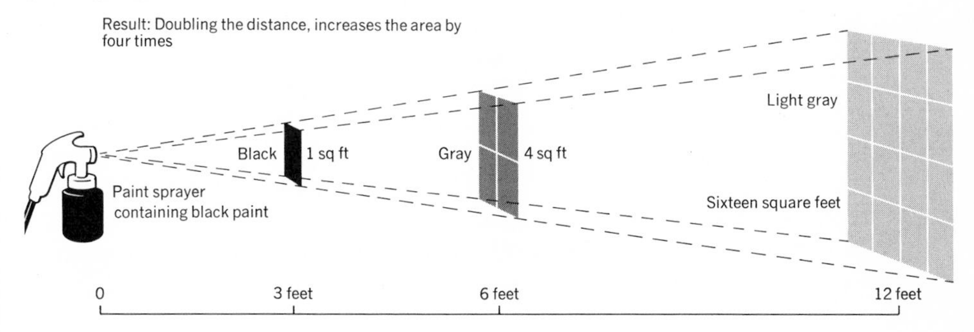

its *period*—is related directly to its luminosity; the most luminous Cepheids have the longest periods (Fig. 19-10).

This *period-luminosity relationship* could be detected because all of the stars in a particular galaxy are about the same distance from the Earth (just as all of the houses in Houston are about the same distance from Miami). Thus differences in brightness are caused chiefly by actual differences in luminosity. At this stage, the shape of the period-luminosity curve was known, but its location on the graph was not. In other words, the actual luminosities of the Cepheids were then unknown; only their relative luminosities had been learned.

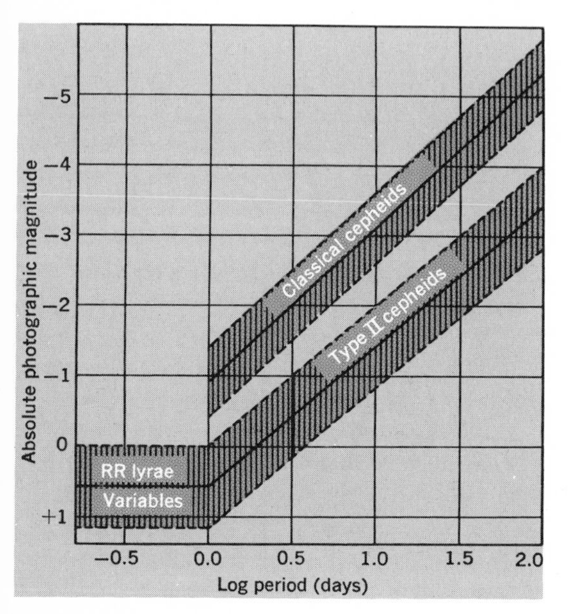

Fig. 19-10. *Period-luminosity curves for Cepheids. The magnitude increases toward the top and the period increases toward the right. The graph shows that (1) the more luminous Cepheids have longer periods than less luminous ones; (2) classical Cepheids are about four times more luminous (1½ magnitudes) than corresponding type II Cepheids. Originally all of the Cepheids were plotted along a single line that extended from the lower left to the upper right. Then it was learned that there should be two lines that parallel each other, shown as the heavy lines in the center of each band. Currently, it is thought that the Cepheids should be plotted in bands about one magnitude wide as shown on the diagram. The presumed positions of the lines (or bands) on the graph have been shifted a number of times and may have to be shifted again. All of these changes cause revisions in the estimates of the distances to stars that are based upon them.* [*From preliminary data supplied by H. C. Arp.*]

The next step involved estimating the distances to relatively nearby Cepheids within our own galaxy by other methods (by an analysis of their proper motions). The apparent brightnesses of these Cepheids were then measured, and the inverse-square law was applied to estimate their luminosities. After the luminosities of these nearby Cepheids in our galaxy had become known, astronomers assumed that Cepheids of similar periods in the Magellanic Clouds or any other galaxies would have similar luminosities. A curve was then plotted on the period-luminosity graph. This relationship was next assumed to apply as far as Cepheids could be observed (about 20 million light-years, since Cepheids are very luminous stars). In other words, all Cepheids with five-day periods were assumed to be equally luminous. Therefore, the faintest ones were farthest away (if gas and dust did not intervene).

The distances to galaxies containing Cepheid stars can thus be determined on the basis of the assumed period–luminosity relationship. Next, the average actual luminosities of these galaxies were calculated and assumed to be representative for other galaxies of similar types that are too far away for individual Cepheids to be distinguished. Thus to obtain the distance to a remote galaxy, the astronomer measures the apparent brightness of the galaxy, assumes that its actual luminosity is similar to that of nearer galaxies of a similar type, and applies the inverse square law of light intensity.

Unfortunately for the accuracy of this method, galaxies vary widely in luminosity, even among those of a similar type. In an attempt to overcome this problem, astronomers have worked with clusters of galaxies. They assume that the 10 or so brightest galaxies in any one cluster will have about the same average luminosity as the average of the same number of brightest galaxies from any other large cluster. The apparent brightnesses of these galaxies are then measured and compared; the faintest ones are assumed to be the most distant.

Cautions Concerning Stellar Distances Calculations of distances in astronomy involve assumptions which pyramid as distances become greater. The distances to nearer stars are obtained by a method which seems reliable but depends upon the validity of certain assumptions. These results are used as a basis for the next method, which likewise involves certain assumptions; e.g., light was assumed to come through space without any diminution of its brightness by intervening dust or gas. If light from a star is dimmed by obscuring matter, our measurement of its apparent brightness will be too small, its calculated distance will be too great, and the magnitude of the error will depend upon the quantity of obscuring matter. Therefore, opportunity exists for the presence of large errors in these successive extrapolations, and caution should be exercised. It would be unwise to accept without question the reliability of all stellar distances.

This need for caution was illustrated in 1952. Until about that year astronomers did not comprehend clearly that two types of Cepheids existed. One type is at least four times as luminous as the other (Fig. 19-10). All of the Cepheids used to obtain distances outside of our galaxy are of the more luminous type. Therefore, if they are actually four times as luminous as was previously assumed, then they must be twice as far away. Distances to other galaxies and the sizes of these galaxies have all been computed on the basis of the Cepheid method, and they are unreliable if the method is not correct.

Prior to 1952 a scale of distances had been calculated for the universe; subsequently, all distances and sizes which had been determined for objects beyond our own galaxy were approximately doubled. The 1952 revision not only resulted in a doubling of the size of the known universe, it also doubled its age (p. 476), and further revisions have been made since 1952. However, sizes and distances within our own galaxy were not affected by this correction. Before 1952 it was estimated that the 200-inch telescope at Palomar could see 1 billion light-years into space; this figure has since been increased to 2 billion light-years, and some astronomers now speak of a 6 billion light-year limit.

Exercises and Questions, Chapter 19

1. A new planet is discovered. It is nearly a perfect sphere. What information, if any, does this give concerning the planet's revolution, rotation, composition, and distance from the sun?
2. Explain why the Earth's axis points approximately at the North Star from all positions in its orbit.
3. Explain the probable seasonal or climatic variations which would result from the following: (1) If the Earth's axis were perpendicular to the plane of its orbit about the sun and remained perpendicular in all parts of its orbit; and (2) If the Earth's axis were parallel to the plane of its orbit and pointed always in the same direction in space (the axis of Uranus is nearly parallel to its orbital plane about the sun).
4. From the equator where would you look to see the North Star? From the North Pole? From your home town? What is the latitude at each place?
5. What changes occur in the time and place of sunrise and sunset in September? December? June?
6. List some of the changes which have been made in the estimates of astronomical distances in the last few decades.
7. Imagine that you are located at about 50° latitude in the Southern Hemisphere. In what direction is the sun at noon? When will the noon shadow be shortest during the year? When longest?
8. Assume that a pilot flies a jet plane high above the equator in a certain direction at a certain rate. He notes that his solar time does not change (e.g., it is always noontime or sunset). What is his direction and rate of flight?

9. Assume that a scholar on a fictitious planet makes measurements similar to those of Eratosthenes. The distance between his two cities (corresponding to Syene and Alexandria) is 1000 miles and the angle between the noon sun and zenith is 20 degrees (the sun is at the zenith at the other city). What is the circumference of this planet? its radius?

10. Saturn's total mass is about 95 times larger than that of the Earth, yet an object at its surface would weigh about the same as at the Earth's surface. Why?

11. What is the approximate latitude of a 100-foot flagpole that casts a noon shadow of 57.7 feet at an equinox?

20 The moon

WITH the exception of the sun, the moon is the most conspicuous of celestial bodies. Old travelers counted on moonlight and the tides, and early man had a calendar in the moon's phases. Lunar and solar eclipses have always stirred wonder. And now, as men prepare to land on the lunar surface, many unmanned spacecraft, Russian and American—some circling, some crashing, some soft-landing—are relaying pictures and data to the Earth (Figs. 20-1 and 20-8 to 20-12). Despite these spectacular advances and thousands of pictures (Surveyor V in September 1967 relayed some 18,000) few questions have been settled. It is said that the photographs are like mirrors: each viewer sees his own theories reflected back to him.

The moon, about 2160 miles in diameter, follows an elliptical path about the Earth at a mean distance of nearly 240,000 miles. An average sample of the moon is approximately 3.4 times as heavy as an equal volume of water (the Earth's specific gravity is about 5.5) and its mass is less than 1/81 that of the Earth. The moon acts as a mirror to reflect light from the sun onto the nighttime side of the Earth. The Earth, at times, likewise reflects light onto the darkened portion of the moon (earthshine).

As measured from the Earth, the diameter of the moon varies as much as 12 percent, an indication that its distance varies. In fact, the moon's elliptical orbit can be calculated on the basis of this apparent change in diameter: the moon is closest to the Earth when its diameter seems largest.

The change in the moon's diameter should not be confused with the apparent increase in size that all celestial bodies undergo near the horizon—an optical illusion, as is proved by photographing an object such as the moon both near the horizon and high overhead.

Phases of the Moon The moon's monthly cycle of phases was possibly the first celestial motion to be explained correctly. As the moon is a nearly spherical body in space, one half is always lighted by the sun, and the other half is darkened and turned away from the sun. The moon's revolution about the Earth brings varying proportions of its lighted half into view and the phases result (Fig. 20-2). No one part of the moon is permanently illuminated; night and day occur, but each lasts about two of our weeks. New moon occurs when the moon is approximately between the Earth and sun and the lighted half is turned completely away from the Earth; full moon occurs a little more than two weeks later when the Earth is approximately between the sun and moon and the illuminated half is facing the Earth.

For the earthly observer, the cycle begins about two days after the invisible new moon when a thin crescent can first be seen low in the western sky at sunset. Almost a week later the half-moon (first quarter phase) is high overhead in the southern sky at sunset (Fig. 20-3). At the end of another week, the full moon rises in the east as the sun sinks in the

Fig. 20-1. *Three pictures of the moon's surface as revealed by the television cameras of Ranger IX spacecraft on March 24, 1965.—Left, the region of the crater Alphonsus (right) as it appears from an altitude of 258 miles at 2 min 50 sec before impact. The area is about 115 miles on a side.—Center, the central peak of Alphonsus and nearby rills and craters seen from an altitude of 58 miles at 38.8 sec before impact.—Right, from a distance of 115 miles the prominent rills show up, lines with dark halo-type craters. Secondary craters are visible in photographs taken by the Ranger spacecraft. These have a different shape from the primary craters (shallower, smoother, elongated, gently sloping walls) and occur only within the rays. Evidently they were produced by the impacts of debris splashed out of the primary craters.* [*Courtesy NASA*.]

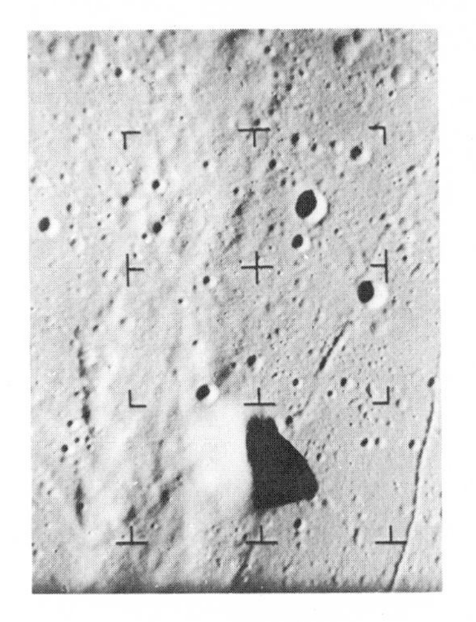

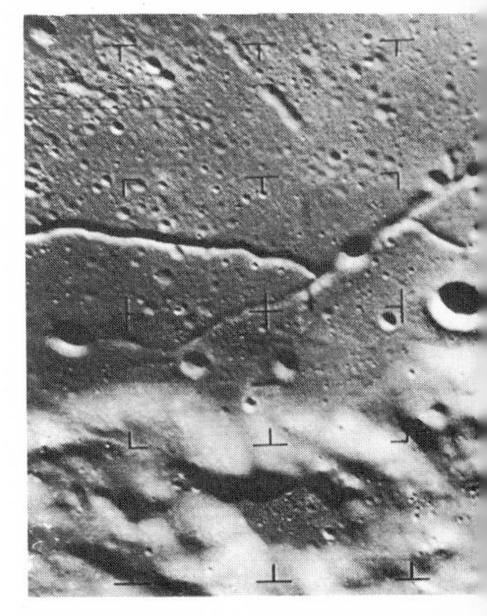

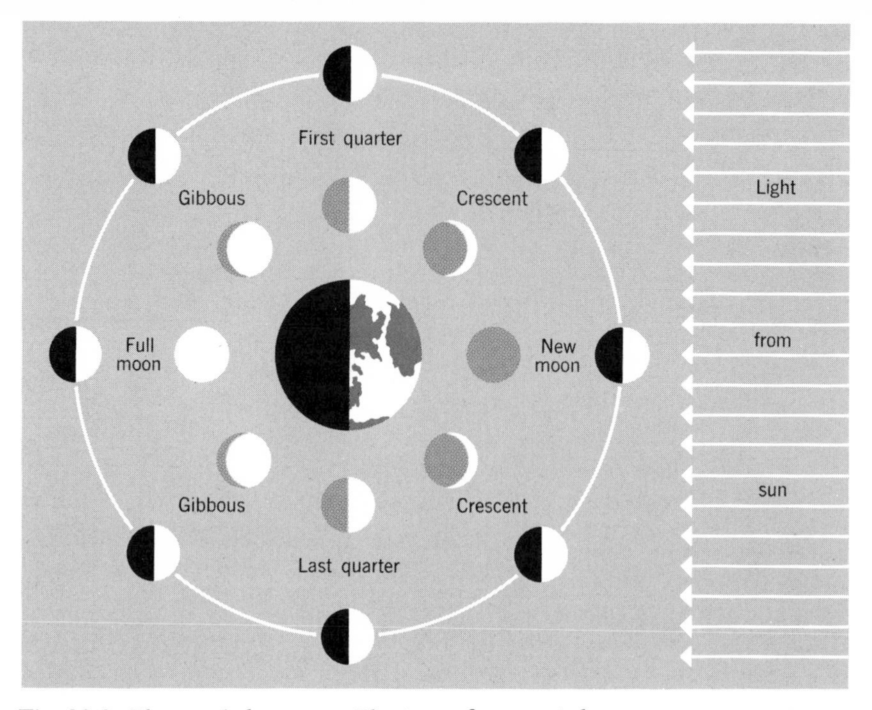

Fig. 20-2. *Phases of the moon. The inner figures of the moon represent its appearance from the Earth.* [*Bowditch,* American Practical Navigator, *1962*.]

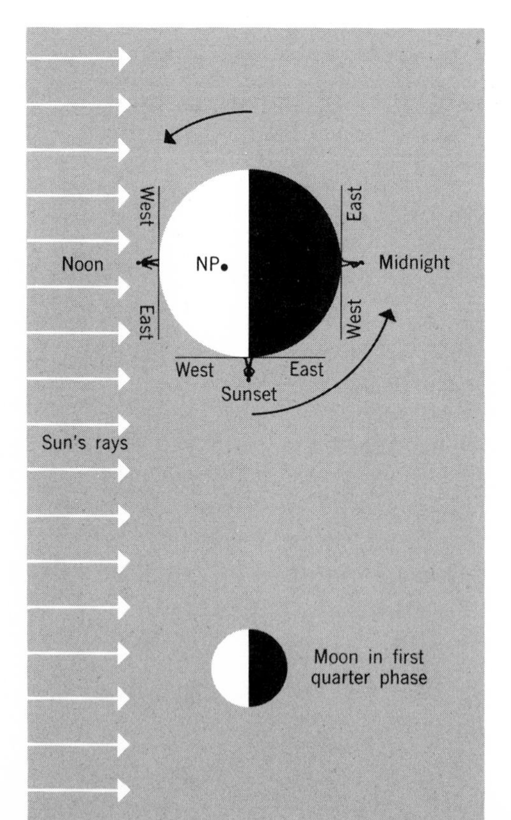

Fig. 20-3. *The moon in the observer's sky during its first quarter. Orientation: in space looking down on the Northern Hemisphere in summer. For one observer on the Earth, it is noon, for a second sunset, and for a third midnight. The moon is on the eastern horizon (rising) for the noon observer, high in the south for the sunset observer, and on the western horizon (setting) for the midnight observer. In six hours the Earth's counterclockwise rotation will spin the noon observer into the sunset position, and six hours more will bring him to midnight. Make other diagrams to show the location of the moon in the observer's sky during its other phases. Sunset occurs at different hours at different seasons of the year.*

west. At the third quarter phase, the half-moon rises above the eastern horizon at midnight. About one month after the beginning of the cycle, the observer again sees a thin crescent low in the western sky at sunset, and a new cycle has commenced.

The length of time spent by the moon above the horizon tends to be inversely proportional to that spent by the sun, a fortunate circumstance for man. During long winter nights, the moon rises earlier, moves higher in the sky, and sets at a later hour than it does during summer months. Note (Fig. 19-3) that when the ecliptic is located $23\frac{1}{2}°$ below the celestial equator at noon (on December 22), it will be located the same distance above the celestial equator at midnight. The moon is always within 5° of the plane of the ecliptic.

The revolution of the moon about the Earth causes the moon to shift eastward against the celestial background more rapidly than the sun. Thus the lunar day—the time for one complete rotation of the Earth relative to the moon—is longer than the solar day and averages 24 hours and 51 minutes (Fig. 20-4).

Fig. 20-4. *The moon rises later each day. Orientation: looking down on the Northern Hemisphere in autumn. The full moon appears above the eastern horizon at sunset on one day for the observer. Next day the Earth's daily rotation has returned him to the sunset position in about 24 hours. Meanwhile the moon has moved forward in its orbit (movement is exaggerated); it is no longer visible at sunset. The earth must spin 51 minutes longer, on the average, to move the observer into position to see the moon rise in the east again. What is the season of the year in the Southern Hemisphere at this time?*

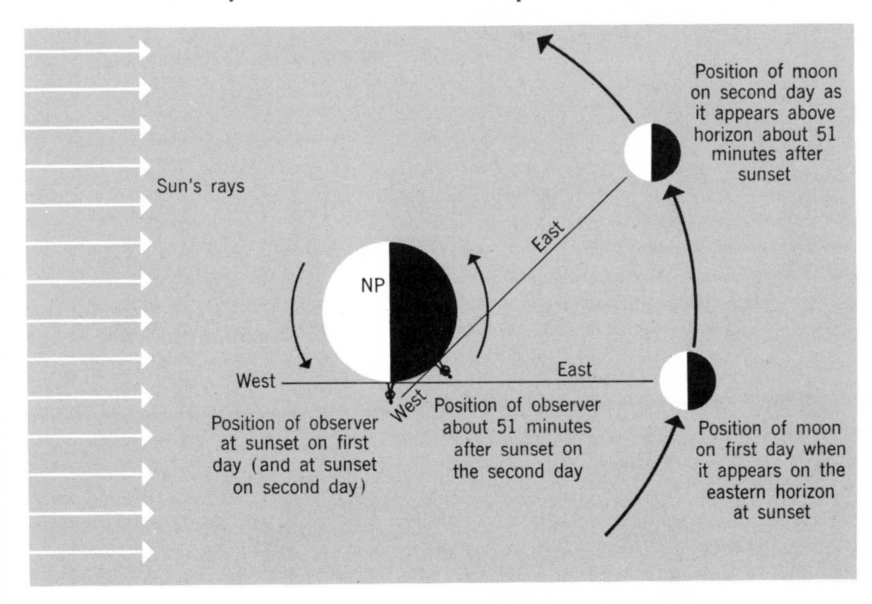

The length of the lunar day varies considerably during the year because the inclination of the ecliptic to the horizon changes; e.g., near the time of the autumn equinox the full moon may rise as little as 20 minutes or so later each night. Therefore, at any one latitude it is visible for a longer time during the early evening hours at this season than it is at other seasons of the year—the so-called harvest moon.

The time that elapses between two full moons or two new moons approximates 29½ days (*synodic month*). On the other hand, the moon completes an eastward swing of 360 degrees relative to the stars in about 27⅓ days (*sidereal month*). The orbital movement of the Earth causes this difference (Fig. 20-5).

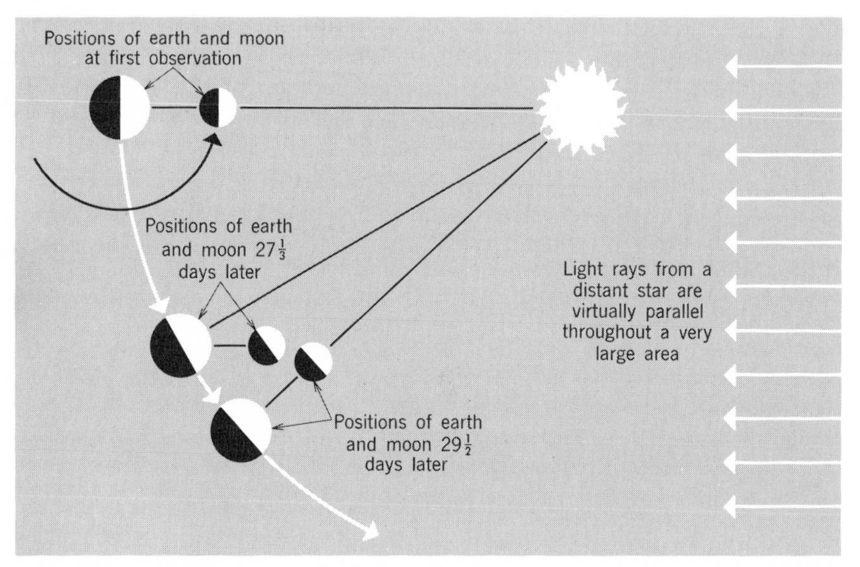

Fig. 20-5. *Sidereal and synodic months. Orientation: looking down on the Northern Hemisphere.* (*1*) *At the first observation, the moon is in its new phase; the Earth, moon, sun and reference star are in line in space.* (*2*) *About 27⅓ days later, after the moon has revolved through 360 angular degrees around the Earth, the sidereal month is completed. Earth, moon, and reference star are once more aligned—but not so the sun. The moon is a crescent.* (*3*) *Approximately two days later, about 29½ days after* (*1*), *the synodic month ends. The moon is again in a line between the Earth and the sun in the new moon phase.*

The Moon's Rotation and Revolution Approximately the same hemisphere of the moon faces the Earth at all times; from the Earth one can see the face of the "man in the moon" but never the back of his head. Two motions cause this: The moon revolves about the Earth, and it also rotates

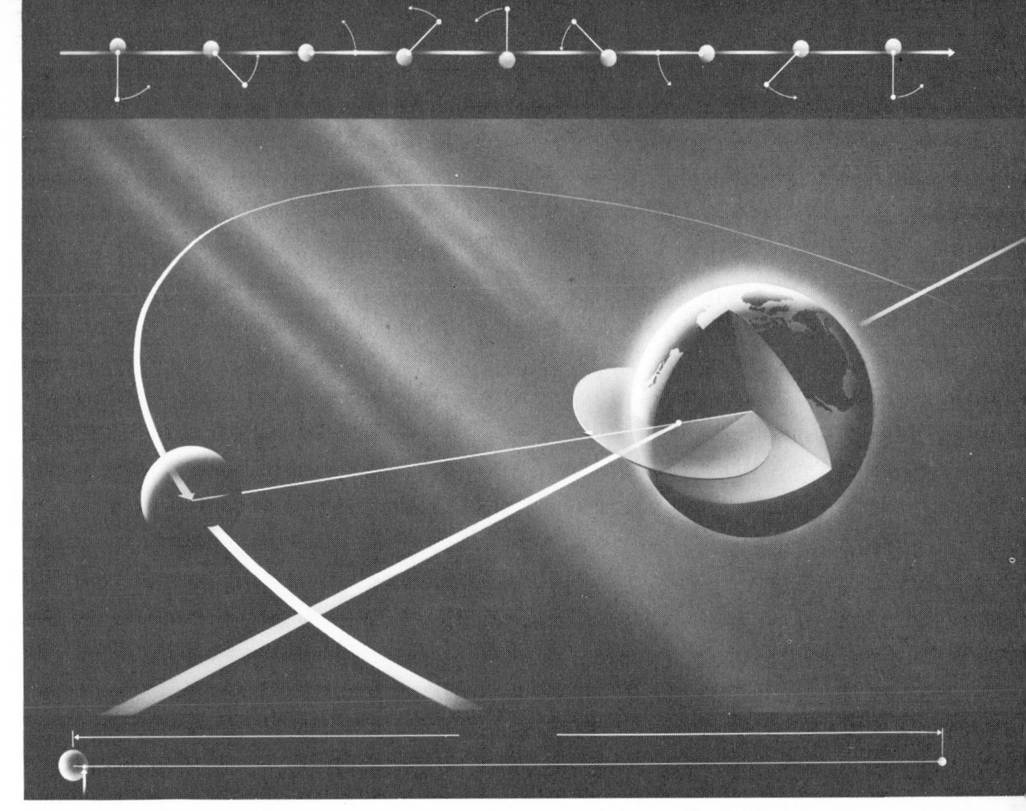

Fig. 20-6. *Center of mass for the Earth-moon system. This is the balance point about which each turns, once each month. The motion of this point around the sun (panel at top) is the "orbit of the Earth." Earth-moon sizes and distance apart are shown to scale at the bottom.* [*American Museum of Natural History.*]

(with respect to the stars) as it revolves; i.e., the moon rotates through 360 degrees in the $27\frac{1}{3}$ days that it takes to complete a sidereal revolution of 360 degrees.

During these motions, a particular point on the moon experiences one period of daytime in which the sun is visible (nearly 14 Earth days) followed by an equal period of nighttime. To illustrate this phenomenon, try following a circular path about a tree, keeping your face turned toward the tree in the center at all times. In order to do this, you will have to rotate once during each complete trip around the tree.

One can speculate that the moon formerly rotated much faster, but has been slowed by tidal effects caused by the Earth's gravitational attraction. Tidal friction would have been particularly powerful if the moon had once been less rigid and closer to the Earth. Apparently many of the satellites of the other planets likewise keep the same hemispheres always toward their primaries.

The moon's motions are actually among the most complex which astronomers resolve. The moon and the Earth are relatively so close to each other and so far from the sun that the moon's path, like that of the Earth, is always concave toward the sun. The moon and the Earth revolve *around each other,* or rather around the center of gravity in the line that joins them. This center is beneath the Earth's surface, 2900 miles from the Earth's own center. As seen from a point in space, the orbits of the Earth and the moon would thus nearly coincide; the moon would periodically overtake the Earth and would later in its turn be passed by the Earth (Fig. 20-6).

When the Earth's gravitational attraction pulls the moon forward in its orbit, as it does at the first-quarter phase, the moon speeds up. The moon is gradually slowed by the Earth's gravitational pull from the full-moon position onward, but its speed is not reduced to that of the Earth's until the third-quarter position is reached, when it begins to fall behind again. The moon's orbit is always concave toward the sun, although this is not shown on most diagrams because of a distortion of the scale.

The distance of the moon from the center of mass of the Earth-moon system can be used to estimate the mass of the moon. If two celestial objects mutually revolve about a point—the center of mass or gravity for this two-body system (Fig. 20-6)—their masses and distances from the center are related as follows. The mass of one body multiplied by its distance from the center equals the mass of the other multiplied by its distance from the center. Since the moon's mean distance from the Earth is nearly 239,000 miles (from center to center), the moon is about 236,100 miles from the center of mass. Thus the moon is about 81 times farther from this center than is the Earth; therefore, its mass is 1/81 that of the Earth.

FEATURES OF THE MOON

Absence of an Appreciable Atmosphere Considerable evidence indicates that no appreciable sea of air envelops the moon. No clouds can be seen when the moon is observed through powerful telescopes, shadows are distinct and black, and observations have not detected signs of twilight or of erosion-produced topographic features similar to those on the Earth. Also, the light of a star is blotted out abruptly as it disappears behind the moon, whereas if the moon had a significant atmosphere, the star would be seen through the moon's atmosphere and its light would be dimmed gradually before being completely extinguished. This lack of an appreciable atmosphere is explained by the moon's small gravitational attraction, about one-sixth that of the Earth. If gas molecules ever surrounded the moon, they have long since had the opportunity to escape the relatively slight gravitational pull by virtue of their rapid movement.

Gas may have been erupted during volcanic activity at the moon's surface in the late 1950's and been detected, because some bright lines were observed in the moon's spectrum at this time. Perhaps these were caused by light emitted by gas molecules of carbon compounds that had been excited by solar radiation.

A hike on the surface of the moon would reveal many conditions which differ from those on the Earth. It would be necessary, of course, to supply one's own oxygen and to overcome the absence of atmospheric pressure. Dangerous ultraviolet radiations from the sun, which are screened out by the Earth's atmosphere, would have to be guarded against on the moon. Normal conversation would be impossible because sound waves need a medium for their transmission. Indigenous life would presumably be absent. The hiker would weigh about one-sixth of his weight on the Earth and would be able to perform startling athletic feats (p. 360). Except in the immediate neighborhood of the sun, the sky would be black both day and night (p. 439), and the stars would be visible at all times. The Earth, as a moon in the lunar sky, would go through phases without rising and setting, and would be about four times as large and many times brighter than the moon appears to an earthly observer. It would be visible from only one side of the moon. Meteoroids should pepper the hiker incessantly but would not be visible as "shooting stars" (p. 412). Large variations in temperature (more than 400°F) would occur between day and night. Weather predictions would be simple: two weeks (as measured on the Earth) of clear and hot weather would always be followed by two weeks that would be clear and cold.

Surface Features of the Moon The most conspicuous topographic features on the moon are its dark areas, craters, mountains, and at times its rays (Figs. 20-7 and 20-8); lesser land forms include cracks, faults, grooves, and small domes. The large dark areas form the so-called lunar seas (*maria*) which make the facial features of the "man in the moon." They are plainly visible to the unaided eye. These relatively smooth, circular regions are confined largely to the moon's northern hemisphere, which appears near the bottom of many lunar photographs. Perhaps such areas are dark because a smooth surface tends to reflect rays of light chiefly in certain directions, whereas a rough surface tends to scatter light in all directions. Therefore, although both types of surfaces may reflect the same amount of light, the smooth surface appears darker than the rough surface unless the observer is favorably located to receive the reflected light.

The maria are generally depressed relative to the rest of the moon's surface, but they occur at different levels. Partially submerged craters, smaller craters, and wrinkled surfaces occur within them. The maria contrast distinctly with other parts of the lunar surface that are higher in altitude, lighter colored, mountainous, cratered, and less smooth. Through the telescope, the maria have the appearance of hardened flows of lava, but they have also been described as dust-filled depressions.

Fig. 20-7. *The moon about 3 days after first quarter. As viewed through a telescope in astronomy, objects are commonly inverted and reversed, and that is the orientation here; the south pole is at the top. Six conspicuous craters are numbered on the photo: (1) Clavius is nearly 150 miles across and has walls that tower as much as 4 miles above its floor. Actually this should be called a walled plain. If one stood at its center, the curvature of the moon's surface might be too great to allow one to see the top of the surrounding rim. Note the smaller (younger) craters within Clavius. (2) Tycho has a striking system of light-colored streaks (rays) radiating outward from it, but they are most conspicuous when the moon's phase is about full. (3) Copernicus—its rays are faintly visible (see also Fig. 20-8). (4) Eratosthenes. (5) Archimedes. (6) Plato has a smooth dark floor.—The large conspicuous dark areas are the maria.* [*Yerkes Observatory photograph.*]

The lunar mountains generally are formed in curved elongated ranges that margin the maria, but isolated peaks also exist. Some project 4 to 5 miles above their surroundings, proportionally a greater altitude than mountains have on the Earth. The mountains are jumbled masses that show little resemblance to the fold and fault block mountains known on the Earth;

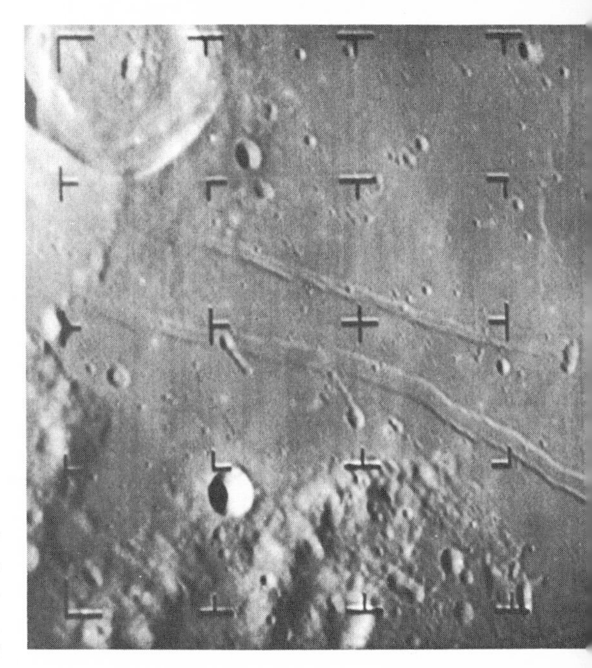

Fig. 20-8. *Above, the craters Copernicus (right) and Eratosthenes (left). The floor of Copernicus is about 40 miles across and contains several small peaks near its center. The walls of the crater rise more than 2 miles above the floor and show a series of concentric, inward-facing steep slopes which suggest an origin by faulting and sliding. The rays are visible and appear to have formed from material splashed or thrown radially outward over all sorts of topography. A string of small craters is located to the left of Copernicus; their alignment is suggestive of volcanic activity along a fault. Note the crater-capped central peak of the crater Eratosthenes.* [*Mount Wilson and Palomar Observatories.*]—*Below, photograph of lunar surface taken from the Ranger VIII spacecraft on 20 February 1965 at an altitude of 270 miles. This is one of about 7500 pictures which were relayed to the Earth before the spacecraft crashed into the moon. The bright crater in the lower left is approximately 3 miles in diameter. Note the two lunar rills which trend diagonally from upper left to lower right. Presumably these are long narrow graben or downfaulted regions of the lunar surface. Note also the depressions and grooves that have scarred the rills in places. To the left of center, one of these shows a line of shadow which suggests that faulting has occurred along this side of the graben since the groove originated. Small secondary craters are numerous in this view and apparently formed by the impact of debris that was splashed out of larger craters at the time of their formation.* [*Courtesy, NASA.*]

perhaps igneous activity and the explosive impacts of space debris have been important factors in their formation.

Thousands of pockmarks occur on the moon's surface. These are easily seen through a small telescope, and probably constitute its most intriguing physical aspect. They can be grouped into primary and secondary craters. A typical primary lunar crater (Fig. 20-8) is circular in outline and is surrounded by a ridge which has a steep inner slope and a gentle outer slope. The floors of most craters are lower than the adjoining terrain, but some are higher. If the material in the ridge encircling a crater were bulldozed into the crater, it would tend to fill the crater about level with its surroundings. A number of the craters contain central peaks and look something like Mexican sombreros. About 30,000 of the craters have been measured; their diameters range from about 140 miles to several hundred feet (for the smallest craters visible through Earth-based telescopes).

The secondary craters (Fig. 20-1) apparently were formed by debris thrown out of the primaries at the time of their formation by the explosive impacts of large meteoroids or small asteroids. Most of the debris that exploded outward from a primary probably moved at a relatively low speed and tended to collect around the primary crater to form a ridge. However, some chunks apparently flew for considerable distances and produced secondary craters where they hit the moon's surface. The secondaries cluster around their primaries and appear to be confined to the areas covered by the lunar rays. The rays are light-colored streaks that radiate outward from certain craters, show no shadows, cross various topographic features without deviation, and have the appearance of material splashed out of the craters. The secondary craters have a smoother appearance than the primaries, are more elongated and shallower, and have more gently sloping walls.

The origin of the craters constitutes an interesting problem. Such features as Meteor Crater in Arizona (Fig. 21-11) suggest that the lunar craters may have formed by the bombardment of large meteoroids and small asteroids. The intense heat developed at impact would vaporize the meteoroid and part of the moon's surface and cause an explosion. This explanation is favored at present. However, lunar craters also resemble certain volcanic features on the Earth, such as Crater Lake in Oregon (Fig. 32-10), and the hypothesis of a volcanic origin has received support. Probably the lunar craters have formed in more ways than one, but collisions with meteoroids and asteroids seem the most important factor.

Shoemaker* has reported his conclusions from a study of Ranger VII photographs (Fig. 20-1) and other materials. There is an inverse relationship between the numbers of craters and their sizes—the smallest craters are most numerous. In fact, it had been estimated that 10-foot-diameter craters would be so numerous that they would overlap and cover the entire surface. However, the Ranger VII photographs indicate that craters about 300 feet in diameter are most numerous. To explain the unexpectedly small numbers of very small craters, Shoemaker has suggested that such craters

* Eugene M. Shoemaker, "The Moon Close Up," *National Geographic,* November 1964, pp. 690-707.

Fig. 20-9. *Moon's surface photographed by Surveyor I's television camera. The rock in the foreground is about 18 inches long by 6 inches high. According to one hypothesis, this is an "instant rock"; it is an aggregate made from powdery material by the shock of a meteoritic impact. Dark spots are pores or cavities. The spacecraft was soft-landed within a crater in June 1966. Thousands of pictures were sent back to the Earth during parts of two lunar days (each equivalent to about 14 Earth days).* [*Courtesy, NASA.*]

once existed but were slowly worn away. Perhaps an average thickness of about 50 feet of the moon's surface has been removed during the $4\frac{1}{2}$ billion years that have presumably elapsed since the moon originated. This very slow wasting away is attributed to very high-speed impacts of very tiny meteoroids. The explosive impact of each high-speed particle probably scattered a quantity of debris much larger than its own mass. Most of this debris probably fell back to accumulate at the moon's surface. However, some of it probably traveled fast enough to escape from the moon, and perhaps some of this debris is eventually swept up by the Earth! Such a presumed wearing away of the moon's surface would have little overall effect upon the larger craters that can be viewed through a telescope from the Earth, but the smaller craters would become shallower and assume gently rounded outlines, and many would eventually disappear. That such small craters still exist, as shown by the Ranger VII photographs, indicates that some have formed rather recently.

Fig. 20-10. *Earth-moon photograph. This picture of the crescent Earth with the back side of the moon in the foreground (never previously seen in this detail and completely invisible to Earth-based observers) was taken by Lunar Orbiter I in August 1966 when the spacecraft was about 75 miles above the moon's surface. The photograph was taken on film and subsequently scanned and relayed to the Earth. The east coast of the United States is in the upper left, and Antarctica is at the bottom of the Earth crescent.* [*Courtesy, NASA.*]

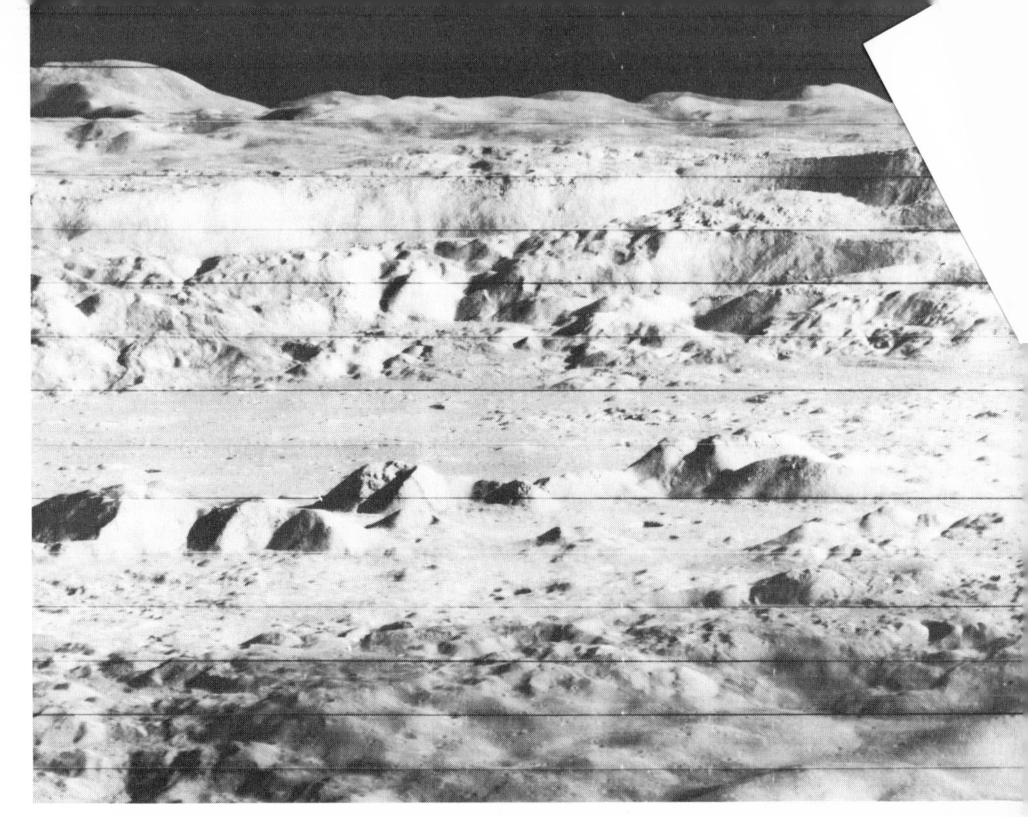

Fig. 20-11. *Low-level telephoto view of the crater Copernicus from Lunar Orbiter II in November 1966. Copernicus is about 60 miles wide and 2 miles deep; a section 17 miles wide is shown in this photograph. The hills showing sharp shadows are about 1000 feet high and occur near its center. The photograph was taken from an altitude of about 28 miles and a distance of 150 miles from the center of Copernicus. It has been described as "one of the great pictures of the century." Orbiter photographs show that cliffs on the rim of the crater are undergoing continual downslope movement of material. Mass-wasting thus seems to occur on the moon as well as the Earth, although the process must be somewhat different in the absence of air and water.* [*Courtesy, NASA.*]

The lunar surface has a granular rough texture on a small scale and is firm enough that spacecraft did not sink out of sight into loose dust as some had feared. However, dust particles in a vacuum might have this appearance; presumably they would stick together in the absence of air and water to produce a spongy structure (possibly somewhat like fresh wet snow). As measured from Surveyor V, the lunar surface material contains oxygen, silicon, and aluminum in about the same proportions as occur in basaltic rock on Earth. In the measurement, alpha particles, radiated by curium, struck atomic nuclei at the lunar surface. The energies with which these particles are back-scattered, and the energies of protons ejected, are characteristic of the atoms struck and can be measured. Other elements were also identified.

The manner of origin of the moon itself is likewise still in doubt. Did it originate as material spun off from the Earth, and did tidal forces then cause the moon to move outward to its present position? Did the moon form as an aggregate of gas and dust as a companion of the Earth (p. 478)? Or did the moon form independently and subsequently approach close enough to the Earth to be captured by it?

LUNAR ECLIPSES

Rays of light from the sun are not quite parallel. Therefore, spherical bodies in space, such as the Earth and moon, cast long, tapering, conelike shadows which always point directly away from the sun. The shadows are invisible unless some object moves through them.

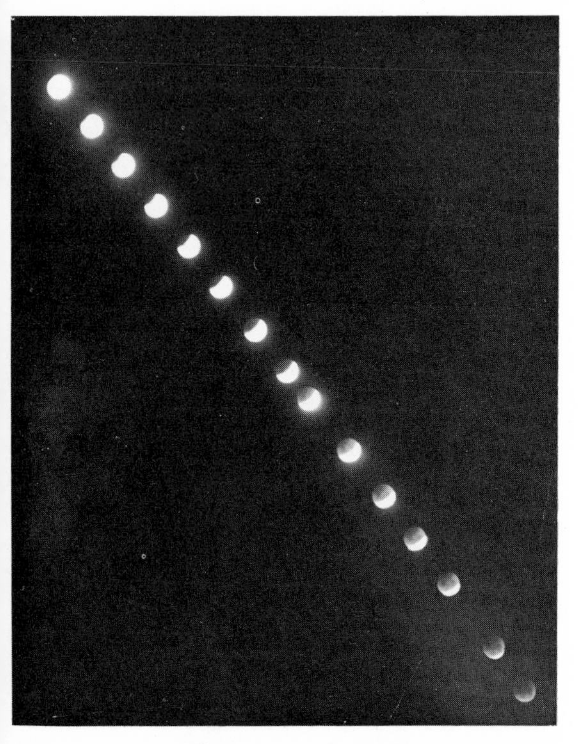

Fig. 20-12. *Lunar eclipse, January 29, 1953. Exposures were made every five minutes when possible as the moon left the Earth's shadow. The moon is completely eclipsed at the lower left. The shadow's curvature indicates that the Earth is round and about four times the size of the moon.* [*Photograph by Neil Croom.*]

The moon is eclipsed whenever the Earth comes directly between it and the sun (Figs. 20-12 and 20-13). The moon is always full at the time of a lunar eclipse, and in a clear sky the eclipse is visible to everyone on the nighttime side of the Earth. For this reason, lunar eclipses are more familiar to the average person than solar eclipses, even though the latter are actually more frequent (about four out of seven). Since the plane of the moon's orbit is inclined approximately 5 degrees to the plane of the Earth's orbit,

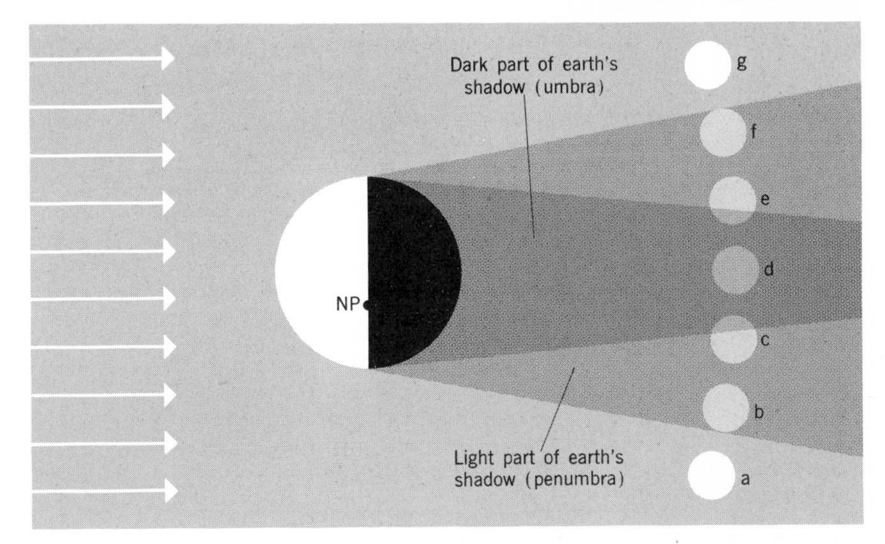

Fig. 20-13. *Top view of a lunar eclipse. In* a *and* g *the moon is full, before and after moving through the Earth's shadow. In* b *and* f *the moon is in the zone from which some of the sun's light is excluded, and the change in the moon's appearance is not pronounced. In* c *and* e *the moon is about half-eclipsed, and in* d *the eclipse is total. Even when totally eclipsed, the moon is usually visible as a reddish disk, because the penetrating reddish waves in the sunlight are bent inward as they pass through the Earth's atmosphere and thus reach the moon, to be reflected back to the Earth.*

an eclipse does not occur each time the moon is full (Fig. 20-14). The longest total lunar eclipses last about $1\frac{1}{2}$ hours. As the moon is moving eastward more rapidly than the Earth's shadow, the eastern side of the moon is always darkened first.

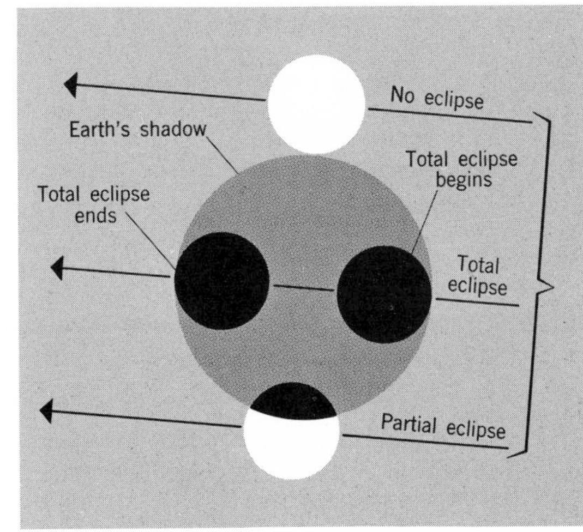

Fig. 20-14. *Lunar eclipse, as if viewed from the sun. The large dark circle represents the Earth's shadow (umbra) at the moon's distance from the Earth. Since the diameter of the shadow at this distance is about 2½ times the diameter of the moon—5700 miles vs. 2160 miles—total eclipses are possible. The different paths shown can occur because the moon's orbital plane around the Earth is inclined about 5 degrees to the ecliptic.*

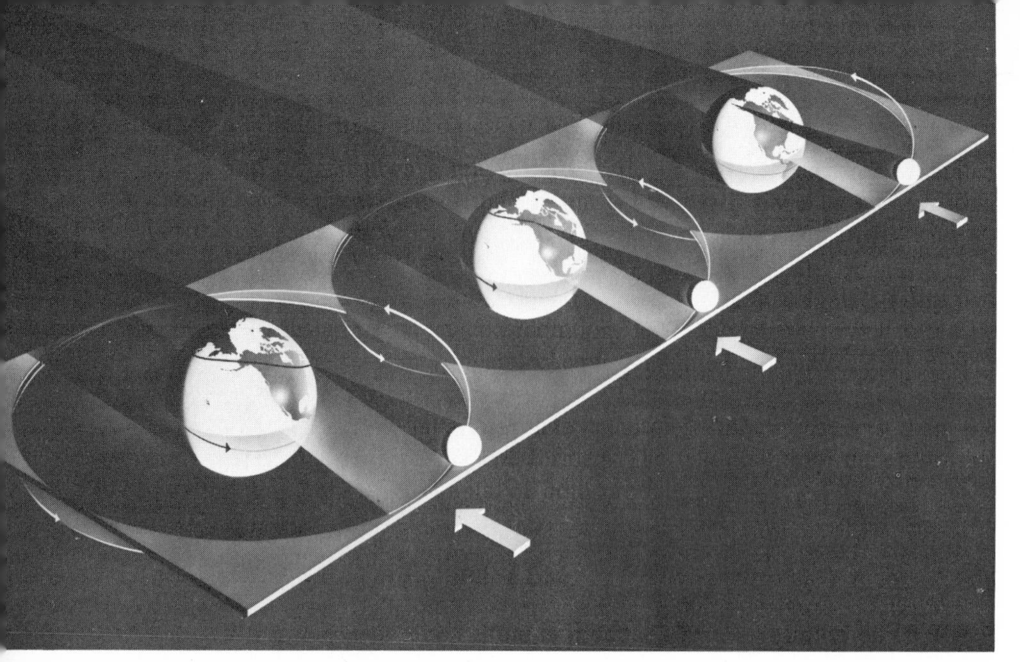

Fig. 20-15. *A total solar eclipse begins in the upper right and ends in the lower left. Only the darkest part of the moon's shadow (umbra) is shown. Note that the moon's orbital plane is inclined to the ecliptic (represented by the plywood sheet) as is the Earth's axis.* [*Courtesy of the American Museum of Natural History.*]

It is reported that Christopher Columbus made good use of his knowledge of eclipses at Jamaica on his fourth voyage to the New World. He was sick and hungry, his crew was mutinous, and the Indians refused to furnish food. Fortunately Columbus knew that a total eclipse of the moon was scheduled for a certain evening. He told the natives that God was angry at them for withholding food and would cause a famine; as a sign of this, the light of the moon would be blotted out. As predicted, the total lunar eclipse took place. Before it was over, however, Columbus reported to the Indians that if they granted the food, the famine would not occur. Columbus obtained the food, and the moon began to shine—certainly a practical application of astronomy!

SOLAR ECLIPSES

A total solar eclipse is a magnificent celestial phenomenon and a rare occurrence in the lives of most of us. Calculations show that an average specific geographic location experiences a total solar eclipse only once in approximately 360 years.

Solar eclipses occur when the new moon comes directly between the Earth and the sun (Fig. 20-15). A solar eclipse does not take place at each new moon for the same reason that a lunar eclipse does not occur at each full moon: the inclination of the moon's orbit to the Earth's orbit causes the tip of the moon's shadow during the new moon phase to sweep above the Earth on some occasions and below the Earth at other times. The moon's shadow has a length of about 232,000 miles. This is about 7000 miles shorter than the mean distance between the moon and the Earth, but along a portion of its elliptical orbit, the moon is closer than 232,000 miles to the Earth. The largest shadow it can cast on the Earth is about 167 miles in diameter; usually the shadow is much smaller than this. The shadow moves eastward and the Earth rotates eastward, but the moon's shadow moves more rapidly. When the shadow overtakes the Earth, it passes across the Earth along a path which extends from west to east. The rate at which the shadow crosses the Earth's surface varies from a minimum of about 1000 mi/hr at the equator to a maximum of about 5000 mi/hr at the poles. The length, width, and location of the path traced by the shadow depend upon the moon's distance from the Earth and its location in space. The longest total solar eclipses last about $7\frac{1}{2}$ minutes. This is the time required for the rapidly moving circular shadow, 167 miles or less in diameter, to pass a particular spot on the Earth.

A total solar eclipse is preceded and followed by a partial eclipse. Partial eclipses also occur without an accompanying total phase and are more frequent than total eclipses and can be seen from a larger area. At least two partial eclipses of the sun must occur each year.

The moon may come directly between the sun and Earth but at too great a distance for the tip of its shadow to graze the Earth's surface. At such times, a total solar eclipse cannot take place. A person situated on the Earth in the area formed by the projection of this shadow sees an annular eclipse: the dark body of the moon does not completely cover the sun, and a thin ring of the sun remains visible. To illustrate this, close one eye and hold a coin between the other eye and something large and round, like a clock. If the coin is close enough to your eye it will completely cover the clock (total eclipse); if the coin is moved farther from your eye, an outer ring-shaped portion of the clock will be visible (*annular eclipse*).

The changes which accompany a total solar eclipse begin with a partial eclipse that lasts for about 1 hour. The moon's dark body causes a circular indentation on the west side of the sun. This indentation grows larger until only a thin crescent of the sun remains visible. The light reaching the Earth is now of a different quality, since it comes entirely from the outer parts of the sun. Just before the sun is completely hidden, the continuous crescentic sliver that remains is subdivided into a string of spots because the sun's light passes only through depressions along the moon's silhouette. With increasing darkness, temperatures drop, dew may form, stars become visible, and animals act as if night were approaching. After totality, these phenomena are repeated in the reverse order.

TIDES

Tides in the Earth's oceans are caused by the gravitational attractions of the moon and the sun, but the effect of the moon is more than double that of the sun (Fig. 20-16). It was realized long ago that the motions of the moon and the magnitude of the tides are directly related. The largest tides occur at times of new and full moon, whereas the smallest tides occur during the quarter phases. The time between two high tides or two low tides tends to approximate 12 hours and 25 minutes, which is one-half of the average lunar day of about 24 hours and 50 minutes. In other words, a point on the Earth directly under the moon at one time is directly opposite it on the far side of the Earth 12 hours and 25 minutes later on the average.

Tides several inches in magnitude are also produced in the Earth's solid body, but they can be detected only with very sensitive instruments. The surface of the ocean may rise and fall from 2 to 5 feet in mid-ocean. However, the funneling effect of bays causes a much greater change in sea level along certain coasts; changes of as much as 70 feet have been reported.

The following explanation of the tides is incomplete and simplified. The Earth's rigid body acts as a unit in responding to the gravitational pull of

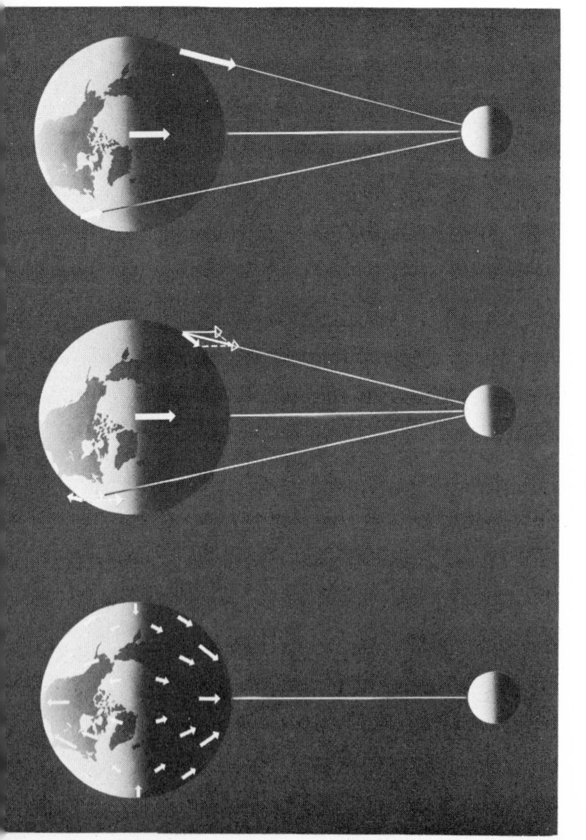

Fig. 20-16. *The moon causes tides on the Earth.* Top diagram: *the gravitational attraction of the moon on the Earth's surface is stronger on the near side and weaker on the far side (arrow length is proportional to the force of attraction).* Middle diagram: *the arrows show components acting along the Earth's surface.* Bottom diagram: *the moon causes high tides of about the same magnitude on both the near and far sides of the Earth. Low tide zones are located midway between the high tide zones. The Earth rotates through these tidal zones. [Courtesy of the American Museum of Natural History.]*

the moon—as if all of the Earth's mass were concentrated at its center. However, water at the Earth's surface is free to move, and its response to the pull of the moon varies with the distance. Because the moon's gravitational attraction on an object varies inversely as the square of its distance from that object, the moon pulls water on the near side more strongly than it pulls the solid Earth (Fig. 20-16). Similarly, the moon's pull on the water on the opposite side of the Earth is less strong than the pull on the rigid Earth. Therefore, with respect to the rigid Earth, the water on the side facing the moon tends to flow toward the moon, and the water on the other side tends to flow away from the moon. Thus the moon simultaneously causes high tides on both the near and far sides of the Earth.

The tidal bulges tend to lie along the line extending from the center of the moon through the center of the Earth (they occur somewhat ahead of this line in the counterclockwise direction). Therefore, as the Earth rotates, any point on the Earth's surface is carried from high to low to high to low tide. Since the moon is circling the Earth, the line joining the Earth and moon is moving, and it requires 24 hours and 50 minutes on the average for a given point on the Earth to complete its tidal cycle.

But successive high tides at a particular locality may not be of the same magnitude. Differences may be caused by the oscillation effect (see below) or by changes produced by the Earth's axial tilt in an observer's position relative to the location of successive high tides. (To see this clearly, draw a side-view diagram showing the moon, the Earth, the two tidal bulges, and a given observer's location on the near and far sides of the Earth.)

The total gravitational attraction between the sun and the Earth exceeds that between the moon and the Earth by 150 times or more. However, the sun's tide-producing effect is smaller because tides are caused by the difference in gravitational attraction between the near and far sides of a body, not by the total pull.

During times of new and full moon, the tidal effects of the moon and the sun reinforce each other and cause exceptionally great tides (*spring tides* —no relationship to the season called spring). During the quarter phases of the moon, the tidal effect of the sun opposes that of the moon and so these tides are weaker (*neap tides*).

The size of the tides is also affected by the moon's varying distance from the Earth; tides are greatest when the distance is least. The greatest tides occur when the moon is closest to the Earth during its new or full moon phases (*perigee tides*).

Tides are also affected by the size, shape, and extent of a body of water and the manner in which the water in it oscillates—the soup in a bowl carried by an unsteady hand shifts back and forth in a somewhat similar manner. If the period of oscillation is such as to reinforce the tides in one area, exceptionally high tides will result; however, in another area, not necessarily far away, the oscillation may act to reduce the magnitude of the tides.

Tidal friction between water and the ocean floors apparently is causing the Earth to rotate more slowly with the passage of time. It has been estimated that the day may now be approximately 1.8 seconds longer than it was 100,000 years ago. At first glance this retardation seems to be too small to be measured. However, a year now would be about 657 seconds longer than a year 100,000 years ago, and each of the 100,000 years would have been slightly longer than the preceding year. If a very precise clock had been started 2000 years ago (based upon the Earth's rate of rotation then), it would be ahead of present-day clocks (based upon the Earth's present rate of rotation) by about 3 hours.

This change is sufficiently great to be checked by the records of ancient eclipses. Modern astronomers can calculate the times and places at which certain ancient total solar eclipses should have occurred (these are based upon the Earth's current rate of rotation). It was found that the times and places did not check; e.g., a total solar eclipse which took place 2000 years ago, occurred about 3 hours earlier and 45° west of the predicted time and place (in 3 hours the Earth rotates $\frac{1}{8}$ of the angular distance around its axis). Such differences are apparent from ancient records. The times and places of ancient eclipses do check, however, if the gradual lengthening of the day is considered.

If this change has been taking place since early in the Earth's history, then notable differences may once have existed in the length of the day and the magnitude of the tides. According to some estimates, there may have been more than 400 days during a year about 400 million years ago (growth patterns in Paleozoic corals seem to provide some supporting evidence). At this time, the daily tidal range on the Earth may have been 100 to 200 ft. If so, the zone between the high- and low-water marks probably spanned many miles.

Exercises and Questions, Chapter 20

1. Make a sketch (orientation: looking "down") which shows the following:
   (a) Location of an observer on the Earth at sunset.
   (b) Moon in the third-quarter phase.
   (c) Mars in the southwestern sky of the observer.
   (d) Winter beginning in the Northern Hemisphere.
2. You see the moon in the sky about 9:00 A.M. one morning. What is the approximate phase of the moon? In what general direction did you look to see it?

3. One evening shortly after sunset you see a bright starlike object in the southeastern sky. Which one of the following might the object be: Polaris, Mercury, Auriga, Venus, Uranus, Neptune, Mars?

4. What measurements could you make over a period of time which would prove that the moon's orbit about the Earth is not a circle?

5. At an early evening hour during a period of about $1\frac{1}{2}$ weeks, observe the moon's different positions against the background of stars as it changes from a crescent to a full moon. Plot these successive locations on a single semicircular star map sketch which is oriented toward the south.

6. Make telescopic observations of the moon when it is in the first quarter and full phases and write descriptions of what you see.

7. Describe the "true" path of the moon in space relative to the Earth and sun. Explain why the moon's orbital speed varies.

8. What changes, if any, would take place in the lengths of the sidereal and synodic months if the moon revolved from east to west about the Earth (assume no other changes from existing conditions)?

9. What would be the ratio of the mass of the moon to the mass of the Earth if their center of mass were located 23,900 miles from the center of the Earth? At 119,500 miles?

10. Why do solar and lunar eclipses tend to occur at intervals of either a few weeks or about six months? How could five solar eclipses occur in the same calendar year?

11. Compare total lunar and solar eclipses relative to each of the following:
    (a) phase of moon at time eclipse occurs.
    (b) maximum length of totality.
    (c) approximate size of area from which eclipse is visible.
    (d) appearance during totality.

12. Explain each of the following concerning tides:
    (a) Why do 12 hours and 25 minutes commonly elapse between two high tides?
    (b) Why may successive high tides be of different magnitude at a certain point?
    (c) Why can predictions be made years in advance that particularly high tides will occur on certain dates whereas other high tides cannot be predicted even a few days in advance?
    (d) Why does a day on the Earth appear to be increasing in length?

(e) Why is the moon more effective than the sun in causing tides on the Earth?

13. In what ways would a hike on the moon's surface probably differ from one on the Earth's surface?

14. An interesting, informative astronomy laboratory exercise was described in *Sky and Telescope* in April 1964 (reprints may be on sale). This involves the determination of the shape of the moon's elliptical orbit by measurements of the angular diameter of the moon on a series of lunar photographs that come with the exercise. The eccentricity of the moon's orbit can be calculated from the data.

21 The sun's family

THE solar family includes the 9 major planets and their 31 (32?) known satellites, the asteroids (planetoids), comets, meteoroids, and vast numbers of dustlike particles. Physical data are given for each planet in Table 21-1 and Fig. 21-1.

The planets all revolve about the sun in the same eastward or counterclockwise direction, and all but Uranus (and perhaps Venus) rotate in this direction. Their orbital planes all lie close to the plane of the ecliptic; seven are tilted 3 degrees or less to it, but the orbital planes of Mercury and Pluto are tilted 7 degrees and 17 degrees respectively. The orbits of the planets are nearly circular, but again Mercury and Pluto deviate the most.

MERCURY

Mercury is the smallest of the major planets, the closest to the sun, and the swiftest: its orbital velocity varies from 23 to 35 mi/sec. It seems also to have the least atmosphere and the hottest areas of any planet.

Mercury is too near the sun to be readily observed from the Earth. In fact, Copernicus is reported never to have seen it. Mercury may be an evening star visible in the west just after the sun sets, or a morning star visible in the east just before sunrise (Figs. 18-11 and 18-12). In both cases, we see the planet obliquely through a great thickness of the Earth's atmosphere, and clouds and dust frequently interfere with clear observation. Some ancient sages believed that there were two planets: Mercury, the evening star, and Apollo, the morning star. Astronomers today obtain their best views of Mercury with a telescope during the day when it is highest above the horizon. Mercury goes through phases similar to those of

Table 21–1 Planetary Statistics

PLANETS	MEAN DISTANCE FROM SUN *(in millions of miles)*	MEAN DIAMETER *(in miles)*	MASS RELATIVE TO THAT OF THE EARTH AS 1
Mercury	36	3000	0.05
Venus	67	7600	0.8
Earth	93	7913	1.0
Mars	142	4200	0.11
Jupiter	483	87,000	318.0
Saturn	886	75,000	95.0
Uranus	1783	30,000	15.0
Neptune	2794	29,000	17.0
Pluto	3670	Small	0.8(?)

Venus. Each year Mercury is an evening star three times and a morning star three times.

In a number of respects Mercury is like the moon. Each is relatively

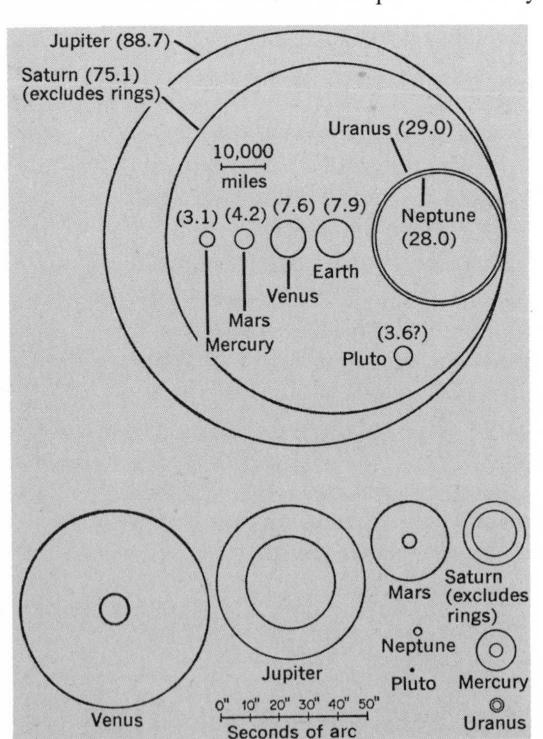

Fig. 21-1. *Actual and apparent diameters of the planets. Above, numbers show diameters in thousands of miles (some discrepancies occur with figures used in Table 21-1). Below, maximum and minimum apparent angular diameters of the planets as observed from the Earth.* [*Alex G. Smith and Thomas D. Carr,* Radio Exploration of the Planetary System, *D. Van Nostrand Company, Inc.*]

SPECIFIC GRAVITY (Water = 1)	PERIOD OF ROTATION	PERIOD OF REVOLUTION	NUMBER OF KNOWN SATELLITES
5.3	88 days	88 days	0
5	About 1(?) month	225 days	0
5.5	23 hours, 56 min.	365.25 days	1
4	24.5 hours	687 days	2
1.3	10 hours	12 years	12
.7	10.2 hours	29.5 years	9
1.6	10.8 hours	84 years	5
2.3	15.8 hours	165 years	2
–	?	248 years	?

small, and they reflect light in a similar manner; thus their surfaces may be similar. Neither is believed to have an appreciable atmosphere.

At one time it was thought that Mercury's period of rotation was equal to its period of revolution of 88 days. The present view, resulting from work with radio telescopes, is that Mercury rotates once in nearly 60 days. Freezing temperatures may occur on the part that is turned away from the sun at a particular time, because the planet lacks an atmosphere that might distribute solar energy from the day side to the night side. Probably little heat energy can be conducted through the planet. On the other hand, temperatures may approach 800°F at times on some sunlighted parts of Mercury. Mercury receives about twice as much solar radiation when it is closest to the sun as when it is farthest away. Life as known on the Earth could not exist on Mercury.

VENUS

Like Mercury, Venus is both an evening and a morning star. The distance between Venus and the Earth varies from about 26 to 160 million miles (Fig. 18-11). When brightest as a thin crescent, Venus is about 15 times as bright as Sirius, the brightest star in the sky. At such times it causes objects on the Earth to show slight shadows at night and can be seen during the day by observers who are aware of its location in the sky. In fact, Venus was once the target for some unsuccessful anti-aircraft fire during World War II!

In both size and mass, Venus is slightly less than the Earth. The surface of Venus has never been seen because it is perpetually hidden by a dense covering of clouds of unknown composition and thickness. The atmosphere above these clouds has a high percentage of carbon dioxide and lacks measurable quantities of free oxygen.

In October 1967 two spacecraft, Venus IV (Russian) and Mariner V (U.S.A.), arrived at Venus after four-month journeys. An insulated capsule from Venus IV was parachuted into Venus's atmosphere and descended to its surface; Mariner V passed within 2480 miles of Venus's surface. Signals from the descending capsule of Venus IV indicated that temperatures increased from about 105°F to 535°F, that carbon dioxide made up 90 to 95 percent of the atmosphere, that oxygen and water vapor made up less than 1.5 percent and nitrogen less than 7 percent, and that the density was 15 to 22 times that of Earth's atmosphere. No significant radiation belts or magnetic fields were detected. Mariner V's results generally tended to confirm those of Venus IV.

The period of rotation of Venus is difficult to determine because permanent markings cannot be observed through optical telescopes. Its nearly spherical shape suggests a slow rate of spinning. Interpretation of data obtained recently by radio telescopes indicates a period of rotation of about 250 Earth days (its period of revolution is 225 Earth days), perhaps in the retrograde or clockwise direction. However, the period of rotation of Venus had previously been estimated as lasting about a few weeks or longer.

If Venus does have a high surface temperature, free oxygen would not be expected in its atmosphere because the oxygen would readily enter into chemical combinations with other elements. Conditions on Venus may be unsuitable for life as known on the Earth, but many uncertainties exist concerning this question.

The orbit of Venus about the sun is very nearly circular. Because of the inclination of its orbit to that of the Earth, Venus transits the sun an average of only two times each century. At such times it can be seen through a smoked glass without a telescope as a round dark spot which moves slowly across the sun's surface. The next transit is predicted for June 8, 2004.

MARS

Mars is the reddish planet named after the bloody god of wars (Fig. 21-2). Its diameter is about half the length of the Earth's. Mars rotates

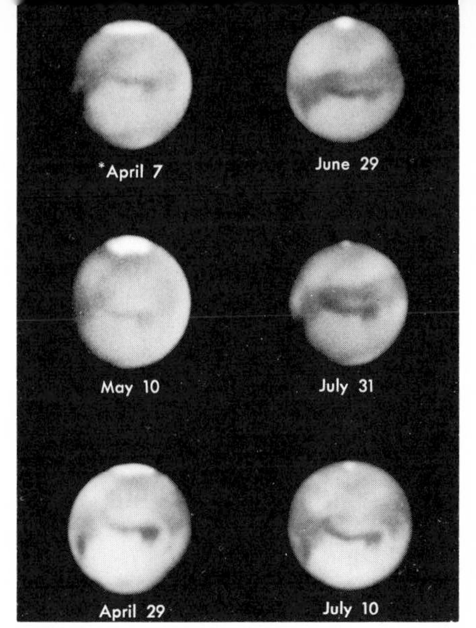

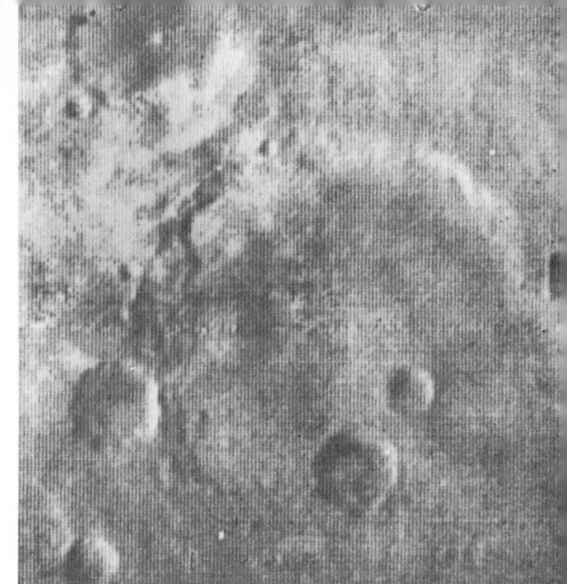

Fig. 21-2. *Left, seasonal changes on Mars. South is at the top. The six photographs are assigned dates in the Martian years and show the shrinking of the south polar cap that accompanies the arrival of its summer season, as well as the gradual darkening of the dark markings.* [*Lowell Observatory photograph.*] —*Right, frame 11 of the* Mariner IV *photographs of Mars made in July 1965. The slant range was about 7800 miles. Note the striking resemblance of the cratered surface of Mars to Ranger photographs of the moon (Fig. 20-1).* The *largest crater is about 75 miles in diameter. The small crater on its upper limb is 3 miles across.* [*Courtesy, NASA.*]

once in about 24 hours, and its axis is tilted constantly to the plane of its orbit at about the same angle as the Earth's. The Martian seasons last nearly twice as long as the corresponding seasons on the Earth, because Mars' period of revolution is nearly 2 years. When closest to the Earth, Mars is about 35 million miles away. Such favorable oppositions (time when Mars is located in the opposite direction from the sun) occur once every 15 to 17 years, although less favorable oppositions take place every other year.

Mars has two satellites, one of which completes a 360-degree, west-to-east revolution in about $7\frac{1}{2}$ hours; however, its synodic period is 10.8 hours. An observer on the more slowly rotating Mars would sometimes see this moon rise in the west and set in the east twice during the same night. No other satellite is known to do this. The effect of Mars on the orbits of these two moons makes it possible for astronomers to calculate the mass of Mars fairly accurately (p. 455).

The surface features of Mars can be seen from the Earth more clearly than those of any other planet because of its relative nearness, its rarefied atmosphere, and its lack of frequent clouds. Photographs taken with film sensitive to atmosphere-piercing infrared light show permanent markings. On the other hand, photographs taken on film which is most sensitive to ultraviolet light show a larger, nearly featureless disk because this light is scattered by the outer part of the atmosphere of Mars and does not penetrate to its surface.

The atmosphere of Mars may contain a somewhat higher proportion of carbon dioxide than does the Earth's atmosphere, but water vapor and oxygen seem to be very scarce. White polar caps are present at the north and south poles of Mars. The caps may be only an inch or so thick because they melt or sublimate so rapidly; perhaps they consist of ice crystals similar to the frost which collects on window panes. As the southern hemisphere is closest to the sun during its summer and farthest from the sun during winter, its seasons are more extreme, and its polar cap may disappear completely during the summer. During its winter, each polar cap advances more than halfway toward the equator, a latitude which would be comparable to that of New York City or Chicago.

The permanent markings are described as blue-green or brown against an orange-red background. The polar caps show the same patterns, as they advance and retreat during each Martian year, probably an indication of topography. Some astronomers report that blue-green markings become more pronounced near a polar cap in the spring, move toward the equator during the spring and summer, and change into brownish colors during fall and winter. Other astronomers claim that this change is one of intensity rather than of color. The change is attributed by some to seasonal growth of vegetation, by others to chemical changes in certain salts which might take up water in the summer and give it off in the winter, and by still others to changes in the wind direction and fluctuations in volcanic activity. The ruddy background color may be caused by the presence of reddish iron oxides at the surface of Mars in desert areas. Huge yellowish clouds, probably very large dust storms, are visible above the deserts at times, and these obscure the surface as they move across it.

Speculation that intelligent life may exist on Mars has captured the fancy of many persons for whom astronomy is only a casual hobby. The Italian astronomer Schiaparelli innocently instigated this controversy at the favorable opposition of 1877 when he discovered new surface features on Mars which he described as fine dark lines crossing the orange areas. He called them *canali,* meaning "channels." Unfortunately the word was transliterated as the English "canals." An American astronomer, Percival Lowell, believed that these lines were straight and arranged in a geometrical manner and hence were artificially produced by some type of intelligent being. According to Lowell, these beings would have to live near the equator where it was warmest. Since water was scarce, they would be forced to use great irrigation ditches to lead the melting ice and snow in the spring from the polar caps toward the equator. However, liquid water presumably cannot flow on Mars (it would freeze or evaporate), and probably the majority of astronomers doubt whether geometrically arranged lines exist, although they agree that some sort of markings, perhaps discontinuous, are present. Some astronomers ascribe these markings to the presence of vegetation along river channels cut through desert areas; they think the Nile might appear like one of these markings from an observation point 35 million miles away.

In observing Mars from the Earth, we must peer through the sea of air about the Earth which is constantly moving and which thus causes features on the surface of Mars to be somewhat blurred. However, if an observer spends enough time, he may see Mars during an occasional few seconds when the Earth's atmosphere is still and details are clear. Much better planetary detail may be obtained from future orbiting telescopes.

Conditions on Mars would be rigorous for life as we know it. Temperatures are always low, and the very low atmospheric pressure might be equivalent to living at an altitude on the Earth a few times greater than that of Mount Everest.

In July 1965 the *Mariner IV* spacecraft passed within about 6000 miles of the surface of Mars after journeying approximately 325 million miles in 228 days. About twenty pictures were taken and relayed to the Earth (Fig. 21-2); these show a cratered surface surprisingly like that of the moon. The nearness of Mars to the asteroid belt had suggested that it would have craters, but these craters are unexpectedly well preserved. Collisions with asteroids large enough to produce the larger craters are probably quite rare, which suggests that the cratered surface of Mars is very old. Perhaps it has never had an atmosphere much denser than the present very thin one, nor has it apparently had enough water to form streams, lakes, or oceans. Mountain chains, ocean basins, and continents are not visible in the pictures, an absence which suggests an internal structure for Mars that is much less dynamic than that of the Earth.

Nearly all hypotheses concerning the origin of life involve processes occurring in a water solution. Thus the apparent absence of abundant water on Mars may indicate that it is a dead planet and has been so for a very long time. Mars seems to be a much less hospitable place for life than was previously thought, although disagreement still exists on this point. Mars resembles the moon, and possibly Mercury, much more than it does the Earth. The results from *Mariner IV* enhance the uniqueness of the Earth as an inhabited planet.

THE FOUR MAJOR PLANETS AND PLUTO

Four giant planets occur in orbits between Mars and Pluto: Jupiter, Saturn, Uranus, and Neptune. Their diameters exceed the Earth's by about

11, 9, 4, and 4 times respectively. All four spin rapidly—their periods of rotation range from 10 to 15 hours—and thus they have flattened polar regions and bulging equators; this is particularly noticeable for Jupiter and Saturn. Their specific gravities are low compared with the Earth's, but their total masses are high.

We see only the outer portions of their atmospheres, which seem similar. Their temperatures are all very low—from about −200°F to −300°F. Spectroscopic studies show that methane (CH_4) and ammonia (NH_3) are present in the atmospheres of Jupiter and Saturn, but the methane bands are stronger for Saturn and the ammonia weaker. This may be due to Saturn's lower temperature—more of the ammonia is frozen and cannot be detected spectroscopically. No ammonia can be detected in the very cold atmospheres of Uranus and Neptune, but methane absorption bands are prominent.

Jupiter and Saturn are very large and are bright starlike objects as viewed from the Earth, although their brightness varies as their distances change. Gaseous bands are distinct in the atmospheres of Jupiter and Saturn, but lacking or very faint in Uranus and Neptune. Perhaps these planets are too remote from us for such atmospheric details to be visible.

The specific gravities of Uranus and Neptune are slightly higher than those of Jupiter and Saturn, even though they are compressed less by their smaller masses. Thus they probably have different and denser materials at their centers (perhaps rock cores). Calculations based upon their shapes and rates of rotation suggest that the masses of all four planets are even more strongly concentrated toward their centers than occurs in the Earth.

Hydrogen and helium probably compose the great bulk of the solid and gaseous portions of the major planets. Only these two elements, apparently the most abundant in the universe, seem light enough to account for their low specific gravities. However, most of the hydrogen and helium probably occurs as solids rather than as gases. The low temperatures make spectroscopic identification of molecular hydrogen in the atmospheres difficult, but it has been identified there for Jupiter and Uranus. Although methane has been identified in the atmospheres of each of the major planets, it probably occurs as a minor constituent of them.

Life as we know it would seem impossible on the major planets because of their low temperatures and poisonous atmospheres.

Jupiter Jupiter (Fig. 21-3) is the giant among the planets, and its volume and mass exceed the combined volumes and masses of the other planets. Next to Venus, and occasionally Mars, Jupiter is the brightest of the planets; rarely is it fainter than Sirius. Its size and satellites make observation exciting with even a small telescope. Of its twelve known moons, four can be seen readily through a small telescope (Fig. 21-4). They form a sort of miniature solar system and were first observed by Galileo. Evidently we see only the outer part of Jupiter's very dense and thick atmosphere. However, certain markings in this, such as the Great Red Spot (Fig. 21-3), remain visible for years. Jupiter's rapid rotation—about 10 hours for one complete rotation at the equator, but different times at different latitudes—apparently

Fig. 21-3. *Jupiter, October 24, 1952. The Great Red Spot and a satellite and its shadow appear near the top of the photograph, which was made with blue light. Prominent dark and light bands of many colors occur parallel to the equator. Jupiter is flattened noticeably at the poles. Bands of various colors—including reds, browns, and bluish white—occur parallel to the equator and can be seen through a small telescope; they are somewhat analogous to wind belts on the Earth. Minor changes in these belts occur constantly, particularly near the equator, but the major zones are rather stable as viewed from the Earth. Evidently we see only the outer part of Jupiter's dense thick atmosphere, but certain markings in this remain visible for many years, e.g., the Great Red Spot has been visible for about 100 years, although its size, color, intensity, and position have all changed somewhat. Its nature is unknown. Jupiter rotates very rapidly, but at different rates at different latitudes. [Mount Wilson and Palomar Observations.]—Jupiter's four largest satellites have diameters that range from about 2000 to 3200 miles. They revolve about Jupiter in the west-to-east direction (direct) and have orbits that are nearly circular and nearly in Jupiter's equatorial plane. Their mean distances range from about 260,000 to 1,170,000 miles and their periods of revolution from about 1¾ to 16⅔ days. Jupiter's outermost four small satellites revolve in the retrograde direction. Their periods of revolution are about 2 years, and their distances approximate 13 to 15 million miles.*

is responsible for the polar flattening, the bands, and the minor disturbances (which would actually be gigantic on an earthly scale (Fig. 21-3).

Jupiter has no significant seasonal changes because its axis of rotation is nearly perpendicular to its orbit.

Saturn Saturn, unique in its possession of a system of rings (Fig. 21-5), is one of the most beautiful of celestial objects viewable through a telescope. To the unaided eye it appears as a bright, yellowish, starlike object in the night sky. It was the most distant planet known to ancient astronomers, and it was given the name of the god of time for its leisurely ($29\frac{1}{2}$ years) revolution through space. Saturn's diameter is somewhat smaller than Jupiter's but about $9\frac{1}{2}$ times as large as the Earth's. Saturn is flattened even more noticeably at its poles than Jupiter, although it spins at approximately the same rate. Saturn's average specific gravity is calculated to be less than 1: i.e., if a sufficiently large ocean could be procured, Saturn would float in it.

Saturn's equator is tilted about 27 degrees to its orbital plane and maintains the same angle of tilt at all times. The rings are in Saturn's equatorial plane. Thus one can see the top of the rings for nearly 15 years and then the bottom (southern) portion of the rings for another 15 years.

Saturn has three concentric rings, paper-thin (an estimated 10 to 20

Fig. 21-4. *In the seventeenth century a Danish astronomer, Roemer, estimated the speed of light by observing Jupiter's moons. Since the period of revolution of any one moon is a constant, this moon should disappear behind Jupiter at constant intervals: assume a period of 24 hours and a time of 9 P.M. for the eclipse. But the eclipse occurs at about 9:17 P.M. when the Earth is on the far side of its orbit. Thus light needs about 17 minutes to move 186 million miles. Details have been oversimplified: e.g., light does not travel directly across the Earth's orbit, Jupiter is moving, and Roemer's result was about 25 percent slow.*

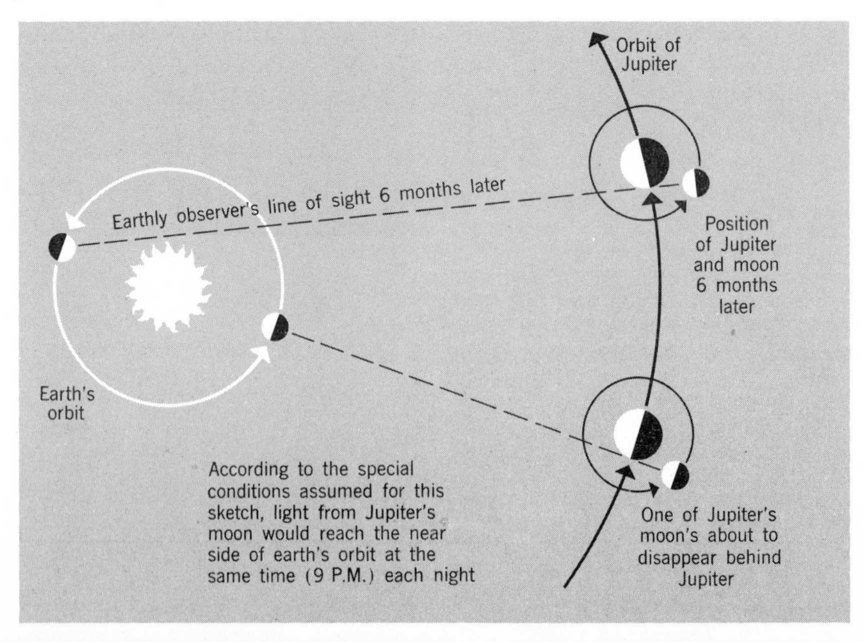

Fig. 21-5. *Saturn as photographed through the 100-inch reflector. Note the flattening of the polar region. The bright middle ring is about 16,000 miles wide and is separated from the outer ring (about 10,000 miles wide) by a gap about 3000 miles wide. The faint inner ring does not show separately.* [*Mount Wilson and Palomar Observatories.*]

miles) in proportion to their widths (about 171,000 miles for the outer ring), and invisible when viewed precisely edgewise. They probably consist of myriads of tiny particles or meteoroids which revolve as miniature satellites in three orbital groupings about Saturn. The rings are not solid: the light of a star, as seen through them, is dimmed but not blotted out; also, the inner parts of the rings revolve more rapidly than the outer parts.

One hypothesis on the origin of the rings suggests that they constitute planetary material which was too close to its parent body to consolidate into a satellite. According to another hypothesis, a former satellite approached too close to Saturn and was torn to pieces by its gravitational attraction.

A tenth satellite of Saturn may have been discovered in late 1966. Tentative calculations indicate that this very faint object may be a few hundred miles in diameter and may orbit Saturn in 18 hours at a distance of nearly 100,000 miles, just outside the outermost ring. Such a nearby object would be most conspicuous with the rings edgewise as they were at the time of discovery, because the rings would then reflect less light toward the Earth.

Uranus, Neptune, and Pluto Uranus and Neptune appear greenish in a telescope, perhaps owing to methane in their atmospheres. Uranus can barely be seen by the unaided eye and was discovered by Herschel in 1781. The discoveries of Neptune and Pluto were discussed in Chapter 18. Uranus is unusual in having its equatorial plane tilted about 82 degrees to the plane of its orbit and also in rotating in the retrograde, westward direction. Uranus's five satellites revolve in the same direction as the planet rotates and in the plane of its equator. Uranus thus has a unique seasonal change; its axis is nearly parallel to its orbital plane and points continuously in a fixed direction in space.

Of all the planets, Pluto is the least well known. Its orbit is the most eccentric of the nine planets and also has the greatest inclination to the plane of the ecliptic. At its closest approach to the sun, Pluto is actually nearer the sun than Neptune is, but Pluto and Neptune are never very close to each other because Pluto's orbit is inclined 17 degrees to the ecliptic. Pluto may once have been a satellite of Neptune. Physical data concerning Pluto are both contradictory and uncertain.

An attempt was made in 1965 to estimate the diameter of Pluto by observing its occultation (eclipsing) of a star. Since the orbital velocity of Pluto is known, the duration of the occultation would be a measure of its diameter. Unfortunately, the occultation could not be seen at any of the participating observatories. However, this indicates that Pluto's diameter is probably less than 4200 miles. This small size accounts for its very low brightness and suggests that Pluto does not have enough mass to have caused the orbital perturbations observed in Uranus and Neptune. Thus one or more planets may exist beyond Pluto (p. 355).

ASTEROIDS (PLANETOIDS)

Asteroids are minor planets which revolve about the sun. The majority have orbits that are located between Mars and Jupiter. The largest of these bodies is a little less than 500 miles in diameter, and the smallest visible ones are 1 to 2 miles across. The number in each size increases as the size decreases; therefore, asteroids too small to be visible from the Earth are probably very abundant, and one type of meteor seems to consist of still smaller fragments. Smaller asteroids seem to have irregular shapes because variations occur in the quantity of light they reflect, but the largest ones appear to be nearly spherical (p. 361).

The orbits of more than 1500 asteroids have been calculated. The orbits of most of the asteroids are not more irregular than those of Mercury and Pluto, but some are much more so. All seem to revolve in the direct eastward direction. The total mass of all the asteroids is probably much less than that of the Earth. Perhaps they represent planetary material which

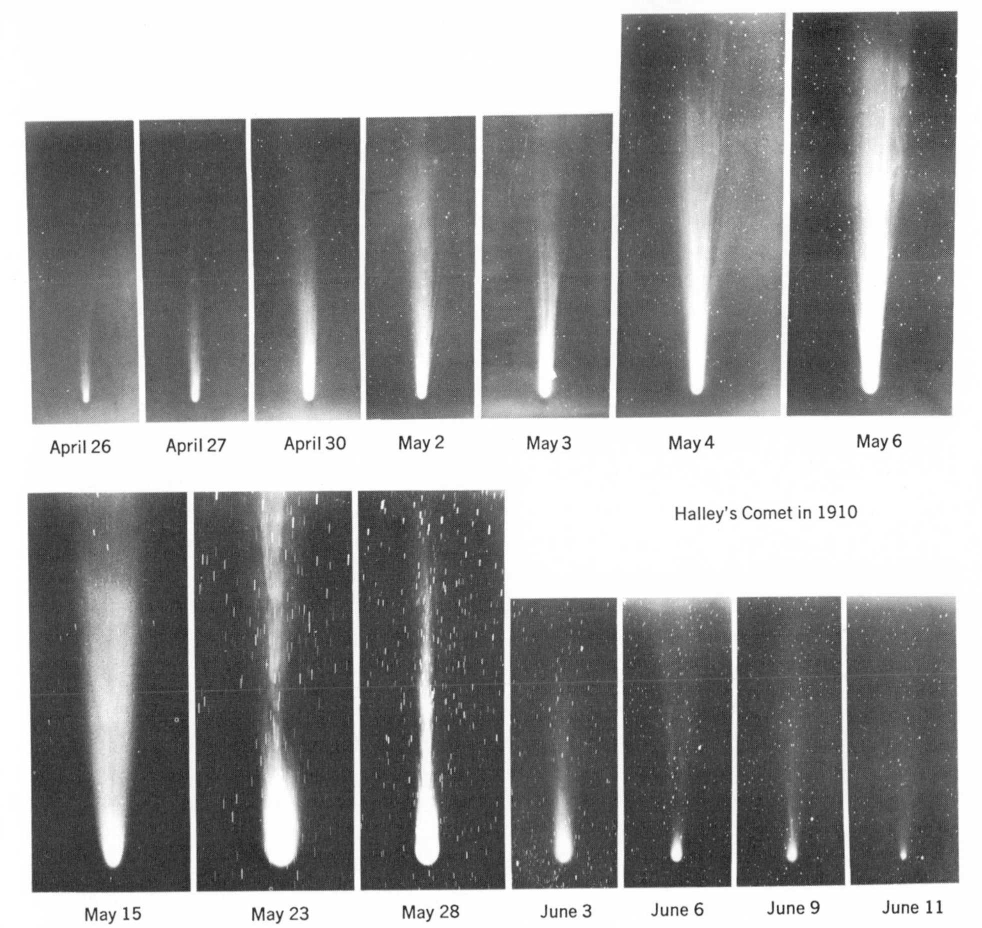

Fig. 21-6. *Halley's Comet. Fourteen successive views from April 26 to June 11, 1910.* [*Mount Wilson and Palomar Observatories.*]

never collected into a major planet, or possibly a former planet may have disrupted into many smaller bodies.

Some asteroids approach to within a million miles of the Earth. At such close distances, their geocentric parallaxes can be determined very accurately, which is of great importance in calculating other distances in astronomy.

COMETS

Comets (Fig. 21-6) probably constitute the most unusual members of the solar system. A well developed comet has a long tail, which always points away from the sun, and a brighter head section (*coma*) that contains a small bright nucleus. However, not all comets have tails, and the nucleus may not be visible. The size, shape, and brightness of a comet vary as it approaches and leaves the vicinity of the sun. Thus the most permanent feature of a comet is its orbit, but this also changes if a comet approaches close enough to a planet to be deflected by its gravitational pull. Comet orbits are oriented at all angles to the plane of the ecliptic, and comets show no preference for the common west-to-east direction of motion.

In size, comets are enormous; the diameters of the heads may average about 80,000 miles, and the tail sections may extend for millions of miles. In mass, however, they are very small. No comet has been known to affect the orbit of a planet or a satellite, although comets themselves have been influenced greatly by these bodies. The light of a star viewed through the tail of a comet is barely dimmed, and apparently the Earth has passed through a comet's tail with no noticeable effect.

Comets are quite different from meteors in appearance. A bright comet covers a large part of the sky, moves very slowly with respect to the background stars, and may be visible on successive nights for weeks or even months. Meteors are short-lived and occur within the Earth's atmosphere; most comets are many millions of miles away.

According to the "dirty snowball" hypothesis, the nucleus of a comet may be a porous structure that ranges in diameter from a few miles to 10 to 20 miles or more. It may consist of stony and metallic fragments embedded in ices of methane, ammonia, and water. When far removed from the sun, a comet may consist only of a nucleus. However, as it approaches the sun, it commonly develops a coma and tail and increases in size and brightness. Increased heat from the sun probably expels gases and dust explosively from the ices and meteoric particles in the head; this material diffuses into space to enlarge the head and form a tail. The comet as a whole is loosely held together by the mutual gravitational attraction of its particles.

Radiation from the sun exerts pressure which is insignificant on large objects but which is greater than the opposing gravitational attraction on very tiny objects. Therefore, as a comet approaches the sun, corpuscular (the "solar wind," p. 458) and electromagnetic radiations exert pressure against tiny particles in the tail and cause it to point away from the sun. Material expelled from the head of a comet to form its tail cannot be recovered because the comet's gravitational attraction is too weak to sweep it up. Each particle must continue to revolve in an orbit about the sun as would a feather or a handful of powdery snow under similar circumstances. Friction is so common on the Earth that the idea of light, diffuse material revolving permanently in orbits about the sun is startling at first, but friction is practically nonexistent in space.

If a comet approaches too close to the sun or to a large planet, it may be disrupted into a number of smaller comets or completely fragmented. In fact, meteoric swarms probably originate in this manner.

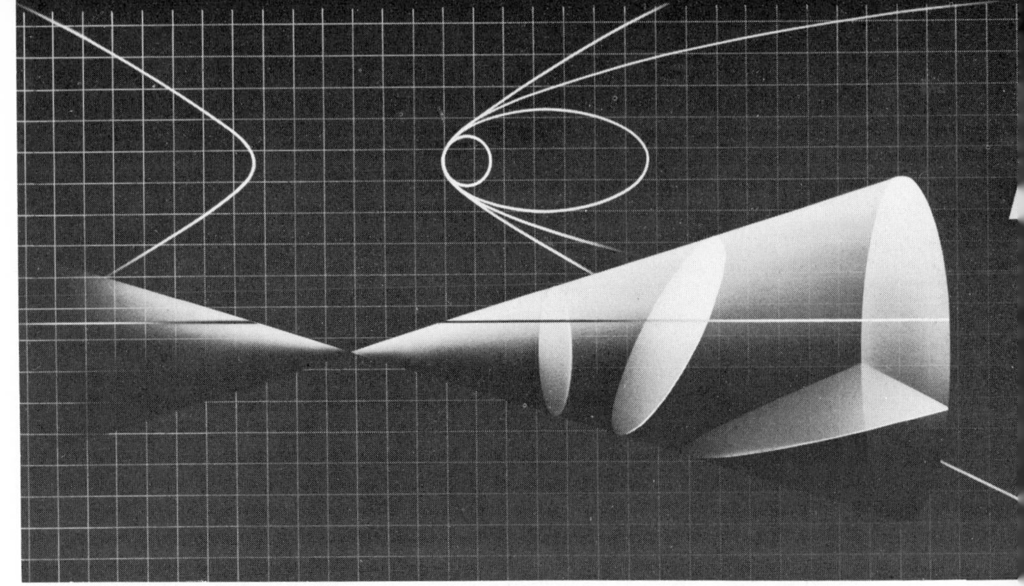

Fig. 21-7. *Four orbits with the same perihelion distance: circle, ellipse, parabola, and hyperbola. Note that each type of orbit can be produced by the intersection of a plane surface with a cone. The size and shape of an orbit are determined by the velocity of an object at perihelion. This could be the orbit of a comet, a meteor, or a planet. A circular orbit results from the least velocity and a hyperbola from the greatest. The parabola and hyperbola are open curves. If a comet leaves the vicinity of the sun along such a curve, it will never return unless its orbit is subsequently changed by the gravitational pull of some other body that it approaches in space. Note that the curves are alike in the relatively small portion near the sun observable from the Earth. No comet is known definitely to follow an open curve.* [*By H. K. Wimmer; courtesy of the American Museum of Natural History.*]

Light from a distant comet is chiefly reflected sunlight, whereas that from a comet within a few astronomical units of the sun is chiefly fluorescent light. According to present estimates, most comets probably belong to the solar system, but many have orbits which cause them to exist for long periods of time far out in space (Fig. 21-7); comets adhere to Kepler's laws and travel most slowly when farthest from the sun. Comets that can be seen by the unaided eye average about one in every ten years. Spectacular comets are much less common. Probably Halley's comet is the most famous. With but one exception, it has been observed at intervals of about 75 years at every perihelion passage since 240 B.C. It is scheduled to reappear about 1986.

METEOROIDS, METEORS, AND METEORITES

Meteors are the familiar objects known as shooting stars, but they are very unlike stars. Three terms are now used: meteoroids, meteors, and meteorites. *Meteoroids* are solid particles, commonly the size of sand grains or pebbles or smaller, that travel in orbits about the sun. However, some meteoroids weigh many tons, and no arbitrary limit separates large meteoroids from small asteroids.

Meteoroids are invisible until they enter the Earth's atmosphere and are heated to incandescence as they collide with molecules in the air. The term *meteor* is used for this luminous phase. Temperatures rise high enough to vaporize most meteors completely and the brighter ones leave trails of gases which may remain visible for a number of minutes. However, most meteors are visible only for seconds. The air through which the meteor travels is ionized and thus made luminous. We see the cylindrical trail left by a meteor, not the meteor itself. Most meteors become visible at altitudes of 60 to 70 miles and disappear at altitudes of 40 to 50 miles; their average speeds approach 26 mi/sec, which is the velocity of escape from the sun at a distance of one astronomical unit. However, the Earth's rotation, revolution, and gravitational pull influence the movement of a meteor and must be allowed for.

Millions of meteors may enter the Earth's atmosphere daily, but most of them are completely volatilized before they reach the Earth's surface. We are fortunate indeed to have an air umbrella.

That stones can fall from the skies has been generally accepted for about 150 years. A notable fall of a few thousand stones in France about 1800 was studied by a commission and contributed greatly to this acceptance. However, when a smaller fall in Connecticut was described by two members of the Yale faculty, Thomas Jefferson is quoted as saying: "I should rather believe that those two Yankee professors would lie than to believe that stones fell from heaven." At that time people failed to understand what force could keep a stone floating in the sky. Today we say that such meteoroids travel in orbits about the sun—they do not float in the Earth's sky—and the laws of planetary motion, universal gravitation, and inertia govern their orbits.

The term *meteorite* refers to the solid object that has reached the Earth's surface. About three-fourths of the known elements have been identified in meteorites, which are metallic or stony or a mixture of the two. Metallic meteorites consist largely of iron (80 to 90 percent or so) with some nickel; stony meteorites are somewhat similar to dark-colored, heavy, igneous rocks found on the Earth. The stony meteors are apparently much more numerous than the metallic ones, but metallic meteorites are more abundant in museums, perhaps because they are recognized more readily.

An interesting speculation is suggested by the composition of meteorites. If this material is a true sample of the type of material that formed the planets, then a sizable fraction of the Earth should be composed of iron and nickel. These heavy elements are not this abundant at the Earth's surface; therefore, perhaps they are concentrated in the core. Other evidence indicates that the Earth has a core about 4300 miles in diameter which is heavier and different from the material above it.

Spectroscopic identification of meteors as they volatilize at high altitudes indicates that the stony variety are actually more numerous than metallic ones. This would be expected if stony meteorites represent the type of material which formed the Earth's mantle (84 percent of the Earth's total volume).

Meteoroids appear to consist of two quite different types. One type apparently had its source in the asteroid belt, and fragments of these form the stony and metallic meteorites. Comets apparently are the source of the meteoroids of the other type, which do not form meteorites. Meteoroids of this second type may consist of particles of friable frozen gases which vaporize completely in the Earth's atmosphere. Swarms of meteors exist and each swarm apparently represents the remains of a comet (Figs. 21-8

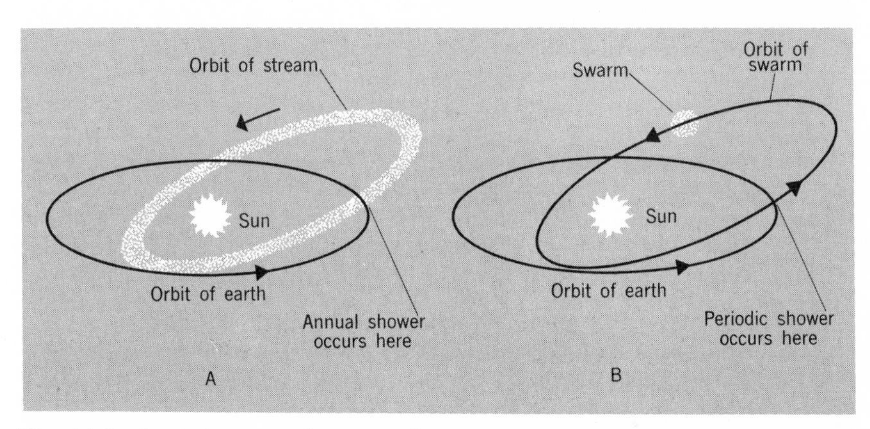

Fig. 21-8. *Origin of annual meteor showers (I) and origin of periodic meteor showers (II). Variations occur in the intensity of an annual shower if the meteoroids (cometary debris) are not distributed uniformly along the orbit.* [*Samuel Glasstone,* Sourcebook on the Space Sciences, *Van Nostrand.*]

and 21-9). Apparently the comet-associated meteoroids travel around the sun in both the direct (eastward) and retrograde directions and approach the sun at all angles to the plane of the ecliptic. Asteroid-associated meteoroids seem to have orbital motions similar to those of the asteroids: the motion is direct, of rather small eccentricity, and occurs near the plane of the ecliptic.

The Leonid meteor shower (Fig. 21-10) of November 1966 was a most spectacular display of celestial fireworks at certain times and in certain places. Observers just before dawn in the western United States were startled and delighted by an intense rain of many thousands of meteors,

Fig. 21-9. *The radiant of a meteor shower. The meteors seem to come from one spot in the sky (radiant) for the same reason that railroad tracks seem to diverge from a distant point. However, meteors in a swarm travel together along parallel paths unless they are deflected.*

perhaps as many as 40 per second during a short interval at the peak of the shower. Some were fireballs or left trails like rockets; others were small. Together they produced a display that will be long remembered. However, clouds prevented observations in parts of the eastern United States, where the most intense part of the shower apparently took place after sunrise.

The Leonids seem to radiate from the constellation Leo and showers take place each November, with spectacular displays tending to occur about every 33 years (Figs. 21-8 and 21-9). The 1966 shower rivaled that of 1833. The Leonids were disappointing in 1932 and 1899.

Characteristically, meteorites have a thin, blackened crust pitted with irregular holes that look something like thumbprints and were caused by differential fusion (some parts were vaporized more readily than adjoining portions). The freshly ground surface of a stony meteorite commonly shows a light grayish color and metallic specks. Surprisingly, most meteorites are not at high temperatures when they strike the Earth's surface. They come from the extreme cold of outer space, and only their surfaces are heated as they penetrate the Earth's atmosphere. If large enough, they pass through the air without volatilizing completely.

No human being is known to have been killed by a meteorite, and the chances that an individual might be hit are extremely slim, although some animals have been killed and buildings have been hit. One example of a near miss occurred in 1924 when a 14-pound stone struck a highway in Colorado just after a funeral procession had passed by. Apparently a

Fig. 21-10. *A display of the Leonid meteors photographed from Kitt Peak in November 1966. "Pouring out of the Big Dipper were 43 Leonids in 43 seconds," wrote the observer.* [*Courtesy,* Sky and Telescope *and David McLean.*]

woman in Alabama was bruised by a meteorite in 1954, and another meteorite penetrated a garage in Illinois in 1938.

Two spectacular meteoritic events have occurred in Siberia during the present century. On June 30, 1908, the Earth collided with a massive body (perhaps the nucleus of a comet) which apparently exploded before reaching the surface. No meteoritic debris has been found in the impact area (central Siberia, Tunquska fall), and depressions once interpreted as meteorite craters are now ascribed to terrestrial processes. According to reports, the blast was heard 400 miles away, windows were broken at a distance of 50 miles, trees were blown over within a radius of 20 or more miles, and several hundred reindeer were killed. If the fall had occurred about 5 hours later, it might have struck Leningrad. A very bright daytime

Fig. 21-11. *Meteor Crater, Arizona. It is 4200 feet in diameter and 570 feet deep. Living cedar trees on the site show that the meteor fell at least 700 years ago.* [*Photograph by John Forrell, Fort Worth, Texas.*]

fireball occurred on February 12, 1947, and devastated the uninhabited side of a mountain in Siberia with iron-nickel meteorites, the largest weighing 1 to 2 tons. Craters dot the impact area.

Meteor Crater in Arizona (Fig. 21-11) undoubtedly was made by the impact of a large meteor or a group of smaller ones, and other large depressions that may have had a similar origin are known. A multitude of

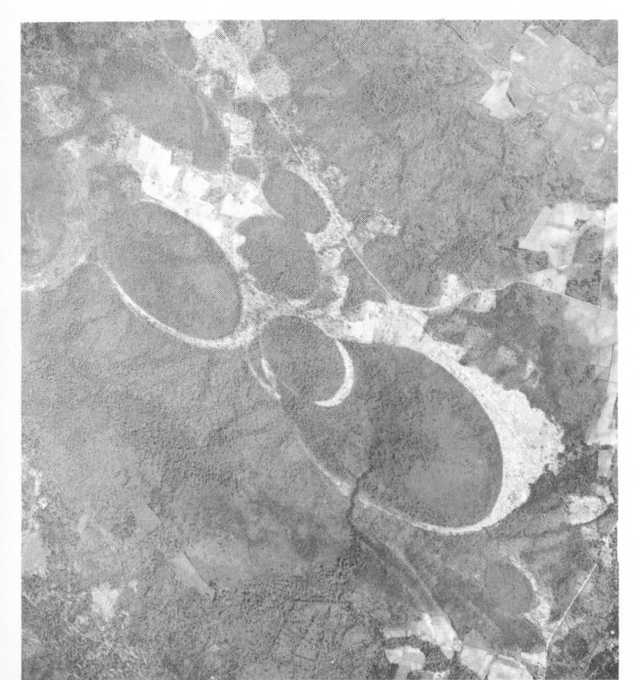

Fig. 21-12. *A few of the Carolina bays. Nineteen bays are recognized here, eighteen in a strip about a mile wide, extending northwest-southeast through picture. Some overlap. Light-colored areas are sand rims which are best developed on southeastern sides.* [*Courtesy, Professor C. E. Prouty, Michigan State University.*]

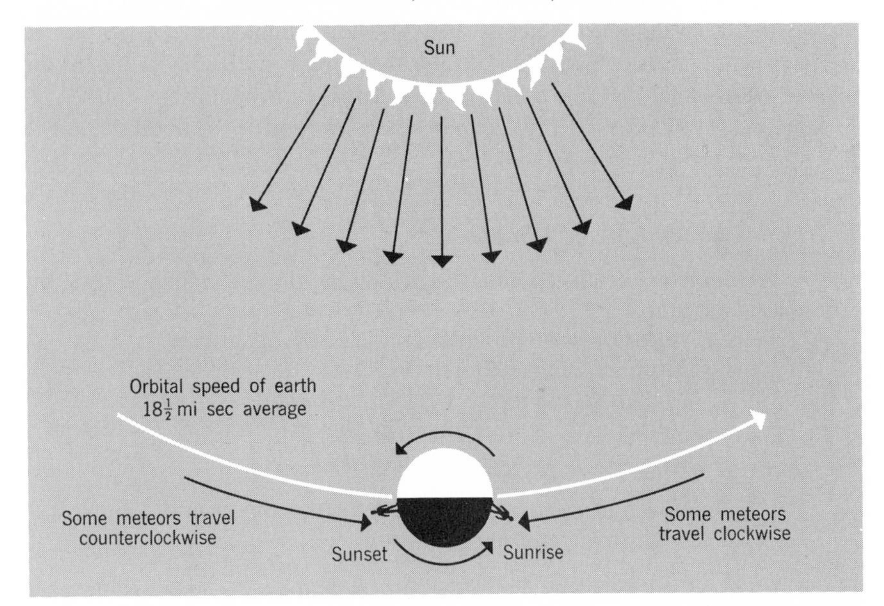

Fig. 21-13. *Meteor colors. Meteors travel both clockwise and counterclockwise in orbits about the sun at rates which may be faster or slower than the speed of the Earth. The rate at which a meteor descends through the Earth's atmosphere determines the amount of friction and heat developed: high speeds produce great friction, high temperatures, and bluish-white colors; slower speeds cause less friction, lower temperatures, and reddish colors. Most meteors noted by an observer in the early evening hours are traveling more rapidly than the Earth and in the same direction. Thus the "front end" of the meteor collides rather gently with the "rear bumper" of the Earth and a reddish meteor results. On the other hand, in the hours immediately preceding sunrise, an observer sees meteors which are traveling in the opposite direction from the Earth. There is head-on collision, and bluish-white meteors result. Meteors moving more slowly than the Earth and in the same direction are exceptional. Their light is reddish. Some meteors, also exceptional, have orbits at angles to the Earth's orbit.*

depressions (Fig. 21-12) in the Carolinas and adjoining coastal areas to the north and south may also have been caused by meteoritic bombardment. Temperatures at the moment of impact of large swift meteors are probably high enough to volatilize completely the rocks, soil, and water at such spots. The great expansion necessitated by the change to a gaseous state causes the explosion. The column of air directly in the path of a meteor does not have time to move completely aside; it is compressed into an air-shock wave that affects a much larger area than the place actually hit by the meteorite.

Usually more meteors are visible in the early morning hours than are visible before midnight. Meteors seen after midnight tend to be bluish-white in color and to have short trails; they are vaporized rapidly by the intense heat generated by their hurried passage through the Earth's atmosphere (Fig. 21-13). Early evening meteors tend to be reddish and to have longer trails.

Exercises and Questions, Chapter 21

1. Make a sketch (orientation: looking down) which shows the following:
   (a) Location of an observer on the Earth at sunrise.
   (b) Moon causing an annular eclipse.
   (c) Mars near observer's western horizon.
   (d) Venus as a morning star in the observer's sky (through a telescope it appears as a crescent).

2. Compare and contrast the phases of the moon and Venus. Can Venus cause an eclipse of the sun?

3. A friend tells you that Mercury is a morning star visible each morning for a slightly longer time before sunrise. What is the approximate phase of Mercury?

4. Make telescopic observations of each of the planets visible this term. If Jupiter is visible, how could you prove that its four easily seen moons are not nearby stars?

5. Make a series of semicircular star map sketches spanning several weeks which prove that planets are "wanderers."

6. Which of the planets other than the Earth has the most favorable conditions for life as we know it?

7. List as many unique features for each of the planets as you can.

8. Does Mars retrograde more or less frequently than Saturn? Explain.

9. What evidence suggests that the smaller asteroids are irregularly shaped objects whereas the biggest asteroids seem to be spherical? Explain.

10. Observations of meteors:
    (a) Explain what one actually sees in observing a meteor in the sky.
    (b) How can the altitude of a meteor be estimated? its velocity?
    (c) Why are meteors more numerous than average at certain times of the year? after midnight rather than before?

11. What significance may meteorites have concerning the composition of the Earth's interior? the age of the Earth?

12. Describe and explain the changes that occur typically as a comet approaches the sun.

13. List as many differences as you can between comets and meteors.

14. How can the speed of a comet or a meteor indicate that it is or is not a member of the solar system?

22 Space, time, and life

LAUNCHING AN ARTIFICIAL SATELLITE INTO AN ORBIT ABOUT THE EARTH

The successful launching of a man-made moon involves Newton's laws of universal gravitation and motion (Chapter 4), Kepler's laws of planetary motion, air resistance, jet propulsion, guidance, and other factors. Two results must be achieved: (1) A satellite has to be lifted above the dense lower part of the Earth's atmosphere, which extends outward for thousands of miles, but which is exceedingly thin at altitudes greater than 100 to 200 miles. (2) When located high enough above the surface, the satellite is accelerated to a very great velocity in a direction nearly parallel to the Earth's surface. As the speeding satellite moves through space, it falls continuously toward the Earth, but its momentum prevents it from being pulled into the Earth (Chapter 18). Eventually air resistance may slow the satellite's forward speed sufficiently to bring it down through the denser air in a meteor-like plunge, unless a braking action has been arranged.

Velocities of 18,000 mi/hr (5 mi/sec) or more are necessary to keep a satellite orbiting the Earth for weeks, months, and years at an altitude of several hundred miles (Figs. 22-1 and 22-2).

The multistage rocket is used to attain such velocities. The first stage lifts a satellite relatively slowly through the dense lower atmosphere. The other stages are then fired in turn to accelerate the payload to greater and greater velocities as the rocket rises higher and higher into thinner and thinner air. If the rocket were given a 5 mi/sec velocity at the Earth's surface, friction would immediately reduce its speed and might be great enough to vaporize it. The rocket would burn up as a meteor. Further-

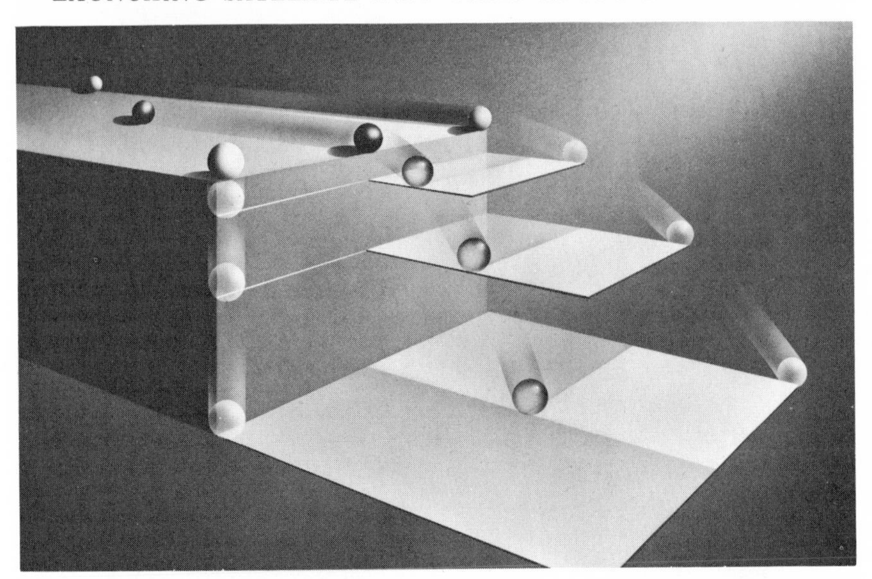

Fig. 22-1. *The effect of the Earth's gravitational attraction on freely falling objects is independent of other motions that they may have. The three balls leave the tabletop at the same instant. The ball on the left merely falls from the tabletop, whereas the other two are moving horizontally at different rates as they leave the tabletop. All three balls strike the floor at the same time.* [*Courtesy of the American Museum of Natural History.*]

more, the waste weight of each burned-out stage can be successively dropped.

The rocket is fired vertically, but its path is so controlled that it turns gradually away from the vertical. By the time the target altitude is reached, the last stage has been slanted into a nearly horizontal path about parallel to the Earth's surface directly beneath it. The last stage is then fired and this accelerates the payload to the necessary velocity to stay in orbit for some time (Figs. 22-3 and 22-4).

The velocity and direction of this last stage are both critical factors. Assume that satellites are launched from the same place (at an altitude of 300 miles or so), in the same direction—parallel to the Earth's surface—but at different velocities. About 5 mi/sec produces a circular orbit. A lower launching velocity produces an elliptical orbit in which the starting point becomes the apogee position (the farthest point in the orbit from the Earth). The perigee point (closest to the Earth) is then exactly halfway around the orbit from the starting point. The center of the Earth is at the more distant focal point in the elliptical orbit. If launched at still lower velocities, the satellite would crash into the Earth. At greater velocities than

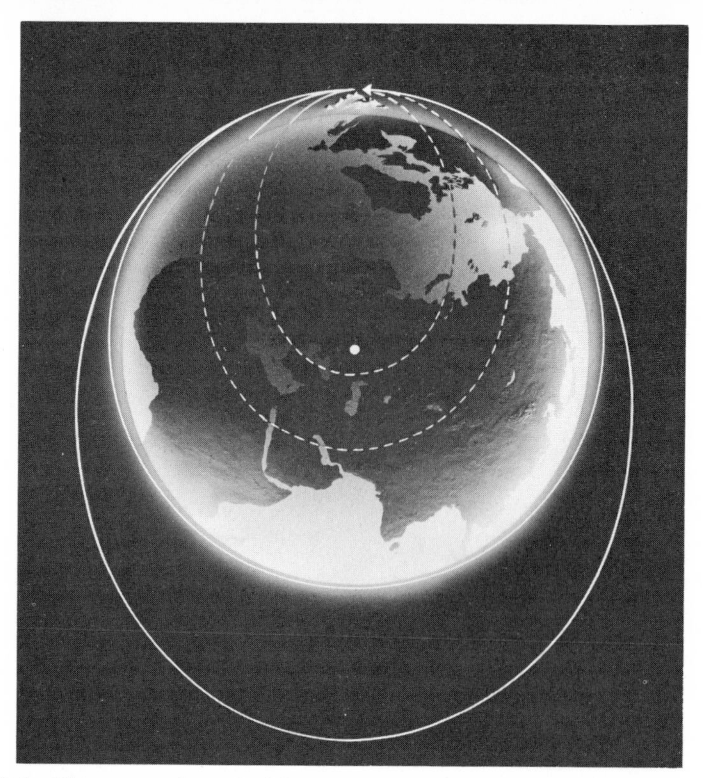

Fig. 22-2. *Objects are launched horizontally from a mountain top at different rates. Slower moving objects "go into orbits" about the Earth's center, but these are interrupted by the Earth's surface. Faster moving objects complete their orbits. Imagine a launching platform located at an altitude of 100 or 200 miles above the Earth's surface; at this altitude the air is very thin and air resistance is slight (but it is not negligible). Next, imagine that bullets are fired horizontally from the platform at different velocities. All fall at the same rate. However, the bullets with the greater horizontal rates travel farther before they strike the Earth's surface. The Earth's surface also curves away from the falling bullets and thus increases their ranges. At a sufficient velocity (about 5 mi/sec or 18,000 mi/hr), the Earth's surface curves away from a bullet at the same distance per mile that it falls. Thus the bullet stays at the same altitude above the surface and goes all of the way around the Earth—it is in orbit. In this case, the acceleration of a bullet revolving in a circular orbit at a constant speed (disregard air resistance for this) involves a change in its direction.* [*Courtesy, American Museum of Natural History.*]

5 mi/sec, the orbit is again an ellipse, but the starting point becomes the perigee position, and the apogee point is 180 degrees beyond. At velocities greater than the escape velocity of 7 mi/sec (25,000 mi/hr), a satellite would go off on a parabolic or hyperbolic path and not return to the neighborhood of the Earth.

Once a satellite is in orbit about the Earth, air resistance gradually reduces its velocity, unless, of course, all parts of the orbital path are so remote from the Earth that atmospheric friction is negligible. When there

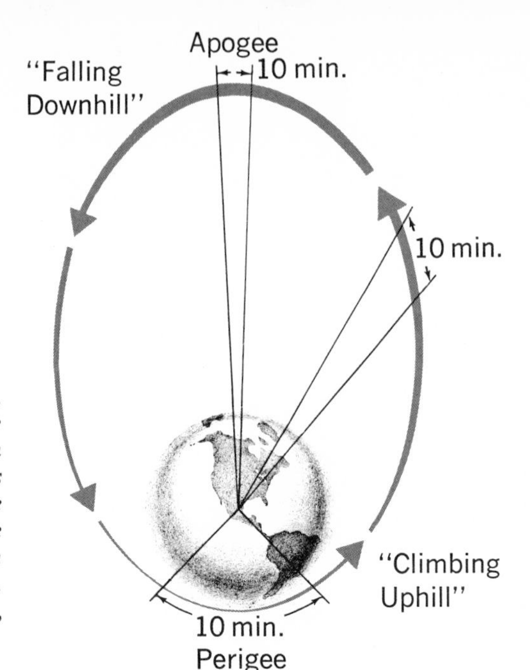

Fig. 22-3. *Elliptical orbit for a satellite. The satellite "falls downhill" when the Earth's gravitational attraction acts in the same direction as the satellite is moving; therefore, it speeds up. The closer a satellite orbits the Earth, the faster it must move in order to stay in orbit. See Table 22-1.* [*Courtesty, H. L. Goodwin,* Space: Frontier Unlimited, *D. Van Nostrand Co., Inc.*]

is any friction, no matter how slight, the orbit changes at each revolution, slowly at first and then more and more rapidly as the shrinking orbit brings the satellite into denser air. At first the perigee position remains about the same, and the altitude of the apogee point is reduced on each revolution. A nearly circular orbit eventually results, and its radius is then gradually reduced (Fig. 22-5). The satellite moves according to Kepler's equal areas

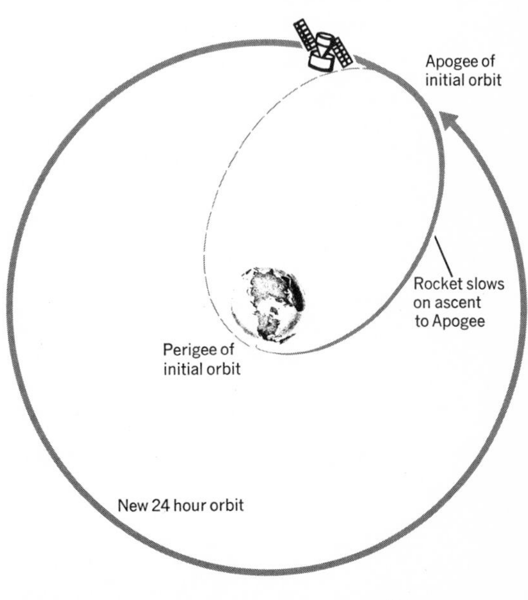

Fig. 22-4. *"Kick in the apogee" to achieve a circular orbit. At the moment the final rocket stage and its attached satellite reach the apogee position of the initial orbit, they are moving parallel to the Earth's surface far below (but too slowly to stay in orbit at this altitude). The final stage is then fired and increases the velocity of the satellite enough to place it in a new circular orbit.* [*H. L. Goodwin,* Space: Frontier Unlimited, *D. Van Nostrand Co., Inc.*]

Table 22–1 Altitude, Velocity, and Period of Orbiting Satellites

APPROXIMATE ALTITUDE OF ORBITING SATELLITE ABOVE THE EARTH'S SURFACE	APPROXIMATE VELOCITY NECESSARY TO KEEP FROM FALLING INTO THE EARTH	APPROXIMATE PERIOD OF REVOLUTION
Several feet	19,000 mi/hr	84 min
100 miles	18,000 mi/hr	90 min
1000 miles	15,700 mi/hr	120 min
22,300 miles	6,900 mi/hr	24 hours
239,000 miles	2,300 mi/hr	About 1 month (the moon illustrates this orbit)

ASSUMPTIONS: The Earth has no atmosphere, the Earth is a true sphere, and the orbits are circular

law; therefore, its greatest velocity is attained when it is closest to the Earth. However, air resistance is also greatest at perigee, and this reduces the velocity at each passage through the perigee position. In other words, the satellite speeds up as it approaches perigee but not so much as it would in the absence of air resistance. Thus the satellite does not gain the momentum necessary to swing out as far on the next revolution, and its apogee point shifts closer to the Earth.

The period of an artificial satellite is directly related to the size of its orbit, but its mean velocity is inversely proportional to the square root of its distance from the Earth (Table 22-1). In other words, more remote satellites travel more slowly and take longer to complete one revolution. This follows from Kepler's three laws. At a distance of about 22,300 miles above the Earth's surface, the period is 24 hours; therefore, a satellite in a circular orbit over the Earth's equator at this altitude remains "fixed" in space directly above a particular spot on the Earth beneath it (Fig. 22-6). The velocity necessary for a circular orbit at this altitude is about 6900 mi/hr.

The direction in which a satellite is launched is an important factor; less energy is needed if the direction is the same as that of the Earth's rotation and revolution. Points on the Earth's surface move in an eastward direction at velocities that vary with latitude; they are greatest at the equator (about

Fig. 22-4A. *The famous walk in space of Astronaut White in June 1965, outside the* Gemini IV *spacecraft. We see the Earth's orb above, the egress door, and parts of the craft at the right and below, and the umbilical cord that sustains White as he maneuvers in weightless space with the aid of a hand-held jet unit. Free floating is not yet an experience familiar to many earthbound men.* [*Courtesy, NASA.*]

1000 mi/hr or 0.3 mi/sec) and decrease toward the poles. All objects at the equator are moving eastward at this rate, and this motion is independent of other motions the objects may have. Thus a ball thrown vertically upward falls back to the starting point. The Earth does not leave such objects be-

Fig. 22-5. *Atmospheric drag changes the apogee position more than the perigee position.* [*Courtesy, NASA.*]

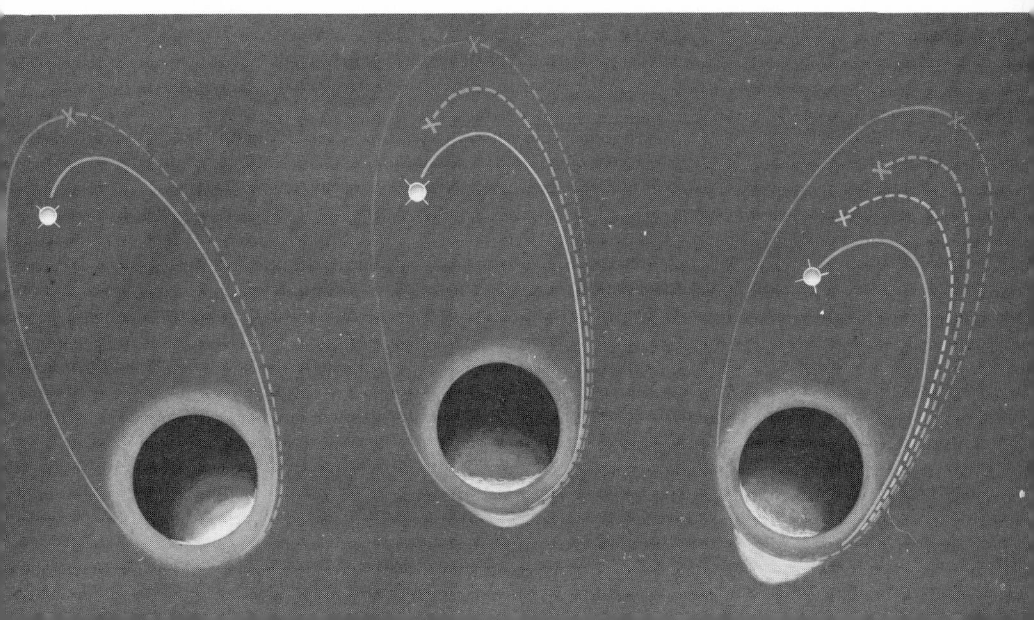

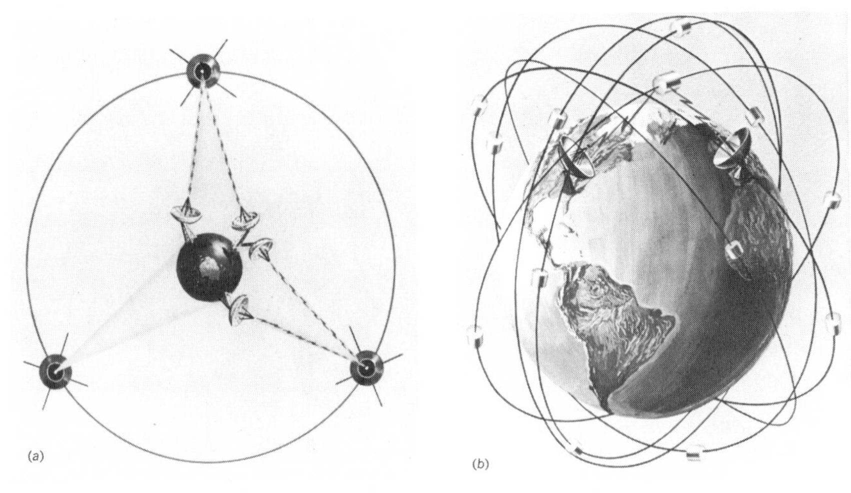

Fig. 22-6. *Communication satellites at an altitude of 22,300 miles in synchronous orbits (A) and at an altitude of about 5000 miles (B). To an observer on the Earth, the satellites in (A) would not appear to move. Either system can be used to provide world-wide television broadcasts and other communication services. The satellites would be equipped with radio receivers and transmitters and other instruments so that they could receive, store, amplify and retransmit messages. They would function as microwave towers in space. Enough "active-repeater" satellites would be placed in orbit so that at least one would always be present in space above any one area.* [*Courtesy, NASA.*]

hind by rotating eastward from under them. However, a bomb dropped from a plane does not strike the Earth directly below because it falls with the motion of the plane superposed on the motion of the Earth. Thus a satellite launched in a general eastward direction at the equator is already moving at a rapid speed. If 5 mi/sec is the velocity the satellite must eventually attain to stay in orbit, it already has 0.3 mi/sec of this; therefore, it must be given an additional velocity of 4.7 mi/sec. If it were launched toward the west, the necessary velocity would be 5.3 mi/sec. The direction of launch becomes even more important in interplanetary travel.

If a satellite were launched in a due eastward direction at the equator, it would move in space directly above the equator. It could not be used to photograph or measure phenomena at higher latitudes. On the other hand, if a satellite were launched along a meridian, it would eventually pass over all parts of the Earth, but much of its time would be spent above the unpopulated polar regions. Thus it is more common from the middle northern latitudes to launch satellites toward the southeast so that they pass back and forth over the heavily populated regions between the middle latitudes and the equator. However, the direction of launch will depend upon the purpose for which the satellite was designed.

OBSERVING SATELLITES FROM THE EARTH

Satellites are seen by reflected sunlight and thus cannot be viewed during the day with the unaided eye against the bright background of diffused sunlight. The moon itself is inconspicuous in the daytime sky. The reflected light of a satellite may fluctuate if it has an asymmetrical shape and spins or tumbles as it revolves. In addition, a satellite has a lighted half and a darkened half and thus it exhibits some phase changes. Most satellites are relatively close to the Earth and nearly half of their orbits lie within the Earth's shadow; thus we can observe satellites only near the times of sunrise and sunset.

For watchers in the middle northern latitudes, satellites may move across the sky from southwest to northeast or from northwest to southeast. However, if a satellite is in a polar orbit, one will see it move from north to south or from south to north. A satellite may be visible at different times on any one night and for a number of nights in succession, but it never follows precisely the same path from one trip to the next. A sinuous line must be drawn on a map to show a satellite's orbital path projected downward to the Earth's surface directly beneath it. This can be readily understood with the aid of models: a globe for the Earth and a hoop for the orbit of the satellite. Once the satellite is in orbit, its orbital plane remains nearly fixed in space relative to the stars (the plane actually shifts slowly but assume it is fixed in space). Place the globe inside the hoop, tilt the hoop at a moderate angle to the equator, keep the hoop motionless, and spin the globe.

Explanations for some of the motions of satellites should now be apparent. A relatively low satellite completes a revolution in about $1\frac{1}{2}$ hours, whereas the Earth rotates once in about 24 hours. Therefore, on successive revolutions at $1\frac{1}{2}$-hour intervals, the satellite moves in about the same path, but the Earth has turned under it. As successive passes are viewed from any one location during a single night, the satellite is seen to follow parallel paths, but each will be higher or lower in the sky than the preceding one. If the direction is from southwest to northeast for an observer on one side of the Earth, on the opposite side of the Earth for another observer less than 1 hour later, the direction will be from northwest to southeast.

If the Earth were a perfect sphere of uniform density or composed of uniform concentric shells, the orbital plane of a satellite would remain fixed in space. However, the Earth's crust is not uniform in density and the Earth has an equatorial bulge. Therefore, the orbital plane of a satellite slowly shifts in space, changing in a manner that is similar to the precession of the equinoxes. The orbital plane maintains the same angle to the equator, but slowly turns in space; the rate may be several degrees in 1 day. The amount of such a shift can be calculated approximately before the satellite is launched and precisely after it has been tracked during several revolutions. If slight deviations from the calculated orbit are observed, they can be interpreted to yield information about the shape of the Earth and the distribution of matter within it. Such data were responsible for the much-publicized "pear-shaped" Earth.

SPACE TRAVEL

Even a very slight acquaintance with astronomical distances indicates that space travel in the foreseeable future will be limited to the vicinity of the solar system. To journey to Pluto at a rate of 5 mi/sec would take more than 20 years. (A higher velocity would be needed to escape from the Earth.) A trip to Alpha Centauri, the nearest star, would last for 150,000 years. Thus trips to even the nearest stars seem entirely unpractical and trips to distant stars or other galaxies are quite impossible. Let us consider some of the problems and principles involved in trips to the moon, Venus, and Mars.

A Trip to the Moon A velocity of about 7 mi/sec (approximately 25,000 mi/hr) must be attained by an object located a few hundred miles above the Earth's surface if it is to move outward from the Earth for a distance of about 240,000 miles and reach the moon. However, this does not mean that a spaceship could travel to the moon in about 10 hours. Two factors increase this time greatly: (1) The Earth's gravitational pull causes a gradual decrease in velocity throughout the journey, because burnout occurs near the Earth, and a spaceship or probe then coasts for the remaining distance in unpowered flight. A spaceship which left the Earth at a speed of 7 mi/sec would be barely moving at the point about 25,000 miles from the moon where the gravitational pull of the moon equals that of the Earth. From this distance onward, the moon's gravitational pull will gradually increase the velocity of the spaceship. If no action is taken to slow down the spaceship (such as firing retrorockets) its velocity will increase and it will strike the moon at a speed of about 1½ mi/sec. This equals the escape velocity from the airless moon. (2) At launching the spaceship is not aimed directly at the moon because the moon will not be at this position when the spaceship eventually arrives—duck hunters may have an advantage in understanding this aspect of space travel. The spaceship is aimed at the point in space where the moon will be located at the time of intersection. However, the Earth, moon, and spaceship are all orbiting the sun. Thus the

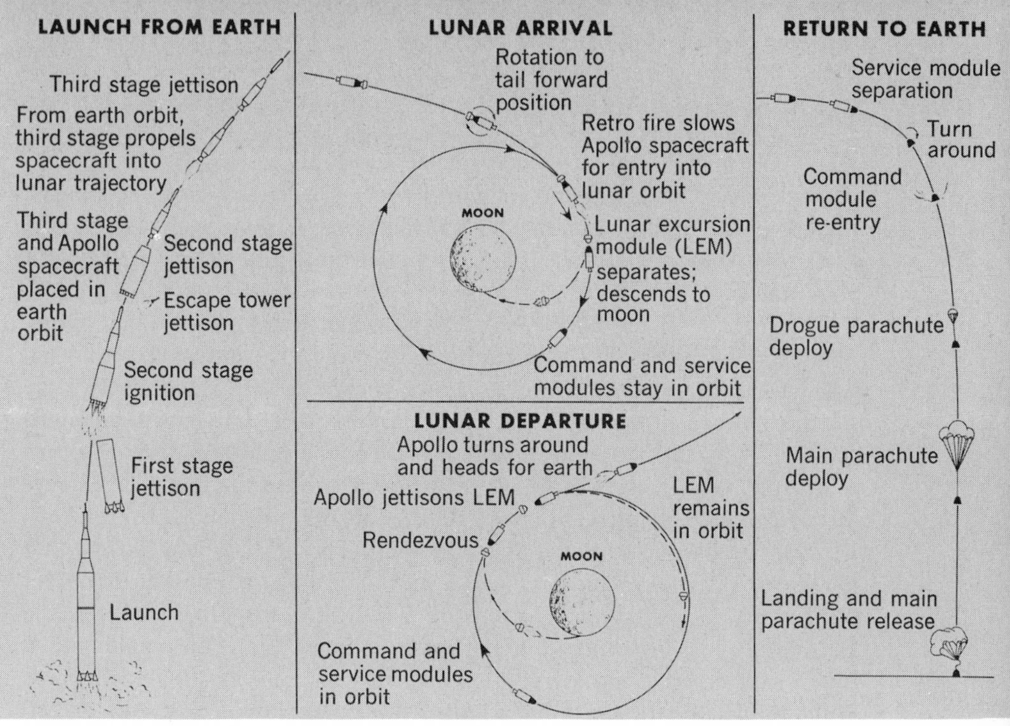

Fig. 22-7. *Sequence of major events in an Apollo lunar exploration mission.* [*Courtesy, NASA.*]

spaceship travels many miles in space to move outward a relatively few miles from the Earth.

Probes and spaceships can be made to approach or land on the moon along different types of orbital paths (Fig. 22-7). The problems, of course, are vastly greater for manned shots. Each added pound of payload means a large increase in the size of the rocket that will propel it through space. (The rocket functions as an expensive truck to move the payload from one place to another.) Thus unmanned probes are much easier in many respects, and observations of widely diverse kinds can be made automatically by instruments in the probes and the results radioed to the Earth. Man would be a nuisance in many probes.

Trips to Venus and Mars At their closest approaches to the Earth, Venus and Mars are still well over a hundred times farther from the Earth than the moon is, but the energy necessary to propel space probes to these planets along minimum-energy flight paths is not much greater than that required for reaching the moon. However, the navigational precision must be of an extremely high order and involves four dimensions: three space coordinates and time. The space probe must arrive at precisely the right point in a vast region of space at precisely the right time (in 1 minute, Venus and Mars move about 1300 miles and 900 miles respectively).

Basic data include three points: (1) The Earth's average velocity is about $18\frac{1}{2}$ mi/sec and all objects on the Earth are moving at this rate independently of other motions they may have. (2) Planets closer to the sun revolve more rapidly, whereas those farther away revolve less rapidly. (3) Changes in velocity must occur to shift a planet into a larger or smaller orbit about the sun.

For example, if the Earth's velocity were increased a few mi/sec by the application of a force for a time, the Earth would move outward from the sun into a larger orbit. However, once the force was discontinued, the Earth would be slowed in its outward journey by the sun's gravitational attraction. When it had slowed sufficiently (at the aphelion position of its new orbit) the Earth would have to fall back toward the sun. The Earth's velocity would now increase until it reached its new perihelion position halfway around the sun from the aphelion point. The Earth would thus be in a new orbit at a greater mean distance, and its average velocity in the larger orbit would be less than its present velocity. On the other hand, if the Earth were to slow down, the sun's gravitational attraction would pull it into a smaller orbit and increase its velocity (once the slowing-down force is discontinued). In the smaller orbit, the Earth's velocity would exceed that in its present orbit.

Thus the general scheme is rather simple. The probe must have a velocity

Fig. 22-8. *Typical 1964 Mars trajectory. A probe launched in November of 1964 would be expected to encounter Mars about eight months later.* [*Courtesy, NASA.*]

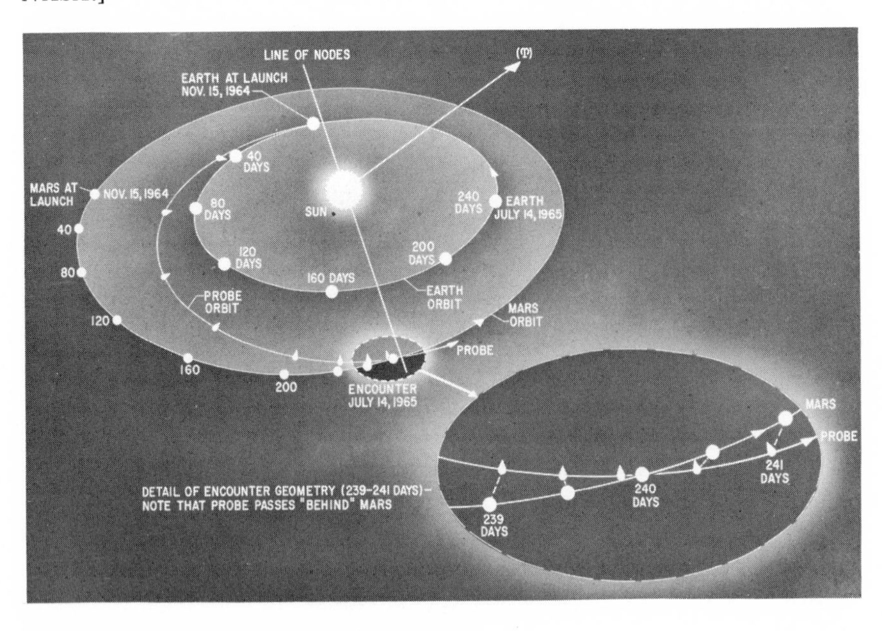

at the start that is faster than the Earth's to spiral outward to the orbit of Mars, or smaller than the Earth's to spiral inward toward the orbit of Venus. If the probe is launched in the direction in which the Earth revolves and rotates, it has an initial velocity of $18\frac{1}{2}$ mi/sec, plus about 0.3 mi/sec, before it is launched. A probe must be given a velocity of about 7 mi/sec at an altitude of a few hundred miles to escape from the Earth. After it has escaped from the Earth, however, it is still in orbit about the sun at approximately the same distance and velocity as the Earth. If the probe is now given an additional velocity of about 1.5 mi/sec, it will gradually move farther outward from the sun until it reaches the distance of Mars (Fig. 22-8). The probe is aimed not at Mars but at the point in space where Mars will be located months later. One must also consider the position of the Earth in space at the time the probe reaches Mars; if the Earth is too far away, data radioed from the probe may not reach us.

To reach Venus the probe must likewise be given an initial escape velocity of about 7 mi/sec, but then it must be slowed down about $1\frac{1}{2}$ mi/sec. This can be accomplished by launching the probe with an initial velocity of about 8.5 mi/sec in the direction opposite to that of the Earth's revolution (Fig. 22-9).

A trip to Venus or Mars thus involves two problems. (1) The probe must first escape from the Earth. An initial velocity of about 7 mi/sec must be attained at an altitude of a few hundred miles to eliminate the effects of air resistance. This takes a space probe out to a distance of a million miles or so where the Earth's gravitational pull is so weak that it does not have an appreciable effect. The probe would slow down continuously as it coasted through frictionless space because of the Earth's gravitational pull, although

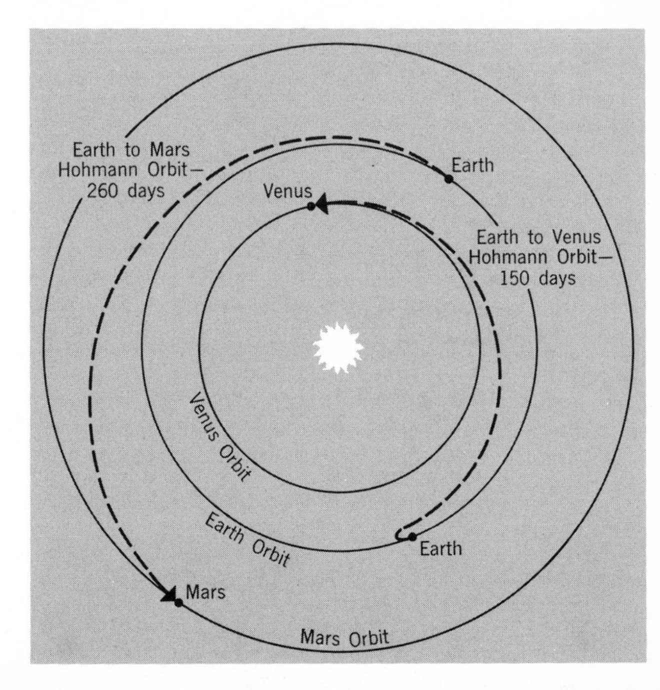

Fig. 22-9. *Minimum-energy flight paths to Venus and Mars (also called Hohmann orbits).* [*H. L. Goodwin,* Space: Frontier Unlimited, *D. Van Nostrand Co., Inc.*]

this pull becomes less and less as the distance becomes greater. The probe uses up the 7-mi/sec initial velocity to escape from the Earth, but it still has the $18\frac{1}{2}$ mi/sec velocity that it had when launched from the Earth, and thus it would revolve about the sun in an orbit about the size of the Earth's. (2) The space probe has now escaped from the Earth but it is still about the same distance from the sun. To change this distance, the velocity of the space probe must be increased or decreased. An increase of about $1\frac{1}{2}$ mi/sec in velocity would send the probe out to the orbit of Mars, a decrease of about $1\frac{1}{2}$ mi/sec would send it spiraling in toward Venus. This is accomplished approximately by launching the space probe with an initial velocity of about $8\frac{1}{2}$ mi/sec—in the eastward direction to reach Mars, in the westward direction to reach Venus. To escape from the solar system, the space probe would have to be launched eastward with an initial velocity of more than 14 mi/sec (to attain the escape velocity at the Earth's distance of 26 mi/sec). In other words, if the Earth's average velocity were increased to about 26 mi/sec, it too would leave the solar system.

Weightlessness has been illustrated in the following manner. Imagine that two satellites are following identical orbits about the Earth or some other celestial body; one would fall toward the Earth at the same rate as the other. If one satellite is placed inside the other, they would still fall at the same rate; if the inner satellite is replaced by a man, the rate of free fall remains the same. The man is weightless inside the satellite unless it is made to spin or change its acceleration.

A satellite might be used to detect the existence of life on another planet (Fig. 22-10).

DOES LIFE EXIST ELSEWHERE IN SPACE?

Uncertainties abound in a discussion of this sort. Conclusions must be based upon whatever assumptions we make concerning the origin of planets, of life itself, and of the conditions under which we think life can exist. We limit the discussion to life as we know it.* The existence of intelligent life elsewhere in the solar system seems highly improbable, although some scientists consider it likely that vegetation exists on Mars. To be a suitable home for life as we know it, a planet should probably have a moderate gravitational attraction. A massive planet such as Jupiter would probably have a sufficient gravitational pull to retain hydrogen and other light elements in its atmosphere; thus poisonous gases such as methane and ammonia might be present. On the other hand, a small planet such as Mercury could not hold an atmosphere. This moderate-sized planet should follow a stable orbit about a star at an appropriate distance so that it is neither too

* Some of the ideas in this unit are based upon an article by Su-Shu Huang, "Life Outside the Solar System," *Scientific American,* April 1960, pp. 55-63.

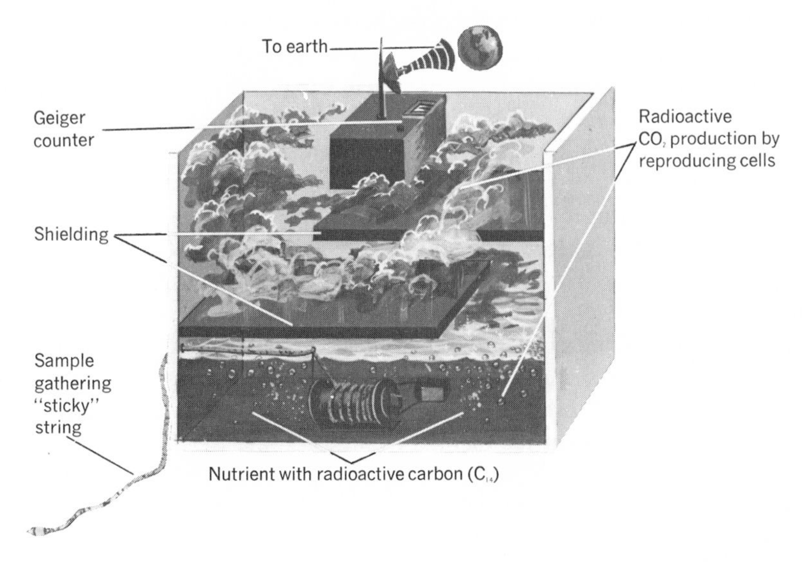

Fig. 22-10. *Device for detection of life on other planets. Planetary material would be drawn by an adhesive into a nutrient solution containing radioactive carbon. Forms of life may ingest the solution and give off radioactive carbon dioxide which would register on a Geiger counter. These data would then be radioed to the Earth.* [*Courtesy, NASA.*]

hot nor too cold; i.e., it must lie within the inhabitable zone that surrounds each star. Temperatures, of course, decrease outward from a star, and if a certain star is hot enough, its inhabitable zone will be quite wide; planets of the proper size might by chance exist within this zone. On the other hand, the inhabitable zone of a small red star, apparently the most abundant type of star, would be rather small, and relatively few planets of the proper size could be expected within such zones.

If the orbits of such planets are to be stable and remain within the inhabitable zones at all times, they probably should revolve about single stars; revolution around one member of a binary star or multiple star group would probably result in unstable orbits. Perhaps half of all stars are double stars or members of multiple-star systems and thus unlikely locations for inhabited planets.

How abundant are planets? How long a time is necessary for life to evolve? Would evolutionary development produce similar end results if it occurred elsewhere in the universe on a planet similar to the Earth? These are questions without definite answers. However, astronomers today favor a manner of origin of stars (p. 459) that should make planets common objects in the universe.

Let us assume for the moment that planets are common by-products in the formation of stars. How long does it take for intelligent life to develop? On the Earth, evolutionary development seems to have taken a few billion years. If this is assumed to be an average, then we eliminate as possible candidates for inhabited planets all hot massive stars that evolve rapidly. It may seem something of an anticlimax now to conclude that intelligent beings would most probably exist on planets like the Earth revolving about stars like the sun—if intelligent life does exist elsewhere in the universe. However, such stars are very numerous in our galaxy, and the universe contains many galaxies.

What are the chances of finding planets inhabited by intelligent beings around any of the sunlike stars within several dozen light-years of the Earth? If they exist, how would we know about it? If life exists elsewhere and has evolved in some such manner as it has on the Earth, it seems highly unlikely that any near stars would have planets on which life and civilization had reached approximately the same stage as on the Earth. For one reason, stars have formed at different times in the past; thus planets and life, if they exist elsewhere, must also have formed at different times. Consider also the technological developments of the past few centuries, of the past few decades, and even of the past few years. It is perhaps more likely that the age of dinosaurs would be occurring on one, whereas amphibians would be evolving from a type of lung fish on another. On the other hand, more advanced civilizations than our own may be present somewhere. In the geological history of the Earth, various organisms have evolved, culminated, and declined—some apparently abruptly and unaccountedly. Will man have a similar fate on the Earth?

The possibility of intelligent beings on a planet located within a few tens of light-years from the Earth and capable of sending radio messages into space seems very, very small. A "listening" program was once undertaken by means of a radio telescope in a project aptly called Project Ozma. It was unsuccessful as expected, but the thinking it provoked probably made the project worthwhile. In fact, the question was raised: If we detect messages from outer space, should we answer? Would other intelligent beings be hostile? Although radio waves travel at the speed of light, communication would be frustratingly slow even with intelligent beings on relatively near stars. One generation might ask a question and their great grandchildren receive the reply. But what question can be asked and in what language? One suggestion is to send the value of pi, a numerical constant, and hope for a continuation of numbers; e.g., QUESTION: 3.14159; ANSWER: 2653589; pi is equal to 3.141592653589. The question of intelligent life elsewhere in space is intriguing, exciting, and profound, but it seems unanswerable for the present at least.

Exercises and Questions, Chapter 22

1. Why are multistage rockets commonly used to launch artificial satellites?

2. State the laws of Kepler and Newton that are directly involved in the launching of an artificial satellite into an orbit about the Earth.

3. Assume the following: a satellite has been launched into an elliptical orbit about the Earth; its perigee position occurs at an altitude of about 100 to 150 miles; its apogee position is at an altitude of several hundred miles. Describe the orbital changes that will probably take place in the future.

4. Why is it difficult to launch an artificial satellite into a precisely circular orbit about the Earth?

5. Why is it common practice to launch an artificial satellite in an eastward direction about the Earth as opposed to a westward direction? Discuss some of the advantages and disadvantages of an equatorial orbit (of a satellite that revolves approximately in the plane of the Earth's equator); of a polar orbit; of an orbit that is inclined about 45 degrees to the Earth's equatorial plane.

6. Artificial satellites that orbit the Earth within a few hundred miles of its surface are visible only at certain hours and dates. Why? Why may variations occur in the brightness of these satellites?

7. Discuss some of the orbital factors involved in:
   (a) A manned flight to the moon.
   (b) Sending an unmanned probe to Venus.
   (c) Sending an unmanned probe to Mars.

8. What are some of the dangers and personal problems that seem likely to confront the occupants of a space vehicle on a flight to the moon?

9. Consider the question of the possible existence of life as we know it on some other planet in the universe and discuss the following factors:
   (a) How common are planets in the universe?
   (b) Why is the size of a planet significant?
   (c) Which types of stars are considered to be most favorable? Why?

10. Assume that intelligent beings exist on a planet orbiting a nearby star (not the sun). Discuss some of the problems of communicating with such beings. What was Project Ozma?

11. Discuss the differences in magnitude involved in each of the following:
    (a) Interplanetary space travel.
    (b) Interstellar space travel.
    (c) Intergalactic space travel.

12. Does it take more energy to send a satellite into an orbit so that it collides with the sun or into an orbit that will take it several light-years from the Earth? Explain.

23 Tools of the astronomer

DURING an ordinary conversation about his favorite subject, an astronomer might make statements such as any of the following.

Carbon dioxide is present in the atmosphere of Venus, and methane occurs on Jupiter. . . . More than sixty of the elements that are known on the Earth have also been identified on the sun. . . . Hydrogen and helium seem to make up about 96 to 99 percent of the matter in the sun, in the other stars, and in the entire universe. . . . Some stars have surface temperatures that exceed 50,000°F, whereas others are less than a tenth this hot. . . . Certain stars rotate more rapidly than others. . . . The distances from the Earth to certain stars are increasing, but to other stars the distances are decreasing. . . . A particular star appears as a single point of light, yet it is really a double star. . . . Millions of galaxies exist in the universe at distances of millions of light-years and all of them seem to be moving away from us.

Such remarks would be commonplace among astronomers, but reaching the ears of the intellectually curious nonastronomer, they cause wonder and skepticism. How have astronomers obtained all of this information about celestial objects—objects that may be so remote that we see them merely as tiny points in space, whose light must travel for many years to reach the Earth? The data have come from a study of the light such objects emit or reflect, and chiefly by means of three instruments—the telescope, the camera, and the spectroscope.

Although light is the source of our information about the stars, the nature of light itself is still something of a mystery. According to the wave theory, the light which we can see consists of electromagnetic waves which move in a straight line at the rate of about 186,000 mi/sec through a vacuum (see Chapter 9). Other portions of the electromagnetic spectrum are also being studied; for example, radio astronomy has been a most important development since the 1940's.

COLOR

Ordinary white light is a mixture of electromagnetic waves of many wavelengths, each associated with a color in the range of the rainbow: red, orange, yellow, green, blue, indigo, and violet. Solar radiation also includes other electromagnetic waves that are too long or too short for our eyes to detect. If our eyes were sensitive to other wavelengths, we would see a different universe: some objects would be larger and more conspicuous and even have a different shape. With a simple prism, sunlight can be separated into its component colors (Figs. 9-1 and 9-2): the wavelengths of red light are longest and of violet light shortest.

The Earth's atmosphere diffuses sunlight, but longer wavelengths penetrate the atmosphere better than shorter wavelengths, which tend to be scattered (i.e., reflected or diffracted without change in wavelength) by dust particles and air molecules. This explains a number of familiar phenomena (Fig. 23-1): daylight, twilight, blue sky, reddish sunsets and sunrises, and a reddish moon during a total lunar eclipse.

If an object is not itself a source of light, it is visible because it reflects or refracts light from another source; the color of such an object depends on which wavelengths it reflects, and hence does not absorb. To state the case simply, an object is yellow because its atomic structure causes the reflection of light of the wavelengths identified with yellow and the absorption of light of all other wavelengths. But our eyes can also respond to a blend of two or more spectral colors with a sensation of a completely different color. Thus the blending of pure blue and pure yellow produces the sensation of green, even though the eye may not be receiving any radiation at all of the wavelength corresponding to a spectral green.

The color of an object that is its own source of light depends chiefly upon the temperature of the object. If a piece of metal is heated gradually, it feels warm before any change in color can be detected; the warmth represents the radiation of invisible infrared electromagnetic waves. As the temperature increases, red light predominates among the radiations, and the metal is seen to glow. At still higher temperatures the radiations become stronger and the color changes in turn to orange, to yellow, and then to green. Highest temperatures produce radiations which are strongest in the blue and violet wavelengths. The relationship between temperature and color enables astronomers to determine the surface temperatures of the stars.

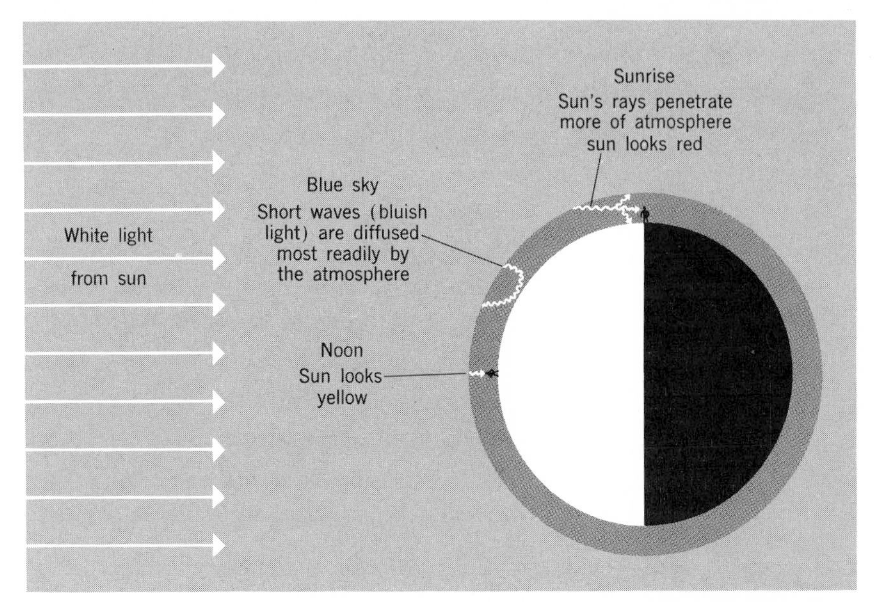

Fig. 23-1. *Blue sky and reddish sunsets. Orientation: looking "down" on the sun and the Earth. One observer is experiencing noon on the Earth, and another is at his sunrise position. The longer red waves penetrate the atmosphere more readily than the shorter blue and violet waves, which are scattered easily by gas and dust in the air.—As the sun's light comes through the Earth's atmosphere, some of the shorter waves are scattered, and hence the sun looks yellowish or reddish instead of white. The scattering is greatest at sunrise and sunset, when the sun's rays come through the Earth's atmosphere at the greatest slant. The scattered light waves, predominantly blue and violet, are reflected by the dust particles in the atmosphere and make the sky look blue.—When there is a great deal of dust in the atmosphere near the horizon, then reddish colors are prominent in the sunset. Occasionally at midday, when considerable dust and haze intervene between the observer and the sun, the sun takes on an orange hue.*

Yellowish stars like the sun have surface temperatures of approximately 6000°C (11,000°F); the surface temperatures of reddish stars are lower, and those of bluish stars are higher.

Daylight is caused by the diffusion of sunlight by gas and dust in the Earth's atmosphere. As seen from a space rocket flying high above the Earth's atmosphere, the sky approaches the black of interstellar space even if the sun is visible (if you looked out of a space rocket at an angle to the sun, sunlight would be streaming by you, but it would remain invisible because nothing exists in space to reflect the sunlight into your eyes). Similarly, seen from the airless moon, the sky would be black.

THE TELESCOPE AND CAMERA IN ASTRONOMY

Telescopes* involve lenses and mirrors and the refraction and reflection of electromagnetic waves. Waves of light from a source such as a star travel outward in all directions. If we consider a minute portion of these waves, we may think of it as a ray of light that travels in a straight line as it passes through a homogeneous medium or through a different medium that it enters at a 90-degree angle. If the ray passes at a slant into another type of medium, its speed and direction are changed, and we say that it is refracted (bent). We are all familiar with the apparent bend in a straight object when part of it is seen under water. The refraction of light waves as they pass through the Earth's atmosphere, causes an apparent increase in the altitude of celestial objects (Fig. 23-2).

If a ray of light meets a polished surface such as a mirror, it is bounced back or reflected. A simple convex lens or concave mirror can be so shaped

* Cf. "The Finest Deep-Sky Objects," James Mullaney and Wallace McCall, in *Sky and Telescope,* November 1965, December 1965, and January 1966; "The Backyard Astronomer," James S. Pickering, in *Natural History:* "Choice of Telescopes" in November 1966, "Setting Up a Telescope" in January 1967, "Finding Celestial Objects" in March 1967, and "The Sun and Double Stars" in August-September 1967; *Consumer Reports,* November 1967.

Fig. 23-2. *Celestial objects are elevated by atmospheric refraction. Atmospheric refraction makes an object appear higher above the horizon than its actual position; this effect increases greatly near the horizon. Therefore, the bottom portion of a large object such as the sun appears to be elevated more than the top portion, and it looks flattened. No distortion occurs in a horizontal direction.*

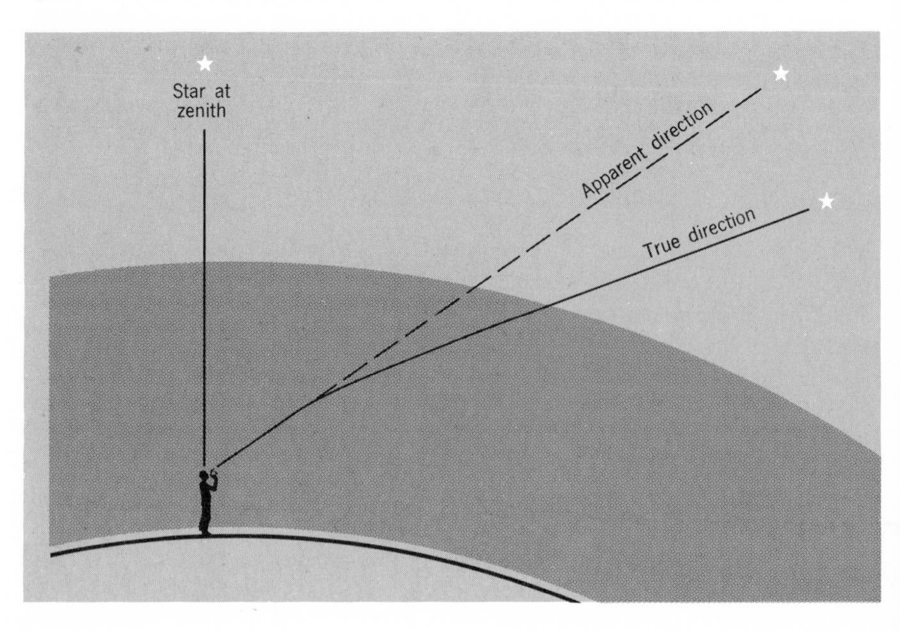

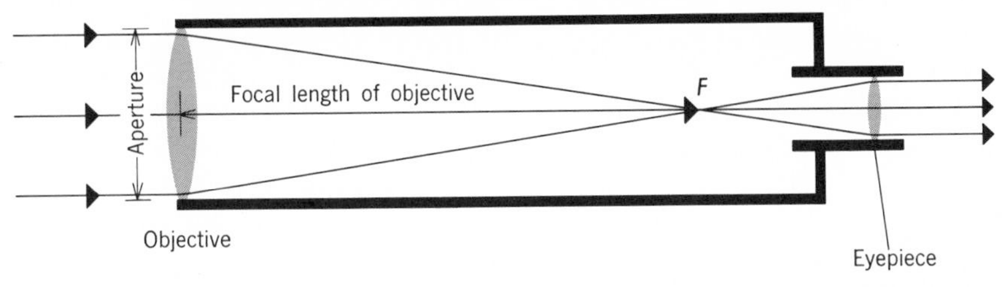

Fig. 23-3. *A simple refracting telescope. Note that all of the parallel rays falling on the large lens also pass through the eyepiece; thus they are concentrated. This makes it possible to observe remote objects too faint to be visible to the unaided eye. The distance from the center of a convex lens to the point where all of the once-parallel rays are converged (its focal point) is called its focal length. The large convex lens is called the objective and the small one is the eyepiece; their distance apart is equal to the sum of their focal lengths when the object being viewed is very far away. In most modern telescopes the objective and eyepiece are both compound lenses involving different shapes and materials. Various eyepieces can be used to achieve different magnifications. Magnification is equal to the focal length of the objective divided by the focal length of the eyepiece.*

that it will concentrate at a point all of the parallel light waves that fall on it from a source. This is the main job of the telescope in astronomy; it acts as a funnel by collecting all of the parallel rays from a star which fall on the objective lens or mirror and focuses them at a point (Fig. 23-3).

The light-gathering power of a telescope varies as the square of the diameter of the objective (i.e., directly as its area). This power can be compared with that of the unaided eye if we choose a suitable figure for the diameter of the pupil. The pupil changes in size with the intensity of the light; but probably about $\frac{1}{3}$ inch is a maximum. Thus the light-gathering power of the 200-inch telescope, the world's largest optical telescope, is at least 360,000 times that of the human eye. Objects much too faint to be seen by the unaided eye are brilliantly visible in this telescope.

The larger telescopes are used for photography much more than for visual observation and might well be regarded as cameras of a special kind. The camera in astronomy has the advantage of providing a permanent record of what is seen, unbiased by vagaries of personal judgment, and it can also photograph objects in both infrared and ultraviolet light that cannot be seen by man. Furthermore, many of the celestial objects which astronomers want to study are too distant for the eye to see even through the 200-inch telescope; here time exposures can extend outward the boundaries of the visible universe. Light from a very distant star funneled through a telescope onto a film, may be too faint to create an image in a minute or an hour or even in a night; but if the time exposure is continued, the star's photograph may eventually be obtained. General fogging of the entire negative limits the length of time a film may be exposed, but a successful 80-hour exposure has been made of the spectrum of a remote galaxy.

To observe objects at even greater distances, astronomers have utilized different techniques. They have increased the amount of light that falls on a photographic plate by building larger telescopes and by making longer time exposures. Faster and more sensitive film has been developed. Most recently, some use has been made of an electronic device—the image converter—which is more sensitive to light than any known film. It is theoretically possible that a 20-inch telescope with this attachment may detect

Fig. 23-4. *Dome of 200-inch telescope on Palomar Mountain in California, seen by moonlight. The entire top part of the dome can move on tracks so that the telescope may be pointed at any part of the sky.* [*Mount Wilson and Palomar Observatories.*]

objects as far away as those that a 60-inch telescope can detect without it. However, the actual achievement to date is considerably less than this.

All large telescopes (more than 40 inches in diameter) are reflectors, for a number of reasons: light does not pass through a mirror, and its quality does not have to be so nearly perfect as in a refractor; a mirror can be supported on the back; light is not separated into its different colors by reflection from the surface of a mirror (chromatic aberration is eliminated).

Large telescopes are oriented on axes and electrically operated so that they can be centered exactly upon any celestial object and follow it, without deviation, on its daily east-to-west journey across the skies (Fig. 23-4).

Astronomers use a system of coordinates similar to terrestrial latitude and longitude to locate celestial bodies. The position of the desired object in the sky is ascertained from a set of tables. The telescope is turned a certain number of angular degrees in one direction and then the designated number of degrees along a second axis at right angles to the first, until it points directly at the target. Except for certain fine adjustments, the astronomer does not need to look in the sky for the object to guide the movement of the telescope.

The Hale telescope is now in operation at the Palomar Observatory in California (Fig. 23-4). In fact, three of the four largest optical telescopes in existence in 1966 were all located in California. (Large observatories are commonly located in areas far from city lights and smoke, in climates that permit good seeing on many nights, and at high altitudes above the dense lowermost atmosphere.) The maximum range of the Hale telescope is probably several billion light-years, and its primary function is to determine the structure and composition of the universe. Plans have been made to triple its focal length, perhaps thereby increasing its range by 50 percent. Several of the photographs reproduced for this text were taken through the Hale telescope. A 61-inch reflector of quartz was installed in the U.S. Naval Observatory near Flagstaff, Arizona, in 1964.

RADIO ASTRONOMY

Celestial objects such as stars, planets, gas clouds, and galaxies produce not only visible light but also long, invisible electromagnetic waves. These waves can penetrate the Earth's atmosphere and are intermediate in length between light waves and standard radio waves. They are being studied at present by radio telescopes (Fig. 23-5). Celestial objects also produce radiation of a wide variety of wavelengths that does not penetrate the atmosphere to the Earth's surface.

A radio telescope is something like an optical telescope, but in place of a lens or mirror it employs a large reflector to focus radio waves of a certain length on a small radio antenna. The antenna corresponds to the photographic plate located in the focal plane of an optical telescope. The current thus induced in the antenna is conducted to a receiver. As in an ordinary radio set, only one wavelength at a time can be utilized, not a range of wavelengths as in visible light. However, the receiver can be tuned quickly from one radio wavelength to another. The reflector is commonly of metal and may be an open fencelike mesh if the openings are kept much smaller than the wavelengths being studied. The strength of the signal on a certain wavelength from a certain part of the sky is recorded on a registering device and then compared with similar signals from other directions in space and on other wavelengths.

Fig. 23-5. *The world's largest radio telescope in Arecibo, Puerto Rico, was dedicated in November 1964. A 2½-ton dump truck illustrates the huge size of the reflector which covers an area of 18.5 acres. The transmitter hangs 435 feet above the center of the reflector.* [*Courtesy, Commonwealth of Puerto Rico.*]

Radio telescopes function as well during the day as at night. They can receive electromagnetic waves that penetrate readily through clouds in the Earth's atmosphere and also through the gas and dust clouds that occur here and there in space, especially in the arms of spiral galaxies. These are advantages over the optical telescopes—full-time, all-weather astronomy and greater penetration of space. However, the target objects are never actually seen and are difficult to locate precisely. Thus the radio telescope will not supplant the optical telescope; each has a useful function, and together they aid in enlarging our knowledge concerning the universe.

Already credited to the radio telescopes are (1) the discovery of peculiar galaxies that may be colliding or exploding (Fig. 23-6), (2) additional information about the size, shape, and number of spiral arms of our own and other galaxies, (3) a navigational sextant which can be used in fog (a minor but practical contribution), (4) new information concerning the amount of matter scattered in clouds of gas and dust in space, and (5) data that have been interpreted as showing that the universe is not uniform in space and time. Such data may have made the Steady State hypothesis untenable (p. 477).

Fig. 23-6. *Colliding galaxies? (NGC 5128). A spiral seen edgewise may be colliding centrally with a large galaxy which is intermediate in type between elliptical and spiral. The discovery of this pair of galaxies is credited to radio telescopes. The galaxies constitute a very powerful source of radio waves. The Hale telescope was called on to photograph this part of space to identify the source of the radio emission.* [*Mount Wilson and Palomar Observatories.*]

THE SPECTROSCOPE IN ASTRONOMY

Spectra and the operation of the spectroscope (Fig. 23-7) are discussed in Chapter 9. Here we deal with the instrument as it finds application in astronomy.

The spectroscope is an instrument designed to separate the components of the different wavelengths present in the light that enters it; thus the presence or absence of light of certain wavelengths can be determined. If a photographic plate is substituted for the eyepiece, which is common practice in astronomy, the spectroscope becomes a spectrograph. The separation of light into component wavelengths can be achieved by means of a glass prism or a diffraction grating.

Spectra Laboratory study has shown that three types of spectra can be produced (Fig. 9-5): *continuous, bright-line* (emission), and *dark-line* (absorption). Solids, liquids, and compressed gases of all known materials, when heated to temperatures high enough to make them luminous, show continuous spectra; i.e., a continuous band of colors like the rainbow is visible. This indicates that such light consists of a mixture of components of all wavelengths from red to violet.

A different spectrum is obtained if light from a luminous noncompressed gas (e.g., from a neon sign) is passed through the slit of a spectroscope—a number of bright-colored lines occur against a dark background. Evidently this light consists of a mixture of a relatively few colors (wavelengths). Most of the wavelengths of visible light are not produced by this gas; hence the black background.

Each element in this state always produces the same characteristic pattern of bright lines; the numbers, locations, and intensities of the lines are different for each element. Thus, elements can be identified in the laboratory on the basis of their bright-line spectra; each has a distinguishing set of "fingerprints." For example, the bright-line spectrum of sodium vapor contains only two lines, both in the yellow part of the spectrum. In contrast, the bright-line spectrum of hydrogen shows five prominent lines in the visible region, and none is yellow.

Elements can also be identified in the sun and the stars by the characteristic spectra they produce, but these are usually dark-line spectra—the characteristic pattern of an element appears as dark lines against a colored background that is continuous except for the dark lines. The number and locations of the dark lines of any one element are identical with the

Fig. 23-7. *Paths of light waves inside a spectroscope. Highly schematic. Careful study of the sketch should show how a spectroscope functions in spreading apart the light waves which enter it so that they may be studied individually. Many shades of each color occur, and the statement that all waves of red light are focused at one point should be interpreted broadly. It should be apparent that dark lines will occur in the spectrum if certain wavelengths* (*colors*) *are missing.*

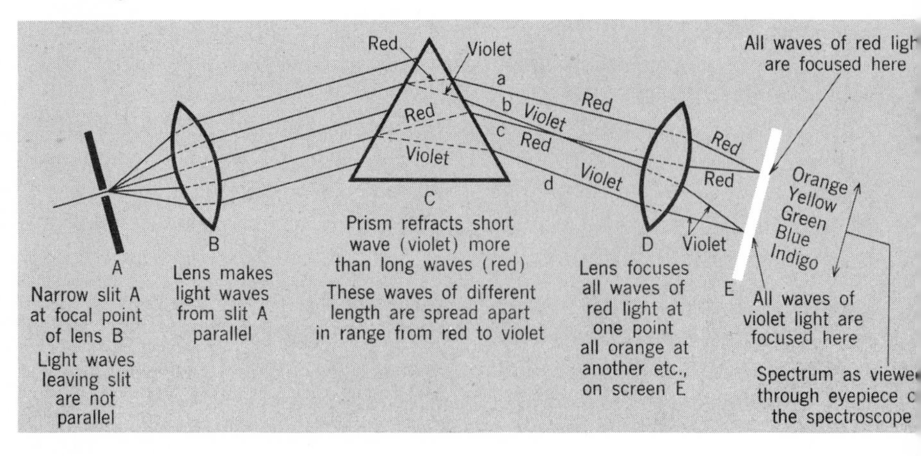

colored lines in its bright-line spectrum, but a dark line stands in the place of each colored line. The two lines of sodium vapor, for example, would appear as dark lines against the yellow part of the spectrum.

There are two prerequisites for the production of a dark-line spectrum: the light must come from a continuous source, and it must pass through a cooler gas before reaching a spectroscope. The cooler gas absorbs from the light the same wavelengths it would emit if it were hot enough to produce a bright-line spectrum (Fig. 9-5). The dark-line spectrum thus shows the composition of the cooler gases located between the source and the spectroscope.

Sunlight is emitted by hot compressed gases at the visible surface of the sun (*photosphere*), and these produce all of the visible wavelengths. However, this light must travel from the photosphere outward through relatively cooler gases in the sun's atmosphere, and these cooler gases absorb certain wavelengths. More than 20,000 dark lines from more than 60 elements have been identified. Note that the spectroscope cannot be used directly to determine the chemical composition of the gases beneath the photosphere.

Fig. 23-8. *The Stratoscope II balloon system a few moments after launch is rising at about 1400 ft/min. The 36-inch-aperture telescope was lauched in Texas and came down in Tennessee. It was lowered gently by balloons, not by a parachute. The object of the initial flight was to make an infrared study of Mars. Total weight to be lifted was about 13,000 pounds; the payload was 6300 pounds of this total. The guidance system of the telescope is designed to track an object in the sky and hold it fixed in the image plane to within 0.02 second of arc for as long as one hour. During a time-exposure photograph, the balloon is floating at an altitude of about 80,000 feet and perhaps is being blown rapidly through the atmosphere while the payload swings back and forth beneath the balloon.* [N.S.F. photograph.]

Cool gases in the Earth's own atmosphere also absorb certain wavelengths, but these are chiefly bands produced by molecules and can be distinguished from lines produced by atoms on the sun. The chemical compositions of the atmospheres of the planets can be partly determined spectroscopically. Light from the sun passes part way through the atmosphere of a planet before it is reflected to the Earth; in this case, astronomers must separate the dark-line spectra produced by the sun's atmosphere from those produced by the atmospheres of the planet and the Earth.

The type of spectrum produced by any given element tends to vary with the temperature, and the spectrum at a very high temperature may be entirely different from that at a low temperature. This seems to account for the main differences observed in the spectra of stars as well as for their colors, which range from blue for the hottest stars to red for the coolest stars.

It is advantageous to study radiations from celestial objects from locations above the Earth's atmosphere, which is opaque to many wavelengths of the electromagnetic spectrum. Spectroscopic studies have been made from equipment suspended from unmanned high-altitude balloons (Fig. 23-8) that are above most of the atmosphere (by mass). If an astronomer studying the atmosphere of Venus or Mars finds water-vapor absorption lines on a photographic plate made at 80,000 feet above the Earth, he can know that these were not produced in the Earth's atmosphere.

An orbiting astronomical observatory is scheduled to go into a 500-mile-high circular orbit in the later 1960's. It will have self-contained stabilization, communications, and power equipment and is designed to point a telescope (as much as 36 inches in diameter) at a given star with an accuracy equivalent to focusing on a dime a mile away.

THE DOPPLER EFFECT

Study of the spectrum of a star gives us information about the chemical composition of its atmosphere, its surface temperature, and its structure (a spherical mass of compressed gases overlain by a cooler atmosphere). We turn now to the technique of determining whether the distance between the Earth and a star is increasing or decreasing and how rapidly this change may be occurring. This involves the phenomenon known as the Doppler effect and red and violet shifts (Fig. 23-9).

To understand the Doppler effect, imagine a motionless plane at an airport. A gun mounted on the plane is fired at uniform intervals at a motionless target, and the bullets strike at uniform intervals. Next, imagine the plane flying at a steady rate toward the target while the gun is fired at the same interval as on the ground. The plane is closer to the target

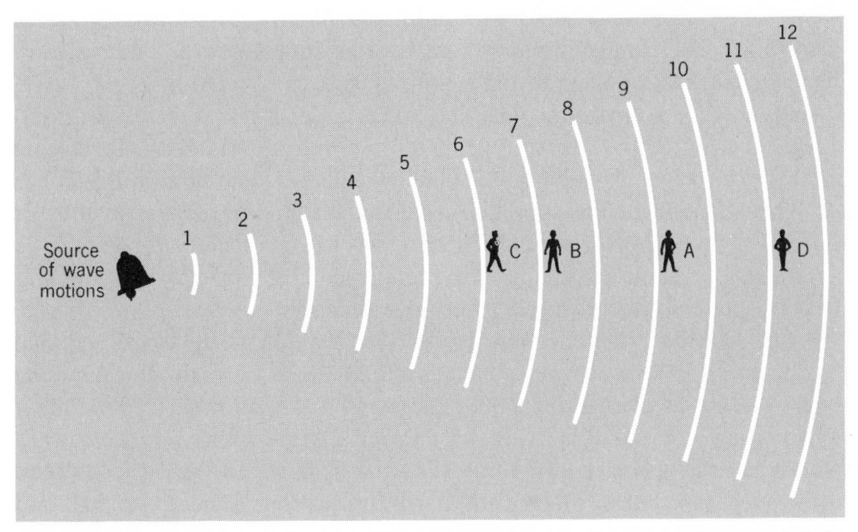

Fig. 23-9. *The Doppler effect explained according to the wave theory of light. Wave fronts are moving out from the source at the rate of six each second. During one second, waves 9, 8, 7, 6, 5, and 4 pass by the observer at A. If the observer during that second moves from A to B, waves 3 and 2 will also pass by him. Thus eight wave fronts move by the observer during this second, and each wave seems to be reduced in length. If the observer remains at B, the rate returns to six per second.—On the other hand, if the observer moves to D during the second, waves 4 and 5 will not reach him, and the rate is reduced to four waves each second. To the observer, each wave seems longer. Thus a decrease in distance (between source and observer) seems to decrease wavelengths, whereas an increase in distance appears to increase wavelengths.*

after each shot, and each bullet in succession has a shorter distance to travel than its predecessor. Therefore, the bullets strike the target at shorter intervals than the rate at which they are fired. The analogy can be carried further by imagining the plane flying away from the target, flying toward or away at different speeds, and by moving the target.

Thus, if the distance between a source of waves and an observer is decreasing, the waves appear shorter to the observer. If the distance is increasing, the waves appear longer. The observer, the source, or both may move; evidence of the Doppler effect cannot be used directly to distinguish among these three possibilities. Finally, the faster the distance changes, the shorter (if decreasing) or longer (if increasing) the waves appear to the observer.

A familiar form of Doppler effect is the drop in pitch of a car horn as the car passes us on a road. The sound waves seem to be shortened as the car approaches, raising the pitch of the horn; as the car recedes, the sound waves seem to be lengthened and there is an immediate drop in the pitch of the horn. The change in pitch should not be confused with a corresponding change in the loudness of the sound.

Astronomers compare the dark-line spectrum of a star with a laboratory-produced spectrum by allowing light from an iron arc to fall on the bottom and top portions of a photographic plate while the middle portion is blocked off. Then light from the star is focused on the middle portion while the top and bottom portions are blocked off. The presence of comparison spectra on the same plate and above and below the spectrum of the star makes it possible to measure wavelengths precisely.

When such comparisons were made for certain stars, it was found that the "fingerprints" of certain elements could be recognized, but that the lines were not located in the same places as on the laboratory spectrum—for a certain star, each line was shifted precisely the same distance toward one end of the spectrum. In some stars the shift was toward the red end of the spectrum, but in other stars the shift was toward the violet end, and the amounts differed from star to star. Evidently these shifts resulted from the Doppler effect.

If a star is moving away from the Earth (upper star spectrum in Fig. 23-10), then fewer light waves reach the spectroscope per unit of time and so each of the waves appears longer. But a change in the wavelength of light means a change in its color. Therefore, each of the dark lines in this spectrum is shifted the same amount toward the red end. Similarly, if the distance separating the star and the Earth is decreasing (lower star spectrum in Fig. 23-10), each wave seen or photographed in the spectrum appears to be shorter, and each of the dark lines is shifted an equal distance toward the violet end of the spectrum. The rate of movement is calculated from the magnitude of the shift. Note that in a red shift the spectral

Fig. 23-10. *Doppler displacements in the dark-line (absorption) spectra of the brighter member of the double star Castor in the constellation Gemini. The top and bottom bands are part of a bright-line spectrum being used for comparison. The other two bands are absorption spectra of the star made on two different occasions. In the upper of these the matching dark lines are displaced to the right—the red end of the spectrum—because the orbital motion of the star is away from the earth. In the lower spectrum the matching dark lines are displaced to the left—the violet end of the spectrum—because the orbital motion of the star is now toward the earth.* [*Lick Observatory.*]

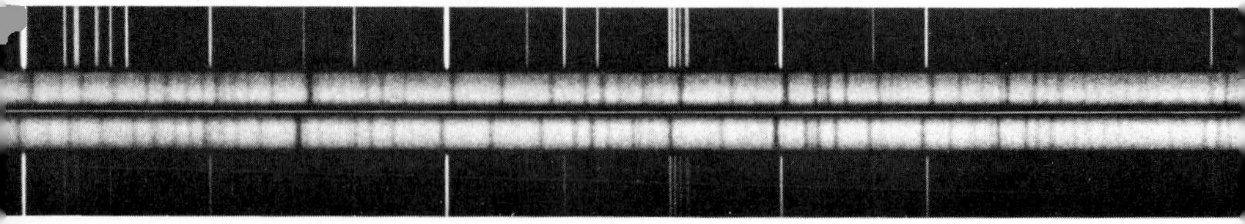

lines do not all become red; in fact, these are shifts in the dark absorption lines toward the red end of the spectrum. Evidence supporting the Doppler interpretation of such shifts is furnished by the rotation of the sun (p. 454) and its movement through space as well as by the Earth's orbital motion.

Exercises and Questions, Chapter 23

1. Three persons wearing white, yellow, and blue shirts respectively enter a room illuminated entirely by yellow light. What colors are their shirts in this room?
2. List some favorable, some unfavorable, and some unique features of reflecting, refracting, and radio telescopes.
3. How does the light-gathering power of a 4-inch refractor compare with that of a 60-inch reflector? How much more light can each of these gather than the human eye (assume a maximum eye opening of $\frac{1}{4}$ inch)?
4. How is the magnifying power of a telescope calculated? How can an astronomer change the magnifying power of a telescope?
5. Describe favorable conditions for the location of a large reflecting telescope; of a large radio telescope.
6. If a moth lands on the lens of your telescope when you are observing the moon, will you see it as a gigantic moth standing on the moon's surface? Explain.
7. Why do stars have different colors?
8. How is a continuous spectrum produced? A bright-line spectrum? A dark-line spectrum?
9. By means of a spectroscope, an astronomer can estimate directly the chemical composition of only part of the sun or of a star. Explain.
10. What is meant by a Doppler effect and why does one occur if the source of the wave motion is moving toward or away from an observer? If the observer is moving toward or away from the source of the wave motion? Why does a larger Doppler effect occur when the distance is changing rapidly than when it is changing slowly?
11. Why have astronomers made spectrographic measurements of the atmospheres of planets from instruments suspended beneath high-altitude balloons?

12. Six spectra are shown schematically in the accompanying diagram. The rectangles at the top and bottom represent comparison spectra produced by a light source in a laboratory. Spectra A, B, C, and D represent the spectra of four stars. Six fictitious spectral lines are shown in the laboratory comparison spectra. These are

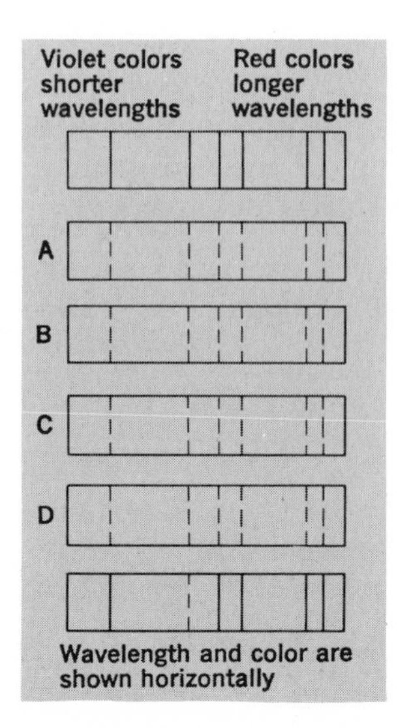

dashed in the spectra of the stars to show their locations in the absence of a Doppler effect. In the spectrum of each star, draw in the six spectral lines in the proper positions to show that the distance between this star and the Earth is:
   (a) Decreasing slowly for star A.
   (b) Decreasing rapidly for star B.
   (c) Increasing slowly for star C.
   (d) Increasing rapidly for star D.

13. Is a star showing a red shift necessarily farther from the Earth than one showing a violet shift?

24 Suns, stars, galaxies–The universe

THE SUN

If a fast jet plane could travel at 750 mi/hr right from the Earth to the sun, a distance of almost 93 million miles, the trip would take about 14 years and 2 months. By terrestrial standards, therefore, the sun is tremendously distant, but by astronomical standards it is our neighbor—the only star close enough to have a visible disk, and close enough to be studied in detail. Fortunately the sun seems to be an average type of star, and information obtained concerning it is probably applicable in varying degrees to other stars. The sun's gravitational attraction controls the orbital motions of the other members of the solar system and keeps them in elliptical orbits about it.

Approximately 699/700 of all the matter in the solar system is concentrated in the sun. Jupiter's mass is about 1/1000 that of the sun and accounts for most of the remainder. The sun's mass is about 333,000 times that of the Earth; its volume is about 1.3 million times as large; and thus its average specific gravity is about a quarter as large (1.4 vs. 5.5). The gravitational pull at its surface is about 28 times that of the Earth. Thus a 200-pound Texan would weigh nearly 3 tons on the sun, but he would, of course, be vaporized instantly. If the Earth were the size of a grapefruit, the sun would be about 50 feet in diameter and about a mile away.

The sun, as a hot gaseous body, has no physical features corresponding exactly to the solid body and atmosphere of the Earth. Direct radiation from the sun comes from the photosphere, the visible surface approximately 864,000 miles in diameter which we see as the disk of the sun. Below this surface, which is not a sharp boundary, gas particles are dense enough to be opaque to visible light. Above the photosphere, a progres-

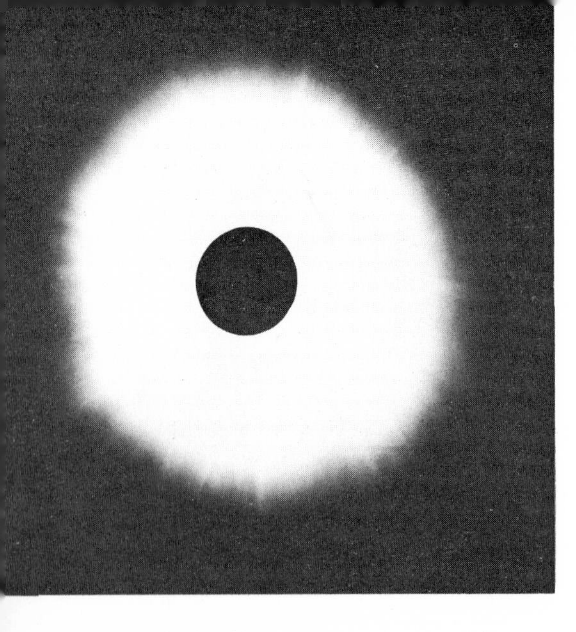

Fig. 24-1. *The total solar eclipse of May 20, 1947, showing the gaseous atmosphere (corona).* [*Yerkes Observatory.*]

sively thinner gaseous atmosphere extends outward for more than a million miles, but it can be seen by the unaided eye only during a total solar eclipse (Fig. 24-1).

About three-fourths of the mass of the sun seems to be hydrogen, about one-fourth is helium, and approximately 1 to 2 percent consists of heavier elements. More than 60 of the elements known on the Earth have been detected on the sun.

The sun rotates on its axis from west to east, but at different rates at different latitudes; thus it cannot be a solid body. The rate is fastest at the equator (about 25 days) and slowest near the poles (about 33 days at latitude 75 degrees). The cause of this variation is unknown. The movement of sunspots (Fig. 24-2) and the Doppler effect show that rotation occurs. When light from opposite sides of the sun is examined in a spectroscope, a red shift shows up in the dark-line spectrum on one side (spinning away from us) and a violet shift shows up on the other side (spinning toward us).

The solar system appears to be moving toward the constellation Hercules at about 12 mi/sec relative to the near stars visible to the unaided eye. Evidence for this motion is based upon a familiar phenomenon: as one drives toward a group of houses, they seem to spread apart; as one drives away, they appear to come together again. Stars in the vicinity of Hercules appear to be diverging, whereas stars on the opposite side of the sky seem to be converging. Furthermore, most of the stars in the vicinity of Hercules show violet shifts in the spectroscope; red shifts are common for stars in the opposite direction.

It is very dangerous to look directly at the sun with the naked eye, though it can be observed safely through heavily smoked glass or two thicknesses of blackened photographic film. The simplest and most convenient

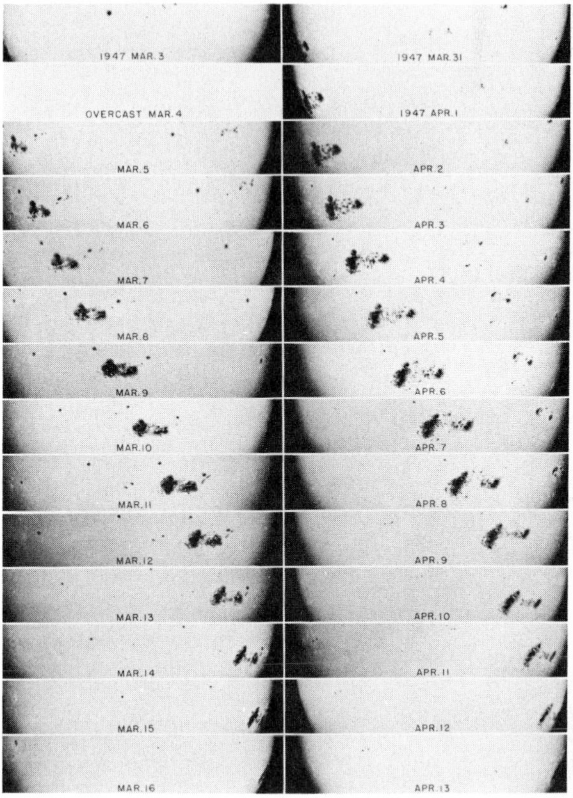

Fig. 24-2. *Sunspots as evidence of the sun's rotation. This group lasted 95 days and made four transits. It attained a very large size and could be seen without a telescope. Note that spots appear, change shape and disappear.* [*Mount Wilson and Palomar Observatories.*]

way of observing the sun with a telescope is to project its image onto a piece of white cardboard held a foot or so behind the eyepiece (with special filters or eyepieces, one can view the sun directly through a telescope). A group of people can view the image on the cardboard at the same time.

Masses of the Sun, Planets, and Stars Newton learned that Kepler's third law should be restated as follows: *The squares of the periods of any two planets, each multiplied by the combined mass of the sun and planet, are proportional to the cubes of their mean distances from the sun.* In fact, any two pairs of mutually revolving bodies may be compared by means of this law. Newton's modification makes only a slight difference when two planets are compared because the mass of the sun is so much larger than that of a planet. It appears in both the numerator and the denominator, and thus it cancels out.

To obtain the mass of the sun, we can use the Earth-moon system as one of the pairs and the sun and any one of the planets—e.g., Mercury—as the other pair. The mass of the planet is so small relative to that of the sun that it can be disregarded. Thus we have the following equation with a single unknown, the mass of the sun (actually the sum of the masses of the sun and Mercury), which is obtained by solving the equation:

$$\frac{(\text{Period of sun-Mercury system})^2(\text{mass of sun} + \text{mass of Mercury}}{(\text{Period of Earth-moon system})^2(\text{mass of Earth} + \text{mass of moon})} = \frac{(\text{mean distance of Mercury from sun})^3}{(\text{mean distance of moon from Earth})^3}$$

In a similar manner the mass of any planet with a satellite can be calculated (the planet-satellite system is substituted for the sun-Mercury system, and the mass of the satellite can be disregarded as was the mass of Mercury).

Likewise the combined masses of a pair of mutually revolving stars (*binaries*) can be found in terms of the sun's mass as the unit. The arithmetic is simplified by comparing the binaries to the sun-Earth system in which the period is 1 year, the distance is 1 astronomical unit, and the sun's mass is taken as 1. If the center of mass of the binary stars can be located, the individual masses of the stars can be determined in the same manner as the mass of the moon was obtained (p. 380).

The Sun's Photosphere and Atmosphere The *photosphere* is mottled by rounded bright areas called granules (Fig. 24-3). Through a telescope the sun's yellowish disk is noticeably darker around its outer margin (limb-darkening). Light from the outer edge of the sun travels to us obliquely through its atmosphere and selective absorption occurs—the shorter wavelengths tend to be absorbed and scattered more readily than the longer wavelengths (sunsets are reddish for a somewhat similar reason). Thus we see deeper and hotter regions of the sun at the center of its disk, and we observe higher-level, cooler, less bright regions near the outer margin.

Sunspots are perhaps the most familiar of solar features (Figs. 24-2 and 24-3). As viewed through a telescope, sunspots appear as relatively small, irregularly shaped dark areas located within 5 to 30 degrees of the sun's equator. The largest of the spots rival Jupiter in size. The temperatures of sunspot centers may be about 4500°C, which is hotter than the surfaces of many stars. However, this is about 1500°C lower than the sun's surface elsewhere; therefore, sunspots appear dark against the brighter, hotter background. The number of sunspots visible on the sun's disk shows a strong periodic variation, with a maximum number of spots occurring about every 11 years on the average. The cause of this cycle is unknown as is the cause of the sunspots themselves. About half of all sunspots exist for a few days or less, but some have lasted for several months.

Prominences are visible during total solar eclipses or in the coronagraph (an instrument that artificially eclipses the sun and allows study of the brighter inner part of its atmosphere at times other than those of total solar eclipses). Prominences consist of great reddish gas clouds that erupt from

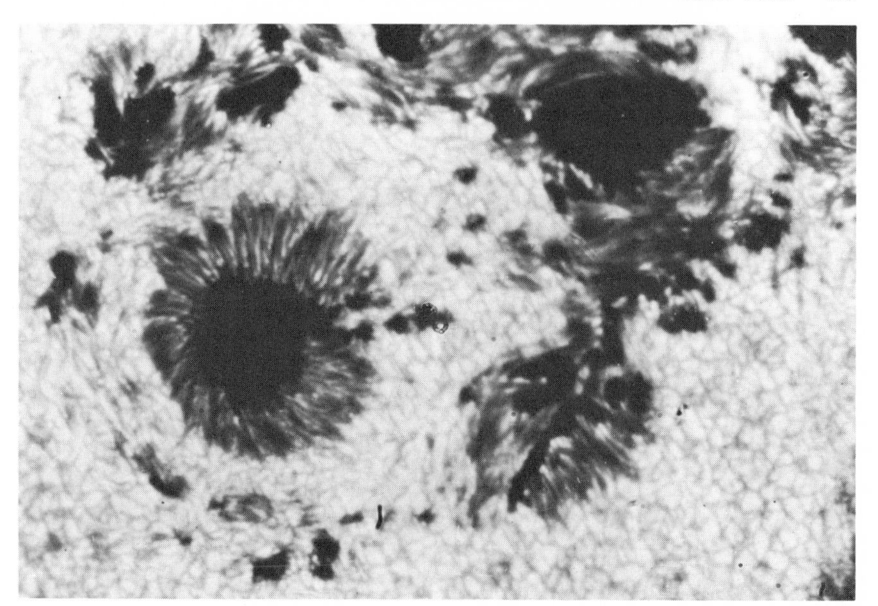

Fig. 24-3. *The sun photographed from an altitude of about 15 miles on August 17, 1957, from an unmanned balloon. The small bright specks may be up to 1000 miles across and exist for only a few minutes. They probably represent rising bubbles of hot gas somewhat like those that form when water boils. Note the radial structure in the less dark, outer portion of the largest sunspot.* [*Courtesy, Perkin-Elmer Corporation.*]

the surface of the sun or stream downward toward it. They are striking features against the white background of the corona, and some extend for several hundred thousand miles. Prominences are widely varied in shape and activity and are puzzling features. Occasionally they seem to form high above the sun's disk, and downward moving material appears to be more abundant than that visibly erupted.

A *solar flare* is a type of solar activity in which an explosive brightening of a small part of the photosphere occurs. Flares are commonly associated with sunspot groups and may extend outward for thousands of miles and last for a few hours, although commonly they exist for a shorter time. They probably affect the Earth by sending outward intense ultraviolet radiations, charged particles, and radio waves.

The *corona* is the exceedingly rarefied but very extensive outer portion of the sun's atmosphere. The inner yellowish part can be viewed or photographed through the coronagraph, but the outer pearly white portion can be seen only at a total solar eclipse (Fig. 24-1). It may extend for millions of miles and merge with the zodiacal light. The shape of the corona varies with the sunspot cycle: a circular outline tends to occur during a sunspot maximum, whereas elongated streamers projecting from the equatorial region are more common at a sunspot minimum.

In addition to the electromagnetic radiation, protons, electrons, and perhaps other charged particles are ejected more or less continuously from the sun. These are particularly intense following a strong solar flare. The charged particles (corpuscular radiation) may cause ionospheric storms which influence radio reception all over the Earth and may last for days. The outward-moving particles take about one day to move from the sun to the Earth and have been described as a *solar wind.* Evidence for its existence includes the auroras, terrestrial magnetic storms, fluctuations in cosmic ray activity, and the behavior of a comet's tail which always points away from the sun. The pressure of sunlight seems inadequate as the agent pushing the tail of a comet outward from the sun, and streams of tiny charged particles presumably exert this pressure. Since comets have been acting as "solar wind socks" for centuries, and in different parts of the solar system, they seem to furnish evidence of the continuous existence of streams of particles that radiate outward in all directions from the sun.

Some Solar-Terrestrial Relations Terrestrial effects of solar activity are known to increase in number and intensity at sunspot maxima. Radio fade-outs may occur throughout the lighted half of the Earth whenever a solar flare occurs near the sun's meridian (electromagnetic waves move vertically outward from the sun at 186,000 mi/sec and strike the Earth). Long-distance radio reception is possible because radio waves are reflected back to the Earth's surface by ionized layers in the Earth's atmosphere (ionization is discussed on p. 272). Ionized layers apparently are created by the action of high-energy solar radiation (ultraviolet and shorter wavelengths) on rarefied gases in the Earth's upper atmosphere. Therefore, the extra amount of this radiation associated with the flare changes the ionized layers sufficiently to produce the radio fade-outs.

Such streams of charged particles from the sun offer at least a partial explanation for the northern and southern lights (Fig. 24-4, aurora borealis and aurora australis). The particles evidently follow lines of force in the Earth's magnetic field (Fig. 10-10) downward through the Earth's atmosphere toward the north and south magnetic poles. As the charged particles pass through the atmosphere, they ionize some of the atoms in their path; these are chiefly nitrogen and oxygen. When the ionized atoms recapture electrons, they give off the varicolored lights observed in the aurora. The magnetic north pole is about at latitude 70 degrees north. Occasionally an aurora can be seen as far away as 40 degrees of latitude from the north magnetic pole. Auroras are most abundant at sunspot maxima.

Origin of Solar (and Stellar) Energy The surface of the sun has an average temperature of nearly 6000°C (11,000°F) and each square inch of the photosphere apparently radiates about 10,000 calories per second; this is equivalent to about 70,000 horsepower per square yard. The total radiation seems to remain about constant and apparently has done so during the past few billion years. Astronomers have developed theories concerning the manner in which the sun and other stars formed and the types of reactions that are the source of their energy.

Stars have formed and seem to be forming today from clouds of interstellar gas and dust in space—from portions of the bright and dark nebulae. Spectroscopic study indicates that these clouds consist chiefly of hydrogen and helium. Gravitational attraction causes a cloud (or a portion of it) to contract, but its mass must be within a certain range for a single star to form. More than this amount of matter results in the formation of multiple stars and of star clusters. Less than this amount of matter probably results in the formation of planets and satellites—central temperatures and pressures never attain the magnitudes necessary to support thermonuclear reactions. Thus planets much larger than Jupiter probably cannot exist.

Fig. 24-4. *The aurora as presented in part of a mural in the Hayden Planetarium. A luminous arch stretches across the northern sky and rays project upward as if giant searchlights were being turned on and off. Other aspects of the aurora resemble luminous curtains or draperies rippling in the sky.* [*Courtesy of the American Museum of Natural History.*]

The source of the sun's energy has long been mysterious, and even now it may be incompletely or imperfectly known. The sun is actually too hot to burn because this requires the formation of molecules, and few molecules can remain intact at such high temperatures. Furthermore, the energy liberated by chemical reactions in combustion is far too small to account for the release of energy from the sun.

Gravitational attraction resulting in contraction was once considered a source of solar energy but was subsequently proved inadequate. Thermonuclear reactions (those involving high temperatures and the nuclei of atoms) are now thought to be the chief source of the energy radiated by the sun and other stars. However, contraction is also important at times, particularly in the initial stages of star formation. Gravitational attraction shrinks a cloud of gas and dust into a sphere and raises the central temperatures and pressures sufficiently for the thermonuclear reactions to begin. The cloud then becomes a star.

Astronomers have concluded that certain types of thermonuclear reactions are most likely to be occurring within the sun and other stars. Each step in these reactions has been reproduced on a small scale in laboratory experiments. The reactions apparently consist in the transformation of hydrogen into helium (the fusion type of thermonuclear reaction). Matter is converted into energy during this transformation ($E = mc^2$—the quantity of energy that results is equal to the amount of matter that is destroyed times the speed of light squared). Thus the amount of matter in the sun and other stars decreases continuously. The sun is estimated to lose 8 billion pounds per second. However, it is also estimated to have enough hydrogen remaining to continue to radiate at its present rate for several billion years.

Astronomers reason somewhat as follows. The interior of the sun must be very hot to keep it from collapsing under the continuous squeezing effect produced by its gravitational attraction. Central temperatures of the order of 10 to 20 million degrees celsius seem necessary to keep the atomic particles moving fast enough to exert an outward counterbalancing pressure. An average sample of this central, high-temperature core may have a density some one hundred times that of water, but it still behaves as a gas. At such extreme temperatures and pressures, many of the nuclei may have lost the electrons that normally occur with them or lose and gain them rapidly. Such matter can have an extraordinarily high density because the nuclei are tiny, and many can be crowded into a small space if their electron shells are missing or incomplete. Yet a nucleus contains nearly all of the mass of an atom.

These thermonuclear reactions presumably can occur only in the central portion of the sun where temperatures and pressures are sufficiently great. One can think of the central core of a star as a gigantic nuclear reactor in which the equivalent of numerous so-called hydrogen bombs are exploding every second. The very powerful self-gravity of a star keeps its core from dispersing violently into space.

Equilibrium exists inside the sun and within most stars between the inward-directed force of gravitational contraction and the outward-directed forces of radiation and gas pressures (more important at lower temperatures). If the sun were to expand, its atomic particles would become separated more widely and fewer nuclear reactions would occur. Thus the outward-directed pressures would lessen, and gravitational contraction would become the dominating force and cause contraction. On the other hand, if gravitational contraction squeezes the central core beyond the equilibrium position, the atomic particles are forced closer together, and the rate of nuclear activity correspondingly speeds up. This increases the outward-directed pressures. Thus relatively slight fluctuations back and forth across the equilibrium position probably occur continuously in stable stars such as the sun.

STARS

Stars are other suns—large hot gaseous bodies shining by their own light. The number of stars visible with the aid of a large telescope is practically limitless; but to the unaided eye, only about 6000 are visible, and an observer will be able to see less than half of these at any one time.

We can speculate that families of planets move about many of these distant suns, with conditions favorable for life. No telescope is powerful enough to prove or disprove this conjecture directly, although current ideas about the origin of stars and of the solar system suggest that planets should be common objects.

In 1963 Peter van de Kamp announced the discovery of a Jupiter-sized planet orbiting a small red star (Barnard's star) which is about 6 light-years from the Earth and invisible to the naked eye. The planet apparently is $1\frac{1}{2}$ times as massive as Jupiter and revolves about Barnard's star once in approximately 24 years at a mean distance of 4 astronomical units. The planet itself has not been seen—it is much too faint. It was detected by the "gravitational wobble" that it produced in the motion of Barnard's star. This deviation from a straight line amounted to a maximum of 1/10,000 of an inch when photographic plates taken decades apart were compared.

Physical Data Concerning Stars Several methods of measuring the distances to stars were discussed in Chapter 19. Stars rotate on their axes, and this shows up as a widening of their spectral lines. Although a spectroscope cannot be aimed first at one side of a star and then at the other as it can for the sun, nevertheless red and violet Doppler shifts occur simultaneously and these broaden the spectral lines. The orientation of a star's axis in space is also a factor in determining the width of its lines; for example, if a star's axis points directly at the Earth, no rotational widening can be observed. In general, the width of a spectral line is proportional to the rate of rotation. The average rotational velocity for white stars may approximate 30 mi/sec, but that of bluish stars tends to be higher, and that of yellow and reddish stars tends to be lower (perhaps unseen planets have absorbed much of their angular momentum).

Stars vary widely in size, density, temperature, and luminosity, but they vary less in mass and chemical composition. Some supergiant stars have average densities that approach the best vacuums that can be produced on the Earth and have diameters that exceed the sun's by several hundred times or more. On the other hand, some stars are as small as Mercury and Mars; but these are amazingly dense objects called white dwarfs. An average cubic inch of the matter in a white dwarf, if it could be moved to the Earth's surface, might weigh a ton or more. Such stars presumably are made of degenerate matter in which atoms have been stripped of their space-consuming electron shells. Thus the mass of a sun can be packed into a volume equal to that of a small planet. However, the great majority of the stars are neither supergiants nor white dwarfs; they are more like the sun in both size and density.

Stars less luminous than the sun are much more numerous in our local region of space than more luminous stars. According to Peter van de Kamp, 55 stars (including the sun) are known to occur within about 16 light-years of the Earth. Only three of these exceed the sun in size. The diameter of Sirius, the largest, is about twice that of the sun, and its luminosity is about 23 times as great. In fact, the luminosity of Sirius exceeds that of the other 54 stars combined. Thus most of the starlight in the universe is probably produced by a relatively small number of very hot massive stars.

Most of the 55 stars are cool enough to have reddish colors. If this is a representative stellar sample, small reddish low-luminosity stars are by far the most numerous in the universe. At greater distances, star counts show higher percentages of the more luminous stars, but astronomers think that this results from our inability to detect the small low-luminosity stars, which are probably very abundant but cannot be observed at such great distances. Many of the 55 nearest stars occur in pairs, and five are white dwarfs.

The colors of stars indicate that their surface temperatures vary from a maximum in the neighborhood of 50,000°C (blue stars) to a minimum of about 2000°C or less (red stars). More stars are closer to the minimum than to the maximum. The radiation from a star whose surface is much less than 2000°C would be too weak to detect visually unless the star were relatively very close. Such stars may exist unknown to us.

The diameter of a star can be calculated indirectly if the temperature and the intrinsic luminosity of the star are known. The temperature of a star determines the amount of light given off by a unit area of its surface; hot stars radiate much more energy per unit area than do cool stars, and a higher proportion of the radiation is emitted as shorter, high-energy wavelengths (p. 484). Therefore, if the temperature and the total amount of light radiated by the star are known, its surface area and its diameter can be determined. Thus a highly luminous red star with a low surface temperature must be very large because a relatively small amount of light is radiated from its surface per unit area. At the other extreme, a hot white star with low intrinsic luminosity must be small because a great deal of light is radiated per unit area.

In comparing other stars with the sun (radius = 1; surface temperature = 6000°K), we find that their luminosities are proportional to the squares of their radii multiplied by the fourth power of their absolute temperatures ($L \propto R^2 \times T^4$). Thus a star that is twice as large as the sun would have four times the surface area. If it is also twice as hot (12,000°K), then each unit area of its surface will radiate 2^4 or 16 times as much light. Therefore, this star is 4×16 or 64 times as luminous as the sun.

If another star has a surface temperature of 18,000°K and is 81 times as luminous as the sun, we can calculate that its radius equals that of the sun ($81 \propto R^2 \times 3^4$).

Stars apparently vary less in mass than in almost any other physical property. Perhaps they are most alike in chemical composition, which is estimated to average nearly three-fourths hydrogen and about one-fourth helium. But these proportions seem to change during the life cycle of a star. Stars are known that have masses more than 50 times that of the sun, but these are exceptional, as are stars that are much less massive than the sun. Stars indeed are rare which are 10 times more or less massive than the sun. However, only the masses of double stars can be calculated directly.

The mass of a star seems to be one of its most fundamental properties and the determining factor in the type of star it is, how much light it radiates, and even apparently how fast it "ages" and how long it "lives." The luminosity of a star is directly related to its mass—i.e., the most massive stars are also the most luminous ones. This is called the mass-luminosity relationship. Perhaps 90 percent or so of all stars conform to this relationship (these are the main-sequence stars, p. 465); the exceptions include the white dwarfs and the low-density giant and supergiant stars.

Multiple and Variable Stars With the aid of the telescope and spectroscope, many, apparently single stars have been shown to consist of two (binary) or more stars that are revolving about a point that represents their mutual center of mass; such stars are relatively close together. In fact, some pairs may be in contact, although others may be many astronomical units apart. True binaries should be distinguished from two stars which appear close only because they are almost on the same line of sight from the Earth.

Novae or "new" stars have been observed a number of times in the history of astronomy. Tycho Brahe's star in Cassiopeia in 1572 is a famous example of an exceptional type called a *supernova.* At its brightest, this star was visible during the day. Apparently such stars were too faint for easy observation before their tremendous, sudden increase in energy output. After a year or so of vastly increased brightness, each of these stars returned to its former inconspicuous place in the universe. It has been suggested that the outer layers of novae become unstable and are blown off explosively.

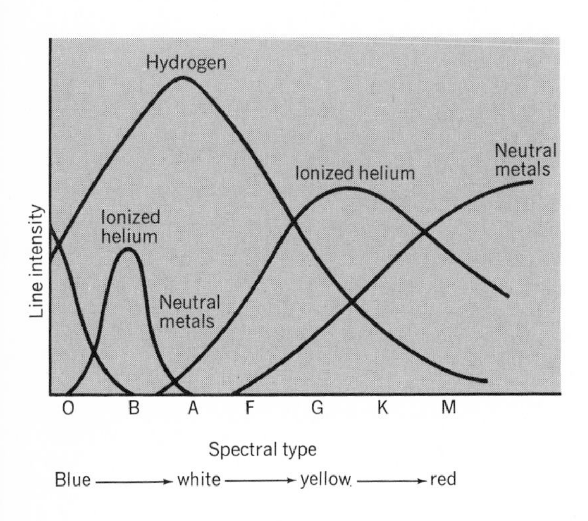

Fig. 24-5. *Temperature differences are the main reason why spectral differences occur among stars. Molecules can be present only in the cooler stars; at high temperatures they dissociate into their constituent atoms. The atoms of a given element will be "excited" enough at a certain temperature to give off light; however, at a higher temperature many of these atoms become ionized and a different set of lines results. Some atoms are excited at lower temperatures than others; thus helium is ionized only at very high temperatures, whereas metals are ionized at lower temperatures. An ionized hydrogen atom cannot produce spectral lines because its single electron has been removed.* [*After Inglis.*]

Stellar Spectra A star may be classified by its spectrum (Fig. 24-5), which depends upon the star's temperature (most important factor), chemical composition, and physical structure. Seven types (O, B, A, F, G, K, and M) include nearly all stars. The following mnemonic sentence keeps these in order: *Oh Be A Fine Girl, Kiss Me!* Members of one type have temperatures, colors, and spectral lines which are transitional to those of neighboring types. The sun is a yellowish G star.

The Hertzsprung-Russell (H-R) Diagram The H-R type of diagram (Fig. 24-6) is constructed by plotting the absolute magnitudes and spectral types of stars on a graph. This can be done for different regions of space,

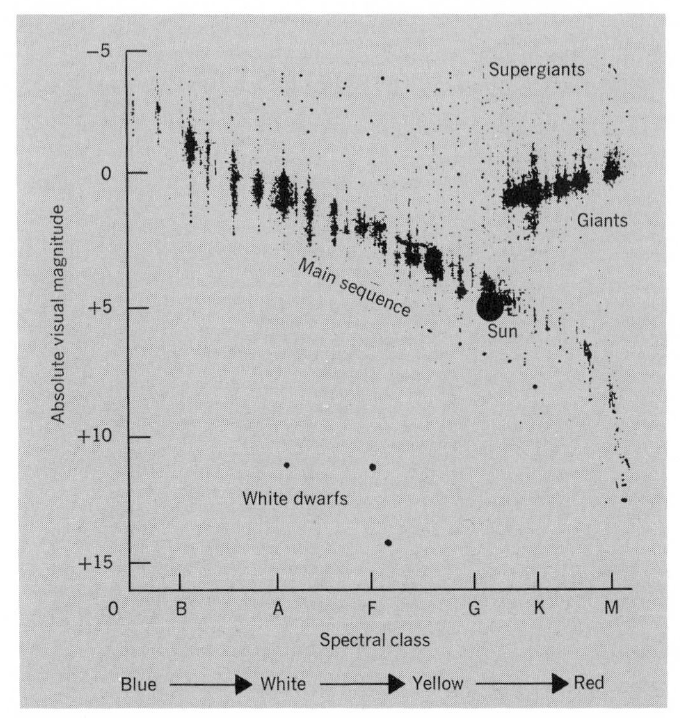

Fig. 24-6. *The Hertzsprung-Russell (H-R) diagram. Stars in our section of the galaxy (part of a spiral arm with Population 1 stars) are plotted on a graph that shows the spectral types along the horizontal axis: type O stars are at the left and type M stars are at the right. Colors and temperatures also change gradually along the horizontal axis from the hottest blue type O stars at the left, through white, yellow, and orange to the coolest red type M stars at the right. The luminosities of the stars are plotted along the vertical axis; the most luminous stars are at the top, and the least luminous ones are at the bottom. The absolute visual luminosity compares the apparent brightnesses the stars would have if they were all located at 10 parsecs or 32.6 light-years from the Earth. Thus the sun is a type G star that would have an apparent magnitude of about +5 if it were located 10 parsecs away. Following the mass-luminosity relationship, the vertical axis also shows the relative masses of the stars on the main sequence: large massive stars are at the upper left, and the stellar masses decrease gradually toward the least massive ones of the lower right. The graph overemphasizes the brighter stars which are the only ones that we can observe at greater distances. In a true sample of all stars in one region of space, stars in the main sequence below the sun would probably be far more abundant than all other stars combined, and white dwarfs would probably far outnumber the giants and supergiants.* [*Modified from a diagram by W. Gyllenberg, Lund Observatory.*]

and thus we have different types of H-R diagrams which reveal certain basic relationships among stars.

The reddish stars on the right side of the diagram are subdivided into two groups, one of much greater luminosity than the other. Since stars of the same color (say type M) emit the same quantity of radiation per unit area, the luminous M stars must be very large in size—giants or supergiants.

Note that most of the stars in Fig. 24-6 cluster along a zone that extends from the upper left downward to the lower right. Large hot bluish massive stars occur at the top and small cool reddish stars of relatively low mass occur at the lower right. These are the end members of a completely gradational sequence. These stars form the *main sequence*—an appropriate term, since 90 percent or so of all stars probably form part of it.

Unusual stars called *white dwarfs* (very small hot stars of low luminosity) occur near the bottom of the diagram. Probably these should be more numerous, but they can be observed only if they are relatively close to the Earth—5 of the 55 nearest stars are white dwarfs. The stars that occur above the main sequence along the upper part of the H-R diagram are appropriately called *giants* and *supergiants*. They are very large.

Life Cycle of a Star The arrangement of stars on H-R diagrams suggests that stars evolve through stages that might be designated as youthful, mature, and old (Fig. 24-6). Thus stars have genetic ages (the present stages in their evolutionary cycles) and chronological ages (the actual number of years since they originated). Similarly the rates at which stars age are also thought to vary widely—from a few million years for large hot massive stars to many billions of years for small reddish stars less massive than the sun.

The study of star clusters has been important in the development of current theories concerning stellar evolution. Star clusters are of two main types. *Galactic clusters* are most abundant along the equatorial plane of our galaxy. They have irregular shapes, and stars do not tend to become more numerous toward their centers. *Globular clusters* (Fig. 24-9) have spherical shapes and stars are more concentrated toward their centers. The members of a star cluster are closer together than average and tend to move as a unit through space. Presumably all of the stars in any one cluster formed at about the same time and from the same cloud of gas and dust; thus they originally had the same chemical composition. Although all stars in any one cluster have the same chronological ages, their genetic ages differ because they are not all equally massive. The more massive stars evolve much more rapidly than the less massive ones.

According to current ideas, a star begins as a huge, cool, diffuse mass of gas and dust whose only heat comes from contraction caused by the mutual gravitational attraction of the entire mass. Temperatures and pressures in the interior of the shrinking mass eventually become high enough for nuclear reactions to begin: at this stage contraction ceases, the mass begins to give off light, and it is said to become a star. An equilibrium is reached between the inward-directed pressure of gravitational attraction and the outward-directed pressure of the hot gases and radiation. When the star reaches equilibrium it is on the main sequence. It began as a cool cloud of gas and dust, and it has shifted about horizontally across the H-R diagram from right to left to reach the main sequence. This contracting youthful stage is relatively quite short; thus few stars are observed in this stage today and few points representing them occur on the H-R diagram.

Stars apparently contract gravitationally at different rates which are determined by their masses. Large massive stars contract rapidly to the main sequence stage, whereas small low-mass stars contract slowly.

Location on the main sequence seems to be determined almost entirely by the mass of the parent cloud of gas and dust: a large mass contracts into a large hot blue star of type O, a medium-sized mass contracts into a yellowish type-G star like the sun, and most abundant of all, a small mass produces a small cool red star of the type M or K. If the mass is too small, contraction cannot raise temperatures to the point at which nuclear reactions can begin; such masses may form planets instead of stars. On the other hand, if the mass is very large, clusters of stars may form.

The mass of a star also determines the rate at which a star consumes its thermonuclear fuel. Large hot bluish stars consume it rapidly (in a million years or so), whereas small cool reddish stars consume it slowly (in several billion years or longer). In the most massive stars, the thermonuclear reactions can occur at very rapid rates and throughout very large parts of their interiors. Furthermore, certain reactions can occur which do not take place at the lower temperatures and pressures existing within smaller stars.

Until a critical portion of the hydrogen in the core has been used up, the star remains in equilibrium and stays in the main sequence. This lasts for a long time for most stars, and thus we find the great majority of stars on the main sequence.

Likely subsequent stages in a star of mass comparable to the sun are shown in Fig. 24-7. When a critical portion of the hydrogen in the core has been transformed into helium, nuclear reactions in the core probably slow down, and it contracts. Central temperatures and pressures are thereby increased, and this initiates nuclear reactions in the shell surrounding the core. As a result the star expands and becomes a red giant. Its outer surface cools as it expands, but its total luminosity increases; thus the star moves upward and toward the right on an H-R diagram. Most of the stars in the

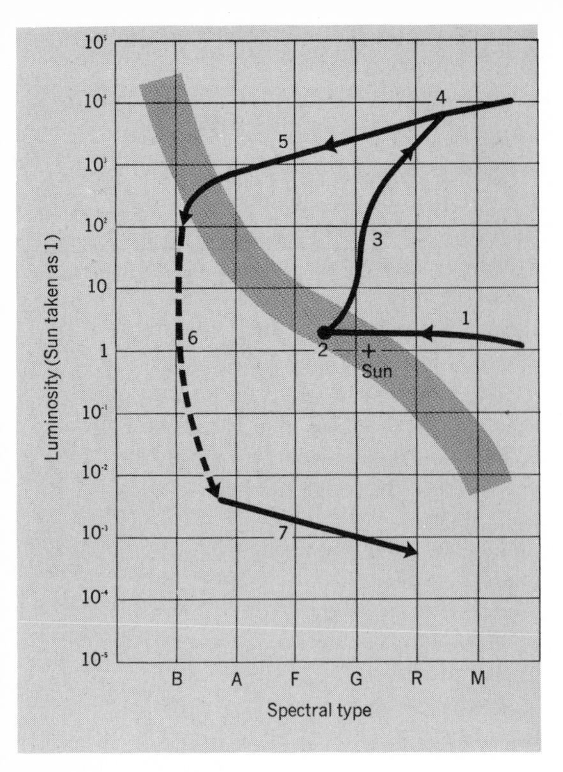

Fig. 24-7. *A possible evolutionary path is traced on an H-R diagram for a star similar to the sun in mass. The vertical scale shows relative luminosities with that of the sun as the unit. Along the horizontal scale are shown the spectral types, colors, and temperatures in the absolute scale (approximately the same as Centigrade in this range). The location of the main sequence is shown by the shaded region. The numbers correspond to stages likely to occur in the evolutionary development of a star of this mass. In stage 1 the star has formed and is in its relatively brief stage of gravitational contraction. It shifts horizontally across the H-R diagram as its temperature increases and its diameter decreases. In stage 2 the star has reached the main sequence and is in equilibrium. Here it spends a long portion of its life cycle. In stage 3 the star is expanding and in stage 4 it is a red giant. The following steps are less certain, but somehow the star probably becomes a white dwarf (stage 7). In Stage 5 it may pulsate somewhat as a Cepheid star, and in stage 6 it may eject matter explosively as a nova. [After Su-Shu Huang,* Scientific American, *April 1960.*]

giant and supergiant stages on an H-R diagram were probably once main-sequence stars.

Following the red giant stage, the star probably changes into a white dwarf, but just how this is accomplished is still quite uncertain. The star may go through an unstable period during which it pulsates as a variable star and an explosively eruptive period during which it ejects matter into space (nova or supernova stage). After the star becomes a white dwarf, it presumably must cool slowly, with corresponding color changes, into a "blackened cinder."

Thus when H-R diagrams for different clusters are compared, any differences should depend primarily upon age differences. Successively larger portions of the upper part of the main sequence are missing in older clusters. Stars that once were present in the upper main sequences of an old cluster are probably represented now by dots for different kinds of stars elsewhere on the diagram. Stars that evolved fastest are now white dwarfs. Stars that evolved less rapidly are now red giants and supergiants. Stars on the lower half of the main sequence evolve so slowly that they have yet to leave it.

Two broad groupings of stars have been recognized: *Population I* and *Population II*. Stars of Population I characteristically occur in the spiral arms of galaxies and in regions where interstellar dust clouds are abundant. The sun belongs to this group. The brightest stars of Population I are the blue supergiants. Cepheids of this group are four times as luminous (or even more) as those in Population II. The stars of Population II are located in regions of space where interstellar gas and dust are absent: in the central nuclei of spiral galaxies, in elliptical galaxies, and in globular clusters. Red supergiants are the brightest stars of Population II. The heavier elements seem to be least abundant in the oldest Population II groups and to be most abundant in the younger Population I stars.

Interstellar Matter This is the parent material of stars and in some regions it acts as a "cosmic smog" by obscuring or altering the light of remote stars and galaxies behind it. However, the interstellar matter is so diffused that even the denser portions constitute better vacuums than man can produce on Earth. The gas is chiefly hydrogen and is estimated to have a total mass exceeding that of the dust (of unknown composition) by 100 or so times.

The bright and dark nebulae are more concentrated portions of the interstellar matter (Fig. 17-3), and the term *nebula* refers to such interstellar clouds of gas and dust. Exceedingly rarefied interstellar matter is also diffused throughout very large regions of space that enclose the bright and dark nebulae. This material is so sparse that it cannot be observed directly. The gas can be detected by the dark absorption lines that it produces in the spectra of distant stars. The dust reddens the light of distant objects by selective scattering. In addition, 21-cm radio waves from neutral hydrogen have been detected by radio telescopes.

GALAXIES AND THE UNIVERSE

Stars are not scattered uniformly through space. The known universe or cosmos consists of vast aggregates of stars called *galaxies*. Stars and star clusters are the main units within galaxies; similarly, galaxies and clusters of galaxies are the main units in the universe.

Our solar system is located in a galaxy whose equatorial plane is marked by the Milky Way (Fig. 24-8), that faint band of light which crosses the sky in the neighborhood of constellations such as Auriga, Perseus, Cassiopeia, Cygnus, Sagittarius, and Scorpio. The light of the Milky Way is produced by the combined radiations of millions of stars, each too faint to be visible to the naked eye. All individual stars seen without optical assistance are part of our galaxy, and most of them are relatively quite near us.

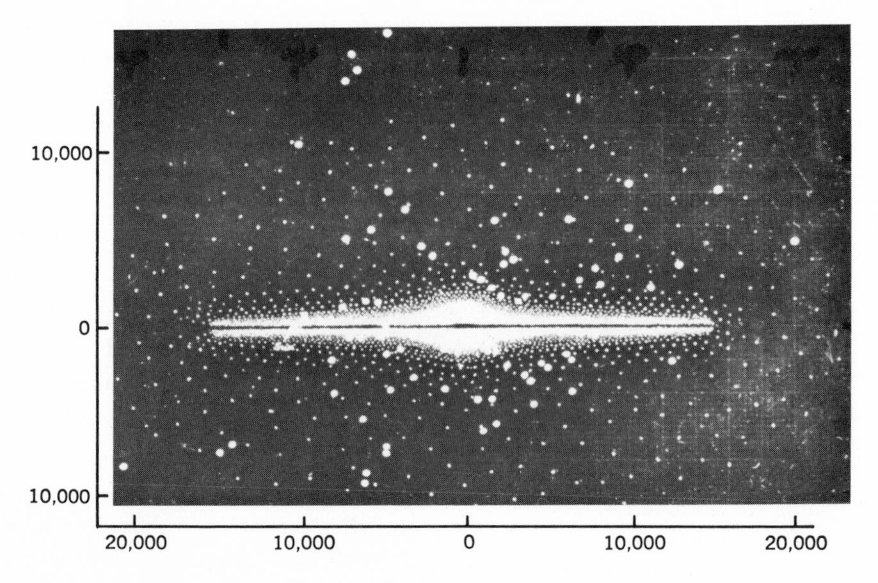

Fig. 24-8. *Schematic edgewise sketch of our galaxy based upon data available at present. Most of the billions of stars in the galaxy are probably grouped along its equatorial plane, although some outlying stars are scattered around the main structure. The stars are most abundant in the spiral arms (not shown in this side view) and central disk. Dark obscuring matter is shown concentrated along the equatorial plane (Milky Way) but is not a continuous uniform sheet. Globular clusters (large white spots) form a nearly spherical halo around the galaxy; they appear to increase in numbers toward the center, and few are more distant than the Earth from the center. Estimated distances are shown along the margins; the unit is the parsec, and 1 parsec = 3.26 light-years.* [*Yerkes Observatory*.]

To determine the shape of our galaxy, star counts may be made at progressively greater distances in various directions, and comparison may be made with what appear to be similar galaxies located far away in space.

If a telescope is pointed toward the Milky Way and a series of photographs are taken with longer and longer time exposures (a way of increasing the distance which can be observed each time), stars continue to be abundant in this direction. In fact, stellar density appears to be greatest in the general vicinity of Sagittarius, although widespread dark nebulae in this region reduce the amount of light that we can observe. However, at right angles to the Milky Way, stars become progressively less numerous at approximately equal distances on either side. Thus, our galaxy is a flattened structure. Within it, stars are more numerous in certain areas (e.g., in the spiral arms and central disk) than in others. It forms a sort of enormous pinwheel with arms spiraling outward from the center.

Our galaxy may be about 100,000 light-years across, and its central disk may have a maximum thickness of 10,000 light-years. One cannot see through the center to the other side because concentrations of interstellar gas and dust occur along the Milky Way and obscure vision. Radio astronomy is not hampered by this cosmic smog and can be used to study the farther reaches of our galaxy. The solar system appears to be located near the equatorial plane within a spiral arm, and about 30,000 light-years outward from the center.

Stars in the equatorial plane of our galaxy seem to revolve about the galactic nucleus in this plane and to maintain a nearly constant distance from the center. Stars nearer the nucleus revolve more rapidly than those out near the edge. Relative to the center of the galaxy, the sun appears to have a velocity of about 100 to 200 mi/sec. At this rate, it completes a revolution in about 200 million years.

Fig. 24-9. *Globular cluster M 13 in Hercules. It has a diameter of about 150 light-years and is some 30,000 light-years away, but these figures are only approximations. The number of stars in it probably exceeds 100,000 (perhaps five times as many) but individual stars cannot be distinguished in the central region.* [*Mount Wilson and Palomar Observatories.*]

Globular clusters (Fig. 24-9), individual stars, and widely dispersed hydrogen gas occur in a spherical region centered on the galactic nucleus. About 100 globular clusters have been discovered forming a spherical "halo" around the central disk of our galaxy, and globular clusters occur also with other spiral galaxies and with some elliptical galaxies. On the average, the stars within a globular cluster are probably less than 1 light-year from their neighbors, but collisions are probably rare. Apparently the globular clusters and other halo objects revolve in highly elongated

Fig. 24-10. *The great spiral galaxy in Andromeda (Messier 31). Two elliptical galaxies can be seen in the photograph: one is directly above the nucleus; the other is to the left of the nucleus. The equatorial plane of the galaxy is inclined about 13 degrees from an edgewise orientation. The galaxy is nearly circular in outline but appears elliptical because of the inclination. The foreground stars are part of our galaxy. Note the bright central disk, spiral arms, and dark nebulae.* [*Mount Wilson and Palomar Observatories.*]

elliptical orbits about the galactic nucleus; the orbits tend to have a high inclination to the equatorial plane and to be elongated ellipses.

The Andromeda galaxy (Fig. 24-10) is visible to the unaided eye on a clear dark night as a faint patch of light located in the direction of the constellation Andromeda. If all of the galaxy were visible to the unaided eye, it would cover about 4 degrees of space. To observe it, we must look through the stars which are part of our galaxy far into space—perhaps more than 2 million light-years away. It is the most remote object visible to the unaided eye. In it astronomers have been able to discover Cepheid variables globular clusters, novae, stars of different types, and interstellar matter. In other words, it seems to be a large spiral galaxy similar to our own.

Other Galaxies Galaxies appear to be distributed throughout the observable universe with no diminution in their numbers outward to the limits of observation. On the largest of scales, this distribution appears to be approximately uniform. On a somewhat smaller scale, galaxies have an irregular distribution and occur in pairs and clusters that range in numbers from several to more than a thousand (perhaps there are also clusters of clusters). Our galaxy is a member of a cluster called the Local Group (Fig. 24-11).

Galaxies have widely different shapes and sizes, but they can be classified into three main types: *spiral, elliptical,* and *irregular*. The relative number of each type is uncertain, and the problem is similar to that of determining which types of stars are most numerous. Rather small elliptical galaxies seem to be most abundant in our neighborhood of space and probably are

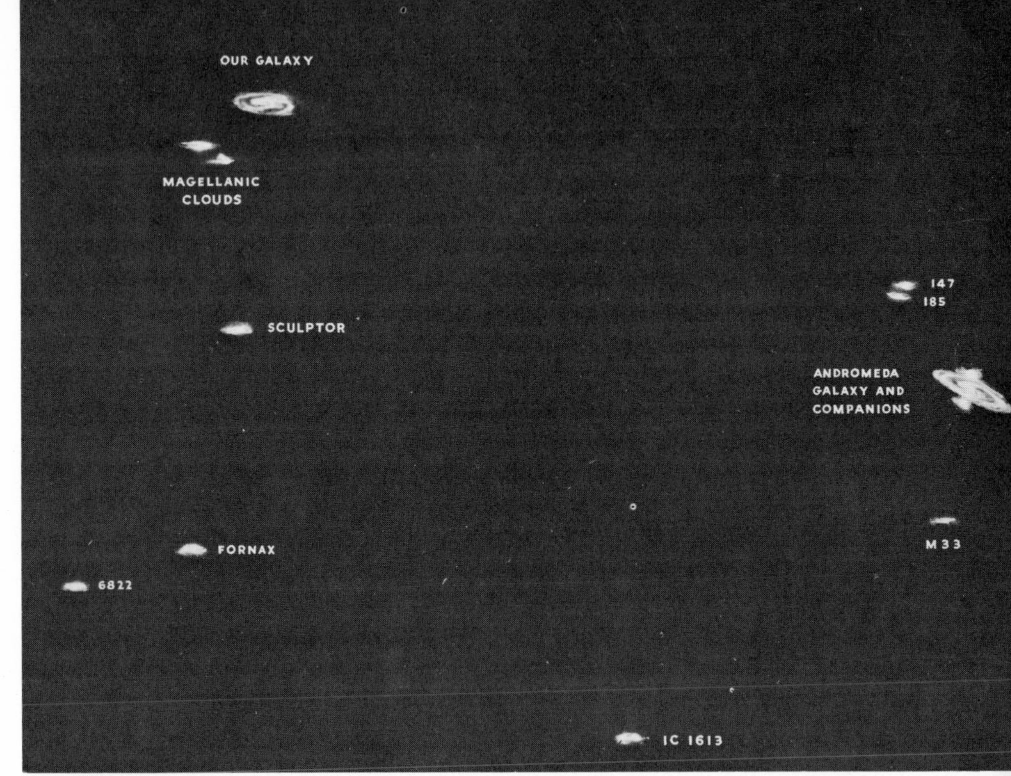

Fig. 24-11. *A model of part of the Local Group of galaxies. Known members total 17: 3 spirals, 4 irregular, and 10 elliptical (including 6 so-called dwarfs). The distance between our galaxy and the Andromeda galaxy is 1½ to 2 million light-years or more. The Andromeda galaxy is apparently the largest of the Group and exceeds 100,000 light-years in diameter. Our diameter appears to be a close second in size and a twin in shape. The larger Magellanic Cloud, diameter about 30,000 light-years, appears to be the fourth largest in the group; it may represent about an average size for galaxies in the universe.* [*Courtesy of the American Museum of Natural History.*]

the most common type of galaxy in the universe, but they are too faint to be detected at great distances.

The spiral galaxies are subdivided into two main types: normal and barred. The members of each type also form a gradational series: some spirals have large central disks and short closely coiled arms, whereas others have small central disks and large loosely coiled arms. All gradations occur between these extremes. Commonly two main arms emerge from opposite sides of a central disk and coil around it in similarly shaped curves, but more than two arms have been observed and branches may extend from the two main arms. The barred spirals are less abundant; their arms project from the ends of a bar that extends through the central disk (Fig. 24-12). The barred spirals also show gradations from tightly coiled arms to widely open arms, and the bar and central disk appear to turn in space as a unit. The cause of the bar is unknown.

Elliptical galaxies contain Population II stars (they lack interstellar ma-

Fig. 24-12. *A barred spiral galaxy (NGC 1300 in Eridanus). Photographed with the 200-inch Hale telescope. This is an intermediate type: the arms are neither tightly coiled nor widely open.* [*Mount Wilson and Palomar Observatories.*]

terial) and range in shape from spherical to flattened disks. They have no spiral arms and appear symmetrical; stars are more concentrated toward their centers. Elliptical galaxies range widely in size. The largest exceed the giant spirals in size, mass, and luminosity, but the smallest measure only several thousand light-years in diameter, considerably smaller than the smallest spirals.

Irregular galaxies appear to be the least numerous of the three types and are unsymmetrical objects that show no definite structures such as central disks and spiral arms. Our two nearest galactic neighbors, the Magellanic Clouds, may belong to this group.

Evolution of Galaxies The gradational nature of spiral and elliptical galaxies suggests that galaxies as well as stars may have life cycles, but other factors must also be considered such as mass, angular momentum, and distribution of Population I and II groups of stars. There is much to be learned about the evolution of galaxies—if indeed they do evolve.

The Expanding Universe The dark lines in the spectra of most galaxies show a shift toward the red end of the spectrum, and the amount of the red shift is roughly proportional to the distance of the galaxy from the Earth:

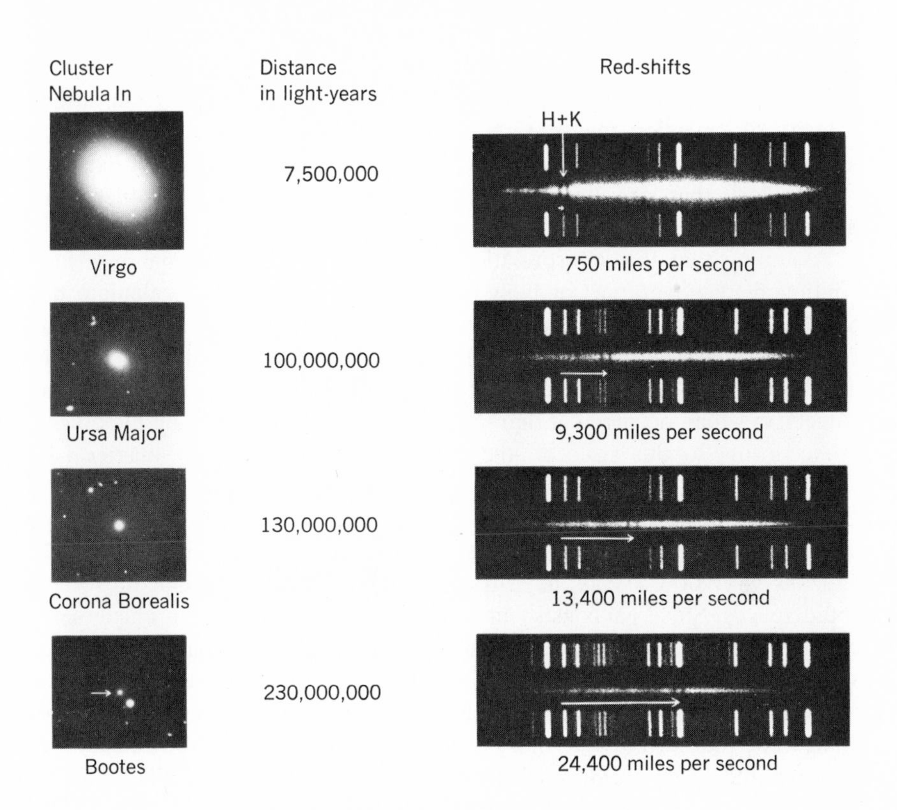

Red-shifts are expressed as velocities, c d λ/λ.
Arrows indicate shift for calcium lines H and K
One light-year equals about 6 trillion miles, or
6×10^{12} miles

Fig. 24-13. *Red shifts in the spectra of galaxies. The records and photographs of four galaxies are shown and their provisionally estimated distances from the Earth are given. The more distant galaxies appear smaller in the photographs but may actually be as large or larger than the closer ones. The spectrum of each galaxy has a comparison spectrum above and below it. The arrows show the shifts of the H and K lines of calcium for each galaxy, but these H and K lines do not appear in the comparison spectra. However, their precise wavelengths are known, as are the wavelengths of the lines that do appear in the comparison spectra. Therefore, astronomers know where the lines would be located if no Doppler red shift occurred. The spectrum of a galaxy is a composite one that shows the lines of its most prominent stars.* [*Mount Wilson and Palomar Observatories.*]

a nearer galaxy shows a small red shift, a moderately distant galaxy has a medium red shift, and a remote galaxy shows a large red shift (Fig. 24-13). This relationship is known as the *law of the galactic red shifts.* It is tentatively interpreted as a Doppler effect which suggests that each galaxy may actually be moving away from the Earth with the magnitude of the shift directly proportional to the rate of recession. A large red shift is thought to show rates measured in thousands of miles per second. In fact, the largest velocity of recession measured to date (excluding quasars, p. 477) may approximate about 46 percent of the speed of light, and the object itself is a radio galaxy that may be about 5 billion light-years away; but these data must be considered extremely tentative.

Assuming that this interpretation is correct and that the galaxies have continued to move apart at their present rates, astronomers calculate that all of the matter in the present galaxies may have begun to move outward from the same general location in space some 10 to 13 billion or so years ago. According to one interpretation, this points to a unique event or cataclysm which occurred at that time in the past—the Big Bang hypothesis.

According to this concept, the material from which the galaxies later developed probably began to move outward at the same time but in different directions and at different rates. Since objects in space should continue to move at uniform speeds in straight lines in the absence of other forces, galaxies with slow speeds have traveled shorter distances than those with rapid speeds. Since galaxies seem to be distributed in space in clusters, the expansion is probably between clusters; the distances separating galaxies within a cluster are probably not increasing.

According to one hypothesis, a Big Squeeze may have preceded the Big Bang. The mutual gravitational attraction of material in space caused it to contract and collapse into a relatively small region of space. In still another modification, a pulsating or oscillating universe is imagined—an endless alternation of contractions and expansions. According to the competing, Steady State universe hypothesis, the universe had no beginning, will have no end, and is uniform in time and space. As expansion occurs, new matter (hydrogen) is created in intergalactic space to maintain a uniform density. New stars and galaxies form and evolve from this new matter and replace those that have expanded outward. Thus individual stars and galaxies change, but the over-all nature of the universe remains uniform. The speculative nature of such ideas should be obvious.

Although the Doppler interpretation of galactic red shifts suggests that most galaxies are moving away from us, the position of our galaxy may not be unique. If a balloon (universe) covered by uniformly spaced dots (galaxies) were inflated, not only would each dot move away from its neighbors, but from any one dot all other dots would appear to be receding. Similarly all galaxies may be moving away from each other.

Observations of certain kinds may eventually make it possible to discard some of these hypotheses. For example, according to the Big Bang account, galaxies must have been closer together billions of years ago. On the other hand, the Steady State hypothesis postulates a universe that is uniform in time and space. Perhaps this discrepancy can be checked by comparing remote regions of space with those that are less remote. We see such regions as they were at some long ago time in the past, and we see the most distant regions as they were longest ago. Radio telescopes are particularly useful in this observational test because they can be used to detect very remote objects. Results to date are probably best regarded as inconclusive.

There is some recent strong evidence against the Steady State view. Cosmic black body radiation surviving from the "primeval fireball" of the Big Bang had been predicted before its apparent discovery in 1965. According to theory, the initial very high-energy gamma radiation produced at the time of the Big Bang should gradually change into lower-energy longer wavelengths during the ensuing expansion. The radiation, billions of years later, should now occur in the radio and microwave regions of the electromagnetic spectrum. Detection is difficult since the radiations are much weaker than other sources of microwave radiation and seem to come toward Earth uniformly from all directions (isotropic). Detection and measurements have now been made at several wavelengths which seem to fit the spectral energy curve produced by black body radiation at about 3°K. In addition, a comparison of remote and less remote regions of the universe supports a lack of uniformity; e.g., more radio signals originate per unit volume of space in very distant regions.

Quasars *Quasi-stellar objects* or *quasars* became a major problem (and opportunity) for astronomers in the 1960's. They can be described as star-like objects of unknown nature, located at uncertain distances, with very large red shifts. Some emit strong radio signals. Relatively rapid light fluctuations have been observed from some quasars and seem to indicate a maximum size of a few light-days—far smaller than galaxies.

If the red shifts are interpreted as resulting from the expansion of the universe, then the most remote of these objects may be about 9 billion light-years away and receding from us at about 80 percent of the speed of light. If the object is relatively small and as remote as this, then thermonuclear reactions will not account for the enormous energy it emits.

On the other hand, quasars may be relatively nearby objects. According to one view, they were ejected at very high speeds from the nucleus of our galaxy some millions of years ago when gravitational collapse occurred there. However, no quasars with blue shifts have been detected; and such should occur if quasars had also been ejected from neighboring galaxies. Thus the quasars' distances must be determined before their role can be elucidated.

ORIGIN OF THE SOLAR SYSTEM

The regularity and uniformity exhibited by planetary motions indicate that the planets originated at more or less the same time and in the same general manner. All hypotheses involve the rearrangement of matter and energy previously in existence, and in many hypotheses the material of the planets was once part of the sun, of material which subsequently formed the sun, or of a companion star of the sun.

Stumbling blocks for most hypotheses have been these: (1) hot gases should disperse quickly into space if they are pulled out of a sun to form planets, and (2) the sun should have most of the angular momentum of the solar system, but it has only about 2 percent. The *angular momentum* of a planet (or other object) revolving in space about a central mass equals the mass of the planet multiplied by its velocity and by the square of its distance from the center. The *principle of conservation of angular momentum* states that in an isolated system in space, the total angular momentum remains constant, although it can be shifted from one part of the system to another. A skater whirls faster by bringing her arms closer to her body and illustrates this principle in operation.

The *protoplanet hypothesis* explains many of the phenomena of the solar system, and we discuss some aspects of it briefly. The starting point is a gaseous disk, perhaps one-tenth as massive as the present sun. This surrounds a large dark central mass that subsequently becomes the sun. As the nebula contracts and flattens, it becomes unstable and divides into a number of separate clouds or protoplanets. Solid particles accumulate into a central core in each protoplanet and are surrounded by a large gaseous envelope. The composition of the nebula is similar to that of the sun: chiefly hydrogen, with some helium, and 1 to 2 percent of heavier elements. At this stage, the protoplanet from which the Earth formed may have been 1000 times more massive than the present Earth. The protoplanets were of somewhat different sizes, but all were far larger and more massive than the present planets.

The satellites may have formed in a similar manner as the protoplanets contracted, but they were relatively much closer to their parent bodies. Their rotation was slowed by tidal friction until they rotated and revolved at the same rate and in the same direction. They remained spherical and did not subdivide further.

The counterclockwise rotation of the original nebula accounts for the counterclockwise revolution of all of the planets. Their common rotation in the counterclockwise direction may have been produced by the tidal attraction of the sun on the protoplanets. This stretched them into elongated shapes and kept their long axes always pointed toward the sun. Rotation was thus in the same direction as revolution and took the same amount of time.

By now the central mass had contracted enough to become a star. As the sun's temperature rose, its radiations and ejected particles ionized the gases in the nebula around it. These gases interacted with the sun's magnetic lines of force to slow its rotation; most of the sun's angular momentum was transferred to the particles in the nebula, and they moved faster as a result of the transfer.

Only a small fraction of the original nebula remains as part of the present planets. The planets nearest the sun lost higher proportions of their protoplanets, and thus they now have smaller masses, greater densities, and larger proportions of heavier elements. As a protoplanet shrank in size and mass, it rotated more rapidly to conserve its angular momentum.

As the masses of the protoplanets decreased, their gravitational attractions became less, and their satellites were able to revolve at greater distances; some eventually escaped. A number of these may eventually have been recaptured, which could account for the orbital irregularities of certain outer satellites. Pluto may have escaped from Neptune in this manner.

CONCLUSIONS AND CAUTIONS

One is quite likely to end a survey of astronomy with mixed emotions. On the one hand, we are amazed at man's ingenuity in learning all that he has about the universe. On the other hand, the inconspicuous role which has been assigned to the Earth, and the fundamental unanswered questions of the origin of the universe, its nature, and its ultimate destiny, fill us with humility, wonder, and awe.

The dynamic and hypothetical nature of many astronomical concepts should be kept conspicuously in mind at all times. In all of the sciences, and in all periods, there have been ardent advocates of concepts which later had to be abandoned as new information became available. Astronomical data, concerned as they are with the structure, composition, life cycle, and motions of very remote celestial objects must necessarily remain tenuous and subject to constant check and revision. Hypotheses and theories in science lay down directions for research, and thus they serve a useful function even when they have to be entirely abandoned soon after they are propounded. But the hypothetical nature of some of these concepts should be given more emphasis than it often gets.

Exercises and Questions, Chapter 24

1. Explain how an astronomer might obtain evidence that a nearby star has a planet revolving about it. Calculate a scale model to illustrate the difficulties of detecting such a planet optically.
2. Why does a so-called critical size (mass) exist between stars and planets?
3. How does an astronomer estimate each of the following for a star?

(a) Surface temperature.
(b) Rate of rotation.
(c) Chemical composition.
(d) Mass.
(e) Distance (distinguish between near and remote stars).
(f) If its distance from the Earth is changing.

4. In what physical properties do stars apparently vary the most? the least?

5. A certain star is compared to the sun and found to have a surface temperature of 3000°K but to be four times as luminous. What is the radius of this star in terms of the sun's radius as 1? The surface temperature of another star is 12,000°K, and it is 256 times as luminous. How does its radius compare with that of the sun?

6. What kinds of stars are most common in the universe? Discuss pertinent data and problems.

7. Why is the mass of a star considered to be one of its most fundamental properties?

8. Describe in general terms some of the differences that occur among stellar spectra. Why do stars have different spectra?

9. Why are studies of clusters of stars very important in a consideration of stellar evolution?

10. Describe the stages that are thought to occur in the life cycle of a star that has a mass about equal to that of the sun. Which stages are least well known? Which stages may last the longest?

11. Discuss some of the problems that are involved in a determination of the shape of the galaxy in which we are located. Describe its shape, size, and general nature.

12. What evidence supports the concept of an expanding universe? Just what is thought to be expanding? What rates are involved?

13. Describe some of the major similarities and differences between the Steady State and Big Bang hypotheses.

14. What observations might be made that could lead to the elimination of either the Steady State or Big Bang hypotheses? If one hypothesis becomes untenable, is the other necessarily proved to be valid?

15. How have changes during the last few decades in the estimates of the distances to galaxies affected estimates concerning the age of the universe according to the Big Bang hypothesis?

25 Weather: Air, heat, and water

BENJAMIN FRANKLIN once said, "Some people are weatherwise, but most are otherwise." There is much talk about the weather, and much adaptation to changes in it; yet to many of us weather changes continue to seem highly mysterious. We can probably never hope to really control the weather. But we can hope to understand it more fully, to be able to predict it more accurately and over longer periods of time, and to take advantage of its good points while minimizing the more unfavorable aspects. Some degree of weather modification has already been achieved.

Meteorology is the science of the atmosphere and of weather phenomena—rain, snow, hurricane, tornado, frost, dew, hail, clouds, fog, wind, and sunshine. Understanding them depends upon a knowledge of air, heat, and water as they function on a rotating, revolving, sun-warmed Earth. Let us consider first the atmosphere.

THE ATMOSPHERE

The atmosphere is the sea of air which envelops the Earth; it has no distinct outer boundary and becomes exceedingly diffused in its outermost parts some thousands of miles above the Earth's surface, where it merges with the even more diffused medium of interplanetary space. This gaseous shell is held to the Earth by gravitational attraction. In it man lives and breathes. Because of it rain falls, vegetation grows, and rocks decay. The importance of the atmosphere to man cannot be exaggerated.

Air consists of a mixture of gases whose nature, density, and physical state change with altitude (Fig. 25-1). An average sample of pure dry air is colorless and odorless and is made up chiefly of molecules of nitrogen (about 78 percent by volume), oxygen (about 21 percent), and argon (nearly 1 percent). These proportions appear to remain about uniform throughout the well-mixed lowermost 70 to 80 miles above the Earth's surface. Overlying this zone are shells composed chiefly of atomic oxygen, helium, and hydrogen. Thus the atmosphere as a whole has a density stratification.

Water vapor (up to 4 or 5 percent by volume), carbon dioxide, and dust particles are present in small and varying amounts in the first 5 to 10 miles or so of the atmosphere (in the troposphere). These three substances are very important in causing weather phenomena and in the heating of the atmosphere.

Atmospheric pressure is produced by the mass of all of the air in the atmosphere being pressed down upon the Earth's surface by the gravitational pull of the Earth. The lower part is greatly compressed by this pressure and contains most of the mass of the atmosphere (Fig. 25-2). For example, 1 cubic yard of air at sea-level pressure may contain about the same quantity of matter as 1 cubic mile of the very diffused air that is located at an altitude of 100 to 200 miles.

Man is not crushed by the atmosphere because the pressure inside his body is equal to the pressure outside. Atmospheric pressure at sea level

Fig. 25-1. *A cross section of the Earth showing the chemical composition of the atmosphere. The sketch is misleading quantitatively because the atmosphere thins out rapidly and is exceedingly tenuous at high altitudes.*

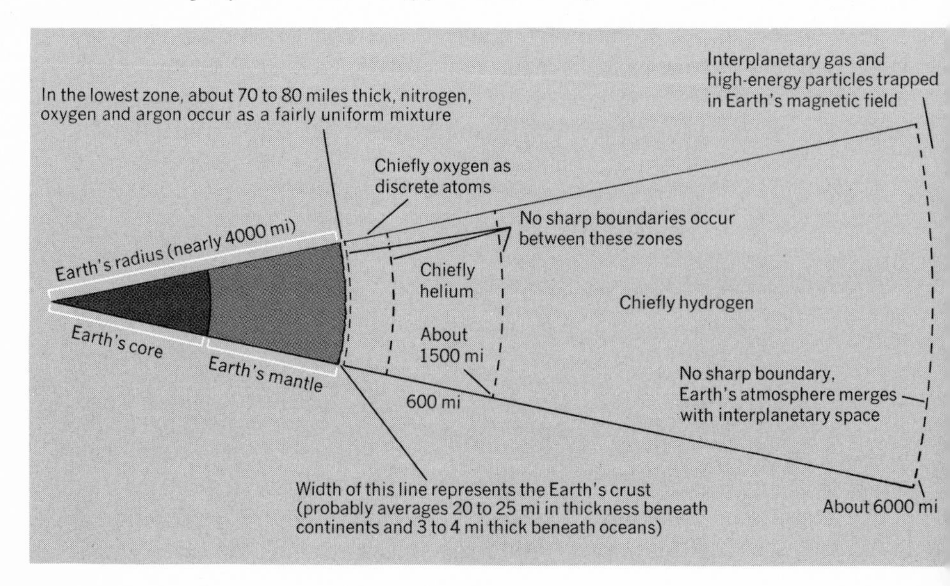

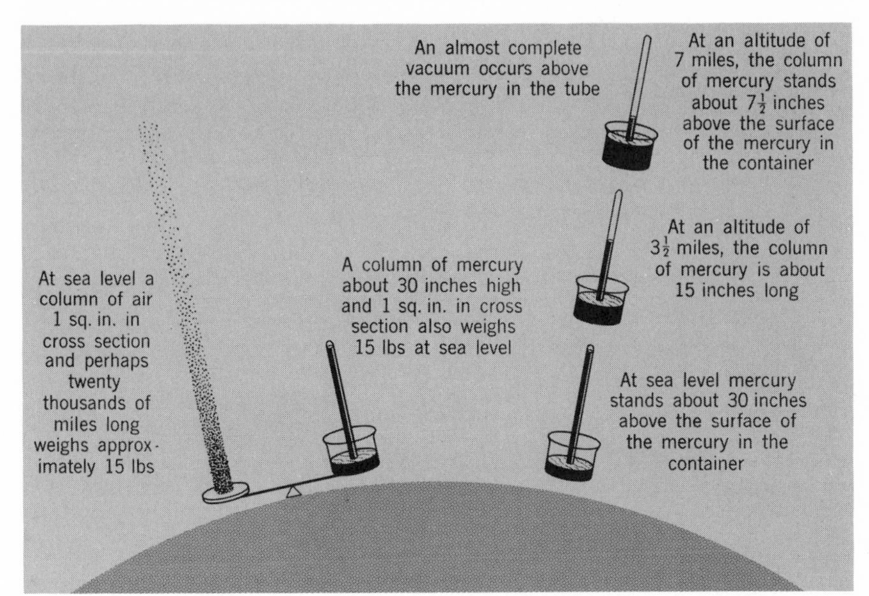

Fig. 25-2. *Atmospheric pressure. About 50 percent of the total mass of the atmosphere is estimated to occur in the lowermost 3½ miles, and about 75 percent may occur in the lowermost 7 miles. Probably 90, 99, and 99.9 percent occur in the lowermost 10, 20, and 30 miles respectively.*

is nearly 15 lb/sq in., or more than 1 ton/sq ft. Atmospheric pressure may be measured by means of a mercurial barometer or an aneroid barometer, an instrument which involves a partially evacuated, airtight metal can, kept from collapsing by a spring inside. When atmospheric pressure increases, the top of the can is depressed, and an attached pointer rises.

The barometer is used to weigh that portion of the atmosphere which occurs above it and is a very important weather instrument. It functions about as well inside a building as outside. Changes in weather are associated frequently with changes in barometric pressure, but no one weather element can be used by itself to make accurate forecasts.

Heating of the Atmosphere The temperature of an object is related directly to the rate of movement or vibration of the atoms and molecules which compose it (p. 70). If the temperature of a body rises, its molecules move more rapidly; if the temperature lowers, the molecular velocity diminishes. In the language of molecules, one desires a drink of slowly moving water molecules on a hot day and longs for rapidly moving air molecules during cold winter months.

The temperature of an object determines the type and quantity of radiation it emits. Any object above the temperature of absolute zero (−459°F, the temperature at which molecular movement ceases) gives off a certain amount of electromagnetic radiation, but our eyes cannot detect the long wavelengths emitted at low temperatures. More radiation of all wavelengths is emitted at higher temperatures, but the proportion of shorter wavelengths increases rapidly. According to Wien's law, *the wavelength in which the maximum amount of energy is radiated from a black body varies inversely as its absolute temperature* (a black body can be defined as one which emits the maximum amount of radiation possible in each wavelength at the existing temperature). *The rate at which energy of all wavelengths is radiated by a black body is proportional to the fourth power of the absolute temperature* (the Stefan-Boltzmann law). Thus an increase in the absolute temperature of 2, 3, or 4 times, means an increase in the total energy radiated of 8, 81, and 256 times respectively.

At the temperatures that exist at the sun's surface, about 41 percent of the energy radiated by the sun is in the form of visible light, about 9 percent consists of invisible ultraviolet and shorter wavelengths, and about 50 percent consists of invisible infrared and longer wavelengths. The maxi-

Fig. 25-3. *Spectral energy curves for a perfect radiator. The intensity of the radiation is shown along the vertical axis and the wavelength along the horizontal axis (1 angstrom is 10^{-8} cm, a very short distance); thus red waves occur at the right and violet ones at the left in the visible region of the spectrum (shaded area). The amount of space beneath any one of the three curves is proportional to the total radiation per unit area at the temperature chosen. Thus the total quantity of radiation and the proportion of shorter wavelengths both increase as the temperature increases. The curves have been calculated for objects that are perfect radiators and are based in part upon laboratory experiments.* [*Baker,* Astronomy.]

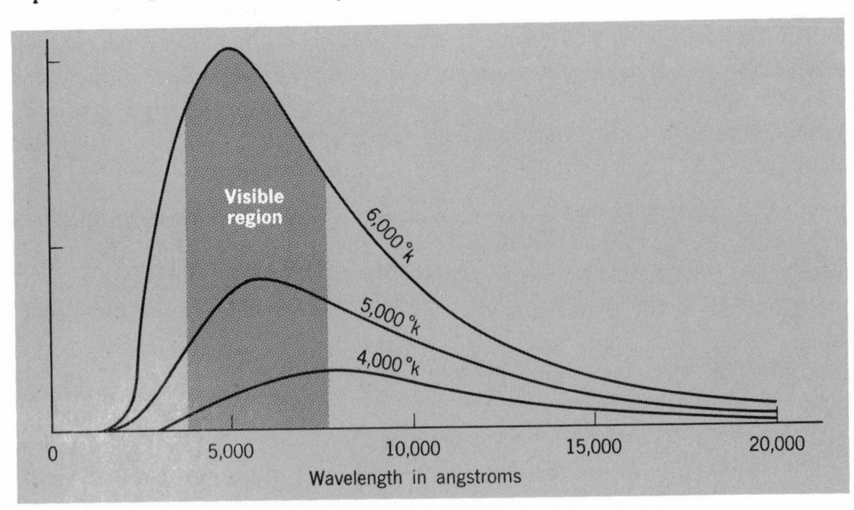

mum intensity is in wavelengths visible to our eyes (blue-green; Fig. 25-3). Solar radiation is called *insolation* (incoming solar radiation) and includes that which comes indirectly from the sky.

The reactions between the insolation and the Earth's atmosphere and surface vary considerably from time to time and place to place. For example, the cloudiness of the atmosphere and its water vapor and carbon dioxide content are factors. Most of the ultraviolet and shorter radiation is absorbed by atoms and molecules of oxygen (particularly ozone) and other gases before it penetrates the troposphere or lowest layer. This absorption increases the temperature of the atmosphere in the regions where it occurs.

Part of the insolation, perhaps some 35 to 40 percent on the average, is wasted because it is reflected back into space by molecules, dust particles, and clouds. Some insolation is absorbed directly by the atmosphere. The remaining radiation, nearly one-half of the original amount, reaches the surface and is absorbed by materials such as water, rocks, vegetation, and buildings; subsequently it is reradiated as wavelengths too long to be visible (Fig. 25-4). Some of these are absorbed on their outward journey by water vapor and carbon dioxide molecules and in this way heat the atmosphere.

Thus, the atmosphere is heated chiefly from the bottom, because the main heat source is the reradiation from the Earth's surface and because

Fig. 25-4. *Heating of the Earth's atmosphere. Solar energy passing through the air is reflected and absorbed so that less than half reaches the ground surface. Percentages vary considerably from place to place and from day to day.*

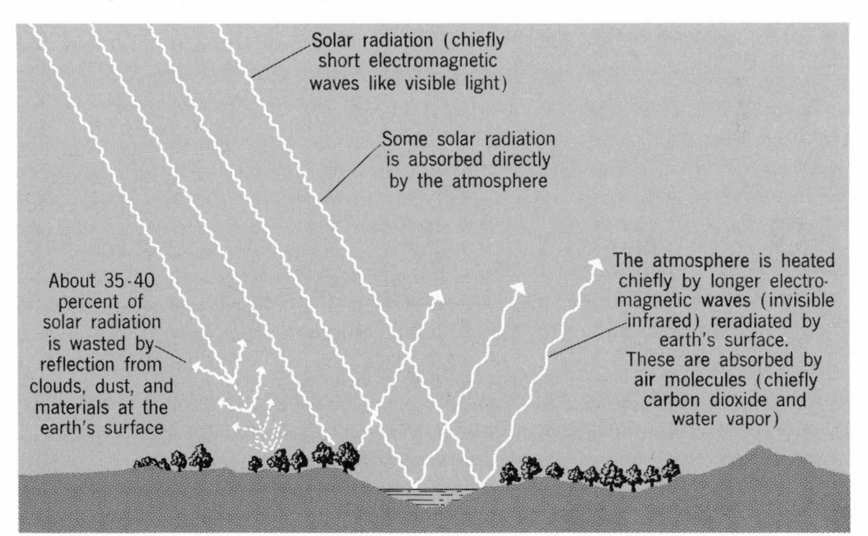

the numbers of water-vapor and carbon dioxide molecules decrease rapidly upward. However, secondary heat sources occur in the ozone region and in the thermosphere, where absorption of high-energy X rays and ultraviolet radiation occurs (p. 488). Thus from the surface upward the atmosphere contains three warm layers separated by two cold layers (Fig. 25-5).

The *greenhouse effect* may be explained in a similar manner. Sunlight can penetrate the glass windows enclosing a greenhouse. Materials inside the greenhouse absorb these high-energy wavelengths and reradiate them as longer low-energy wavelengths which cannot readily penetrate the glass. These are trapped inside where they are absorbed by molecules such as water vapor and carbon dioxide and increase the temperature. This phenomenon is familiar to anyone who has entered a car which has been parked for some time in sunshine with its windows closed.

A number of weather phenomena can now be explained readily: temperatures commonly decrease at higher altitudes, and extreme temperature changes from hot days to cold nights are common in desert areas because little water vapor is present in the atmosphere to make the air an efficient blanket for trapping the Earth's heat. Frosts are more likely on clear nights than on cloudy ones for a similar reason. The ground steadily loses heat by radiation until it reaches a lower temperature than the air immediately above it. Subsequently this bottom air loses heat to the ground and becomes cooler and heavier than the air above it. The cool heavy air may drain down slopes and collect in depressions and produce frosts. For this reason citrus orchards in California and coffee plantations in Brazil are located on sloping land rather than in depressions, where frosts are more common.

Temperature Variations During the night in the absence of solar radiation, the ground continuously loses heat by reradiation, and the coolest part of the day normally occurs around sunrise. On the other hand, energy received from the sun is at a maximum at noon when the rays of sunlight are most direct. However, radiations from the sun continue to add more energy than is reradiated into space for approximately three hours after the noon maximum. The warmest part of the day normally occurs at about 3:00 P.M. for a land area (about 5:00 P.M. over the oceans).

For similar reasons the coolest month of the year for a land area is normally January in the Northern Hemisphere and the warmest month is July. Over the oceans the moderating effects of the water shift the coolest and warmest months to February and August. The greatest known temperature variation during the year is experienced in the huge land mass of northern Siberia, whereas the least temperature range during the year occurs over the oceans. A record high temperature of 136°F (not in direct

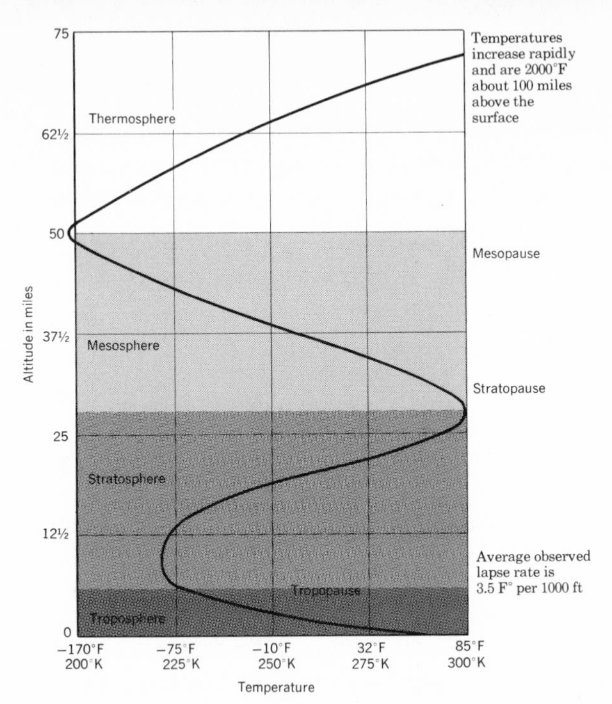

Fig. 25-5. *Main units of the atmosphere based upon variations in temperature with altitude (generalized). The atmosphere is heated chiefly from the bottom but higher temperatures occur at certain altitudes because ultraviolet and shorter radiations from the sun are absorbed at these altitudes. The altitude of the tropopause varies with the latitude and the season; it ranges from about 4 to 12 miles. Note the uniform temperatures in the lower stratosphere; they vary with latitude and range from about −50° to −100°F.* [*Temperature-altitude graph after Butler.*]

sunlight) was recorded in the Sahara in September 1922. Record low temperatures occur in Antarctica (below −100°F). Temperatures sometimes differ considerably at locations that are not far apart. One record showed a variation of 28°F within a distance of 225 feet on a steep hillside.

Layers of the Atmosphere Temperature variations at increasing altitudes provide one criterion for subdividing the atmosphere into layers (Fig. 25-5). Nearly all weather phenomena—storms, clouds, convectional circulation, and the like—are confined to the layer which rests upon the Earth's surface and is called the *troposphere.* In the troposphere, temperatures tend to decrease upward at the average rate of about 3.5F°/1000 ft (2C°/1000 ft), which is called the *observed lapse rate.* In other words, if one were lifted by a balloon and measured the air temperature at 1000-foot intervals, he would be determining the observed lapse rate. This applies to air that is not rising or sinking appreciably and varies considerably. Vertical currents result in expansion or compression of the air, and the resulting temperature changes are known as adiabatic (p. 508).

The troposphere is thickest at the equator (some 11 to 12 miles) and thinnest at the poles (some 4 to 5 miles). Its thickness at any one latitude also varies seasonally (particularly in the middle latitudes). It tends to be thickest during the warmest times of the year, which seems to correlate with an increase in turbulence produced by more intense solar heating. It is interesting that lower temperatures occur near the top of the troposphere above the equator (lower than −100°F) than above the poles (perhaps an average of −65 to −75°F). Winds (air moving more or less horizontally) tend to increase in velocity upward in the troposphere, partly because more friction and turbulence occur along the Earth's surface.

The top of the troposphere is called the *tropopause* (i.e., the zone separating the troposphere and stratosphere). It is located at the altitude where temperatures stop decreasing and remain about uniform. At times and places, the tropopause is a sharp boundary only a few hundred feet thick, but at other times and places the troposphere and stratosphere merge gradually.

The tropopause apparently is not a single continuous zone that decreases gradually in altitude with increasing distance from the equator. Rather, it appears to consist of a number of zones that occur at different altitudes at different latitudes; they may overlap to some extent, and gaps may occur between them. One important gap seems closely related to the location of the subtropical jet stream and is located in each hemisphere above a latitude of approximately 25 to 30 degrees.

Temperatures tend to remain about the same with increase in altitude in the lower *stratosphere,* the next layer above the troposphere. Above the lower stratosphere, temperatures tend to increase upward to a maximum of about 80°F at an altitude of nearly 30 miles. This marks the top of the stratosphere (the *stratopause*). Higher temperatures in the upper stratosphere apparently result from the selective absorption of high-energy ultraviolet radiation by oxygen molecules to form ozone, the three-atom molecule of oxygen which has a faint bluish color and a distinctive odor. Temperatures decrease upward for a distance from the 30-mile altitude because fewer oxygen molecules occur at higher altitudes. On the other hand, temperatures decrease downward for a distance because less ultraviolet radiation is present (some was absorbed at higher altitudes). If this ultraviolet radiation were not absorbed by atoms and molecules in the atmosphere, our skins would be burned and our eyes would be blinded.

Above the stratopause, temperatures decrease again to a minimum of about −175°F at an altitude of approximately 50 miles. Above this, temperatures increase again, probably because oxygen atoms (not molecules) occur at higher altitudes and absorb ultraviolet and other high-energy radiation. The terms *mesosphere* and *thermosphere* have been suggested for these regions.

The Ionosphere. Solar radiation, particularly its high-energy shorter wavelengths, produces electrified particles in the upper atmosphere by detaching electrons from some atoms. This region of electrically charged particles is called the *ionosphere,* and it extends from an altitude of 30 or 40 miles outward for several hundred miles. These free electrons and ions (atoms that have gained or lost electrons) act as mirrors at certain altitudes and for certain wavelengths (frequencies). Thus they reflect radio waves back to the Earth's surface and make long-distance radio reception possible (Fig. 25-6). The number, altitude, and effectiveness of the reflect-

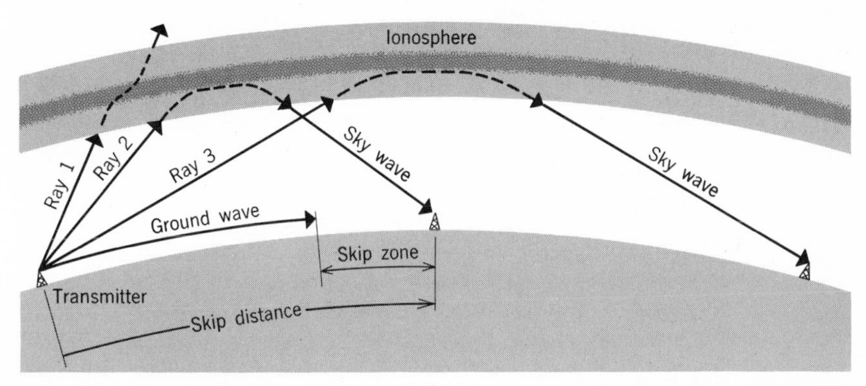

Fig. 25-6. *Effect of the ionosphere on radio waves.* [*Courtesy, U.S. Navy Hydrographic Office.*]

ing layers vary from day to night and during a year. Fewer and weaker reflecting layers occur on the night side of the Earth because some free electrons unite with positively charged ions to form electrically neutral atoms. However, the reflecting layers reappear during the next day when solar radiation again causes ionization. The reflecting layers constitute relatively small differences within the ionosphere, and its overall unity should be emphasized.

The concentration of electrons and ions tends to increase upward to a peak at an altitude of 200 miles or so. Ultraviolet radiation is greater at still higher altitudes, but relatively few atoms and molecules occur. Therefore the percentage of ionized atoms is high, but their numbers are comparatively low. In denser air at lower altitudes, more frequent collisions occur among atoms and electrons, and ions tend to be transformed into electrically neutral atoms.

The Magnetosphere. The magnetosphere (Fig. 25-7) forms an extension of the Earth's atmosphere and consists of a huge doughnut-shaped halo of charged particles that envelops the Earth. It begins at an altitude of about 600 miles and extends outward to 30,000 or 40,000 miles and

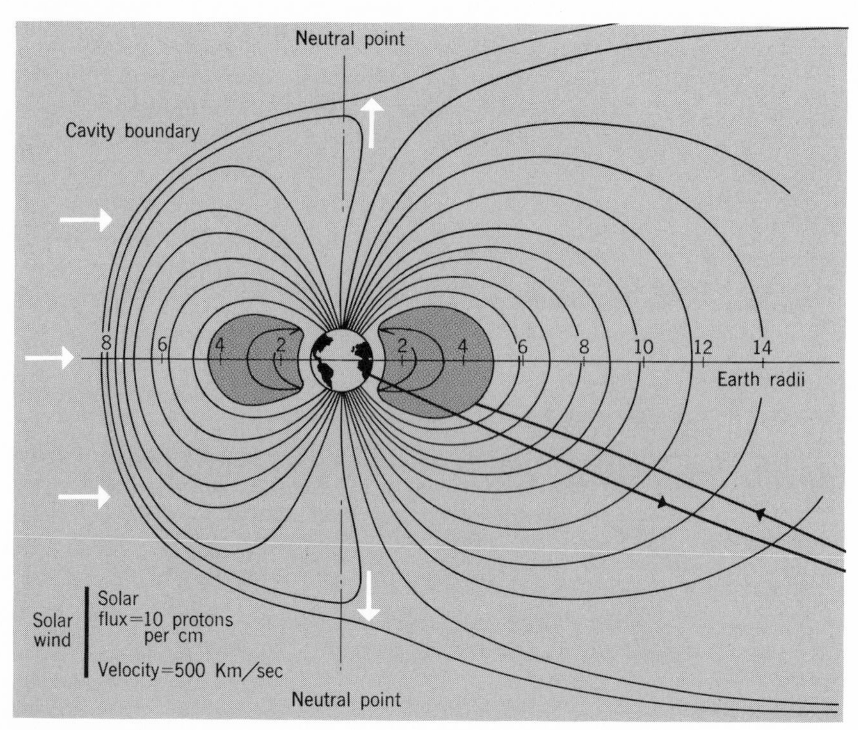

Fig. 25-7. *Interaction between the Earth's magnetic field and the solar wind. The solar wind consists of a stream of low-energy charged particles that are ejected continuously by the sun. The particles vary in number with the rate of solar activity and may travel at a few hundred miles per second. The solar wind distorts the Earth's magnetic field, which presumably would be symmetrical in its absence. As the solar wind moves past the Earth, a wake is produced, somewhat similar to that formed by a boulder in a stream. A shock wave also forms in front of the magnetosphere boundary on the sunward side of the Earth, and a zone of turbulence occurs between this shock wave and the outer boundary of the magnetosphere. The heavy lines in the lower right show the outward and inward paths of an orbiting satellite.* [*Courtesy, NASA.*]

more. Ions and electrons are trapped for a time by the Earth's magnetic field, spiral around lines of force, and move quickly between the Northern and the Southern Hemispheres.

The magnetosphere was first called the Van Allen radiation belts when it appeared to consist of two discrete regions of high-energy particles. By-products of cosmic radiation are probably the main source of the particles in the inner belt (chiefly protons; the belt is relatively stable), whereas solar radiation produces those in the outer belt (chiefly electrons; the belt fluctuates with solar activity). These zones of intense radiation are hazards for manned flights at very high altitudes. Subsequent discoveries have shown that low-energy protons and electrons are far more numerous than the high-energy ones both within the belts and throughout the magnetosphere. Thus the magnetosphere is best regarded as a single large radiation belt containing inner and outer zones of high-energy particles with peak intensities at about 2000 and 10,000 miles respectively. Furthermore, the magnetosphere is distorted by the solar wind; it is compressed on the side toward the sun and forms a very long magnetic "tail" on the opposite side.

Particles are constantly being discharged into the lower atmosphere, particularly at the horns of the belts. Such discharges are most pronounced after a solar flare and may produce auroras (p. 458).

Circulation of the Atmosphere On a global scale, and with emphasis upon mean conditions, three factors are important in understanding the circulation of the Earth's atmosphere.*

(1) Heated air expands and becomes more buoyant; therefore, it is pushed upward by nearby heavier, cooler air. At any given altitude, cold air is denser than warm air because its molecules move less rapidly and are closer together. To air a room quickly in cool weather, one opens

* Pointing up the world-wide dispersion of atmospheric gases and the dangers of contamination, Shapley (*Beyond the Observatory*) estimates that today you will breathe 15 argon atoms from every single breath of every person in the world a year ago. Your every breath contains 400,000 of the argon atoms breathed by Gandhi in his lifetime.

Fig. 25-8. *The relationship of surface winds to pressure patterns. Surface winds tend to make angles of 10 to 50 degrees with the isobars (at higher altitudes, they tend to move parallel to the isobars). Cyclones and anticyclones are synonymous with lows and highs. A cyclone is not a hurricane. Note the counterclockwise circulation around the low and the clockwise circulation around the high. These directions are reversed in the Southern Hemisphere and result from the Coriolis deflection.* [*AF Manual 105-5,* Weather for Aircrews.]

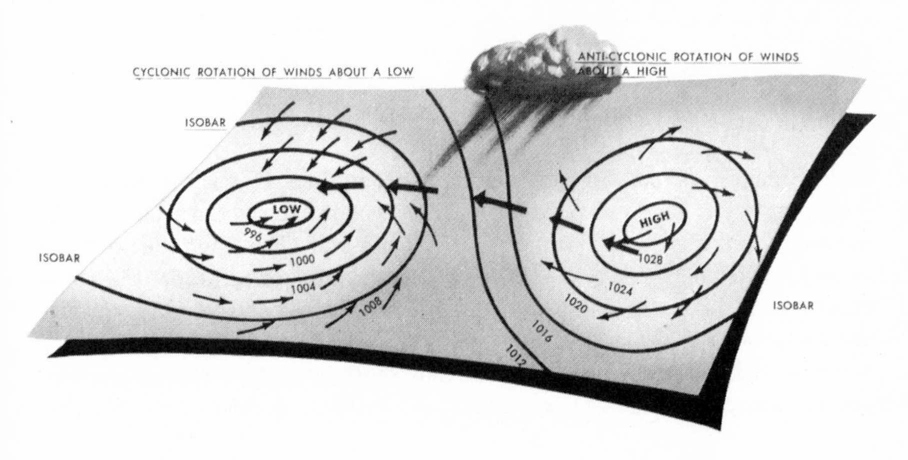

windows from both top (warm stale air leaves) and bottom (cold fresh air enters).

(2) Air has a tendency to move horizontally from higher to lower pressure areas (i.e., down the pressure gradient), and the greater the difference in pressure the more rapidly the air moves. However, air moves indirectly from high to low pressure areas (Fig. 25-8).

(3) Moving objects such as bullets, planes, and air masses are deflected to the right in the direction of movement in the Northern Hemisphere but to the left in the Southern Hemisphere. This is the Coriolis effect (Fig. 25-9). The Coriolis effect is not caused by the existence of a force that

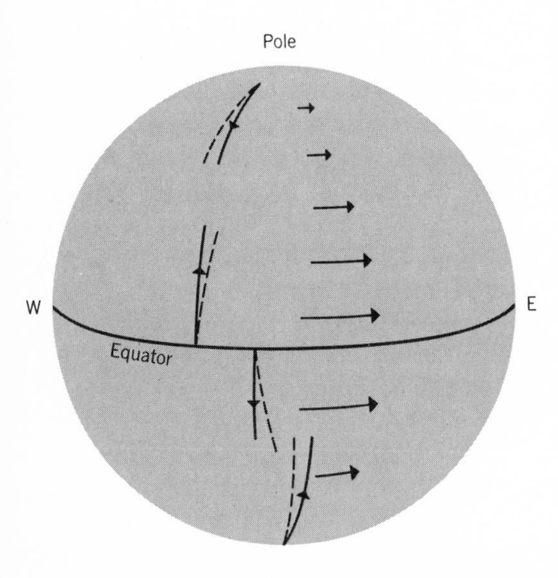

Fig. 25-9. *Deviation of projectiles produced by the Earth's rotation.* The *heavy lines show the direction of firing; the broken lines, the direction of flight as modified by the Earth's rotation. Arrows indicate the difference of the Earth's rotational velocity at various latitudes. Whether the projectile is fired north or south, east or west, it always deviates to the right in the Northern Hemisphere and to the left in the Southern Hemisphere. If an object such as a bullet or an air mass moves northward or southward, it tends to retain the rate of eastward rotation that it had at the latitude at which it began moving. To a space observer not rotating with the Earth, the bullet or air mass would appear to follow a straight line, and the target spot on the Earth, because of its different rate of eastward rotation, would shift away from its path. [Courtesy, Cecilia Payne-Gaposchkin,* Introduction to Astronomy, *Prentice-Hall, 1954.*]

pushes moving objects to the right in the Northern Hemisphere. Rather, a moving object tends to proceed in a straight line (Newton's first law of motion), and the Earth rotates underneath it. The magnitude of the Coriolis effect depends upon latitude, friction, and velocity: it is zero at the equator and increases in magnitude toward the poles; it is less where friction is great; it is less at lower velocities. Therefore, wind directions at the surface tend to be different from those at higher altitudes where greater deflection occurs (less friction and higher velocities).

The sun is the engine of the Earth's global circulation and the rotation of the Earth is its steering mechanism. Air moving about parallel with the Earth's surface beneath it is called wind, and winds are designated according to the directions from which they blow (e.g., a north wind blows from the north). Although a generalized explanation of the wind systems of the Earth is relatively simple, air movements are not precisely known at high altitudes and latitudes, and the overall pattern may not be apparent at any one location or time.

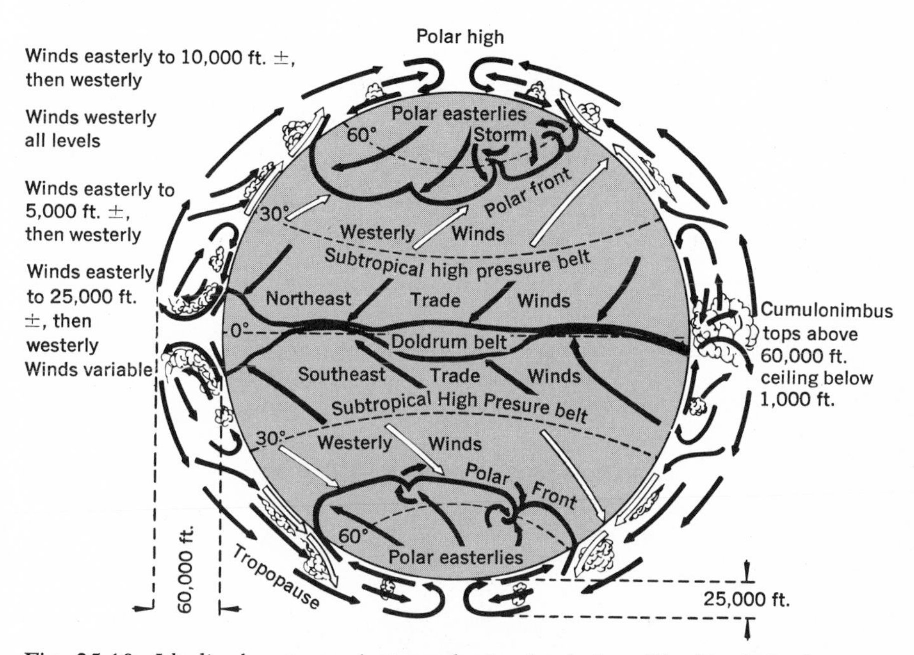

Fig. 25-10. *Idealized pattern of atmospheric circulation. The Earth is shown without continents for simplicity. The subtropical high-pressure belts are not continuous. Winds from the west predominate in the upper troposphere and presumably result from the Coriolis effect on the general poleward movement of air that occurs at higher altitudes.* [*TM 1-300,* Meteorology for Army Aviation.]

Circulation begins at the equator where solar radiation is most intense and air tends to rise (Fig. 25-10). However, this air cools gradually during its ascent, and in the upper troposphere it spills out of the equatorial region; some heads northward, some southward. Along the Earth's surface, winds move toward the equator to replace the rising air. If the Earth were not rotating and if complicating continental masses were not present, a huge simple convection current would presumably develop. Air would rise at the equator, sink at the poles, and move horizontally between these two areas. However, the spinning of the Earth deflects to the east much of the air that rises at the equator and starts northward or southward in the upper troposphere. By the time this air arrives at a latitude of about 30 degrees, it is moving almost due eastward (but high above the Earth's surface). Gradually the air descends to the Earth's surface to form high pressure zones whose centers are located at a latitude of about 30 degrees (the *horse latitudes*). However, these zones are not continuous; they are subdivided into large cells centering over the oceans.

In the Northern Hemisphere, air returns from these cells along the Earth's surface to the equator. On the return journey, it is deflected to the right (west) to form the *northeast tradewinds.* Air also moves from the horse latitudes along the Earth's surface toward the north. Deflection produces the belt of *westerlies.* According to one view, cold air moves southward along the Earth's surface from the high-pressure polar area and is deflected to the right (west) to form a belt known as the *polar easterlies.* The southward moving air of the polar easterlies meets the northern margin of the belt of westerlies at different latitudes at different times (perhaps at 50 to 60 degrees on the average, but ranging from 30 to 70 degrees). The line of convergence is known as the *polar front* (Fig. 25-11). Along it, warmer air in the belt of westerlies tends to be lifted above the colder heavier air from the north. The polar front is quite irregular and fre-

Fig. 25-11. *The polar front. Air from the westerlies moves upward and across the polar front and accumulates at high latitudes. Air pressures thus increase enough to cause occasional outbreaks of polar air toward the equator.* [*TM 1-300,* Meteorology for Army Aviation.]

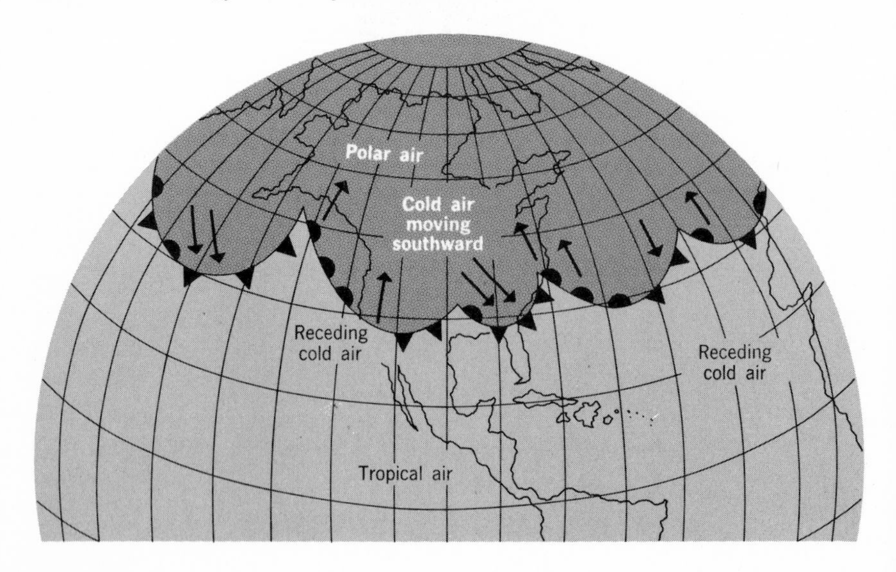

quently develops huge bulges that extend southward into the middle latitudes. It is a major factor in causing weather changes at such latitudes. A discontinuous zone of low pressure tends to occur at latitudes of 55 to 65 degrees.

Similar circulation in the Southern Hemisphere, combined with a deflection to the left, produces comparable belts of trade winds, westerlies, and perhaps polar easterlies. All of the pressure belts shift northward or southward with the seasons (up to several hundred miles toward a polar region during summer) and are influenced by local topographic conditions. The pressure belts and wind systems are better developed in the Southern Hemisphere than in the Northern Hemisphere because of the greater proportion of water to land.

Part of the preceding explanation concerning the global circulation of the Earth's atmosphere is theoretical, especially that concerning the movements of air masses high above the Earth's surface. Evidence for a convectional cell on either side of the equator seems reliable. Such cells are needed to transfer heat energy away from the equator. Some evidence is available for convectional cells in the belts of westerlies. However, polar cells may not exist.

Atmospheric circulation aids in causing certain kinds of climate at certain places on the Earth's surface. At the equator warm moist air rises, expands, and cools. Heavy precipitation, high temperatures, and gentle variable winds are common. At the horse latitudes, relatively dry air descends and is warmed. Its capacity to hold moisture is thereby increased, and little precipitation results. The world's desert areas are situated at latitudes of about 25 to 35 degrees. In the trade-wind belts, winds tend to blow more steadily in one direction, temperatures are high, and little precipitation takes place unless air is forced locally to climb over mountain barriers.

Weather in the belts of westerlies is noted for the regularity with which it changes from fair to stormy and back again as storm systems (lows or cyclones) form and move along the polar front. Mark Twain had these middle latitudes in mind when he said, "If you do not like our weather, wait five minutes."

During the cooler months when the polar front and its accompanying storm systems are best developed, they shift toward the equator, and the lower middle latitudes experience more frequent periods of precipitation and changeable weather.

Thus global circulation of the Earth's atmosphere tends to cause convergence of different air masses at certain latitudes. This results in upward movement, expansion, cooling, and precipitation along the line of convergence (unless temperatures are so low that the air can hold relatively

Fig. 25-12. *Schematic cross section of a jet stream. A jet stream may encircle the Earth. Note the core of very fast winds (1 knot means 1 nautical mile per hour; 100 knots approximates 115 mi/hr).* [*AF Manual 105-5,* Weather for Aircrews.]

little water vapor). At other latitudes, subsidence and divergence tend to occur, and the resulting compression and warming allow very little precipitation.

The Jet Streams The jet streams are powerful high-altitude winds which commonly blow eastward in great sweeping curves near the top of the troposphere (Fig. 25-12). They may extend all the way around the Earth at certain altitudes, or they may occur as discontinuous elongated belts. Furthermore, a jet stream may subdivide over some regions into two or more parts. Their locations shift from north to south considerably from season to season and even in the same season from one year to the next.

Three jet stream systems may occur in the Northern Hemisphere, especially in winter. A *subtropical jet stream* marks the poleward limit of the trade wind cell of the general circulation. A *polar front jet stream* is associated with the principal frontal zones and cyclones of middle and subpolar latitudes. A *polar night jet stream* is situated high in the stratosphere in and around the Arctic Circle. A similar set of systems presumably occurs in the Southern Hemisphere.

Jet streams, which are not well understood, seem to be the concentrated portions of the westerly winds that are common in the upper troposphere. Although jet streams have widths up to 300 miles or more, they are not more than a few miles thick. If one were to stretch a ribbon around a globe at the 30-degree latitude and bend the ribbon into sinuous meandering loops, he would have a model with some resemblance to the subtropical jet stream. Speeds up to 200 and 300 mi/hr have been measured at their centers, but these are much higher than average, and velocities decrease rapidly away from the centers. The winds blow in gusts. A jet stream tends to vary in velocity along its trend as well as outward from the central core, and these zones of higher velocity may shift from one region to another. Thus irregularity in velocity and location is characteristic of a jet stream.

The importance of the jet streams in high-altitude flying is obvious. Planes which fly eastward in the jet streams may have their speeds increased by 100 mi/hr or more. The exact relationship between the jet streams and weather changes is still uncertain.

HEATING OF LAND AND WATER

A number of weather phenomena can be explained by the fact that land heats up and cools off more quickly than water (Fig. 25-13). The heating or cooling of the soil is limited to a thin surface layer, whereas circulation causes the temperature of the water to be nearly uniform through-

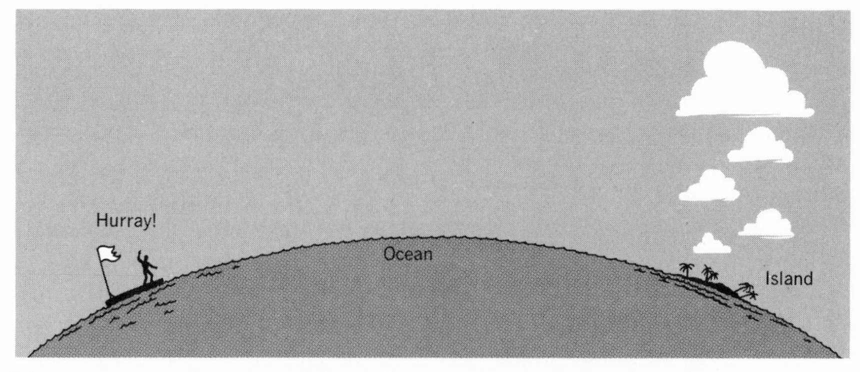

Fig. 25-13. *Land gains and loses heat more rapidly than water. The isolated white clouds hailed with joy by the shipwrecked sailor are interpreted by him as signs of a land mass. The island gains heat more rapidly than the surrounding water. Thus, during the day the strongly heated air above the island expands and rises. Air moves in horizontally from the water to replace it. The air is cooled during the upward movement; its capacity to hold water vapor decreases; and condensation occurs to form clouds. [Modified from Schneider.]*

Fig. 25-14. *The sea breeze. Cause: land gains heat more quickly than water. It is best developed when the ocean water is cold and the land is warm. A lake breeze is a similar phenomenon on a smaller scale. Sea breezes are most pronounced (speed, depth, and inland extent) where temperatures are high: during the hottest time of the day and nearer the equator. In middle latitudes, sea breezes tend to be shallow (up to 1000 feet or so) and gentle (up to about 12 mi/hr) and affect a narrow belt along the coast (as much as 10 miles or so). In the tropics, comparable maxima might be 4000 feet, 25 mi/hr, and 100 miles.*

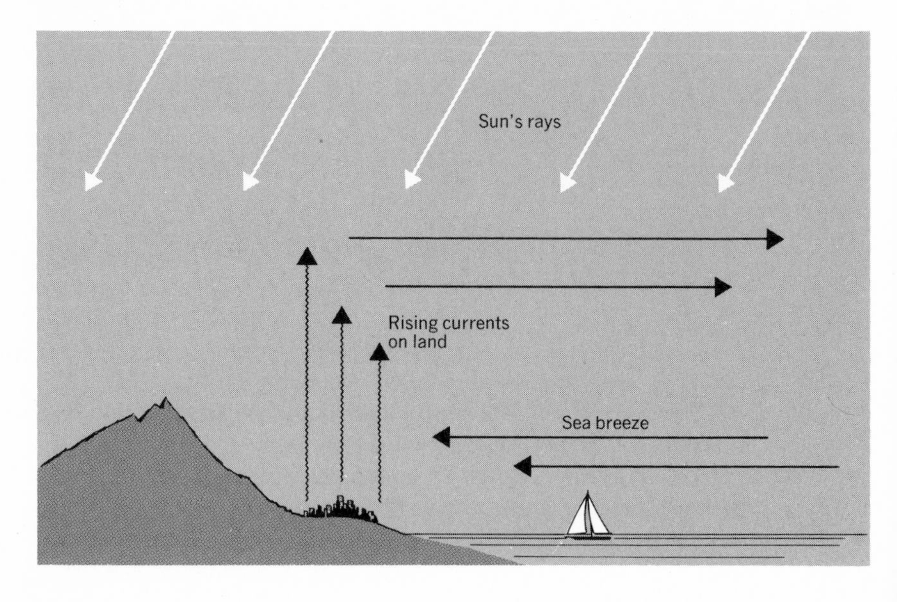

out a much thicker zone. In addition, evaporation cools the surface, and more heat energy is required to raise the temperature of a given mass of water a specific number of degrees than is needed to raise the temperature of an equal mass of soil an equal number of degrees (the specific heat of water is high, about 3 to 5 times that of dry soil).

In coastal regions a sea breeze commonly blows from the sea onto the land during the day (Fig. 25-14); at night the wind direction may be reversed and the wind may blow from land to sea (land breeze). During the day, air over coastal land areas commonly becomes warmer than air above the adjoining water. Cooler, heavier air above the surface of the ocean then moves inland to push upward the warmer, lighter air over the land. A more rapid cooling of the land at night chills the air above it, and this cooler, now heavier land air then moves seaward as a land breeze. Exceptions to this general process occur locally.

The monsoon seasons of India and of other areas are similar but much larger phenomena that occur seasonally rather than diurnally. During hot

Fig. 25-15. *Valley breeze. In the morning, sunlight may strike a mountain slope at a steep angle and thus be concentrated upon it; at the same time sunlight may strike an adjacent flat surface at a gentle angle. The slopes of a mountain are heated on a sunny day to higher temperatures than occur at comparable altitudes above adjacent valleys and lowlands. Air in contact with the mountain slopes is thus warmed and rises up the slope; it appears to come from adjacent lowlands as a valley breeze. These temperature differences tend to be reversed at night because radiational cooling occurs along the mountain slopes.* [*AF Manual 105-5,* Weather for Aircrews.]

Fig. 25-16. *Radiosonde equipment being released. At the right is radar equipment which automatically follows the flight of the balloon and gives the direction and velocity of the upper winds. During its rise and descent by parachute after the balloon has expanded and burst, the radiosonde automatically transmits signals which indicate the temperature, pressure, and humidity of the air it is moving through. Radiosondes are considered expendable.* [*U.S. Army photograph.*]

summer weather, air above India and the great land mass of central Asia is heated strongly and is pushed upward by moisture-laden air which moves northward from the Indian Ocean across the land. Altitudes increase gradually in the direction of the massive Himalaya Mountains. The moist northward-moving air, therefore, must rise higher and higher, and it cools adiabatically (p. 508) as it rises. Chilling of the air at higher altitudes results in enormous quantities of precipitation—more than 1000 inches of water per year have fallen on Cherrapunji, India. However, Cherrapunji averages only 1 inch of rainfall per month during December and January. In contrast, the average rainfall for the United States is approximately 30 inches per year.

Mountain and valley breezes also occur. Cold heavy air may move down a mountain slope at night, and a narrow valley between high mountain slopes will experience a pronounced mountain breeze. During the day, on the other hand, air may move up a mountain slope (Fig. 25-15).

At the surface, wind direction and speed are measured by weather vanes and anemometers. Winds at higher altitudes may be measured by tracking a balloon that is precisely inflated so that it rises at a known rate. Materials may be attached to a rising balloon so that it can be tracked by radar through clouds (the radiosonde, Fig. 25-16).

WATER IN THE AIR

The amount of water in the air as a gas, liquid, or solid is of tremendous importance to man. It affects the temperature by absorbing heat energy radiated from the Earth, and it determines the quantity of precipitation. However, the total amount of water in the atmosphere at any one time is surprisingly small, and it is apparently replenished once every two weeks or so. If all of it were precipitated as rain onto the surface of the Earth at one time, the average accumulation would approximate 1 inch. However, many inches of rain have fallen on an area in a day or so, the result of a continuous flow of warm, moist, precipitating air across the area. The precipitation occurs only over this area, and thus the moisture content of a very large volume of air can be concentrated upon it.

Water has a number of remarkable properties. It reaches its greatest density at 4°C (about 40°F) and expands upon freezing. Therefore, lakes freeze from the top rather than from the bottom as they would if ice did not float. The amount of heat energy necessary to raise the temperature of 1 gram of water 1 C° is called its specific heat and is very high relative to that of other substances; e.g., it is several times greater than that of dry soil or aluminum and about ten times that of iron, zinc, and copper. Thus, a thick soup boils more readily than clear water, but a cup of coffee stays warm longer than the soup. Large bodies of water have an important moderating influence upon temperature changes in the air over them and over nearby land areas, especially if prevailing winds are from the water onto the land. The thermal capacity (mass times specific heat) of the oceans is enormous; therefore, they absorb huge quantities of heat energy in the summer and release them during the winter.

The heat of fusion and of vaporization of water are both very great. To change 1 gram of ice at 0°C into 1 gram of water also at 0°C requires the addition of 80 calories of heat energy. This is called the *heat of fusion* (1 calorie is the amount of heat energy needed to raise the temperature of 1 gram of water 1 C° at 15°C). The same quantity of energy, known as the *heat of crystallization,* must be released whenever water changes into ice. Thus the formation of ice on water bodies in the winter, and the melting of ice and icebergs during the warmer months, limit the temperature changes of large water bodies during a year.

The *heat of vaporization* is nearly seven times the heat of fusion. Thus to change 1 gram of water at 100°C to 1 gram of water vapor (steam) at 100°C requires the addition of about 540 calories of heat energy. On the other hand, when the change is from the gas to the liquid, this same quantity of energy must be liberated (the *heat of condensation*)—e.g., whenever condensation produces dew, fog, or clouds. When the change is directly from the solid to the gas, or vice versa, the process is called *sublimation,* and even larger amounts of heat energy are involved. Thus the absorption or release of heat energy when water changes phase is a most important process in meteorology—in the transfer of heat energy from one latitude to another, for the development of storms, and in the circulation of the atmosphere.

The solvent power of water is large, both in the numbers of substances it can dissolve and in their quantities. On the other hand, its compressibility is less than that of steel. The surface tension of water is very great—important in the formation of raindrops and in various life processes involving capillary action (p. 676).

Water: Its Condensation and Evaporation Water vapor enters the air because evaporation and sublimation occur. Evaporation requires the addition of energy (sunlight is the chief source) because water molecules must move faster to escape into the air as a gas. The transpiration of plants is also a very large source of water vapor. Almost as much water vapor is added to the air over a field of lush vegetation on a hot sunny day as is evaporated from the surface of a nearby lake; perhaps $\frac{1}{2}$ to $\frac{3}{4}$ of an inch of water may be evaporated from a lake on a hot dry day.

If water is placed in the bottom of a container and the top is sealed, certain changes occur in the water and air in the container. Water molecules leave the surface of the water and evaporate into the air as a gas. Since these are the fastest moving molecules, the average rate of molecular motion in the liquid is decreased, and evaporation is a cooling process. Occasionally a water-vapor molecule returns to the liquid, but for a time more water-vapor molecules are added to the air than are subtracted from it. This raises the vapor pressure of the water vapor, which depends upon the number of water molecules in the air and upon their average rate of motion (their temperature); it is independent of the pressure exerted by other gases in the air. As evaporation continues, the vapor pressure of the water vapor increases, and more water molecules return to the liquid state. Eventually an equilibrium condition is reached, and as many molecules return to the water surface as leave it; the air is then said to be saturated at this temperature and pressure. If the temperature is raised, however, the average rate of molecular motion is increased, and less tendency exists for the faster moving molecules to return to the liquid state. Thus more water vapor must be added to the air to make it saturated.

Here we have a most important relationship in meteorology (Fig. 25-17): the capacity of air to contain water vapor—the amount of water vapor that can occur in a given volume of air—increases as its temperature increases; i.e., warm air can hold more moisture than cold air. This capacity increases very rapidly at higher temperatures. On a warm summer afternoon, the air has approximately five times the capacity to hold water vapor in its invisible gaseous state as air does at freezing temperatures.

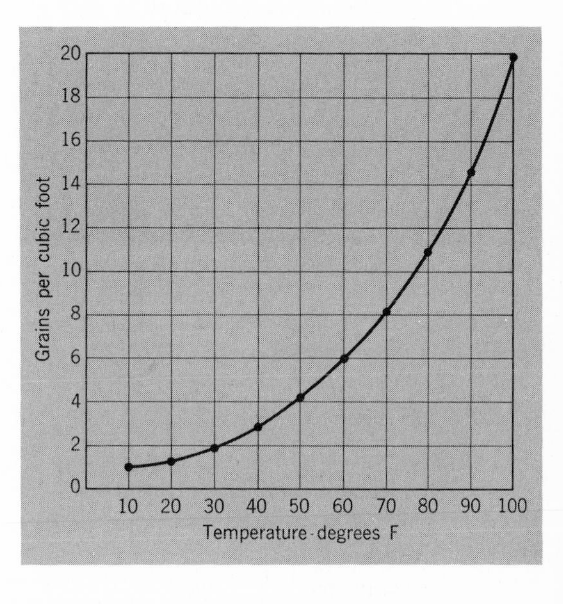

Fig. 25-17. *A most important relationship in meteorology involves the temperature of a mass of air and the amount of water vapor necessary to saturate the air. This amount is quite different at different temperatures.*

Condensation involves the joining together of many invisible water-vapor molecules to form a tiny visible liquid droplet. Condensation commonly does not begin until the air has become saturated with water vapor. Even then it may not begin, and supersaturation results. Saturation may be produced by the cooling of a parcel of air or by the addition of water vapor to it. The *dew point* refers to the cooling process; it is the critical temperature at which saturation occurs when a certain parcel of air is cooled at a constant pressure without the addition or subtraction of water vapor. At any given air temperature, therefore, the dew point is higher in moist air than in dry, and a smaller drop in temperature causes saturation. At temperatures above 32°F, condensation produces a change from the gaseous state to the liquid state; at temperatures below the freezing point, the change tends to be a direct one from the gaseous to the solid state. However, supercooled clouds are common, and these contain liquid droplets at temperatures below 32°F.

Water vapor molecules need a surface to condense on: the windshield of a car, a blade of grass, or a tiny invisible particle (nucleus) in the air. But some nuclei are much more effective than others in causing molecules to aggregate into condensation droplets. Nuclei are always present in the air, and the smaller nuclei are much more abundant than the larger. However, the very smallest and most numerous of the nuclei are not effective as condensation nuclei, and many of the larger ones are not effective until conditions of supersaturation have been attained. Nuclei are unnecessary at very low temperatures (p. 520). Nuclei are several times more abundant over continents than over oceans and are most abundant over large industrialized areas or immediately downwind from them.

Cooling causes a decrease in molecular motion; thus molecules cohere more readily to each other and adhere to some nucleus upon collision. If many water-vapor molecules are present in a certain amount of air, chances are opportune for numerous collisions and for the development of clusters. On the other hand, a rise in temperature increases the rate of molecular movement and tends to disrupt visible clusters of water-vapor molecules into invisible individual particles. Factors favorable for condensation, therefore, are the presence of abundant water molecules, suitable condensation nuclei, and a decrease in temperature.

Ready explanations should now be possible for such commonplace phenomena as the fogging of the bathroom mirror when one takes a shower, seeing one's breath on a cool morning, the presence of tiny water droplets on the outside of a cool glass of liquid on a hot humid day, and the "sweating" of pipes in the summer. Jet planes leave vapor trails at high altitudes where the air is very cold because water vapor is added to the air as fuel is consumed.

Dew, frost, fog, and clouds result from condensation, commonly because air has been chilled until saturation occurs. If air is cooled below the dew point by contact with a cool surface, water vapor condenses as dew directly upon that surface—rock, grass, or tractor; the dew does not fall. If the temperature is below 32°F, delicate feathery white frost crystals tend to develop instead. *Precipitation* in the form of drizzle, hail, rain, sleet, and snow involves the falling of moisture condensed from the air and is discussed later (cloud droplets are too small to fall rapidly).

FOG

A fog is essentially a cloud that occurs at or near the Earth's surface and reduces visibility to less than 1 kilometer (about 0.62 mile). To a man at a low altitude, the peak of a mountain appears covered by clouds; for a mountain climber standing on the same peak, fog blankets the surface. Commonly fogs consist of very small droplets of water (about 0.001 inch in diameter) which condensed from and remain suspended in the air, but some fogs consist of tiny ice crystals.

Two processes account for the formation of most fogs: air along the ground may be cooled below its dew point, or enough water vapor may be added to the air to saturate it. In advection, radiation, and upslope fogs, the cooling process predominates; in evaporation (steam) and frontal fogs, the addition of water vapor predominates. However, the two processes may go on simultaneously.

Radiation fog may develop along the ground during a "C" evening—clear, cool, and calm. After sunset, the ground cools off as it continues to radiate heat energy into the air. As a result, the temperature of the air near the ground may drop below its dew point, and condensation may occur to produce a fog. (Fig. 25-18). A radiation fog is thus associated with a local, shallow, and temporary *temperature inversion* (Fig. 25-19). The absence of strong winds and clouds are favorable factors. Water molecules in clouds absorb some terrestrial radiation and reradiate some of this downward to warm the surface. Turbulence and mixing are associated with winds; thus

Fig. 25-18. *Early morning fog at Norwich, England. Tree tops and other tall objects project above this shallow fog that was produced by radiational cooling of the ground during the night. Air in contact with the ground was chilled, became denser, and drained into depressions. Air along the ground is thus stable (shallow surface inversion) and smoke emitted at the surface is prevented from rising. However, warm gases emitted from the tall power station chimney are above the stable layer and rise further before spreading out horizontally. [Photograph by Charles E. Brown, from F. H. Ludlam and R. S. Scorer,* Cloud Study.]

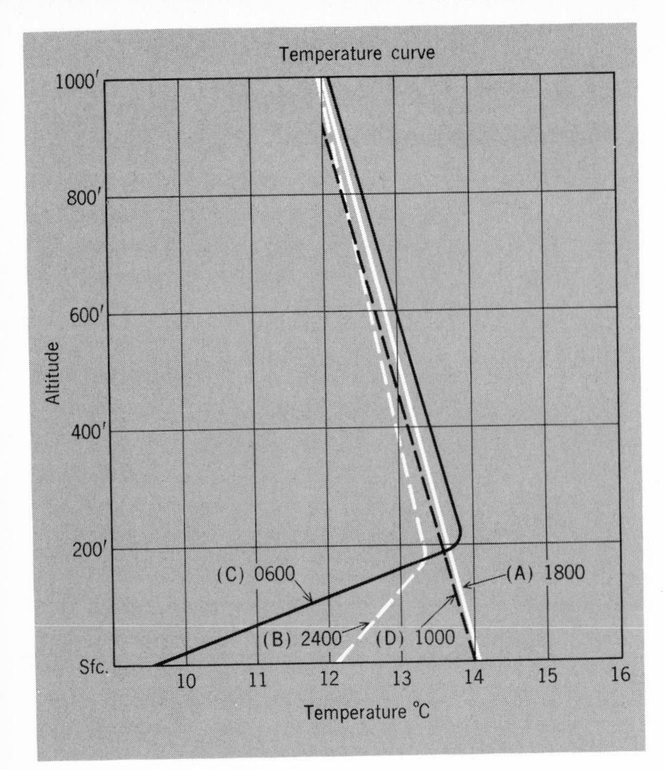

Fig. 25-19. *Surface inversion produced by radiational cooling during the night. An inversion involves a part of the atmosphere in which temperatures increase upward rather than decrease. From left to right the times are 6 A.M., midnight, 10 A.M., and 6 P.M. Note that the inversion becomes more intense from midnight to 6 A.M.* [*AF Manual 105-5,* Weather for Aircrews.]

no one parcel of air may remain near the ground long enough to be chilled to the dew point. However, a slight wind will tend to thicken a fog.

Fogs tend to develop over low-lying areas because cool heavy air at night moves down slopes to collect along the bottoms of depressions. In addition, air obtains water vapor by evaporation more readily from low-lying areas. During the morning, energy from the sun warms the air (chiefly by radiation from the surface), evaporation occurs, and the fog disappears. Radiation fogs are associated with stable air and clear weather.

Upslope fogs form in areas where stable moist air moves along a part of the Earth's surface that slants upward to a higher altitude. The upslope movement results in adiabatic cooling at the rate of $5\frac{1}{2}$ F°/1000 ft (p. 508). In a region of hills and valleys, upslope fog may form wreaths about each of the hills at a certain altitude, whereas lower sections remain free of fog. Upslope fogs are not restricted to nighttime formation and may develop in winds of 20 to 30 mi/hr or more.

Advection means "being carried in from elsewhere" and *advection fogs* may form over land or sea wherever relatively warm moist air moves across a cooler surface. The frequent and persistent fogs off Newfoundland where the cold Labrador Current and the warm Gulf Stream meet are of this type.

Evaporation fogs result if sufficient water vapor is added to a parcel of air to saturate it without a drop in temperature. These conditions may be illustrated by placing a pan of hot water in cold air or in another pan of cold water, or by running the hot shower in a cold bathroom. (According to reports, evaporation fog is a familiar sight in unheated English bathrooms.) An evaporation fog (also called steam fog) will form if cool air moves across a warmer water body, and rapid evaporation from the warmer water saturates the colder air (the temperature difference probably should be of the order of 20 to 25°F). The frontal variety of evaporation fog involves the forward margins of air masses, especially warm fronts (Fig. 26-9). Under these conditions rain may fall from a high-altitude warm air mass downward through a cooler stable air mass along the Earth's surface. Sufficient evaporation may occur to produce stratus clouds or fog in the cooler air.

Smog (smoke and fog) has been called man's "sin of emission," and it constitutes a growing nuisance and hazard. Tiny particles of various solids, liquids, and gases are added to the air by the diverse activities of a large city, and some of these act as condensation nuclei for the development of tiny fog droplets. The abundance of the nuclei seems to make the size of the fog droplet over an industrialized area smaller than average, and fog may develop before saturation occurs. Such small droplets fall more slowly, and the fogs persist longer. Smogs also persist because they prevent some

Fig. 25-20. *Effect of increase in altitude on air and water.* [*After Neuberger and Stephens.*]

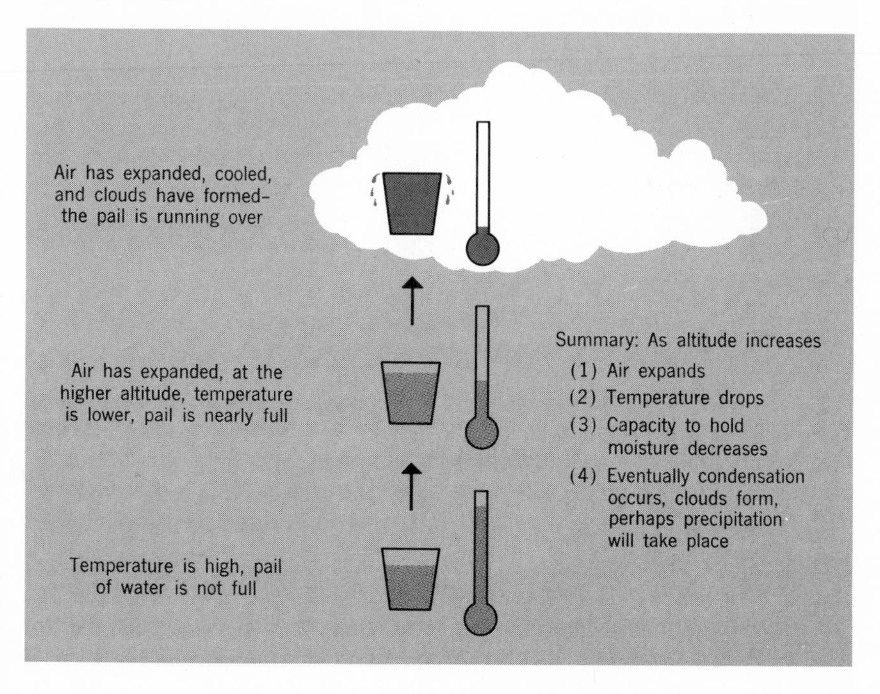

insolation from reaching the surface and thus from warming the ground and the air above it. Therefore, relatively cool, heavy, stable, smog-filled air may blanket an area and accumulate industrialized effluents until moved along by strong winds or washed out by heavy rains.

ADIABATIC COOLING AND HEATING

Clouds consist of moisture which has condensed from the air as tiny water droplets, of the order of $\frac{1}{500}$ of an inch in diameter, or sublimated as small ice crystals. Very slight upward currents prevent them from falling, or they rise and fall gently. Clouds commonly form because air containing water vapor and suitable condensation nuclei has been cooled below the dew point. This cooling frequently results from upward motion which produces expansion and adiabatic cooling (Fig. 25-20). Clouds may also form if enough water vapor is added to a parcel of air to cause saturation. Some cloud droplets fall out of a cloud into unsaturated air below and evaporate, but additional droplets form within the cloud and thus it is maintained for a time.

The cooling that occurs as air rises and expands, and the heating that it undergoes as it descends and is compressed, constitute a most important relationship in meteorology—adiabatic cooling or heating, each of which may occur under either dry or moist conditions. In the adiabatic process, no heat energy exchange occurs between a rising or sinking air parcel and the air it moves through. The change occurs within the parcel and results entirely from the change in volume that accompanies the change in altitude.

Fig. 25-21. *Vertical currents caused by unequal surface heating.* [*AF Manual 105-5,* Weather for Aircrews.]

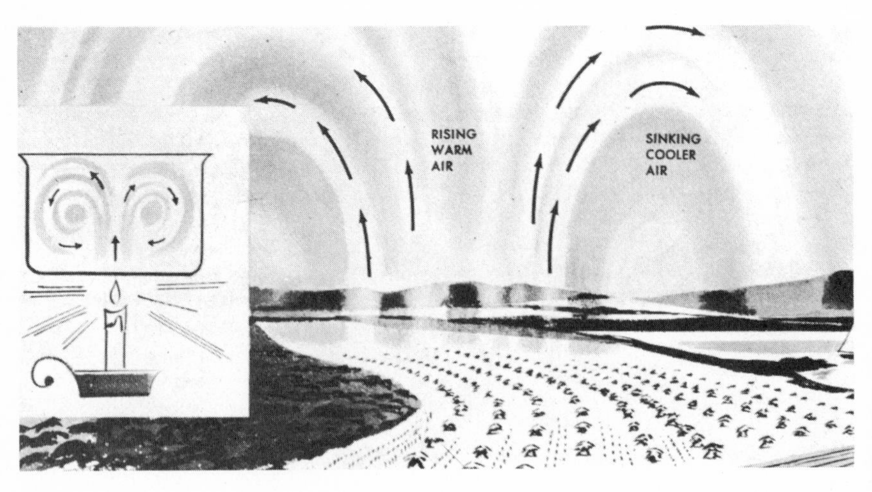

Fig. 25-22. *Cumulus cap cloud on top of smoke column rising from forest fire, Bitterroot National Forest, Montana, 5 August 1961.* [*Courtesy, U.S. Forest Service.*]

As air rises, it must push upward or aside the air which it replaces. This requires energy and results in a drop in the temperature of the rising air. Rising air cools at the nearly uniform rate of $5\frac{1}{2}$ F°/1000 ft (about 1 C°/100 meters) of upward movement which is known as the *dry adiabatic lapse rate*. As air descends, it is warmed adiabatically at the same rate.

As a parcel of air rises and cools, its capacity to hold water vapor decreases, and its relative humidity is thus increased. Eventually the saturation condition is reached. Condensation commonly occurs at this point, and the latent heat energy of condensation is released (540 calories per gram of water condensed). The addition of this energy to the cloud decreases the rate at which cooling occurs and provides energy which may aid continued upward movement. If the heat energy given off during condensation makes this rising parcel of air warmer than the air that surrounds it at any one level (and thus more buoyant), the parcel will continue to rise until it be-

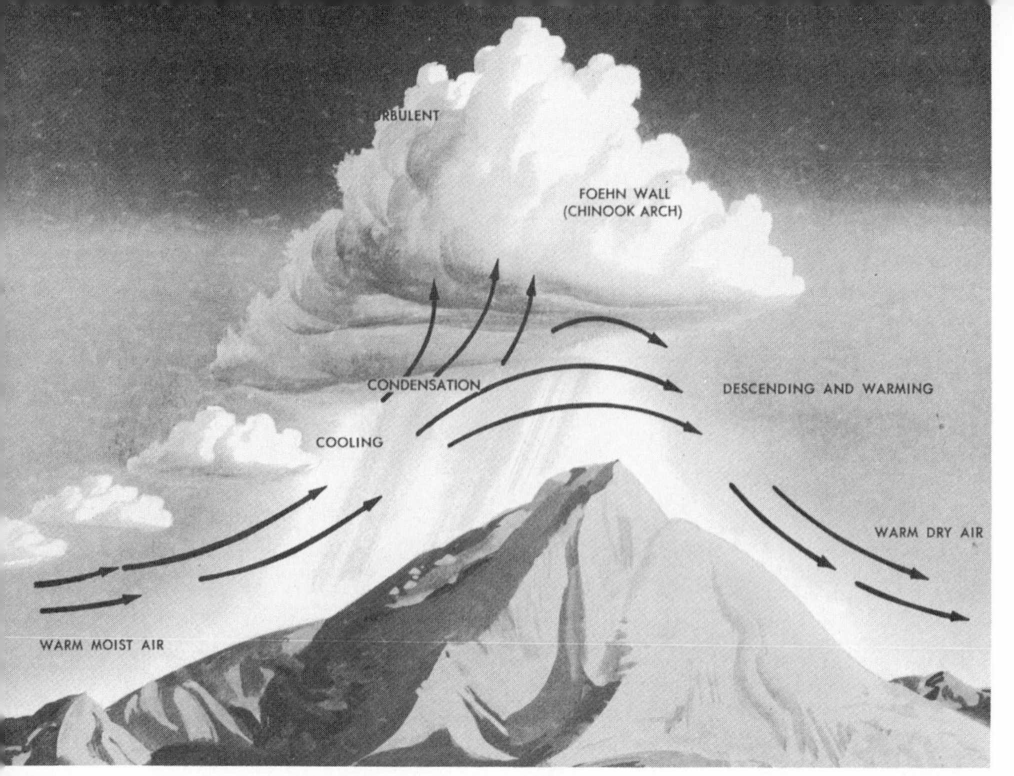

Fig. 25-23. *Warm dry downslope (foehn or chinook) winds may be produced on the lee side of a mountain.* [*AF Manual 105-5,* Weather for Aircrews.]

comes equal in density to the air around it. The mushroom clouds of nuclear blasts are caused by uplift, cooling, and condensation in this manner (Fig. 14-6).

The *wet adiabatic lapse rate* is quite variable. In warm moist air, it may be less than 2 F°/1000 ft since much condensation occurs, and this releases large quantities of heat energy. At low temperatures, the air can hold very little moisture even when saturated. Therefore, very little heat energy is returned to the air by condensation, and the wet adiabatic lapse rate closely approximates the 5.5 F°/1000 ft of the dry adiabatic lapse rate.

The dew point within a rising parcel of air decreases with increasing altitude at a rate of about 0.5 to 1 F°/1000 feet, and thus we have a *dew point lapse rate*. As air rises and expands, there is less water vapor per unit volume. Therefore, for condensation to occur, the temperature must be lower than was necessary at a lower altitude where more water vapor molecules occurred per unit volume.

Upward movements of air have several causes.

(1) One area of the Earth's surface may be heated more strongly than adjoining sections. The air above this area becomes warmed, it expands, and it is then pushed upward by adjoining cooler heavier air (Fig. 25-21). Bare soil, or a paved runway surrounded by vegetation, or a land area

adjoining a water body illustrates suitable conditions for initiating such a convectional circulation. A forest fire may produce a cloud above it by upward movement of the air that it heats (Fig. 25-22), and precipitation may fall from such a cloud. Convection may be on a local or global scale.

(2) Prevailing winds may move air along the Earth's surface and upward over mountain barriers (Fig. 25-23).

(3) Upward movements are associated with low-pressure systems, fronts, and the convergence of air masses.

CLOUDS

Clouds are of two general types: stratus (layered, blanketlike) and cumulus (globular, heaped masses). Other terms used in naming clouds are cirrus (curl), alto (high), nimbus (rain cloud), and fracto (broken). Relative to local observations, clouds constitute one of the best indicators of future weather changes.

Clouds are classified into four families based upon their altitudes and into ten main types (Fig. 25-24). The highest clouds consist chiefly of ice crystals, whereas the others consist chiefly of liquid droplets.

The standardized classification of clouds with abbreviated definitions

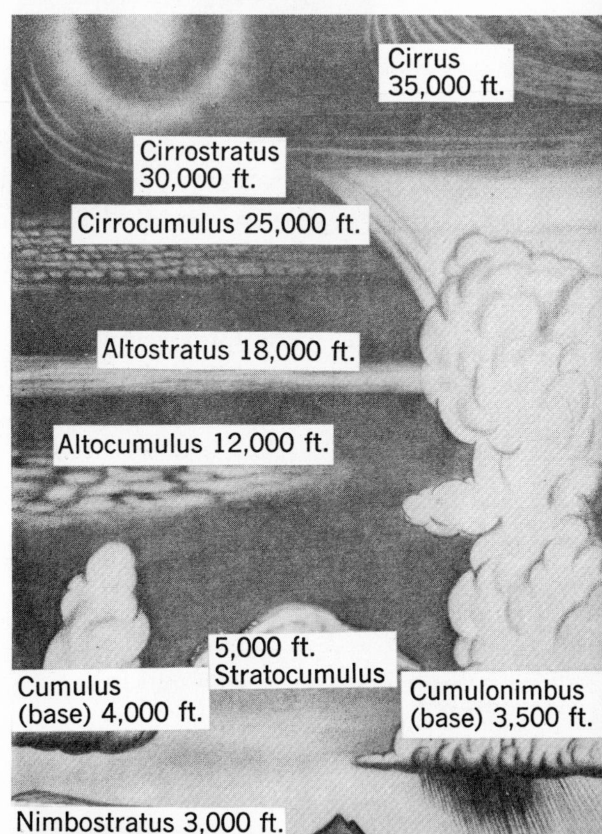

Fig. 25-24. *Basic clouds at average altitude levels.*

which appears below has been reproduced from a weather map published by the U.S. Weather Bureau on January 24, 1946:

> In the International System there are ten principal kinds of clouds. Their names, classification and mean heights are shown in the following table. The mean heights are for temperate latitudes and refer not to sea level but to the general level of land in the region. There is nearly always some variation from the mean height, and in certain cases there may be large departures. Thus, cirrus clouds may sometimes be observed as low as 10,000 feet in temperate regions and at lower levels in higher latitudes.
>
> *Family A:* HIGH CLOUDS (mean lower level, 20,000 feet)
>
> 1. Cirrus (Fig. 25-24A)

Fig. 25-24A. *Cirrus.* [*Courtesy, U.S. Weather Bureau.*]

> 2. Cirrocumulus
> 3. Cirrostratus (Fig. 25-24B)
>
> *Family B:* MIDDLE CLOUDS (mean upper level 20,000 feet; mean lower level, 6500 feet)
>
> 4. Altocumulus (Fig. 25-24C)
> 5. Altostratus
>
> *Family C:* LOW CLOUDS (mean upper level 6500 feet; mean lower level, close to surface)
>
> 6. Stratocumulus (Fig. 25-24D)
> 7. Stratus (Fig. 25-24E)
> 8. Nimbostratus

Fig. 25-24B. *Cirrostratus showing a halo.* [*G. A. Clarke.*]

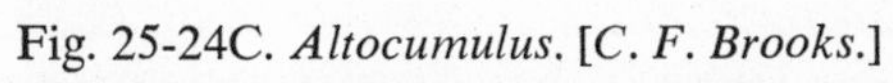

Fig. 25-24C. *Altocumulus.* [*C. F. Brooks.*]

Fig. 25-24D. *Summer afternoon photograph of clouds taken from a plane at an altitude about 3000 feet above a lake in Sweden. Cumulus clouds occur beneath an extensive layer of shallow stratocumulus. The base of the stratocumulus clouds is at the bottom of a very stable layer of air (upper air inversion). The cumulus clouds rise to the stable layer and spread out. No cumulus are forming over the cooler waters of the lake, except for those that rise intermittently above the islands. The lower portions of two of these (in center and to the right) have disappeared, leaving cut-off, mushrooming heads. [Photograph by L. Larsson, from F. H. Ludlam and R. S. Scorer,* Cloud Study.]

Family D: CLOUDS WITH VERTICAL DEVELOPMENT (mean upper level, that of cirrus; mean lower level, 1600 feet)

9. Cumulus (Fig. 25-24F)
10. Cumulonimbus (Figs. 25-24G and 25-24H)

Cloud Definitions *(Abbreviated)* *

1. CIRRUS—Detached clouds of delicate and fibrous appearance, usually without shading, generally white in color, often of silky appearance. They

* Material in parentheses has been added by the authors.

Fig. 25-24E. *Stratus.* [*Courtesy, U.S. Weather Bureau.*]

are always composed of ice crystals. (Cirrus clouds have a greater variety of forms than other types: isolated tufts, lines resembling strands of hair and forming a curl at one end, branching featherlike plumes, and bands.)

2. CIRROCUMULUS—A layer or patch composed of small white flakes or of very small globular masses, which are arranged in groups or lines, or more often in ripples resembling those of the sand on the seashore. (Cirrocumulus clouds are rare. They should be associated with cirrus and cirrostratus clouds and are smaller and higher than altocumulus clouds with which they are commonly confused.)

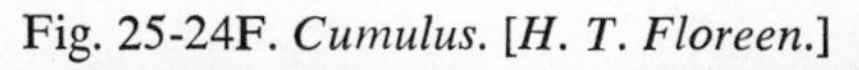

Fig. 25-24F. *Cumulus.* [*H. T. Floreen.*]

Fig. 25-24G. *Cumulonimbus with rain shower at base.* [*U.S. Navy.*]

Fig. 25-24H. *Cumulonimbus clouds showing anvil tops.* [*Courtesy, U.S. Weather Bureau.*]

3. CIRROSTRATUS—A thin whitish veil which does not blur the outlines of the sun or moon, but usually gives rise to halos. (A halo is a circle, sometimes incomplete, of light that is commonly colored red on the inside and occurs 22 degrees from the sun or moon. It is produced by the refraction of sunlight passing through ice crystals. Cirrostratus clouds may give the sky a milky look. They are thicker and lower and form a more uniform sheet than cirrus clouds. If cirrostratus clouds are observed a few hours after cirrus clouds, precipitation will probably occur within the next 24 hours. Objects at the surface cast shadows, a feature which distinguishes cirrostratus from altostratus.)

4. ALTOCUMULUS—A layer (or patches) composed of rather flattened globular masses, the smallest elements of the regularly arranged layer being small and thin. These elements are arranged in groups, in lines, or waves, following one or two directions and are sometimes so close together that their edges join. (Altocumulus clouds may or may not show shadows and are commonly associated with fair weather. They grade toward altostratus clouds when they are packed closely and toward cumulus clouds when they show vertical development. Altocumulus clouds may form a dense layer showing definite relief on its lower surface. Most of the regularly arranged heaps range from 1 to 5 degrees in angular diameter. Those of cirrocumulus tend to be smaller and of stratocumulus larger.

5. ALTOSTRATUS—Striated or fibrous veil, more or less gray or bluish in color. This cloud is like thick cirrostratus, but without halo phenomena; the sun or moon shows vaguely, with a faint gleam as though through ground glass. (If altostratus clouds are observed soon after cirrus and cirrostratus clouds, precipitation is probable within the next 12 hours. Altostratus clouds commonly form by a lowering of cirrostratus clouds; they hide the sun and the moon only in their darker portions. Nimbostratus clouds are darker than altostratus clouds and hide the sun and moon in all parts.)

6. STRATOCUMULUS—A layer (or patches) composed of globular masses or rolls; the smallest of the elements are fairly large; they are soft and gray with darker parts. The elements are arranged in groups, in lines, or in waves, aligned in one or two directions. Often the rolls are so close that their edges join. (Stratocumulus clouds may cover the entire sky and are most common in winter. They may form from stratus clouds by increased convectional circulation or vice versa.)

7. STRATUS—A low uniform layer of cloud resembling fog but not resting on the ground.

8. NIMBOSTRATUS—A low, formless, and rainy layer, of a dark color, usually nearly uniform; when it gives precipitation it is in the form of continuous rain or snow.

9. CUMULUS—Dense clouds with vertical development; the upper surface is dome shaped and exhibits rounded protuberances, while the base is nearly horizontal. (These are the white cottony puffs that enhance the

beauty of any landscape. They are visible evidence of columns of air that are rising above the condensation level.)

10. CUMULONIMBUS—Heavy masses of cloud with great vertical development, whose cumuliform summits rise in the form of mountains or towers, the upper parts having a fibrous texture and often spreading out in the shape of an anvil. (The following are associated with cumulonimbus clouds: thunderstorms, lightning, hail, strong updrafts and downdrafts, sudden heavy showers, and tornadoes.)

In attempting to identify clouds, one should remember that all types of gradations occur, that more than one kind of cloud may be in the sky at one time, and that clouds tend to form and disappear rapidly. Identification may also be difficult unless one has been observing the development of the clouds and is aware of the physical conditions involved in their formation.

Clouds may disperse for several reasons: precipitation may remove surplus water; the clouds may be warmed by radiations from the sun or Earth which decrease the relative humidity; surrounding drier air may mix with the clouds, especially near the margins; if downward air currents develop, the air will be warmed by compression.

A cloud will appear white to a distant observer if it reflects all of the sunlight falling on it. However, the same cloud will be gray or black to an observer beneath it from whom the sunlight is largely or entirely obscured.

ORIGIN OF PRECIPITATION

Probably raindrops form in different ways under different conditions. The two main processes apparently involve the melting of snowflakes and the rapid coalescence of cloud droplets. However, details are still puzzling, and the electricity of a cloud may be an important factor. Rain has fallen from clouds that had above-freezing temperatures throughout; such raindrops were not once snowflakes. On the other hand, radar has provided evidence that some rain was once snow and ice.

A raindrop of average size may be a million times larger in volume than a cloud droplet (Fig. 25-25). A basketball has a similar size relationship to a BB shot. Thus cloud droplets grow by condensation too slowly to produce raindrops in the short time intervals in which some clouds may be seen to form and produce rain. An exception to this involves condensation on giant salt nuclei (from 1 to 10 microns in diameter) which can quickly produce large droplets up to 100 microns or so in diameter (about the size of a human hair).

Coalescence can take place in a cloud containing a mixture of large and small droplets because the large droplets rise and fall at rates different from those of smaller droplets. Thus they tend to grow larger by amassing the smaller droplets that collide with them. Collision and collection may occur at the bottom of a descending large droplet which falls faster than smaller droplets and also at the top; at the same time, air resistance is reduced in its wake, and adjacent droplets tend to fall faster and overtake it. However, droplets occasionally bounce apart from a collision without coalescing. On the other hand, if droplets in a cloud have a uniform size, they tend to rise and fall at approximately equal rates and relatively few collisions result.

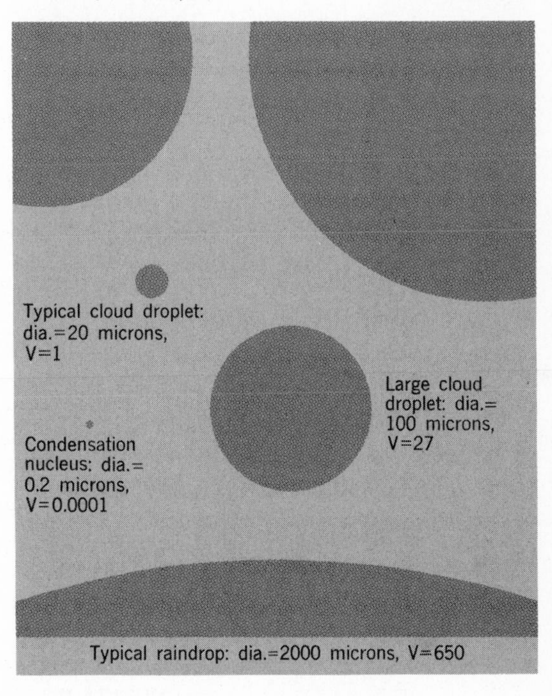

Fig. 25-25. *Comparative sizes and terminal falling velocities of some particles involved in condensation and precipitation. Diameters* (d) *are given in microns* (*1 micron equals 1/1000 mm*) *and the terminal velocities* (V) *in cm/sec. The diameter of a typical raindrop probably is about 100 times larger than that of a typical cloud droplet and its volume is about 1 million times greater. As compared with cloud droplets, raindrops tend to fall faster, evaporate more slowly, and reach the ground.* [*After McDonald.*]

If strong updrafts are present, as in cumulus clouds, coalescence may occur in a different manner. After a few raindrops form in a cloud, they remain suspended or fall slowly, and tiny cloud droplets stream upward past them. Collisions and coalescence follow. When a raindrop reaches about $\frac{1}{4}$ or $\frac{1}{5}$ of an inch in diameter, surface tension can no longer keep it as a single drop, and it subdivides into two or more smaller drops, which in their turn grow larger and subdivide again. This type of chain reaction may produce many raindrops in a short time. However, warm-temperature clouds need a depth of nearly a mile to produce precipitation. Not enough coalescence occurs in thinner clouds. Some raindrops evaporate before falling all of the way to the ground; thus the height of the cloud base is also a factor.

The second method of raindrop formation occurs when scattered ice crystals develop in a saturated, supercooled cloud and subsequently grow by sublimation as liquid droplets in the cloud evaporate. Water molecules can be detached more readily from liquid droplets than from ice crystals; therefore, the water-vapor pressure over water at below-freezing temperatures is greater than that over ice. Thus when air is saturated relative to water, it is supersaturated relative to ice. Under such conditions of supersaturation, water molecules in the gaseous state attach themselves directly to the ice crystals. To replace these molecules and maintain the saturated equilibrium condition relative to liquid water, other water molecules evaporate from the liquid droplets. In a rather short time the ice crystals grow large enough to fall through the cloud. In so doing, they collide with and assimilate some of the cloud droplets in their paths; they also encounter other ice crystals and unite with them to form snowflakes. Thus coalescence becomes the most important growth process in the later stages of the ice crystal–snowflake origin of raindrops. For this reason, snow may fall on a mountain slope at the same time that rain falls in an adjoining valley.

Radar (radio detection and ranging) has aided greatly in the study of precipitation as well as in the location and tracking of hailstorms, hurricanes, and tornadoes. Most radar sets can detect only the larger precipitation particles within a cloud, not the tiny cloud droplets themselves. Thus the shapes and locations of the precipitation regions within clouds can be located and their development and movement traced (Fig. 25-26).

Although nimbostratus clouds have a uniform appearance as seen from below, radar study has shown that some consist of three main layers: rain clouds in the lower part, clouds of snowflakes in the middle portion, and ice-crystal clouds in the upper and coldest zone. Where precipitation is heaviest, these zones merge into one.

WEATHER MODIFICATION

A number of factors are involved in modern attempts at rainmaking. Supercooled clouds occur in middle and high latitudes. In such clouds, liquid water droplets are common at temperatures down to about −20°C and occasionally occur as low as −35°C or so. In fact, very tiny droplets of pure water have been cooled in laboratory experiments to −41°C before they crystallized spontaneously. Thus liquid cloud droplets freeze over a range of about 40 C° and are common in the warmer portion of this range. However, if ice crystals or other suitable ice-forming nuclei such as silver iodide particles (geometrically similar to ice crystals) form or are intro-

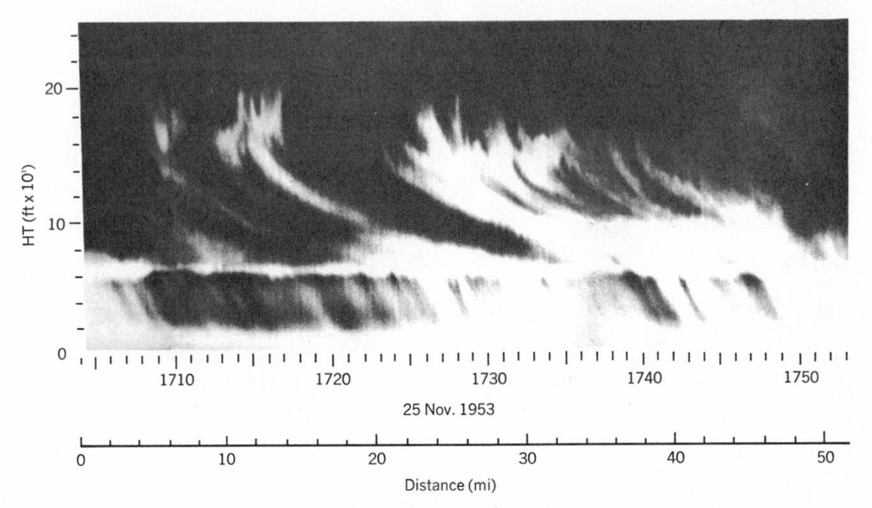

Fig. 25-26. *Height-versus-time record of radar echoes from a cloud system that passed above Montreal, Canada. A vertically pointing 3-cm radar set was used. In calculating the distance scale, the weather pattern was assumed to move at a velocity of about 60 mph, which was the wind velocity at 18,000 feet. Note the bright-echo band located at an altitude of about 6000 feet; this occurs just beneath the melting level of 32°F. Echo streaks slant more steeply below the bright-echo band than they do above it. Experimentation has shown that small reflecting particles (i.e., smaller than the wavelength being used) are about five times more reflective as liquids than as solids. Thus the radar record of this storm seems to show that snowflakes formed in cold air at high altitudes and fluttered slowly downward. They melted at an altitude of about 6000 feet as they fell into warmer air. The raindrops thus formed gave a much stronger radar reflecting signal, fell more rapidly, and produced steeper echo streaks.* [*Courtesy, Stormy Weather Group, McGill University.*]

duced into such supercooled clouds, the ice crystals grow larger at the expense of the liquid droplets.* Eventually they fall at different rates, and collisions at lower altitudes and warmer temperatures cause numerous snow crystals to fuse together into snowflakes. At still lower levels, these melt to form raindrops.

Evidently, suitable ice-forming nuclei are absent in effective quantities in such supercooled clouds. Therefore, the addition of suitable nuclei to the clouds by man might initiate precipitation. The main attempts to date have involved adding nuclei such as silver iodide (Fig. 25-27) or pellets of Dry Ice (solid carbon dioxide at a temperature of about −79°C). The Dry Ice does not act directly as condensation nuclei. Rather, a falling pellet chills the

* The addition of monomolecular layers or patches of iodine on the surfaces of submicroscopic particles of lead oxide from automobile exhausts transforms them into nuclei on which water vapor molecules sublimate to produce ice crystals. One gram of iodine may activate 10^{18} lead oxide particles. It is as if (Schaefer, *Science*, 12 December 1966) a golf ball swelled to the size of the Empire State Building in a few seconds. The discovery may have great significance for weather modification.

Fig. 25-27. *Development of a seeded cloud on August 20, 1963, over the Caribbean Sea.* Top left, *time of seeding.* Top right, *9 minutes later just before maximum vertical growth.* Bottom left, *19 minutes after seeding. The horizontal "explosion" is under way.* Bottom right, *38 minutes after seeding the cloud has attained giant proportions. The pyrotechnic silver iodide generator shown in the center was developed by the Naval Ordnance Test Station. A number of these generators are dropped at intervals from a plane and in seconds fill a cloud with a vast number of large silver iodide particles.* [*Joanne S. Malkus and Robert H. Simpson, "Modification Experiments on Tropical Cumulus Clouds,"* Science, *7 August 1964, Vol. 145, pp. 541-548.*]

air in its path enough to produce ice crystals, and these spread through the cloud and grow larger at the expense of the supercooled water droplets. Each pellet of Dry Ice may cause the development of millions of tiny ice crystals.

Silver iodide and Dry Ice are effective only in supercooled clouds. In convective clouds at above-freezing temperatures, two techniques have been used. Each involves the formation of a number of large cloud droplets that initiate the coalescence process because they fall at rates different from those of the bulk of the cloud droplets. Large salt nuclei (much larger than silver iodide particles) have been added to clouds, and condensation on these can produce large droplets up to 100 microns in diameter. Water droplets have also been sprayed directly into warm-temperature cumulus clouds from a plane. In experiments performed on Mt. Washington, New Hampshire, handfuls of Dry Ice were tossed into supercooled clouds moving across the summit, and snow was observed to fall from the clouds immediately downwind.

Cloud seeding is commonly done under conditions in which precipitation is possible. If precipitation does result, one must still wonder whether the rain or snow would have fallen naturally without artificial stimulation by man. The extreme variability of precipitation—from one area to another, from one day to the next, and from year to year—makes it difficult to assess results.

The first successful experiments received much more publicity than the failures, but enough tests have been made to show that man can increase precipitation to some extent, although many uncertainties remain. For example, certain experiments have cast doubt upon the durability of silver iodide particles as nucleating agents and upon their chances of actually rising up into clouds that are being seeded from the ground. The most propitious circumstances involve the movement of moisture-laden prevailing winds across mountain barriers. Such clouds may be seeded from the ground by silver iodide generators that are at times actually within the clouds. Such conditions exist in the western third of the United States in which the Coast Ranges and Rocky Mountains are located. Perhaps precipitation can be increased by 10 to 15 percent under these conditions.

Hail suppression, fog dispersal, and severe storm modification are other areas in which experimentation is under way to modify the weather.

TYPES OF PRECIPITATION

Together with the processes of evaporation, sublimation, and condensation, precipitation forms part of the water cycle. The major forms of precipitation are drizzle, rain, sleet, hail, and snow.

Sleet According to the definition used in the United States, sleet consists of frozen raindrops. It can form if raindrops fall through a subfreezing layer of air near the Earth's surface and are frozen into ice beads. These bounce when they hit the ground and may form a white layer.

Glaze, on the other hand, is an ice coating and not a type of precipitation. It forms when supercooled raindrops strike cold objects at the Earth's surface and freeze into clear ice. Glaze makes walking and driving treacherous and occasional ice storms cause a great deal of damage. The ice may coat wires, twigs, trees, cars, and other objects, and the mass of the ice may at times be some twenty times greater than that of the wire or twig it surrounds. Thus trees bend, branches snap off, wires and power poles break, and much damage and inconvenience occur.

Rain Raindrops range in diameter from about $\frac{1}{100}$ to $\frac{1}{4}$ inch (Fig. 25-25). Their rate of fall in still air varies directly with the drop size, and the largest attain speeds of about 20 to 25 mi/hr. Surface tension pulls the smallest raindrops into spherical shapes, but the largest tend to be flattened at the bottom and rounded at the top. Raindrops cannot exceed a certain size because surface tension becomes less effective as the size increases, whereas the disrupting effects of the increased rate of fall become greater. (Surface tension varies with the surface area and thus with the square of the diameter, whereas volume varies with the cube of the diameter.) The largest drops tend to fall from cumulonimbus clouds because these have strong upward currents that support the raindrops until they attain a maximum size. Heavy downpours come from such clouds when the upward currents stop or shift locations. Rain tends to wash impurities such as dust, pollen, and soot from the air and may occasionally be colored by these materials.

Drizzle droplets range from about $\frac{1}{50}$ to $\frac{1}{500}$ of an inch and fall very slowly or seem to float in the air. They are produced in stratus clouds which are relatively thin and located near the Earth's surface. Apparently the droplets are small and numerous because the cloud depth is shallow and the droplets fall at approximately similar rates—too slowly for coalescence to produce larger drops.

Some of the heaviest observed amounts of rainfall reported by the U.S. Weather Bureau follow: 12 inches in 42 minutes at Holt, Missouri, in June 1947; nearly 31 inches in about $4\frac{1}{2}$ hours at Smethport, Pennsylvania, in July 1942; 102 inches at Cherrapunji, India, in four days in June 1876; and 884 inches at Cherrapunji during six months in 1861, April to September. On the other hand, in one location in northern Chile rainfall averaged only 0.02 inches per year over a 43-year period.

Since the total amount of water vapor in the air at any time is relatively small and is probably replenished about once every two weeks, it follows that total worldwide precipitation and evaporation must be about equal. Thus if one area has a drought, another area must simultaneously have an excess of precipitation. A similar simultaneous balancing out occurs with temperature differences; thus a hot summer in one area does not mean that this same area will subsequently experience a colder-than-average season.

Hail Hail consists of rounded particles of ice which fall out of cumulonimbus clouds. Larger hailstones commonly exhibit an onionlike structure of alternating shells of clear and opaque ice. The clear ice tends to consist of larger crystals in thicker layers (cooled more slowly) and appears to represent frozen water. The smaller ice crystals of the thinner opaque layers contain air bubbles and apparently cooled more quickly. The opaque layers may also result from the accumulation of snow crystals. Larger hailstones apparently require several round trips through a cloud and are shifted about between ascending and descending currents. The smaller stones may form during a single descent.

Successive concentric shells may be added to a hailstone until it becomes too large to be supported by the moving air and falls to the ground. A hailstone about $5\frac{1}{2}$ inches in diameter and weighing $1\frac{1}{2}$ pounds fell in Nebraska in July 1928. However, perhaps only one hailstone out of 100 to 1000 exceeds 1 inch in diameter. Many of the larger ones are flattened lumps formed by the coalescence of two or more smaller hailstones. Disklike masses of ice 6 to 10 inches across and 2 to 3 inches thick are reported to have fallen in Kansas in June 1917. Hailstones do not last long, cannot be predicted precisely, and are most capricious in their extent—they may destroy crops in one field and miss those in a neighboring field.

Snow Snow crystals consist of ice crystals which developed in the air by sublimation: water vapor changed directly into solid particles at temperatures below freezing. Most snow crystals are six-sided in design, but some are three-pointed. Thousands of variations on this general design have been

Fig. 25-28. *The six types of snow (ice) crystals.* Top left to right: *hexagonal plates, plates with extensions, stellar.* Bottom left to right: *ice needles, hexagonal columns, capped hexagonal columns.* [*Courtesy of the American Museum of Natural History.*]

observed (Fig. 25-28). Most snowflakes consist of aggregates of ice crystals. If flakes fall through warmer air they may melt together to form large clots. According to Tannehill, huge clots of snow, 15 inches across and 8 inches thick, fell in Montana in 1887! Temperatures are never too low for snow to form, but very cold air contains little moisture. On the average, about 10 inches of snow will melt to 1 inch of water, but the range is from about 6 to 30 inches.

Although the term *blizzard* is popularly applied to any heavy, somewhat windy snowfall, a blizzard is officially defined as "a violent, intensely cold wind, laden with snow mostly or entirely picked up from the ground." It may or may not snow during a blizzard.

According to the U.S. Weather Bureau, the greatest recorded seasonal snowfalls include the 884 inches that fell at Tamarack, California, during the winter of 1906-07 and the 1000 inches that piled up on Mount Rainier in 1955-56 (at an altitude of 5500 feet). During the New York snowstorm of December 26-27, 1947, 26 inches of snow fell in 24 hours, a record for the city. More than $8,000,000 were spent in removing the snow from the metropolitan area. Additional heavy snowfalls include the 87 inches which fell at Silver Lake, Colorado, in 27½ hours in April 1921 and the 108 inches which accumulated at Tahoe, California, in four days.

HUMIDITY

Absolute humidity refers to the mass of water vapor actually present in a unit volume of air. *Relative humidity* refers to the ratio between the amount of water vapor actually present and the amount that would have to be present to saturate the air at the existing temperature and pressure. Humidity may be less than 100 percent during a rainstorm if rain is falling from a saturated cloud situated a few thousand feet above the surface through unsaturated air nearer the ground. High humidity is uncomfortable on warm days because the rate of evaporation, which is a cooling process, slows down when the air is nearly saturated with water vapor.

A breeze aids evaporation by preventing saturated air from stagnating. A windy cold day feels particularly chilly because the rate of evaporation of moisture from one's skin is increased. A swimmer feels chilly as he steps out of the water and rapid evaporation occurs. However, he feels quite comfortable a moment later after a brisk rub with a towel has removed excess water.

In the heating of buildings, lower temperatures and higher humidity are less expensive and more healthful than higher temperatures coupled with low humidity. The air in some homes during the winter is so dry that rapid evaporation occurs and one feels cool despite rather high temperatures.

Cold winter air can contain little moisture even if its relative humidity is high. Therefore, when the air is heated to summer temperatures inside a house, its relative humidity becomes very low.

Humidity may be measured by a hygrometer which uses strands of human hair as an essential element. The hairs become shorter in dry air and longer in humid air. (Girls—and boys—who curl their hair know what happens to curls when hair becomes wet.) The instrument is relatively simple and inexpensive but is not particularly precise and must be calibrated occasionally.

Humidity may also be measured by a pair of thermometers one of which has a wet cloth around its bulb. If the humidity of the air is 100 percent, evaporation does not occur, and the temperature readings of the two thermometers will be the same. However, if humidity is not 100 percent, evaporation and cooling take place. Dry air causes a considerable drop in the temperature around the bulb of the wet-bulb thermometer. Thus the difference in the readings of the wet- and dry-bulb thermometers can be calibrated in such a manner that the humidity of the air is directly indicated (the greater the spread, the drier the air).

The *infrared absorption hygrometer* is a newer and superior method of measuring the moisture content of the air. It utilizes a beam of light containing two wavelengths. One of the wavelengths is absorbed by water vapor in the air, whereas the other is unaffected. Therefore, the ratio of energy transmitted in the two wavelengths indicates the quantity of water vapor present in the air sample in the path of the waves.

Exercises and Questions, Chapter 25

1. Describe the main temperature changes that occur at increased altitudes in the atmosphere and explain why these occur.
2. You blow up and seal a balloon on a calm day and release it. Will the balloon rise more readily on a cold day or a warm day? Explain.
3. When driving through a fog, special yellow lights are often more effective than regular headlights. Why?
4. Assume that the Earth's average surface temperature is about 300°K and that the Earth radiates more energy at a wavelength of 10 microns than at any other wavelength.
   (a) What is the temperature of the surface of a star that radiates the maximum amount of energy at a wavelength of 0.5 micron?
   (b) What is the temperature of a planet that radiates the maximum amount of energy at a wavelength of 5 microns? At 30 microns?

(c) Calculate these three temperatures in degrees Fahrenheit.
(d) State the radiation law that you used in your calculations.

5. A low-pressure system occurs at 45 degrees south latitude. Describe the associated winds (relative to the isobars) near the surface and at an altitude of a few thousand feet. Explain why differences occur.

6. Describe in a generalized way the Earth's global rainfall pattern and explain why more precipitation occurs at some latitudes than at others. Construct a graph that approximately relates rainfall to latitude.

7. List at least two criteria that could be used in distinguishing between each of the following pairs of clouds:
   (a) Cirrocumulus vs. altocumulus.
   (b) Cirrostratus vs. altostratus.
   (c) Altostratus vs. nimbostratus.
   (d) Nimbostratus vs. stratus.
   (e) Cumulus vs. cumulonimbus.
   (f) Stratocumulus vs. altocumulus.

8. Identify the clouds in the sky every day of the week at some particular time of the day—say, just prior to a class meeting or at noon. Record your identifications and check their accuracy periodically.

9. If a pail of water were emptied from the top of a skyscraper, what would be experienced by persons on the street directly below?

10. How does rain originate? What evidence has radar provided concerning this problem?

11. Describe and discuss some of the key ideas and data involved in weather modification. Why is evaluation of weather modification experiments controversial?

12. How many calories of heat energy must be added to 10 grams of ice at 0°C to change it to 10 grams of steam at 100°C? Assume that the ice does not sublimate and that the water (after the ice has melted) does not evaporate until it reaches 100°C.

13. A thunderstorm cell 3 miles in diameter may contain 500,000 tons of water that have condensed (some of this would probably be snow and ice, but assume it is all water). How much latent heat energy was liberated as a result of this condensation? How many tons of water could be raised from the freezing point to the boiling point by this quantity of heat energy? (There are 453.6 grams per pound, but use 450 in your calculations.)

14. Assume that 1 inch of rain falls on an area. How many tons of water per square mile does this amount to? (One cubic foot of water weighs 62.4 pounds and there are 63,360 inches in a mile —you may use 60 and 60,000 in your calculations.)

15. Get an old phonograph record, lay a ruler across it, and while the record spins, try to draw a straight line on it, either toward or away from the center. Does this aid in understanding the Coriolis effect?

26 Weather: Air masses, pressure systems, and forecasts

AIR MASSES

Until the technique of air-mass analysis (developed after World War I) clarified weather phenomena greatly, explanations of weather changes seemed confusing and inadequate to most amateur meteorologists. But nowadays it seems obvious that the kind of air blanketing an area will be most important in determining the kind of weather experienced within this area. Air masses normally are not stationary, and they are especially mobile in the belt of westerlies where the dominant direction of air movement is from west to east. Therefore, if the type of air above an area one day moves eastward, it will probably be replaced by a noticeably different kind of air, thus causing a rather sudden change in the weather conditions in that locality—in temperature, humidity, pressure, wind, clouds, and precipitation.

An *air mass* comprises a huge section of the lower troposphere and may extend horizontally for hundreds of miles. At any one altitude above the surface, an air mass has nearly uniform temperature and moisture conditions, or these change gradually from one portion of the air mass to another. However, temperature and moisture conditions commonly are different at different altitudes.

Four types of air masses are of particular importance in causing weather changes in the middle latitudes: *continental polar, maritime polar, continental tropical,* and *maritime tropical.* These terms indicate the chief physical properties of the air masses: hot if tropical, cold if polar, moist if the air accumulated over water (maritime), and dry if the source region was a land area.

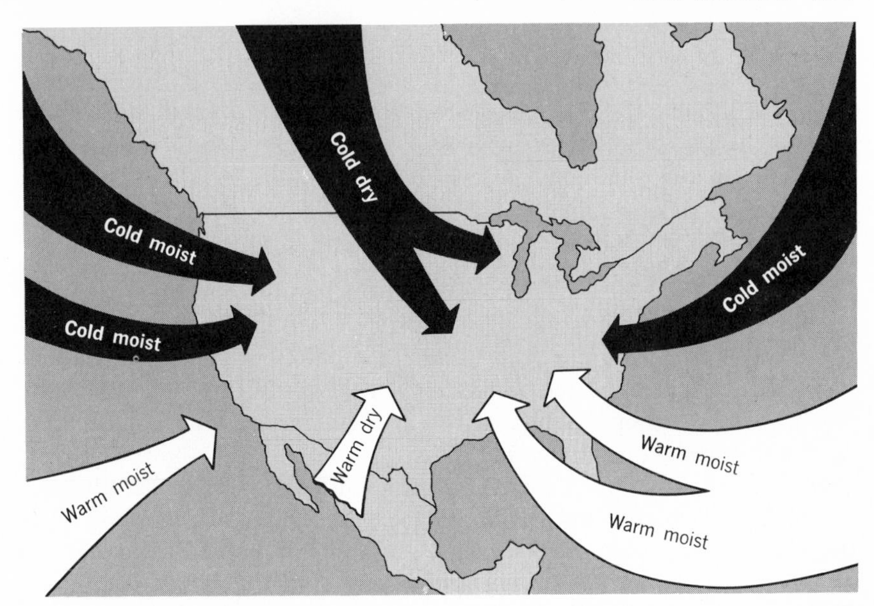

Fig. 26-1. *Air mass source regions for the United States.* [*AF Manual 105-5,* Weather for Aircrews.]

Large sections of the troposphere are more commonly "air-conditioned" in certain regions than in others (Fig. 26-1). This occurs most readily if air stagnates above a large uniform surface or moves slowly across it, whether land or sea. Favorable areas are those of high pressure where air sinks and diverges slowly along the surface: along the horse latitudes and poleward from the polar front. Thus air masses are of two main types—polar and continental.

If air moves across a large area with uniform surface conditions, it may reach fairly complete equilibrium with the surface beneath it in a few days or a week or two. Further changes in the air mass are very slow. Equilibrium develops between the air and the surface by exchanges involving turbulence, convection, evaporation, condensation, and radiation (from the air as well as from the sun and Earth's surface). Equilibrium is attained more rapidly over a warm surface than over a cold surface because of mixing produced by convection. Such air masses may subsequently move hundreds and even thousands of miles from their source regions and still retain enough of their original characteristics to be recognizable.

For example, in North America during the winter, a source of continental polar air is the entire area north of 55 degrees latitude, from Labrador in the east to Alaska in the west. Because of the prevalent snow and ice, this nearly stationary air becomes intensely chilled, particularly near the ground. As the air is coldest and heaviest along the ground, it is stable; little tendency for upward movement exists, and condensation and clouds are not common.

When continental polar air leaves its source area in Canada to travel southward into the United States, its arrival can be recognized readily even by the nonmeteorologist. It produces a cold wave in the winter and sends fuel trucks scurrying quickly from one house to another. In the summer, its vigorous gusty arrival from the northwest may herald a welcome relief from a preceding heat wave. Occasionally tongues of continental polar air reach southward into Florida. The Earth's surface in the United States is commonly warmer than such air, and some parts are commonly warmer than others. As a result, the bottom air is heated and rises, not uniformly, but in great ascending bubbles or columns (thermals) something like the boiling of water on a Gargantuan scale.

A typical spring day in New York inside a mass of polar Canadian air would dawn with cloudless sparkling blue skies. During the morning, solar radiation becomes increasingly intense and initiates the upward movement of large blobs of air. These upward-moving air currents become visible when they reach altitudes at which condensation occurs; white cottony puffs of cumulus clouds develop. By late afternoon the sky may be nearly covered by such clouds. However, after sunset, the upward movement decreases, the clouds disappear by evaporation, and a brilliant, cool, starlit evening results.

Pilots and their passengers know when they fly through such currents by the upward bump as they enter and the sudden drop as they leave. Glider pilots can attain great altitudes by circling within these rising masses of air. Dust is carried upward and a clear sparkling day results. Precipitation is not common from such clouds, for little moisture is present. However, cumulonimbus clouds and thunderstorms may develop.

A different reaction is involved when maritime tropical air moves northward and northeastward from the Gulf of Mexico into the United States (Fig. 26-1). The ground in winter is colder than such an air mass, the bottom air becomes chilled, stability develops, and moisture readily condenses into clouds of the stratus type. Precipitation is common along the forward edge of the advancing air mass if it is forced to climb over some barrier (e.g., over a mountain or a denser air mass).

The shape of the North American continent—wide in the north and narrow in the south—explains why continental polar and maritime tropical air masses are the two most important to affect the United States, especially its central and eastern portions. However, maritime polar air is also a frequent visitor.

The type of weather an air mass brings to a certain location depends upon several factors: conditions in the source area that produced the air mass, conditions in the areas over which it has traveled since leaving the source region, and finally the rate at which it has moved. As an air mass moves away from a source area, it may be modified by heating or cooling from the surface, by the addition or subtraction of moisture, and by mixing resulting from turbulence, subsidence, and uplift. Vertical movements tend to occur if the bottom portion of an air mass is heated, if it crosses a mountain, or if two unlike air masses converge. Cloudiness and precipitation are likely if an uplifted air mass contains abundant moisture.

TEMPERATURE CHANGES IN VERTICAL AIR CURRENTS

Measurements made at different altitudes and seasons and over both land and sea show that temperatures commonly decrease upward in the troposphere at an average rate of about 3.5 F°/1000 ft. This is called the observed lapse rate because temperatures tend to lapse or decrease upward (Fig. 26-2), and it relates to air that is not actively rising or falling. A steep lapse rate is one in which larger-than-average differences in temperature occur at increased altitudes.

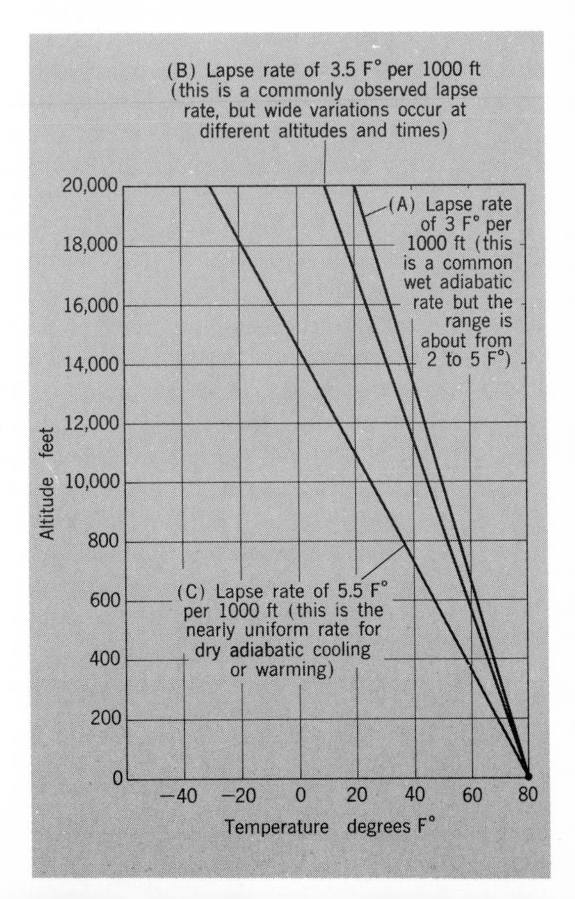

Fig. 26-2. *Three typical lapse rates plotted on a temperature-altitude graph. Relative to the horizontal axis, the so-called steepest lapse rates (i.e., those that show the largest temperature changes per 1000 feet) actually have the least steep slopes, a paradox which may lead to some confusion. On the graph, (C) shows the steepest lapse rate, and (A) has the least steep lapse rate.*

If a parcel of air is rising, it expands and cools at the nearly uniform dry adiabatic lapse rate of 5.5 F°/1000 ft (p. 508). However, when condensation begins in this rising parcel, very large quantities of heat energy are released, and the air cools at a slower, less steep rate—at the wet adiabatic lapse rate, which varies with the temperature and moisture content. The wet adiabatic lapse rate may average about 3 F°/1000 ft in warmer air, but it ranges from about 2 F° to 5 F°/1000 ft; it is steepest in very cold and therefore dry air, in which it approximates the dry adiabatic lapse rate.

When a parcel of air subsides, it is warmed at comparable dry and wet adiabatic lapse rates. If evaporation occurs as air descends, large quantities of heat energy are absorbed, and thus its temperature increases less rapidly. Because of these adiabatic processes, air on the lee side of a mountain may have a higher temperature than that on the windward side at the same altitude (Fig. 25-23).

Rising masses or parcels of air are called thermals, and they attain peak velocities in thunderstorms where speeds may at times approach 150 to 200 mi/hr. They form as part of a convectional circulation because of localized heating at the bottom or localized cooling at the top (less common).

The buoyancy of a rising parcel of air depends upon the difference in density at any particular level between it and the air that it is rising through —that surrounds it at this level (Archimedes' principle, Chapter 1). The buoyancy tends to be greatest when a parcel of rising warm moist air, cooling at a slow wet adiabatic lapse rate, moves upward through air that has a steep observed lapse rate. The differences in density are caused chiefly by differences in temperature, but a difference in moisture content is also a factor. At comparable temperatures and pressures, moist air is lighter than dry air because each water-vapor molecule in the air takes the place of a more massive oxygen or nitrogen molecule (Avogadro's principle, p. 126).

The stability or instability of the atmosphere is an important factor in determining the type and magnitude of the weather changes that occur in air that rises or falls. The processes involved can be illustrated by imagining what would happen under different conditions to a small parcel of air that is made to move upward or downward through the atmosphere. Such vertical movements do not start spontaneously. They must be initiated by a push of some kind (e.g., as part of a convectional circulation, because of turbulence caused by winds moving across a rough surface, or because of convergence or divergence). Under different conditions, the air is said to show absolute stability, neutral stability, absolute instability, and conditional instability (Fig. 26-3).

If the air is in a state of *absolute stability,* a parcel, upon being displaced,

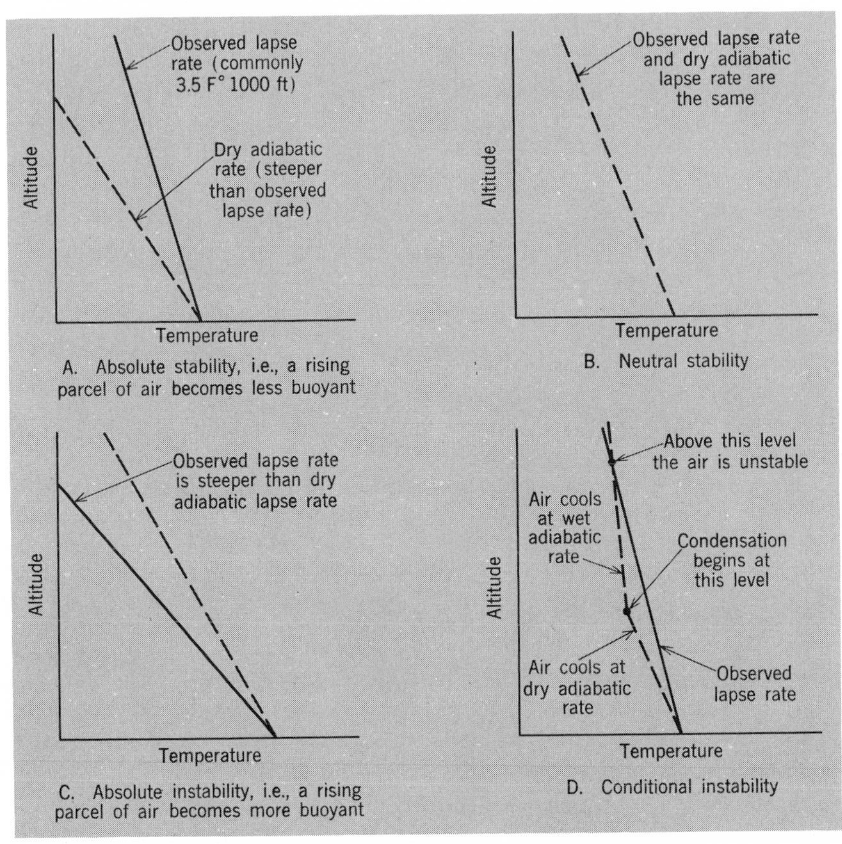

Fig. 26-3. *Conditions illustrating stability and instability are plotted on four temperature-altitude graphs. The steep observed lapse rate shown in (C) is common during the warmer months in desert areas but uncommon elsewhere.*

tends to return to its original level (Fig. 26-3A). Thus stable air moved downward becomes more buoyant and returns to its former level.

If upon being displaced a parcel remains at the level to which it was moved by a push of some sort, then it is in a state of *neutral stability* (Fig. 26-3B). After the pushing effect stops, the parcel neither returns to its original level nor continues in the direction toward which it was being displaced.

The air is in a state of *absolute instability* if upon being displaced a parcel continues (after the force is removed) to move in the direction toward which it was being displaced, either up or down (Fig. 26-3C). Thus unstable air, moved downward, becomes less buoyant and continues to sink. A steep lapse rate favors instability.

Conditional instability exists when the observed lapse rate in the air is less than the dry adiabatic lapse rate and more than the wet adiabatic lapse rate (Fig. 26-3D). A common observed lapse rate is 3.5 F°/1000 ft. Rising unsaturated air tends to cool at about 5.5 F°/1000 ft, and thus it becomes less buoyant as it rises. However, if it has been given a strong upward push, the air may rise high enough so that condensation begins and heat energy is released. The wet adiabatic rate of cooling upon ascent is frequently less than the observed lapse rate that exists in the air surrounding the rising parcel. Thus the air becomes relatively lighter and more buoyant as it moves upward, and its velocity is increased. This is known as conditional

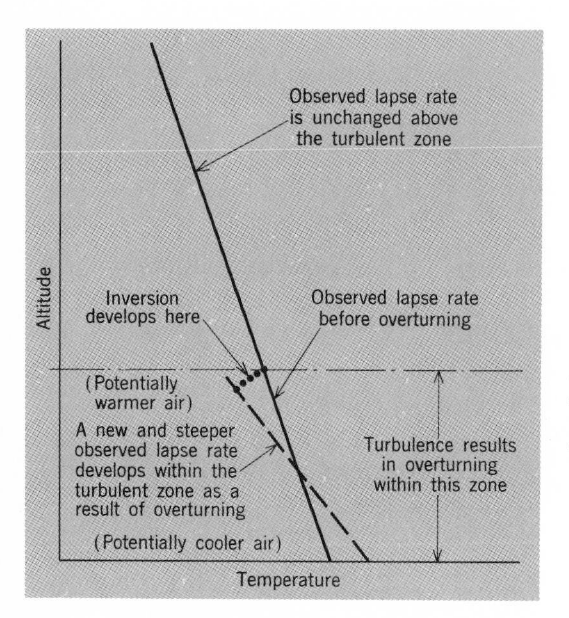

Fig. 26-4. *Development of an upper-air inversion. Since the dry adiabatic lapse rate is usually steeper than the observed lapse rate, air at the surface is potentially cooler than air that occurs at a certain altitude above it—say at 4000 feet. Assume an observed lapse rate of 3.5 F°/1000 ft. Therefore, if a parcel of surface air rises to 4000 feet, its temperature will drop 22 F° (assume no condensation). However, air already at 4000 feet will be only 14 F° cooler than the surface air.*

instability because it depends upon (is conditioned by) the cooling of a rising parcel at a wet adiabatic lapse rate that is less than the observed lapse rate. Under favorable circumstances (steep lapse rate in the surrounding air and much moisture in the rising parcel) such an upward movement, as part of a giant cumulonimbus cloud, may penetrate for 1, 2, or even 3 miles into the stratosphere. However, the buoyancy decreases rapidly as the air moves upward through the stable stratosphere (cooler and heavier at the base).

Although temperatures generally decrease upward through the troposphere, exceptions are rather common. These occur locally, temporarily, and at different altitudes and are called *inversions*—temperatures increase upward for a certain distance before they resume the normal lapse in temperature at still higher altitudes (Fig. 26-4). To bring about a temperature inversion, either the lower part of a mass of air must be chilled or an upper portion must be warmed. Shallow temperature inversions are associated with the formation of some types of fog (p. 506).

An inversion acts as a lid because it hinders the upward movement of parcels of air through it. This occurs because the upward moving parcel is cooling off, whereas the temperature of the air it moves through is becoming higher. This rapidly eliminates any buoyancy the parcel might have at the base of the inversion, and its upward movement stops. The effects of air pollution are thus greatly intensified beneath an inversion, especially if little wind occurs. The pollutants become concentrated in the layer beneath the inversion instead of being diffused throughout a very large volume of air.

AIR-MASS vs. FRONTAL WEATHER

Most of the United States occurs within the belt of westerlies and has prevailing winds that blow from the southwest and west. Polar air masses tend to develop north of this belt and tropical air masses south of it. These converge along a zone that is called the polar front (Fig. 25-11). Pressures north of the polar front occasionally become great enough to send a huge blob of continental polar air into the United States, thus shifting the polar front far to the south at this time and place. At other times, pressures in the vicinity of the horse latitudes build up sufficiently to push maritime tropical air northward across the United States and into Canada. Thus the polar front advances and retreats. Once these contrasting air masses move into the United States, they are shifted across it in a generally eastward direction, and a few days of continental polar air-mass weather tend to alternate with a few days of maritime tropical weather.

Because air masses are very large—sometimes they span more than 1000 miles—a few days may be needed for any one air mass to pass a given locality. During this passage, one experiences air-mass weather; the same general weather conditions are repeated each day, although gradual modifications occur from one day to the next as the physical conditions of the air mass change slowly. Major weather changes occur as the rear section of one air mass moves past a certain location and is replaced by the leading edge of a different air mass. Thus the more sudden weather changes occur along the margins of air masses, and we may speak of air-mass weather vs. frontal weather.

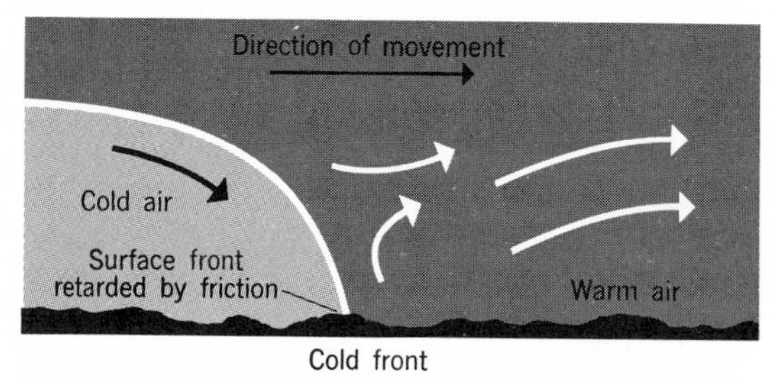

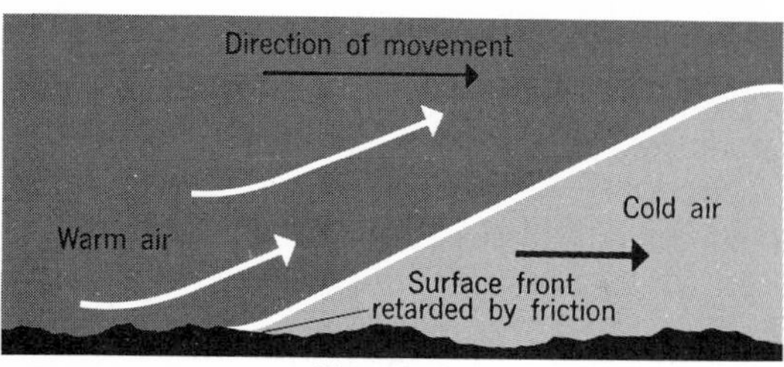

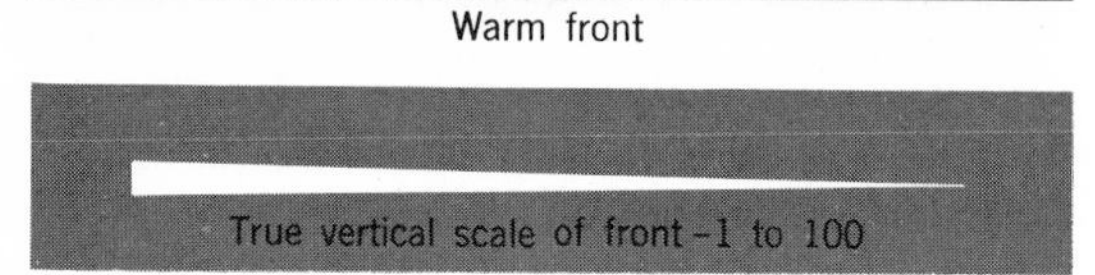

Fig. 26-5. *Vertical cross sections showing frontal slopes. Note that surface friction acts to steepen a cold front but makes a warm front less steep. Cold fronts and warm fronts have slopes that range from about 1 to 40 through 1 to 400.* [*TM 1-300,* Meteorology for Army Aviation.]

FRONTS

Fronts are the transitional zones that occur between two contiguous air masses (Fig. 26-5). Although shown on weather maps as lines, fronts are mixed zones that range from about 3 to 50 miles in width. Temperature and moisture conditions tend to be noticeably different on opposite sides of

Fig. 26-6. *Photograph of a cold front. This dust storm occurred at Manteer, Kansas, in April 1935. The dust was picked up by the advancing cold air.* [*Courtesy, U.S. Weather Bureau.*]

a front. Two air masses do not mix readily, and the transition zone between them is relatively narrow. Lines with appropriate symbols are drawn on weather maps to show where the base of a front touches the surface. Commonly these lines are convex in the direction of movement and may extend for hundreds of miles. The front may slope upward from the surface for 1 to 2 miles or less, or it may extend to the top of the troposphere.

Different types of fronts have been recognized: cold, warm, stationary, occluded (two types), and upper air fronts. Each of these may be prominent and persistent or weak and evanescent. An important factor here is the stability or instability of the uplifted air mass.

The forward margin of an advancing mass of polar air is aptly termed a *cold front* (Fig. 26-6) and certain weather changes are characteristically associated with it (Figs. 26-7 and 26-8). Because cold air is heavier than warm air, it stays near the surface and wedges under warmer air that occurs ahead of it. This warmer air is shoved upward along the frontal zone and it rises, expands, and cools. Therefore, towering cumulus and cumulonimbus clouds tend to develop along the forward margin of the cold air, and sudden heavy precipitation is characteristic of the passage of a cold front. From its line of contact with the Earth's surface, a cold front slants back-

Fig. 26-7. *Fast-moving cold front.* [*AF Manual 105-5,* Weather for Aircrews.]

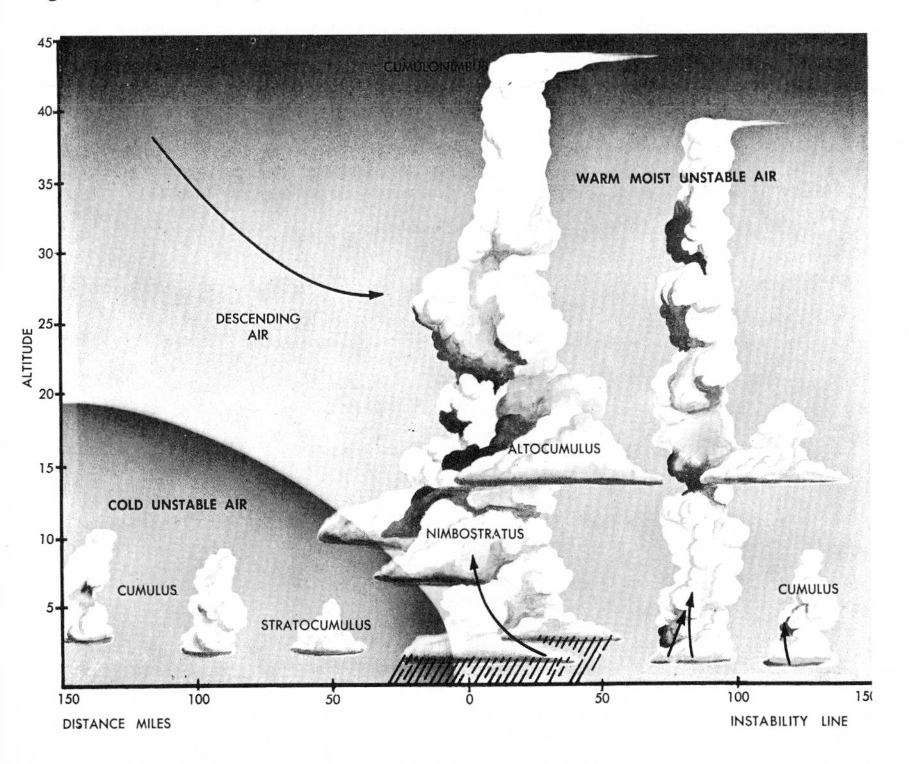

ward and upward over the cold air mass at a slope that may average about 1 mile in a vertical direction for each 40 to 80 miles along the Earth's surface. Cold fronts tend to move southeastward into the United States.

The amount of precipitation and turbulence associated with the arrival of a cold front depends upon the nature of the two air masses involved and the velocity of the front. Extreme conditions result if very warm moist unstable air is underrun by fast-moving very cold air. If a front moves rapidly (up to 60 mi/hr on occasions, but commonly less than half of this rate), the slope at its forward margin is steepened because of friction with the ground. Thus the warm air is shoved upward vigorously in a relatively narrow zone located chiefly in advance of the cold front (Fig. 26-7). On the other hand, if a cold front moves slowly, its slope is less steep, and the warm air ahead of it can slide upward gradually along the frontal surface. Thus layered clouds can form and extend for many miles behind the base of the front (Fig. 26-8). Although clouds tend to develop all along a cold front, they may be absent in places if the air is very dry or is shoved upward less vigorously. Thus precipitation tends to occur in association with thunderstorms scattered here and there along the front.

In a *warm front* (Fig. 26-9) the warm air mass advances more rapidly than the colder air ahead of it. The warmer lighter air glides gradually upward above the colder air along a gently sloping boundary zone (approximate range: 1 to 100 / 1 to 300). Surface friction tends to slow the motion of the base of a warm front, and this decreases the slope. Because a cold front slopes in the opposite direction, friction with the ground tends to steepen its slope.

Cirrus clouds at the top of a warm front may be at an altitude of several

Fig. 26-8. *Slow-moving cold front.* [*AF Manual 105-5,* Weather for Aircrews.]

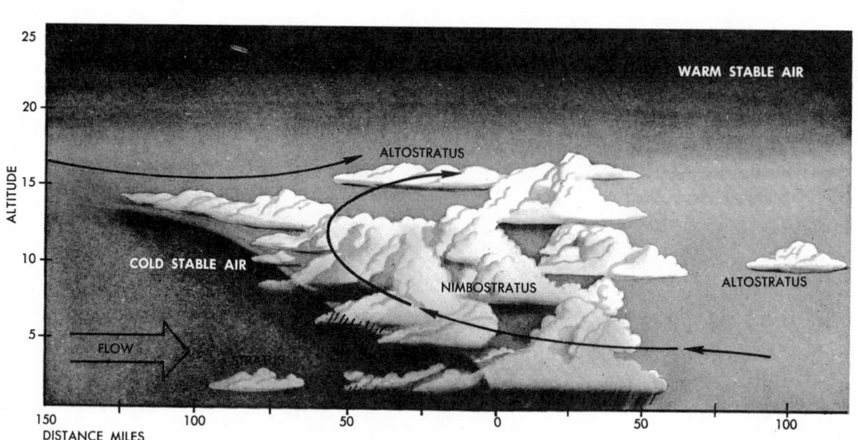

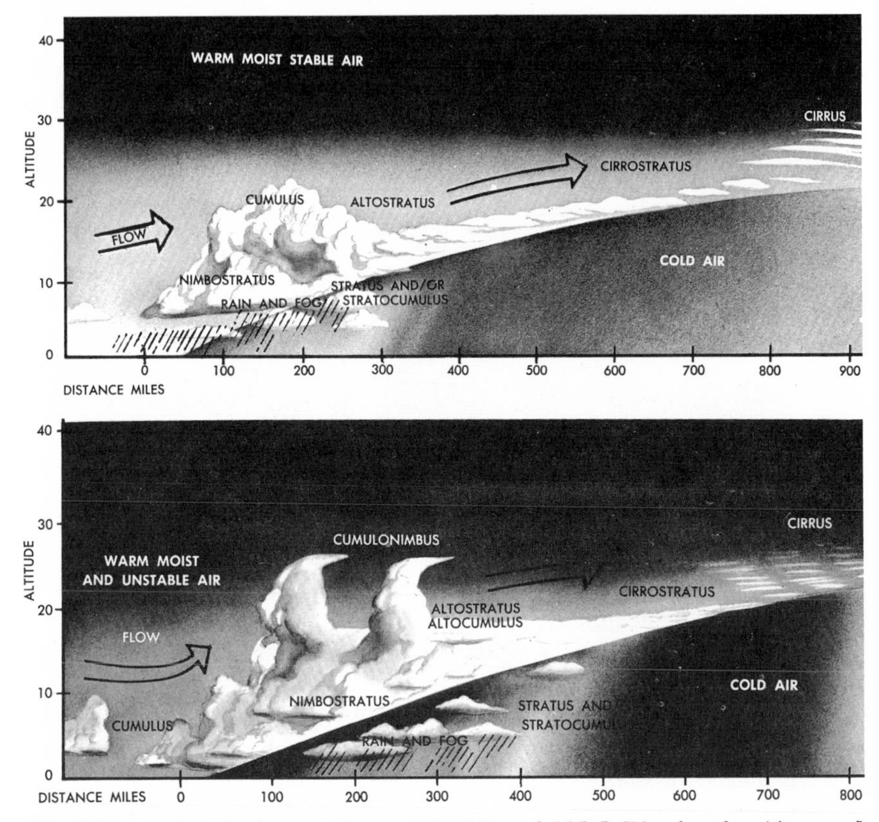

Fig. 26-9. *Two typical warm fronts.* [*AF Manual 105-5,* Weather for Aircrews.]

miles and 500 to 1000 miles ahead of the base of the warm front. Layered clouds are commonly associated with a warm front. As a warm front approaches, one observes that thinner, higher types of clouds are gradually replaced by lower, denser varieties. Cirrus is followed by cirrostratus and perhaps by cirrocumulus—the mackerel sky that may be a sign of approaching precipitation. Alto-type clouds come next and precipitation may begin from altostratus. Precipitation continues from nimbostratus clouds that are next in the procession and that may extend a few hundred miles ahead of the base of the warm front.

Rain falling through cooler air beneath a sloping frontal zone may add enough moisture by evaporation to saturate the cold air in places and produce fog and stratus or stratocumulus clouds. Warm front precipitation generally covers a wider area, lasts longer, and is less intense than that associated with a cold front. If warm moist unstable air moves upward along a warm front, cumulus and cumulonimbus clouds may also form; thus thunderstorms occasionally occur in association with warm fronts. As with

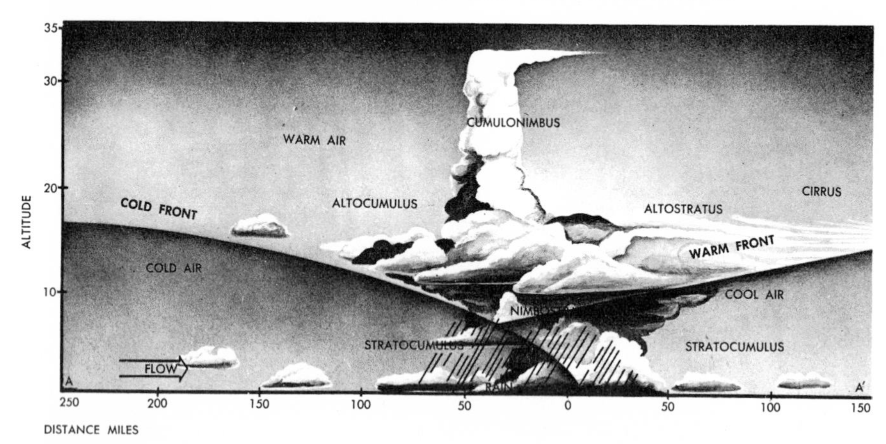

Fig. 26-10. *A cold-front type of occlusion.* [*AF Manual 105-5,* Weather for Aircrews.]

cold fronts, no two warm fronts are identical. Weather conditions depend upon the rate at which the air masses move and the extent of their temperature and moisture differences. Warm fronts in the United States commonly move eastward and northeastward.

At times a warm air mass may be in contact with a cold air mass, and yet neither air mass actively advances or retreats. In such instances the frontal surface is called a *stationary front.* On a weather map, a maritime tropical air mass may be separated from a continental polar air mass by a long sinuous line. This line shows warm front symbols in an area where the warm air is advancing; the same line shows cold front symbols in an ad-

Fig. 26-11. *A warm-front type of occlusion.* [*AF manual 105-5,* Weather for Aircrews.]

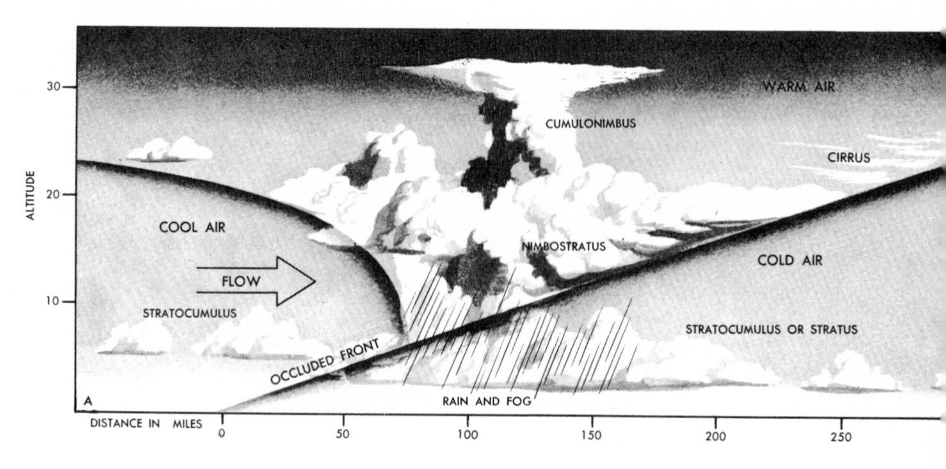

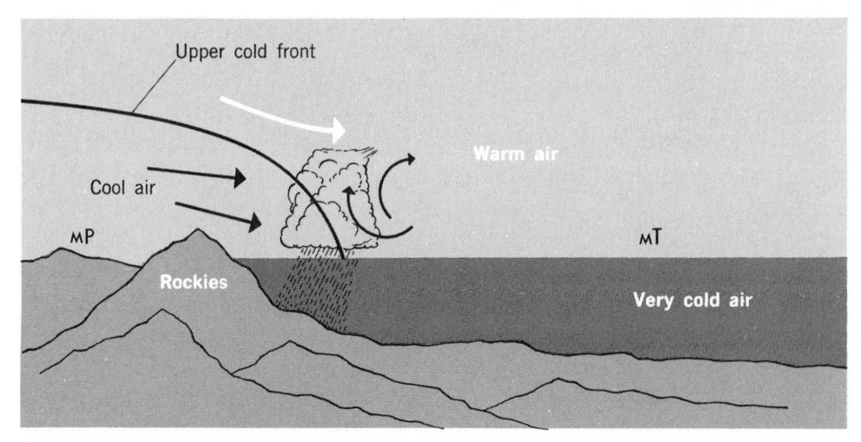

Fig. 26-12. *Upper air cold front. A mass of maritime polar air moves eastward across the Rocky Mountains but cannot descend through very cold and dense continental polar air that blankets the surface eastward from the mountains. Therefore, the maritime polar air continues eastward along the upper boundary of the layer of continental polar air and interacts with warm air in its path (overlying the continental polar air) in a typical cold-front fashion.* [*Courtesy, U.S. Weather Bureau.*]

joining area where the cold air is advancing; and it is marked by stationary front symbols in still another area where neither air mass is advancing.

Occluded fronts form because cold fronts tend to move eastward faster

Fig. 26-13. *Prefrontal squall line. TM 1-300,* Meteorology for Army Aviation.]

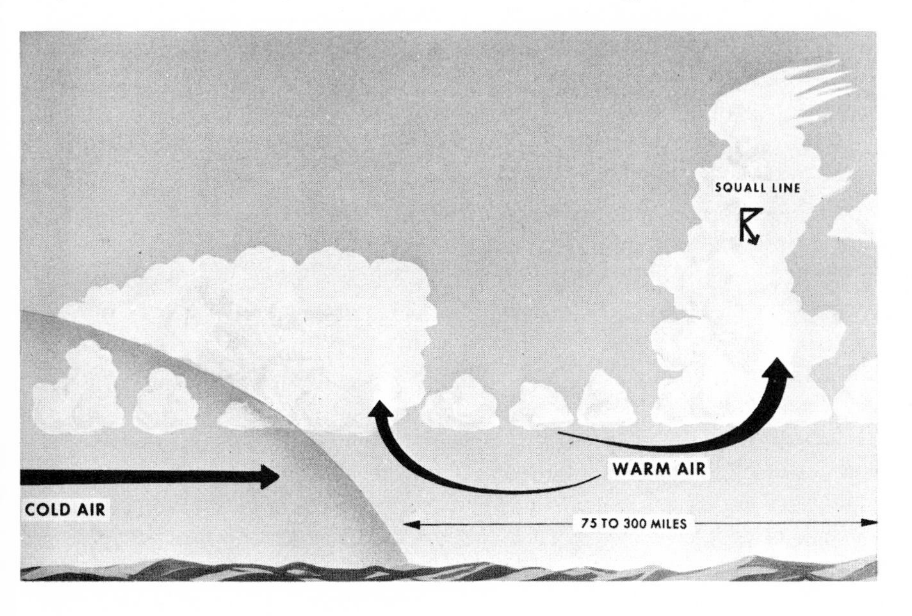

than warm fronts (cold air can displace warm air along the surface more readily). Thus a cold front may overtake and lift above the ground a large part of the northern portion of a warm front. On a weather map, a smooth curved line frequently connects a cold front with the occluded front which is an extension of it. Two types of occlusion occur: a *cold-front type* (Fig. 26-10) and a *warm-front type* (Fig. 26-11). The type is determined by differences in density (due primarily to temperature differences) between the cold air masses that are situated ahead of and behind the occluded front. A plane flying at a certain altitude through an occluded front will thus pass through three different air masses and two frontal surfaces.

An *upper air front* is shown in Fig. 26-12 and a *prefrontal squall line* in Fig. 26-13. The cause of squall lines is uncertain, but they are often accompanied by violent thunderstorms and occasionally by tornadoes. A squall line may be several hundred miles in length and is likely to parallel a cold front that occurs behind it. A squall line may occur some 50 to 300 miles ahead of a cold front (in the United States, to the east of it). A typical squall line lasts less than 24 hours and moves at 20 to 25 mi/hr.

LOWS AND HIGHS

Barometric readings from widely scattered stations show that atmospheric pressure varies considerably at any one time at different stations and at any one station over a period of time. However, the pressure differences between any two cities are commonly less than those between the top and bottom floors of a tall building. When lines (*isobars*) are drawn on a map through the locations of stations which have the same barometric pressure, definite centers of low or high pressure commonly appear. Variations in pressure caused by altitude differences are eliminated by corrections which reduce all of the readings to sea level. These pressure systems are large: diameters range from a few hundred to more than a thousand miles. The *lows* (cyclones) and *highs* (anticyclones) have characteristic conditions of wind, temperature, clouds, and precipitation associated with them.

Lows probably form in more than one way, perhaps most commonly in association with the polar front (Fig. 26-14). Initially a stationary front may separate continental polar air in the north from maritime tropical air in the south. The air masses move parallel to the front and do not mix readily. However, at some irregularity along the boundary, the warm air may begin to flow toward the cold air and upward. This northward bulge develops into a warm front. The cold air on the north side of the irregularity pushes southward forming a cold front. The motion is somewhat similar to the passage of people through a revolving door. A current of warm air rises in the center because of the convergence of the two different air masses and causes low barometric pressure.

At locations away from the center, air moves inward along the Earth's surface and is deflected to the right of the direction of movement in the Northern Hemisphere until it is blowing at an angle to the isobars. Thus a counterclockwise circulation originates about the low-pressure area, and winds spiral inward and upward toward the center. Lows tend to be elliptical in shape and elongated in a northeast-southwest direction. Eastward and northeastward movement of lows in the United States averages about 20 mi/hr in the summer and 30 mi/hr in the winter (about 500 to 700 mi/day). Because warm moist air moves upward near a low-pressure center, clouds and precipitation are common.

It was once thought that lows developed because of convectional circulation in a warm air mass. However, this hypothesis was discarded when it was learned that lows are more common in winter than in summer and that they form over the oceans as well as over the lands; neither condition is favorable for the heating of the bottom of an air mass to produce convection. Also, two or more air masses and two or more frontal systems (warm, cold, and occluded) are associated with mature lows.

Fig. 26-14. *Development of a typical middle-latitude low. Six stages are shown. As the cold front moves faster than the warm front, the area of warm air in the system becomes narrowed until occlusion occurs. The whirling counterclockwise circulation then weakens in the cold air. Variations of this manner of formation are common.* [*Courtesy, U.S. Weather Bureau.*]

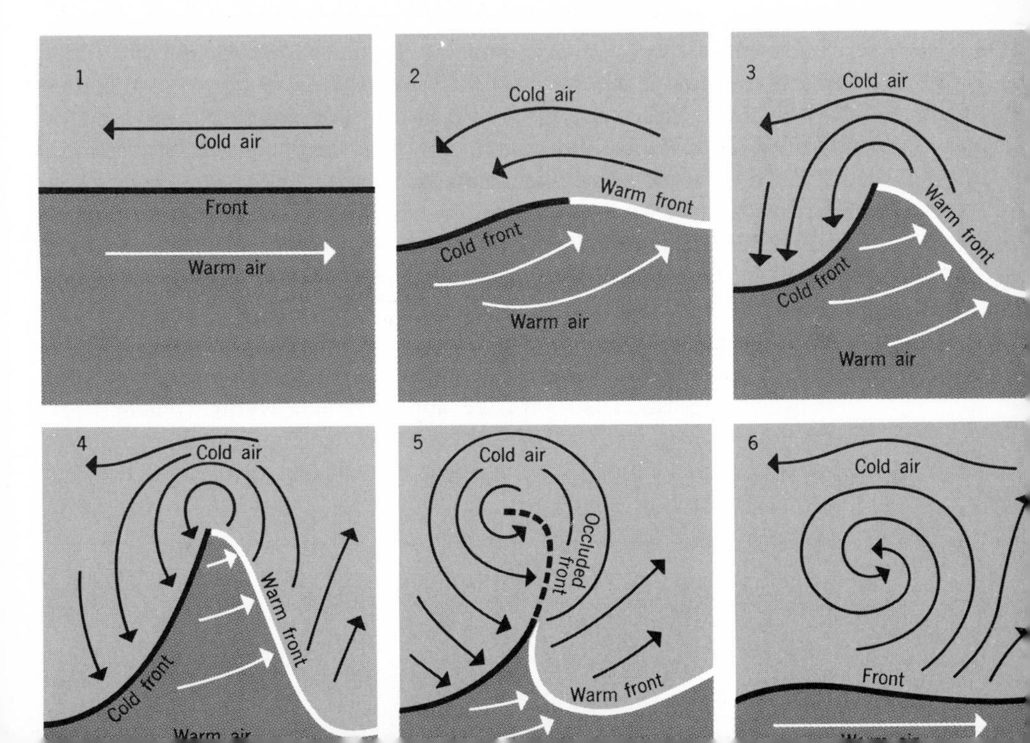

High-pressure areas are commonly much larger than low-pressure areas. In a high-pressure system, air subsides at the center and diverges along the Earth's surface outward from the center; it is deflected to the right and spirals outward and around the center in a clockwise circulation in the Northern Hemisphere. A high-pressure system commonly forms as part of a continental polar air mass that moves southeastward into the United States.

As air moves horizontally from high- to low-pressure areas because of the pressure differential, winds are strong when the difference in pressure is great. If the pressure lines (isobars) are close together on a map, winds are strong, and the pressure system is well developed. On the other hand, if the pressure lines are far apart, winds tend to be gentle, and the pressure system is not intensively developed.

Along the surface as a result of friction, winds tend to blow at acute angles to the pressure lines (about 10 to 40 degrees). At a higher altitude, winds tend to parallel the isobars—friction is negligible, velocities are greater, and the Coriolis effect is more pronounced. Winds also tend to blow more nearly parallel to the isobars over water than over land (less friction, higher velocity, and more pronounced Coriolis effect).

Highs and lows are less common and less intense during the summer months than in the winter months and move at a slower rate and along different paths. In the Southern Hemisphere, movement of air in the low- and high-pressure systems is reversed—clockwise in a low, and counterclockwise in a high.

Fig. 26-15 relates pressure systems, air masses, fronts, and characteristic weather conditions as they tend to develop in the United States. An observer is located in the eastern part of the United States. A low-pressure center has developed along the boundary separating continental polar from maritime tropical air. On Monday the low-pressure center is situated about 600 miles west of the observer. A warm front extends southeastward from the center, and a center of high pressure is located to the northwest in the continental polar air mass. The air masses, fronts, and pressure systems are all moving to the east and northeast. Weather data collected during preceding years suggest that the center of the low will pass north of the observer's location. A movement of about 600 miles per day for the pressure systems is typical.

A person at the observation point commonly experiences the following weather changes during the warmer months. First, certain phenomena take place which are associated with the approach of a warm front and low-pressure system: on the first day, say Monday, the barometer falls slowly; temperature and humidity remain about the same or rise gradually; winds are from the east or southeast; and a few high harmless-looking clouds

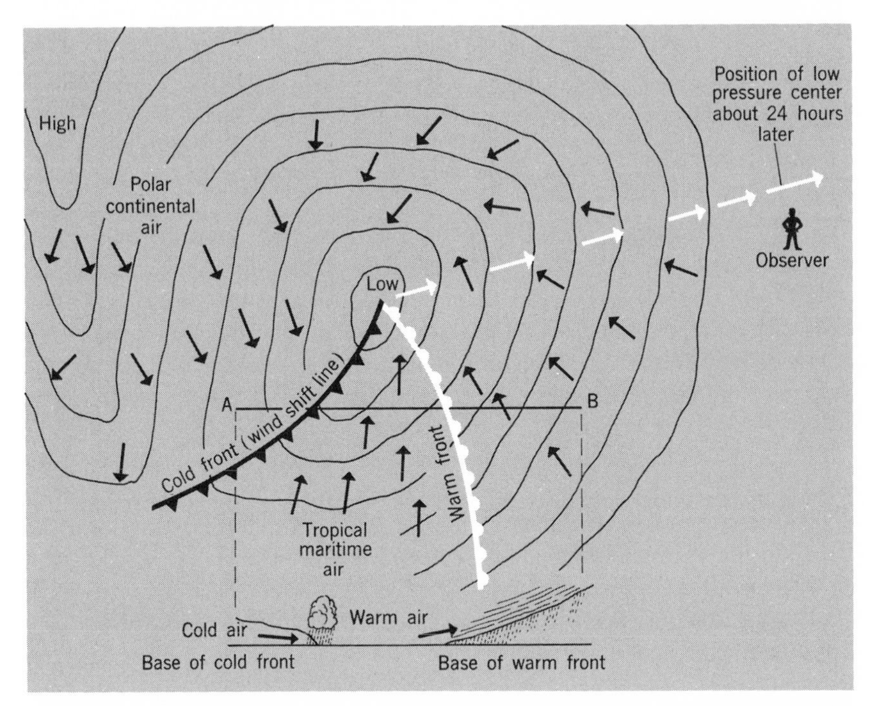

Fig. 26-15. *The diagrammatic weather map above shows continental polar and maritime tropical air masses associated with well-developed high- and low-pressure areas with accompanying cold and warm fronts. The future path of the low pressure center is plotted, and its probable location the next day is marked. The high-pressure center will move southeastward and then follow the low to the east. Note the counterclockwise circulation which has developed around the low-pressure system and the clockwise movement around the high. AB is a cross section through the cold and warm fronts. The cold front commonly overtakes the northern part of the warm front, thereby forming an occlusion.*

(cirrus) appear in the west. Several hours later the clouds are thicker and lower, and the wind blows from the southeast or south. The sky next becomes completely overcast (altostratus), then the nimbostratus clouds arrive, and a light rain falls steadily for a number of hours.

On Tuesday the lower part of the warm front passes over the observation point. It is warm and humid, and the sky may clear. The barometer is low and steady, and the wind blows from the south or southwest.

At a later hour on Tuesday, the cold front approaches the observer. Its presence is heralded by a high cloud bank capped here and there by huge towering masses (cumulonimbus) which appear in the west and move eastward. Thunderstorms and brief heavy downpours are common as the cold front passes over an area, although the storms occur here and there along the front.

By Wednesday the sky has cleared, and by afternoon it contains white puffs of fair-weather clouds (cumulus) which sparkle against the blue background. The barometer is rising, and humidity is low; it is cool, and the wind blows vigorously from the northwest. The observer can now predict with some confidence that he will experience these same general weather conditions (air-mass weather) for another day or so as the continental polar mass moves by, although each day should be a little warmer than the previous one. However, a watchful eye should be kept on the barometer, the wind, and on the southwestern sky for the first signs of the approach of the next low.

Many variations centering about these general weather changes are possible, and the changes may be quite different locally from those described. The fronts may move at different rates of speed in different directions; the low- and high-pressure systems may be weakly or strongly developed; and the air masses themselves may vary in physical properties. For example, in the Kansas flood of 1951, the front part of a cold air mass remained nearly motionless over Kansas for about four days instead of moving eastward as

Fig. 26-16. *Principal tracks of lows in the United States. The lows tend to form in the southeast, the southwest, or the northwest. Their paths tend to converge on the northeastern part of the United States.* [*Adapted from U.S. Weather Bureau map.*]

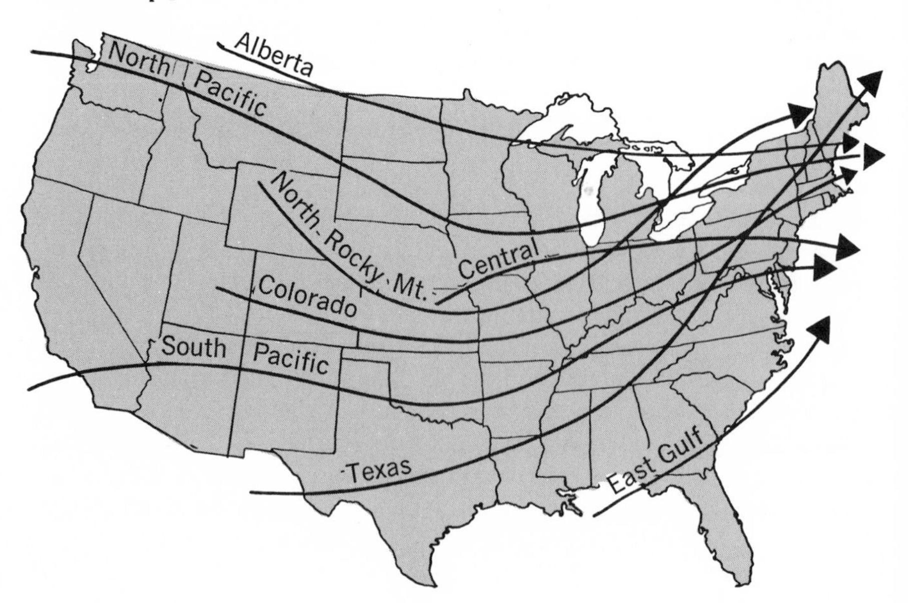

normally happens. During this time, warm moist air from the Gulf of Mexico moved northward and upward over the cold air, shedding torrents of rain on the surface below.

The rate at which a storm moves is likely to be an important factor in the quantity of precipitation that falls on a given location (a slow rate usually means a greater accumulation). Furthermore, a relatively small change in the direction of a storm during the colder months can determine whether a certain city receives snow, sleet, or rain; e.g., an inch of rain may fall instead of 10 inches of snow. Thus not all weather forecasts are accurate.

The belt of westerlies has been likened to a huge river which flows in an easterly direction across the United States. In this river, low- and high-pressure systems are like gigantic moving eddies with their counterclockwise and clockwise circulations. In fact, the eddies are so numerous that they obscure the general eastward movement of the river itself. The eddies develop somewhat periodically and follow each other eastward, some persisting much longer than others. This eastward procession of lows and highs and the passage of large masses of different kinds of air cause the alternation of weather conditions common to the middle latitudes. For this reason, a telephone call concerning the weather to a friend located some distance to the west will often aid one in predicting the type of weather which is headed eastward and which is due to arrive in one's local area several hours later. Although the general trend of movement is eastward, from the Mississippi basin the most frequented tracks of the weather disturbances are northeastward across the New England states or northeastward to the Great Lakes and then along the St. Lawrence Valley (Fig. 26-16). Some storms cross the Atlantic and travel eastward as far as Siberia.

WEATHER FORECASTING

Weather forecasting is based upon information concerning the whereabouts of different air masses, pressure systems, fronts, and their movements at any one time (Fig. 26-17). The most important elements of the weather are (1) temperature of the air, (2) atmospheric pressure, (3) direction and speed of the wind, (4) humidity, (5) type and amount of cloudiness, and (6) amount of precipitation. This information is then interpreted in the light of past experience which indicates the paths followed most frequently by low- and high-pressure systems and the rates at which they commonly travel at different times of the year. Here electronic computers are of great value. In addition, the characteristic reactions of different air masses to various ground conditions must be known by the weather forecaster if he is to predict future weather changes with some accuracy.

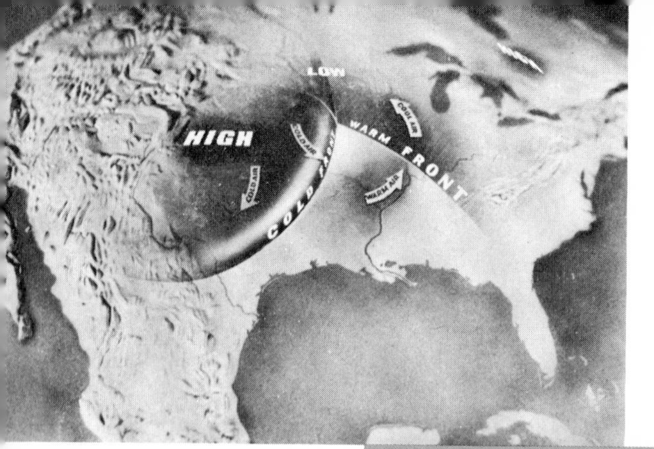

Fig. 26-17. *Weather map. If air masses were visible, the frontal system drawn on the small map might appear as shown in the three-dimensional illustration. A cold-front type of occlusion has been created north of the surface junction of the two fronts.* [*Courtesy, U.S. Weather Bureau.*]

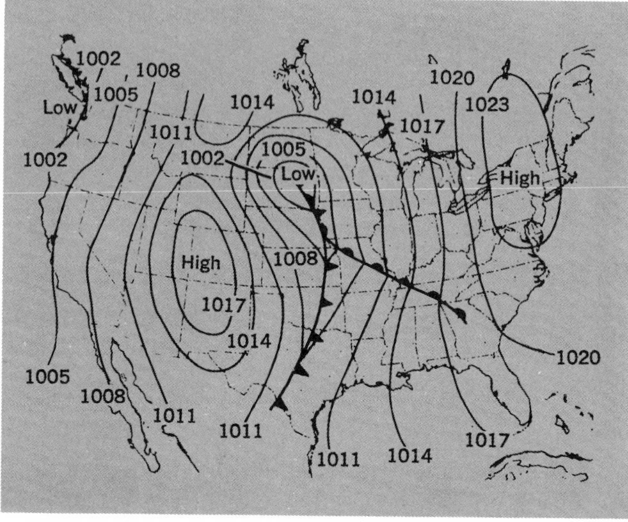

In general, accurate predictions depend upon speedy interpretations by experienced observers of a large amount of weather data recorded simultaneously by weather observers distributed here and there over a very large part of the Earth's surface and then collected rapidly and analyzed at a central office (the synoptic weather map). Weather records are made from stations on land, ships at sea, planes and balloons in the air, and more recently from various types of automatic devices and specially designed weather satellites orbiting the Earth. Since the weather that affects any one country tends to approach it from some other country or from the oceans, adequate forecasting demands the availability of nearly global weather data and its study on the same worldwide basis.

Weather forecasting as a public service began about 100 years ago. The invention of the telegraph (about 1840) and radio (about 1900—needed for data from the oceans) were necessary before modern forecasting methods could advance. The first radiosonde weather records were made in the 1930's (Fig. 25-16). A balloon lifts the radiosonde instruments to altitudes of 15 to 20 miles before it bursts. Such a balloon can be tracked to provide information concerning winds at higher altitudes (a balloon can be designed and inflated to rise at a known rate).

An interesting weather instrument is a self-contained automatic weather station called the Grasshopper. This instrument may be parachuted from a plane to inaccessible areas. An automatic clock is started before the weather station is dropped, and this clock controls all future activity. A small explosive disengages the parachute after the landing. Six other charges cause legs attached to springs to stick out and place the weather station in an upright position. Another explosive sends up a 20-foot antenna. Weather-responsive devices then function to turn a radio transmitter on and off. A receiving station can change the radio signals into temperature, pressure, and humidity readings. The automatic weather station can also be used as a radio beacon. Similar automatic weather stations are being stationed on buoys located here and there on the oceans.

Tiros I (Television Infrared Observation Satellite—Fig. 26-18) was launched in April 1960, proved to be highly successful, and has been followed by other Tiros satellites. Thousands of television pictures of clouds, weather systems, and the Earth's surface have been made (Fig. 26-19). A Tiros satellite may follow a nearly circular orbit about the Earth at an altitude of about 500 miles and complete a revolution in about 100 minutes. Pictures may be taken in areas remote from readout stations at the surface.

Fig. 26-18. *Major orbital and ground elements of the Tiros weather system.* [*Courtesy, Robert Jastrow, Goddard Space Flight Center.*]

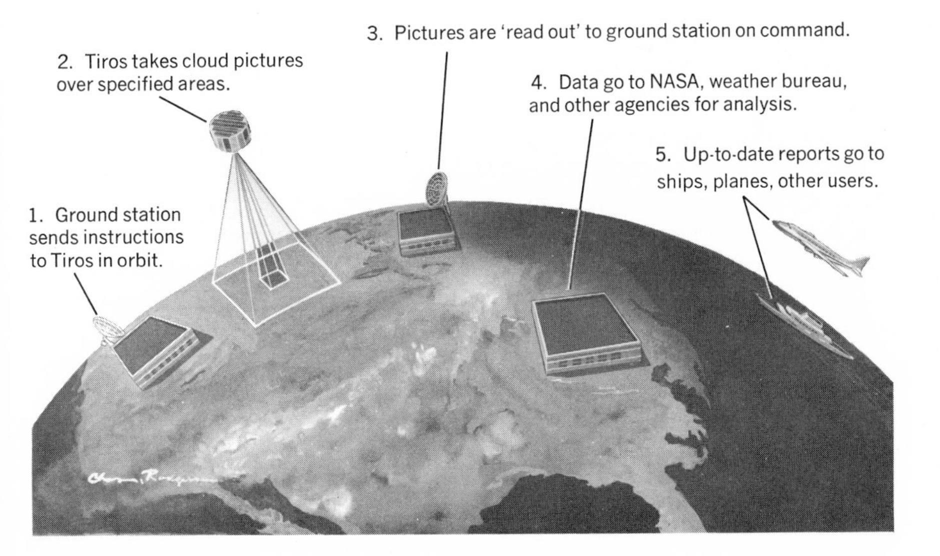

These are stored on magnetic tape and subsequently transmitted upon command to the ground as a satellite passes above a receiving station. Signals can be sent to a satellite and cause it to photograph the area beneath it at a certain time. The pictures are received at the surface on a television-type tube which is photographed. Prints so made can then be projected onto maps and the data correlated with surface weather records (Fig. 26-20).

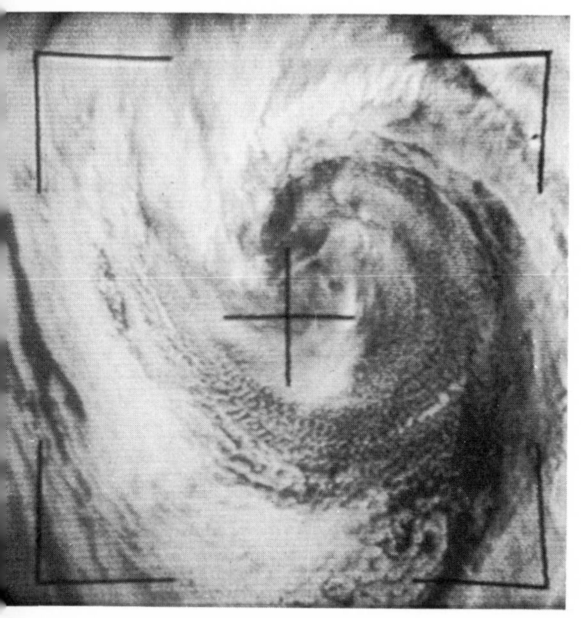

Fig. 26-19. *Tiros photograph of a vortex (a mass of fluid with a whirling or circular motion) in the Central Pacific (Tiros VII, April 6, 1964).* [*Courtesy, U.S. Weather Bureau.*]

Weather maps show the development and movement of weather systems and are among the principal tools used in weather forecasting. Some maps show conditions near the surface of the Earth, while others portray conditions at various heights in the atmosphere. Symbols of various types are used so that a large amount of weather data can be plotted in a small space.

Great advances seem to lie just ahead in meteorology. A truly global system of weather-data gathering seems feasible and will involve weather satellites, balloons that float at designated levels, automatic weather stations attached to ocean buoys, and an enlarged network of ground-based stations. To handle these data adequately, faster computers and better mathematical models are needed. All three of these factors seem capable of achievement in the years immediately ahead.

THUNDERSTORMS

A thunderstorm is an intense, local, rainstorm of short duration which is accompanied by thunder and lightning and sometimes by hail. More than

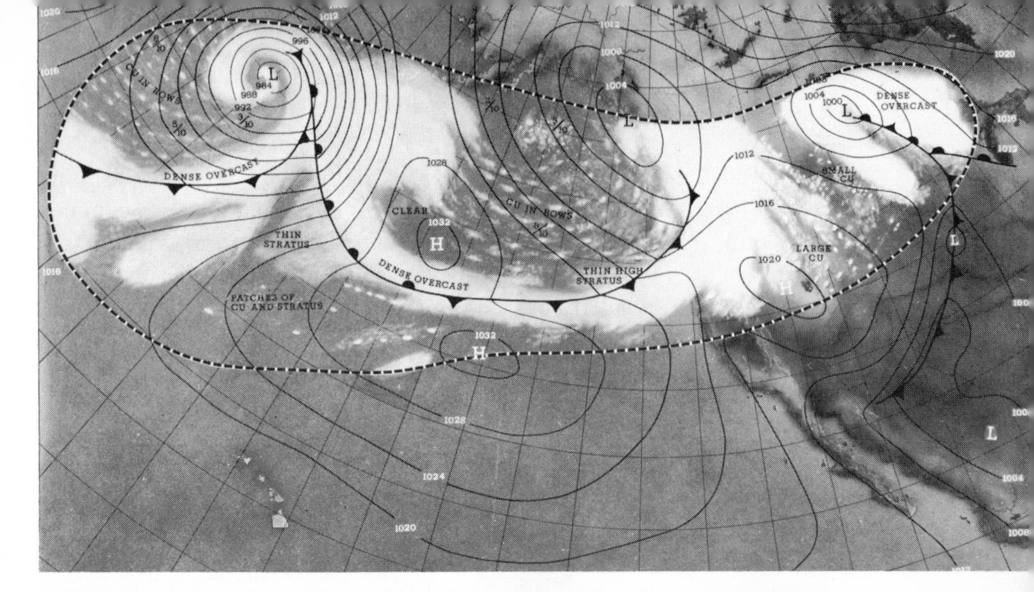

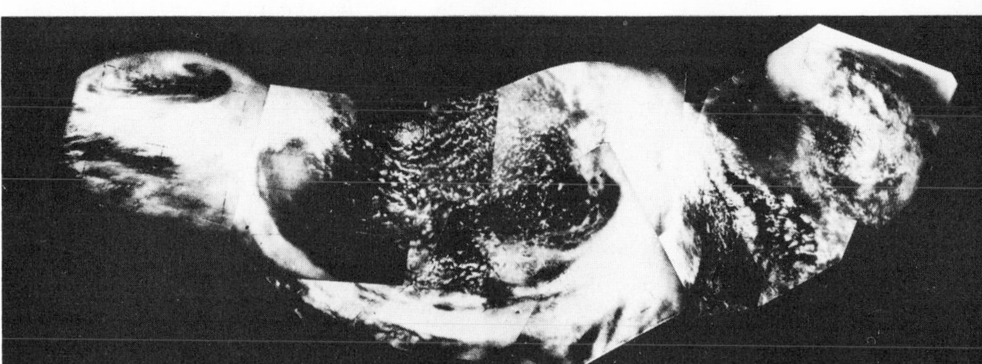

Fig. 26-20. *On May 20, 1960, photographs taken by the Tiros I weather satellite showed a series of storms extending from north of Japan to the eastern portion of the United States. The picture at the top shows the surface weather map for the eastern Pacific Ocean and western North America on May 20. Superimposed on the weather map is an artist's drawing of the clouds shown in the actual Tiros photographs below. Few routine weather observations are available from the vast areas of the Pacific Ocean. Without the Tiros photographs, weathermen might have been unaware of the position, or even the existence, of these storms.* [*Courtesy, U.S. Department of Commerce, Weather Bureau.*]

40,000 thunderstorms may occur during an average day on the Earth. Thunderstorms commonly result from the rapid upward movement of a parcel of warm moist air. Thus they tend to be associated with high temperatures. The latent heat energy released during condensation is a main factor in their formation, and only warm air can contain large quantities of water vapor. Thunderstorms are most frequent in the afternoon over land areas in the lower latitudes. They occur under similar conditions in the middle latitudes during the warmer months of the year, and they are uncommon at high latitudes. Thunderstorms occur in association with cumulonimbus clouds (Figs. 25-24 and 26-21) and are characterized by local extent, short lifetime, gusty winds, thunder and lightning, and heavy rains. Hail falls only from thunderstorms, but many thunderstorms do not produce hail that falls to the surface.

Fig. 26-21. *Tops of cumulonimbus clouds viewed from an altitude of 8 miles near the Shetland Islands (distant clouds on the right are over mountains in Norway). The anvil tops of the two nearer cumulonimbus clouds are unusually symmetrical. A third anvil is located just beyond the second one. [Photograph by RAF, from F. H. Ludlam and R. S. Scorer,* Cloud Study.]

Thunderstorms tend to occur under conditions of two different types. Airmass thunderstorms are more common and develop within a warm moist air mass by localized convectional circulation. These form most frequently on hot humid afternoons. Frontal thunderstorms may occur at any hour during any season as a result of the uplift of relatively warm moist air along cold and occluded fronts. Such storms may be scattered here and there along a front because locally the air is lifted more vigorously, contains more water vapor, or is less stable.

A thunderstorm probably consists of one or more units called cells which range from about 1 to 5 miles in diameter (Fig. 26-22). According to this model, each cell passes through three main phases: *cumulus, mature,* and *dissipating* (Fig. 26-23). Thus at any one time a thunderstorm may contain several different cells, and these may be in different stages of their development. The entire life cycle of any one cell is commonly played out within an hour or so. However, the storm itself may last much longer because new cells form and replace those that have dissipated.

In the cumulus stage of a thunderstorm cell, the chief movement of air is upward. Within a rising parcel the air at any one level tends to be warmer than the air outside of it because condensation is occurring and it is cooling at the less steep, wet adiabatic lapse rate (note the upward bulge in the isotherms). Probably this stage lasts for 10 to 15 minutes after the top of a cloud has reached the freezing level. Velocities are greater in the upper parts of such cumulus clouds because conditional instability has developed. However, velocities tend to vary from one part of a cloud to another and also to change with time. Large quantities of air are drawn into an upward moving current from the sides (entrainment).

Fig. 26-22. *A thunderstorm commonly contains a number of cells in different stages of development. Radar echoes are obtained from raindrops, ice crystals, and hailstones within a cloud, but not from cloud droplets (with radar wavelengths commonly used). Precipitation and strong vertical air movements tend to be absent from portions of a thunderstorm cloud outside of the cells, but enough raindrops are commonly present to produce a weak radar signal.* [*After Byers.*]

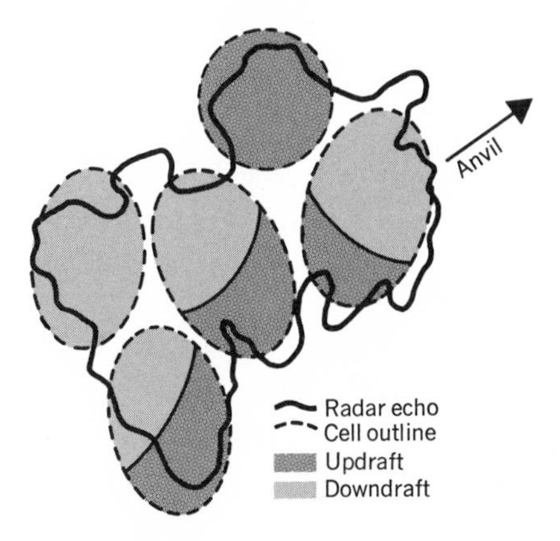

Precipitation develops in the upward-moving air but does not fall immediately. Ice crystals and snowflakes tend to form at the below-freezing temperatures that occur in the upper parts of a cumulus cloud, whereas raindrops form in the warmer air at lower altitudes. However, some of the raindrops may be blown upward across the freezing level and remain as a supercooled liquid. Cumulus clouds of this type are quite common, but only a small percentage continue to grow into full-fledged thunderstorms. Some cells fail to develop further because they are "diluted" by additions of dry air or because they rise upward into stable layers of dry air.

Fig. 26-23. *Three stages in the life cycle of a thunderstorm.* [*AF Manual 105-5,* Weather for Aircrews.]

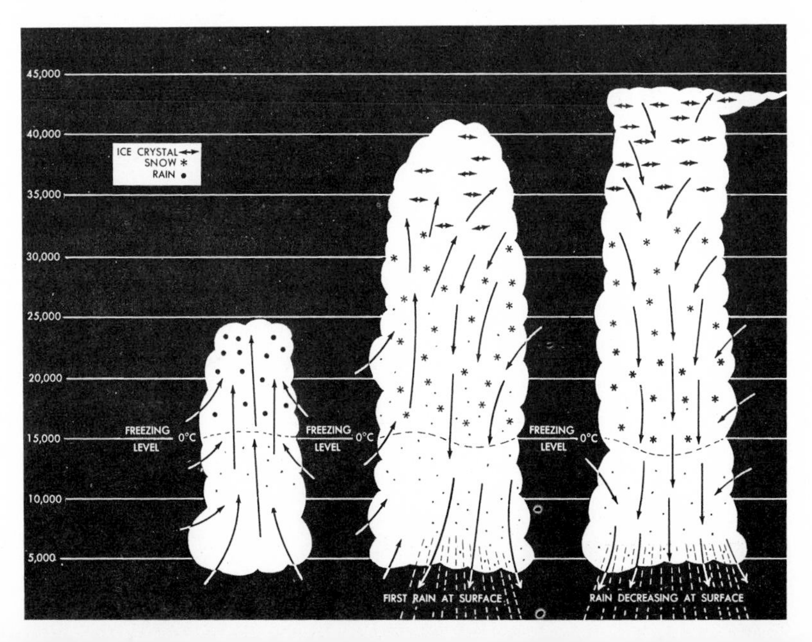

When the precipitation particles become too large to be kept suspended in the updrafts, they begin to fall. When they reach the ground, the mature stage of the cell is said to begin. Friction between the air and falling raindrops causes some of the air in the cloud to be dragged downward. The downdrafts occur first in the lower sections of the cloud but extend upward and increase somewhat in velocity as time passes. Thus the most powerful currents of air develop during the mature stage. These are upward in one part of a cell and downward in another part. At this stage the top part of the cloud (now a cumulonimbus) may extend all of the way to the stratosphere, or even penetrate for some distance into it.

Cumulonimbus clouds are the sources of some very intense rainfalls and of maximum-size raindrops—a direct result of the strong updrafts that occur within them. Large quantities of precipitation can be supported within an upward-rushing air current, and the particles can grow to maximum sizes. However, when a downward movement begins in one part of a thunderstorm cell, the enclosed raindrops fall quickly upon the surface below—the rate at which drops would fall in still air must be added to the downward velocity of the air itself.

The downward moving air current in the precipitation zone spreads out as it reaches the surface, particularly in the direction in which the entire storm is moving. This cool air functions as a miniature cold air mass, and its forward portion resembles a small cold front (Fig. 26-24). It produces

Fig. 26-24. *Movement of air beneath a thunderstorm cell in the mature stage.* [*TM 1-300,* Meteorology for Army Aviation.]

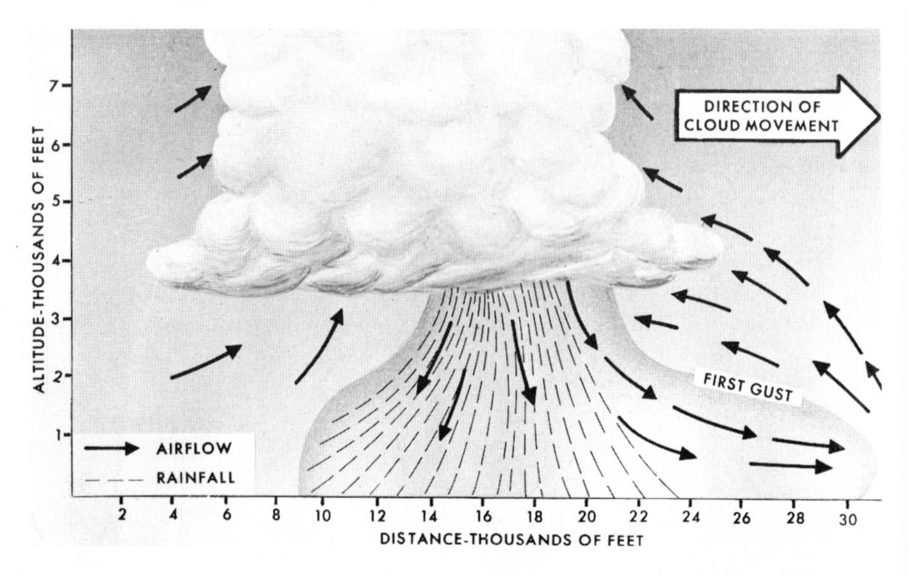

Fig. 26-25. *Location of electrical charges inside a typical thunderstorm cell.* [*TM 1-300,* Meteorology for Army Aviation.]

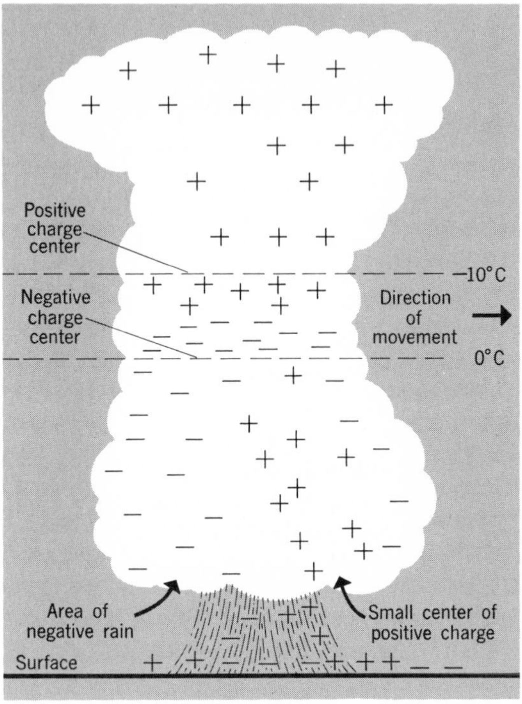

one of the characteristic features of a thunderstorm—the strong, gusty, cool wind that precedes the arrival of the precipitation and may attain velocities greater than 50 mi/hr.

The mature stage may last for 15 to 30 minutes or so. During it, downdrafts become more extensive as air is dragged into a cell from all sides. When the downdrafts have spread throughout the entire lower region of a cell, the third or dissipating stage in its life cycle has been reached. The supply of upward moving air has now been cut off, and condensation and precipitation stop. Relatively little vertical movement occurs in the uppermost portion of a cell during its dissipating stage. Strong winds commonly occur at such high altitudes, and these spread the cirrus clouds into the characteristic anvil top, especially in a downwind direction.

Lightning is the flash in the sky caused by a very rapid flow of electric current; thunder is the sound produced by the intensely heated, rapidly expanding air in the vicinity of a lightning flash. Lightning discharges may occur from one part of a cloud to another, from one cloud to another, and from a cloud to the Earth's surface. Centers of electric charge are generated or induced within clouds and on the Earth (Fig. 26-25). When they become powerful enough, the air between two nearby centers of unlike charge can no longer act as an insulator, and the discharge occurs. As sound travels about 1 mile in 5 seconds, the distance to a lightning flash can be calculated by measuring the time which elapses between the flash and the arrival of the sound.

TORNADOES

Tornadoes are smaller and more intense than either thunderstorms or hurricanes. In fact, they are the most violent storms known on Earth. They are more abundant in the United States than in most regions of the Earth and are most common in the afternoon in the spring and early summer in the Mississippi Valley region (Kansas, Iowa, Texas, and Oklahoma have had the greatest number). However, every state in the contiguous United States has had a tornado.

Tornadoes form in association with violent thunderstorms and consist of a center of low pressure (a vortex) around which air swirls violently (commonly counterclockwise in the Northern Hemisphere). From a distance a tornado appears as a long, dark, funnel-shaped or ropelike cloud which hangs like an elephant's trunk from the base of a cumulonimbus cloud (Fig. 26-26). The funnel appears to form in the cloud and to descend toward the ground, but not all funnels reach the surface. The outer boundary of a funnel may be sharp or indistinct. Commonly the diameter of a funnel cloud is less than $\frac{1}{4}$ mile, but its havoc-wreaking trail along the surface is generally wider than this because winds are also very strong around the cloud. The cloud is commonly dark because of dust and debris picked up from the surface.

Cloud droplets constituting a funnel form because moist air is cooled very rapidly. Barometric pressure inside a funnel is extremely low, air is drawn into it from all sides, and this air expands suddenly. The air is cooled well below its dew point; the result is rapid condensation of some of its water vapor. Violently uprushing air within a funnel cloud is likewise cooled adiabatically as it ascends. On occasions the middle portion of a tornado may not be visible: dust and debris have darkened the lower portion, but condensation to form a cloud has occurred only in its upper portion. The top part of a tornado may move faster than the portion near

Fig. 26-26. *Four photographs of a destructive tornado at Gothenburg, Kansas. [Photographs by Mrs. Ray Homer. Courtesy, U.S. Weather Bureau.]*

the surface, and thus after a time it does not hang vertically downward from the cumulonimbus cloud that spawned it.

The average tornado lasts for less than 1 hour, and the length of its path approximates 10 to 40 miles or less. However, some have traveled for more than 100 miles. In the United States the common direction is toward the northeast, and the average rate seems to be some 25 to 40 mi/hr, but faster and slower rates have been recorded. The average time that a tornado exists above any one spot is less than 1 minute. The twisting funnel may rise here and there above the surface and spare the buildings and people in its path.

Probably some 200 or more tornadoes occur in the United States in an average year, but nearly 600 tornadoes occurred in the first half of 1957. The destruction produced by a tornado is caused by the power of its winds, by the explosive effect of the sudden drop in atmospheric pressure outside buildings, and by the tremendous lifting effect of its updrafts. As a unit, a tornado may move at a rate of a few tens of miles per hour, but winds whirling around its low pressure center may blow at rates up to 200 to 400 mi/hr or even more. No anemometer has survived the passage of a tornado, but certain events indicate the extreme speeds involved: e.g., straws have been blown into tree trunks, and a piece of wood 2 inches by 4 inches in cross section was once blown through a solid iron sheet $\frac{5}{8}$ of an inch in thickness.

Updrafts associated with a tornado may attain speeds of 100 to 200 mi/hr; thus it acts as a giant vacuum cleaner. The drop in barometric pressure that occurs as the center of a tornado passes a given location has been estimated as equivalent to a decrease in the length of a mercury column by 1 to 5 inches. Barometric pressure shifts from normal to minimum and back to normal in less than 1 minute. Thus within 10 to 20 seconds, pressures may change as much as 400 lbs/sq ft. Pressures inside a closed building cannot adjust so quickly; the buildings may explode outward, and roofs may be blown off. Debris whirled along by the wind adds to the danger, and observers have reported loud roaring sounds.

Just how tornadoes originate is uncertain, and their precise locations cannot be forecast. Many occur (as do squall lines) an average of 150 miles or so ahead of a cold front at the surface and along a line parallel to this cold front. Several conditions are known to be favorable for the development of tornadoes, and area forecasts can be made. Three of these conditions in the central United States are these:

(1) A layer of warm moist air occurs along the surface, has a thickness of about 1 to 2 miles, and is moving northward from the Gulf of Mexico.

(2) Above this is a thick layer of drier air that has crossed the Rockies. The lower part of this air was warmed adiabatically as it descended east-

Fig. 26-27. *A waterspout (tornado over water) over Subic Bay in the Philippines on October 5, 1962. The lower portion of the waterspout's funnel-shaped cloud consists of spray. Such a tornado-waterspout builds downward from a massive threatening cloud. A different and relatively harmless type of waterspout is similar to the dust devils that form occasionally over land. These whirls begin at the surface and extend upward and may form beneath a cloudless sky. A small cloud sometimes forms above this type of waterspout at the altitude where condensation occurs in the whirling, upward-moving air.* [*U.S. Navy Official Photograph.*]

ward from the mountains; its relative humidity decreased, and thus it was "dried out." This air is still cold in its upper part, and it has a steep lapse rate because its lower part has been warmed.

(3) A jet stream blowing eastward occurs a few miles above the surface.

Many weird events happen in tornadoes. Horses and cars have been picked up by these violent storms and carried some distance before being lowered to the ground more or less undamaged. Such objects are whirled upward by violent updrafts near the center of a funnel and lowered slowly through lesser upward-moving currents surrounding the center. In other cases, fences may be rolled into giant balls and cars tumbled into shapeless masses. Chickens and turkeys may have their feathers stripped away, and some have lived through this experience. Reduced pressures may cause corks to pop from bottles.

Tornadoes may form over water and are then called waterspouts, but these tend to be less violent (Fig. 26-27). At the center of a waterspout, the water level may rise by a foot or two. The funnel cloud apparently forms in the same manner as over land and consists of moisture condensed from air that has been suddenly cooled, but it also sucks up large quantities of spray from the water.

The storm cellar of the midwesterner is the safest place in a tornado. In the United States, the storms usually move northeastward. Thus a place on the floor along a southwest wall and under a bed or table provides some protection. Windows and doors should be opened on the northeast side. If a tornado occurs at night, the first warning may be a loud roar. This means the tornado is about to strike, and one has very little time to do anything about it. A prone position in a hole offers the best chance of survival.

HURRICANES

Hurricane, typhoon, and tropical cyclone are synonymous names used in different countries for a particularly violent type of circular storm associated with torrential rains, thunder and lightning, and powerful winds that spiral inward and upward around a calm center of low barometric pressure (the eye of the storm). A hurricane is quite thin (a few miles or

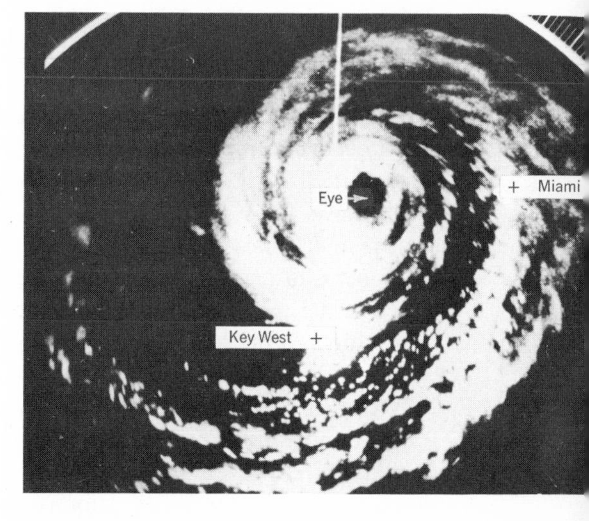

Fig. 26-28. *Photograph of radarscope showing a hurricane. Note the eye and the thunderstorms and regions of intense rainfall that occur in spiral bands around it. The concentric circles on the radarscope are spaced 20 miles apart. [Courtesy, U.S. Weather Bureau.]*

so from bottom to top) relative to its diameter (of the order of 100 to 500 miles), and thus it somewhat resembles a whirling phonograph record. The exact manner of their origin is uncertain, but hurricanes tend to have certain characteristics and to follow certain paths after they form. A tropical

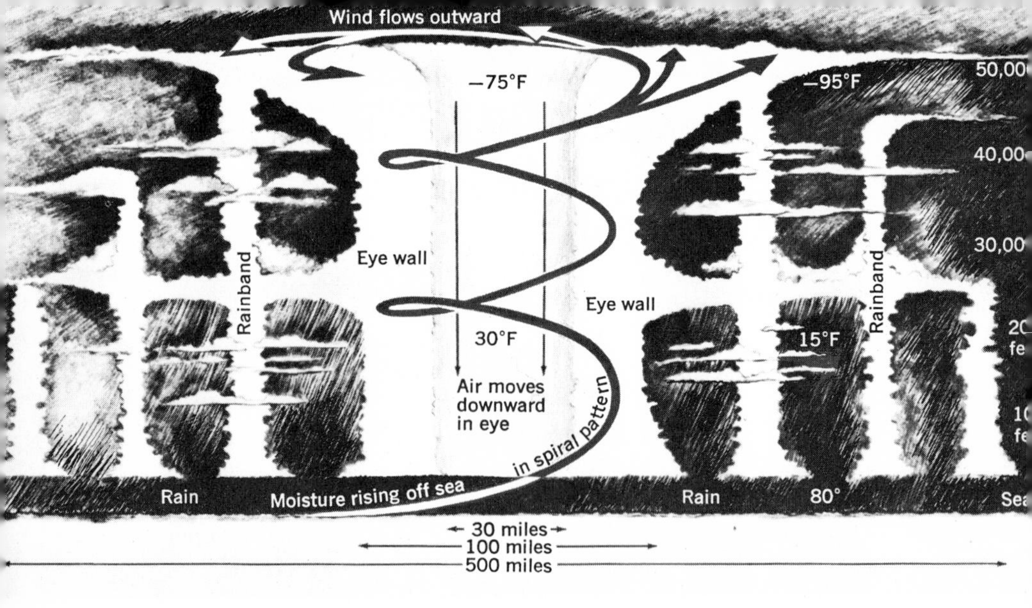

Fig. 26-29. *The hurricane model. The primary energy cell (convective chimney) is located in the area enclosed by the broken line. Seeding experiments with silver iodide on tropical hurricanes are part of the current Project Stormfury. Latent heat energy is released as a supercooled cloud is converted to ice. It is hoped that this increased amount of heat energy at a critical location may cause the eye-wall cloud to move outward, thus reducing the counterclockwise hurricane-force winds.* [*Courtesy, Robert H. and Joanne Simpson.*]

storm is not called a hurricane until its winds attain a speed of 75 mi/hr. Partly because of the conservation of angular momentum, wind velocities increase inward toward an eye. Maximum velocities may exceed 200 mi/hr, although this is quite exceptional.

In striking contrast to the violence that rages around it, relatively little wind occurs within the *eye* of a hurricane, and clouds are scattered or absent (Fig. 26-28). Among the various cyclonic storms, only the hurricane has an eye. This may average some 10 to 20 miles in diameter, but eyes that span 40 miles or more have been recorded. An eye increases in diameter at higher altitudes; convergence occurs at the bottom, and divergence takes place at the top. A solid wall of clouds surrounds an eye and extends upward for several miles. Although air spirals upward around an eye, air within it apparently moves downward.

Weather conditions tend to be symmetrical about the eye of a hurricane: temperature, pressure (nearly circular isobars), cloudiness and precipitation. However, in the Northern Hemisphere the right side of a hurricane (facing in the direction of its movement) is more severe than the left side because the forward motion of the storm is added to the counterclockwise movement of the winds. Furthermore, radar observations show that thunderstorms and areas of most intense rainfall are not uniformly distributed about an eye but are arranged in spiral bands (Figs. 26-28 and 26-29). These may also form complete circles near an eye. Any one

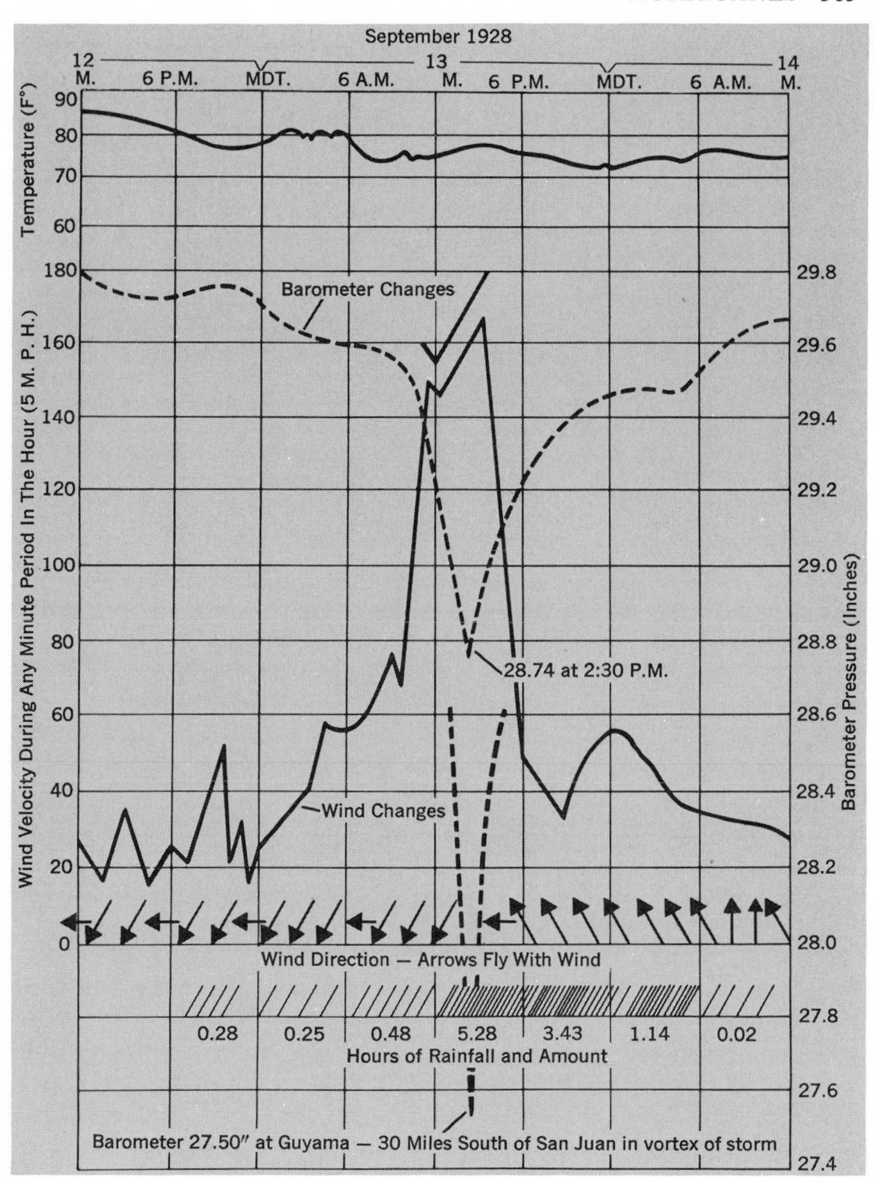

Fig. 26-30. *Changes in barometric pressure and other conditions at San Juan, Puerto Rico, during the passage of a hurricane in September 1928. Fig. 26-31 shows Puerto Rico in the lower right. Winds blew first from the northeast and then from the southeast. They did not diminish, because the eye passed south of San Juan.* [*TM 1-300,* Meteorology for Army Aviation.]

thunderstorm cell exists for only a short time, but others form elsewhere along a line, and thus the spiral structure is preserved.

The heaviest rainfall at any one location tends to occur when a slowly moving hurricane passes above it, and millions of tons of water may be dumped on the surface below. For example, a 1-inch rainfall adds about 100 tons of water to a square plot 200 feet on a side, and several inches to a foot or so of rain commonly accumulate at one location from one hurricane. In fact, a rainfall of 6 feet 4 inches was once recorded in 24 hours at Baguio in the Philippines.

The arrival of the eye of a hurricane at a land location means a brief respite from the fury of the storm (for 1 to 2 hours or more) but then the winds resume their violent movement, although now they blow in the opposite direction. For a ship at sea, however, there is little relief within an eye because monstrous waves occur there. Apparently waves cannot grow very high in the region surrounding an eye because the tops of the waves are blown off by the strong winds.

The length of the mercury column in a barometer has been measured at less than 27 inches at the center of a hurricane, but pressures below 28 inches are not common. In contrast to the abrupt drop of pressure that occurs during the passage of a tornado, the changes in barometric pressure associated with a hurricane occur relatively slowly. Therefore, closed buildings do not explode outward.

Hurricanes resemble the low-pressure systems of the middle latitudes in some ways. Winds converge and spiral inward and upward about a center of low barometric pressure; widespread areas of cloudiness and precipitation develop, and the direction of motion is counterclockwise in the Northern Hemisphere. However, hurricanes differ from middle latitude lows in other ways. They are smaller, more intense, and have eyes at their centers. Hurricanes occur within one type of air mass (maritime tropical), whereas a low is usually associated with two or more air masses and with two or more frontal systems and is followed by a center of high pressure. Hurricanes lack these phenomena. Hurricanes are more symmetrical; their isobars are circular and closer together. Precipitation tends to be much heavier from a hurricane. Hurricanes are late summer and early fall developments, whereas lows are best developed during the colder months of the year. However, hurricanes commonly change into extratropical cyclones as they weaken and encounter cold fronts in the middle latitudes.

Hurricanes originate over warm-water regions of the oceans at latitudes of about 5 to 15 degrees (but not at the equator) during the late summer and early fall seasons of each hemisphere, when the surface waters of the oceans attain their highest temperatures (the seasonal lag tends to be

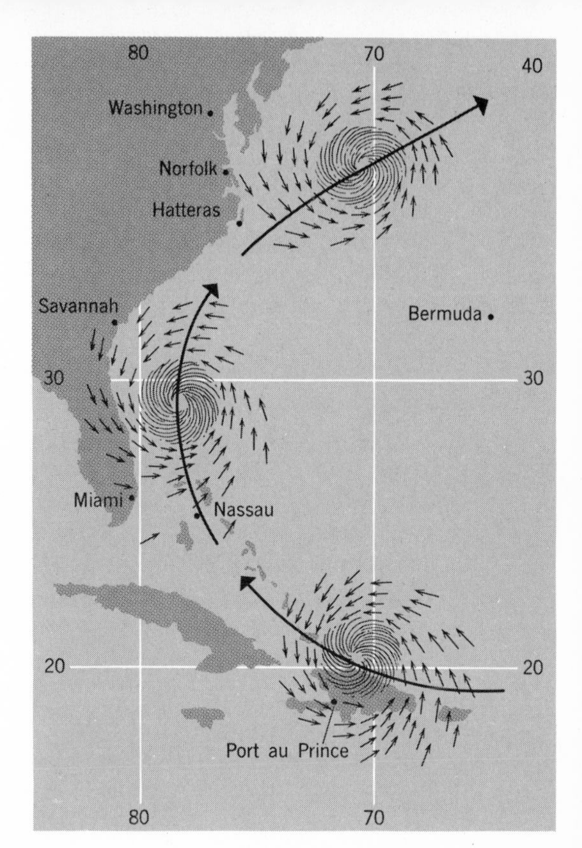

Fig. 26-31. *Typical track and wind system of a hurricane in the North Atlantic.* [*Courtesy, U.S. Weather Bureau.*]

greater over water than over land). At such dates, the intertropical convergence zone may have shifted as much as 15 degrees of latitude away from the equator—far enough from the equator for the Coriolis effect to occur. As tropical storms develop and grow into hurricanes, they usually move westward and slightly away from the equator across the trade-wind belts (Fig. 26-31). Next they are likely to cross the horse latitudes and circulate in a clockwise direction in the Northern Hemisphere around the western ends of the semipermanent high-pressure cells that occur over the oceans at a latitude of about 30 degrees.

Hurricanes that affect the United States form over the eastern Atlantic near the Cape Verde Islands and move westward toward the West Indies. Some hurricanes then continue westward across the Gulf of Mexico, whereas others curve northward toward Florida and thence northeastward along the coast of the United States (Fig. 26-32). Most of them continue curving northeastward across the Atlantic and do not damage the northeastern United States, but some do, and these are long remembered. Irregular looping movements also develop here and there and make precise prediction difficult. The entire hurricane tends to move at 10 to 30 mi/hr, but rates of 60 mi/hr have been observed, and 12 mi/hr may be closest to the average rate, particularly at lower latitudes. A full-fledged hurricane (after the winds exceed 74 mi/hr) commonly lasts from several days to more than one week.

Men and animals have been killed by hurricanes and, on certain tragic occasions, in very large numbers. Most of the deaths have occurred in low-lying coastal areas and have been caused by drowning. Onshore winds may pile up water along a coast, and irregularities in the coastline may concentrate this upon certain areas. The large drop in barometric pressure at the storm center may then cause the water level to rise several feet higher. Huge waves, superimposed upon these high waters, batter the shore and add to the destruction. Hurricanes can now be tracked by means of planes, radar, and other equipment, and their paths can be predicted with some precision. Thus hurricanes seldom arrive without warning. People have time to protect themselves and their possessions. However, property damage is still very great, and that caused by a single hurricane in the United States has totaled as much as an estimated billion dollars (Donna in September 1960).

Fig. 26-32. *Tracks of some devastating North Atlantic hurricanes.* [*TM 1-300,* Meteorology for Army Aviation.]

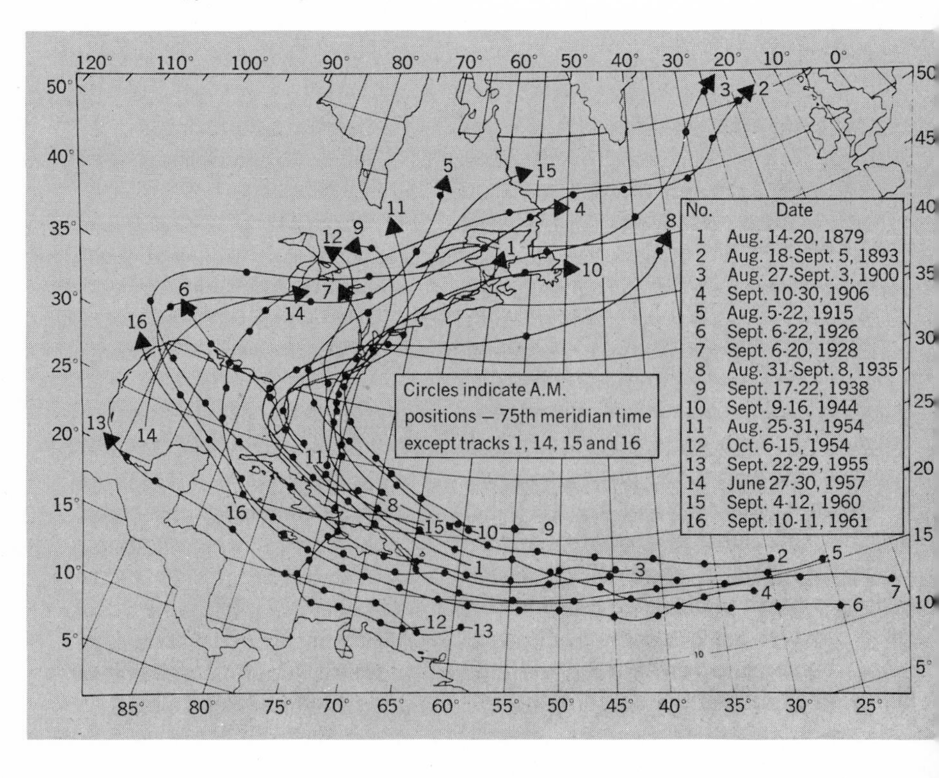

The first sign at a coastal location of an approaching hurricane may be the swell (very long waves coming from the direction of the storm center). The swell was produced far out at sea by the violent hurricane winds and it has traveled much more rapidly than the storm itself (perhaps at a rate of 1000 mi/day). Thunder and lightning are associated with many hurricanes but are obscured by noises such as the roar of wind and rain and the crashing of buildings. Occasionally a tornado may form and add to the destruction. The violent part of a storm at any one place may last from several hours to a day or so.

The main source of energy of a hurricane is the latent heat released during condensation. Thus a hurricane has an almost unlimited source of water vapor as it moves across warm ocean water, and it can grow to a maximum intensity (friction is one limiting factor). A hurricane tends to weaken as it moves across a land area or colder ocean water because its supply of moisture is reduced. Furthermore, friction is greater over irregular land surfaces.

Hurricanes can be studied in different ways. Specially reinforced and equipped planes fly into and out of a storm, locate its eye, and report upon the intensity and direction of motion of the storm. Special balloons carrying weather instruments may be dropped in an eye and drift with it; these make vital observations automatically and report them by radio. Radar is most useful. Perhaps of greatest value are the weather satellites that keep critical areas under almost continuous watch. A hurricane occurs within a large air mass and moves with it. Therefore, meteorologists need data from a very large area to determine the factors that can influence the movement of the air mass and thus of the hurricane contained within it. Results of attempts to modify hurricanes and their paths by cloud seeding are inconclusive (Fig. 26-29).

The annual frequency of *tropical storms* (wind velocities of 25 to 75 mi/hr) in the North Atlantic from 1879 to 1956 was 7.4 (range 1 to 22). About half of these tropical storms subsequently reached hurricane intensity. In recent years the U.S. Weather Bureau has given such hurricanes girls' names: the first hurricane of the season in the North Atlantic receives a name beginning with the letter A, the second one with B, etc. A new set of names is made up for each year.

Exercises and Questions, Chapter 26

1. What types of exchanges occur between objects and materials at the Earth's surface and the overlying air to effect an equilibrium and produce an air mass? Is the equilibrium attained more rapidly over a warm or cold surface? Why?

2. Which has a higher specific gravity, a gallon of dry air or a gallon of moist air? Assume that pressure and temperature are the same. Hint: the same number of molecules occur in each gallon.

3. Assume that the air in a certain region at a certain time has an observed lapse rate of 3.5 F°/1000 feet and a surface temperature of 60°F. Assume further that a small parcel of this air near the surface is heated to 72°F. Calculate how high it will rise and the temperature it will have at the altitude at which the upward movement stops. Assume that the air is too dry for condensation to occur.

4. Air moves from west to east across an island in the ocean. The island is 12,000 feet high. Assume: (1) that the air temperature is 70°F on the west side of the island at sea level; (2) that condensation begins at 5000 feet and continues to the summit (considerable precipitation occurs); (3) that evaporation occurs during descent on the east side of the mountain from the summit to an altitude of 10,000 feet; (4) that no additional evaporation occurs from the 10,000-foot level down to sea level; and (5) that the wet adiabatic lapse rate is 3 F°/1000 feet. What is the temperature at sea level on the east side of the island? Why was it unnecessary to give you data about the dry adiabatic lapse rate? Plot the pertinent data on a temperature-altitude graph.

5. Assume that the surface air temperature is 70°F and that a parcel of this air is forced to rise. Assume that condensation begins at an altitude of 5000 feet and the wet adiabatic lapse rate is 2 F°/1000 feet. The observed lapse rate at this time and place is 3.5 F°/1000 feet. At what altitude would the air reach a state of absolute instability? Plot the pertinent data on a temperature-altitude graph.

6. The types of clouds and weather associated with the passage of one warm front may be quite different from those associated with the passage of another warm front at a different time. Explain why this is possible.

7. How do low-pressure systems form?

8. How does a high-pressure system commonly differ from a low? Shape? Size? Direction and rate of movement? Place of origin? Associated air mass or air masses? Commonly associated weather conditions?

9. What special influence, if any, does your local topography have on weather phenomena?

10. Check the accuracy of weather forecasts for a time and assign an accuracy percentage.

11. Arrange a number of weather maps in chronological order with the most recent on the bottom. Study the top map and predict the changes which you think are likely to occur in the next one, then check, and repeat the procedure. You will probably make many mistakes and unpredictable weather changes may have occurred, but you should learn a good deal about weather changes. Daily weather maps can be obtained from the U.S. Weather Bureau.

12. Observe the weather phenomena that occur in your area and attempt to explain their occurrence and to predict changes in the immediate future.

13. If a barograph is available, mark on each week's record the weather changes which have occurred. Is there a correlation between highs and lows and certain weather changes? How reliable are weather predictions based solely upon changes in barometric pressure?

14. The accompanying weather map is for the central or central-eastern part of the United States. Seven weather stations are shown by numbers.

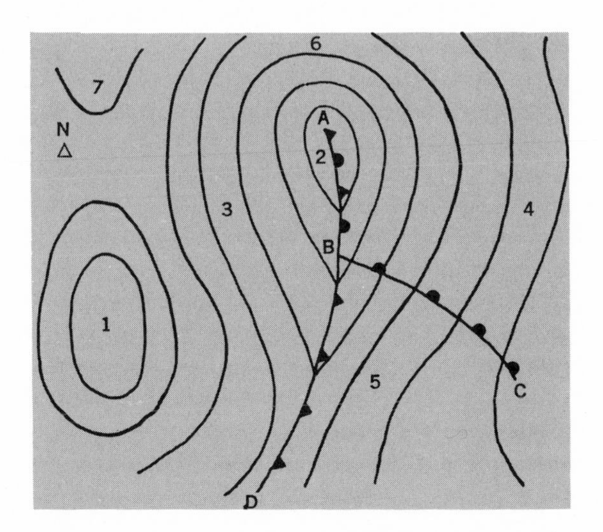

(a) Name an occluded front, a warm front, a cold front.
(b) Locate the weather stations with the highest and lowest air pressures.
(c) Which station most probably has southeast winds? Northwest winds?
(d) Which station or stations most probably has (have) precipitation?

(e) Which station most probably has nimbostratus clouds? Cumulus clouds? Altostratus clouds?
(f) Which station probably has the highest temperature? The lowest?

15. Where and when are thunderstorms most frequent? Why? What exceptions occur?
16. What causes the downdraft in the mature stage of a thunderstorm? Why does it tend to spread across the lower part of the cell to form the dissipating stage?
17. What causes the funnel cloud of a tornado?
18. Compare and contrast a low-pressure system with a hurricane.
19. Describe the place of origin, paths, and rates of movement followed by hurricanes that affect the United States.
20. Compare and contrast the destructiveness of a tornado with that of a hurricane.

27 Sermons in stones

And this our life, exempt from public haunts,
Finds tongues in trees, books in the running brooks,
Sermons in stones, and good in everything.
—*As You Like It: Act II, Scene I*

GEOLOGY (Greek: "earth study") is concerned with the Earth, its origin, its surface features (Fig. 27-1) and physical conditions, with the organisms which once inhabited it (Fig. 27-2), and with the countless changes that have occurred in these during the 4 to 5 billion years that the planet Earth has orbited the sun.

Geology is a fascinating science for inquiring individuals who want to understand certain facets of the planet on which they live: e.g., the eruption of a geyser or volcano, the nature of its hidden interior, an earthquake, a landslide, a natural bridge, an artesian well, a towering mountain, a glacier, the stalactites in a cavern, a meandering river on a wide floodplain, a steep-sided canyon, and the diverse fossils found in rocks. Scenery and landscapes, the commonplace even more than the spectacular, provide a mature lasting satisfaction when observation is coupled with an understanding of their origin and nature. Geology makes a major contribution to the intellectual stimulation and satisfaction of all who come in contact with the great out-of-doors.

Mountains, hills, plateaus, and valleys are not permanent features of the Earth's surface, but only temporary forms in an ever-changing pattern. Rocks are not dead, inert, and unchanging; they are alive with messages about a geologic history that stretches backward into time for many hundreds of millions of years. There really are "sermons in stones" for anyone who knows how to decode the records.

A hike along trails close to home will mean much more if one can stop here and there to win from the rocks tales of events that occurred long ago. Perhaps exposed on a slab of sedimentary rock will be raindrop prints which record the impact of the last drops of a storm which occurred millions of years ago; nearby may be birdlike three-toed footprints informing the observer that a dinosaur once walked by (Fig. 27-3).

Sandstone layers encountered along the way may contain abundant fossils of animals whose present relatives live only in the ocean. According to this "sermon": many years ago the area was covered by marine water; sand was then carried by rivers from nearby lands to this sea and deposited; next, shelled animals were buried in the accumulating sand and later fossilized; subsequently, the sediments became hard rocks and

Fig. 27-1. *Lightning in a volcanic cloud over Surtsey Island, Iceland, on 1 December 1963 (90-second exposure). The altitude of the top of the picture is about 5 miles. Eruptions occurred where water was about 400 feet deep. In 10 days enough volcanic debris had piled up to form an island about ½ mile long with its summit 350 feet above sea level.* [*Sigurgeir Jónasson photograph.*]

Fig. 27-2. *A fossil fish 14 feet long was discovered in limestone strata in Kansas* (Portheus; *Late Cretaceous.*) *Another fossil fish* (Gillicus), *6 feet long, occurs within the remains of Portheus.* Portheus *swallowed* Gillicus *head first a short time before it died and sank to the bottom. Burial in limy muds followed. Eventually the area became land, erosion took place, and overlying rocks were removed.* [*Courtesy of the American Museum of Natural History.*]

were eventually uplifted to their present positions; finally, erosion* exposed them to view (Fig. 27-4).

Thomas Henry Huxley underscored strikingly the intellectual importance of geology to man when he wrote: "To a person uninstructed in natural history, his country or seaside stroll is a walk through a gallery filled with wonderful works of art, nine-tenths of which have their faces turned to the wall."

Utilitarian Aspects of Geology In today's civilization, which relies so heavily upon the natural resources of the Earth, geology is an exceedingly utilitarian science. For such essential materials as uranium, iron, copper, lead, zinc, tungsten, manganese, oil, and coal, geology must furnish answers for important questions. Where can suitable deposits be found? How can these best be exploited? What are the reserves of these "one-crop" materials whose accumulation is too slow for new deposits to form in the centuries which lie immediately ahead?

World consumption of many mineral resources during the past two or three decades has exceeded that of all the preceding years in the history of man. Such consumption makes deposits that once seemed inexhaustible

* Erosion includes all the processes of loosening, removal, and transportation which tend to wear away the Earth's surface.

Fig. 27-3. *Dinosaur tracks in Cretaceous rocks near Glen Rose, Texas. The large footprints are those of a herbivorous dinosaur, and the three-toed tracks were made by a carnivorous dinosaur that walked on two legs. Did the carnivore catch his prey?* [*Courtesy of the American Museum of Natural History.*]

now seem quite inadequate. Furthermore, the necessary mineral deposits are concentrated in relatively few places. No nation is self-sufficient in all the minerals that it requires, and yet industrial power is based upon ample supplies of a very great variety of mineral resources. Thus problems arise that involve international trade, politics, transportation, exploration, development, exploitation, have and have-not nations, and war or peace.

The engineering applications of geology are many and essential and seem destined for a large and rapid increase because of the multitude of pressures being exerted upon the lands and their resources by the startling increase in world population. Examples include the location of satisfactory sites for dams and buildings, flood control and water supply, tunnel construction, determination of the best locations for roads, pipelines, and undersea cables, and protection of coastal areas from erosion. Military applications of geology are likewise numerous and diverse.

Probably more than half of all geologists in the United States are employed in the petroleum industry, but a number aid in the search for deposits of more prosaic materials: clay or shale for bricks, limestone for cement or building construction, and sand, gravel, and traprock for road-building materials. On the other hand, the search for glamorous minerals such as diamonds and rubies is also part of geology.

The availability of almost limitless supplies of good-quality water is assumed as a kind of natural heritage by many—about in the same category as the air we breathe. However, the task of finding adequate supplies and of achieving suitable pollution control during the next few decades may involve a greater public investment than any other field of natural resource development or conservation. Adequate knowledge of the sediments and

Fig. 27-4. *What happened from the time an animal was buried during the Cenozoic era until its discovery.—A. A flood carries sediment which is deposited on the floodplain. The line of trees locates the stream channel.—B. The flood recedes and a drowned animal's bones lie on the ground.—C. Sediments from successive floods bury the skeleton, which becomes fossilized.—D. The fossil is buried far below the surface as the channel shifts back and forth across the floodplain.—E. Uplift of the region results in erosion of the area.—F. After deep erosion, a man finds the fossil projecting from a rock layer. [Copyright, Chicago Natural History Museum. Drawings by John Conrad Hansin.]*

rocks at and below the surface is an important factor here, and thus geology is directly involved.

The Earth and Its Major Subdivisions The Earth is a nearly spherical planet, slightly flattened at the poles and bulging at the equator (Chapter 19). It has a circumference of nearly 25,000 miles, a polar diameter of about 7900 miles, and an equatorial diameter of approximately 7927 miles. On a similarly shaped spheroid 100 feet across, the 27-mile difference in diameters would shrink to 4 inches, too small to be noticeable to the eye.

The specific gravity of the Earth as a whole is about 5.5: an average sample contains $5\frac{1}{2}$ times more matter than does an equal volume of water. This is about double the specific gravity of rocks at the Earth's surface and indicates that materials in its interior are quite dense. The total mass of the Earth is estimated as 6×10^{21} tons (6 followed by 21 zeros).

Materials constituting the outer part of the Earth form three spheres: the atmosphere, the hydrosphere, and the lithosphere—the spheres of air, water, and stone. According to one definition, geology is the science of the lithosphere, including its relationship to the hydrosphere and atmosphere. The three spheres interpenetrate: gases (atmosphere) and solids (lithosphere) are dissolved or suspended in the waters of the hydrosphere; dust and water vapor occur in the air; and gases and water occur in openings in the rocks of the lithosphere.

The *atmosphere* (Chapters 25 and 26) is an important geologic agent. Wind causes erosion directly by blowing sand grains and other fragments against obstructions and rock surfaces. Moving air transports rock fragments from one area to another—witness migrating sand dunes and dust storms. Winds produce waves and currents at sea and indirectly cause much of the erosion that occurs along a coastline. Water vapor is evaporated from the oceans by solar energy, and some is transported inland where condensation and precipitation occur. Subsequently some of this flows to the oceans to complete the water cycle. The atmosphere is thus indirectly responsible for the eroding and transporting activities of running water. As a direct agent in the weathering of rocks (i.e., in their disintegration and decay), some gases in the atmosphere react chemically with rock materials.

The *hydrosphere* is the discontinuous envelope of water covering parts of the Earth's surface. It includes the oceans—which submerge about 71 percent of the Earth's surface—lakes, rivers, and water located in subsurface cracks and other openings. If irregularities such as continents and ocean basins did not exist, water would completely cover the outer part of the Earth to an estimated depth of more than $1\frac{1}{2}$ miles.

The *lithosphere* is the solid outer part of the Earth, but the precise meaning of the term has not been agreed upon. The crust on which we live (Fig. 27-5) forms the outer part of the lithosphere, but only its upper part is more or less accessible to direct study. "Crust" is a useful but misleading term. At the time it came into use, the whole interior of the Earth was believed to be molten except for a thin outer rind.

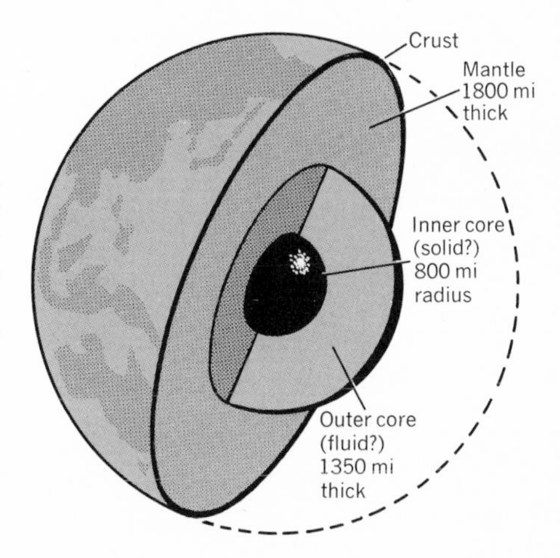

Fig. 27-5. *The three main units of the Earth's interior: core, mantle, and crust. The mantle makes up about 84 percent of the volume of the Earth; the core makes up about 16 percent. Thus the crust forms a very small part of the Earth as a whole.*

On the average, the crust is probably about six times as thick beneath the continents as beneath the ocean floors (approximately 20 to 25 miles vs. 3 to 4 miles). The composition of the continental crust likewise differs from that beneath the oceans (p. 775). The upper surface of the crust may be covered by water, by unconsolidated sediments tens or even thousands of feet thick, and by soil and vegetation, or it may be exposed at the surface as an outcrop of bedrock (Fig. 29-4). Wherever soil or loose surface debris is penetrated deeply enough, commonly a few tens of feet or less, bedrock is found beneath.

The Earth's surface can be subdivided into three main units: the ocean basins, the system of mid-ocean ridges, and the continents. The continental shelves and slopes form an extensive marginal zone between the continents and ocean basins (Chapter 34). The recently discovered system of mid-ocean ridges extends for about 40,000 miles and covers nearly as much area as the continents. The Mid-Atlantic Ridge is its best-known portion.

The Earth's surface has two fairly well defined levels (Fig. 34-2): the upper level is associated with the continents which have an average altitude of about $\frac{1}{2}$ mile above sea level; the lower level, about $2\frac{1}{2}$ miles below sea level, coincides with the average depth of the ocean basins. Presumably these levels occur because the thinner, denser oceanic crust is in isostatic balance with the thicker, less dense continental crust (Fig. 28-19).

THE EARTH'S CRUST IS MADE OF MINERALS AND ROCKS

By direct observation, man has access to only the outer part of the crust: the deepest mines penetrate less than 2 miles beneath the surface, and the deepest wells about 5 miles. However, certain rocks now exposed at the surface may once have been buried several miles below it.

An intriguing research project, the Mohole, has been proposed and may one day be carried out. It involves an attempt to drill a hole from a ship at sea, in water about 3 miles deep, downward through the sediments on the sea floor, through the oceanic crust, and into the upper mantle.

Chemical analyses have been completed for various types of rocks, and an estimate has been made (Clarke and Washington) of the relative proportions of the elements in the outer 10-mile zone of the lithosphere. Eight elements—oxygen (most abundant), silicon, aluminum, iron, calcium, sodium, potassium, and magnesium (least abundant)—probably constitute more than 98 percent by weight of this 10-mile zone. Of these, oxygen and silicon together may make up about three-fourths of the total. The following mnemonic expression puts these eight elements in the order of their relative abundance: "Only Silly Artists In College Study Past Midnight." If the materials in the atmosphere and hydrosphere are added to those of the 10-mile zone, percentages are changed only slightly.

Minerals Minerals (see the Appendix) are natural inorganic substances that have characteristic physical properties and chemical compositions. They are the basic homogeneous units which have been combined in various ways and under different conditions to form rocks, and minerals and rocks are the materials which make up the crust of the Earth. As aggregates of minerals, rocks tend to be heterogeneous materials.

Most minerals consist of elements combined as chemical compounds, although a few may occur as native elements: e.g., gold, silver, copper, and carbon (diamond and graphite). Each mineral possesses physical characteristics and a chemical composition which distinguish it from all other minerals. One mineral may be different from another because it breaks in a special manner, develops with a distinctive shape, or has some unique property such as a peculiar taste, feel, or magnetism (Figs. 27-6 and 27-7). Although hundreds of mineral species are known, only two dozen or so are important as rock-making or ore minerals. As examples of minerals we may cite feldspar, quartz, mica, pyrite, and galena.

The atoms and groups of atoms which make up minerals are arranged

Fig. 27-6. *Mineral crystals have different shapes. Top left, a large quartz crystal from Auburn, Maine. Top right, a 10-pound garnet crystal, found within a stone's throw of Macy's Department Store in New York City. Bottom left, fluorite crystal. Bottom right, a group of feldspar crystals (microcline).* [*Courtesy of the American Museum of Natural History.*]

Fig. 27-7. *Minerals may fracture and/or cleave when they break. To cleave, a mineral must break along one or more smooth plane surfaces. Variations occur in the number of cleavage directions and in the angles at which the cleavage surfaces intersect.* Top left, *mica illustrates nearly perfect cleavage in one direction.* Top right, *feldspar (orthoclose) cleaves in two directions (top and front sides) that intersect at about 90 degrees. Feldspar also fractures, and an uneven fracture surface is shown on the right.* Bottom left, *calcite cleaves in three directions, but the intersections do not form 90-degree angles.* Bottom right, *fluorite cleaves in four different directions. The specimen on the left has been cleaved into a nearly symmetrical form. The large faceted gem is 1 5/16 inches in length. [Neil Croom; Ward's Natural Science Establishment, Inc.; Sinkankas,* Gemstones of North America.]

according to a definite three-dimensional pattern. Certain minerals always have certain shapes or break in a certain way because they are always made of the same types of atoms joined together in the same kind of structural pattern. The electrical and other forces holding such atoms together within a mineral are very strong in some directions but weak in others. Moreover, these directions and patterns extend throughout an entire mineral specimen. Thus a large crystal of a certain mineral tends to have the same basic shape as a small crystal of this mineral, or it cleaves in the same way when broken.

Some minerals originate by precipitation from solution, in much the same way that rock candy is made or the way that salt crystals form from a water solution. The amount of a substance which may be dissolved in a solvent depends of course on what it is and what the solvent is, but also on physical factors like temperature and pressure. Ordinarily, the lower the temperature, the smaller the amount of a substance that can be dissolved. Thus as high-temperature solutions rise from the depths along cracks or other passageways in rocks, they reach zones of lower temperatures nearer the Earth's surface, and precipitation may occur. If several materials are dissolved, the least soluble substance crystallizes first and the most soluble substance is precipitated last. In this manner the walls of a crack in bedrock may be lined with minerals of various kinds (Fig. 27-8). Molten rock-making material like that erupted at volcanoes is a

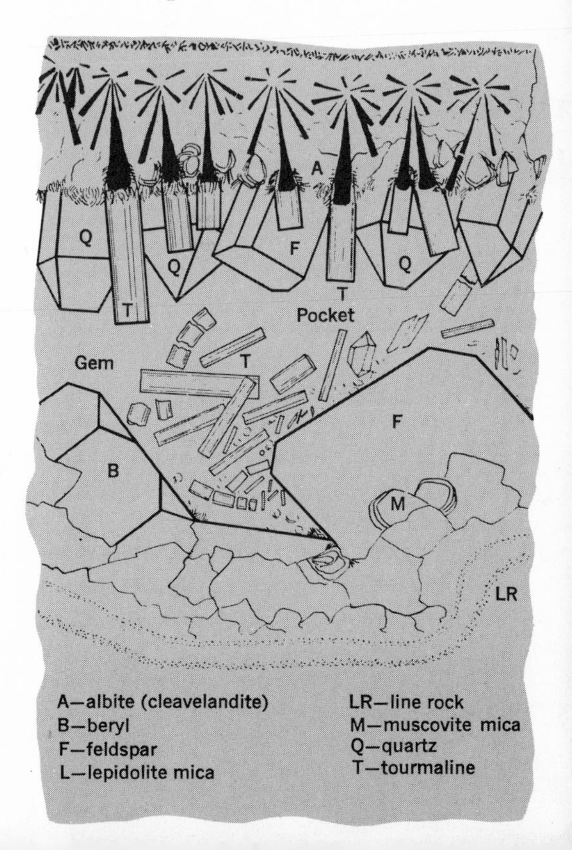

Fig. 27-8. *A cross section through a gem pocket. The cavity was filled with water when the different crystals formed.* [*John Sinkankas,* Gemstones of North America.]

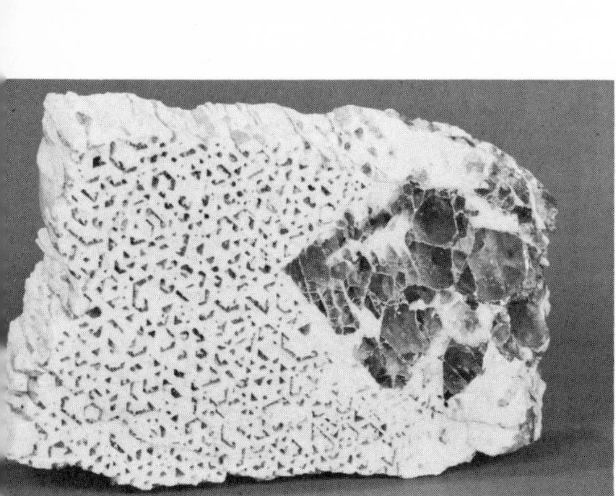

Fig. 27-9. *Two buried lava flows in Yellowstone National Park. Sedimentary rocks such as conglomerate and volcanic debris occur above and below the flows and glacially transported debris (till) occurs at the top. The following events are recorded: (1) deposition of sediment, (2) lava flow, (3) deposition of additional sediments, (4) another lava flow, (5) sedimentation again, and (6) erosion which results in a valley and the exposure of these rocks along one side of the valley.* [*J. P. Iddings, U.S. Geological Survey.*]

type of solution from which many minerals have formed, and hot-water solutions are another type. Minerals have also been formed by metamorphism and sublimation.

Rocks The majority of rocks (Figs. 27-9 and 27-10; see the Appendix) are heterogeneous aggregates of more than one kind of mineral, but some rocks consist largely of one mineral. Some geological processes make rocks; others break apart and destroy them. Products formed by the destruction of one kind of rock may later be combined into a new rock type, and different rock cycles result (Fig. 27-11).

All-inclusive, brief, simple definitions are difficult to formulate, and

---

Fig. 27-10. *Igneous, sedimentary, and metamorphic rocks.* Top left, *conglomerate is a sedimentary rock that was deposited as gravel.*—[*Ward's Natural Science Establishment, Inc.*] Top right, *micaceous schist containing garnet crystals. Schist is a metamorphic rock which tends to break into thin slabs; the parallel orientation of mica flakes is the cause in this specimen.* [*The Smithsonian Institution*]—Center left, *a polished slab of graphic granite, a type of igneous rock. The lighter-colored mineral in the rock is feldspar and the darker-colored one is quartz.* [*Ward's National Science Establishment, Inc.*]—Center right, *an igneous rock (trachyte porphyry) with a porphyritic texture (Appendix IV). The large light-colored minerals (phenocrysts) are feldspar. The darker part of the rock consists of much smaller grains of feldspar and other minerals.* [*Ward's National Science Establishment, Inc.*]—Bottom left, *a volcanic bomb that formed when a clot of lava was hurled high into the air by a volcanic explosion. The outside cooled and hardened, thus trapping bubbles of gas inside; the lava later solidified around the gas bubbles. The shape of the bomb resulted from its twisting and turning passage through the air. When the bomb hit the ground, it broke and the interior was exposed.* [*Neil Croom.*]—Bottom right, *obsidian, a natural glass that forms when lava cools very quickly.* [*Neil Croom.*]

the term "rock" is no exception. A rock is commonly defined as an aggregate of one or more minerals which make up essential parts of the Earth's crust. Exceptions are coal and the natural glasses which are not made of minerals. What constitutes a large enough mass to be an essential part of

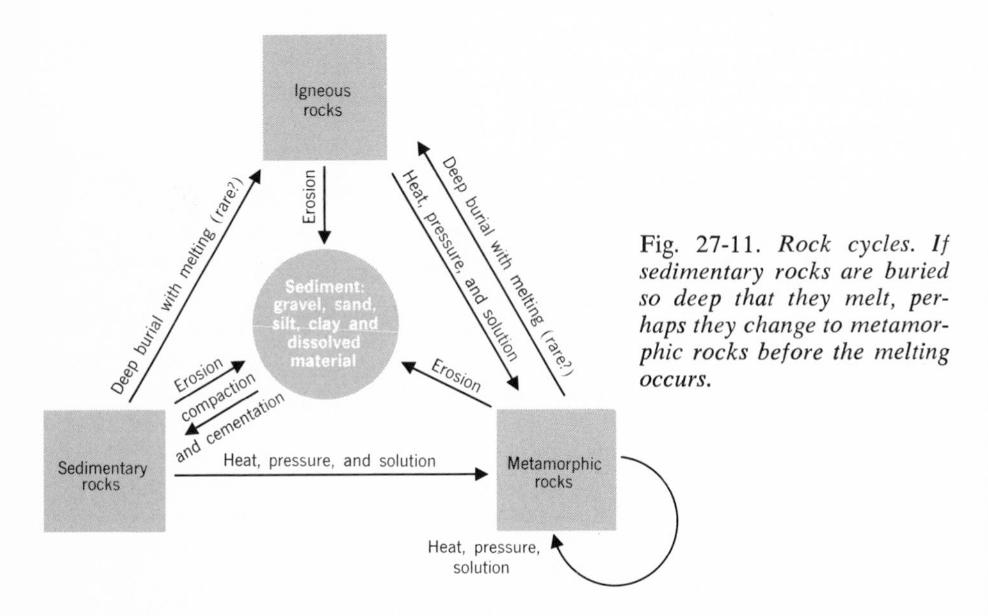

Fig. 27-11. *Rock cycles. If sedimentary rocks are buried so deep that they melt, perhaps they change to metamorphic rocks before the melting occurs.*

the crust is sometimes debatable. Rocks are more variable than minerals in their physical and chemical properties, but any one type of rock (e.g., a granite or a sandstone) does have certain mineral associations, textures, and other properties which are characteristic of it.

Fig. 27-12. *Interlocking texture and order of crystallization in an igneous rock.* [*After Knopf.*]

The following analogy may be useful. A building is constructed of wood, bricks, and glass just as a rock is composed of different minerals. Different buildings result from varying proportions of such building materials and from different architectural styles. Similarly, the same types of minerals (e.g., feldspar, quartz, and mica) are common in many different kinds of rocks, but they occur in different proportions and arrangements.

Based upon their manner of origin, rocks are commonly subdivided into three groups: *igneous, sedimentary,* and *metamorphic* (see the Appendix).

Igneous Rocks. Igneous rocks are produced by the cooling and crystallization of molten rock-making material called magma or lava. Granite, basalt, and obsidian are familiar examples of this group. Dissolved gases are important constituents of this material but tend to be excluded from the rock-making minerals when the hardening process occurs. This may take place within or upon the Earth's crust, and thus the igneous rocks may be subdivided into intrusive (from magma) and extrusive (from lava) groups. Intrusive rock masses are younger than the rocks they intrude. Intrusive rocks are exposed in places at the surface today because erosion has removed older rocks which once were around and on top of them.

The minerals in igneous rocks occur in characteristic assemblages of certain mineral types. Individual mineral grains (particles) are shaped

Fig. 27-13. *Concentration of sand as seen under the microscope:* (a) *loose sand from an Oregon beach;* (b) *partially cemented sandstone from a Brazilian coral reef; and* (c) *completely cemented sandstone from Ohio.* [*From* Principles of Geology *by Gilluly, Waters, and Woodford, 2d ed., 1959. W. H. Freeman and Company.*]

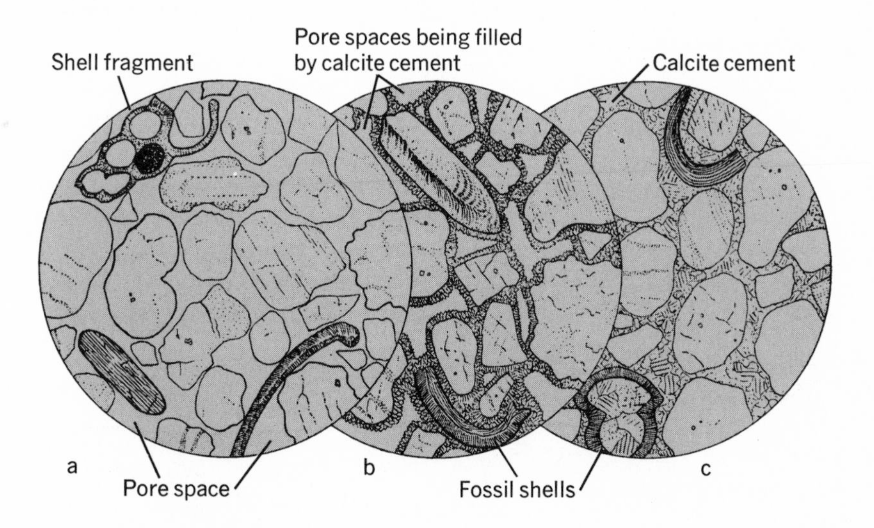

and oriented so that they mutually interlock and leave few, if any, openings (Fig. 27-12). Single grains are large enough to be visible in granite but too small to be readily distinguished by the unaided eye in basalt. Feldspar, pyroxene, amphibole, quartz, and mica are common minerals in igneous rocks. Variations in the proportions of such minerals produce light, medium, and dark colors.

Sedimentary Rocks. Sediments are produced by the weathering and erosion of minerals and rocks of all kinds. Sediments range in size from huge boulders to microscopic particles carried in solution. They are transported and dropped or precipitated by such geologic agents as running water, ocean waves and currents, wind, and ice. Some of the fragments may be rounded; others are angular. Such features depend upon the length of transportation and other factors. Loose sediments may be changed into solid sedimentary rocks by pressure from overlying strata which are deposited later and by the precipitation of cementing material as a binder around individual grains (Fig. 27-13).

Fig. 27-14. *Skeleton of the oldest known fossil bat. The picture shows shadowy, theoretical restoration of the wings and feet extended as they might have been when the animal was flying in Wyoming some 50 million years ago. This vertebrate fossil, from Eocene Green River Formation, is remarkably complete; e.g., small remnants of carbonized wing membranes and fragmentary residues of ingested food have been preserved.* [*Courtesy, Princeton University Museum of Natural History and Professor Glenn L. Jepsen.*]

The presence of different layers, beds, or strata constitutes an outstanding feature of most sedimentary rocks and serves to distinguish them from many igneous and metamorphic rocks. Layers result from changes in conditions as deposition takes place. For example, velocities of the transporting medium may be increased or decreased so that larger or smaller particles are dropped at a certain spot.

Fossils (Fig. 27-14) are the recognizable remains or traces of prehistoric animals and plants that have been preserved in sediments, rocks, and other materials such as ice, tar, and amber. Fossils are abundant in some sedimentary rocks and readily distinguish them from most specimens of igneous and metamorphic rocks. Fossils "label" the rocks in which they occur and yield much information about the past; but not all sedimentary rocks contain fossils.

Conglomerate, sandstone, and shale—formerly gravel, sand, and mud or clay, respectively—are common types of sedimentary rocks. The size of the average particle is the basis for the classification. On the other hand, limestones are classified on their mineral content, which is chiefly calcium carbonate as the mineral calcite. Limestones are likewise common. The

Fig. 27-15. *Size gradation is from coarse to fine away from the source area. The thickness of individual beds is greatly exaggerated in this highly diagrammatic sketch. One type of rock grades gradually into the next as the formations are followed from west to east across the area. The conglomerate and sandstone are composed chiefly of well-rounded particles of quartz.—Several events in geologic history apparently can be interpreted from the rocks. Since the particles become finer grained toward the east, streams which deposited them probably flowed from a source area in the west. Evidently the site of the source area was some distance away, because the particles are well rounded and almost entirely quartz. Thus larger pieces of feldspar and other minerals, which must originally have been more numerous than those of quartz, were reduced during transportation to fine particles. These were then carried farther eastward than the larger, more durable quartz pebbles and sand grains. Furthermore, the size of the pebbles in the conglomerate implies a moderately high land mass such as could give streams flowing down its slopes sufficient velocity to transport the pieces. The source area has subsequently been destroyed by erosion. The sedimentary rocks in the sketch are the only evidence of its former existence.*

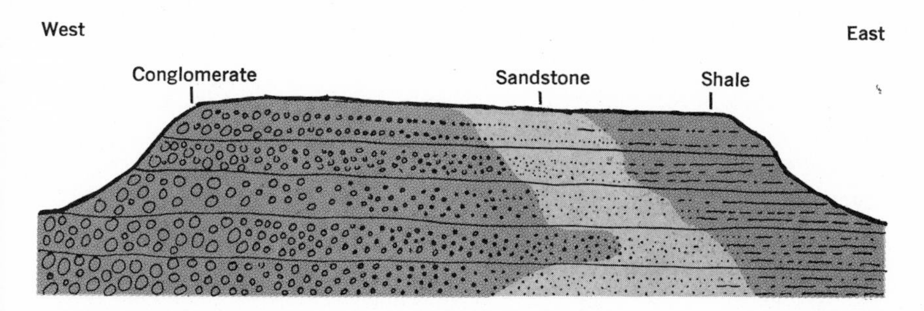

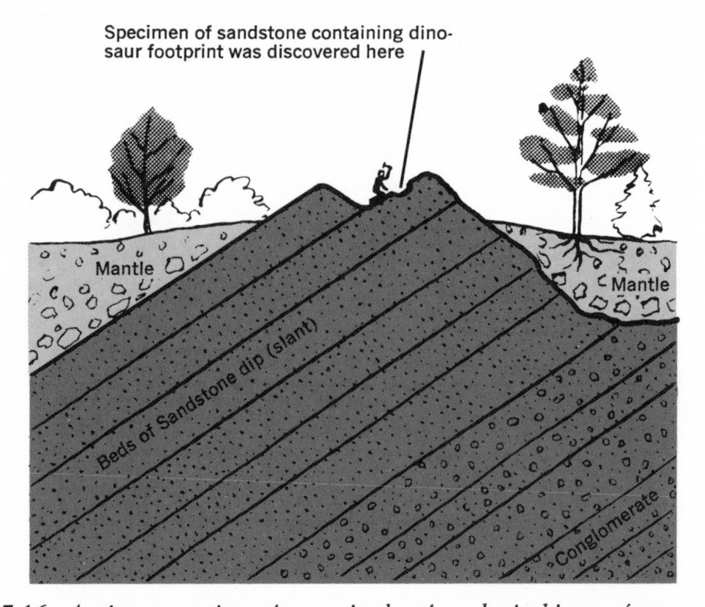

Fig. 27-16. *An interpretation of an episode of geologic history from rocks (a "sermon in stones"). (1) The conglomerate formed first and the standstone later; i.e., layers of gravel and then sand were transported to this area and deposited. Presumably the layers were nearly horizontal because this is the way such layers form today. (2) On one occasion during the deposition of the sand, a dinosaur walked across the surface. Enough moisture was present in the sand for footprints to form. The surface subsequently was consolidated (baked in sunshine), and the footprint was not destroyed when the next layer of sand was deposited. This occurred during the Mesozoic era when dinosaurs existed. (3) The layers of sand and gravel were compacted and cemented into sandstone and conglomerate. (4) The rock layers were tilted by a crustal movement (e.g., by a fault or fold). (5) Erosion removed overlying layers and eventually exposed the dinosaur footprint. (6) Loose debris (mantle) was deposited on top of the bedrock in places. Lack of sorting in the mantle and the abrupt contact between the mantle and bedrock indicate that the mantle consists of debris transported to this area and deposited—probably by a glacier.*

calcite may have formed by precipitation from solution or by the accumulation of shell fragments previously produced by various organisms from calcium carbonate dissolved in water.

Observation of the manner in which sediments are produced, transported, and deposited today yields information which is useful in interpreting the geologic history of ancient sedimentary rocks. Sediments come from a source area and are carried to a basin of deposition (Figs. 27-15 and 27-16). However, this may be only a temporary resting place; these sediments may again be eroded and transported to yet another basin of deposition in a process which may be repeated many times. The sea floor is the ultimate destination of most sediments, but even here the cycle does not stop—the sea floor may subsequently be uplifted to become land, and the materials may again be eroded, transported, and deposited.

Although the factors influencing the formation of sedimentary rocks are many and diverse, a few may be mentioned here. The type of rock or rocks eroded in a source area is an obvious factor. The environment of a source area and its crustal stability or instability are other factors. Different processes of weathering and erosion in different climates result in different types of sediment. Warm humid climates produce decayed, chemically altered rock debris which differs greatly from that produced by the freeze-thaw effect in colder regions. Glacially eroded sediments differ from those produced by wind. The size and nature of sediments will be affected by crustal movements in their source area: up or down, slow or relatively rapid, large or small. The length and type of transportation are other obvious factors. Finally, the nature of a basin of deposition and its crustal stability or instability have important effects. Deposition on a slowly subsiding, shallow sea floor permits waves to break less resistant fragments into clay-size particles and to sort, round, and smooth the larger pieces. Various chemical and physical changes may occur in sediments after they have been deposited.

Thus in studying a sedimentary rock, a geologist attempts to interpret

Fig. 27-17. *Tops vs. bottoms in steeply dipping sedimentary rocks. Costly mining exploratory work may fail or succeed depending upon the accuracy of the interpretation. Some of the signs to be noted are shown in the sketch below.—Another method involves the matching of the sequence of beds at one site with undisturbed layers some distance away. Graded bedding can be illustrated by tossing a number of handfuls of mixed sand, silt, and gravel into a deep pool of water. The coarser pieces commonly reach the bottom first.*

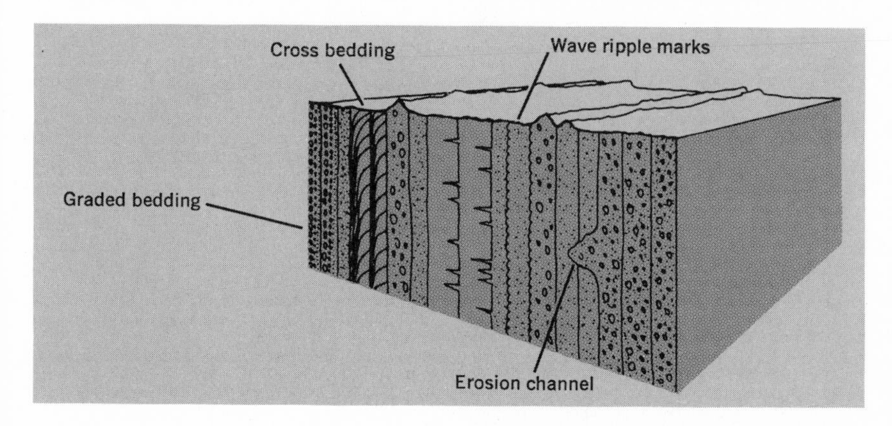

as much information as possible concerning the nature and environment of its source area, its deposition area, and its transporting agents. This might be thought of as a Sherlock Holmes, detective-searching-for-clues type of approach. It follows that a successful geologist must be a keen, imaginative observer with the ability and desire to deduce and interpret, and he must be thoroughly trained in biology, chemistry, physics, and mathematics. However, such interpretations must be based upon evidence and must be subjected to the strict, nonbiased limitations imposed by the methods of science.

Present-day observation shows that sediments tend to pile up in layers that are nearly parallel to the floor of deposition (*principle of original horizontality*). Younger layers are piled on top of older layers (*principle of superposition*). Yet many ancient sedimentary rocks are tilted (Fig. 27-17). Thus crustal movements occurred after deposition which raised, lowered, twisted, crumpled, broke, or folded the rocks. Sedimentary layers tend to thin out gradually either within a basin of deposition or near its margins. Thus thick sedimentary rock strata which stop abruptly along a cliff face today must once have extended farther (*principle of original continuity*). The strategy of the geologist, therefore, is to study the Earth and its processes as they function at present so that he can determine the manner of its functioning during the long, long past.

Sedimentary rocks constitute the most readable portions of the diary

Fig. 27-18. *A varve is a sedimentary deposit that has accumulated in one year. Varves deposited in lakes near glaciers tend to show alternations of thicker, lighter-colored, coarser-grained laminae (summer deposition) and thinner, darker-colored, finer-grained laminae (winter deposition). Each light-and-dark couplet represents the deposition of one year (nails mark off the years in the varves shown in these photos). Varves can be correlated sometimes from one area to another. Varves may vary in thickness from a fraction of an inch to more than a foot, but thicknesses ranging from 1 to 2 inches or less are most common. Pollen grains have been found in some varves and show that a couplet was deposited during a single year.* [*Courtesy of the American Museum of Natural History.*]

that nature has recorded in rocks for geologists to decipher (Figs. 27-18 and 27-19). However, some of the language has yet to be made intelligible, and many pages and even whole groups of pages have become badly tattered or are entirely missing.

Some sedimentary rocks have a conspicuous minor layering oriented at an angle to the main trend of the beds. This feature is called *cross-bedding* (Fig. 27-17). It may form in a stream bed where a depression is being

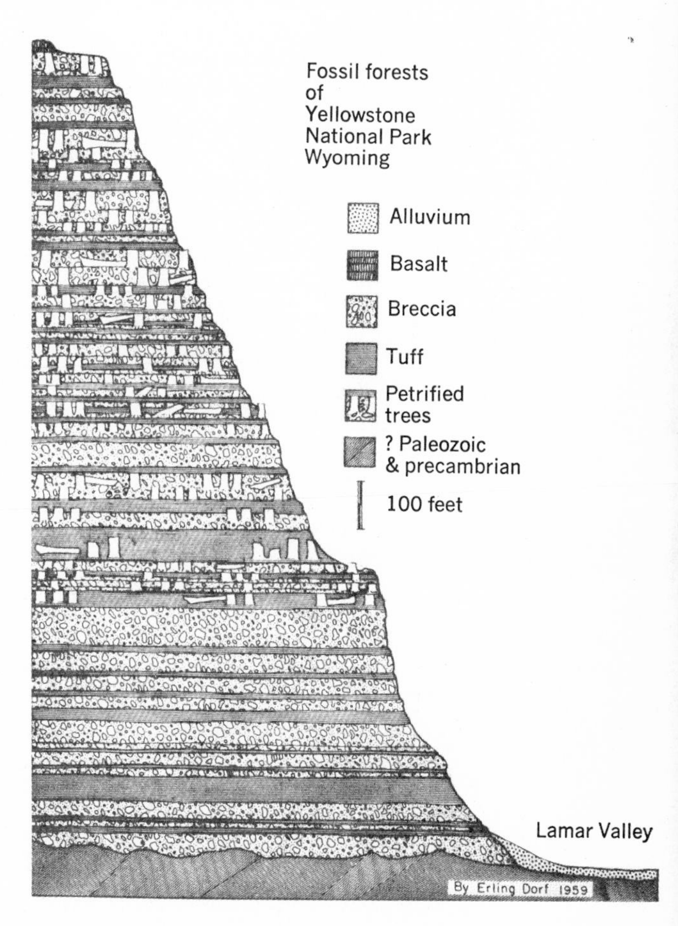

Fig. 27-19. *The fossil forests of Yellowstone National Park, Wyoming. The petrified trees in the twenty-seven buried forests resemble living redwoods. Tree rings show that one of the petrified trees probably lived for about 1000 years, and many trees show up to 500 annual growth rings. Some of the organic matter once present in the trees has been removed by solution and replaced by the mineral quartz, but in most cases the quartz fills cellular cavities in the wood. Each of the forests was buried by a single pyroclastic deposit* (*i.e., fragmental debris erupted during a volcanic explosion*). *Studies of volcanic areas in Mexico show that new trees can begin to grow in an area about 200 years after pyroclastic material has fallen upon its surface. However, the fossilized trees in any one of the forests were not necessarily the first generation of trees to grow after an eruption. The following sequence was repeated twenty-seven times: soil formed on the surface of the volcanic debris following an eruption; eventually a forest developed; an eruption occurred and buried part of the forest.* [*Courtesy, Erling Dorf.*]

filled rapidly by deposition from upstream. Although the main volume of sand dropped by the stream in this channel has a horizontal trend, it is deposited along a sloping surface. Cross-bedding is well developed in some dune sands. It may form wherever sediments are deposited rapidly over fairly steep slopes. Cross-beds tend to be concave upward, and any one layer tends to thin out gradually toward the bottom and to stop abruptly at its top. Thus cross-bedding can sometimes be used to tell the top of a steeply tilted layer from its bottom (Fig. 27-17).

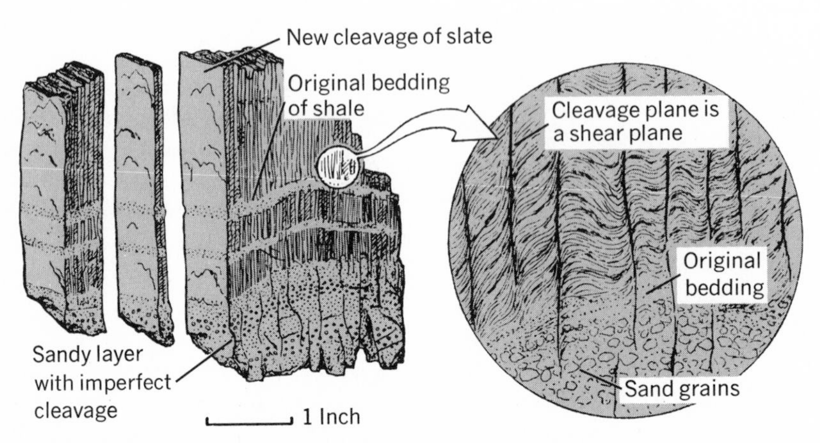

Fig. 27-20. *Slate vs. shale. Relics of original bedding are shown in the slate. Before metamorphism the rock would have broken into flat pieces parallel to the stratification. After metamorphism the rock breaks parallel to the foliation. The enlargement shows small offsets along the foliation (rock cleavage) surfaces.* [*From* Principles of Geology *by Gilluly, Waters, and Woodford, 2d ed., 1959, W. H. Freeman and Company.*]

Metamorphic Rocks. This group has been formed by the transformation of older sedimentary or igneous rocks into new, notably different types. New kinds of minerals, or mineral particles with different shapes or orientations, have been produced in the crust by heat, pressure, and the chemical action of solutions. Some metamorphic rocks are characterized by the segregation of different kinds of minerals into bands (gneiss) and others (schist and slate) by a tendency to break along closely spaced parallel surfaces into flat slabs (Fig. 27-20: foliation or rock cleavage). Marble is a recrystallized limestone which lacks foliation unless impurities are present.

The study of metamorphism has been aided greatly by observations in the field (out-of-doors in contrast to the laboratory or classroom) of rocks at stages that represent gradual changes between nonmetamorphosed and completely metamorphosed rocks. Recognizable features in the original rocks are seen to become fainter and new minerals or structures to become better developed as the metamorphic rocks are approached.

Exercises and Questions, Chapter 27

1. List some of the ways in which geology is of economic and intellectual value to man. Why is geology likely to be of greater economic importance to man in the future than in the past?
2. Grow artificial crystals in the laboratory or at home. Vary conditions (the rate of cooling) so as to produce large crystals at one time and small ones at another time. A simple experiment can be performed by dissolving sugar in hot water until a thick syrupy solution results. Allow this to cool. Suspend a string in the solution so that some of the crystals will attach themselves to it. They will be better developed than those which form on the bottom of the container. Why?
3. Explore your neighborhood or be on the lookout as you travel for geologic features that seem to tell a story: e.g., rock outcrops, hills, stream valleys, clay pits, volcanoes, mines, lakes, and coastal areas. Can you develop the geologic history that each such feature records? If you are unsuccessful now, try later on when your knowledge of geology has increased.
4. Visit a nearby museum and study the minerals, rocks, fossils, and other exhibits.
5. Examine the material from a salt shaker and describe the shapes of the salt particles. What mineral property is illustrated? A magnifying glass is useful but not essential.
6. A certain town is located on a hill near a steep cliff which has been used as the town dump—waste materials have been transported to the cliff and pushed over the edge to pile up on lower ground below. Included among the waste products have been automobiles made in the 1920's; 1930's, 1949, 1954, 1959, and 1965. Make a sketch (do you have a sense of humor?) which illustrates what an observer would see in the sides of a valley subsequently eroded through this dump by a stream. If the observer were acquainted with ancient model automobiles, could he date the different zones in the dump? Do you detect any similarities between this illustration and a pile of sedimentary rocks containing fossils?

7. On a scale of 1 centimeter = 1 mile, calculate each of the following for the Earth.
   (a) Polar diameter: in cm________; in inches________; in feet________.
   (b) Difference in length between polar and equatorial diameters: in cm________; in inches________.
   (c) Approximate diameter of the core: in cm________; in inches________; in feet________.
   (d) Approximate thickness of the mantle: in cm________; in inches________.
   (e) Average thickness of the continental crust: in cm________; in inches________.
   (f) Average thickness of the oceanic crust: in cm________; in inches________.
   (g) Highest mountain: in cm________; in inches________.
   (h) Greatest oceanic depth: in cm________; in inches________.

8. Summarize important data concerning the Earth's three spheres.

9. Questions concerning minerals and rocks.
   (a) What is the difference between a mineral and a rock?
   (b) What accounts for such physical properties of minerals as cleavage, crystal shape, and hardness?
   (c) Describe a number of ways in which minerals are useful to mankind and describe the physical properties which account for these.
   (d) Why is it possible to identify many rocks more readily in the field than as small specimens in a laboratory (this assumes that each identification is made without the use of a microscope)?
   (e) What is meant by the rock cycle?

10. How can a geologist locate the source area of the sediments that are contained in a certain sedimentary formation?
    (a) Assume that the source area for a thick pile of sedimentary rocks was uplifted a considerable distance while they were accumulating. How might this be deduced from a study of the sedimentary rocks?
    (b) Why are sediments deposited in layers?
    (c) How do sediments become rocks?

28 Crustal movements and geological time

THE GRAND CANYON of the Colorado River is some 200 miles long and has an average depth of about 5000 feet. Its width from rim to rim varies from about 4 to 18 miles (Figs. 28-1, 28-2, and 28-3). As for the spectacle from the brink of the north or south rim, it is one of the wonders of the world, and the feelings that overwhelm a visitor have never found adequate expression in words. Many miss completely the significance of this vast valley and the layered rocks which make cliffs and slopes of red, white, and gray along its walls. Others understand it partially; e.g., the farmer from a soil-eroded farm in the semiarid west who exclaimed: "Golly, what a gully!"

Among the feelings that arise is the desire to know the causes of this vast opening and its history. Here the geologist is aided by the record in the rocks. In the long geologic history of this region the canyon is only the most recent of six immense chapters of geologic change.

THE GRAND CANYON

Chapter One of the story begins more than a billion years ago with the origin of the ancient metamorphic and igneous rocks (Precambrian) now exposed in the inner gorge. Similar rocks are found today only in mountainous regions or in areas that presumably were once mountainous. Although erosion has removed all but the "stumps" of these mountains, yet their former existence—but not, of course, their precise shapes and altitudes—seems as certain as does that of the forest which the imagination can so easily reconstruct for a freshly lumbered hillside. Thus the presence of such rocks implies the former existence of a range of mountains in this

Fig. 28-1. *The Grand Canyon region.*

area long before the origin of the Colorado Plateau and Colorado River (Fig. 28-3A). Since some of these ancient metamorphic rocks seem once to have been sedimentary, and since sedimentary rocks are made of fragments of still older rocks, the history of the Grand Canyon begins after geologic processes had been in operation for some time.

The upper parts of these Precambrian rocks everywhere are beveled to a remarkably flat surface (Fig. 28-3B), apparently the end product of a very long period of erosion during which the mountains were worn down to an approximately level surface (a *peneplain*) near sea level.

Chapter Two in the geologic history of the Grand Canyon deals with the tilted layers of sedimentary rocks (younger Precambrian) which rest unconformably upon the older metamorphic rocks (Figs. 28-3C and 3D). Such a relationship between overlying and underlying sets of rocks, where a gap occurs in the rock-recorded history of the Earth, is called *unconformity* (p. 739). These tilted sedimentary beds likewise are very ancient. They originated with the gradual subsidence of the area after it had been peneplained. Upon this slowly sinking, nearly horizontal surface was deposited a thickness of more than 12,000 feet of sediments which later hardened into sedimentary rocks (the ultimate cause of the subsidence, of subsequent major crustal movements, and of similar movements in other parts of the Earth, is essentially unknown). Geologists assume that the strata were deposited originally in a nearly horizontal position because they can observe similar beds forming in that manner today.

Later, this 2-mile thick pile of rocks was broken into separate blocks along great fractures in the Earth's crust (faulting, Fig. 28-3D). One side of each block was pushed upward to form a mountain, whereas the other side was tilted downward. This second generation of mountains in the Grand Canyon area was then destroyed by erosion during another inconceivably long stretch of time (Fig. 28-3E). The uptilted, cut-off, younger Precambrian rocks are evidence for these steps.

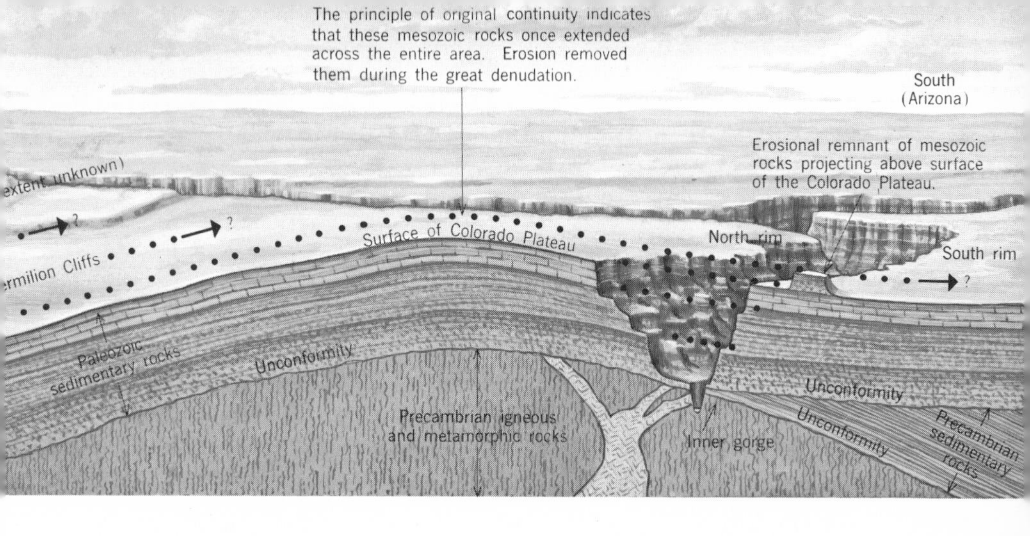

The third great chapter (Paleozoic) in the geologic history of the Grand Canyon begins with the formation of the horizontal strata, about 4000 feet thick, which today rest unconformably upon the older rocks below (Fig. 28-3F). In a number of places, these Chapter Three rocks rest directly upon the eroded, peneplained igneous and metamorphic rocks of Chapter One. In such places, all of the events of Chapter Two—the deposition of the sediments, the faulting that made them into mountains, and the erosion that wore them down—have been wiped completely from the geologic

Fig. 28-2. *The Grand Canyon and the Colorado River. The unconformity between horizontal Paleozoic sedimentary rocks and Precambrian igneous and metamorphic rocks is located near the top of the inner gorge (about 1000 feet deep).* [*Photograph George Grant, U.S. Geological Survey.*]

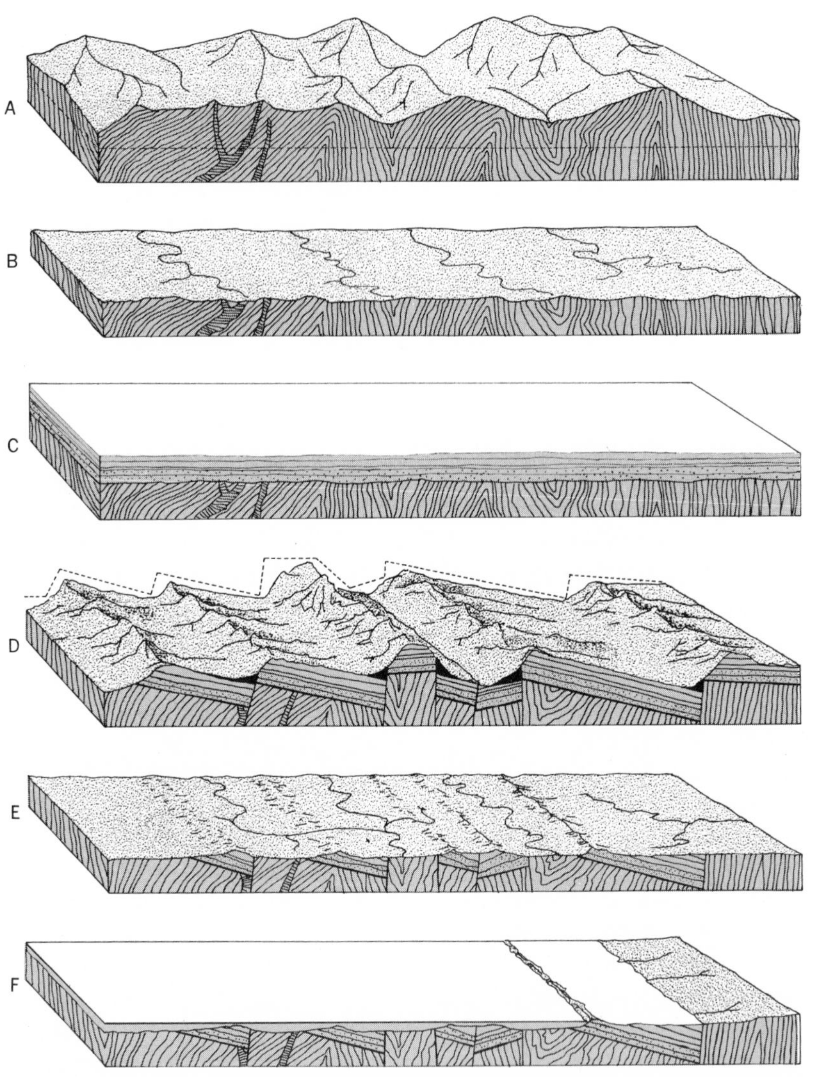

Fig. 28-3. *Five stages in the Precambrian history of the Grand Canyon region. The view is northward, and the sections represent an east-west distance of about 15 miles. The solid black shading in block D represents alluvium.* (*A*) *Precambrian rocks are formed by sedimentation, folding, metamorphism, and igneous activity. Mountains result.* (*B*) *Peneplanation leaves only stumps of the mountains.* (*C*) *Precambrian sedimentary rocks are formed.* (*D*) *Mountains are produced by faulting.* (*E*) *Peneplanation occurs and again only the stumps of the mountains remain.* (*F*) *Sedimentation is resumed at the start of the Paleozoic.* [*Courtesy, C. O. Dunbar,* Historical Geology, *1949, John Wiley & Sons, Inc.*]

record. Impressed geologists have called this a profound unconformity, and certainly it is difficult to exaggerate the immensity of this gap in the Earth's history.

A number of these Chapter Three strata contain abundant fossils of marine origin; other layers include fossils of land organisms; still others—the white sandstones exposed in the 300-foot cliff near the top of the canyon—resemble somewhat the wind-blown sands which accumulate in dunes in desert areas today. Apparently, therefore, the Grand Canyon area was covered by the sea during several long intervals. Dry land occurred at other times when the sea receded from the area, and once the region may have been a desert.

Chapters Four and Five of our story must be compressed here. They tell of the origin of the Mesozoic and Cenozoic sedimentary rocks now exposed in the spectacular scenery of Zion and Bryce Canyons (Fig. 28-1).

The present-day topographic features of the Grand Canyon region were produced by erosion and uplift during Chapter Six. Mesozoic rocks like those in Cedar Mountain now rest as erosional remnants on the Colorado Plateau. They indicate that the entire region was formerly blanketed by these Mesozoic strata and perhaps by some of the Cenozoic rocks also (principle of original continuity, p. 622). During a time picturesquely named the Great Denudation these strata, hundreds of feet in thickness, were stripped from above the Paleozoic rocks that now form the surface of the Colorado Plateau.

As erosion wore away the rocks toward the north, the edges of resistant beds formed cliffs and the steplike topography shown in the Pink, White, and Vermilion Cliffs today. Furthermore, the downward tilt of the sedimentary strata toward the north explains why one encounters successively younger rocks in each cliff without much overall increase in altitude. The quantity of rock debris transported from the area during the Great Denudation exceeds that removed from the region during the origin of the Grand Canyon itself.

The river now called the Colorado began to flow across this region when the surface was still at a low altitude. As the region was gradually uplifted a mile or more to form the present dome-shaped plateau, the river flowed more swiftly. Like a saw cutting through a board, the river eroded downward in its channel through the central portions of the rising dome. As evidence, the river now flows from lower ground in the north, through the elevated plateau region, into lower ground to the southwest. If the plateau had been in existence first, the river would have flowed around and not across it. Downcutting by the river and its tributaries enabled weathering and various types of mass-wasting (p. 648) to widen the canyon by moving disintegrated, decayed rock debris down the canyon walls to the river channel where it could be carried toward the sea.

As geologic time is measured, the uplift and canyon cutting occurred only a short time ago. The Grand Canyon itself is still a youthful valley in terms of the amount of erosion yet to be done to wear the area down near sea level. As one stands on the brink of the canyon, he finds it difficult to realize that the tiny-appearing, muddy stream in the bottom of the canyon has performed such a prodigious amount of erosion. Yet it is estimated that the Colorado River carries an average of about a million tons of sediment through the canyon every 24 hours.

In volume, this tonnage is approximately equivalent to 12 million cubic feet of the average kind of rock making up the Earth's crust: e.g., a solid that is 300 feet long, 200 feet wide, and 200 feet high. If erosion were to continue at this rate for about three centuries and a half, the Colorado River would have carried the equivalent of 1 cubic mile through the canyon. Of course, this material comes from the entire drainage area of the river, not from the Grand Canyon alone.

Many of the Paleozoic, Mesozoic, and Cenozoic strata bear fossils; e.g., trilobites (extinct members of the phylum Arthropoda, p. 796) and certain kinds of primitive fish and plants are found only in Paleozoic rocks, dinosaur fossils are confined to Mesozoic beds, and certain mammals occur only in Cenozoic strata. Thus the rocks of the Grand Canyon region also furnish a thick picture book of the life of the past.

PROBLEMS AND PRINCIPLES

Two disturbing questions are sharply pointed up by the story of the Grand Canyon. First, does the Earth's crust actually move this much? Have areas that once were beneath the surface of the ocean actually been lifted upward for 1 or 2 miles? What have geologists discovered about crustal movements? Certainly no one living has seen the crust move such great distances. Presumably such a large movement is the end result of a long period of gradual upwarping or sinking or of a succession of much smaller but abrupt shifts of the crust (of the type that produce earthquakes). Second, is the Earth really as old as the history of the Grand Canyon seems to indicate? What is the evidence for the great antiquity of the Earth?

In summary, a number of fundamental geologic ideas are lodged in the story of the Grand Canyon:

(1) Rocks have formed at different times, in different places, and in different ways.
(2) Rocks are the written records of geologic history ("sermons in stones").

(3) Fossils furnish much information about the Earth's history—about physical changes such as the advances and retreats of ancient seas as well as about biological changes.
(4) The Earth is very old.
(5) Forces within the Earth cause the crust to move.
(6) As sediments pile up parallel to the Earth's surface in nearly horizontal layers, they constitute a gigantic rock calendar. The youngest rocks occur at the top of a pile, the oldest rocks at the bottom.
(7) Surface and climatic conditions in any given area have varied greatly at different times in the past.

INSTABILITY OF THE EARTH'S CRUST

The notion of a *terra firma* is incompatible with geology and with the experiences of persons living in earthquake and volcanic regions. Evidence for crustal movements can be separated into three groups: recent movements such as that which produced the Alaskan earthquake of March 27, 1964; movements shown by records kept by man within historic times; and movements that are attested to by evidence found in rocks.

The term *diastrophism* encompasses all movements of the Earth's crust which involve changes of position and rock deformation, whether large, small, quick, slow, up, down, or sideways.

Sea level is somewhat satisfactory as a reference surface for the measurement of vertical movements because the oceans are interconnected, and mean tide is nearly level throughout the world. Sea level may be projected inland as an imaginary surface; e.g., Mt. Washington in New Hampshire is 6288 feet above sea level. However, sea level is not a fixed measuring surface. The total volume of water in the oceans has changed throughout the geologic past and is changing slowly today. Furthermore, vertical movements of the ocean floors and the addition of sediments and lavas to the ocean basins have caused changes in sea level in the past. Moreover, tides, winds, and changes in atmospheric pressure produce distortions in the ocean surface. However, such changes are thought to be small, slow, or nearly uniform throughout the world.

Recent Movements of the Earth's Crust A group of elastic shock waves is produced whenever an abrupt movement of a part of the Earth's crust or upper mantle occurs, and most commonly this takes place along a fault surface. These waves radiate outward in all directions from the disturbed zone; some go through the Earth, whereas others are confined to the crust and go around it. As the waves pass any one place, they cause the ground to shake and vibrate; they produce an *earthquake*. The Earth reacts as if it had been violently jarred by a blow from a Bunyan-sized hammer. According to the *elastic rebound theory* (Fig. 28-4) the following sequence may be common. During the years preceding an earthquake, slow movements in the crust—in opposite directions on opposite sides of a stressed zone—bend the rocks gradually until they reach and exceed their breaking point. Their sudden rupture and abrupt return approximately to their pre-stressed positions produce an earthquake.

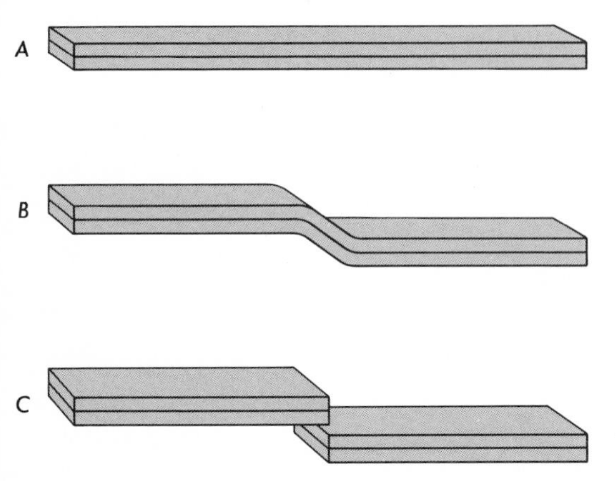

Fig. 28-4. *Elastic rebound theory. A represents part of the crust and is unaffected by stresses. B shows bending in response to vertical stresses which have been applied slowly; stresses may be applied at any angle. In C the strain has been relieved by a sudden movement along a fault surface. This causes an earthquake.*

No part of the Earth's surface is immune to earthquakes, though certain regions have many (p. 713), whereas others have few. The Tibet earthquake of 1950 apparently had one of the greatest magnitudes yet measured and was described as follows in the *New York Times* for August 27, 1950:

In reporting a great earthquake twelve days ago, seismologists used such words as "terrific," "dizzy," and "colossal." But the region hit by the quake —Northern India and Tibet—was so remote that just what happened there was unknown. Last week the Indian Air Force disclosed the results of a survey it had made of the Indian territory affected. This is what the survey showed: The entire geography of the area was changed. Roads and rivers have disappeared and sections of railway have been left hanging in midair. Hundreds of square miles are covered by new lakes. Thousands of acres of land have been laid waste by yawning fissures. Sulphur has polluted the water of some rivers and fish are piling onto the crumbled banks as if trying to escape. More than 100,000 homes have been wholly or partly destroyed, 50,000 head of cattle killed and 5,000,000 persons in all affected. It will be weeks or months before all the destruction can be measured.

Earthquakes can produce notable topographic changes. In 1899 in the Yakutat Bay region of southern Alaska, several great earthquakes resulted from the sudden lifting of part of the coast a vertical distance of nearly 50 feet. Simultaneously another part of the same coast was moved downward.

After an earthquake in Chile in 1882, coastal towns were found to be located 3 to 4 feet higher above sea level than before the shock. Earthquakes of very large magnitude also originated in Chile in 1960 and in the Anchorage, Alaska, area in 1964.

In the greatest quakes, the ground is reported to have moved up and down in waves, causing fissures to open and close in the soil and trees to tilt at various angles. Accounts of such fissures in earthquakes are very commonly exaggerated; the terror, fright, and emotional excitement experienced by individuals during an earthquake are readily conducive to such distortions. Fissures and cracks may be produced in surficial debris during the passage of earthquake waves and also in loose debris which may be jarred loose from a nearby mountain side to slide violently down its slope. However, the Earth's crust is not pulled apart to leave gaping chasms which later are closed.

In April 1906 the San Francisco Bay region of California experienced a violent earthquake. Sudden movements occurred on opposite sides of a great vertical crack (fault) in the Earth's crust. Most of the movement was in a horizontal direction. Roads, fences, pipes, and houses were offset a maximum distance of 21 feet along the fracture, and the ground was torn apart. Movement occurred throughout a distance of some 250 miles along the fault. The greatest destruction in San Francisco was caused by fire which spread quickly from damaged buildings. A second destructive earthquake, less violent than that of 1906, occurred in 1957, and additional movements will probably occur in the future.

A powerful earthquake shook the Hebgen Lake region of southwestern Montana about 11:40 P.M. on August 17, 1959. Small cliffs up to 20 feet high were produced by the faulting, which represented renewed movement along old faults. The crustal movement tilted the surface in the vicinity of Hebgen Lake; one shoreline was submerged about 10 feet, and the opposite shoreline emerged about 10 feet.

The most disastrous effect of the earthquake was a gigantic landslide in the canyon of the Madison River 6 miles below Hebgen Dam (Fig. 29-10). In a minute or so, 35 million cubic yards of broken rock slid into the canyon and covered a 1-mile span of the river and highway to a depth of 100 to 300 feet. Three weeks later a lake 175 feet deep and nearly 6 miles long had formed upstream from the landslide area. The sliding occurred along cracks which slanted downward toward the river and were nearly parallel to one steep wall of the canyon. The slide started about 1300 feet above the river, swept downward and across the valley, and extended as much as 400 feet above the river on the opposite side. From this unstable position, some debris subsequently slid back into the valley. A number of people lost their lives in this disaster.

The presence of many marine shells on land surfaces bordering the coast in the Baltic area had suggested to some people that uplift was taking place, and marks were placed along the coast to measure this. Some stakes, once at mean tide level, are now several feet above sea level and more than a mile inland from the present shore line. A maximum uplift in northern Sweden of about 7 feet in the last 175 years has been measured. Movement is so slow that inhabitants in the area cannot detect any change.

Many similar examples could be cited showing how the Earth's crust has moved in both vertical and horizontal directions. To be sure, the magnitudes are quite small when compared with movements recorded by rocks in the Grand Canyon, but some happen so quickly and cause so much destruction that they make us keenly aware of crustal movements.

Evidence of Crustal Movements from Historical Records Ancient docks built on the island of Crete in the Mediterranean about 2000 years ago are today as much as 27 feet above sea level. However, other structures are now below sea level. Thus Crete shows evidence of local crustal warping within the last 2000 years—some parts up, other parts down. If uplift continues at the rate of 27 feet per 2000 years for a million years (geologically a short time) it will amount to 13,500 feet, a calculation which illustrates an important principle in geology. If relatively small movements or apparently tiny forces act through a very long period of time, they produce major results. More rapid uplift has been measured elsewhere by precise surveys that span several decades: e.g., in one part of the San Joaquin Valley of California the rate is about 4 feet per 100 years.

Fig. 28-5. *Surf-eroded benches, recently uplifted, on San Clemente Island, California. A surf-eroded bench formed when the crust was relatively stable in this area, and an uplift then followed. This sequence has been repeated several times.* [*R. S. Dietz,* Geol. Soc. of Am. Bull., *Vol. 74, August 1963.*]

Rocks Show Evidence of Crustal Movements In interpreting evidence of crustal movements from rocks, we assume that the present is the key to the past. Along a seacoast, we find certain features of the landscape associated closely with sea level: beaches, cliffs cut by the surf, and smooth beveled rock surfaces (surf-eroded terraces) sloping gently out to sea (surf-eroded benches or terraces, Fig. 28-5). Erosional remnants (stacks) may project above the surface of a surf-eroded bench. Such topographic features form only along a coast line near sea level. Their marks on the lands are somewhat analogous to the dirt line on a tub that shows the former depth of water (Junior's after a Saturday morning scrimmage on a muddy football field). If such features are found today above sea level, their present location implies changes in the relative positions of land and sea. If the sea surface actually rises or falls, then the effect is worldwide instead of local.

A comparison of shells and animal remains found in the ocean with others found inland shows that recognizable differences exist today between marine and nonmarine forms. A check with other observers in different parts of the world substantiates the conclusion that certain types of animal life live only in the sea.

Unconsolidated sands, silts, and muds containing the shells and bones of marine organisms occur today a hundred or more feet above sea level in some areas. If such deposits are located near existing coastal areas, one does not hesitate to conclude that the land was elevated after deposition of the sediments. Consolidated sedimentary rocks containing similar marine fossils can be found at elevations of hundreds and even thousands of feet above

Fig. 28-6. *An eroded structural basin in Africa. The youngest exposed rocks are located at the center of the basin, and successively older rocks occur outward from the center.* [*Photograph, U.S. Air Force: N. 32° 30′, W. 03° 35′.*]

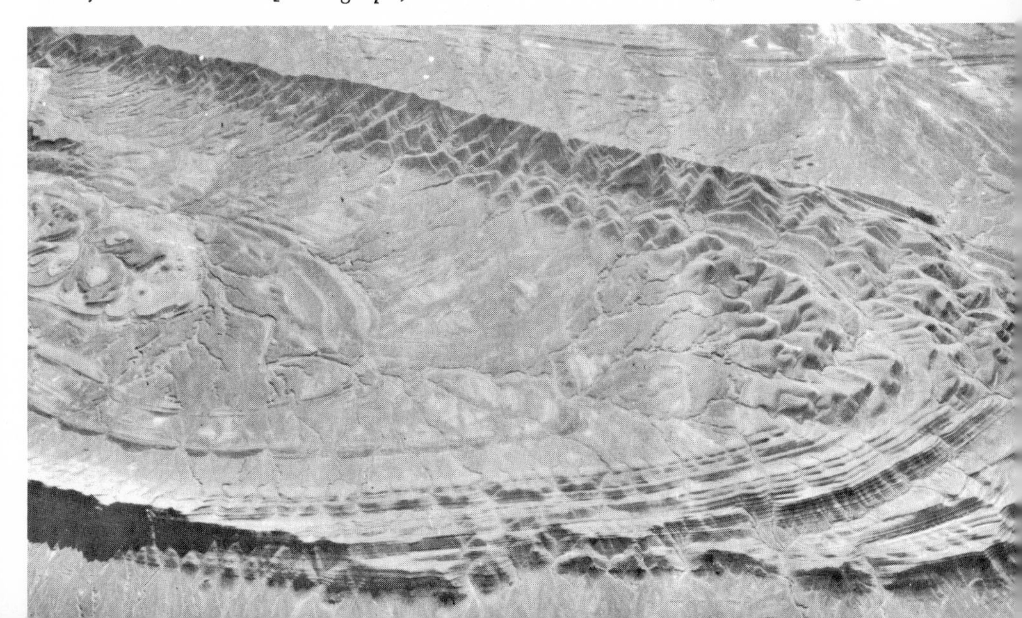

Fig. 28-7. *Air photograph of a fault. The view is toward the southwest along the south side of Macdonald Lake near Great Slave Lake. The fault separates ancient Precambrian granite on the left from less ancient Precambrian sedimentary rocks on the right.* [*Courtesy, Dept. of Mines and Technical Surveys, Canada.*]

sea level (3 to 4 miles in the Himalayas). Were such areas also formerly beneath the surface of the ocean? The difference between the two examples is one of magnitude only. The evidence is conclusive that some movements of the Earth's crust have been very great.

The discovery of once-horizontal sedimentary rocks which have been bent and folded, is striking and furnishes convincing evidence of crustal movements (p. 732 and Fig. 28-6). Furthermore, rock formations have been broken and offset along great fractures in the crust (Figs. 28-7 and 28-8). Previous examples have dealt chiefly with vertical movements, but some folds and faults result from great compressive forces in the crust.

Another type of evidence indicates that the Earth's crust has moved greatly in the past. If erosion has been taking place unendingly upon the Earth's surface since the first rains fell, why have the continents not been worn down to a relatively smooth surface near sea level? Forces must exist which oppose those of erosion and act to increase and elevate the land areas.

THE PROBLEM OF GEOLOGIC TIME

The time and manner of the Earth's origin pose questions upon which men of all centuries have expended much in the way of speculation and ingenuity. Estimates of the age of the Earth have ranged from a minimum of about 6000 years to eternity. Unfortunately, the 6000-year figure gained authority in the seventeenth century as the actual age of the Earth. For 150 years or so, attempts were made to explain all of the Earth's features in terms of 6000 years or less. The past had to be credited with a succession of great catastrophes which had no counterparts in the present. Streams flowed in valleys formed by the sudden opening of great fissures in the Earth's crust; mountains were formed by quick cataclysmic upheavals; great floods covered large portions of the Earth, leaving behind huge boulders which the raging waters had picked up and carried. As these catastrophes were not occurring during the present age, men believed they could explain the Earth's history best by speculations conceived while sitting at their desks rather than by actual study of the Earth in the field.

Near the end of the eighteenth century, James Hutton, a Scottish farmer-physician, introduced a point of view which radically changed all geologic thinking. Hutton was an observer of nature. It was only after years of careful, patient field work that he published in 1785 the book embodying his

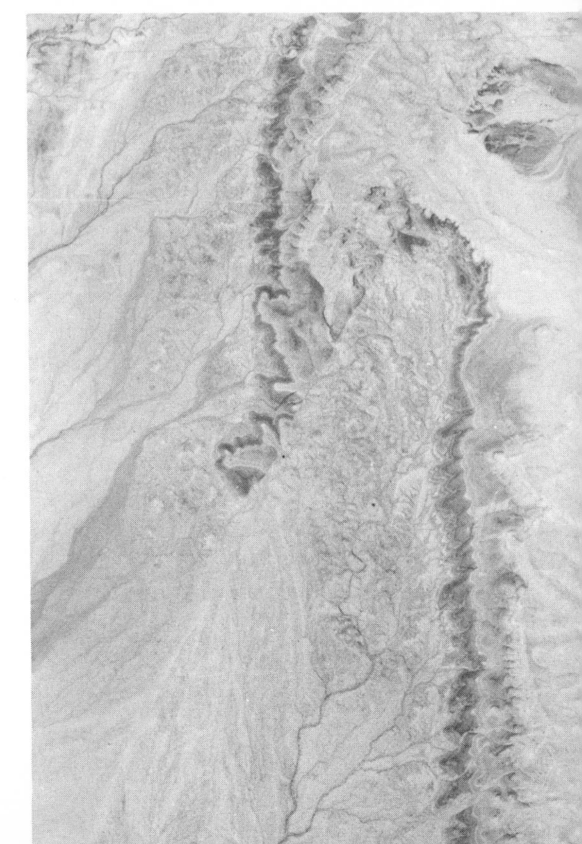

Fig. 28-8. *Gently dipping sedimentary rocks are offset by a near-vertical fault in Utah. The dark-toned layers are shale. The fault can be traced across the alluvium in the center and left (note difference in drainage patterns on opposite sides of the fault).* [*Figure 83,* U.S. Geological Survey Professional Paper *373*.]

principle: *The present is the key to the past.* Hutton argued that the study of processes now operating on the Earth would yield information for unraveling the mysteries of the Earth's past. According to this principle, agents now at work shaping the Earth's surface have been functioning throughout the Earth's history. The fundamental laws of physics, chemistry, and biology apply now as they did in the past. Today gravitational attraction operates, light travels at 186,000 mi/sec, winds blow, rains fall, water flows down hill, volcanoes erupt, sediments are deposited, shells are buried, earthquakes occur, rocks crumble and decay, snow changes to ice, and coastal areas are eaten away by wave action. Of course these processes have not always functioned at their present rates. Volcanic activity, for example, was probably more pronounced in some periods of the past than it is today. Yet the study of today's volcanoes is our guide in interpreting ancient volcanic activity. Given time enough, processes of the kinds that we are familiar with in the present can account for most of the Earth's features.

Fig. 28-9. *Three methods of estimating the age of the Earth, illustrated schematically.*

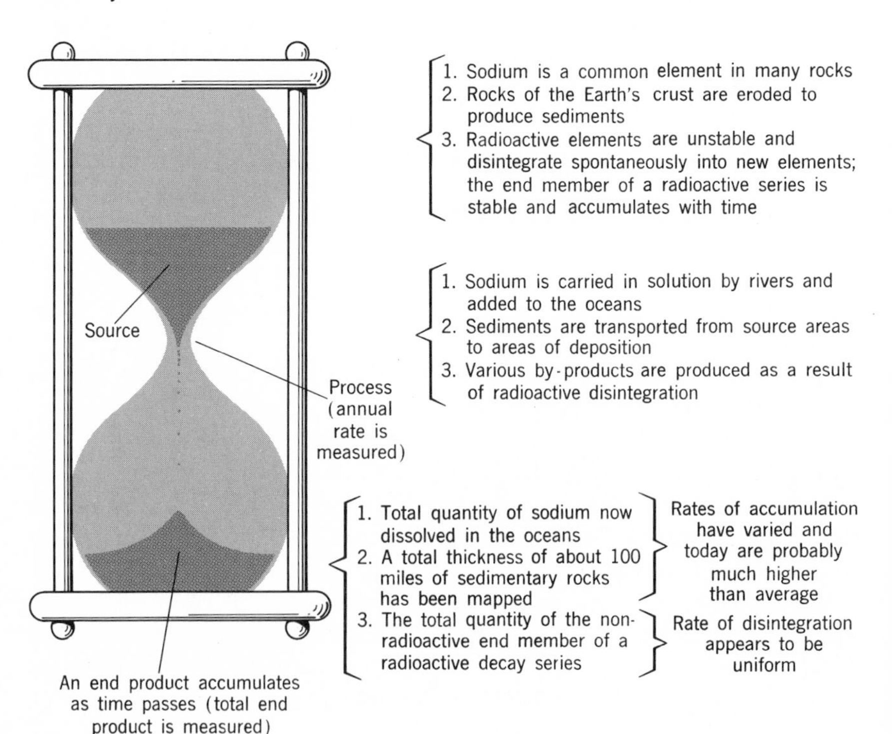

Key Concepts in Geology Hutton's inductive reasoning was a far cry from the preceding years of speculation. He reasoned from the particular to the general; i.e., he used the facts gained from field observations as a basis for working out general principles that then could be applied widely to new situations. Further observation was needed to check the validity of each principle. Perhaps Hutton's greatest contribution to geology was this emphasis upon painstaking field observation as opposed to the armchair deduction practiced by many of his contemporaries. Deduction involves reasoning from the general to the particular. Some principle or generalization is assumed to be valid, and the specific consequences that should follow are predicted. Conclusions reached in this manner involve reasoning alone, and some of Hutton's contemporaries did not verify theirs by observation in the field. Unfortunately, a number of their basic generalizations were erroneous.

Concerning the Earth itself, Hutton stated that he could "find no vestige of a beginning . . . no prospect of an end." Hutton's motto—*the present is the key to the past*—was later championed by the great British geologist Charles Lyell and remains fundamental in geology. It is now known as the principle or assumption of uniformitarianism, and it is a specimen of what may be called *key concepts.* These are principles that guide research and give a structure to the separate facts of science. They have been called part of the tactics and strategy of geology and will be met later in this chapter and are emphasized throughout the chapters on geology.

THE MEASUREMENT OF GEOLOGIC TIME

If topographic and geologic changes are to be accounted for chiefly by the familiar processes that occur today, they must have come about slowly, over immense stretches of time. This immensity was for many years a stumbling block to the acceptance of Hutton's principle, but the evidence in favor of it has been overwhelming.

Methods of estimating geologic time have certain features in common (Fig. 28-9): attention is fixed upon some change or process which is taking place on the Earth at present; this change is assumed to have been going on uniformly since early in the history of the Earth, or an average rate is estimated; the total change is measured; and the annual quantity of the change is determined. Dividing the total change by the annual change then gives the age of the Earth.

Unfortunately for age determinations, the rates of accumulation of sodium and sediments have varied, and average rates are not known. Thus these two methods show only that the Earth is very old, though sediments can also be used for their indications of relative ages.

Determination of geologic time by radioactivity emphasizes the close interrelationship of geology with chemistry and physics. The nuclei of a number of elements are radioactive naturally; they spontaneously emit certain radiations and are thereby transformed into the nuclei of atoms of different elements (atomic structure and radioactivity are discussed in Chapters 12 and 15). Nothing known to man can alter the rates of these spontaneous disintegrations, and thus the radioactive decay of a particular isotope proceeds at a uniform rate; its half-life period remains constant. Here is a geologic process which apparently has not varied in rate during the long past. Therefore, radioactive disintegration apparently provides an accurate means of measuring the ages of rocks. The Earth's age is greater than that of the oldest rock measured by the radioactive method.

Meteorites provide an additional clue. Their age determinations apparently average about $4\frac{1}{2}$ billion years. If meteorites are representative of the material from which the planets of the solar system were formed—if meteorites are the unused remnants of this material—then the age of the Earth should be about the same as the ages of the meteorites.

Radioactive isotopes of uranium, thorium, rubidium, and potassium are important in determining the ages of rocks, but we shall concentrate upon the uranium-238 isotope as an example. In nature, uranium is found as a constituent of certain kinds of minerals which have formed in different ways, e.g., by precipitation from solutions such as magma and hot water (Fig. 28-10). A uranium-238 isotope spontaneously undergoes a series of changes which result in the formation of fourteen new isotopes, the fourteenth of which is stable lead-206. To start the decay series, a uranium-238 isotope emits an alpha particle and becomes an isotope of thorium; this in its turn emits a beta particle and becomes an isotope of protactinium, which in its turn . . . until the fourteenth isotope, stable lead-206, is reached. The number of uranium-238 isotopes decreases at a known uniform rate, and stable end-product lead-206 isotopes accumulate at a known uniform rate. Therefore, the proportion of uranium-238 to lead-206 in a rock measures the age of that rock. The oldest rocks have the highest proportion of lead.

The rate of disintegration can be determined in several ways. A statistical approach and the half-life period of an isotope are involved. In life insurance it is impossible to predict exactly which men in a group of 60-year-old men will die in the next year. However, on the basis of past experience, one can predict quite reliably, if the total group is large enough, the number of deaths to expect within the coming year. In a similar manner, one cannot determine which of the isotopes of a radioactive element will change to another isotope or when this will occur. However, one can predict the length of time necessary for half of any large number to change. Thus in approximately $4\frac{1}{2}$ billion years, one-half of the uranium-238 isotopes in any given quantity will have disintegrated into the isotopes of other elements; a large portion will be lead, but not half because a dozen in-between members occur in the uranium-238 decay series. One-half of the remainder will be changed in another $4\frac{1}{2}$ billion years and so on. This relationship is known as the half-life period.

Let us illustrate its meaning with a fictitious example. Imagine that a certain radioactive isotope has a half-life period of 10 years and that 100 pounds of this was placed in a safe in 1900. How much was left in 1950?

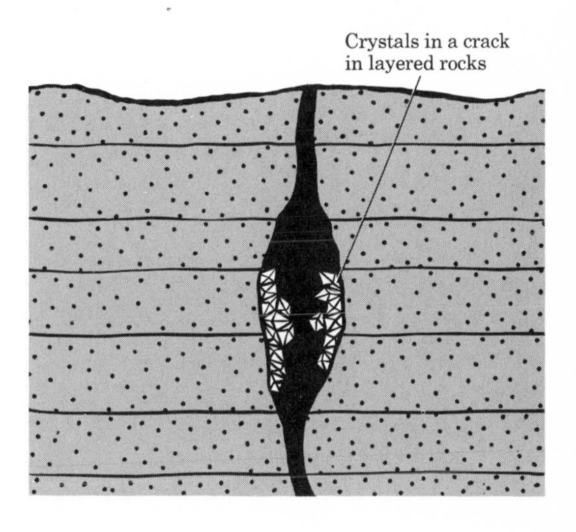

Fig. 28-10. *The radioactive method of dating minerals and rocks. After the layered rocks had formed, a crack developed across them and was later filled by hot water solutions which rose along the crack (forced upward by the great pressures that exist at depth). The solutions contained uranium, oxygen, and other elements, and some of these combined and were precipitated to form the uranium-bearing crystals shown in the crack. Since precipitation, the number of uranium atoms in the crystals has steadily decreased, whereas the number of lead atoms has steadily increased. Thus the age of the crystal (i.e., the time when it was precipitated from solution) can be determined. The sedimentary layers are older than the crystals, but how much older cannot be determined from the limited data included in this illustration.*

In 1910, 50 pounds would remain; in 1920, 25 pounds; in 1930, $12\frac{1}{2}$ pounds; in 1940, $6\frac{1}{4}$ pounds; and in 1950, $3\frac{1}{8}$ pounds.

The original quantity of uranium in a rock formation can be determined by adding the number of uranium atoms in the rock being tested to the total number of atoms in the specimen which result from the disintegration of uranium. Each of these atoms was once a uranium atom. Of course, one must take care at this point. Has the specimen remained a closed chemical system? We might go wrong if we included in our count "common" lead which did not form via the radioactive disintegration of uranium. Such common-origin lead may subsequently be added to lead of radioactive origin. Fortunately, if contamination occurs, it can commonly be detected. On the other hand, some of the radioactive lead isotopes that formed in the

Table 28–1 Major Methods in Geochronometry*

NUCLIDES	HALF LIFE (YRS.)	λ (YRS.$^{-1}$)
U^{238}-Pb^{206}	4.50×10^{9}	1.54×10^{-10}
U^{235}-Pb^{207}	0.71×10^{9}	9.72×10^{-10}
Rb^{87}-Sr^{87}	4.7×10^{10}	1.47×10^{-11}
K^{40}-Ar^{40}	1.30×10^{9} (total)	$\lambda\beta$ 4.72×10^{-10} λe 5.83×10^{-11}
C^{14}	5710 ± 30	0.21×10^{-4}

*T_0 = age of the earth, i.e., 4.6×10^9 yrs.
**For paleogeographic studies.
†Under certain favorable conditions the lower limit of this method can be extended to approximately 10^4 yrs.
λ = decay constants.

specimen may have left it during weathering. However, if this occurred, it can generally be detected and allowed for. To avoid the effects of contamination and solution, the best specimens to use for age determinations are single unweathered crystals of uranium-bearing minerals. The lead iso-

EFFECTIVE RANGE (YRS.)N	MATERIALS
$10^7 - T_0$	Zircon, uraninite, pitchblende
$10^7 - T_0$	Zircon, uraninite, pitchblende
$10^7 - T_0$	Muscovite, biotite, lepidolite, microline, glauconite, whole metamorphic rock
†$10^5 - T_0$	Muscovite, biotite, hornblende, phlogopite, glauconite, sanidine, whole volcanic rock, sylvite (arkose, sandstone, siltstone)**
0-50,000	Wood, charcoal, peat, grain, tissue, charred bone, cloth, shells, tufa, ground water, ocean water

topes which form by disintegration of the uranium are trapped in these crystals.

Tests of various uranium-bearing minerals from different locations have given results which range from a few million years to a few billion years.

This spread indicates that uranium-bearing minerals have formed at many different times in the past. The oldest rocks yet measured by this method are about 3 to 3.5 billion years old. Perhaps somewhat older rocks will be found in the future, but present theory suggests that the $4\frac{1}{2}$ billion year age for meteorites is a limit.

Radioactive isotopes of rubidium and potassium have been added to those of the uranium isotopes and thorium in age determinations (Table 28-1). The hydrogen isotope, tritium, with a mass number of 3 (two neutrons and one proton in its nucleus) and a half-life period of $12\frac{1}{2}$ years, can be used to determine the age of water—i.e., to tell the time since the water was removed from direct contact with the atmosphere to become part of the ground-water supply. The range is limited to less than 100 years by the short half-life period.

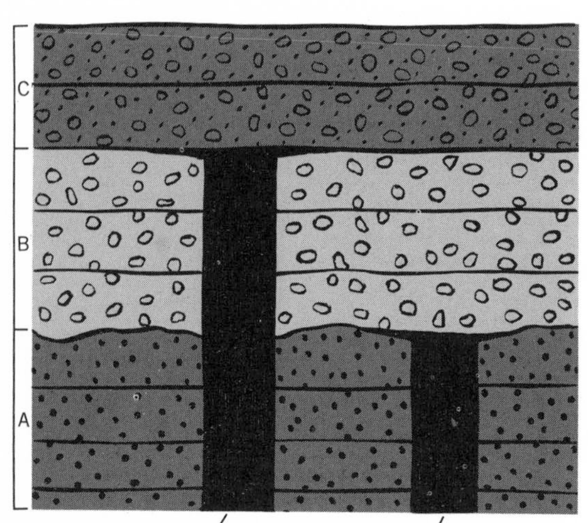

Fig. 28-11. *Absolute ages of sedimentary rocks are determined by their relationships to igneous rocks. Three sequences of sedimentary rocks (A, B, and C) are shown in the structure section. They are separated by two unconformities and have been intruded by two dikes. The relationships between the igneous and sedimentary rocks show that sequence A is older than 200 million years; sequence* B *is between 150 and 200 million years, and sequence C is younger than 150 million years.*

There are checks on the accuracy of the radioactive method of measuring the ages of rocks; e.g., different radioactive isotopes disintegrate at different rates into different end products, and several may be found in a single specimen. This is analogous to finding several clocks which all give the same time. One assumes the time is correct. Thus the methods, half-life periods, and results seem to be generally reliable but only, it should be emphasized, in terms of millions of years.

Although the ages of many igneous and metamorphic rocks can now be determined, sedimentary rocks cannot generally be dated so readily by the radioactive method. Although a pebble in a sedimentary rock may contain suitable radioactive elements, its age determination would give the time that

Table 28–2 Examples of Concordant and Discordant Isotopic Ages*

LOCALITY	MINERAL	METHOD	ISOTOPIC AGE (M.Y.)
Black Hills, S.D.	Uraninite	U^{238}-Pb^{206}	1610 ± 20
		U^{235}-Pb^{207}	1615 ± 20
		Pb^{207}-Pb^{206}	1620 ± 15
	Microcline	Rb^{87}-Sr^{87}	1630 ± 45
	Muscovite	K^{40}-Ar^{40}	1590 ± 40
Wilberforce, Ont. (Grenville Province)	Uraninite (Pegmatite)	U^{238}-Pb^{206}	1020 ± 10
		U^{235}-Pb^{207}	1025 ± 15
		Pb^{207}-Pb^{206}	1035 ± 30
	Zircon (Pegmatite)	U^{238}-Pb^{206}	1900 ± 10
		U^{235}-Pb^{207}	930 ± 15
		Pb^{207}-Pb^{206}	1000 ± 30
	Biotite (Pegmatite)	Rb^{87}-Sr^{87}	970 ± 30
		K^{40}-Ar^{40}	940 ± 30
	Biotite (Gneiss)	K^{40}-Ar^{40}	850 ± 25
Pikes Peak, Colorado	Zircon (Granite)	U^{238}-Pb^{206}	625 ± 25
		U^{235}-Pb^{207}	705 ± 20
		Pb^{207}-Pb^{206}	980 ± 40

Data from G.R. Tilton *et al.* 1960 and 1957
Eckelmann and Kulp, 1957.

*Courtesy of Professor J. Laurence Kulp.

the source rock formed and not the more recent time when the sediments were deposited. However, geologists can obtain the approximate ages of many sedimentary rocks by their relationship to igneous rocks which may have intruded into them or which may be correlated with them in some other way (Fig. 28-11).

Furthermore, in some sedimentary rocks the cementing material may contain a radioactive isotope (e.g., potassium-40) that makes an age determination of the "cement" possible. If the cement formed shortly after the sediments were deposited, this dates the sedimentary rock.

The immensity of geologic time is quite unimaginable, but the relative length of its epochs can be grasped with the help of analogies. Imagine that you are walking into the past at the rate of 100 years for each 3-foot step. Walking a mile (1760 steps) on this scale would be the equivalent of going back 176,000 years into time. The first step takes you back to about the Civil War, the second step approximately to the Revolutionary War. Eighteen more steps would total 2000 years and carry you back to the time of Christ. Now to walk into time far enough back to see a dinosaur (at the end of the Mesozoic Era) would require a 360-mile hike! For the next 950 miles or so you would be passing through the Mesozoic Age of Reptiles. A total of about 3000 miles, roughly the distance from New York to San Francisco, would have to be walked to arrive at the beginning of the Paleozoic era, some 600 or so million years ago. On the scale we have chosen, 4 billion years is the equivalent of about four round trips from the Atlantic to the Pacific—each 3-foot step equivalent to 100 years!

Again, suppose that all of the Earth's 4 billion or so years of conjectured history is compressed into 1 year. On this scale, the Paleozoic era, which contains the oldest rocks with abundant fossils, would not even begin until $10\frac{1}{2}$ months had passed. Mammals would appear on the scene about the second week in December. Apparently man would arrive around 11:45 P.M. on December 31, and all of recorded history would be represented by the last 60 seconds.

Thus the remarkable discoveries made since the turn of the century in the fields of atomic structure and radioactivity have had their impact on geology as well as on other sciences. The first radioactive age determinations of minerals were made shortly before World War I. For the first time in the history of geology, ages of minerals could apparently be measured accurately in actual millions of years. Before this, geologic time had been measured chiefly in a relative sense: this fossil or rock was older or younger that that one, but exactly how much older or younger was not known. The Earth was found to be much older than had previously been believed.

Prior to the discovery of radioactivity, it had generally been thought that the Earth was cooling off by the gradual loss of heat residual from its supposed fiery origin. Now, the heat which is known to be given off during radioactive disintegration has made it uncertain whether the Earth is cooling off or warming up. This problem has an important bearing on fundamental geologic processes.

The Carbon-14 Method of Measuring the Age of Dead Organic Matter Akin to the methods of measuring geologic time, an exciting technique has been developed to measure the age of dead organic matter. This method is applied to once-living matter and indicates the approximate number of years which have elapsed since the source organism stopped living. The technique was discovered somewhat incidentally rather than as the result of a planned search (such unexpected fruitful results occur commonly in scientific research). It provides an excellent illustration of science in action: the careful unbiased observation of natural phenomena and accumlation of data; the coming of ideas not necessarily visualized when the work started; the formulation of hypotheses to be verified, modified, and expanded or discarded; and the development of new instrumentation. All of these are merged into a closely knit, intimately interwoven pattern of activity that is not readily organized into a preset, one-two-three series of steps which can be called the scientific method. Its goal is the development of principles, concepts, and laws. A complete understanding of natural phenomena by the formulation of theoretical schemes which make a meaningful whole out of seemingly unrelated, uninteresting factual knowledge.

The carbon-14 method began as an outgrowth of research by Libby and others on cosmic radiation. It was learned that cosmic rays, during their bombardment of the upper atmosphere produce secondary neutrons which react with some of the nitrogen in the atmosphere to make a carbon isotope which is radioactive. This carbon isotope has an atomic weight of 14, whereas common carbon has an atomic weight of 12 and is not radioactive (a carbon-13 isotope exists but can be ignored here). Although a single carbon-14 atom does not exist long (it disintegrates back into nitrogen), additional carbon-14 atoms are constantly forming in the air.

Libby theorized that all living organisms should contain some carbon-14 and that a constant ratio of carbon-14 atoms to carbon-12 atoms should be present in the atmosphere and in all living organisms including man. He reasoned somewhat as follows: carbon-14 is forming continuously everywhere in the Earth's atmosphere; some of it unites with oxygen to form carbon dioxide; plants assimilate carbon dioxide during their life functions; and animals eat plants. The hypothesis was at once tested. Samples of such diverse organic matter as sewage, trees, seal oil, and sea shells were obtained from all parts of the Earth and checked. The proportions of carbon-14 to carbon-12 were found to be approximately the same in all instances.

A carefully measured quantity of the sample is processed chemically (which destroys the sample) before the amount of carbon-14 in it can be measured with a Geiger counter (p. 221). The radioactivity is directly proportional to the quantity of carbon-14 atoms present in the specimen. The level of activity in living organisms is about 14 disintegrations per minute per gram of carbon. This is exceedingly small when compared with the normal background radioactivity—for example, the researcher is more radioactive than the sample of dead tree that he may be checking—and necessitates a very sensitive measurement and ingenious shielding devices.

A second hypothesis was now formulated. In living organisms, a constant proportion of carbon-14 to carbon-12 is maintained because new supplies of carbon-14 are constantly being taken in to replace those lost by disintegration. But after death, replenishment ceases, whereas disintegration continues. Recent measurements indicate a half-life period for carbon-14 of about 5600 years (5710 $\pm$ 30 years may be more accurate, but a figure of 5568 years has been used in most measurements to date). The proportion of carbon-14 left in a dead organism, therefore, should indicate the length of time that has gone by since the organism died. A small proportion of carbon-14 to carbon-12 would show that the organism died long ago. If only half of the original quantity of carbon-14 remains, the specimen is about 5600 years old; if one-fourth remains, the age is about 11,200 years, etc.

This hypothesis was tested against known historic records. For example, tests were made on wood taken from coffins that contained Egyptian mummies of known age (presumably the wood for a coffin was cut at about the time the death occurred). The carbon-14 age determinations corresponded closely to the dates based on recorded history. The oldest checks that historical evidence permits were in the neighborhood of 5000 years.

To be sure, the method requires some precautions particularly with the oldest specimens. Cosmic radiation may have varied in intensity in the past. Young carbon may contaminate old samples; e.g., it has been estimated that the addition of only 1 percent of modern carbon to a 57,000 year old specimen would reduce its age to about 35,000 years. Such contamination is difficult to detect.

Carbon-14 has a relatively short half-life period, and thus this method has a time limitation of about 40,000 years, although further refinements may lengthen this somewhat.

Imagine the eagerness with which an archaeologist, for example, might wait for age determinations of his samples. Perhaps he submitted some charcoal from an ancient campfire around which he had found the bones of a few extinct animal species and the arrowheads and artifacts of prehistoric human culture. Before the discovery of the carbon-14 method, an exact age determination would have been impossible. Now the date when the campfire was built (actually when the firewood died) is determinable within a small percentage of error. Or the archaeologist may have cut a trench into the floor of a cave inhabited by man for many centuries, and he may have collected samples of charcoal and artifacts at different levels. Now these levels can be dated; proportionally less carbon-14 occurs in specimens from lower and lower levels.

Some time ago, about 300 pairs of sandals made from sagebrush bark were found in an Oregon cave. The sandals were woven out of grass rope and are attractive in design and shape. Of course, one is curious about the time they were made. Carbon-14 tests give an age of about 9000 years. They provide evidence of an American cobbler of 7000 B.C. or so.

A few years ago some herdsmen camped for the night near a spring in Palestine. According to one story, they noticed a small opening in the ground nearby. Investigating, they found a cave and in it a number of sealed earthen jars. In their rage at finding rolls of parchment in the jars rather than gold coins, the herdsmen destroyed most of the material. Fortunately they carried away a few rolls and these passed from hand to hand until at last one of the parchments reached a scholar capable of recognizing its importance. It proved to be a scroll almost 25 feet long which contained a nearly complete version of the Book of Isaiah. Since the present text of Isaiah has been taken from a copy of a still older version in the 1600's, the date of this new discovery was quite significant to biblical scholars. Would the two versions be similar, and how old are the scrolls? The amount of carbon-14 present in the linen wrappings of the parchment indicates that the scroll was written about 1900 years ago. The ancient and modern versions proved nearly identical. Fortunately, many more of these Dead Sea scrolls have since been discovered in other caves and are being studied.

THE TACTICS AND STRATEGY OF GEOLOGY

Nearly all who come in contact with geology find the subject interesting, and many find it fascinating. Geology probably has more to offer the nonscientist than almost any other science because the background necessary to appreciate and understand many of its fundamental aspects is relatively limited and rather easily attained. Geological science is, of course, complex in many of its aspects and is rapidly becoming more so. The modern trend emphasizes laboratory studies and experimentation, the use of complex instruments, the application of mathematics, statistics, and computers, and the quantitative studies of many geologic processes. Geology has been defined as the application of the other sciences to the study of the Earth and is, therefore, fully as complex as these sciences. However, a number of its key concepts can be readily and rewardingly grasped. Such concepts can be the active, illuminating, lifetime companions of any one who has been properly exposed to geology and who observes the Earth around and under him.

We have already met a number of the basic, wide-reaching concepts, principles, and conclusions of geology—its tactics and strategy—and others will be taken up in the chapters that follow. However, we shall collect and emphasize them here.

- *The present is the key to the past (uniformitarianism).*

Unlike the laboratory proofs possible in sciences such as chemistry, geologists must assume that the present is the key to the past (p. 608)—that the physical laws and processes that function on the Earth today have done so throughout the past, although their rates and geographic locations have varied widely. Despite this lack of laboratory confirmation, everything known to scientists today strongly supports this assumption and warrants its key position as the foundation upon which geology has been built.

- *Rocks have formed in different ways, at different times, and in different places. Fossils are the remains or traces of organisms that were preserved in sediments and other materials prior to historic times.*

After a brief introduction to the meaning of igneous, sedimentary, and metamorphic rocks and to the rock cycle, one is probably quite astonished to learn that the true nature of rocks has been generally understood only for the past two centuries or so. It seems evident that all rocks could not have formed simultaneously early in the Earth's history and have subsequently remained unchanged. Yet this was the generally accepted explanation until the latter part of the eighteenth century. Now that it has been established, today's interpretation seems quite obvious and illustrates one of the helpful aspects to the nonscientist of many of the basic conclusions of geology. Upon meeting them, we feel fully acquainted.

One of the first important milestones in the development of geology was the recognition of the true nature and significance of fossils, particularly of marine fossils exposed on the upper slopes of present-day mountains or in areas far from the sea. Much imagination is needed to reconstruct these as former sea floors alive with organisms of many kinds. In the fifteenth century, Leonardo da Vinci stated the modern view: floods carry mud; the mud is deposited over animals that live in the sea near its shores; the animals die; the mud within and around their shells turns to stone; the sea withdraws; finally, erosion exposes the fossils, which in many places show the original shells encased in stone.

- *There are sermons in stones.*

The history of the Earth, and of the organisms that have lived upon it, is preserved in rocks. The history is incomplete; it is imperfectly known and preserved; but it is there for the interpreting. The more ingenious man can become in finding and understanding the clues and data recorded in the rocks, the more completely will we know this history.

- *Superposition occurs as sediments are deposited.*

In many places on the Earth today, sediments are being deposited layer upon layer, and presumably sediments accumulated in a similar manner in the past. Thus, the strata in a pile of sedimentary rocks are arranged in chronological order; the oldest layer is at the bottom, and the youngest layer is at the top. This may appear too obvious to be considered as a key principle, yet it has been generally recognized and accepted for only the past 200 years or so. Exceptionally, crustal movements have overturned layered rocks, and the oldest stratum is no longer on the bottom. However, intense rock deformation found in such areas warns the geologist to expect the unusual.

- *Strata have a nearly horizontal orientation when they are deposited.*

At present, sediments are deposited in layers which tend to parallel the Earth's surface at the places of deposition and to be nearly horizontal. Presumably they were deposited in a similar manner in the past. However, many ancient sedimentary rocks are found in a crumpled or tilted condition today. According to the principle of original horizontality, these ancient strata were approximately horizontal when they formed, but subsequent crustal movements have changed their original attitude. Acceptance of this

Fig. 28-12. *Truncation by erosion or dislocation. The sketch is highly diagrammatic. The layer of conglomerate exposed in the hill at the left once extended unbroken toward the right where it changes to sandstone. Erosion has truncated the strata along the slopes of the hills and removed the large volumes of rock which once existed in the spaces separating the hills. Truncation by dislocation (faulting) is shown at the right. Even if mantle obscured the actual trace of the fault at the Earth's surface, a geologist could deduce by matching the offset layers that a fault existed.*

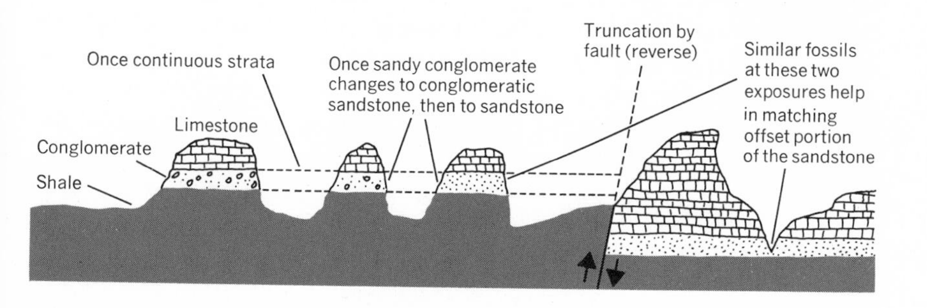

principle was postponed for some time by the sheer immensity of the crustal forces necessary to produce the deformation.

- *There is an original continuity of strata.*

Sedimentary layers forming today commonly do not end abruptly as, for example, do the beds exposed on opposite sides of the Grand Canyon (Fig. 28-2). Instead they tend to thin or pinch out gradually. If strata are found to terminate abruptly today, it is assumed that erosion or offsetting by faulting has caused this truncation (Figs. 28-12 and 28-13). In the Grand Canyon, the series of sedimentary beds which are now exposed along the north and south rims must formerly have extended unbroken from one side to the other. Erosion has produced a great gorge where part of a plateau once existed.

- *Fossil succession results from superposition and biological evolution.*

Fossils can be found in sedimentary beds that have been piled one on top of the other and thus represent organisms that lived at different times during the past. The oldest organisms must have been in existence when the bottom layer was deposited, but the youngest organism could not have come into

Fig. 28-13. *Monument Valley in Utah. Thick resistant sandstone beds form the upper parts of the monuments, and thinner beds of sandstone and shale form the lower portions.*

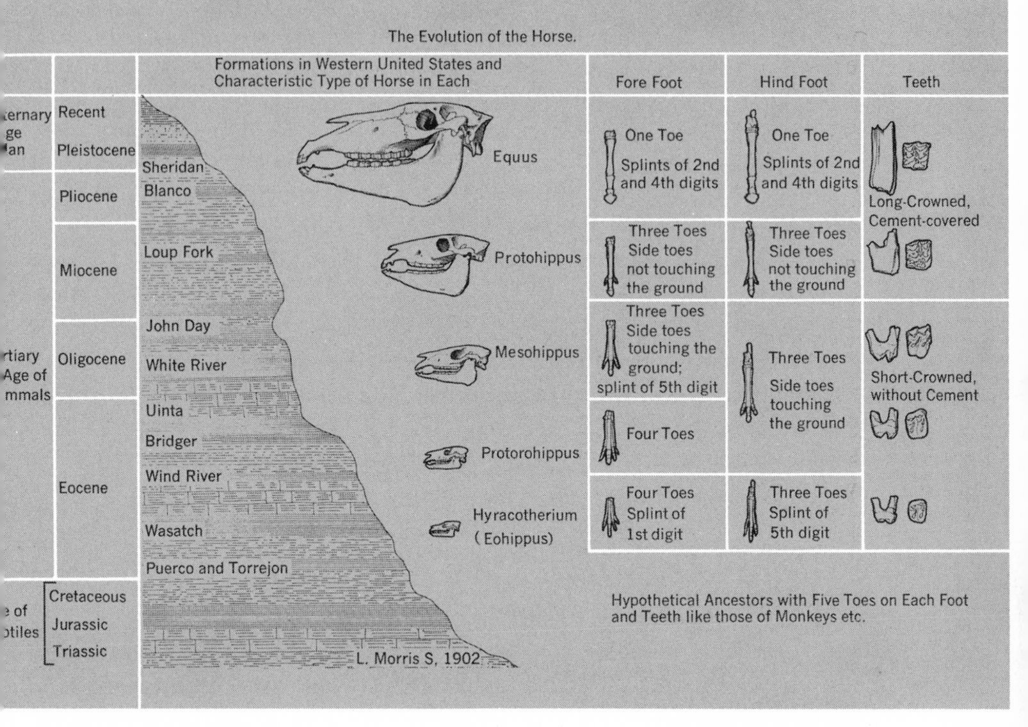

Fig. 28-14. *Evolution of the horse. Contrary to popular belief, the evolutionary development of the horse was not a straight-line, undeviating, progressive development which began with a small, four-toed, browsing animal that gradually became larger, lost all but one of its toes, and developed teeth suitable for grazing. Several varieties of horses apparently existed simultaneously, and of these, all but one type subsequently became extinct. For example, three-toed browsing horses lived at the same time as one-toed grazing horses; the one-toed horse evolved toward the present-day horse, whereas the three-toed horse became extinct. Some varieties of horse became smaller rather than larger. Furthermore, rates of evolutionary development apparently differed at different intervals during the Cenozoic.*

Tertiary and Quaternary refer to periods in the Cenozoic era and are terms, along with the name of the epoch called Recent, which some geologists would like to eliminate. [*After W. D. Matthew,* American Museum of Natural History Journal.]

being until much later, when the top layer was formed. When fossils are collected from such rocks and arranged chronologically from oldest to youngest, a striking feature is noted. The youngest fossils resemble today's organisms more closely than do the oldest fossils, and gradations occur between them. In fact, many of the older forms long ago became extinct. In general, the changes are from more simple to more complex forms. The horse series is a well-known illustration (Fig. 28-14).

Such fossil collections strongly support the concept of evolution (p. 790). Each type of animal and plant in existence during any one interval of geologic history apparently had reached a more or less similar stage in its development in most parts of the Earth. Each type possessed characteristic features which differed from those of its ancient relatives and from those of the generations which followed it. Some forms evidently evolved slowly through millions of years, whereas others evolved at a faster pace. However, at any one time in geologic history a certain stage in development tended to exist throughout the Earth, although exceptions are known. Relatively rapid and widespread migration, dispersion, and mixing are factors in accomplishing this. Therefore, sediments deposited more or less simultaneously (as a geologist regards time) in various parts of the world should contain fossils which are somewhat similar. Sedimentary rocks are thus tagged by the fossils which they contain, and scattered outcrops may be matched or correlated on the basis of these labels.

William "Strata" Smith, an English surveyor, is generally given credit for this discovery. However, as with many other scientific discoveries, the labors and thoughts of other men had also been contributed. If Smith had not discovered the principle of fossil succession at this time, the discovery might well have been made by someone else in the relatively near future—the season was ripe for the harvesting of this idea. Smith observed layers of sedimentary rocks in excavations and natural exposures in the areas in which he worked. As a hobby he collected the fossils which were so abundant in these strata, and near the end of the eighteenth century he arrived at two important conclusions. In a certain area individual layers always occur in the same succession at each outcrop; e.g., a gray cherty limestone would always underlie a cross-bedded white sandstone. Each kind of sedimentary rock has a distinctive set of fossils which distinguishes it from all other kinds (oversimplified: some sedimentary rocks, in whole or in part, do not contain fossils). Smith amazed his fossil-collecting friends by his ability to examine their collections and tell them the locations and strata from which each of the fossils came.

Correlation of sedimentary rocks on the basis of their fossil content is complicated by the fact that most living organisms are confined to certain distinctive environments; e.g., certain types of marine animals dwell only on a sandy bottom, a different group lives on muddy bottoms, and only a few species live in both environments. Identical fossil forms, therefore, cannot be found in sedimentary beds which formed simultaneously in different environments. However, the age equivalence of the different fossil types can be determined by tracing the strata involved from one area to another. A series of sandstone beds may crop out in one area, and layers of shale may be exposed a few miles away. As a geologist walks from one area to the other, he may be able to observe that the sandstone changes gradually to shaly sandstone, then to sandy shale, and finally to shale. The age equivalence of fossils found in the sandstone and the shale thus is demonstrated. A geologic chronology based upon worldwide fossil collections has enabled investigators to put the Earth's history more or less into chronological order.

- *Shallow seas were once more extensive.*

Marine waters at one time or another in the past have covered most areas which are now dry land, commonly more than once, and some areas now covered by the sea were formerly land. This does not mean that continents at times become ocean basins or vice versa. The different structure of the continental and oceanic crust and the concept of isostasy (p. 631) argue against this. Rather, local sinking of the land or a worldwide rise in sea level causes ocean water to advance across some sections of a continent.

Geologically speaking, the coast line separating land from sea has proved to be an extremely flexible and movable boundary. Local sinking of the land or worldwide rise in sea level causes ocean water to advance inexorably on continental areas. A rise in sea level is caused by the displacement of sea water by sediments, by eruptions of lava on the sea floor, by upward movements of the sea floor, and by the melting of glaciers. If the floor of a shallow sea subsided more or less continuously, space was thereby made available for thick accumulations of sediments.

- *Remnants of the sedimentary record may be correlated.*

Sites of deposition have shifted from one area to another throughout geologic time, and erosion commonly has begun in an area as soon as deposition ceased. Both erosion and deposition do take place in the same area, but one or the other tends to predominate. As erosion occurs in a region, previously formed rocks are destroyed, together with their record of the Earth's history. In addition, in that region no sediments are deposited which can be preserved as permanent records of the events then taking place. Thus determination of the Earth's geologic history must be based upon collecting and matching many scattered fragments into a more complete composite record. Arranging the numerous isolated sequences of sedimentary beds into their chronological order and deciphering the messages they contain have proved to be monumental tasks made feasible only by worldwide cooperation (Fig. 28-15).

The total thickness of known sedimentary rocks is nearly 100 miles. Of course, this is a composite record; only a small fraction has accumulated at any one place. This 100-mile column is composed of shales, sandstones, limestones, conglomerates, and similar rocks. If an average annual rate of deposition could be determined, the length of time involved in the formation

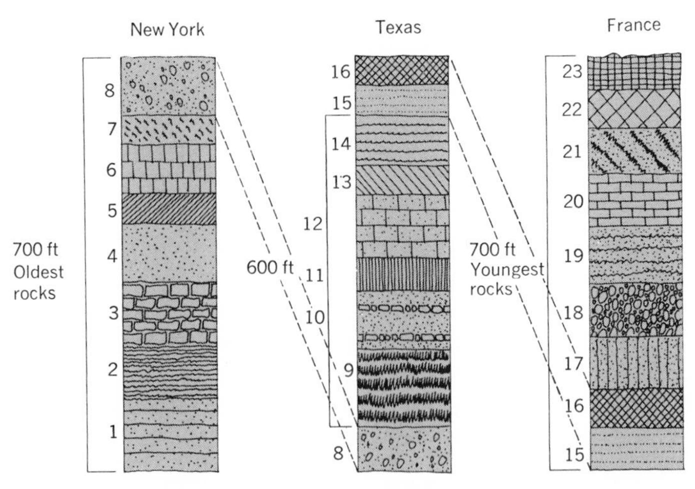

Fig. 28-15. *Composite nature of the sedimentary record. The sketch is highly diagrammatic. Layers of sedimentary rocks are exposed along the slopes of mountains in New York, Texas, and France. Strata numbered 1 through 8 in New York are shown by fossils to be the oldest of the three groups of rocks illustrated.—Stratum 8 in Texas contains fossils which are nearly identical with those found in bed 8 in New York. Thus the organisms probably lived at about the same time in the past and the strata in which they are found were probably deposited about simultaneously in terms of geologic time. (This does not, of course, indicate that layer 8 was once continuous from New York to Texas.) Therefore, beds numbered 9 through 16 in Texas apparently are younger than the rocks exposed in New York.—In a similar manner, layers 15 and 16 in France seem to have been deposited at the same time as beds 15 and 16 in Texas. Therefore, beds numbered 17 through 23 in France form the youngest group of sedimentary rocks in the three areas. During deposition of the youngest strata in France, erosion may have been taking place in New York and Texas.—Thus the composite thickness of sedimentary rocks for the time interval represented by the three groups of rocks equals the combined thicknesses of the three groups minus the overlapping layers, a total of 2000 feet. Of two layers deposited at the same time, the thicker is counted in the composite geologic record. Thus layer 8 in New York and layers 15 and 16 in France were included in the total thickness.*

of these rocks could be calculated—a minimum age for the Earth. Unfortunately, this is difficult or impossible. A foot of sand may accumulate in one place in the same interval that 20 feet of gravel and one-tenth of an inch of calcareous material pile up elsewhere.

Since land areas today appear higher and more extensive than they were in most of the past, the present rate of accumulation of sediments is far too rapid for an all-time average. High lands speed up the processes of erosion and deposition. However, no accurate estimate of an all-time average rate seems possible. In addition, great breaks (unconformities) still occur in the geologic record; they represent long periods of time for which no deposits have yet been found. Furthermore, many of the oldest known rocks are metamorphic ones which originally were sedimentary strata. No accurate estimate is possible for their thickness. Therefore, the bulk of the 100-mile column is made up of sediments deposited since the Precambrian, a relatively small fraction of total geologic time.

Thus the rate of accumulation of sediments like that of sodium indicates that the Earth is very ancient, but it does not provide an accurate estimate in terms of millions of years.

Correlation refers to the determination of the age equivalence or contemporaneity of fossils, strata, or events in different areas. In other words, is this particular fossil (layer, event) older, younger, or about the same age as that one? Fossils are the most useful means of correlating sedimentary rocks from one area to another and from one continent to another. However, the radioactive dating of igneous and metamorphic rocks, and thus indirectly of the sedimentary rocks associated with them, provides another tool for intercontinental correlation. Over shorter distances, geologists may match similar types of rocks or similar sequences of rocks, or utilize other techniques.

- *The age of the Earth is apparently about 4 to 5 billion years.*

The conclusion that the Earth is very old is important because it provides time enough for processes such as stream erosion and deposition—processes which seem exceedingly slow in terms of human lifetimes—to accomplish the results of very large magnitude which are shown by the geologic record.

- *Movements of the Earth's crust are common and diverse.*

In one type of crustal movement, very gradual upwarping or subsidence occurs and affects large regions of the Earth, either land or sea; little or no crumpling and folding of the rocks occur, although some tilting may take place. A different type of movement is known as *orogenic* (Greek: origin of mountains). It affects elongated belts of the crust and results in the formation of mountains by folding, faulting, and thrusting. Both horizontal and vertical crustal movements are involved in orogenies, and the crust in the elongated belts appears to be severely compressed or squeezed at times.

- *The Earth's present topographic features, its scenery and landscapes, are only transient forms. They have evolved from the different shapes of the past and will pass slowly and inexorably into the still different forms of the future.*

The evolutionary development of land forms depends upon factors such as the following: the kind or kinds of rocks underlying a surface; the rock structures present—folds, faults, domes, or flat-lying beds; the types of

Fig. 28-16. *Landforms of the United States.* [*Erwin Raisz.*]

erosional processes which are occurring; the time during which these various factors have been interacting; and the stability or instability of the crust. The Earth's land forms (Fig. 28-16), although differing widely in detail, are not haphazard. They can be classified genetically with some success, and their shapes and histories can generally be understood, at least in part.

Differential erosion is evident in the Earth's surface features in all gradations of scale from the very large to the very small. The relative resistances or weaknesses of the different types of rocks to erosion are soon made evident. On a large scale the locations of mountains, ridges, and valleys commonly depend upon the locations of resistant and weak rocks. However, rocks which are resistant in one climate may not be so under the different weathering conditions that may occur in another.

The rock structures of the crust are key factors in controlling the shape of the Earth's surface in many areas. A flat-topped mesa may be sculptured from flat-lying strata in a semiarid region. This contrasts sharply with the elongated ridges which develop upon tilted strata and with the elliptical outline of an eroded structural dome or basin. Mountains formed by faulting differ in shape from those produced by folding, particularly the loop-shaped mountains which are sculptured from folds whose axes plunge (Fig. 33-20).

Each erosional process has its diagnostic features. An area in a climate favoring stream erosion differs in appearance from a colder region where glaciers predominate. Characteristic land forms are produced by winds in an arid region, by waves and currents along a coast, and by the action of subsurface water in a humid region underlain by soluble rocks.

The Earth's surface undergoes an orderly sequential development as erosion proceeds in an area, although details of the sequences in different areas vary because of differences in the controlling factors listed above (Fig. 28-17). However, these are variations of a limited number of basic

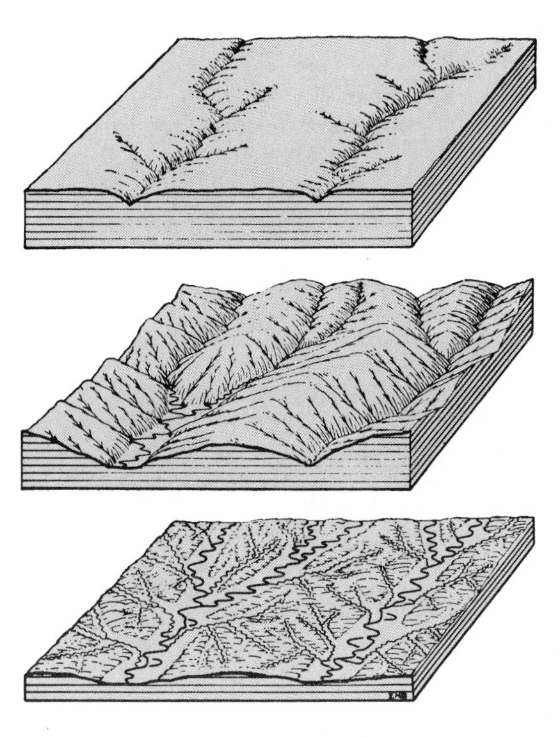

Fig. 28-17. *The normal cycle of erosion in a moderately elevated region of essentially uniform, stratified rocks devoid of important structures. Top, region in youth; middle, in maturity; bottom, in old age. A region in typical maturity is one of slopes with a rolling hill-and-valley topography. Drainage is good, relief is at its maximum, and erosion is most effective. Major streams have begun to develop floodplains and meanders.*

sequences. Because erosion is a slow process, it is necessary to construct a sequence in theory, and then to check this against the actual features present here and there on the Earth. Such features represent different stages in the development of different sequences. This has led to the concept of a geomorphic cycle and to the application of the terms *youth, maturity,* and *old age* to its three main stages. Each cycle, however, is a continuous gradual development. It can be subdivided only arbitrarily, and any one cycle may be interrupted before it has been completed. A geomorphic cycle is not defined in terms of an actual number of years (time is relative only) and its three main subdivisions are not of equal length. The term cycle is somewhat misleading because a complete sequence of events does not occur at regular intervals. Furthermore, so much diversity exists in land forms that no single geomorphic cycle can account adequately for all of the features observed. Despite these limitations, the concept of a geomorphic cycle remains a useful one.

Most landscapes are not the result of a single process but of a combination of processes. Similarly most cycles are interrupted before completion by a change in climate or by crustal movement (Fig. 28-18).

Fig. 28-18. *The entrenched meanders or goosenecks of the San Juan River in southeastern Utah. The canyon exceeds 1200 feet in depth. Through it, the winding river flows six miles to travel an airline distance of one mile. The meanders developed when the river flowed on a wide floodplain. Subsequent uplift occurred, and the river entrenched its channel into the horizontally layered sedimentary strata.*

The landscapes and scenery of today are likely to be relatively youthful features. Many of the rocks, folds, and faults found in mountain belts such as the Appalachians and Rockies are quite ancient, but each of these resulted from widespread extensive uplift and subsequent differential erosion during the Cenozoic.

- *Pleistocene glaciation and climatic changes have had recent and worldwide influence.*

A succession of glacial and interglacial ages occurred during the recent Ice Age and was accompanied by widespread changes that extended far beyond the regions covered directly by glacier ice. Some of these changes involved crustal movements, the volume of ocean water, and the migrations of plants and animals (p. 695). In addition, other features originated within the glaciated areas as a result of glacial erosion and deposition. Thus we may observe features today that were formed under different climatic conditions during parts of the Pleistocene. Therefore, the assumption that the present is the key to the past must be modified somewhat: e.g., to understand certain land forms located at a latitude of 40 degrees, it may be necessary to study current processes located some 10 to 20 degrees closer to the poles—a modern glacier or a permafrost region.

• *Minor catastrophes are allowed for in uniformitarianism.*

The cumulative effects of relatively small, more or less continuous processes such as stream erosion and deposition are emphasized in uniformitarianism. However, it is now realized that in a number of processes, the bulk of the erosion and deposition is accomplished during brief, infrequent storms or floods—during minor catastrophes. During an exceptionally powerful flood—a type that occurs about one, two, or three times per decade in some valleys—much more erosion and transportation may be accomplished by a river than during the intervening months and years when low, normal, and flood stages alternate. Perhaps more wave erosion is produced along certain coasts when a tsunami (p. 744) strikes than is accomplished by all of the wave action that takes place in the intervening years. Hurricanes, some landslides, movements along some faults, volcanic eruptions, and the occasional rain squall in a desert all indicate that one must be cognizant of infrequent minor catastrophes in assessing the relative roles of various geologic processes. Of course, in a geologic sense, these are not catastrophes at all, but merely times of increased erosion and deposition which occur occasionally.

• *Isostasy and isostatic adjustment are continuing.*

Isostasy (Greek: "equal standing") is a balance apparently present in the outer part of the Earth (Fig. 28-19). Isostatic balance is upset if enough matter is transferred from one area to the other at the Earth's surface. This happened during the Pleistocene when water was evaporated from the oceans and fell as snow to make glaciers on the lands. It also apparently occurs when a large mountain range is worn down and its eroded debris transported to an adjoining ocean area. In these circumstances, isostatic adjustment occurs. At depth, rocks apparently flow slowly outward from an overloaded area, which sinks, to an underloaded area, which thus rises. This continues until balance is restored.

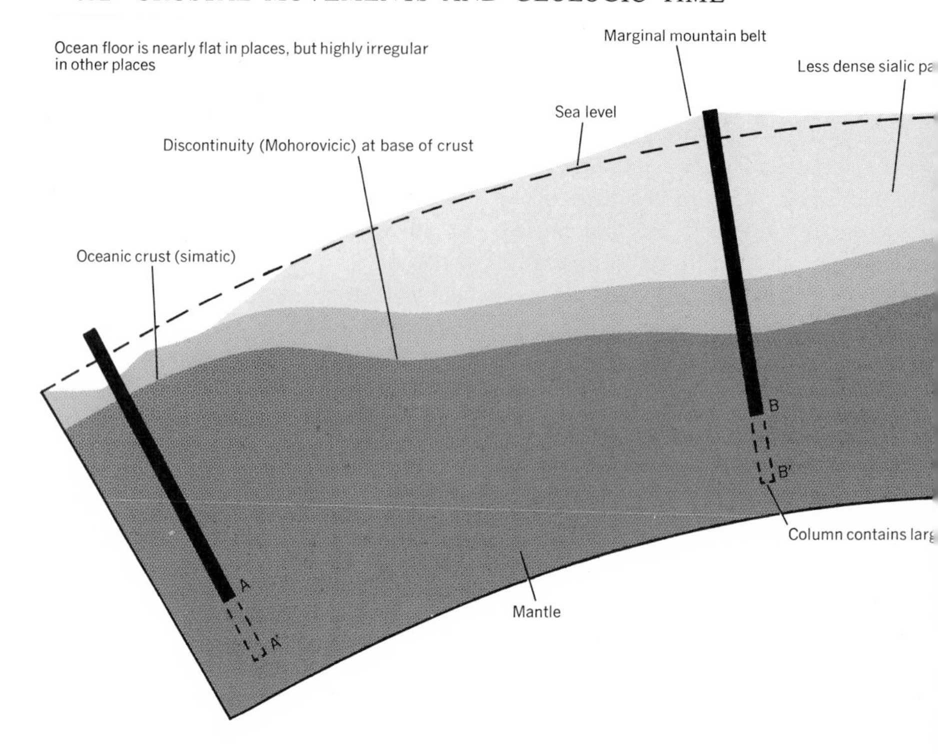

Fig. 28-19. *Schematic structure section across a continent illustrating isostatic balance and the apparent increase in thickness of the crust beneath a great mountain belt. Three hypothetical columns* (A, B, *and* C) *are imagined to be equal in length, cross section, and mass. They contain different proportions of sial (least dense), sima, and mantle (most dense). Although column A is formed partly of air and water, it has the largest proportion of the densest type of rock and thus has the same weight (mass) as columns* B *and* C. *If the columns were extended equal distances to* A′, B′, *and* C′ *their masses would still be equal because the same quantity of mantle would be added to the base of each column.*

- *The geosynclinal theory helps account for complex mountain belts.*

These are very important, complex topics which are discussed briefly in Chapter 34.

- *A sequence of geologic events may be interpreted in chronological order from geologic maps and structure sections.*

This provides a framework for the inclusion of a number of important ideas and extends the concept of "sermons in stones." The significance of an unconformity is included here (Fig. 28-20 and p. 738). For example,

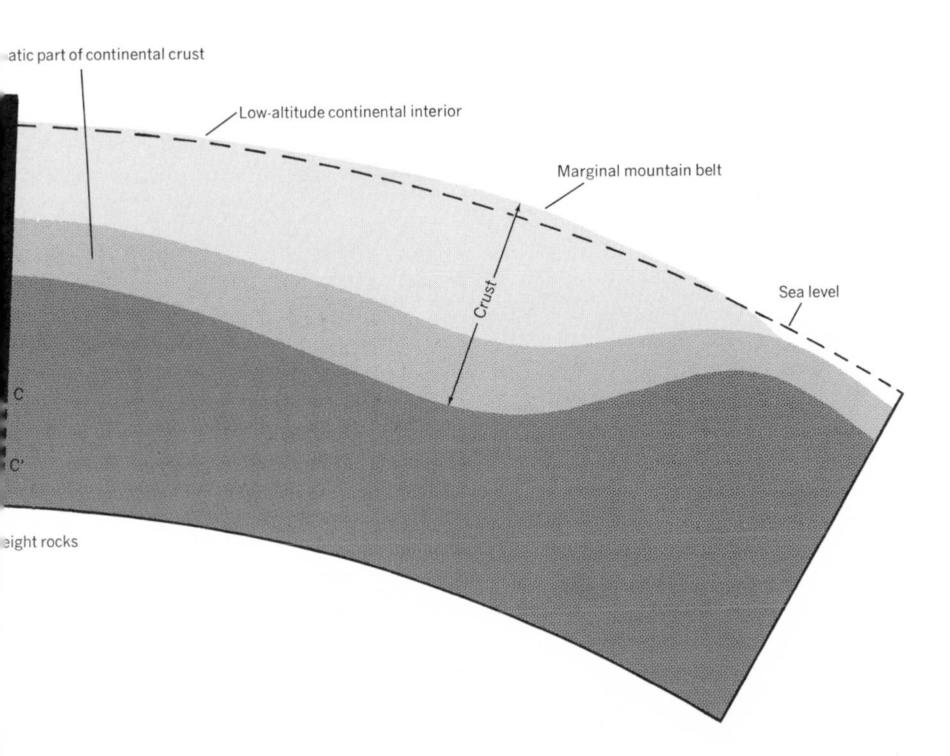

assume that a marine limestone is overlain unconformably by a marine shale, and that a buried erosion surface occurs between the two formations. The following sequence of events is indicated. Calcareous sediments accumulated on a sea floor and eventually were compacted and cemented to form limestone. Next, the area became land, either because the area was uplifted, or because sea level fell; erosion then carved out gullies and small valleys and removed some of the limestone. Next, the sea again covered the

Fig. 28-20. *Unconformities. Older rocks beneath an unconformity may or may not be affected by crustal movements before younger rocks form on top of them.*

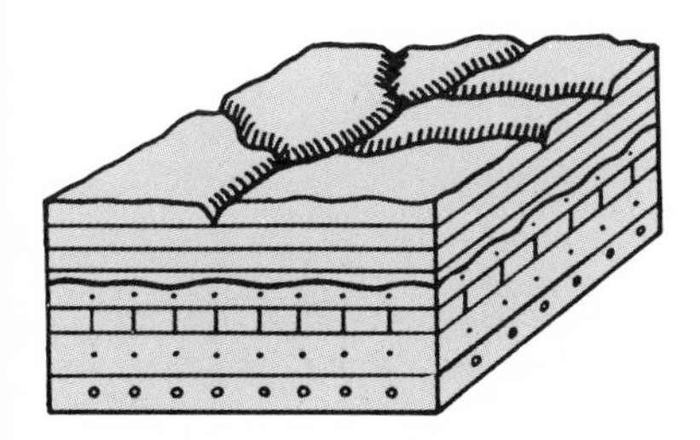

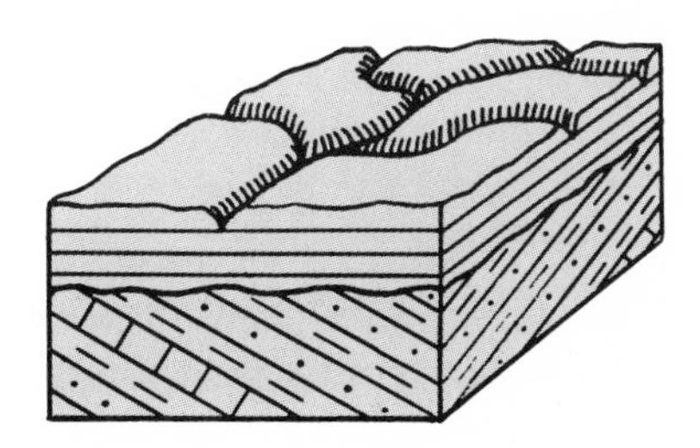

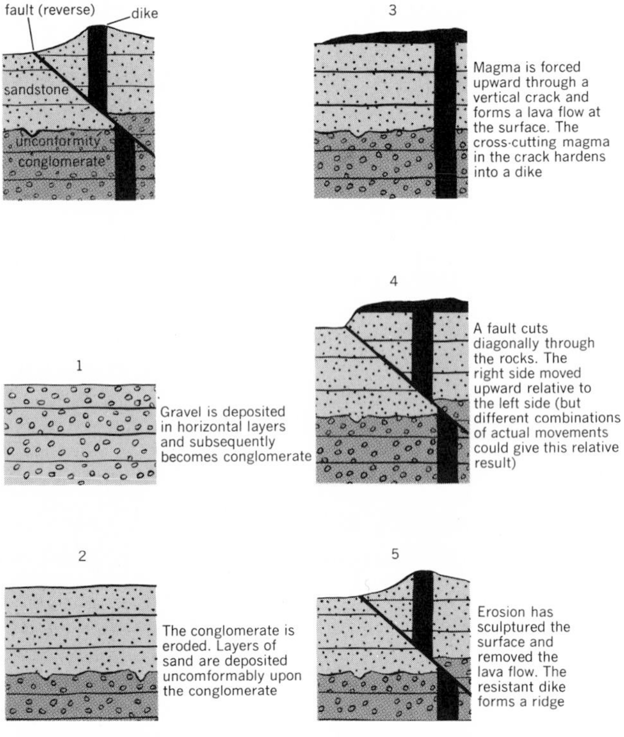

Fig. 28-21. *Interpretation of a geologic structure section.*

area and mud and clay were deposited on top of this erosion surface and eventually buried it. Finally, uplift and erosion occurred again to expose the rocks as they are observed today.

The interpretation of structure sections involves relationships such as the following. If an igneous mass is intruded into the Earth's crust, the igneous material is younger than the rocks it has intruded. If a fault extends upward and stops abruptly at an unconformity, we know that the faulting occurred before the erosion which produced the unconformity. On the other hand, a fault is younger than any of the rocks it offsets. In general, in working out the sequence of events observed in a structure section, one begins at the bottom, following the principle of superposition, and works toward the top (Figs. 28-21 and 28-22). Intrusive igneous masses are exceptions here.

- ***Interpretations of geologic events are guided and limited by the evidence available.***

Geology has its share of assumptions, hypotheses, and unsolved problems. Imagination is needed to reconstruct the partially hidden events of the

Fig. 28-22. *The distinction between a sill and a buried lava flow. A sill is a tabular-shaped intrusive igneous rock body that is parallel to the layers of rock that enclose it. In the sketch a series of sedimentary strata contain a buried lava flow of basalt and a felsite sill. The sill can be distinguished from the flow thus: (1) The sill has altered both its upper and lower contacts; the flow has baked only the rocks beneath it.—(2) Blocks of the overlying rock (conglomerate) have fallen into the felsite; on the other hand, boulders of basalt are found in the first stratum to be deposited above the flow.—(3) Tongues of felsite have intruded the enclosing rocks.*

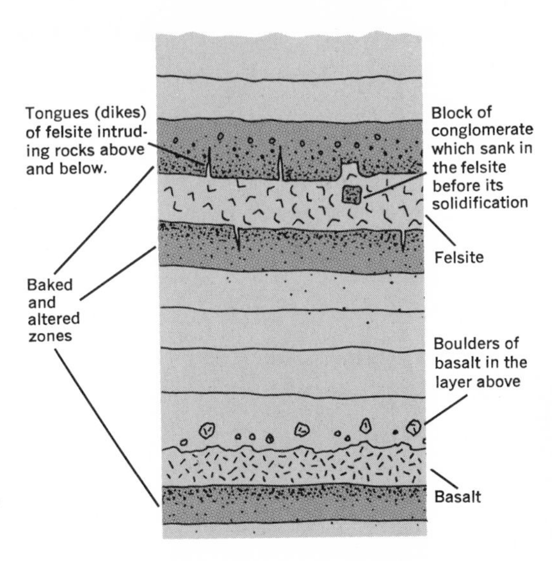

past. However, in all of this, a distinction is made between fact and hypothesis; interpretations are based upon the evidence available and must be changed when new evidence so demands. Here is the place for a healthy curiosity and for numerous probing questions: What is your evidence for that statement? Why do you interpret the problem in that manner? How do you know that uplift occurred then?

- ***The methods and viewpoints of geology are varied.***

Geology is primarily a field study, although laboratory experimentation has increased greatly since the 1940's. Data painstakingly collected from thousands of sites are recorded on geologic maps which show kinds, distribution, and attitudes of rock formations at the Earth's surface. Often a study of scattered outcrops will justify a confident prediction that a certain rock formation will be found buried beneath loose mantle here, or another one there.

Change is the keynote of the Earth's history. Changes and processes which now act on the Earth are studied, and the information thus obtained is applied to the past. The aim is to understand the Earth's history by reading all of the sermons inscribed in the rocks. Rocks originate and break up, and their remains go into the formation of new rocks. Erosion functions to wear down the continental areas, but some crustal forces and igneous activity renew the lands. This interplay apparently has continued since the Earth formed.

But this account omits some modern aspects of geology which emphasize a quantitative approach, laboratory experimentation, and the application to geological problems of nongeological techniques taken from fields such as biology, chemistry, physics, and mathematics. Quantitative studies and the mathematical manipulation of geological data have become possible in recent decades because of the large amounts of observational data now available. These data can be correlated with the rapidly multiplying results of laboratory experiments.

The geologist is concerned with the processes that shape the Earth as well as with the Earth itself. He can study these processes as they operate in that part of nature that is accessible to him and other aspects under controlled conditions in the laboratory. In this manner, he may view the results of chemical reactions that occur at temperatures and pressures simulating those that exist at depths of several miles and more below the Earth's surface. He may then examine these end products by means of X rays, polarized light, the electron microscope, and the mass spectrometer. The techniques of physics and chemistry have been emphasized in recent decades and the new disciplines of geochemistry and geophysics have resulted. Discovery and progress are most rapid today in such overlapping zones between two related sciences.

Exercises and Questions, Chapter 28

1. Questions relating to the geologic history of the Grand Canyon region:
   (a) Describe a number of basic geologic concepts illustrated by this history.
   (b) If you studied the rocks on the north and south rims, what evidence could you find that an uplift of about 1½ miles had occurred?
   (c) What evidence indicates that mountains were formed at two different times and subsequently were worn away?
   (d) What is meant by the "great denudation" and what is the evidence for it?

2. How can the relative ages of a number of sedimentary rocks occurring in different areas be determined? How can their absolute ages be determined? What assumptions or concepts are involved? How accurately are such ages known?

3. List a number of events that might occur in the geologic history of an area. Next, draw and label a structure section that illustrates this sequence of events and give it to a classmate to interpret.

4. Assume that a tree was buried by glacial debris about 22,400 years ago. How should its carbon-14 content compare with that of a tree living today? Assume the tree has remained a closed chemical system and use 5600 years as the half-life period of carbon-14.

(a) Why is there a time limit of about 40,000 years on age determinations made using the carbon-14 technique?
(b) Why has it been said that age determinations exceeding this limit would probably represent inaccurate results achieved with great precision?
(c) The development of the carbon-14 method has been described as an instructive case history illustrating scientific discovery and the methods of science. Discuss.

5. In what way is the sodium-in-the-sea method of estimating the age of the Earth similar to that of the radioactive method? In what significant way do they differ?

6. Choose five of the fundamental principles of geology and give the evidence and/or supporting argument on which each is based.

7. Make your own illustration to show the enormous extent of geologic time.

8. Describe briefly three possible causes of a rise in sea level.

9. Interpret the "sermons" in each of the following:
(a) A bed of coal occurs in a mine at a depth of 2000 feet below the surface. It is part of a thick sequence of horizontal Paleozoic strata. Brachiopods and trilobites occur in some of the layers above the coal.
(b) In another region, sandstone beds are steeply tilted and contain dinosaur footprints and ripple marks.

10. You study a certain outcrop and eventually are able to identify it as a coarse-grained intrusive igneous rock (e.g., granite or diorite). What does this tell you in a generalized way about the geologic history of the area?

11. Assume that the rocks in an area consist chiefly of two types: granite occurs in one part of the area and limestone occurs in the other part.
(a) Describe three different types of contact which can occur between such rocks.
(b) Describe criteria that might be used to distinguish one type of contact from another.

12. How did the viewpoint of the catastrophists differ from that of the uniformitarianists concerning the history of the Earth?

13. How does the deductive method of solving a problem differ from that of the inductive method?

14. Give an example of the integration (overlapping) of several different sciences.

29 Weathering, mass-wasting, and stream erosion

There rolls the deep where grew the tree.
  O earth, what changes hast thou seen!
  There where the long street roars, hath been
The stillness of the central sea.

The hills are shadows, and they flow
  From form to form, and nothing stands;
  They melt like mist, the solid lands,
Like clouds they shape themselves and go.

—TENNYSON: *In Memoriam*

ONE feature of Figs. 29-1 and 28-13 will strike the eye of any observer: the layers of sedimentary rock in the scattered columns and monuments match exactly. Evidently we observe today the remnants of once-continuous layers that have been partly eroded away; and perhaps former overlying layers have been entirely removed.

An important geologic concept is thus illustrated: the "everlasting hills" are really not everlasting at all. Weathering causes rocks to break apart, decay, and crumble; the resulting fragments and dissolved materials then move from higher to lower ground (mass-wasting); and next, the sediments are transported from the area by some agent such as running water or wind. Eventual burial in the sea is the fate of most sediments. How such vast quantities of rock can be removed from an area and how the Earth's surface can be sculptured into mountains, valleys, cliffs, and badlands is the theme of this chapter.

As a "team," *weathering, mass-wasting,* and *stream erosion* form the dominant erosive force on the Earth today, although distinctive land forms are also produced by subsurface water, wind, and ice. If upward movements of the crust and igneous activity did not take place, the continents

Fig. 29-1. *Bryce Canyon in southern Utah. The horizontal sedimentary rocks are cut by systems of vertical cracks (joints) which separate the strata into columns. Erosion has worn away the edges of the columns more rapidly than the sides and produced the rounded shapes seen in the photograph.*

would be worn down to almost featureless surfaces located near sea level (in about 10 million years according to one estimate).

WEATHERING

From the time someone first bounced a stone off a tender shin, most people have been thoroughly impressed by the hardness of rocks. Rock outcrops in one's neighborhood show no readily discernible changes during a lifetime, and stone monuments and buildings are accepted as permanent features. Yet the action of the weather in changing shiny iron nails to rust is familiar to all, and careful observation will show that rocks do crumble and decay in buildings and in nature. The weathered surface of a rock may be strikingly different in color from that of a fresh surface; e.g., a reddish exterior may hide the true white color of a rock. The outer parts of some rock exposures can be crumpled by the fingers, but like the rusty nails they too were once solid.

Fig. 29-2. *Prominent, widely spaced joints in sandstone. A joint is a fracture or break in a rock along which no appreciable displacement has occurred parallel to the joint surface. In this way it differs from a fault. Joints tend to occur in sets that are separated by distances commonly ranging from a few inches to several feet. The joints in any one set are approximately parallel. Many rocks exhibit more than one set of joints, and two mutually perpendicular sets are common. In igneous rocks, many joints are caused by the decrease in volume that occurs during cooling from a liquid to a solid. In other rocks, many joints probably result from crustal movements that have raised, lowered, squeezed, extended, and twisted the crust at one time or another.* [*Fig. 42,* U.S. Geological Survey Professional Paper 373.]

Weathering is the static part of the general process of erosion; it is a name for all processes which combine to cause the disintegration and chemical alteration of rocks at or near the Earth's surface. Weathering is furthered by the presence in all surface or near-surface rocks of cracks or joints (Fig. 29-2 and 29-3).

Two general kinds are recognized: *chemical weathering* causes the decomposition and rotting of rocks; *mechanical weathering* causes larger

Fig. 29-3. *Delicate Arch, Arches National Monument, southeastern Utah. Arches form in the Entrada sandstone (Jurassic, about 150 million years old). Note the gently dipping layers in and below the arch. The rocks are cut by nearly parallel vertical joints. According to geologists at the monument, water enters the joints and dissolves some of the cementing material. This frees some of the sand grains, which are then removed by wind and water. The vertical cracks slowly become wider and develop into narrow steep-walled canyons separated by narrow steep-sided ridges (fins). More rapid weathering of one section of a fin then forms an arch. Delicate Arch is thus an isolated remnant of a fin; the rest of this fin, as well as the fins and arches that once were its neighbors, have been entirely removed by erosion. In the monument, one can see fins and arches in all stages of development. Natural bridges form in a different manner; stream erosion is involved).*

pieces of rock to disintegrate into smaller pieces. Although one type of weathering may predominate under certain conditions, the two are intimately related, and one aids the other; e.g., the disintegration of rocks furnishes additional surfaces for attack by chemical action. To illustrate, a 1-inch cube has 6 square inches of surface area. Subdivision into eight $\frac{1}{2}$-inch cubes doubles the surface area, and further subdivision into $\frac{1}{4}$-, $\frac{1}{8}$-, and $\frac{1}{16}$-inch cubes results in surface areas of 24, 48, and 96 square inches respectively. Since chemical reactions tend to occur at rock surfaces, disintegration to tiny particles greatly increases the rate of chemical weathering. On the other hand, some chemical reactions result in an increase in volume which thus causes disintegration.

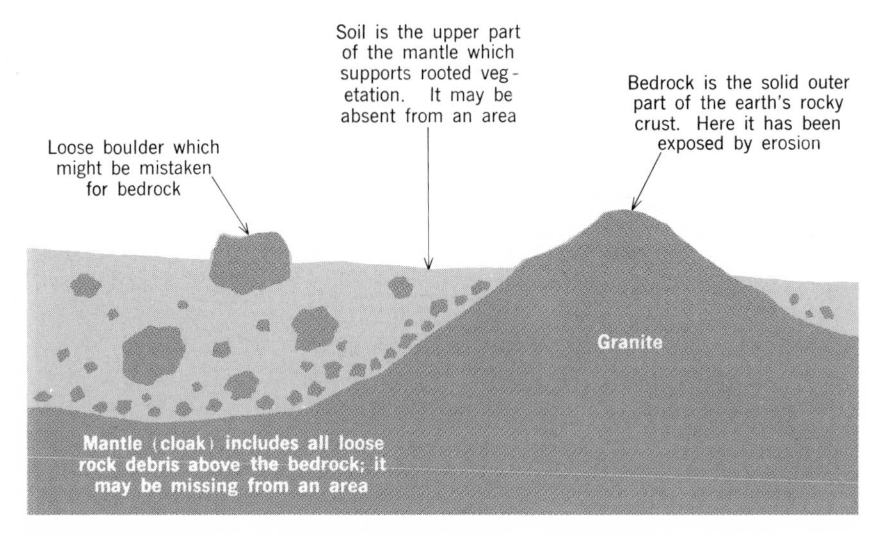

Fig. 29-4. *Bedrock, mantle, and soil. Mantle may be either transported (as shown in this sketch) or residual (accumulated essentially in its present location by the disintegration and decay of bedrock). Bedrock may consist of igneous, sedimentary, or metamorphic rocks. According to one report, a drill hole was once made in Ireland to test a site for the construction of a dam. The drill bit encountered granite; this was interpreted as solid bedrock and a firm foundation for the dam. After construction had begun, it was learned that the granite was only a boulder. A very large amount of unplanned excavation was necessary to get down to a solid foundation for the dam.*

Weathering produces three types of materials: broken rock and mineral fragments, residual decomposition products, and dissolved substances. The terms *bedrock, mantle,* and *soil* are explained in Fig. 29-4.

Different rocks weather at different rates under different conditions—one rock vs. another, different portions of the same rock, and even one mineral grain vs. its neighbors. Thus *differential weathering* etches out much of the finer detail of the Earth's scenery. Similarly, *differential erosion* shapes the larger units.

Chemical Weathering Warm moist climate, gentle slopes, and abundant vegetation are most effective for chemical weathering. Water is important as a medium in which chemical reactions can occur that take place slowly, if at all, between dry substances. Water combines with carbon dioxide in the air and with the humus of decayed vegetation to form carbonic and humic acids, which can dissolve and remove many rock materials (leaching). An increase in temperature speeds up most chemical reactions and causes some which cannot take place at lower temperatures. The commonplace process of rusting involves the chemical union of oxygen and water with iron to form limonite, a hydrous iron oxide group of minerals. The presence of cracks and other openings in rocks is important because they allow penetration by air and water.

Since feldspars, pyroxenes, amphiboles, quartz, and mica are estimated to constitute over 90 percent of the minerals of the Earth's crust, the manner in which they tend to weather is important. Feldspars decompose chiefly to clay minerals (hydrous aluminum silicates); pyroxene, amphibole, and black mica may yield clay and various iron oxides; quartz and white mica are resistant to chemical weathering.

Mechanical Weathering Freezing and thawing of water confined in rock openings is the most important purely mechanical process involved in the breaking of rocks. On freezing, water expands about 9 percent by volume and exerts a pressure which may exceed many hundreds of pounds per

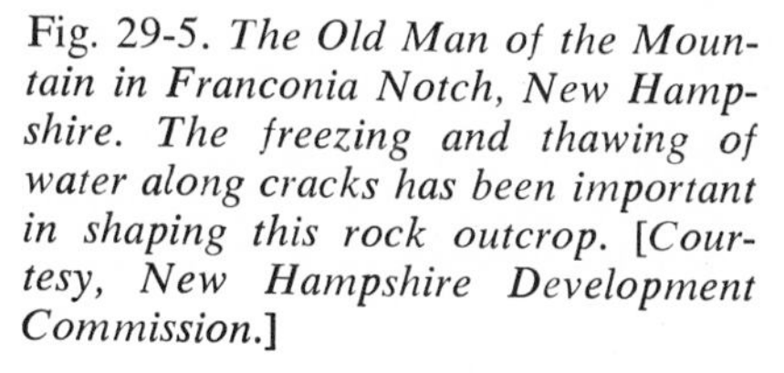

Fig. 29-5. *The Old Man of the Mountain in Franconia Notch, New Hampshire. The freezing and thawing of water along cracks has been important in shaping this rock outcrop.* [*Courtesy, New Hampshire Development Commission.*]

square inch. Confinement can occur after surface water freezes in a crack and seals off the water beneath it. Steep slopes with little vegetation in regions where temperatures fluctuate frequently back and forth across the freezing mark promote this freeze-thaw type of weathering; mountain tops in cool moist climates are favorable locations (Fig. 29-5).

Rock decay may involve an increase in volume and thus cause disintegration, e.g., by the addition of water, oxygen, or carbon dioxide. This process seems to cause some types of *exfoliation,* i.e., the spalling or flaking off of concentric shells from massive types of rock such as granite. The shells are a few inches or less in thickness, and are illustrated by certain spheroidal boulders (Fig. 29-6) which form at or near the surface. The

Fig. 29-6. *Spheroidal weathering in gabbro.* [*W. T. Schaller, U.S. Geological Survey.*]

rate of weathering of a rectangular block is least along its sides, more rapid along its edges, and fastest of all at its corners where the rock is attacked from three sides simultaneously. If evenly spaced joints occur in a massive rock, the initial blocks will have approximately cubical shapes, and nearly spherical exfoliated shells result.

Much thicker exfoliated shells (Fig. 29-7) and sheet structures in massive, relatively unweathered rocks may be caused by release of pressure due to unloading. Fractures divide such rocks into sheets or shells and tend to be parallel to present surfaces. According to one view, such exfoliated rocks formed far below the surface. Therefore, as erosion removed the tremendous masses of rocks that once rested upon them, they expanded, and the concentric shells formed.

The wedging action of rooted vegetation growing in cracks likewise reduces rocks to smaller pieces. Frost heaving causes rock disintegration and is most effective in fine-grained materials in moist regions with frequent frosts. Lenses and layers of ice form at different levels below the surface by drawing up water from below like a blotter. Frost heaves up to 18 inches have been measured, and sometimes the amount of heave has equaled the combined thicknesses of the ice lenses. Roads may be damaged severely by this process which also accounts for the presence of some of the rock fragments that may be encountered in a garden each spring.

Flaking and disintegration are produced by the extreme temperature changes which occur when a forest fire burns an area and is extinguished by a heavy rain. Individual minerals in a thin zone near the surfaces of such rocks expand and contract under the alternate heating and cooling. As different minerals swell and shrink at different rates, strains are set up which can cause eventual disintegration. It is uncertain whether lesser daily temperature fluctuations in desert regions likewise cause disintegration. Specimens have been alternately heated and chilled through many cycles in the laboratory without noticeable effect, although disintegration did occur when water occasionally was sprayed on the specimens.

SOILS

Soils consist of decomposed rock debris and decayed organic matter (humus) which have been produced by weathering. Thousands of years

Fig. 29-7. *Thick exfoliated shells exposed on Half Dome in Yosemite National Park, California.* [*U.S. Geological Survey.*]

are normally required for the formation of a thick fertile topsoil from a naked rock outcrop. However, soils several inches or more in thickness have developed in a few centuries or less in Japan and Mexico by the weathering of volcanic ash deposits; the eruption dates of such deposits are known.

Abundant humus colors soils dark, whereas iron oxides impart yellowish, brownish, or reddish colors. Mature soils commonly have three distinct layers (Figs. 29-8 and 29-9).

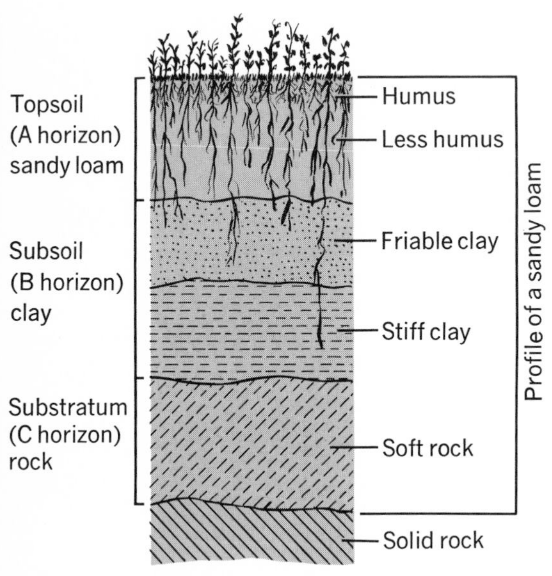

Fig. 29-8. *Vertical section through a residual soil. Loam is a soil composed of sand, silt, and clay. Three major horizons or zones are recognizable in most mature soils, designated the A (surface), B, and C horizons. Organic matter is most abundant in the topsoil. The original minerals in the topsoil of the A horizon have been entirely weathered to new minerals, except the most resitant ones such as quartz. Some of the material from horizon A has been leached out and deposited in the underlying B horizon. The finest particles also tend to be moved downward out of the A horizon by percolating water. Materials in the lowest C zone are only partially weathered and merge gradually with the bedrock if the soil is a residual one. With the passage of time, each zone becomes thicker at the expense of the underlying one.* [*Soil Conservation Service.*]

Some important factors in the development of soil are kind or kinds of parent rock material, climate, types of animals and plants present, shape of the surface, and time. Formerly it was thought that each type of parent rock would produce a distinctive type of soil, but now it is known that climate and time are perhaps even more significant. Weathering for a long enough time under one type of climate may produce similar soils from widely different parent rocks.

Worms are important soil makers. Earthworms extract vegetable matter from the soil by literally eating their way through it. As the soil passes through their bodies, mechanical and chemical modifications occur. In humid temperate regions it has been estimated that earthworms completely

Fig. 29-9. *Profile of a very fine sandy loam in Walsh County, North Dakota. The dark surface soil is fairly deep. Charles E. Kellogg has made the following comments concerning the ideal arable soil: "The farmer makes his arable soil from a natural soil or old arable soil. He develops and maintains a deep rooting zone, easily penetrated by air, water, and roots. It holds water between rains, but allows the excess to pass through it. It has a balanced supply of nutrients. It neither washes away during rains nor blows away with high winds. The combination of practices to use depends on what is necessary to develop and maintain a soil as nearly as possible to the ideal on a sustained long-time basis. They vary widely among the many kinds of soil. Successful farmers choose the practices for their fields according to two primary considerations: What practices do I need to come near the ideal? How will the costs and returns fit into my farm budget?"* [*U.S. Dept. of Agriculture photograph.*]

work over a soil layer from 6 to 12 inches thick every 50 years. Multitudes of smaller organisms and microscopic bacteria also live in soil and are important in its formation.

Practical Aspects of Weathering Without question, the formation of soil is the most important single result of weathering to mankind, but weathering has also produced many valuable mineral deposits. Either an originally useful mineral is concentrated by removal of waste products, or valuable minerals are produced by weathering and then accumulate; e.g., ore deposits of iron, copper, and aluminum have formed in this way.

One must know how various rocks weather under different climatic conditions to make wise choices of building stones. In a warm moist climate, a sandstone naturally cemented by calcium carbonate or iron oxide would be unsatisfactory, because the cement would either dissolve or stain the rock with rust spots. A badly fractured silica-rich sandstone would not be suitable in a cool moist climate, because water would enter cracks, freeze, and disintegrate the rock. On the other hand, a soluble limestone makes an excellent building stone in dry areas.

MASS-WASTING

Large masses of material move downslope as units in mass-wasting, an essential process in the general wasting away and leveling of the lands. The processes of weathering loosen, disintegrate, decompose, and dissolve rock debris, and the processes of mass-wasting then move this debris to lower altitudes where streams, ice, and wind can transport it away (Figs. 29-10 and 29-11). Weathering and mass-wasting are nearly ubiquitous on

Fig. 29-10. *The gigantic landslide in the canyon of the Madison River below Hebgen Dam, a part of the earthquake of August 17, 1959. The lake which was forming upstream from this slide is not yet visible in this picture taken on August 22.* [*Courtesy, Montana Highway Department.*]

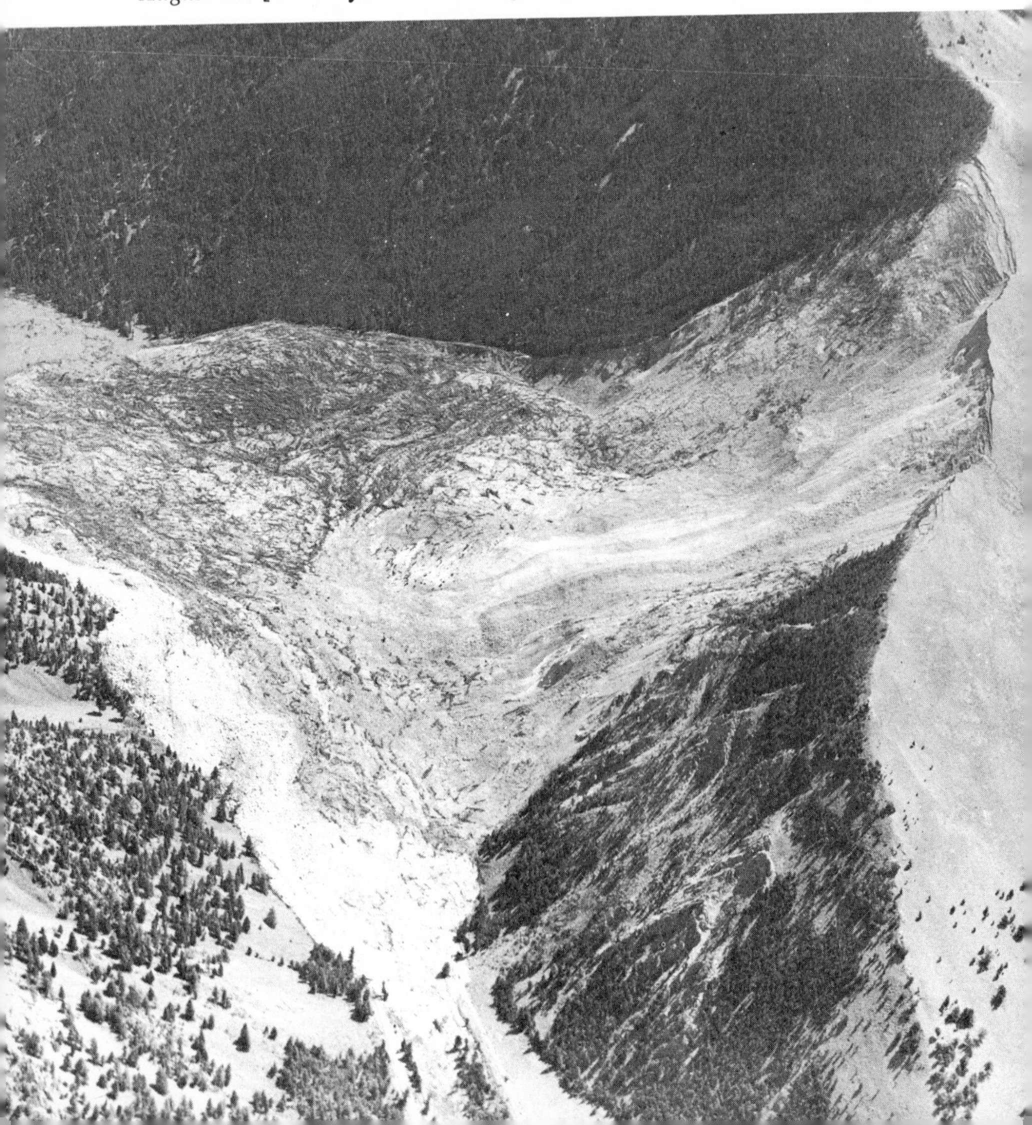

the lands, whereas streams are concentrated in gullies and valleys and cover a much smaller area. The great effect of mass-wasting in wearing down the lands has been appreciated only recently; the almost imperceptible rate at which some types occur was partly responsible. Thus a valley is deepened by stream erosion, but the bulk of valley widening results from weathering and mass-wasting.

Gravity is the direct controlling agent of movements which may be slow or fast. Rock falls, landslides, soil slump, creep, mud flows, and slopewash are examples of mass-wasting. Perhaps the different varieties of mass-wasting may be regarded as gradational members of a more or less continuous series from landslides (steep slopes, large load, and little water) to streams (much water, little load, gentle slopes, and commonly more rapid movement).

Two general types of movement are involved, either of which may be very fast—fast enough to bring death and destruction—or almost imperceptibly slow. One type is represented by the familiar landslide in which a relatively dry mass of bedrock and mantle slides as a unit upon some

Fig. 29-11. *A landslide over Sherman Glacier in Alaska. Debris fell from the high peak at the right during the Alaska earthquake of 27 March 1964 and slid at high speed across Sherman Glacier, 1.5 miles wide here. The heat-insulating effect of the debris will probably cause the glacier to advance. According to Shreve, such landslides may acquire so much momentum that they leave the ground at a break in slope and override and trap a cushion of compressed air upon which they can then slide at high speed with little friction.* [*Photograph by Austin Post, U.S. Geological Survey.*]

sort of lubricated surface. Landslides tend to occur on steep slopes underlain by strata or structures (e.g., joints or foliation) that dip parallel to the surface. Water adds weight and tends to reduce cohesion and internal friction among particles in surficial materials. Thus landslides may be more common in the spring in some areas because of rain and melting snow.

Creep The cloak of loose material above bedrock may creep slowly downward along even gentle slopes. This is the other general type of mass-wasting in which irregular movement occurs throughout the slowly moving mass. A covering mat of vegetation decreases the rate but does not halt the movement. Several kinds of evidence indicate that the mantle is moving downward. Roads, tunnels, and railroad tracks may be shoved out of line; fence posts, monuments, and buildings may be tilted or disrupted by the irregular movements. As the surface part of the mantle moves to lower ground more rapidly than the deeper part, a downslope tilt develops in fence posts and power-line poles which have been sunk for some depth into the loose material. The trunks of many trees are distinctly curved (convex downslope) as the result of a compromise between this downward tilting and their tendency to grow vertically (Fig. 29-12).

The causes of creep are numerous. Frost heaving tends to push loose

Fig. 29-12. *Soil creep. The upper part of the mantle moves slowly downslope, but more rapidly than the lower part. Objects which project downward into the mantle for some distance tend to be tilted.*

pebbles at the surface outward at right angles to a slope. However, when the ice beneath a pebble melts, the pebble falls vertically downward, and thus it has moved a tiny distance downslope. Weathering produces loose fragments which may fall or roll down the slope. Openings which are formed by the dissolving of soluble materials, by the decay of tree roots, or by the burrowing activities of animals are all eventually filled by downward movement of upslope material. Constantly repeated, these forces, all of them tiny, combine to move material downslope. Movement at the rate of 1 foot every 5 to 10 years has been measured and may be representative.

Rounded vs. Angular Topography Topography tends to be rounded in humid areas and angular in arid and semiarid regions. In humid temperate zones, a thick protective mat of vegetation may cover the mantle and prevent gullying. The entire mantle creeps downslope, filling in irregularities and hiding differences between resistant and weak rock formations. Furthermore, chemical weathering so weakens the rocks that steep slopes cannot commonly be supported. On the other hand, mechanical weathering predominates in drier areas; rock falls, landslides, and mud flows tend to leave slopes steep. Insufficient creeping mantle exists to mask differences in bedrock resistance by filling in low areas, and resistant rocks stand out as ledges or ridges.

WORK OF RUNNING WATER *

From the atmosphere water falls upon the Earth's surface as rain, snow, hail, and sleet. Some water evaporates or is taken up by plants, some runs off immediately into streams, and the remainder sinks into the ground. Much groundwater later emerges at the surface at a lower altitude and becomes runoff. Streams carry excess water from the land to the sea. In doing so, they erode valleys and help shape the Earth's surface. They transport rock debris and dissolved materials, and eventually they deposit most of the sediment in the oceans.

Streams are important to man whether he uses them as sources of drinking water, irrigation, or electric power, as scenic inspiration, or as places in which to fish or swim. Valleys furnish the most convenient courses for many roads and railroads. The location of a number of important cities depended upon the navigability of large rivers. Civilization flourished first

* Little drops of water,
Little grains of sand,
Run away together
And destroy the land.
—ROBERT E. HORTON.

on fertile floodplains. Bridges, dams, and reservoirs have to be built. Frequent floods with appalling loss of life and property emphasize the importance of streams to man.

Valleys as a Result of Stream Erosion With few exceptions, streams have excavated the valleys in which they now flow. Different kinds of evidence indicate that this is so. A branching treelike pattern is common in river systems (dendritic: Fig. 29-13); small tributaries enter larger streams at angles which are acute upstream, and this relationship tends to hold

Fig. 29-13. *Soda Canyon, Colorado. This area illustrates the youthful stage in the erosion cycle in a region of flat-lying rocks, dendritic drainage pattern, canyons, plateaus, and mesas. This relief map-model is one of a set of twelve which have been selected to illustrate typical topography studied in geology.* [*Courtesy, A. J. Nystrom & Co.*]

true from the tiniest gully to the chief river in the area. A rectangular stream pattern may also develop (trellised: Fig. 33-20) where sedimentary rocks of differing resistances to erosion have been folded and eroded. Ridges develop on the uptilted beds of resistant rocks, whereas valleys form along the trends of weaker rocks. Such orderly patterns would be unlikely if valleys originated primarily because of faulting.

In general, a definite proportion exists between the size of a valley and the size of the stream flowing in it. However, did the valley determine the location and size of the stream, or did the stream determine where the valley would occur? Each stream tends to flow on a steep slope near its head and on a gentle slope near its mouth; thus its profile is concave upward. Commonly a tributary joins a larger stream at the same level as the larger stream, and this accordance at intersections is maintained throughout the system. The sediments being transported by many rivers are readily observed and indicate that excavation has gone on somewhere upstream. Extensive gully systems have been observed to develop in some farming regions within a few generations. Thus most valleys have been produced and shaped by the streams which flow in them. What happens can be traced in the miniature replica of a branching river system that can be produced by spraying water from a garden hose for a few minutes on a pile of loose soil.

Stream Erosion A stream carries part of its load of sediments in suspension and in solution, but larger fragments make short jumps or roll and slide along the bottom. Under certain conditions, this bedload makes up half of the total, but quantitative data are lacking. A study of about 70 rivers in the United States showed that approximately 20 percent of the total measured load was carried in solution (range: 1 to 64 percent).

Stream erosion is achieved by abrasive impacts of transported fragments on the beds and sides of channels, by the solvent action of water (relatively small), and by the lifting effect of running water. Without sediment, streams cannot scratch and scour their channels, but a turbulent river readily picks up the smaller sizes of sediments. A sediment-laden river, the muddy Missouri, was aptly described by Mark Twain as "too thick to navigate, but too thin to cultivate." Potholes are produced by stream abrasion (Fig. 29-14).

A stream normally originates as a tiny gully in a depression at the Earth's surface where runoff is concentrated. This miniature valley grows deeper and wider, and it becomes longer by headward erosion because more water generally enters at the head of a valley than at any one place along its sides. However, most large rivers have not grown throughout their entire length by headward erosion; rather, smaller segments formed at different times and places and later united. At first, water flows down a valley only after a rain, but eventually a stream cuts into the surface deeply enough to reach the zone in which all open spaces are filled by groundwater (see p. 675). It then becomes permanent, shrinking in size during dry spells and enlarging greatly after heavy rains.

Steep slopes, high velocities, great volumes of water, and weak bedrock all increase the erosive capacity of a stream. In its mountainous headward portion, where a stream commonly must transport large boulders, steep slopes are required. However, sediment sizes have decreased near its mouth, and the volume and velocity of the stream have increased. Friction reduction caused by deeper channels more than compensates for the decrease in gradient. Therefore the stream can carry its load on a gentler gradient; its long profile is commonly concave upward.

Base Level A stream cannot cut downward indefinitely. Where a river enters the sea, it may scour its channel a little below sea level. Headward from its mouth, however, the channel must rise above sea level in order to furnish a slope down which the water can flow. Sea level projected inland as an imaginary surface is thus the deepest approximate level to which a stream can lower its channel; it is called *base level.* Lakes, dams, and resistant rock formations may form local, temporary base levels.

Fig. 29-14. *Potholes in the granite bed of the James River, Virginia. Eddies in the stream whirl gravel and sand around in a small circular area and gradually bore a cylindrical hole in the bedrock. New supplies of gravel and sand replace older pieces as they are worn out. Potholes may form rapidly. One 10 feet deep was observed to form in limestone in 18 months; one 5 feet deep formed in 75 years in granite. Resistance of the rock is only one of the factors involved.* [*Courtesy, U.S. Geological Survey.*]

Graded Streams At any one time a stream has just so much energy for its work of erosion and transportation. The quantity of energy depends upon the volume of water and its altitude above sea level. Most of the energy is wasted by friction produced by turbulent flow and by contact with channel margins. Near its head a stream tends to flow on steep slopes and to carry a relatively small load of sediments. Some of its energy is available to saw vertically downward into the Earth's surface. However, as the stream cuts downward, its gradient decreases, whereas its load increases. Thus the rate of down-cutting diminishes because an increasingly larger proportion of the total energy supply is used in transportation. Increased volumes of water at lower altitudes do not offset this tendency. Eventually an approximate balance is reached between the load a stream carries at any one point and the volume, channel shape, and gradient at that point. To reach this condition, a stream has cut downward in many sections but has deposited sediments to build up its floor in other sections. Such a stream is said to be *graded.* If the equilibrium is disturbed by a change in any one of the many factors involved, the other factors then change in an offsetting combination to restore the equilibrium.

After a river has become graded, conditions may change; e.g., the load of sediment transported by a river may be increased, or the amount of water available to it may be decreased. In each of these instances, the river would deposit some sediment all along the floor of its channel thereby steepening its gradient (the mouth of the river remains at the same altitude). At the greater velocity thus produced, it once more carries a full load and is again graded. On the other hand, if the load of a river is decreased, the excess energy thus made available is applied to lower the channel by downcutting. In this manner, a graded condition is once more attained. When a stream is graded, minor fluctuations occur through a norm; the stream deposits material at one place and picks up sediment at another. Downcutting continues after the graded condition has been attained, although it then becomes exceedingly slow.

Meanders and Valleys with Wide Flat Floors Wide flat-floored valleys are common at lower altitudes and contrast sharply with the V-shaped cross sections (includes asymmetrical valleys with irregularly sloping sides) that we observe in mountainous regions. However, all gradations occur between these two types, which suggests a genetic connection between them; i.e., the wide valleys with flat floors were once V-shaped, had steeper gradients, and falls, rapids, and lakes occurred along their paths (Fig. 29-15).

As a stream cuts downward, it would leave vertical banks if weathering and mass-wasting did not combine to widen the valley. The upper part of a valley, where the widening process has occurred longest, is thus broader

Fig. 29-15. *Three stages in the sequential development of a stream valley. [Courtesy of the American Museum of Natural History.]*

than the lower part. In unconsolidated or less resistant materials, the process is relatively rapid; in resistant rocks, the process is slow, and the valley walls remain steep. Continued deepening by a stream is thus an important factor in determining the shape of a valley. The long profile of a stream at this time is irregular; a relatively smooth, concave-up profile develops later.

As erosion proceeds in an area, numerous small streams form on slopes and flow downward into depressions, creating lakes. The lakes subsequently rise high enough to spill over. Initially a number of small gullies might develop along any one slope, but certain ones are favored and grow larger by capturing the drainage of neighboring gullies. These grow longer by headward erosion, and the capture process may be repeated again and again on a larger and larger scale. This causes additional irregularities in the long profiles of master streams which form eventually by union of numerous segments. Where streams flow from more resistant to less resistant rocks, rapids and waterfalls develop. As streams cut downward, lakes are drained, and rapids and waterfalls slowly disappear. By erosion here and deposition there, a master stream eventually becomes graded. Commonly this occurs first near its mouth and is slowly extended upstream. A fairly smooth, concave-up profile is thus produced in a valley that may still be relatively narrow at the bottom.

Irregularities in the path of a river cause the deepest and most turbulent part of the channel to impinge against the outer part of each bend. Deposition takes place in the slack, less turbulent water on the inside of a bend and decreases the cross-sectional area of the channel. This causes a stream to undercut the outer and downstream side of a bend. Flat crescent-shaped areas develop on the insides of each bend. They coalesce eventually to form a continuous floodplain, because each bend migrates slowly downstream as well as laterally. The winding sinuous course which develops eventually is called *meandering* (Fig. 29-16).

Experiments with models show that meanders tend to form in streams with banks consisting of easily eroded sediments. Bank caving produces irregularities in a channel which initiate the formation of meanders (Fig. 29-17).

Fig. 29-16. *A meandering river and its floodplain. The meander of Crooked Creek, California, in the left foreground shows an undercut outside bank and gently sloping inside bank.* [*Courtesy, U.S. Geological Survey.*]

Fig. 29-17. *Laboratory development of river meanders. Test materials used in experiments include Mississippi River sands and haydite (used in this series), a porous granular material which is hard and inert and has a specific gravity of 1.85. The sign in each photograph shows the lapse of time since the experiment began. A fixed condition of slope and discharge was maintained during this sequence. Material was added to the upper stream flow during testing operations.* [*U.S. Army photograph.*]

This wide, flat, laterally planed, valley floor has a relatively thin covering of river-deposited sediment and is called a floodplain. The sediment has accumulated in two different ways: deposition has occurred at the inside of each bend, and finer sediments have been spread over most of the floodplain during overbank floods. As meanders have migrated both laterally and downstream, older deposits have been reworked and the channel has shifted so widely that at some time or other it has been present on all portions of the floodplain.

For such a floodplain to develop by lateral planation, the crust in an area must remain stable for a very long time, so long in fact that this seems not to have occurred frequently. Many major existing floodplains have formed in a somewhat different manner; they are fill-type floodplains (Fig. 29-18). Sedimentation has filled the entire bottom portions of such valleys. Perhaps this is a direct result of the addition to such streams of huge volumes of glacially derived sediment during the Pleistocene.

As meanders migrate down a valley and shift laterally, the neck portion of a loop tends to narrow and may eventually be cut through. A stream thus temporarily straightens and shortens its course, but new loops then form. Sediment may be deposited at the entrances to a former channel, and a crescent-shaped *oxbow lake* may result (Fig. 29-19). If an abandoned loop does not contain a standing body of water, it should be called an *oxbow*. Terraces may form along the sides of a valley (Fig. 29-20).

Erosion Cycle in a Single Valley If erosion is not interrupted by some geological change, such as an uplift, a valley tends to evolve through a cycle of gradual changes: vigorous youth, powerful maturity, and sluggish

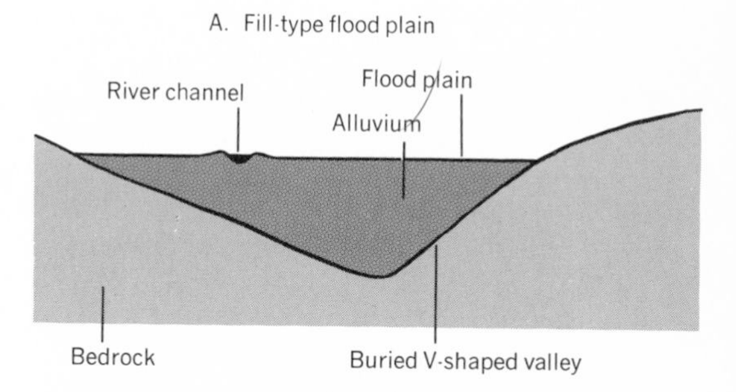

Fig. 29-18. *Fill-type floodplain (A) and erosional type floodplain (B). Alluvium refers to sediments deposited relatively recently by streams, excluding those that accumulate in lakes and seas. In B the buried, beveled, bedrock floor of the valley has been formed by side-cutting (lateral erosion) by the river. During a flood (shown in B), a river may deepen its channel and erode its bedrock floor. During normal-water stages, the floor of the channel is formed by alluvium.*

Fig. 29-19. *Erosion and deposition on a wide flat-floored valley in Alaska. Note the meander scars, oxbows, and oxbow lakes. [Photograph, U.S. Air Force: N. 65° 55′, W. 156° 30′.]*

old age. These stages are not marked by absolute numbers of years. The time it takes to reach a given point in the sequence depends upon the composition and structure of the rocks, the altitude above sea level, the climate, and other factors. The stages are determined by the amount of

Fig. 29-20. *Stream terraces. Nearly level benches or terraces occur along the sides of some wide-floored valleys, and these may be at the same altitude on each side or at different altitudes. Each terrace is an erosional remnant of a former higher floodplain that once extended all the way across the valley. Such terraces form in a number of ways, e.g., by an increase in the transporting ability of a stream.*

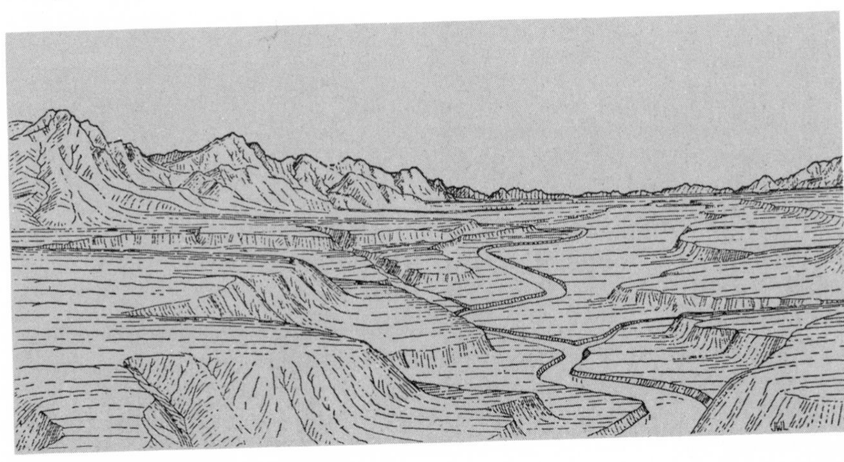

erosion which a river has already accomplished relative to the amount of work that remains for its drainage area to be reduced to base level (Fig. 29-21).

The change between youth and maturity is more marked than that between maturity and old age (chiefly an exaggeration of features associated

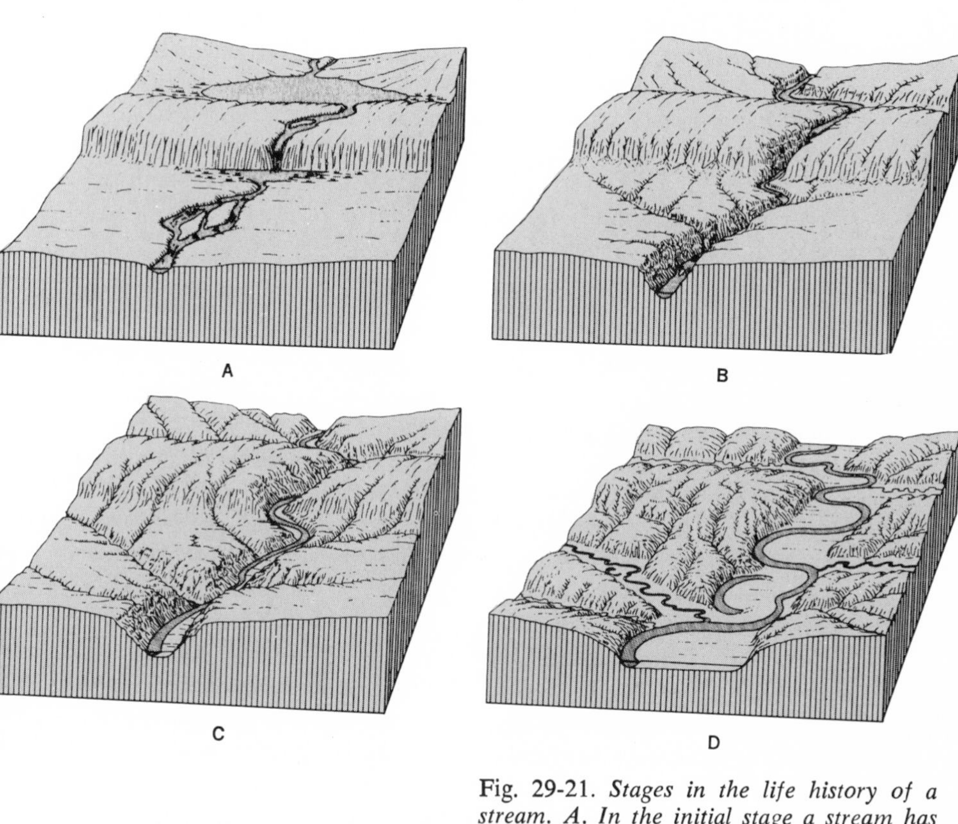

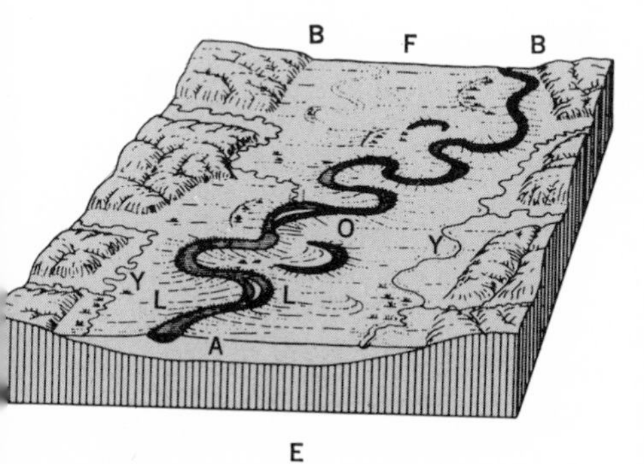

Fig. 29-21. *Stages in the life history of a stream. A. In the initial stage a stream has lakes, waterfalls, and rapids. B. By middle youth the lakes are gone, but falls and rapids persist along the narrow incised gorge. C. Early maturity brings a smoothly graded profile without rapids or falls, but with the beginnings of a floodplain. D. Approaching full maturity, the stream has a floodplain almost wide enough to accommodate its meanders. E. Full maturity is marked by a broad floodplain and freely developed meanders. L = levee; O = oxbow lake; Y = yazoo stream; A = alluvium; B = bluffs; F = floodplain. [After E. Raiz. Courtesy, A. N. Strahler,* Physical Geography, *1951, John Wiley & Sons, Inc.*]

with maturity). The continuous, gradational nature of the sequential development should be emphasized. Although this concept of an erosion cycle has been criticized (p. 629), sequential development of landscapes does occur, and the concept of an erosion cycle aids the imagination in considering the changes involved.

A river may be youthful in its upper reaches, mature in its middle portions, and old near its mouth where erosion has continued for the longest period of time and less downcutting had to occur. Variations occur if a river passes through rocks of different resistances or through areas which have been affected by crustal movements during the development of the valley; e.g., a wide valley could develop in weak rocks both upstream and downstream from a region underlain by resistant rocks in which the valley is still V-shaped.

Erosion Cycle in a Region An entire region may develop slowly through stages which can be characterized as youth, maturity, and old age like the stages of a single valley as discussed above.

If uplift of an area of subdued topography should take place rather rapidly (in geologic terms), the youthful stage in the erosion cycle would be featured by broad flat-topped divides between a few V-shaped valleys. The area is chiefly upland. As streams enlarge their valleys and become more numerous, the amount of upland decreases, slopes (sides of valleys and mountains) predominate, and the area is mature. As erosion continues and a region is lowered closer to base level, the amount of bottom land becomes greater, and the old-age stage is eventually reached.

The land forms which are produced during an erosion cycle are influenced by factors such as the following: differences in rock structures, crustal movements or igneous activity which may occur after initiation of a cycle, variations in original altitude above sea level and distances from the sea, climatic dissimilarities, the length of time involved, and the effects of other erosive agents, such as wind, waves, and ice (p. 706). The stage

Fig. 29-22. *Idealized structure section through a small delta. Top-set beds* (*T*), *fore-set beds* (*F*), *and bottom-set beds* (*B*) *are shown.* [*Modified after G. K. Gilbert.*]

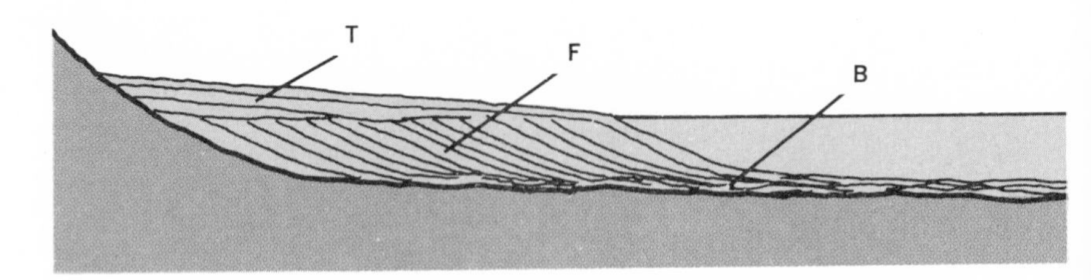

in the erosion cycle attained by a single valley may not be the same as that of the entire region; e.g., numerous youthful tributary streams may exist in a region which has reached maturity.

Peneplain is the term for the nearly smooth erosion surface of low relief and low altitude which covers a large area and which develops in late old age. Its surface may be dotted here and there by an occasional small, rounded hill underlain by resistant rock. A peneplain bevels different kinds of rock formations and geologic structures as if a gigantic carpenter's plane had been pushed back and forth across the region. Peneplaned surfaces may subsequently be uplifted and dissected by erosion.

Alluvial Fans and Deltas *Alluvial fans* are common in drier regions where streams flow abruptly from steeper to gentler gradients, as at the base of a mountain or ridge. A stream drops most of its load of sediments near the base of a mountain, and eventually a fan-shaped mass of debris is built up as the channel shifts laterally back and forth across the fan; the apex of a fan is located near the mouth of a valley. Fans may be large or small. Deposition is rapid, and sorting and stratification are not well developed. A number of fans along the base of a mountain may coalesce to form a continuous alluvial slope (bajada).

A *delta* may form near the junction of a river with a standing body of water in a manner analogous to the formation of an alluvial fan on land. Sorted, stratified sediments accumulate up to and even slightly above the water level. Small deltas may exhibit a characteristic stratification (a type of cross-bedding: Fig. 29-22) not present in many large deltas built into the ocean.

Prevention and Control of Soil Erosion Soil is one of man's most important natural resources, an irreplaceable heritage which he has abused in the past and continues to misuse. According to some estimates, approximately one-half of the total quantity of agricultural land in the United States may have lost much of its topsoil in the last two centuries, is in danger of doing so, or has been destroyed. Perhaps 100,000 acres are being lost annually at present. Fertile lands are also lost when buildings, roads, and air fields are located on them.

Stated in another way, topsoil in the United States may have averaged about 9 inches in depth before settlement. Today this thickness is estimated at 6 inches. These 6 inches of topsoil stand between us and starvation (the current surplus of agricultural products in the United States does not lessen the importance of topsoil when one takes a long-range view). Although a number of remarkable soil conditioners have been introduced, no economic method is known which produces rich topsoil from infertile mantle.

In addition to the direct loss of crop-producing capacity, soil erosion increases the destructiveness of floods and decreases the storage capacity of water in reservoirs; e.g., about one-half of the water supply reservoirs in the United States may be useless within 100 years because they will be filled by sediment.

A protective mat of vegetation prevents or greatly retards soil erosion by wind and water. Of first importance, therefore, is the restoration or maintenance of such a cover on slopes which have been deforested, over-cropped, overgrazed, or burned over. Submarginal soils, especially those on steep slopes, should not be plowed. A double loss occurs when good soil is washed down a slope, mixed with infertile material, and then dropped on top of fertile soils at lower altitudes.

Fig. 29-23. *Soil pedestals caused by raindrop splash. Raindrop splash may remove as much as two inches of top-soil in one heavy rain of high intensity. Plant cover controls splash erosion by intercepting the falling raindrops and absorbing their energy. The pebble-capped pedestals show that the force of the raindrop was applied from above. The soil not protected by pebbles was detached and washed away.* [*U.S. Dept. of Agriculture.*]

Raindrops striking soil on sloping land unprotected by vegetation have an important erosional effect. Tiny soil particles may be moved a few inches or a few feet upon each impact, and the violent disruption and pounding of a heavy rain may result in the shifting about of more than 100 tons of soil per acre. On level ground struck by vertically falling drops, the soil is splashed around, but the effects are canceled by the random motions. However, on sloping ground, raindrops combine with sheetwash to cause much soil erosion. Detachment of tiny particles, in which raindrop impact is most important, and transportation where sheetwash predominates, are the two main steps involved.

Fig. 29-24. *A remarkable example of terracing of a steep valley, in Lebanon.* [*U.S. Dept. of Agriculture.*]

As raindrops batter a surface like tiny bullets, finer particles in a soil may be shifted between larger ones to form an impervious crust. In addition, a soil may be compacted by raindrop impacts. In this manner, an impervious surface crust may form when evaporation occurs after a heavy rain. This crust increases surface runoff and thereby the development of gullies. As erosion breaks through this crust, the underlying unconsolidated particles may be splashed away leaving a tiny soil pedestal (Fig. 29-23).

Various techniques can be used to prevent or decrease the downward movement of soil along a slope. Drainage may be controlled by a system of ditches or small dams. Small gullies may be healed by the planting of quick-growing vines and shrubs. Row crops may be planted parallel to the contours of a slope, not up and down the incline. Crops planted in any given field may be varied from year to year. Strip cropping on a slope intersperses areas of row crops such as corn with protective strips of thick-growing grains and grasses. Terracing (Fig. 29-24) of a slope permits the tilling of gently sloping or flat areas which are separated by more steeply sloping sections that are kept permanently covered by a mat of vegetation or held in place by retaining walls. Materials such as hay or straw can be scattered about between rows of corn or among other crops and prevent much soil erosion. Such mulches have the added advantage of slowing the growth of weeds.

Perhaps the U.S. Conservation Service has best summarized the elements of good practice in this recommendation: Use each acre according to its capabilities; treat each acre according to its needs.

Floods and Flood Control Streams perform most of their work of erosion, transportation, and deposition during floods. In fact, in the few days of a powerful flood a river may accomplish more in these respects than it does during the rest of the year or perhaps even for several years. Some large rivers have been known to deepen their channels more than 100 feet when flood waters are exceptionally high. Such large rivers flow upon unconsolidated sediments that bury the bedrock floors of their valleys and can erode the bedrock only during floods (Fig. 29-18). As a flood subsides, other sediment is deposited, refilling the bottom part of the channel.

When a river overflows its banks during times of high water, its velocity is checked abruptly beyond the margins of the channel where the water becomes shallow. This sudden decrease in velocity causes deposition; coarser materials are deposited at once along the banks of the channel, and finer sediment is spread over the rest of the floodplain. Numerous repetitions of this process may produce low ridges along each side of a channel. Such ridges are called natural levees (Fig. 29-21). Commonly they rise only a few feet above a floodplain, yet during extensive floods they may form long, low islands which parallel a channel and constitute the only dry land in the floor of a valley.

As floodplains are exceedingly fertile areas for farming and furnish flat land for roads and building purposes, many farms, towns, and cities are located on them, even though they are subject to periodic flood. In a sense, these works of man are located in rivers, not just near them, because floodplains are integral parts of a stream system. To prevent floods, artificial levees were built to confine floodwaters to river channels. Unfortunately, this construction inaugurated a vicious cycle. During a flood a river carries an extra load of sediment, part of which is normally spread over a floodplain as a flood subsides. If a river is confined to its channel, the extra sediment is deposited on the channel floor instead of the floodplain, and the level of the water is thus raised. The levees subsequently are not adequate everywhere and must be built higher. Such construction provides only temporary protection, because a river continues to build up the floor of its channel by deposition of sediment during each flood. In some valleys the tops of levees are above the roofs of houses built on adjacent floodplains. A break-through under these circumstances is very destructive, and large areas of a floodplain are inundated. The first levees built along the Mississippi River more than 100 years ago were only about 4 feet high and relatively short. Since then, hundreds of miles of levees of increasingly greater heights have been constructed.

Floods are the most disastrous of natural forces to mankind and are becoming increasingly destructive year by year, primarily because of greater use of the floodplain by man. Floods and flood control affect the lives of all citizens today either directly or indirectly. According to the U.S. Department of Agriculture, the total average annual damage from floods and sediment in the United States is about 1 billion dollars. This damage may be separated into upstream and downstream parts: upstream areas lie above existing or proposed major flood control structures; downstream areas lie below such structures. More than half of the total flood damage probably takes place in upstream areas from frequent, small, unpublicized but destructive floods. None of the existing or proposed major flood-control structures aid in preventing these floods. Estimates suggest that more than $60 has been spent on downstream structures for each dollar spent for upstream flood control. Obviously this is not a proper balance. More than two-thirds of the upstream damage is agricultural: damage to crops, roads, and buildings, erosion of good farmland, deposition of unfertile sediment on productive soil, and indirect losses such as delays in planting or marketing.

Flood control is a many-faceted complex problem. A large river system which extends throughout several states is a natural unit and must be treated as a unit—omissions or commissions in one part of the system can affect remote areas vitally. The multiple uses of a stream system must be considered: floods, power, water supply, navigation, irrigation, recreation, sewage disposal, fish, and wildlife. Sometimes these interests conflict; for example, reservoirs need to be kept nearly empty for maximum efficiency in preventing floods, but they need to be full to serve as adequate water supply.

One reason for the more extensive damage caused by recent floods (other than greater construction of property along the floodplain) is the destruction by man of the cover of vegetation which formerly protected slopes throughout the entire drainage area of a river. When water falls on slopes which are covered by thick grass and numerous trees, much of it sinks into the ground, and very little runs off immediately—almost none from a forested area. On a slope not held together firmly by a mat of vegetation, much water at once washes down the slope, carrying large quantities of soil with it. Therefore, anything which reduces the amount of immediate runoff after heavy rains and melting snows must help to prevent high waters downstream and deposition of sediment in confined channels and reservoirs. Thus reforestation and soil conservation throughout the area drained by a river system appear to constitute an important portion of a satisfactory long-range solution of the flood problem. Rapid runoff must be halted. Straightening and dredging channels and strengthening levees are temporary expedients. Less valuable areas along a river may be selected as temporary reservoirs to be flooded at times of exceptionally high water in order to lessen the pressure on more valuable areas downstream.

Although reforestation is important in preventing or limiting numerous small upstream floods, it apparently has much less effect in preventing major disastrous downstream floods. Reforestation throughout an entire drainage basin would not "soak up" these major floods before they started. Such floods occurred before man entered the scene and cut down the forests. They occur when heavy rains fall on saturated or frozen ground (melting snow may also be an important factor).

A system of dams and reservoirs, therefore, is likewise essential in the control or prevention of floods. If it can prevent the many tributaries which feed a large river from flooding simultaneously, the danger may be averted, and water can be released from reservoirs later at times of drought. Yet agitation by the population of a large city for the building of dams and reservoirs upstream may arouse the hostility of the people who live and farm in the upstream areas which would have to be condemned and flooded to make room for the reservoirs. A number of well-informed people believe that the construction of a few large expensive dams across major rivers will provide less storage volume at greater cost than would the building of many small dams across numerous minor streams. Furthermore, water which collects behind the small dams may not flood large areas of valuable land. In a number of instances, further study has shown that the annual income from the area which is to be flooded by a proposed reservoir far exceeds the total benefits to be derived from its construction.

Settlement control should play a vital role in any long-range plan for controlling floods. To keep flood waters entirely confined to a river channel is extremely difficult and expensive. However, if valuable properties could be zoned to higher altitudes or to areas somewhat removed from a river channel, much property damage would be prevented. Dikes could be located far enough from a channel to provide space for overflow in time of floods. Only temporary or less valuable buildings would be permitted in the zone between the dikes and river channel. Somewhat similar zoning regulations could be applied to farm buildings constructed on the floodplain. Such zoning would tend to eliminate the two most important reasons for the construction of many huge dams and reservoirs.

Exercises and Questions, Chapter 29

1. Assume that a cube 1 meter on a side is subdivided into smaller cubes each 1 centimeter on a side. How much greater is the total surface area of all the 1-cm cubes than that of the original cube? What application does this have to weathering?

2. Describe briefly as many processes as you can that act to reduce the altitude of the Earth's surface. Describe other processes that act to build up the surface.

3. Assume that you are in an area containing a number of large, glacially transported boulders. Some of these have been partly exposed by erosion and project from the ground in much the same manner as does a small outcrop of bedrock. How could you distinguish these boulders from outcrops of bedrock?

4. Weathering:
   (a) Describe the important processes involved in chemical and mechanical weathering.
   (b) What conditions are most favorable for chemical weathering? for mechanical weathering?

5. Soil:
   (a) How does soil originate?
   (b) Why is a mature soil zoned? What are the distinguishing features of each of the three main zones?
   (c) What kinds of data might be collected in a volcanic region that would lead to estimates concerning the length of time necessary for soils to form? Would these rates necessarily apply to other areas and other types of rocks?
   (d) Describe some methods of soil conservation.

6. Describe the roles played by weathering, mass-wasting, and streams in the sculpturing of the lands. Is the analogy of a stream with a conveyor belt a useful one?

7. What is meant by differential weathering and differential erosion? On what scales do these processes operate? Look for examples in your neighborhood and in your travels.

8. Why is topography commonly rounded in humid areas and angular in dry areas?

9. List some of the names applied to different types of mass-wasting.
   (a) What do these tend to have in common?
   (b) In what ways do these tend to differ?

10. Stream flow:
    (a) How does a stream transport its load of sediments?
    (b) Describe some of the factors that affect the velocity of a stream.
    (c) How does stream flow tend to differ when flow around a bend is contrasted with flow in a straight stretch?

11. Describe the changes that tend to occur in the development of a stream valley as it gradually evolves through the stages of youth, maturity, and old age.
    (a) Describe some of the features commonly associated with a stream that flows on a wide floodplain.
    (b) Describe some of the similarities and differences between deltas and alluvial fans.

12. A certain river is in the following stages of the erosion cycle from its head to its mouth: (1) youth, (2) maturity, (3) youth, (4) maturity, and (5) old. Give at least two different explanations for the stage of youth in number 3.

13. How will an area in the mature stage of the erosion cycle in a region of folded rocks differ from an area in the youthful stage of the erosion cycle in a region of nearly horizontal sedimentary rocks?

14. Discuss some of the factors involved in floods and flood control.

30 Subsurface water

WATER in the ground beneath the Earth's surface is our most important mineral resource (Fig. 30-1). People living in humid areas with ample supplies of surface water commonly fail to appreciate the utter dependence of very large numbers of men, women, and children and their animals and crops upon supplies of groundwater.* In the United States in 1960, nearly 50 billion gallons of groundwater were probably used each day (about one-fifth of the total amount used). More than half of all the families in the nation may have used some groundwater. The availability of this water is taken for granted until it is gone. Only then is its true value fully appreciated.

Subsurface water has for thousands of years been a vital need in many areas of the world. Like so many vital matters, the search for it was surrounded with magic and mystery. The superstition of the water diviner's forked stick is still with us, and so is an agelong ignorance that condones the close intimacy of the outhouse and the well in many farming areas, and supports belief in great underground rivers, or in springs occurring on mountain tops. Even educated people are liable to know little more about their water supply than that a turn of the faucet brings forth clear cold water. Yet questions of groundwater supply have gained rather than lost importance in modern civilization, and conservation has become a vital matter that all need to understand.

THE WATER CYCLE

The oceans are the great reservoirs for the world's water supply. Energy from the sun causes evaporation at the surface; some of the resulting water

* Groundwater and subsurface water are used synonymously here, although technically groundwater refers only to water beneath the water table.

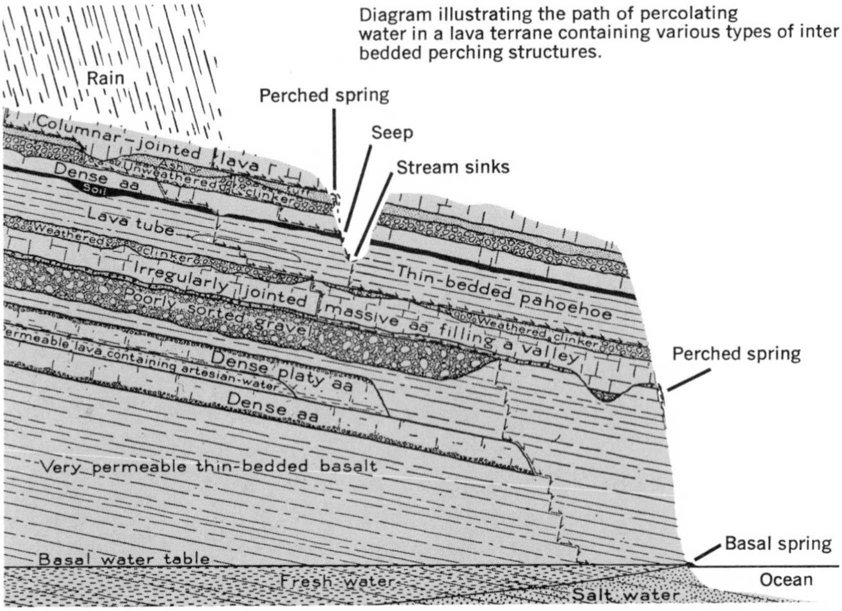

Fig. 30-1. *Typical occurrence of subsurface water in the Hawaiian Islands. Thus some springs, swamps, and intermittent streams may occur above the main water table, especially if impervious barriers occur below them.* [*From Stearns and McDonald, Division of Hydrography, Hawaii.*]

vapor is carried by the atmosphere over the lands where it may be precipitated or condensed as rain, snow, hail, frost, or dew (Fig. 30-2). Part of this may be evaporated again directly, or indirectly by the transpiration of plants (e.g., tree roots extract water from the soil; this water moves up the trunk as part of the sap and is later evaporated from the foliage). Part runs off immediately to join streams and lakes, and part sinks downward into open spaces in sediments and rocks. Much of this groundwater emerges at the surface again at lower altitudes.

The average annual precipitation in the United States is equivalent to some 30 inches of rainfall. Of this, about 21 inches (70 percent) returns directly to the atmosphere by evaporation and transpiration. A field of vigorously growing plants on a summer day probably furnishes nearly as much moisture to the air as is evaporated from the surface of a water body of the same size. The remaining 9 inches is the source of all surface and groundwater. This would be an ample supply if distributed uniformly in time and space. Problems arise because too much water or too little water occurs at a certain place at a certain time. These problems are increasing in number and magnitude each year.

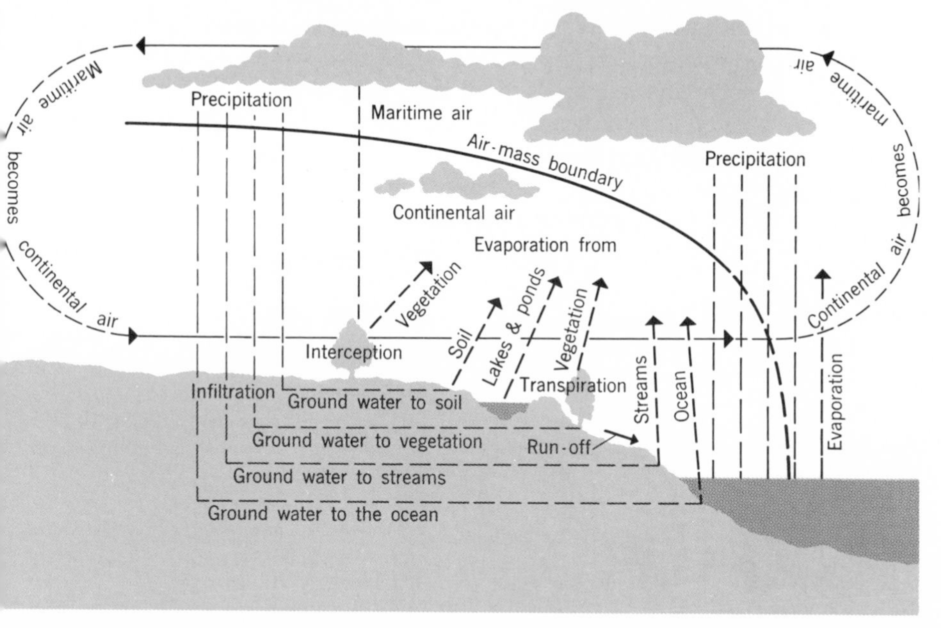

Fig. 30-2. *The hydrologic cycle. A drink of water in a city such as Boston or Chicago may consume water molecules that were evaporated from a maple tree, a tobacco patch, or the Atlantic or Pacific. This may have occurred recently or centuries ago because the water may have been stored in the ground. Before this, the water molecules may have participated in countless phases of the water cycle, on land and sea, and in the northern and southern hemispheres; they certainly may be described as well traveled and aged.* [*Courtesy, U.S. Dept. of Agriculture.*]

Near the end of the seventeenth century, experiments were performed to study the origin and nature of groundwater. A French scientist (Perrault) measured rainfall in the Seine drainage basin for a period of three years and the volume of water discharged by the river during that time. His results showed that about six times as much rain and snow fell in the drainage area as left it via the Seine River (the life functions of organisms and evaporation accounted for the five-sixths that seemed to disappear). Another scientist (Halley) explained the water cycle correctly; he showed experimentally that evaporation from the oceans was sufficient to account for all surface and subsurface water.

DISTRIBUTION OF SUBSURFACE WATER

When water is precipitated from the atmosphere onto the lands in areas where permeable sediments or rocks occur, some of it sinks below the surface into cracks and pores (spaces in rocks and sediments not filled with solid mineral matter). However, heavy rains on steep slopes underlain by relatively impervious materials result in a high percentage of runoff.

The porosity and permeability of the bedrock and mantle are important factors in determining the amount of infiltration. *Porosity* refers to the percentage of interstitial space (openings such as pores and cracks) in a given volume of rocks and sediments; *permeability* refers to the relative ease with which water may move through these interstices (Fig. 30-3). A high poros-

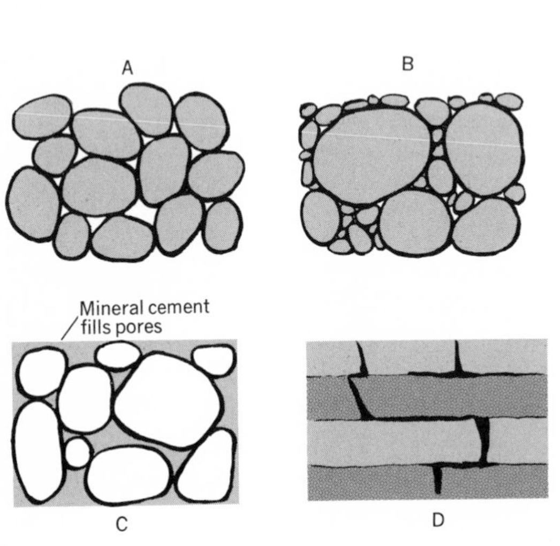

Fig. 30-3. *Some factors determining the porosity of a rock or sediment. (A) Spaces of different sizes and shapes may occur among particles. (B) If sorting is poor, then smaller particles fill in spaces among larger ones and reduce the porosity. (C) Mineral cement reduces the porosity. (D) Water can occur in joints and along bedding surfaces. The amount of porosity in a sediment depends upon the manner in which the particles are packed together as well as upon the shapes and sizes of the particles. If the particles are spheres of a uniform size, and if they are packed in the same manner, then large spheres (lower diagram) have the same porosity as small ones (porosity range: about 48 to 26 percent). Quicksand is sand in open packing (maximum porosity), and is generally caused by water moving through and lifting the sand.*

ity does not assure a high permeability. The size of the interstices is a controlling factor; larger openings are associated with higher permeabilities because less surface area is exposed to cause friction (p. 641). Thus sand is more permeable than clay, although clay may have a higher porosity than sand. Unconsolidated, well-sorted gravels are highly permeable, as are some vesicular lavas and limestones with numerous cracks that have been enlarged by solution.

Below a certain variable depth, which is near the surface in humid regions, all interstices are filled with water, and the upper surface of this saturated zone is called the *water table* (Fig. 30-4). The water table rises in wet seasons and falls in dry seasons, but the cracks and pores below it are always filled with water. (Locally, interstices may be filled with oil or gas.)

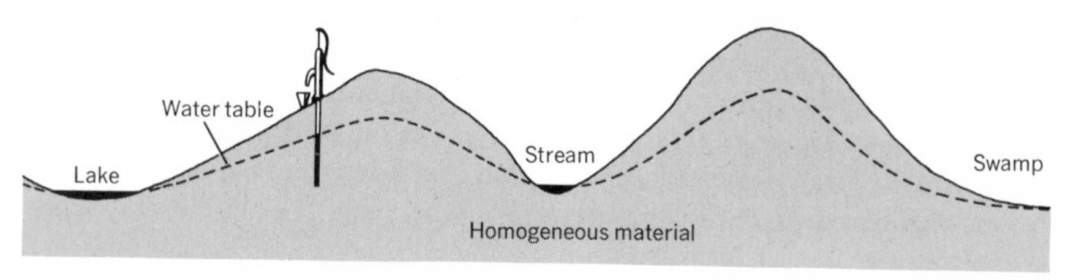

Fig. 30-4. *The water table in homogeneous rocks or sediments. As many impervious materials exist below the Earth's surface, subsurface water does not constitute a vast interconnected underground reservoir.*

Between the water table and the Earth's surface is a zone in which open spaces are alternately filled with air and water. In this zone chemical weathering and leaching take place; below this zone cementation of rock particles and precipitation are prominent. If the water table intersects the Earth's surface, the result is seepage (Fig. 30-1), a swamp, a spring, a stream, or a lake.

An ordinary well dug or drilled below the water table is actually a tiny pond. If the well is not deep enough, it goes dry during a drought. However, if it is then drilled to a greater depth, it again contains water. Nothing mysterious has occurred. The water table merely fell to a lower level during the dry period. In a similar manner, many shallow lakes or ponds disappear during dry seasons. In addition, great industrial demands on large deep wells may cause shallow domestic wells in the vicinity to go dry (Fig. 30-8). Thus the water table is not a fixed or level surface. In homogeneous rocks in humid regions, it forms a subdued replica of the topography; it is farthest from the surface under hills and nearest the surface in valleys.

The water table is not level as is a lake surface or the surface of a pail of water, for a number of reasons: the Earth's surface is irregular, and thus water enters the ground at different altitudes in different areas; differences in rainfall and in permeability cause uneven infiltration; water moves slowly through pores and cracks and a complete leveling out of the water table would take a very long time. New supplies of water are added frequently enough to prevent this from occurring.

Estimates indicate that enough water is present beneath the surface in the United States within the outer 100 feet of rocks and mantle to completely cover the whole area to an average depth of approximately 17 feet. For example, if a sandstone formation has a thickness of 100 feet and a 10 percent porosity and is located beneath the water table, it contains the same amount of water as a lake 10 feet deep. Porosities from 5 to 15 percent are considered average in rocks, and higher porosites occur in unconsolidated sediments. Unfortunately, a considerable volume of subsurface water is in material with very small pore spaces and cannot be obtained readily, or its quality is unsatisfactory. Streams commonly do not flow continuously until they have cut below the water table.

Water is a serious problem in many mining operations, and a number of mines have been closed before exhaustion because pumping costs became too great, e.g., one in cavernous limestone was abandoned after two years of constant pumping had failed to lower the water table appreciably. However, deep mines tend to be dry and hot (the deepest mine extends about $1\frac{3}{4}$ miles below the surface). Pressures increase with depth, and few open spaces can exist in rocks below one-half mile or so.

Movement of Subsurface Water As a result of the Earth's gravitational attraction, rain falls and water moves down slopes into streams. Gravity is likewise the force which causes movement of groundwater, but capillarity influences movement. Capillarity is the tendency of a liquid to cling to a solid surface. Surface-tensional forces are involved, and in very tiny tubes or interstitial openings these may cause a liquid to rise against the pull of gravity. The movement may be up, down, or sideways. Capillarity is an important phenomenon, and illustrations of it are numerous: excess ink rises into a blotter, and the liquid fuel in a cigarette lighter moves up a wick.

In the movement of water below the ground, capillary forces cause a thin film of water to adhere to the surface of each grain and to coat the surface of each crack. Such films have a top-priority claim on the supply of water available. Moisture is pulled from the wet grains to the dry ones downward and laterally. Water cannot collect and fill the larger openings and move downward under the pull of gravity until all of these films have been formed. Gravitational attraction cannot pull capillary water away from the grains.

Evaporation and the roots of plants, however, can remove some of this capillary water. Shortly after a rainstorm, the soil is moist near the surface. However, it may be relatively dry below this because insufficient rain fell to produce capillary films to a greater depth. As plant roots take up capillary water near the surface, this uppermost subsurface zone dries out in the next few days; and as this zone dries out, more water is brought up from below by capillary action. The next rain must replace all of this capillary water before any water can sink deeper. Consequently, if considerable intervals of time elapse between rains, little or no water can sink to the water table.

Thus, in humid regions with seasonal climates, the groundwater supply is replenished chiefly in the spring, owing to a combination of rains and melting snow, relatively little evaporation, and dormant vegetation. Some water may also be added in the fall as the growth of vegetation slows down and less evaporation occurs. Little water is added in the summer, even though more precipitation may occur, because evaporation and transpiration are too great.

Movement from the surface to the water table is dominantly straight down—a slow seepage in which the soil acts as a natural sponge. When water reaches the saturated zone beneath the water table, it oozes slowly downward and laterally, generally in much the same direction as runoff at the surface. According to Meinzer the rate of movement varies greatly: 420 feet per day is one of the fastest by field test; 1 foot in 10 years is one of the slowest by laboratory test. Less extreme rates range from a few feet per day to a few feet per year. A drop of water may move from the surface through the ground and return to the surface at a lower altitude, and this journey may take a few days or a few centuries; it may cover a few yards or a few hundred miles. Of course what is said here about the water table does not apply generally to impermeable rocks or to confined water.

WELLS

There is a familiar problem in areas in which many inhabitants rely upon individual wells. Why do neighboring successful wells vary in depth? One's interest in the location of wells is sharpened by a drilling cost of $8 per foot or by weeks of hauling water during a dry spell (Fig. 30-5).

In areas of impermeable rocks, groundwater circulation is confined mainly to cracks, and a successful well depends upon the more or less chance intersection of an adequate network of cracks and fissures. Since openings tend to decrease with depth, it is sometimes more practical to start a second well at a new location than to extend a first unsuccessful one ever deeper.

Ordinary Wells The ordinary well is simply a hole dug or drilled below the water table. A miniature lake forms in the hole; its surface is at the water table. The level of this tiny lake rises in wet seasons and falls in dry seasons. Pumping is necessary to move the water upward to the surface. To be successful, the well must extend below the water table into permeable materials (aquifers) in areas where the quality of the water is satisfactory. An *aquifer* may be defined as any sediment or rock from which supplies of groundwater may be obtained.

Artesian Wells Any drilled well, especially a deep one, is popularly thought to be an artesian well, but this usage is incorrect. An artesian well is one in which confined water under pressure rises above the aquifer which

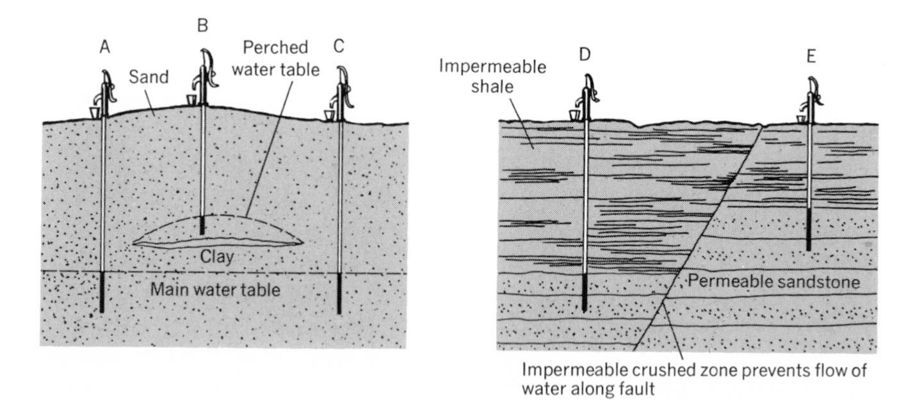

Fig. 30-5. *Successful wells vary in depth. At the left is a perched water body. A local water table may occur above the main water table in an area and cause variation in the depths of successful wells. A lens of impermeable clay is located in permeable sand in the sketch at the left. The main water table is situated a number of feet below the clay. However, the clay traps any water that enters the ground above it and starts to move downward. Thus a local saturated zone forms above the clay. This perched water table would flatten out if frequent replenishment did not occur. At the right, successful wells on either side of a fault. Water enters the sandstone in adjoining areas where it crops out at the surface.*

contains it (Figs. 30-6 and 30-7). Commonly this is above the local water table. Artesian systems are less common in igneous and metamorphic rocks than they are in sedimentary rocks.

When an artesian system is first tapped, prodigious quantities of water may be made available. However, in dry areas the annual addition of water

Fig. 30-6. *A town water supply illustrates some of the conditions necessary for an artesian circulation.* [*After Leopold and Langbein.*]

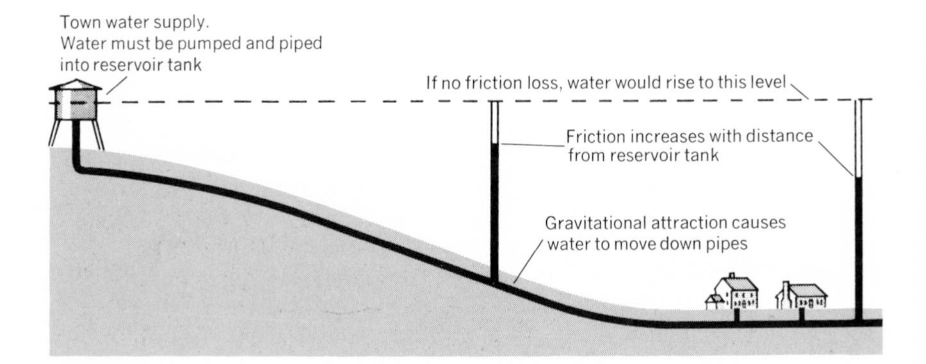

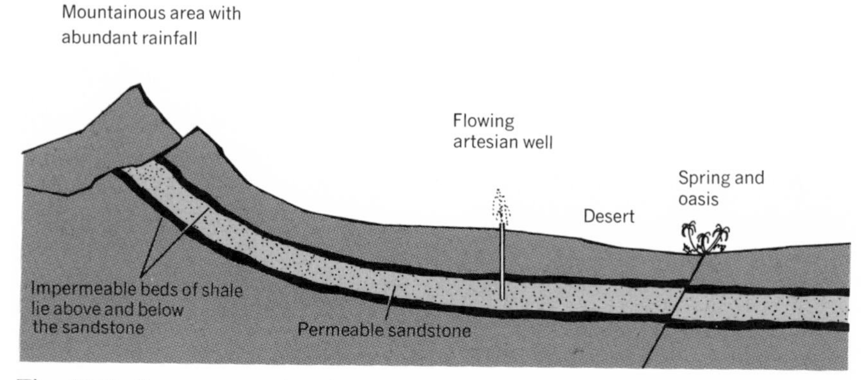

Fig. 30-7. *One type of Artesian system. Necessary conditions are (1) a permeable bed (aquifer), (2) impermeable beds above and below the permeable one, (3) a dip (slant) to the beds steep enough to establish a hydraulic gradient, and (4) exposure of the permeable bed at the surface so that it receives supplies of ground water.—Rain falls on the outcrop of the permeable sandstone which is at a high altitude (at the left in the sketch). Some of this rain sinks into the sandstone and oozes slowly down the bed under the pull of gravity. When the sandstone is tapped by the well, pressure forces water up the hole. In this instance, pressure is sufficient to cause the well to flow.—At the oasis, water has seeped upward along the fractured fault surface. If the water table is located above the permeable bed, no impermeable layer is needed below it.*

to the system from intake areas may be relatively small, and the water in an aquifer may be virtually irreplaceable—almost a one-crop resource. If we use the slowly accumulated supply rapidly, we are like the man inheriting a large sum of money who spends his fortune quickly and recklessly and is poor for the rest of his life. Conservation is necessary to maintain a balance between the natural supply and the total yield.

The important Illinois-Wisconsin artesian system formerly supplied the city of Chicago with a large portion of its water requirements. Permeable sandstones dip southward toward Chicago from their intake area in Wisconsin. A drop of water is estimated to require about 200 years to travel the several hundred miles from the intake area to the Chicago wells. Under such conditions, Chicagoans were drinking "fossil" water.

To find out the rate of movement of groundwater in such a system, we may put a colored dye down one well. When the water in a second well downdip first shows the color, we register the time and measure the distance between wells to get the rate of travel. Several such measurements made at strategic locations give an average for the entire artesian system.

Oases in desert areas may cause wonder. Why should vegetation grow in small areas scattered here and there throughout a hot desert of sand and bare rock surfaces? Many oases are located at places where an artesian water system locally cuts the Earth's surface (Fig. 30-7).

WATER CONSERVATION

Groundwater forms a great reservoir for the storage of water supplies. It commonly extends downward from the water table to depths of 2000 to 3000 feet below the surface. However, much of this water is in impervious beds, the quality is not everywhere desirable, and the supply is not inexhaustible. Conservation must thus be practiced.

Of the 240 billion gallons of water estimated to have been used each day in 1960 in the United States (this total excludes that used in the generation of electric power and thus merely diverted through turbines), an estimated 46 percent was used for irrigation, another 46 percent by self-supplied industries, 1 percent for rural use, and 7 percent for public water supplies, a major portion of which was also industrial. For example, 300 gallons of water are needed to make a barrel of beer, 10 gallons must be used to refine 1 gallon of gasoline, and one large paper mill uses more water than a city of 50,000 inhabitants. Thus a very small amount of the total water used in the United States actually goes for the multitude of familiar personal and household purposes such as bathing, drinking, and washing dishes and cars. The average cost of this water at the faucet for municipal supplies was about 10 cents per ton in 1960, which makes water cheaper than almost anything else except the air we breathe. The average cost of irrigation and industrial water is much less than this.

Total resources of surface and subsurface water in the United States are enormous. However, shortages exist in some areas and for some special uses. Not all areas have the desired amount (neither too much nor too little) at the desired time. These problems will be aggravated as population and per capita use increase. Air conditioning and supplemental irrigation are new uses requiring enormous volumes of water.

Mining vs. Sustained Yield In some areas water is being withdrawn from the ground much more rapidly than it is being replenished. Water tables have dropped alarmingly and continue to fall in such areas. Both artesian and nonartesian sources are affected. Water is virtually an irreplaceable natural resource in these areas, and its withdrawal from the ground can properly be called mining (e.g., mining occurs in Texas, Arizona, and California). There are plainly social and political aspects to any decision to mine such stored water—entirely, in part, or not at all. Decisions require an accurate knowledge of the facts involved.

Long-term figures are needed. A certain part of Kansas experienced falling water tables for a number of years, enough to suggest that the water was being mined. But subsequent heavy rains raised the water tables higher than they had been twenty years earlier. There is some evidence that most recharge in arid and semiarid areas occurs during an occasional wet year.

If groundwater is removed more rapidly than recharge takes place, the Earth's surface may sink as compaction occurs in an aquifer. The permeability of the aquifer may be seriously and permanently damaged. An extreme example is Mexico City. Some 9000 water wells were developed between 1910 and 1952 in unconsolidated materials underlying the city. In some places the surface has subsided as much as 16 feet. Some buildings are sinking into the ground—the former entrance at ground level is now a basement reinforced against collapse. The maximum rate of sinking in 1953 was 20 inches per year. New buildings are located on firmer ground, on piles, or have a floating type of foundation; but the problem remains very serious. Every time a person has a drink of water in Mexico City, they say, the city sinks a little. Withdrawals of oil may likewise cause subsidence.*

Withdrawal and Recharge Groundwater cannot be conserved simply by non-use. Some of it will drain out of underground reservoirs as part of the

* "Geological Subsidence," S. S. Marsden, Jr., and S. N. Davis, *Scientific American,* June 1967.

Fig. 30-8. *Withdrawal and recharge of ground water. In the area illustrated, the water table originally was near the surface. A swamp was located in one place, and a shallow domestic well yielded water when pumped. A large industrial concern moved to the area and drilled a deep well for a large water supply. Pumping was almost continuous and resulted in the development of a large cone of depression. The swamp and domestic well dried up.—The problem was solved when the company was persuaded to return its uncontaminated water to the swamp, where it seeped into the ground and raised the level of the water table.*

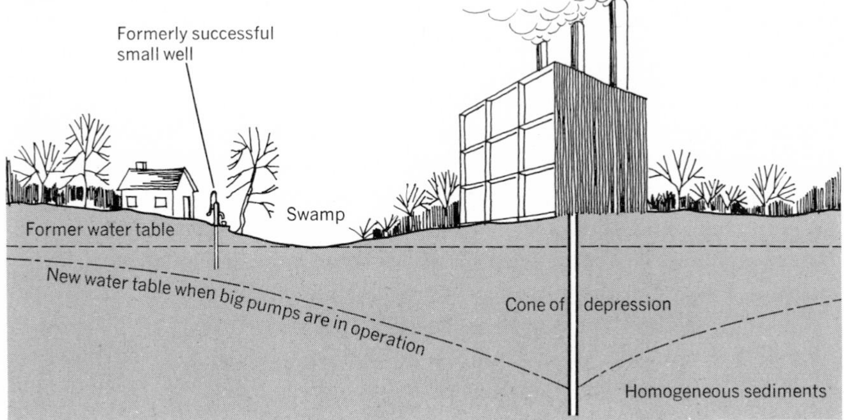

water cycle. When a well is pumped in homogeneous materials, a *cone of depression* develops as the water table is lowered around the well (Fig. 30-8). A balance occurs when the recovery by inflow during pumping equals the amount withdrawn. This is considered a safe yield.

The size and shape of a cone of depression depend upon permeability and rate of pumping. As water is pumped from a well, additional water enters it from the sides and bottom of the hole. As water leaves interstices near a well, it is replaced by water from adjoining and overlying interstices. In their turn, these are filled. This transfer eventually reaches the surface of the water table, which is thus depressed in the area encircling the well. Water moves down the inward-sloping sides of this cone of depression toward the well, and it moves more rapidly as the slope steepens. This steepening continues until equilibrium is attained. As pumping slows down and stops, the cone of depression gradually flattens out and eventually disappears. A rapid withdrawal, or withdrawal from a less permeable aquifer, causes a steep-sided cone of depression. At the other extreme is a wide, shallow cone of depression associated with a slow rate of pumping and permeable materials. Commonly a cone of depression does not extend outward more than a few hundred yards; however, neighboring cones may overlap.

Natural recharge occurs in areas of abundant rain and snow, especially where the percentage of runoff is limited, but replenishment of groundwater by natural recharge is not uniform throughout an area. Just as there are cones of depression, there are the reverse—water table mounds. The mounds tend to flatten out if new supplies are not added frequently. Lakes and ponds do not cause recharge in humid areas because they themselves are fed by ground water. The floors of such lakes do not need to consist of impervious materials because they occur below the water table and all interstices are filled with water.

However, in an arid region the water table is commonly highest beneath a stream channel, and recharge occurs when water occasionally flows down the channel and filters into the unsaturated materials below.

Artificial recharge has been practiced successfully in many areas. In one method, water is forced under pressure down a well from which it slowly oozes into adjacent rocks. However, the local water supply should not be contaminated (Fig. 30-9). Polluted water can be purified by percolation through rocks and sediments. At one city, the slow lateral filtration of sewage-polluted river water through about 500 feet of sediment purified the water.

A second method of artificial recharge is water spreading in which waste water is directed into permeable ground, ditches, and shallow basins. Occasional cleaning operations are necessary to prevent clogging. Stream flow

Fig. 30-9. *Contamination of wells. The shallow dug well A was unwisely located downslope from a cesspool C_1 and therefore received contaminated drainage from it. The owner then drilled a deeper well at B. This well tapped layers of soluble limestone inclined toward it from the direction of the lower cesspool C_2. Water dissolved openings in the limestone and flowed, unpurified by slow percolation, to where it was drawn out through well B. The owner of the two wells must either relocate his cesspool or dig a shallow well somewhere near B.*

may also be regulated by dams. During floods water may rush through an area without much opportunity to sink into the ground for future use. Stream flow may also be diverted from place to place.

Increasing Usable Water Supplies The volume of water stored in underground reservoirs is many times greater than the capacity of all surface reservoirs. Usable supplies of water, both surface and subsurface, may be conserved in many ways. Water may be used and re-used with very great savings if circulation is arranged according to purity requirements. Following such a plan, a California steel plant consumes about 1400 gallons of water per ton of steel instead of the 65,000 gallons more commonly consumed elsewhere.

Reclaimed water may be used for some purposes. For example, Bethlehem Steel uses great quantities of water available after wastes from the city of Baltimore have been treated at sewage disposal plants. In Santee, California, reclaimed water fills a number of man-made lakes used for swimming, boating, and fishing. According to city officials, there is no more reason for throwing away dirty water than there is for throwing away a dirty shirt—each can be laundered and used again.

The presence or absence of vegetation has great and controversial effects upon water supplies. The water used by waste vegetation in seventeen western states may equal $1\frac{1}{2}$ times the annual flow of the Colorado River, but how much of this could be saved by destruction of the waste vegetation without causing harmful erosion is uncertain.

Contrary to many opinions, forestation does not always increase usable supplies. In some experiments deforestation actually increased the total quantity of surface and subsurface water available in the area, because the amounts used by the growth of vegetation each year are saved. Snow may also fall to the ground rather than collecting on trees and sublimating. In some tests, usable trees have been removed in strips in a forest. Snow reaches the surface in the cut-over strips and may then be blown under trees in adjoining tree belts. Shade from these trees causes the snow to melt later in the spring when more of the water can sink into the ground following the rapid runoff produced by spring rains. Coniferous forests may be converted to hardwoods, and dense growths may be thinned to increase snow-water supplies. In the test basins, as natural reforestation occurred, the total runoff decreased. Of course, the soil must not be exposed to dangerous erosion.

The Problem of Seabrook Farms Some years ago Seabrook Farms in southern New Jersey faced a daily problem of getting rid of as much as 10 million gallons of water polluted with vegetable scraps. Although foreign matter was removed by filtration and chlorination, objections were made to the addition of this material to the local water supply. The farmers then tried to return the water to the earth in unfarmed acres, but a spraying operation on an unused field soon changed the surface to a sandy soup. Next the sprayers were moved to the edge of a scrubby white-oak forest, and for two days the ground soaked up water as fast as it was sprayed on. Then 50 inches of water were poured on in the next 10 hours. Still the earth acted like a sponge and absorbed all of the water. The forest floor had the capacity to soak up almost limitless amounts of water, which filtered down through alternating layers of sand and gravel to some black muck at a depth of about 200 feet.

Seabrook Farms then solved its water disposal problem by scattering rotary sprayers strategically through the woods. Their precipitation averaged more than 50 inches per week and thus made this one of the wettest forests in the world. Yet the forests seemed no wetter than before except for droplets of water glittering on the leaves. Readings at observational wells showed an upward movement of the water table, but enough springs developed at the surface to establish an equilibrium. Pure potable water is flowing out of the woods as fast as the waste water is being pumped in.

In 1956, after six years of increasingly successful operation, the infiltration capacity and permeability of the soil had actually increased. Purification by filtration is complete, and there is no contamination or offensive odor from surface or subsurface waters in the area.

The effect on the growth of vegetation of the tremendous increase in the amount of water received by the forest within the range of each nozzle has been startling, and an important discovery has been made. For years it was thought that water in excess of ground saturation was unusable by plants. However, the amazing junglelike growth of plants—both rates and sizes have increased enormously—seems to disprove this idea. Preliminary experiments with food plants such as peas and potatoes indicate favorable results.

Long Island's Water Supply Intensive use of groundwater occurs on Long Island, which is about 115 miles long by 20 miles wide; maximum altitude is 420 feet. The island is underlain by clays, sands, and gravels which dip gently seaward and contain three water-bearing permeable beds. As these beds do not extend to the coast, groundwater is supplied to them only by precipitation, which averages 40 inches or more per year. An estimated one-third to one-half of this rainfall sinks to the water table. About one-half the water used on Long Island is pumped from the mainland.

Fresh groundwater extends for considerable depths below sea level (in

Fig. 30-10. *Fresh vs. salt water on a small island. The area is assumed to be underlain by permeable homogeneous materials. As the area became settled (left), many shallow wells were drilled, and great volumes of water were used. The water table thus fell enough throughout the island to let salt water enter the deep well on the opposite side of the island. The cone of depression commonly does not extend more than a few hundred yards from a well, but measurable lowering of the water table has occurred several miles from a well during heavy pumping in highly permeable sediments.*

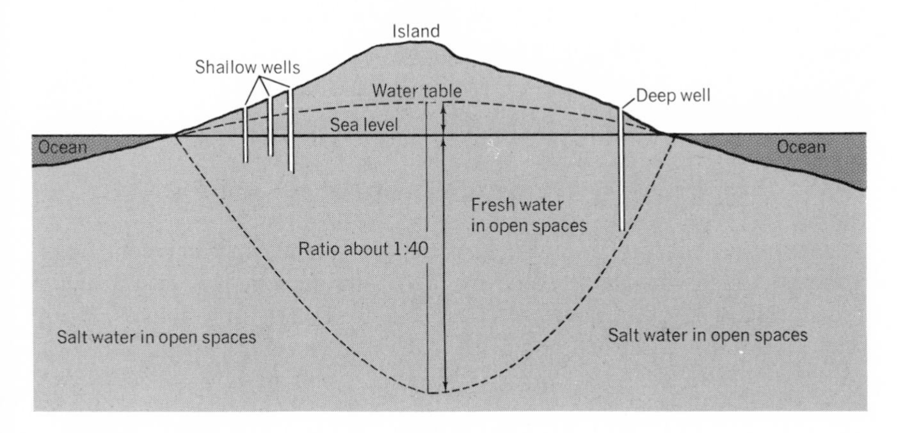

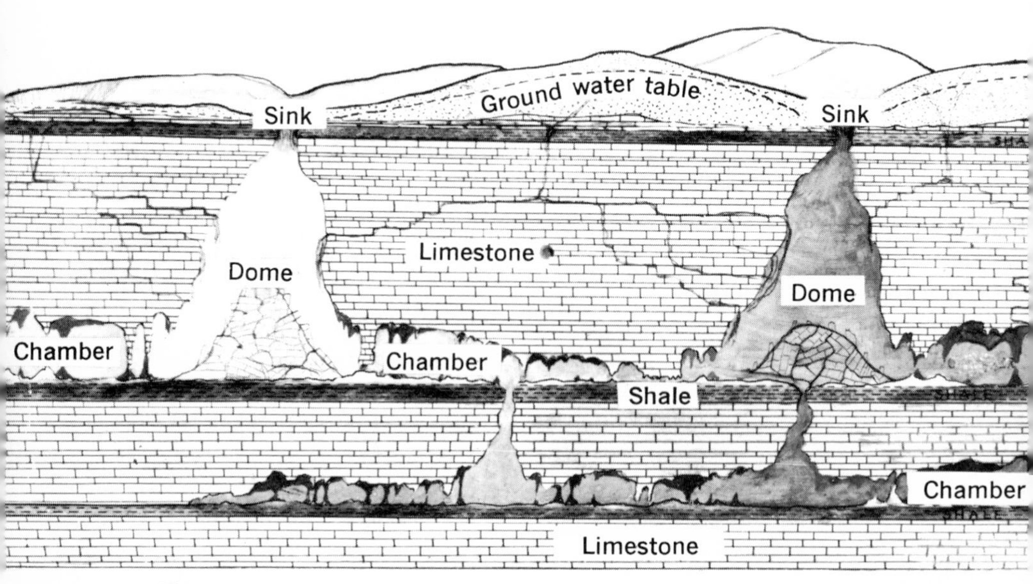

Fig. 30-11. *Some features of a limestone area. Shown are caverns and sinkholes forming and a natural bridge, once the roof of a cave.* [*Courtesy of the American Museum of Natural History.*]

spaces among the sediments) and rests upon the heavier salt water beneath it. Difference in densities is slight. Every foot of groundwater above sea level may indicate an extension of approximately 40 feet beneath sea level (Fig. 30-10). For example, if the water table is 20 feet above sea level at one point, fresh water may extend to a depth of 800 feet below sea level at that point. However, if the water table is lowered 1 foot by pumping, the bottom of the groundwater zone rises 40 feet. Therefore, excessive withdrawal results in contamination by salt water of the deepest wells. This occurred on Long Island. Conservation practices have since corrected this situation, but new problems have also developed.

SUBSURFACE WATER IN SOLUBLE ROCKS

Limestones, marbles, and related rocks are soluble in water containing carbon dioxide, which may come from the atmosphere or from decaying organic material. Commonly rocks are cut by cracks or fractures along which subsurface water can move. Thus the more soluble parts of the rock are slowly dissolved, and the cracks widen. Solution is greatest at the surface, where carbon dioxide in the water is most abundant. Thus cracks

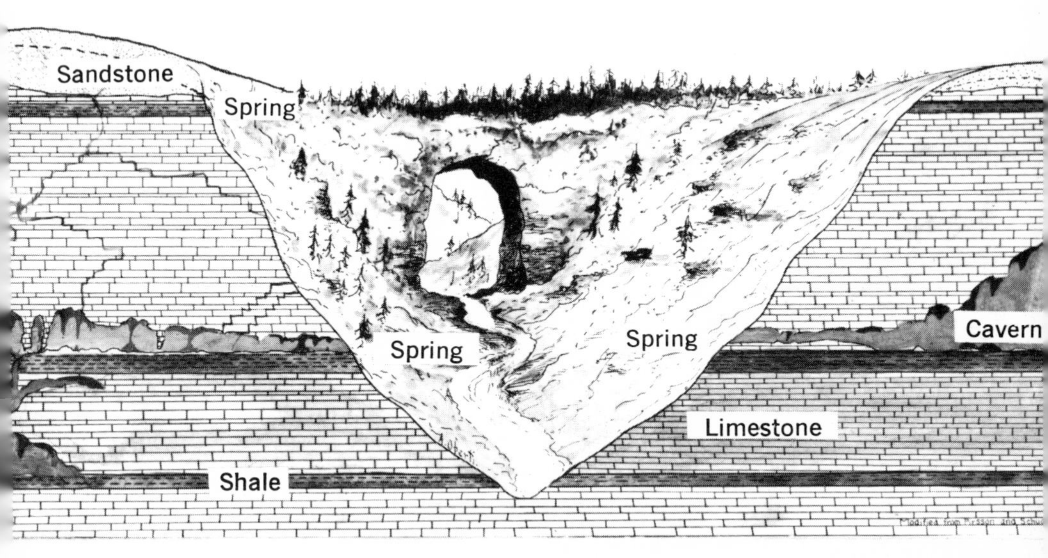

tend to widen near the surface. Where two cracks intersect, a funnel-shaped opening may develop and grow larger. Finally the roof may collapse to form a sinkhole (Figs. 30-11 and 30-12). Sinks may be tiny pits or large holes more than a mile in diameter. If floored with impervious material or cut below the water table, sinkholes may contain lakes or ponds. They may be circular in outline or quite irregular in shape, and several sinks may coalesce to form a very large one.

Most of the groundwater that sinks below the surface in limestone areas tends to move downward and laterally; eventually it joins streams in the deepest valleys of the area. The soluble rocks may be cut by two sets of parallel cracks which intersect at about right angles. Although limestone is commonly dense, individual layers are separated by bedding surfaces along which water can move slowly. Thus a three-dimensional network of cracks and fissures comes into existence to carry groundwater through the rocks in the area. The dissolving action of the water along these cracks and bedding planes gradually enlarges them into a series of interconnected passageways; later these enlarge into caves and caverns by continued solution and roof collapse. Surface streams may enter such a system and further cave formation. Undermined parts of the system may collapse and enlarge the caverns or make holes at the surface. Remnants of such collapsed roofs may form natural bridges.

The dissolving action of water may take place above or below the water

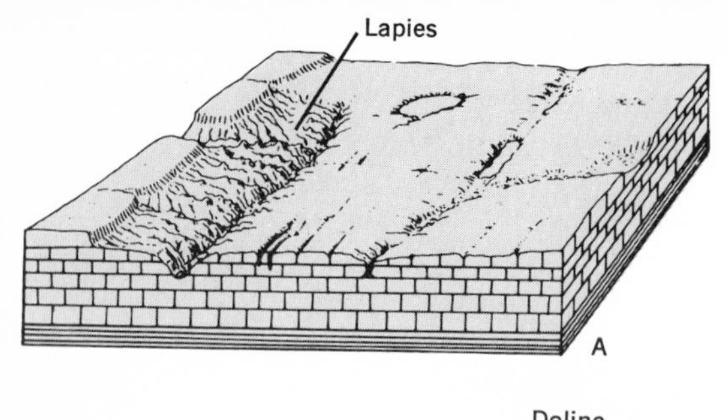

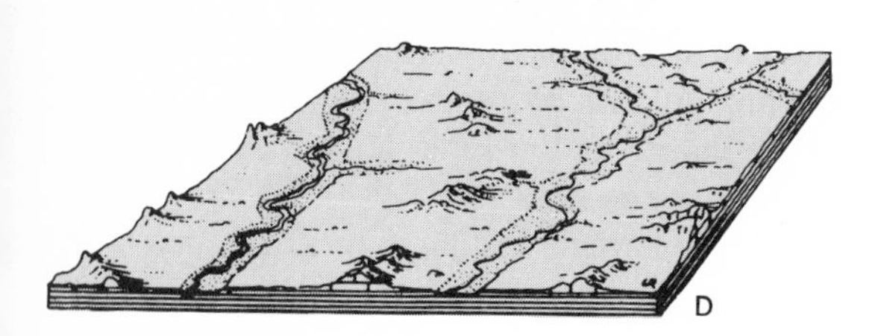

Fig. 30-12. *Possible stages of evolution of a karst landscape. Increasing relief and cavern development are followed by decreasing relief and the removal of the limestone mass. Note the sinkholes and natural bridge. Exposed limestone surfaces in A are deeply grooved and fluted into lapies. The doline in B is a deep, steep-walled, funnel-like sinkhole. In C, a polje is an open, flat-floored valley formed by the coalescence of several sinkholes. (Karst is the name of the type region in Yugoslavia.) [From A. N. Strahler,* Physical Geography, *John Wiley & Sons, Inc.]*

table. As the master stream in an area cuts downward, it lowers the water table. Tributary streams follow suit, and many caves lose their underground streams. Beautiful caverns, visited each year by interested tourists, have thus been formed by the slow, unspectacular action of groundwater (Fig. 30-13).

Sometimes water containing calcium carbonate in solution descends along a crack to the roof of a cave. At this place it partly evaporates and a very little calcium carbonate is precipitated out of solution. Centuries of this process result in iciclelike projections from the roof of the cave (stalactites). Part of the water may fall to the floor and there evaporate a little, thus precipitating more calcium carbonate. Eventually another iciclelike body is

Fig. 30-13. *An immense stalagmite (Giant Dome) in the Big Room of Carlsbad Caverns, New Mexico. Smaller stalactites in great numbers descend from the roof, here at a depth of 750 feet.* [Santa Fe Railway photograph.]

formed which grows upward from the floor (stalagmite*). Stalactite and stalagmite may meet, and form a column. Precipitation may also occur in layers to form terraces in many weird shapes.

Solution in limestones in northern Kentucky may remove an average of about 1 foot of rock every 2000 years, but solution is much slower in other areas and in other rocks. Nevertheless, in a North Carolina area underlain by igneous rocks (diorite), solution may be removing about 1 foot every 28,000 years.

* It has been said that stalactites and stalagmites have a certain resemblance to "ants in the pants." The "mites" go up and the "tites" come down.

Exercises and Questions, Chapter 30

1. Water table:
   (a) What is meant by the water table? Why does its depth beneath the surface vary both areally and seasonally?
   (b) In a particular location in an area with a climate similar to that of New York, when would you expect the water table to be farthest from the surface during a year? Why?
   (c) Discuss capillarity as a factor in the movement of water beneath the surface.

2. Porosity and permeability:
   (a) Describe some of the types of openings that may occur in rocks.
   (b) Some materials have low permeabilities and yet are quite porous. Why?

3. Describe the shapes of the cones of depression that will develop under each of the following conditions.
   (a) A well located in permeable materials is pumped at a moderate rate.
   (b) A well located in materials of a rather low permeability is pumped at a rapid rate.

4. Artesian systems:
   (a) Describe the necessary conditions.
   (b) How does an artesian well differ from an ordinary well?
   (c) What is meant by the statement that one drinks "fossil water" from a certain artesian well?
   (d) How can tritium be used to measure the rate of movement in an artesian system?

5. Describe two different types of conditions that could cause neighboring wells to vary considerably in depth in order to obtain adequate supplies of groundwater.

6. Water conservation:
   (a) When the water supply of a town or city becomes low during a drought, a common practice to conserve water is that of not serving a glass of water with a meal in a restaurant. Discuss the usefulness of this practice.
   (b) What is meant by the "mining" of groundwater? What effect does this tend to have upon an aquifer?
   (c) Describe ways in which usable supplies of groundwater may be increased.
   (d) Polluted river water moves laterally through sand in one area, through fractured limestone in another area, and through well sorted pebbles in a third area. In which area is the water likely to become drinkable? Why?

7. Why are dowsers commonly able to locate supplies of groundwater?
8. Precipitation of minerals from water solutions:
   (a) Why are deposits around hot springs commonly thicker than those around cold springs?
   (b) Why are veins of calcite common in limestones?
9. Discuss some of the problems involved in obtaining a supply of fresh groundwater on a small island in the ocean or in a coastal area.
10. Describe some of the topographic differences that can develop in two regions in warm humid climates if one area is underlain directly by soluble rocks and the other is not.

31 Glaciation

THE story of the recent Ice Age is one to capture the imagination. It is a gift to us from the combined labors of many geologists who have traced the courses of the glaciers that covered almost a third of the Earth's land surface during their maximum extent. At this time about half of the states in the United States were covered by ice—some completely, others partially. Evidence of the former presence of the ice is abundant, widespread, and clear; it is little affected by geologic forces, such as weathering, erosion, burial, and metamorphism. Approximately 10 percent of the Earth's land surface is still covered by glacier ice, most of it in the Antarctic ice sheet (about 5,000,000 square miles) and the Greenland ice sheet (about 670,000 square miles). Neither of these ice sheets was much larger when glaciers were most widespread during the Ice Age. Therefore, glaciers have shrunk the most or disappeared entirely in the upper middle latitudes, especially in the Northern Hemisphere.

A *glacier* has been defined as a mass of ice that has formed from compacted, recrystallized snow and refrozen meltwater, which is moving or has moved, and which lies entirely or partly on land. A *valley glacier* (Fig. 31-1) moves down a channel previously eroded by a stream. It may be wide and thick and tens of miles long, with numerous tributary glaciers, or it may be quite narrow and short. Commonly the ice fills the entire lower part of a valley from wall to wall. In contrast, an *ice sheet* is a very extensive mass of ice that spreads radially outward from a central area and rests like a blanket upon the surface; it is not confined to a single channel. Continental ice sheets may be a mile or more thick and completely bury millions of square miles of the Earth's surface, except for an occasional projecting mountain peak. The term ice cap may be used for a small ice

Fig. 31-1. *Spectacular glacial scenery. Note the long straight U-shaped valley (fjord) that formerly contained a valley glacier.* [*Photograph, Royal Canadian Air Force; courtesy, Dept. of Mine and Technical Surveys.*]

sheet. A *piedmont glacier* (Fig. 31-2) is gradational between a valley glacier and an ice sheet and forms along the base of a mountain by the coalescence of a number of valley glaciers. If growth continues, both areally and vertically, and if a number of piedmont glaciers coalesce, an ice sheet may eventually be formed.

The Ice Age may have begun approximately $\frac{1}{2}$ to $1\frac{1}{2}$ million years ago. The magnitude seems to be of this order. Glacier ice covered the northern part of the United States and Canada as recently as 10,000 years ago. During the Ice Age, great changes took place at and below the Earth's surface both in the glaciated areas and outside of them. The changes involved climates, animals and plants, crustal movements, sea water, erosion, and deposition. The Pleistocene (Greek: "most recent") epoch is the name given by geologists to the time in which these changes occurred. It is more or less synonymous with the popular term, Ice Age.

Fig. 31-2. *Malaspina glacier in Alaska.* [*Photography, U.S. Air Force: N. 59° 57′, W. 140° 33′.*]

Evidence of Glaciation Louis Agassiz, a Swiss naturalist (1807-1873), is often credited with the development of the glacial concept about a hundred years ago; but a number of Agassiz's contemporaries, scientists of earlier days, and many observant individuals deserve to share the credit, as is often true of key principles in science. A basic geologic assumption—the present is the key to the past—was involved in the discovery.

The basic idea came from people who lived in areas of existing glaciers. Erosion and deposition by the ice created features which could be seen and compared readily with similar features located beyond the present margins of the glaciers. From the plain hint that glaciers once extended farther down their valleys than they now do, there slowly grew a realization that great sheets of ice actually once covered large parts of the Earth's surface. This idea was accepted after much argument and discussion.

Agassiz himself was at first skeptical of the concept of widespread glaciation. He studied existing glaciers and supposedly glaciated areas with the intention of disproving the whole idea. However, the field evidence was so convincing that Agassiz became the leading figure in spreading and developing the glacial concept.

There were rival theories. British geologists had noticed large foreign boulders scattered widely, some at higher altitudes than their distant sources. They had observed scratched and polished bedrock surfaces and had studied widespread nonstratified deposits of boulders, gravel, sand, and mud (Fig. 31-3). These were quite unlike the sorted layers of sand

Fig. 31-3. *Till. This term is used for material that has been transported by glaciers and dropped directly from the ice into unsorted piles; it is unstratified glacial drift. Although till is unsorted and has no obvious structures, measurements do show in some instances that the longest dimensions of boulders are aligned parallel to the direction of ice movement. Relatively few large boulders will be present in tills eroded from such weak rocks as shales.* [*Photograph, G. Termier.*]

and silt they could see being deposited by streams. Furthermore, these deposits occurred on hilltops as well as in low areas. Such unusual phenomena called for an unusual answer. Being familiar with the sea and not with glaciers, they theorized that a great flood had submerged the entire area; icebergs had carried the huge boulders and had scratched the bedrock surfaces. The unsorted sedimentary deposits, which they termed "drift," had been picked up and dumped by a great rush of waters—perhaps the Noachian deluge. Similar ideas had developed in other countries.

WORLDWIDE EFFECTS OF PLEISTOCENE GLACIATION *

Glaciation has occurred several times during the Earth's geological history, but that of the Pleistocene is the most recent and provides the clearest record. Still extant glaciers were then larger, and regions which today are without ice were then covered by it. Three huge ice sheets formed more or less simultaneously in the Northern Hemisphere. (1) At its maximum size, the North American (Laurentide) Ice Sheet covered about 5 million square miles and extended southward to Long Island and the channels of the Ohio and Missouri Rivers (Fig. 31-4). It joined the glacier complex that covered parts of the Rockies and the West Coast. (2) A smaller

* R. F. Flint, *Glacial and Pleistocene Geology,* Wiley, New York, 1957, Ch. 1.

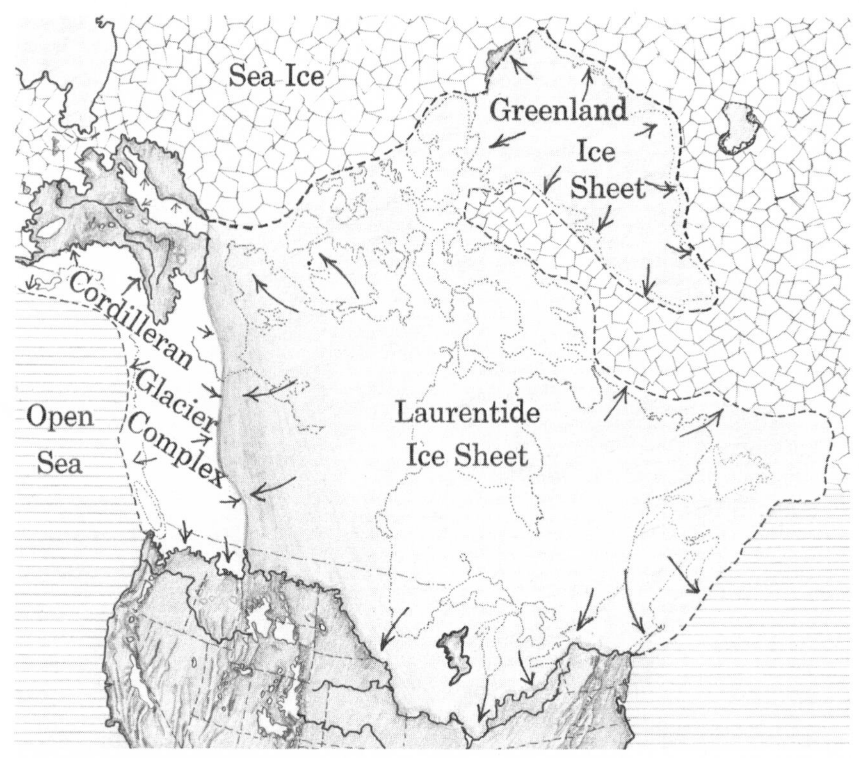

Fig. 31-4. *The maximum extent of glaciers in North America during Pleistocene glaciation. Details are somewhat generalized because of insufficient information, especially in the north. Boundaries between glacier ice, sea ice, and open sea are conjectural but are based upon modern analogies. Arrows show the general directions of ice movement. A winding east-west line that begins on Long Island, continues along sections of the Ohio and Missouri Rivers, and extends to the West Coast marks approximately the southernmost advance of the glaciers into the United States. Smaller isolated glaciers occurred south of this line, especially in the west.* [*After Longwell, Knopf, and Flint,* Physical Geology, *John Wiley & Sons, Inc.*]

Scandinavian Ice Sheet once extended from northwestern Europe, through Denmark, to the northern parts of Germany and European Russia. (3) A Siberian Ice Sheet also existed, but less is known about it. With the exception of Antarctica, comparable ice sheets were not present in the Southern Hemisphere.

Glacial and Interglacial Ages The ice sheets and other glaciers probably advanced and retreated four times during the Pleistocene (Fig. 31-5), but fluctuations of lesser magnitude were superposed on the four main ones and have led to a difference of opinion concerning the total number. A

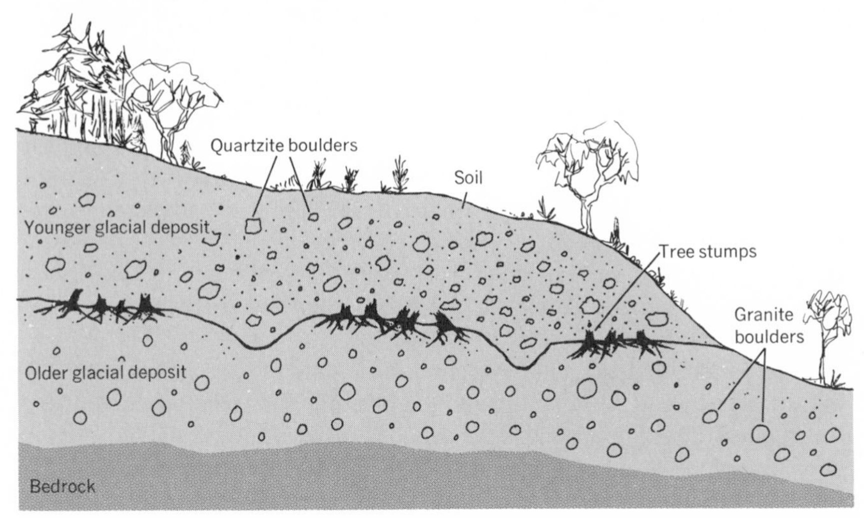

Fig. 31-5. *Evidence indicating multiple glaciation: (1) Granite boulders predominate in the older glacier material, whereas boulders of quartzite predominate in the younger glacial material. Evidently the glacier came from two different directions (bedrock in source areas is different) on its two trips to this area. (2) Stumps of trees at the top of the older deposit indicate that sufficient time elapsed between glaciations for soil to develop and a forest to grow. (3) Two V-shaped valleys are shown eroded in the older glacial deposit. These are filled by the younger material. (4) The soil is thickest on the right where the younger material does not rest upon the older. In this location weathering has been producing soil for a longer time: burial under the younger glacial deposit stopped the process elsewhere. [Modified from Leet and Judson.]*

glacial age occurred near the beginning of the Pleistocene epoch and was followed by an interglacial age. Then in succession came a second glacial age, a second interglacial age, a third glacial age, a third interglacial age, a fourth glacial age, and finally the present, which may or may not be a fourth interglacial age.

The interglacial ages probably lasted much longer than the glacial ages, and conditions during parts of the interglacial ages were probably somewhat similar to present conditions. In fact, during at least one interglacial age, temperatures were higher than they are today. The latest glacial maximum apparently occurred about 18,000 years ago, but this was followed by a number of lesser fluctuations. Temperatures apparently increased rather abruptly about 11,000 years ago, and glaciers began to shrink relatively rapidly.

Cores extracted from sediments that accumulated on the deep-sea floor also provide evidence of multiple glaciations (p. 700).

Volume Changes in Sea Water Glacier ice is metamorphosed snow which has fallen as part of the water cycle. Because the total amount of water at the Earth's surface in the Pleistocene was probably about constant, the volume of ocean water must have been decreased by the amount that was frozen on the lands as ice. Thus, sea level fluctuated as the volume of glacier ice fluctuated during the Pleistocene, and it may have dropped about 300 to 500 feet below its present level during the maximum extent of the glaciers. The effects of local crustal movements must be eliminated in making such estimates, and this is sometimes difficult to do. On the other hand, if all glacier ice on the Earth today were melted, sea level might rise approximately 100 to 200 feet.

Various kinds of geologic evidence show that sea level fluctuated during the Pleistocene, although exact amounts are somewhat uncertain; e.g., the channel of the Hudson River can be traced seaward on the Atlantic floor for some 90 miles to a depth of about 240 feet (pp. 766-767). Presumably such a channel was cut when this area was land instead of sea floor. Similar channel continuations occur in association with other rivers. Furthermore, near their mouths, rivers such as the Hudson flow on thick piles of sediment; their bedrock floors are now as much as 200 feet or more below sea level.

An erosional bench associated with shallow-water fossils has recently been discovered off the eastern United States in water nearly 500 feet deep; presumably it was formed by the surf along a coast at sea level.

During one or more of the interglacial ages in the Pleistocene, sea level was also higher than it is now. Like the dirt line in a tub, certain features furnish evidence of the level where the water formerly intersected the land (e.g., surf-eroded cliffs and benches, deltas, beaches).

Crustal Movements Caused by Growth and Shrinkage of the Ice Sheets The Earth's crust sags beneath the tremendous weight of an ice sheet. This downwarping probably occurs simultaneously with a slow plastic flow of rock materials outward from the overloaded glaciated areas. At the depths where this flow occurs (probably within 100 miles of the surface), temperatures are high enough for rocks to be deformed slowly without breaking. Apparently the weight of a growing ice sheet squeezes rock material from beneath it, and the space thus made available permits the Earth's surface to bend downward. As the ice sheet melts, the crust warps upward, and rock material at depth returns slowly under the glaciated areas. Similar warping of the crust was associated with each advance and retreat of the ice sheets. Because the ice retreated only recently in terms of geologic time, the Earth's surface is still rising slowly in these areas.

Evidence of crustal movement as a direct result of glacial loading and unloading is especially well exposed in Scandinavia. First, the elliptically shaped area known to have been glaciated coincides closely with the area in which uplift has occurred and continues today. Second, the maximum rate of uplift occurs in the central portion of this area where the ice apparently was thickest and the crust was depressed the most. The rate is about 3 feet per century at present, but it seems to have been faster in the past. This is also the location where the greatest amount of uplift has already taken place—perhaps 900 feet or more—and where the largest upward movement is still to come—possibly another 900 feet or so. Thus the center of a dome-shaped upwarp coincides closely with the center of the former ice sheet.

Estimates concerning the amount of uplift still to come depend upon assumptions concerning the strength of the crust, the density of the displaced rock far below the surface, and the thickness of the ice sheet. Although these are not known with any precision, the relative specific gravities of the ice (less than 1) and the displaced rock (probably about 3.4) indicate that several thousand feet of ice are needed to depress the surface a thousand feet.

Pluvial Lakes in Nonglacial Areas Climatic changes also occurred during the Pleistocene in nonglaciated areas because climatic belts were shifted equatorward and narrowed. Today areas such as Utah and Nevada are semiarid and arid; but during the glacial ages of the Pleistocene they were cooler and moister and had less evaporation. Accordingly, present-day lakes were much larger during the glacial ages, and lakes existed in basins that are now dry. Such lakes have been called *pluvial lakes* (Latin: "of rain"). Apparently the pluvial lakes grew and shrank several times during the glacial and interglacial ages.

Migrations of Animals and Plants As the Pleistocene ice sheets gradually advanced, animals and plants apparently migrated to warmer climates; but this must have been a slow process, each generation inhabiting an area a little nearer the equator than its predecessor. As evidence, fossil reindeer and the woolly mammoth have been found in southern New England, walrus along the Georgia coast, white spruce in Louisiana, and musk-oxen in Arkansas and Texas. However, some of these organisms may have lived under climatic conditions different from those of their present-day descendants, and plants are more reliable indications of climate than large mammals. Musk-oxen in Texas do not imply arctic conditions there.

As the ice fronts retreated, the animals and plants could return to their former habitats, again a process stretching across the centuries. Therefore, in sediments deposited at any given location, assemblages of fossil animals and plants indicating a cool climate may be found above or below fossil assemblages indicating a warm climate.

Conclusions concerning the migrations of flowering plants and climatic changes can be drawn from a study of fossil pollen grains and other spores (palynology). Pollen grains are the microspores produced in flowering plants as part of the reproductive process. Countless pollen grains are distributed widely during the spring and summer by air and water currents, and they form a veritable pollen rain upon the Earth's surface that is perhaps most familiar as a yellowish scum on ponds and lakes. These tiny, nearly indestructible organic particles eventually settle to the bottom of a pond along with other sediment and organic matter in a process that is repeated year by year. Since the pollen grains tend to remain intact and can be identified, they provide a method of determining which flowering plants were abundant in a certain area at a certain time in the past.

A core is taken of the sedimentary layers that piled up year by year in the bottom of a bog or elsewhere. The pollen is separated from the mineral matter at different horizons, and the individual grains from different plants are identified and counted. The ages of the pollen grains can be determined by using the carbon-14 method on the enclosing organic matter in each horizon. Correlations and comparisons can then be made among cores taken at different places. The advance or retreat of a particular type of forest across a wide region as the climate of this region gradually changed can be mapped and dated.

At any one level in any one area, certain types of pollen grains predominate, and it is generally assumed that the parent plants of these grains were most abundant in that area at that time in the past. However, some plants are more prolific producers of pollen than others, and the pollen grains of some plants are better airborne travelers than others.

Pleistocene history may be elucidated by study of sediments deposited at great depths on the sea floor. As erosion is limited in this environment, a nearly complete record should be furnished by the accumulated sediments. Instruments have been devised which can extract cores more than 70 feet in length from the ocean floor at depths of thousands of feet. These cores have shown alternations of the remains of warm- and cool-water animals (Foraminifera are particularly useful). Such tiny organisms lived near the surface of the ocean, but when they died their shells sank to the bottom. The shifts suggest changes in the temperature of the water near the surface that are related to the glacial and interglacial ages. Fossil pollen and spores have also been found in clays deposited in the deep ocean.

Summary of Major Worldwide Effects of the Ice Age Pleistocene glaciation originated when continents were more extensive and mountainous than they had been for much of the geologic past. At this time temperatures probably dropped (8°C is the estimated decrease in temperate areas), more precipitation occurred as snow, and less snow melted each summer.

Because a large ice sheet requires many years to disappear, a glacial age can be said to end at different times in different places; e.g., the North American Ice Sheet retreated northward from the Great Lakes region about 10,000 to 11,000 years ago. After that date, temperatures apparently rose gradually until some 4000 to 6000 years ago, at which time it may have been warmer than today. Lesser fluctuations have followed and continue today; e.g., some areas which were inhabited during the Middle Ages have since been covered by valley glaciers. In the 1900's the ice has been retreating from these areas and exposing buildings that had remained buried for several centuries. However, a new cooler trend may even now be taking place.

What of the future? The fluctuations of the last several millennia certainly emphasize the difficulty or impossibility of long-range forecasting.

GLACIERS AND GLACIER ICE

In certain areas on the Earth's surface, some of a winter's snowfall does not melt or sublimate during the succeeding summer. In the polar regions such places occur at every altitude, but in equatorial areas they occur only near the tops of high mountains. Thus each year in certain parts of these areas, the snow piles higher and higher. The total weight of many snowfalls, combined with the moisture that is present from melting snow near the surface, gradually cause the snow at the bottom of a pile to take on a granular texture similar to that of many snowdrifts during the spring. Each snowflake is changed to a tiny granule of ice. Continued compaction by more snowfalls changes the granular material to solid ice. Commonly this ice is layered, evidence that it was formed from many separate falls of snow.

When sufficient ice has piled up on a slope, gravity tends to move it gradually to a lower altitude (Fig. 31-6). Commonly a thickness of 100 feet or more of ice is needed before movement occurs; this depends upon the steepness of the slope, temperature of the ice, and other factors.

Glacier ice can be considered a rock—in fact, two kinds of rock: snowflakes and tiny ice granules are sediments; the stratified ice that has undergone little downslope movement is a sedimentary rock; and glacier ice, recrystallized into an interlocking texture, is a metamorphic rock.

Besides low temperatures, abundant snowfall is needed for glaciers to develop. Northern Siberia is extremely cold, but the small amount of winter snowfall wastes away during the short cool summer, and glacier formation is prevented.

Glacier ice consists of a mass of interlocking grains which can flow

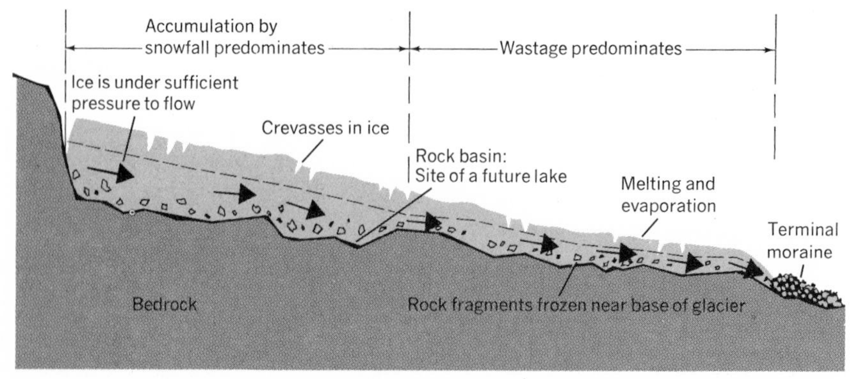

Fig. 31-6. *Nourishment and wastage in a valley glacier. Glacial gouging will be most pronounced in parts of the valley underlain by less resistant or badly fractured rocks. These are sites for future lakes.*

under pressure, apparently by more or less continuous changes in the individual particles. Stresses set up by the weight and motion of the ice cause individual ice grains to grow in size, to twist and rotate, and to slide along planes of weakness in the crystals much as a pack of cards slides when pushed by the hand. Perhaps skating illustrates how part of the flow takes place. A skater actually glides on a thin film of water which pressure forms momentarily beneath the blades of the skates. When ice changes to water, the volume is reduced; therefore, pressure on ice promotes liquefaction. The water freezes again immediately after the pressure is removed, as when a snowball is made. Stresses in glacier ice thus cause momentary liquefaction, the transfer of water downslope, and immediate freezing.

A type of movement akin to faulting also occurs within a glacier and is prominent near its terminus: thin plates of ice are shoved forward and upward, one above the other, along surfaces that are concave upward. In addition to this internal movement, a glacier also moves as a unit by sliding along the surface beneath it. Ice at the bottom of a glacier moves plastically, but the upper part, some 100 feet or more in thickness, is too brittle and rigid to flow under the lower pressure there. It is carried by the flowing ice below. The rate of flow of glacier ice varies from a maximum measured rate of over 100 feet a day to a general average of a few feet or less per day. The surface of a valley glacier moves more rapidly in the center of the valley than along its sides because less friction occurs there. This fact was first determined by placing a straight line of stakes across a valley glacier and measuring the movements of individual stakes.

A glacier tends to shrink in volume more by overall thinning than by upslope retreat of its terminus. When the ice has thinned sufficiently, the brittle zone extends to the bottom, and flow stops.

GLACIAL EROSION AND DEPOSITION

Three short words succinctly describe the erosional and depositional work of glaciers: plow, file, and sled. The rock-studded bottom of a glacier is an effective file or rasp and polishes, scratches, and abrades the surfaces over which it moves. A glacier gathers debris from the mantle or bedrock in its path and from rock materials that slide upon it from valley walls, a process that is less important for ice sheets. Glaciers also pluck out blocks of bedrock loosened by the freezing and thawing of water in fractures beneath the ice. Like a plow, a glacier pushes and shoves loosened debris ahead of it.

In arctic areas, where some glaciers end at the water's edge in cliffs, the transported materials are exposed, evidence that a glacier acts as a

Fig. 31-7. *A Pleistocene glacier left this grooved, striated, and polished rock wall at the side of its valley in Glacier Creek, Flathead National Forest, Montana.* [*Photo by Asahel Curtis. Courtesy, U.S. Forest Service.*]

sled. Large blocks of ice may break off the end of such glaciers to form icebergs which later melt and deposit their rock loads far from land in places ordinarily reached by only the finest of materials. The origin of such deposits, fine-grained strata containing an occasional large boulder, puzzled geologists for some time.

Blocks of rock frozen in the basal ice of a glacier are dragged under great pressure over the underlying bedrock and polish, scratch, and groove its surface (Fig. 31-7). The scratches (striae) are nearly parallel in any one area. The finest scratches are probably formed by silt and sand grains held like an abrasive between a boulder and the bedrock. Very large boulders have been transported by glaciers.

Drainage changes are numerous within glaciated areas and outside them. The Great Lakes and Niagara Falls (Figs. 31-8 and 31-9) formed during the Pleistocene and owe their existence partly to glaciation. Lakes and swamps are very abundant in glaciated areas; the depressions in which they occur may result from glacial gouging or deposition, or by a com-

Fig. 31-8. *The relationship between rock structures and topography in the Great Lakes region. Note that most features in the eastern half have equivalents in the western half: e.g., Saugeen and Door Peninsulas, Green and Georgian Bays, Lake Michigan and Lake Huron. Stratified rocks form a broad shallow structural basin and are arranged somewhat like a stack of saucers.* [*A. K. Lobeck, Things Maps Don't Tell Us, The Macmillan Company, 1956.*]

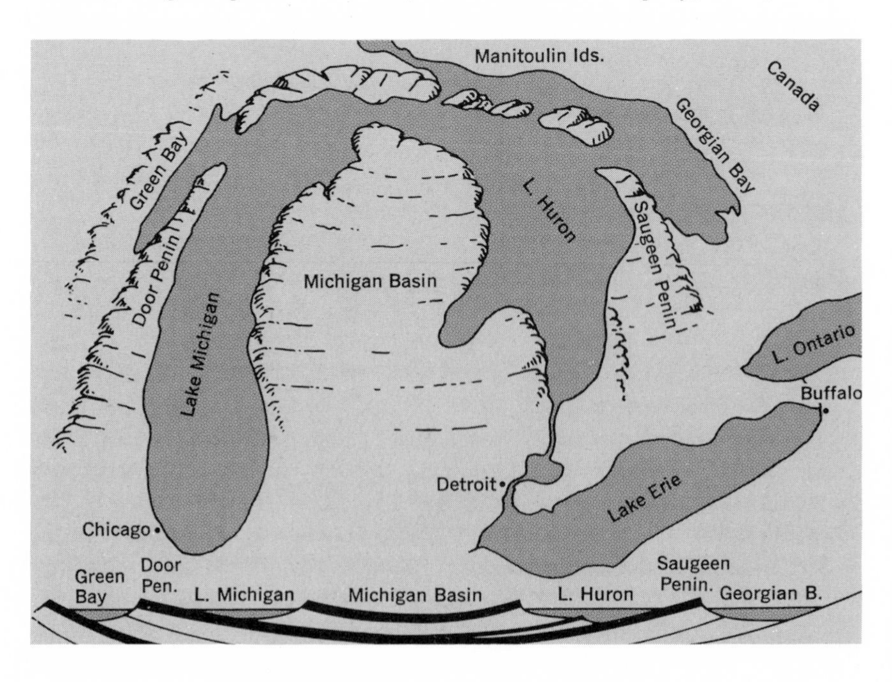

Fig. 31-9. *Niagara Falls and part of Niagara Gorge. After the ice sheet retreated northward from this area, the Great Lakes formed, and water flowed from Lake Erie to Lake Ontario across the Niagara escarpment to produce Niagara Falls. The falls when first formed were about 7 miles downstream from their present position. During the past century the falls have retreated about 4 feet per year. The 7-mile gorge that has been formed as the result of this retreat indicates that this process has been going on for thousands of years. The bedrock in the area consists of sedimentary rocks that dip gently toward the south. A resistant formation forms the lip of the falls at present. It is undermined by the turbulent waters at the foot of the falls and breaks off occasionally in large blocks. As the falls retreat upstream, the dip of this resistant formation takes it to a lower altitude. Thus the falls should eventually change into a rapids before Lake Erie is reached (in an estimated 25,000 to 30,000 years).* [*Niagara Falls Chamber of Commerce.*]

bination of the two. Many rivers or parts of rivers (e.g., the Ohio and the Missouri) owe their locations to the former presence of an ice sheet which blocked and ponded their drainage. Such rivers then eroded new channels in low areas located along the edge of the ice sheet.

Valley glaciers reshape their valleys by widening, deepening, and straightening them. The typical V-shaped cross section of a youthful mountain valley is altered to a steep-sided, flat-floored U-shape (Fig. 31-10). Lakes are common in glaciated valleys. Constricted parts may be deepened more than wider sections, and excessive gouging also occurs in sections of a

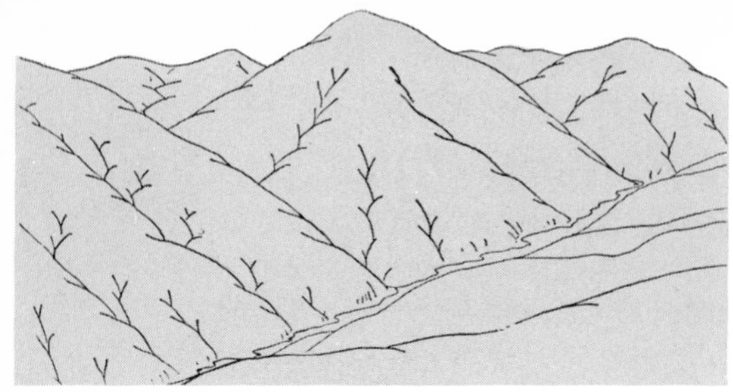

A. Before glaciation sets in, the region has smoothly rounded divides and narrow, V-shaped stream valleys.

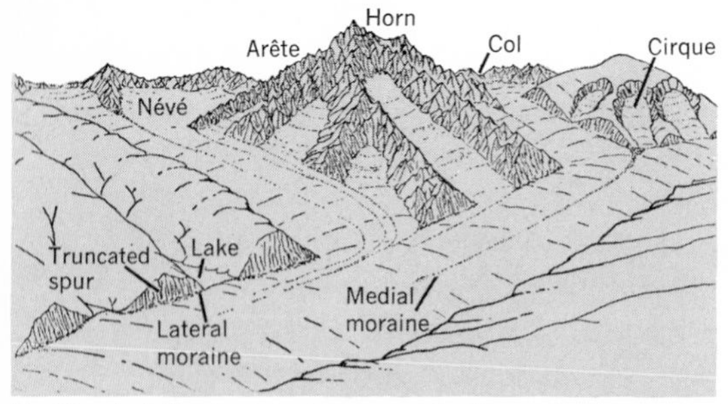

B. After glaciation has been in progress for thousands of years new erosional forms are developed.

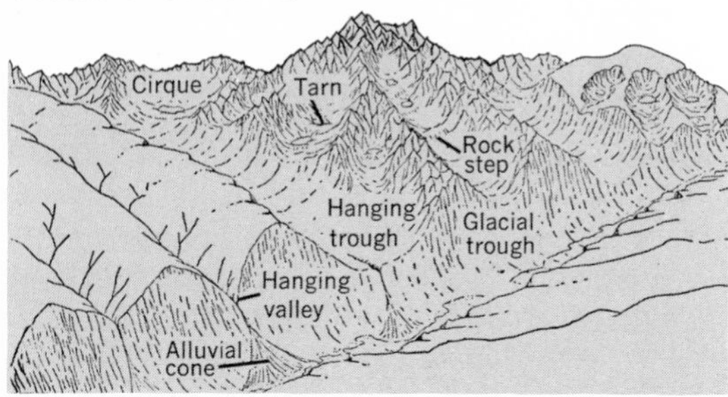

C. With the disappearance of the ice a system of glacial troughs is exposed.

Fig. 31-10. *Land forms produced by valley glaciers.* Top, *before glaciation sets in, the region has smoothly rounded divides and narrow, V-shaped stream valleys.*—Middle, *after glaciation has been in progress for thousands of years new erosional forms are developed.*—Bottom, *with the disappearance of the ice a system of glacial troughs is exposed. [After W. M. Davis and A. K. Lobeck; reprinted with permission from A. N. Strahler,* Physical Geography, *John Wiley & Sons, Inc.*]

bedrock floor that are cut by abundant cracks or are underlain by weak rocks.

Tributary glaciers contain a small amount of ice relative to the main valley glacier into which they flow. Therefore, they are unable to cut downward as rapidly. When the ice eventually wastes away, such tributaries may be left hanging above the main valley, and beautiful waterfalls may occur at their mouths.

Glaciers deposit their loads in a number of different ways, each with a characteristic shape and appearance. Rock materials rich in clay may be plastered onto the surface over which the ice is moving and produce low, streamlined hills called *drumlins* (Fig. 31-11). These commonly occur in swarms. New York State alone is estimated to have over 10,000 drumlins.

Moraines are topographic features ranging from rather thin, irregular

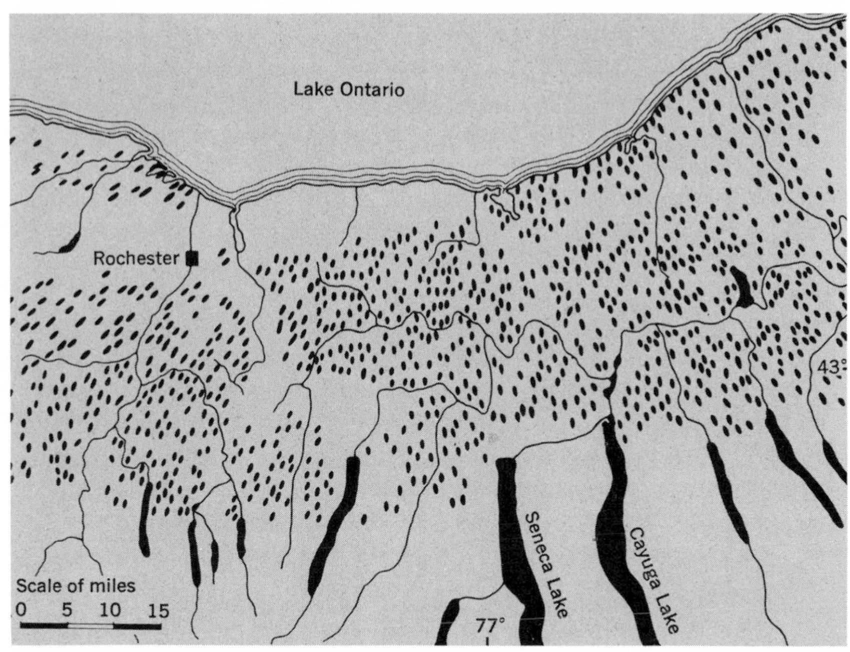

Fig. 31-11. *Drumlins are very numerous south of Lake Ontario. The Finger Lakes are situated in valleys that were eroded by northward-flowing streams. These valleys were then widened, deepened, and straightened by southward-flowing ice. Bedrock consists chiefly of thin-bedded shales that dip gently toward the south. Thus ice flowed across the upward slanting edges of these beds and blocks were plucked out readily. The bedrock floor of Seneca Lake is 1000 feet or more below the lake surface. The mouths of tributary streams "hang" above the bedrock floors of some of these lakes.* [*Monnett and Brown,* Principles of Physical Geology, *Ginn and Company.*]

sheets to low ridges. They consist of rock debris that has been transported and deposited by glaciers, and commonly the material has not been reworked much by streams. *Lateral* and *medial moraines* are associated with valley glaciers (Figs. 31-12 and 31-10) but may also form near the margin of an ice sheet where the ice flows outward through valleys. Glacial erosion steepens valley walls and thus increases the amount of rock debris that tumbles, slides, or is avalanched onto the margins of the ice to form low ridges or lateral moraines. The lateral moraines of tributary glaciers subsequently become the medial moraines of a main glacier.

Terminal and *recessional end moraines* may be deposited either by valley glaciers or by ice sheets. Glaciers advance and retreat, but a retreating glacier is not actually flowing backward toward its source. The forces of nourishment which make a glacier larger oppose the forces of wastage

Fig. 31-12. *Convict Lake area along the eastern front of the Sierra Nevada. The lake (altitude 7583 feet above sea level) is located in a basin formerly occupied by a valley glacier which flowed out upon the valley floor for about 1½ miles as shown by the moraines: terminal moraine T, lateral moraines L,L, and recessional moraines* R_1, R_2, *and* R_3. *The outlet creek has cut through the moraines on the left.* [*Fairchild Aerial Surveys.*]

which make it smaller. If nourishment gains the upper hand, the terminus of a glacier advances. When wastage exceeds nourishment, the front of a glacier recedes. If forward flow and backward wastage are in equilibrium, the front of a glacier remains in the same location, even though the ice continues to flow forward from the source region. New supplies of rock debris are continuously brought by the moving ice to this location and are dumped in a process that reminds one of debris piling up at the end of a conveyor belt. This material is augmented by that previously shoved ahead of the ice. A ridge of material—an end moraine—thus forms along the margin of an ice sheet or terminus of a valley glacier. At any one point the ridge is oriented at an angle of about 90 degrees to the direction of flow. In a valley glacier, the end moraine is in part a continuation of the lateral moraines and forms a crescent because ice flows fastest in the central part of the valley.

End moraines tend to be discontinuous ridges; gaps may develop during formation and streams flowing outward from a glacier may subse-

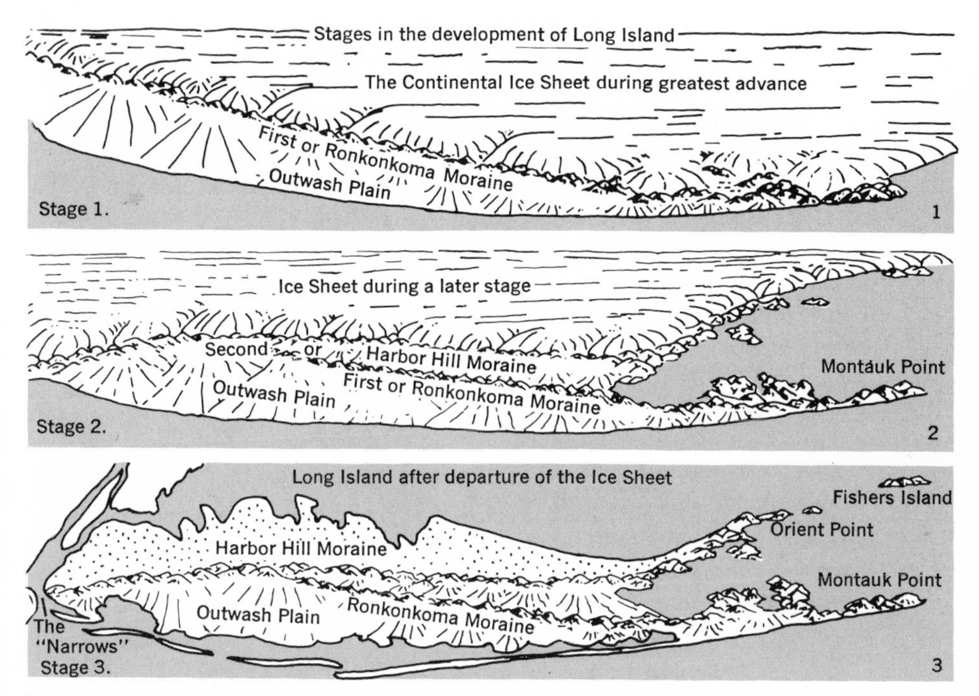

Fig. 31-13. *End moraines on Long Island. The Ronkonkoma end moraine formed first (stage 1). Lake Ronkonkoma is located in a kettle hole in the middle of this moraine. An outwash plain is shown. The Harbor Hill end moraine is shown in stage 2. Its formation caused the destruction of the western end of the Ronkonkoma moraine (stage 3). These two moraines form the two "tails" of Long Island: the Ronkonkoma moraine forms Montauk Point; the Harbor Hill moraine forms Orient Point and some of the islands toward the east and northeast. [After A. K. Lobeck, Example 27,* Things Maps Don't Tell Us, *The Macmillan Company, 1956.*]

quently remove parts of a ridge. A widespread outwash plain of stratified drift may extend from it beyond the ice margin (Fig. 31-13).

After the terminus of a glacier has retreated some distance, equilibrium between nourishment and wastage may again be achieved, and a second end moraine thus forms (recessional). Thus several end moraines may form; the terminal moraine is the oldest.

Depressions are common in glaciated areas, and some form at locations where a block of ice becomes separated from the crest of a glacier as it thins and retreats. Outwash may then be deposited around the block or even bury it completely. When the block subsequently melts, a depression called a kettle hole is formed.

Glaciers, even thick ice sheets, have less effect on the surfaces over which they move than is popularly credited to them. They tend to modify the shapes of major topographic features such as mountains and valleys rather than create them.

The direction in which an ice sheet moved can be determined in a number of ways. Scratches, grooves, and drumlins tend to be aligned parallel with the direction of ice movement, whereas end moraines are elongated about at right angles to this direction.

Indicator stones provide additional evidence. Unique rock types may occur as ice-transported boulders. If their distant sources have been located, the transported boulders are called *indicator stones*. Elongated fan-shaped areas which contain numerous ice-transported boulders of a unique kind are known as *boulder trains*. Such boulders were picked up by a glacier as it moved over a relatively small source outcrop. Later they were scattered in a fan-shaped area by irregular forward movements of the ice. For example, one of Finland's most important copper deposits was found by tracing the source of some ice-transported copper-bearing boulders, and a dozen or so diamonds have been found in glacier drift in Wisconsin, Michigan, Indiana, and Ohio. The source is located somewhere north of the Great Lakes.

ORIGIN OF PLEISTOCENE GLACIATION

In the past, glaciation seems to have occurred only when lands were high. At the present time the land areas of the Earth apparently are more extensive and higher than they were throughout much of geologic history. Yet extensive glaciation has not occurred every time that lands have been high and widespread. Moreover, glaciers advanced and retreated several times during the Pleistocene while the lands remained high.

According to Flint, a drop in mean annual temperature at a time when lands are high may bring on a period of glaciation. A rise in the mean annual temperature would cause recession of the glaciers. This seems to furnish a possible explanation of the origin of Pleistocene glaciation as well as of some glaciations of the remoter past. However, other hypotheses explain the onset of Pleistocene glaciation in a quite different manner. Furthermore, the widespread glaciation that occurred near the end of the Paleozoic era cannot be explained in this manner: at this time, ice sheets apparently formed in Australia, India, Madagascar, South Africa, and

Brazil. We must list the question of origin with the unsolved problems of geology.

Exercises and Questions, Chapter 31

1. What kinds of evidence indicate that an ice sheet once covered an area?

2. What kinds of evidence support the concept of multiple glaciation during the Pleistocene (i.e., the alternation of glacial and interglacial stages)?

3. Discuss some of the major effects of glaciation during the Pleistocene:
   (a) Changes in sea level and in the volume of ocean water.
   (b) Crustal movements (isostasy and isostatic adjustment).
   (c) Migrations of animals and plants.
   (d) Growth and shrinkage of pluvial lakes.
   (e) Erosional and depositional changes produced directly by glaciers—such effects are chiefly modifications of topographic features already in existence.

4. Has Pleistocene glaciation ended?

5. How can sediments transported and deposited by streams commonly be distinguished from those carried and dropped by a glacier?

6. Discuss some of the factors involved in the movement of glacier ice.

7. How do glaciated mountain valleys tend to differ from nonglaciated valleys?

8. End moraines:
   (a) What is an end moraine?
   (b) Why are many end moraines curved like crescents?
   (c) Distinguish between a terminal end moraine and a recessional end moraine.

9. What is the evidence that parts of the Missouri and Ohio Rivers have resulted from glaciation?

10. Why are lakes common in glaciated areas?

11. How can pollen grains, in combination with carbon-14 age determinations, be used to study plant migrations?

12. How can cores of sediments extracted from the ocean floor aid in a study of Pleistocene glaciation?

13. How can the direction of flow of an ice sheet be determined?

32 Volcanos and related phenomena

ON 20 FEBRUARY 1943 a volcano originated in a cornfield near the village of Parícutin about 200 miles west of Mexico City. Earthquakes in increasing intensity and numbers had occurred during the preceding two weeks. Out of the cornfield spouted steam, ashes, and rocks, and nearby trees caught fire. The debris fell to the surface and built a cone-shaped pile around the opening. It was 30 feet high by the next morning. In one week the volcanic cone grew to an altitude of nearly 500 feet, and within a year it towered about 1100 feet above the surrounding lands (Fig. 32-1). During its growth the volcano erupted gaseous, liquid, and solid material, and lava flowed out from its base and flanks. At first the explosions were like cannon fire and occurred every few seconds. Thuds of falling fragments punctuated this background of explosive sounds. As volcanic debris fell back to the earth, it occasionally smashed against upthrown pieces to add to the dust in the air. Clouds of condensed steam and other gases boiled upward to heights of three miles or more and varied from white to black as their ash content changed. At night the spectacle attained its dramatic peak and became a display of fireworks as glowing clots of erupted lava flew through the air and later bounced and rolled down the flanks of the volcano to streak the slopes with red and orange colors. The cinder cone grew wider as it piled higher, although the slope of its flanks remained at about the same angle—the angle of repose for chunks of volcanic debris.

As the months went by, Parícutin's surroundings were buried beneath dust and lava. Flow after flow issued from the base of the volcano, moved outward for distances up to six miles, and overwhelmed nearby villages. As the flows cooled into dark rock, they were covered by new flows until a superposed pile several hundred feet thick had accumulated around the base of the cone. The dust made a much larger area uninhabitable and was thickest near the volcano (up to 10 feet or more). Occasionally it drifted down on areas more than 100 miles away. In 1952 the volcano Parícutin became inactive, and probably it will not erupt again. Its present height above the former cornfield is about 1400 feet.

Fig. 32-1. *Parícutin volcano, October 9, 1944, showing the horseshoe-shaped remains of a parasitic cone (Sapichu in left center) and erupted material in the foreground. Sapichu originated in October 1943 at the base of Parícutin. It grew to a height of about 350 feet before lava flows destroyed part of the cone. Subsequently it was buried by additional lava flows. Parícutin remained dormant while Sapichu was active. [Fred M. Bullard,* Volcanoes: In History, In Theory, In Eruption, *University of Texas Press, 1962.*]

A volcano consists of two parts: an external hill or mountain which commonly is heaped on the Earth's surface as a conical mass; and a more important internal part consisting of a fissure or cylindrically shaped opening (vent or pipe) or openings which lead from the surface to a source of magma far below (Fig. 32-2). Rock debris and gases may also erupt from the sea (Figs. 32-3 and 27-1).

DISTRIBUTION OF VOLCANOES

Some 500 volcanoes are active now or have erupted during historic times, and thousands of others have been active in the recent geologic past. Erosion has had time to modify them only slightly. These volcanoes tend to be grouped in two great belts, the more pronounced of which rings the Pacific in a "girdle of fire" on the east, north, and west. The other great belt extends from the Alpine-Mediterranean area eastward to the Indonesian states. Earthquakes and young mountainous zones occur in close association with these volcanic belts. Many volcanoes have a linear arrangement on a small scale also, apparently because they are aligned along crustal fractures.

Volcanoes originating in historic times have been confined to volcanic areas. However, geologic history indicates that almost all parts of the Earth's surface have been visited by some kind of igneous activity in the past. Thus no area is immune to volcanic action, but some places are much more susceptible than others, and the locations of such volcano-prone places have varied throughout the Earth's history.

VOLCANIC ERUPTIONS

Volcanoes may explode violently, or lava may flow out at the surface rather quietly. A single volcano may erupt explosively at one time and quietly at another; but in many volcanoes, one type of activity or the other predominates. The explosive activity of a volcano depends upon the amount of gas in the magma beneath it and the manner in which that gas escapes. An explosion cannot occur unless gas is abundant and has been trapped to build pressures to the bursting point. Much of this gas was formerly dissolved in the magma below the volcano.

As magma cools, different minerals crystallize (are precipitated) from the molten solution, but dissolved gases tend to be excluded from such mineral grains. Hence, the volume of liquid material in a magma body dwindles steadily as crystallization proceeds, and the relative proportion of

Fig. 32-2. *Structure section of a volcano. Explosive eruptions alternated with lava flows. Flows issue from the slopes and base of a volcano as well as from its crater.* [*Courtesy of the American Museum of Natural History.*]

Fig. 32-3. *The eruption of the Myojin Reef Volcano, 170 miles south of Tokyo.* [*Official U.S. Navy Photograph.*]

gas in the remaining liquid increases. When the gas content becomes too great, gas bubbles separate from the liquid solution and stream upward in the magma chamber to zones of lower temperature and pressure near its roof. In a fluid magma—one that is very hot or that has a relatively low silica content—such gas may be able to escape readily and continuously without creating explosive pressures. As the gas expands, it forces the magma quietly upward to the surface through cracks. If the magma is viscous (nonfluid) or if the volcanic vent has been plugged by solidification of once molten material, the gas cannot escape and pressures increase until an explosion occurs.

Most major volcanoes may be underlain at a relatively shallow depth (some 3 to 4 miles or less) by a magma chamber in which magma accumulates prior to an eruption (Fig. 32-4). The magma, however, apparently has its origin much deeper below the surface and flows upward to the chamber before the eruption occurs. This magma chamber is probably lenticular in shape. Evidence for its existence is the swelling and shrinking observed before and during an eruption. The entire top of a volcano such as Kilauea (Hawaii) commonly bulges upward for several feet and tilts outward during the months before an eruption; the amount of tilting increases toward the summit. Presumably this occurs as the chamber is filled and inflated by accumulating magma. Shrinkage follows swelling, and the two may alternate somewhat prior to an eruption. The rate of swelling may increase rapidly immediately before an eruption (however, the inflation is not constant), and small earthquakes may become very numerous.

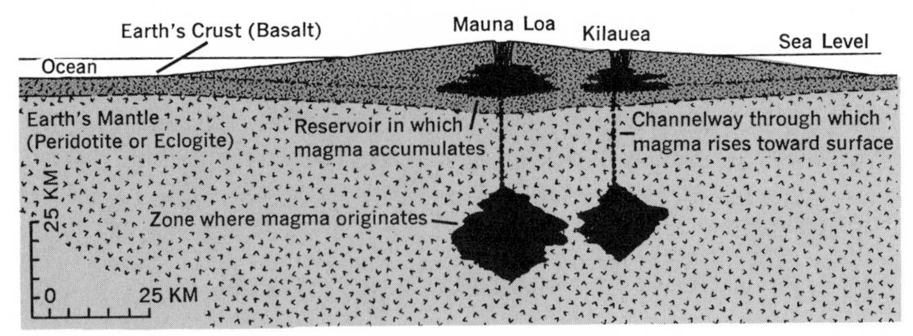

Fig. 32-4. *Hypothetical structure section through Mauna Loa and Kilauea volcanoes, Hawaii.* [*Gordon A. Macdonald, Science, Vol. 133, p. 677, 10 March 1961.*]

This suggests that the times of future eruptions can be predicted fairly accurately for volcanoes of this type, although the level of tilt has also varied from one eruption to the next.

The eruptive action of a volcano may change during its lifetime because the compositions, temperatures, and amounts of the dissolved gases and magma are all variable factors. Furthermore, if the magma moves upward through calcareous rocks like limestone, it may enhance its explosive power by melting the limestone and obtaining an abundant supply of carbon dioxide gas. In addition, the heat of the igneous material may change ground water into steam.

Steam is far the most abundant material given off. Other gases include carbon dioxide, nitrogen, sulfur dioxide, hydrogen, carbon monoxide, chlorine, and fluorine. Heavy rains are commonly associated with volcanic eruptions: huge quantities of steam may condense; and air above a volcano may be blown violently upward causing condensation of much of the moisture in it. Torrents of water sometimes pour down on loose fragments resting on steep volcanic slopes and result in disastrous mud flows.

A particularly destructive and explosive type of volcanic eruption is that known as the fiery cloud (*nuée ardente,* Fig. 32-11C) in which huge quantities of gas and incandescent volcanic debris are blasted upward or laterally through the side of a cone if the volcanic vent is plugged securely. This material forms a dense, seething, turbulent mass that hurtles down a volcanic slope at speeds that may exceed 100 mi/hr. The extreme mobility of such a turbid cloud is caused by the presence of vast quantities of gas that bubbles, froths, and expands continuously and explosively at the reduced pressures at the Earth's surface. Rock fragments are readily transported within the lower part of a fiery cloud, which is much denser than air. Thus it flows rapidly down a volcanic slope. Such a cloud from

Mount Pelée overwhelmed the town of St. Pierre on the island of Martinique in the West Indies in 1902 and killed most of its 28,000 inhabitants.

Lava flows may have a blocky, jagged, highly irregular surface produced by the fragmentation of a hardened crust by still-flowing lava beneath it or by the explosive escape of gases through this crust. On the other hand, the surfaces of fluid flows tend to be relatively smooth and become wrinkled into ropy, rounded, corded shapes.

The surfaces of lava flows commonly contain many spheroidal holes (vesicles) produced by the cooling and hardening of lava around bubbles of gas that had formed and expanded in it. Some lava flows show columnar jointing, a feature that is produced by shrinkage during cooling. Such joints also form in some intrusive igneous masses (Fig. 32-8).

Volcanic bombs (Fig. 32-5) and pumice (Appendix IV) may be produced during some eruptions.

Lava may also well upward through long fissures and pour out quietly without building cones at the surface. The lava flows to lower elevations and fills in depressions. Flow after flow may pile up and from a distance

Fig. 32-5. *Volcanic bombs and impact trails photographed at Parícutin volcano in Mexico. Each bomb was heaved violently into the air as a mass of liquid lava during a volcanic eruption. The turning of the mass and its solidification during flight produced the rounded shape with pointed twisted ends. Some bombs weigh many tons, whereas others are as small as pebbles. If the mass is still plastic when it strikes the surface, its shape is distorted; in fact, some bombs have been molded around objects at the surface such as the limb of a tree. Cracks or holes in a bomb result from the expansion of gas that was once dissolved in the molten material at great pressures.* [*Fred M. Bullard,* Volcanoes: In History, In Theory, In Eruption, *University of Texas Press, 1962.*]

resemble a stack of very thick-bedded sedimentary layers. Commonly the lava is the highly fluid basaltic type. At times in the past, enormous volumes of such material have flooded entire regions and built plateaus.

VOLCANIC TOPOGRAPHY

Characteristically, a volcano has a conical shape, consists of layers of volcanic materials that dip outward from its center, and has a crater at its top (Fig. 32-2). Eruptions throw masses of molten lava and rock fragments high into the air because volcanic vents commonly are vertical. The bulk of the material falls around the opening and forms a cone. In a single eruption large pieces fall first and smaller particles later, thus producing a crude stratification. Successive eruptions vary in magnitude, and there are corresponding changes in the sizes of the fragments thrown out. Between explosions, lava may flow down the surface of a volcano from its top or, more frequently as it grows larger, from openings on its flanks or near its base. The continuous upward movement of volcanic debris from the central opening during an eruption prevents most of the

Fig. 32-6. *The snow-covered summit of Mauna Loa in Hawaii. Its oval base is about 3 miles below sea level (Fig. 32-4). Some 10,000 cubic miles of lava, in flows that may average 10 feet or so in thickness, were probably extruded to form this gently sloping shield volcano which extends nearly 6 miles above its base. Mauna Loa may be the largest single volcano on the Earth. It is capped by an oval caldera about 3 miles long by 1½ miles wide which formed by collapse of the summit portion of the volcano. Smaller craters appear in the foreground.* [*Photograph, U.S. Air Force.*]

material from falling back directly into this central opening; thus a crater forms at the top. When the explosions cease, some material falls into the crater and other debris slumps in from the sides.

Volcanic cones differ widely in shape and size, but three types are most common: *cinder cone, shield,* and *composite.* Cinder cones consist chiefly of erupted fragments and are steep-sided.

Shield volcanoes (Fig. 32-6) have been built largely by the piling up of flow after flow of fluid lava. They form large, dome-shaped, gently sloping masses like Mauna Loa in the Hawaiian Islands (Fig. 32-4). Instead of rising up a central pipelike vent, as is common in continental volcanoes, the lava at Mauna Loa apparently rises through numerous narrow cracks that radiate outward from the center like the spokes in a wheel.

A composite volcanic cone (Fig. 32-2) is formed partly by explosive eruptions, which are chiefly responsible for increasing its height and steepening its summit slopes, and partly by lava that flows forth quietly and widens its base. The symmetry and beauty of some cones of this type are admired throughout the world: Fujiyama in Japan, Mayon in the Philippines, Mount Rainier in Washington, and Vesuvius in Italy (Fig. 32-9).

The circular depressions that cap some volcanoes are much larger than the average crater; in fact, the entire top sections of these volcanoes are

Fig. 32-7. *Shiprock, New Mexico. This volcanic neck and radiating dikes (only one is visible here) are remnants of a once-majestic volcano. The neck extends more than 1000 feet above the surrounding countryside.* [*W. T. Lee, U.S. Geological Survey.*]

Fig. 32-8. *Devil's Tower, Wyoming. According to one hypothesis, this structure is a volcanic neck. The towerlike, erosion-resistant mass of igneous rock stands about 700 feet above its surroundings. Note the excellent columnar jointing which forms under certain conditions when igneous rocks cool and shrink. The columns average about 10 feet across.* [*N. H. Darton, U.S. Geological Survey.*]

missing. Such depressions are called *calderas,* and Crater Lake (Figs. 32-10 and 32-11) is a well known and strikingly beautiful example. The diameter of a caldera is many times wider than that of its volcanic vent.

Many volcanoes contain a system of vertical fractures, now filled by hardened magma (dikes), which lead radially outward from central vents. Deep erosion has exposed these in some areas (Fig. 32-7). Other objects may have been erroneously called volcanic necks (Fig. 32-8).

DORMANT OR EXTINCT?

A volcano may be active, dormant, or extinct. The distinction between a dormant and extinct volcano is difficult or impossible to determine in terms of human lifetimes. Vesuvius near Naples, Italy, illustrates the problem (Fig. 32-9). It is situated on the remnants of an ancient volcano called Mt. Somma. Mt. Somma, with its forest-clad slopes and fertile soil, was believed to be an extinct volcano by the Romans—Vesuvius was not yet in existence. However, in A.D. 79, following several years of frequent earthquakes, Somma exploded mightily, blew off much of its top, and buried two towns, Pompeii and Herculaneum, lying between it and the sea. Little lava was extruded at this time, but enormous volumes of debris and gases were hurled high into the air by the powerful eruptions. Gases killed some of the inhabitants. Others were buried beneath falling debris or under mud flows which formed during the accompanying torrential rains and moved rapidly down Somma's steep slopes.

The buried towns were forgotten until their accidental discovery and

Fig. 32-9. *Mount Vesuvius, partially encircled by the remains of Mt. Somma, and the Harbor of Naples. The upper part of the magma chamber beneath Vesuvius apparently occurs at a depth of about three miles, but the ultimate source of the magma is probably much deeper than this. The three-mile estimate is based upon a knowledge of the layered rocks that occur beneath Vesuvius (broad synclinal structure). These are penetrated by the vent that connects its cone at the surface with the immediate source of molten matter within the Earth. Blocks of these layered rocks are carried upward and occur in lava flows. The blocks estimated to come from a three-mile depth are much altered and indicate proximity to the magma chamber; blocks of overlying rocks are less altered.* [*Courtesy, F. M. Bullard,* Volcanoes: In History, in Theory, in Eruption, *University of Texas Press, Austin, Texas, 1961.*]

subsequent excavation that began during the eighteenth century. Molds had formed of many of the victims, some in restful poses and others in the act of fleeing. Casts have been made from some of these molds. Vesuvius, which now exceeds 4000 feet in altitude, has since been built in the huge hole which formed at the top of Mt. Somma during this eruption.

In the seventeenth century, following a long period of dormancy during which its crater had become overgrown, Vesuvius was again believed to be extinct. However, in 1631 it erupted, and thousands of people were killed. Since that time Vesuvius has been almost constantly active.

Crater Lake Crater Lake (Figs. 32-10 and 32-11) in Oregon lies on the crest of the Cascade Range amid the great volcanic mountains of the Pacific northwest. This is one of the world's most scenic spots, and its

Fig. 32-10. *Crater Lake in Oregon. The circular lake is about 5 miles in diameter and completely enclosed by cliffs which rise 500 to 2000 feet above the water surface, about 6000 feet above sea level. Maximum depth is 2000 feet.* [*Photograph U.S. Air Force.*]

geologic history illustrates some of the events which may take place during the life of a volcano.

Crater Lake occupies a huge circular depression located at the top of a mountain. Bedrock exposed in the cliffs consists of volcanic debris and former lava flows in layers that everywhere dip away from the lake. Glacial deposits are interbedded with the volcanic rocks, and U-shaped valleys can be seen in cross section where they have been truncated by the crater walls. Evidently a radial glacier system existed when the volcano was much higher, perhaps resembling that now present on nearby Mt. Rainier. If the dips of the volcanic strata are projected toward the center of the lake and upward, they suggest that this beheaded volcano once towered approximately 12,000 feet above sea level. However, the top was south of the present lake center. The geologic story of Crater Lake is told in several chapters (Fig. 32-11).

DIKES, SILLS, AND BATHOLITHS

Intrusive igneous rock bodies are classified chiefly on the basis of their shapes and relationships to surrounding rocks (Fig. 32-12); dikes, sills, and batholiths are three common types. Dikes and sills are tabular-shaped masses which formed when magma was squeezed by pressure into cracks

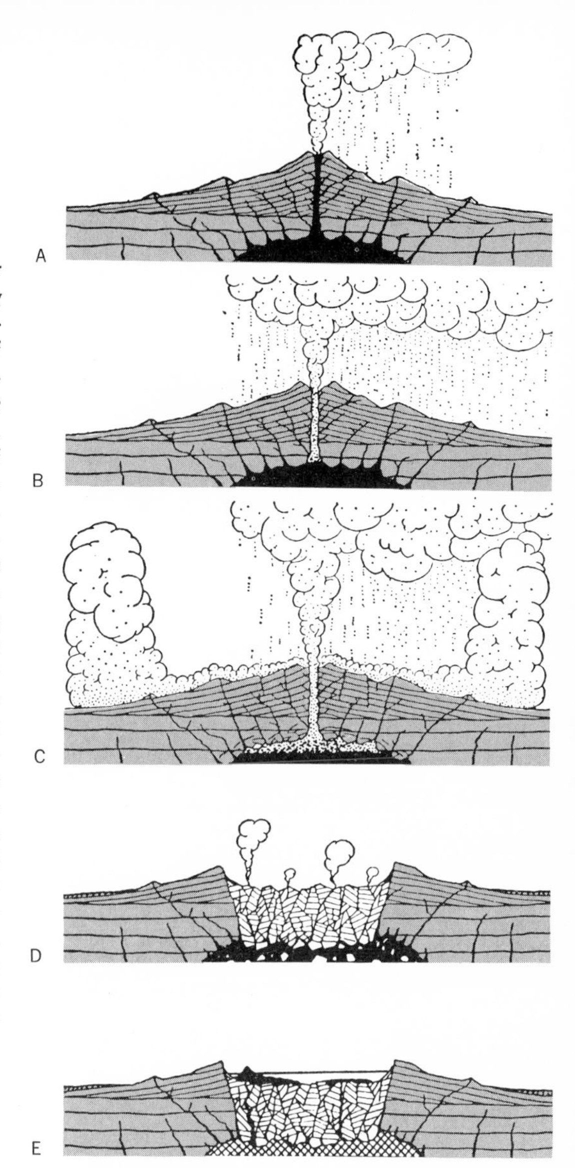

Fig. 32-11. *The evolution of Crater Lake.—A. Beginning of culminating eruptions. Magma is high in the conduit and there is a mild eruption of pumice.—B. Activity increases in violence. Showers of pumice are more voluminous and the ejecta are larger. The magma level lowers to the top of the feeding chamber.—C. Activity approaches the climax. There is a combination of vertically directed explosions and glowing avalanches (nuées ardentes). The chamber is being emptied rapidly and the roof is beginning to fracture and founder. Magma is also being drained from the chamber through fissures at depth.—D. Collapse of the cone as a jumble of enormous blocks, some of which are shown sinking through the magma. Fumaroles appear on the caldera floor.—E. Crater Lake today. Post-collapse eruptions have formed the cone of Wizard Island and probably have covered parts of the lake bottom with lava. Magma in the chamber is largely crystallized. Carbon-14 age determinations of trees buried by the climatic eruptions show that they took place about 6500 years ago.*

or weak zones in rocks. They are shaped like sheets of plywood—two large dimensions and one small. They may be quite large and extend for miles or be very small. If the magma squeezed between two layers in a pile of stratified rocks, it formed a *sill;* if it cut across the layers or squeezed into a crack in nonstratified rocks, it formed a dike. The intrusive rocks may be less resistant to erosion than the materials they intruded, or more so.

A *batholith* (p. 725) is a huge intrusive igneous rock body which tends to become larger at increasing depths; it exceeds 40 square miles in exposed surface area and tends to cut across the structures of the enclosing

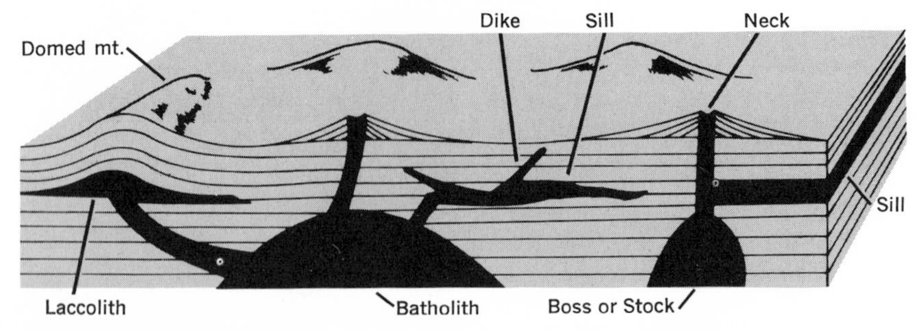

Fig. 32-12. *Types of igneous intrusives. A laccolith resembles a sill, for the magma is squeezed between rock layers; but the laccolith is lens-shaped and domes the rock above it. A stock is a small batholith.*

rocks. Batholiths pose important problems for geologists. They formed far below the surface and are located only on the continents in association with great mountain belts.

Identification of a certain rock outcrop as part of an intrusive igneous rock body indicates that the following events have occurred: (1) the rock was once hot molten magma, (2) the igneous rock is younger than the surrounding rocks into which it intruded, and (3) erosion has removed the rocks which once covered this outcrop.

ORIGIN OF IGNEOUS ROCKS

That some rocks had hardened from a previous molten state was recognized by certain men long ago. They observed lava flowing copiously down volcanic slopes and watched as it cooled into hard slaggy rock. Establishing the igneous origin of rocks no longer connected with active volcanoes was more difficult.

That coarse-grained igneous rocks, such as some granites, were once molten is indicated by several lines of evidence. Certain types of minerals are present, the grains interlock, and the minerals crystallized in a definite order. Similar features can be seen in rocks found on volcanoes, except that individual mineral grains in the granite are larger. Is this merely a coincidence?

In studying the margins of a granite batholith in the field, geologists have noted that along its outer boundaries the granite commonly grades into fine-grained rocks that are nearly identical with the rocks on the volcanic slope. Beyond the margins of the granite are rocks that have been metamorphosed; the effects of the alteration fade out gradually from the granite mass. In addition, granite dikes or sills extend outward from the main body.

However, other granite bodies are known which show gradation: unquestioned granite occurs in their central portions; this grades outward into rocks which look like granite yet show traces of sedimentary stratification; these in turn grade outward into metamorphic rocks and finally into sedimentary rocks. Exposures of this type suggest another origin for granite—that it has formed by the transformation of older sedimentary and metamorphic rocks. This process is called *granitization.* It involves the replacement of pre-existing rocks by emanations from below; wood is petrified in a somewhat similar manner. Thus some granites are metamorphic rocks rather than igneous, and one kind cannot be distinguished from the other in hand specimens. Most geologists agree that granite has formed in both ways and that the chief problems lie in determining which manner of origin is the more common and how each process operates. The source of the emanations is quite mysterious; yet the source of the granitic magma is also uncertain.

Granites are far more voluminous than all other kinds of intrusive igneous rocks, and basalts have a similar position among the extrusive group. This is anomalous in the eyes of many geologists because they believe that the extrusive igneous rocks should correspond in composition to the intrusive group. Basaltic magma is much more fluid than granitic magma, and thus it commonly remains fluid long enough to reach the surface. However, this can be only a partial explanation of the discrepancy.

Perhaps granitic and basaltic magmas originate in two quite different ways. Granitic magma may form by the melting of sedimentary rocks which have the same approximate bulk chemical composition as granite (metamorphism may precede melting).

Thick accumulations of such sediments are associated with the great mountain belts of the Earth and seem to have been buried far below the surface during times of mountain building. Granitic batholiths are located only in the cores of such mountainous areas.

A major problem involves the emplacement of granitic batholiths. Huge volumes of preexisting rocks seem to have been displaced by the granite. How did the granite bodies find or make room for themselves? If the batholiths have been formed by grantization, the space problem may be largely eliminated, although other difficulties remain.

Granitic magma may also originate by magmatic differentiation (p. 727), but very large quantities of granitic magma have probably not been produced in this manner. Thus granites and similar rocks may form by granitization and by crystallization from magma. Granitic magma may form by differentiation and by the melting of sediments and other rocks having the same bulk chemical composition as granite.

LOCALIZED FUSION (MELTING) MAY PRODUCE BASALTIC MAGMA

According to one hypothesis, all basaltic magma was originally solid material somewhat similar to basalt and gabbro in chemical composition (low in its silica content, high in iron, calcium, and magnesium). Part of the supporting evidence consists of basaltic extrusions at the Earth's surface at numerous times in the past, in different parts of the world, and in enormous volumes.

Measurements in wells and mines in the outer 5 miles of the Earth's crust show that temperatures increase downward an average of approximately 1 F° every 60 feet. Hot springs, geysers, and volcanic phenomena also show that temperatures rise with increasing depth. If the average increase in temperature is continued downward to the postulated worldwide source rock, this material should be above its melting point. However, for most solids, melting necessitates expansion. Since pressures at such depths are very great, melting probably cannot occur. Now if pressures are reduced locally, expansion can take place, and these deep-seated source rocks (probably part of the upper mantle but possibly the lower part of the crust) can then melt. The resulting magma would subsequently move upward to zones of lesser pressure. Expansion caused by the melting might crack overlying rocks to produce passageways toward the surface.

Reduction in pressure conceivably might be brought about in several ways: by faulting in earthquake areas, by erosion of a mountainous region, or by upward bending of part of the crust. The intimate association of earthquakes and volcanoes might be explained in this way. However, some geologists hold that local reduction in pressure is unlikely at great depths where hydrostatic equilibrium presumably exists (i.e., where the confining pressures on any one point are the same in all directions; *lithostatic* is a better term).

Dissolved gases may aid in the upward movement of the magma, which is less dense than the surrounding rocks; in the same way, when a warm bottle of coke is uncorked, release of pressure permits dissolved carbon dioxide gas to expand and carry some of the coke out of the bottle with it. Furthermore, the weight of the crust pressing down on either side of a weak zone may be sufficient to force magma upward through it. Analogously, in the winter cracks may occur in ice which covers a lake, and the weight of the ice pushing downward on each side of a crack may force water up the crack and onto the surface. The upward-moving magma may cool and harden before it reaches the Earth's surface, thereby producing intrusive igneous rocks.

Fusion may be complete or selective—basaltic magma might form by the complete fusion of rock with the same chemical composition as basalt, or by the selective melting of the more readily fusible minerals in a source rock of somewhat different chemical composition. Laboratory experiments have shown that a rock does not melt as a unit at any one temperature; rather, there is differential melting throughout a range of a few hundred degrees before complete fusion has taken place. Moreover, if fusion of a homogeneous source rock occurs at different depths, different magmas will probably form because the melting points of some minerals do not change uniformly with increases in pressure.

If localized fusion occurs because of an increase in temperature, geologists can only speculate as to the cause. Local concentrations of radioactive elements could produce the necessary heat energy, yet tests with Geiger counters on flowing lava show no such concentrations. Friction along fault surfaces at depth has been proposed, but this seems inadequate for the generation of large volumes of magma.

Magmatic Differentiation Continuous gradations exist between low-silica magmas that harden to basalt or gabbro and high-silica magmas that produce granitic rocks when they cool slowly. Apparently some magmas were originally basaltic in chemical composition but were later transformed into higher silica magmas through a process called *magmatic differentiation.*

Imagine that a large volume of low-silica basaltic magma moves upward through a weak zone in the crust, but stops several thousand feet below the surface. Here the top and outer margins of the basaltic mass cool. The first minerals to crystallize contain large quantities of iron, calcium, and magnesium, but relatively little silica. These solids tend to sink. Thus the percentage of silica within the upper part of the magma gradually increases, and eventually the molten matter may become granitic in composition. At any stage in this process, part of the magma may resume its upward movement. If the resurgent magma comes from the upper portions of the igneous chamber after crystallization has been going on for some time, its silica content will be higher than that of the original basaltic material. Therefore, any gradation between a high- and low-silica magma is theoretically possible.

Presumably the first-formed low-silica minerals sink in the molten matter until they are melted or come to rest on the floor of the igneous chamber. Therefore, an intrusive igneous body, such as a thick sill, should show a crudely layered structure: lower-silica minerals should be concentrated at the bottom; higher-silica minerals should be more abundant at the top. The Palisades along the Hudson River in New York is a well-known sill which illustrates this phenomenon. The concept of magmatic differentiation grew out of studies of similar structures.

Furthermore, field study has shown that, in a series of intrusive or extrusive igneous masses in an area, the lower-silica members tend to be older than the higher-silica ones. This general order of intrusion or extrusion likewise supports the concept of magmatic differentiation.

We can conclude that magmatic differentiation has probably produced some granitic magmas from basaltic ones, but the process seems to be one of limited volume that can account for only a small percentage of the total volume of granitic rocks. Otherwise one would expect to find a still greater volume of igneous rocks that occur between granite and basalt in chemical composition, and these we do not find.

ORIGIN OF SOME PRIMARY ORE DEPOSITS

According to one explanation, gases and rare elements such as gold and copper that are dissolved in magma do not tend to enter into the growth of the rock-forming minerals. Therefore, the volume of magma shrinks as crystallization proceeds, and these materials are concentrated in the dwindling portion that remains. Near the end of the process of crystallization, pressure may force the gases, together with the rare elements dissolved in them, to leave the igneous chamber and to move along zones of weakness toward the surface. Such fluids may begin their journey upward as liquids, or as gases which later become liquids. Water is their chief constituent.

The hot water solutions rise into zones of lower pressures and temperatures where they can no longer hold in solution all of the dissolved materials. The substances in solution are precipitated, perhaps along fissures to form veins of ore minerals (Fig. 27-8). Many ore deposits (primary) are believed to have formed in this way. Such deposits represent tremendous concentrations of materials once widely dispersed. According to this theory, magma is the parent of such ore deposits.

This concept of the origin of ore deposits developed partly from studies made at volcanoes. During observations, cracks were filled by minerals such as galena and hematite which were deposited from gases that carried lead, iron, sulfur, and other elements in solution.

GEYSERS, FUMAROLES, AND HOT SPRINGS

Volcanoes and intrusive igneous material may give off hot gases which produce a number of interesting phenomena as they emerge from openings in the ground. Geysers, fumaroles, and some hot springs originate in this manner. Steam is the chief gas involved. Some of the steam has a magmatic source, but much more probably originates by vaporization of groundwater upon contact with upward-moving hot magmatic gases. Not all hot springs are related directly to igneous activity: groundwater has percolated downward considerable distances to zones of higher temperature, and later the heated waters may rise along a fault or other channelway and emerge at the Earth's surface.

Fumaroles are holes or fissures in the ground from which issue steam and other gases. Hot springs and fumaroles are of practical importance to man. In a number of areas, hot springs have been used for many years as laundries and as baths. Natural steam has been used to heat buildings and for the generation of electricity.

Geysers (Fig. 32-13) are hot springs which occasionally erupt columns of hot water and steam. Most geysers occur in three areas: Yellowstone National Park, New Zealand, and Iceland. Geysers hold a great deal of interest for many people because on a small scale they illustrate some of the spectacular effects of volcanic activity.

Old Faithful in Yellowstone National Park is justly famous as the tourist's friend. An average of approximately 66 minutes elapses between eruptions that throw thousands of gallons of hot water higher than 100 feet into the air. Old Faithful has been performing at these intervals winter and summer since its discovery by white men in 1871. Other geysers have since become extinct and new ones are being formed. Several thousand

Fig. 32-13. *Old Faithful in Yellowstone National Park.* [*Northern Pacific Railroad Co.*]

geysers, hot springs, and fumaroles exist in Yellowstone National Park. Some geysers are transformed into hot springs during the wet season, just as certain hot springs become fumaroles during dry seasons.

The action of a geyser probably depends upon the relationship between the boiling point of water and pressure. An increase in pressure raises the boiling point; a decrease lowers it. Familiar examples are the ease with which water in a radiator boils at high altitudes (low pressure allows a low boiling point) and the short time interval in which foods can be made edible in a pressure cooker (high pressure produces a high boiling point).

Below a geyser vent, according to this explanation, a series of interconnected fissures and openings are filled with hot water, chiefly ordinary groundwater. Heat is supplied by steam rising from below. At first the steam is assimilated, and the temperature of the water rises until it boils. However, since pressure increases downward from the surface, the boiling point is reached only at higher temperatures at greater depths. Thus water may boil at 212°F at the surface but at 290°F some 200 feet below the surface. If the passageways are tortuous enough to prevent convection, water in a geyser tube will everywhere be heated to its boiling point. When all of the water in a geyser tube is boiling (at different temperatures at different depths), steam must accumulate. When sufficient steam has accumulated, its expansive force causes overflow at the surface and the consequent upward movement of water throughout the system. This upward movement transports water into zones of lower pressures where it can boil at lower temperatures. It flashes instantaneously into steam and erupts.

Exercises and Questions, Chapter 32

1. Volcanic cones and eruptions:
   (a) What causes volcanic eruptions? Why do eruptions differ in violence from one volcano to another? from one time to another at the same volcano?
   (b) Why do volcanic cones differ in shape?
   (c) Why do craters tend to occur at volcano tops? How does a crater differ from a caldera?
   (d) How might one account for a cinder cone that has an elliptically shaped ground plan?
   (e) What is the relationship of some dikes to volcanic cones?
   (f) Can volcanic eruptions be predicted?
   (g) Why are volcanoes likely to occur along a line in any one region?
   (h) Why are the surfaces of some lava flows conspicuously different from those of other flows? Where does lava usually flow out of a volcanic cone?

2. Make a topographic map of an island that is formed by the top portion of a volcanic cone. The top of the island is 910 feet high and has a crater 70 feet deep at the top. Use a 20-foot contour interval and choose a convenient horizontal scale.
3. How does a dike differ from a sill? from a vein?
4. What problems are involved in the emplacement of a granitic batholith?
5. Discuss some of the factors involved in the origin of basaltic magma and describe the evidence that supports your statements.
6. Discuss the origin of granite and describe some of the evidence that supports each of the three main hypotheses involved.
7. Why have magmas been called the parent materials of some ore bodies?
8. Do fossils occur in igneous rocks?
9. What problem occurs in the classification of rocks that form from debris erupted from volcanoes?
10. Describe one way in which the eruption of a geyser may be accounted for.

33 Rocks bend, break, and flow

POWERFUL forces have acted upon the Earth's crust many times in the geologic past and are acting upon it today. Parts of the crust have been squeezed together or have been stretched apart; areas have been uplifted and depressed; movements have been slow or fast, large or small. Depending upon conditions of pressure, temperature, composition, solutions, and time, rocks have reacted to these forces by flowing, cracking, faulting, folding, and warping, or by developing a foliation (metamorphism). A sudden rupture of a section of the Earth's crust or upper mantle produces an earthquake, and some of the waves from a powerful earthquake pass through the Earth. Their records at different stations provide the most illuminating evidence available concerning the nature of the interior of the Earth. These aspects of diastrophism are discussed below.

FOLDS

Most of us think of rocks as hard, brittle substances incapable of any change of shape without fracturing. Yet even at the Earth's surface, rocks have been known to bend a little without breaking; e.g., in some monuments, limestone slabs, supported only at their ends, have sagged slightly under their own weight in a few hundred years. Under powerful confining pressures in laboratory experiments, certain kinds of rocks have been made to flow plastically (i.e., to change shape by slow continuous movement without breaking). When crustal forces act slowly, layered rocks under great pressure tend to be deformed by folding (Fig. 33-1). In contrast to warps, folds commonly show a pairing of crests and troughs, more uniformity of structure, and a greater amount of bending.

Fig. 33-1. *Land forms produced by the erosion of folded rocks in Africa. The sedimentary rocks differ in their resistances to erosion, and ridges are formed where upturned resistant beds crop out. The ridges and valleys in the lower right outline a plunging anticline cut by water gaps. A prominent water gap appears near the center of the photograph, and a braided stream is located to its left.* [*Photograph, U.S. Air Force: N. 28° 20′, W. 09° 35′.*]

Folding may take place by plastic flow, by microscopic fracturing and slipping, and in layered rocks by the sliding of one layer over another. Any one stratum slides upward toward an anticlinal crest over the bed lying beneath it. What happens can be illustrated if a package of file cards is bent by pushing inward on the ends. The top layer slides the greatest distance.

Major folds result from compression of an area by powerful forces which act more or less parallel to the Earth's surface upon rocks already compressed by the weight of overlying rocks. Crustal shortening occurs in such areas, but this may affect only the uppermost part of the crust.

Upfolds forming arches are called *anticlines,* and downfolds forming troughs are termed *synclines*. Following prolonged erosion, a syncline may form the crest of a mountain (Fig. 33-2) and an anticline underlie a valley. Anticlines and synclines, the two most common kinds of folds, tend to occur together like waves on a lake surface (Fig. 33-20).

Folds vary in size from tiny ones measured in fractions of an inch to huge

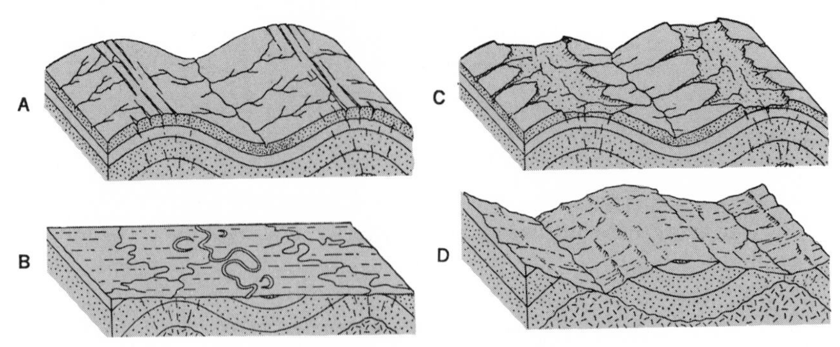

Fig. 33-2. *Origin of a synclinal mountain.*

ones several miles across. Folds may be upright and symmetrical, or one side may be steeper than the other.

A series of ridges and valleys is produced by the erosion of a sequence of tilted strata that differ in their resistances to erosion. The strata may be tilted on the flanks of folds, or they may have no association with folds. The

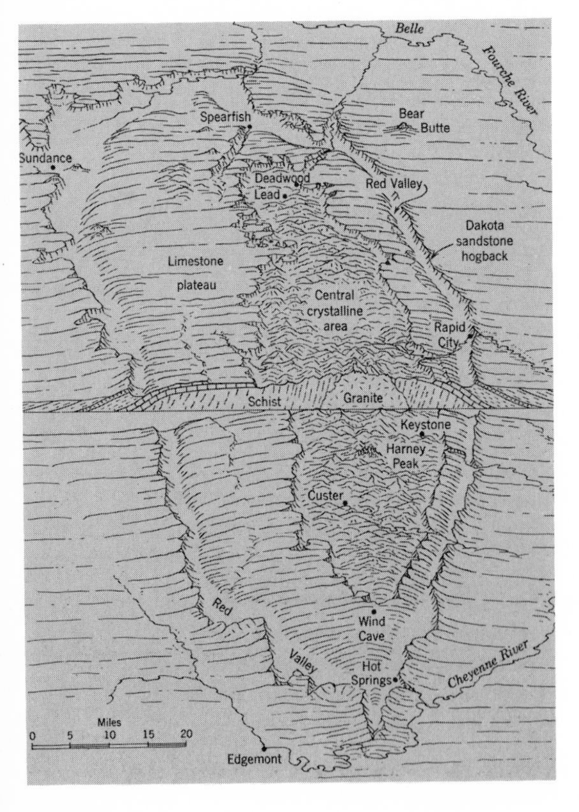

Fig. 33-3. *The Dakota Black Hills consist of a broad flat-topped dome deeply eroded to expose a core of crystalline rock. Ridges occur where resistant rocks crop out, and their steeper slopes face inward. Intervening low lands occur where less resistant rocks crop out. [Reprinted with permission from A. N. Strahler,* Physical Geography, *1951, John Wiley & Sons, Inc.]*

ridges are located where the resistant rocks project out at the surface, and the valleys develop along the zones underlain by the weaker rocks.

The shape of a ridge may be determined primarily by the dip of the strata that underlie it, and this can range from the horizontal to the vertical. Erosion of horizontal beds in a dry region tends to produce flat-topped, steep-sided hills or plateaus. Erosion of gently dipping strata commonly results in the formation of asymmetrical ridges (*cuestas*). The gentler slope of such a ridge parallels the dip of the underlying rocks and is an example of a *dip slope:* a land surface whose slope conforms approximately to that of the underlying rocks. On the other hand, erosion of steeply dipping strata tends to produce more symmetrically shaped ridges. Ridges likewise occur in humid regions along belts of resistant rocks, but the differences in resistance to erosion tend to be less conspicuous.

The axis of a fold is a line along the crest of an anticline or along the trough of a syncline; it may be horizontal or it may slant or plunge. The orientation of the axis has an important influence upon the landscape where a pile of sedimentary rocks of different resistances to erosion have been folded and eroded. If the axis is horizontal, the ridges and valleys tend to be parallel, but if the axis plunges, loop-shaped or zigzag patterns of ridges and valleys result (Fig. 33-20).

Smaller folds can be seen in their entirety in a single outcrop, and larger ones show up on aerial photographs. However, the structure of a very

Fig. 33-4. *Strike and dip. Strike is the direction of a horizontal line in the plane of the bedding. Dip is the angle between the horizontal and the plane of the bedding. If an inclined sheet of plywood (representing a sedimentary bed) is partly inserted into a pool of water, the line of intersection between the water and plywood is the strike. The dip, measured at 90 degrees to the strike, is the angle between the surface of the water and the plywood sheet. The direction of dip is that followed by a ball rolling down the sheet of plywood.*

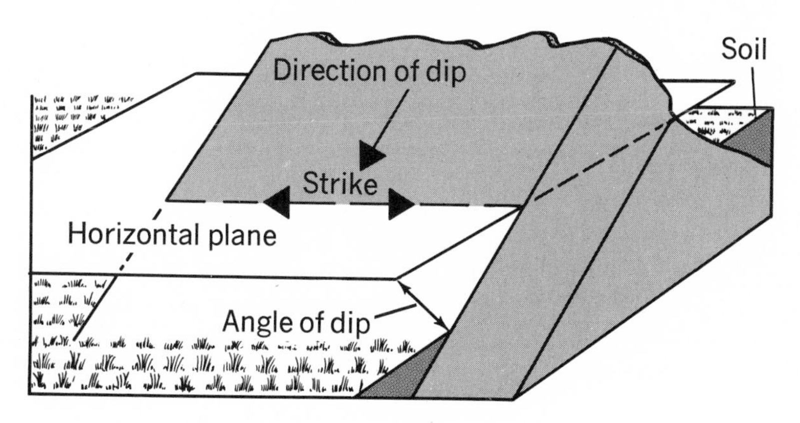

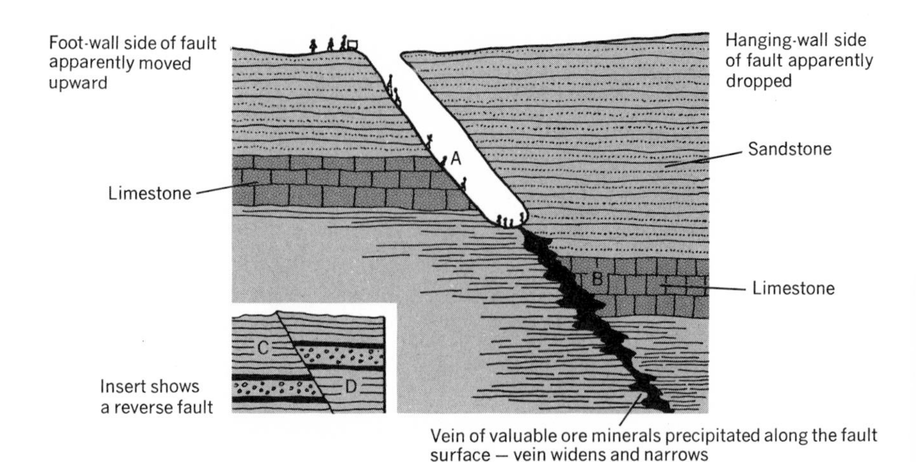

Fig. 33-5. *Normal fault. In the sketch, mining operations are proceeding downward along the fault surface. The miners walk along the foot-wall side of the fault; the other side hangs over their heads. In a normal fault, the hanging-wall side apparently moved downward. Miners found that this was the common kind of fault they encountered. The opposite direction of movement along the fault surface is called a reverse fault. Commonly normal faults are caused by tensional (pulling apart) forces. Formerly contiguous rocks have been separated so that a gap exists between their offset edges (e.g., points A and B above). Reverse faults commonly result from compression which produces an overlap—CD in the insert. In each fault, erosion has worn down the uplifted side.*

large fold or dome can sometimes be determined only by mapping and matching outcrops scattered over many square miles (Fig. 33-3).

Strike and *dip* (Fig. 33-4) are terms used to describe the orientation in the crust of a layered structure, a fault surface, or a crack.

FAULTS

Rocks tend to break or rupture rather than fold when crustal forces are applied relatively rapidly, when confining pressure is insufficient for plastic flow, or when rocks are of a type resistant to flow. A fault is a fracture in rocks along which one side has moved relative to the other; formerly contiguous points along the crack are offset by the movement. Like folds, faults may be measured in inches, yards, or miles; they may be visible in a single outcrop or require careful mapping over large areas for detection. The two most common kinds of faults, normal and reverse, were originally defined by miners (Figs. 33-5 and 33-6).

Fig. 33-6. *A normal fault in sandy shale showing a displacement of several feet. The hanging-wall side of the fault is at the left. Note the drag folds. These bend in a direction opposite to the direction of movement along a fault. The hanging-wall side at the left apparently went down and the dragfold on that side bends up.* [*A. Keith, U.S. Geological Survey.*]

The word "apparent" in the definition of a normal or reverse fault is necessary. Although the hanging-wall side of a normal fault may seem to have dropped, the actual movement may have been quite different. (1) Both sides may have moved downward, the hanging-wall block moving a greater distance. (2) Each block may have moved upward, the foot-wall block covering the greater distance. (3) The hanging-wall block may have remained motionless while the foot-wall block moved upward.

The development of a large fault undoubtedly requires much time. Movements of thousands of feet along some fault surfaces probably are the end results of a number of smaller movements each involving a few tens of feet at the most; each quick movement perhaps caused an earthquake. However, movement may also be slow and not produce earthquakes; an example is movement along a fault surface which is causing the bending of pipes in drill holes in a California oil field.

Fault surfaces commonly are curved, and movement along them may be sideways as well as vertical, or any combination of the two. Such faults are given special names. A very large, low-angle reverse fault is called a *thrust* fault (Fig. 33-7). In some thrust faults, formerly contiguous rocks are now miles apart.

Fault surfaces may be polished, scratched, or grooved by friction be-

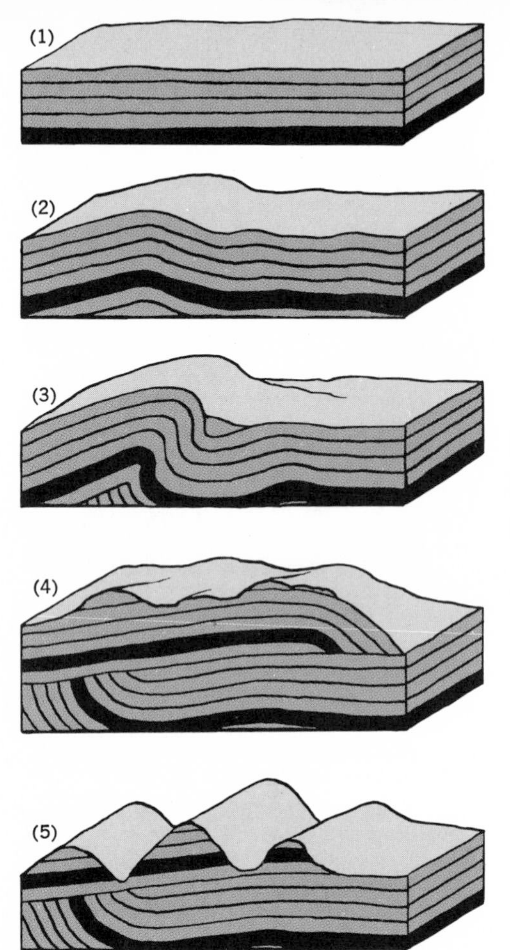

Fig. 33-7. *Development and erosion of a large thrust fault. (1) Horizontal sedimentary rocks are shown; the oldest rocks are on the bottom. (2) Compression has produced an anticline. (3) The anticline has become asymmetrical. (4) Continued compression has caused the anticline to rupture and older (once deeper) rocks from the left have been shoved upward toward the right where they now rest upon younger rocks—an exception to the principle of superposition. (5) Erosion has produced ridges and valleys.* [*After Hussey.*]

tween the moving blocks. The scratches and grooves are aligned in the direction of movement. Such surfaces are said to be *slickensided.*

Joints are fractures in rocks along which little or no faulting has occurred (Fig. 29-2). They are produced by crustal movements of various kinds, by the contraction caused by the hardening of magma or lava, and by other means.

UNCONFORMITIES

An unconformity records a significant story about Earth history. Sediments are conformable when they accumulate layer upon layer without important delay or change occurring during deposition. As strata pile up, fossils may be buried with them; thus they record information concerning events of their times.

However, conditions may change in an area—perhaps an uplift occurs.

Strata are numbered in chronological order-number 1
is the oldest rock exposed here

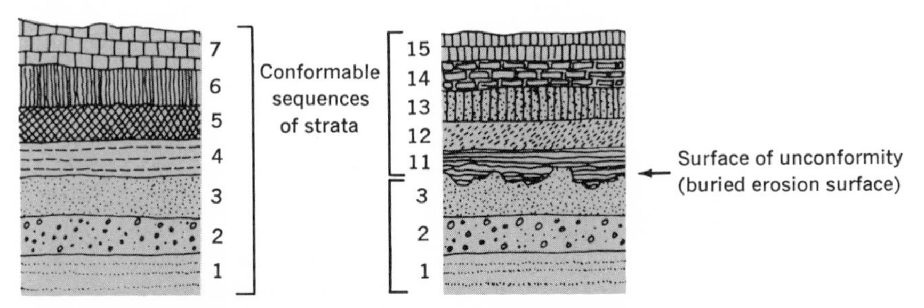

Fig. 33-8. *Unconformity. After layers 1 through 7 had been deposited as shown in the sketch, erosion proceeded to remove layers 7, 6, 5, and 4 from the area. During this interval of erosion, beds 8, 9, and 10 were deposited in an adjoining area. Bed 11 was deposited when sedimentation began again in this area. Thus, the unconformity below shows a local gap in the geologic record represented elsewhere by beds 4 through 10. There are two causes for the break: (1) erosion has removed rocks which had formed earlier, and (2) deposition did not occur in this area while erosion was going on.*

Deposition then becomes less important, and erosion becomes the dominant process. Tiny valleys are carved out of the recent products of sedimentation; these valleys then grow larger and become more numerous. In a later

Fig. 33-9. *Striking unconformity. Horizontal sands and gravels rest upon tilted and eroded shales (Cenozoic). [G. W. Stose, U.S. Geological Survey.]*

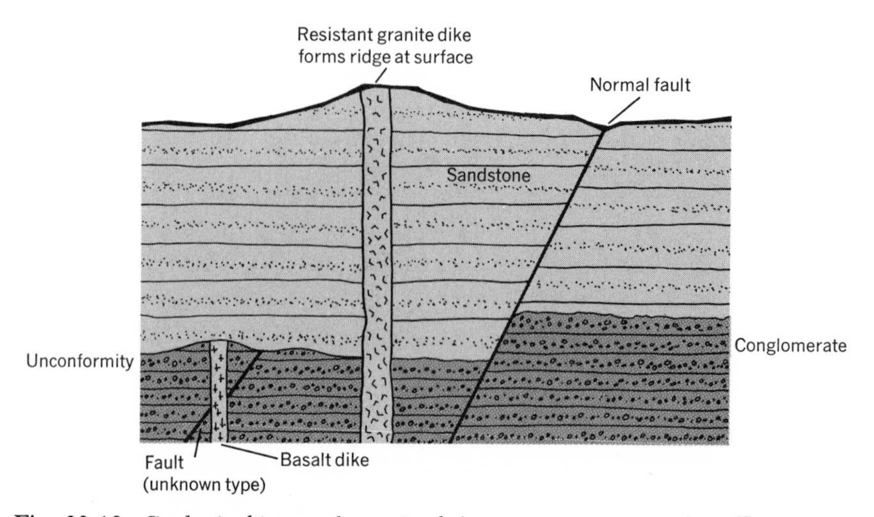

Fig. 33-10. *Geologic history determined from a structure section. Events occurred in the following chronological order.—(1) The conglomerate was deposited first.—(2) A fault cut across the conglomerate. As the layers cannot be matched on opposite sides of the fault, its type cannot be determined.—(3) A basalt dike was intruded into the conglomerate and across the fault. If the fault had come after the dike, the dike would be offset by it.—(4) Erosion occurred.—(5) Sandstone was deposited on the erosion surface forming an unconformity.—(6) and (7) (The order of the next two events cannot be determined)* a. *A granite dike was intruded into the conglomerate and sandstone.* b. *A normal fault intersected the conglomerate and sandstone.—(8) Erosion occurred at the surface. A small valley has developed along the fault surface because erosion has proceeded more rapidly in the crushed-rock zone along the fault.*

age the Earth's crust may again subside, and deposition may again become dominant in this location. These new strata are said to be unconformable upon the older ones beneath them because a lack of continuity occurs between the two groups of rocks. Accordingly, an *unconformity* is a relationship between older and younger rocks that involves a break in the geologic record, and a *surface of unconformity* is commonly a buried surface of erosion (Fig. 33-8).

Crustal movements such as folding or faulting may occur before the younger, overlying layers are deposited. Older rocks beneath the surface of an unconformity are thus deformed, whereas younger rocks above this surface are not (Figs. 33-9 and 28-20). The lost time intervals represented by unconformities in the geologic record are both numerous and long.

INTERPRETATION OF A SEQUENCE OF EVENTS FROM ROCKS

Part of the fascination of geology lies in the attempt to interpret from rock outcrops the sequence of events which occurred long ago in an area (Figs. 28-21 and 33-10).

A practical problem frequently encountered in mining involves the abrupt termination of an ore body by a fault. If the sequence of strata enclosing the ore body has been worked out carefully, then identification of the bed exposed on the opposite side of the fault surface will indicate whether the offset ore body is above or below, and about how far. *Drag folds* (Fig. 33-6) may also be used for this purpose at times.

As someone once said, geologists cannot see into the ground any farther than anyone else. Yet a knowledge of geologic structures and processes permits shrewd interpretations of subsurface conditions (Fig. 33-11 and 33-12).

GEOLOGY AND THE SEARCH FOR PETROLEUM

Geology plays an important role in the search for oil. Four prerequisites are necessary for petroleum to accumulate in commercial quantities in an area. (1) The oil originates in a source bed, and a marine shale, once a

Fig. 33-11. *Discovery of a hidden ore deposit. The outcrops of conglomerate in the valley provide the key clue to the anticlinal structure. The following geological events are recorded by the structure section: (1) Origin of the older sedimentary rocks and the ore body.—(2) Folding.—(3) Erosion.—(4) Deposition of horizontal beds of sandstone which probably once covered the entire area.—(5) Erosion to the present surface.*

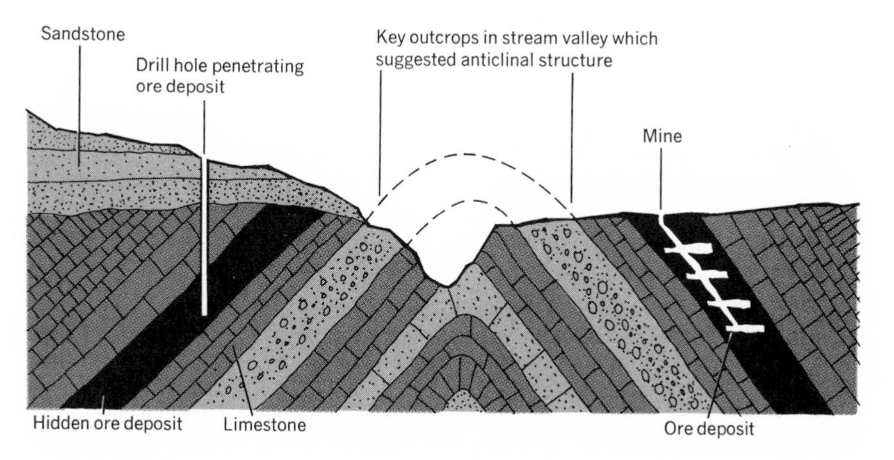

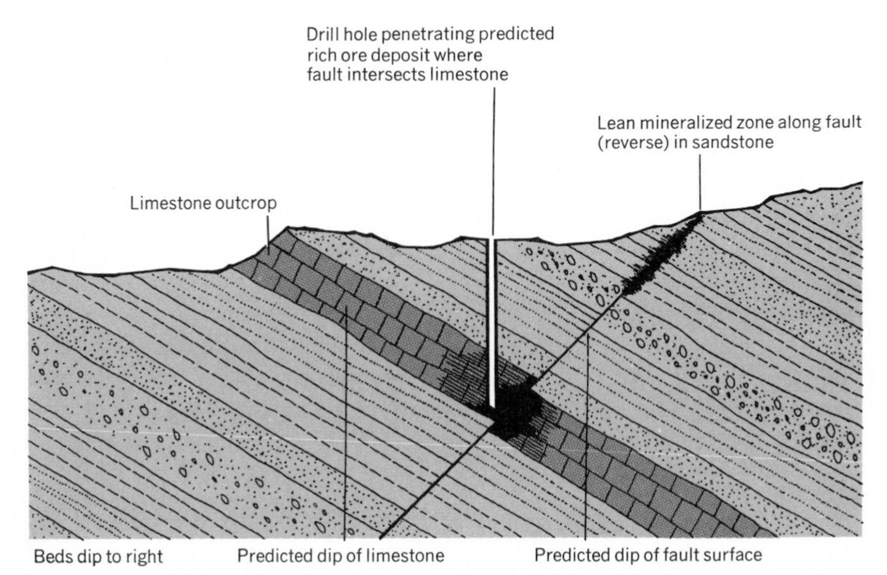

Fig. 33-12. *An ore deposit predicted on the basis of surface outcrops. Most of the sedimentary rocks shown below are insoluble in water, and the mineralized zone is lean along the fault in them. However, limestone is readily soluble in water. The geologist who found the outcrop of limestone hoped that it had been replaced by the ore minerals along the fault to form a rich deposit. The drill hole proved his hypothesis correct. Another ore deposit occurs on the foot-wall side of the fault in the offset part of the limestone.*

black mud rich in organic compounds, is thought to be a common source rock. (2) The oil then migrates to a permeable reservoir rock, and to do this it may travel for long distances both vertically and horizontally. The source beds tend to lack the permeability necessary for profitable extraction of the oil. (3) A nonpermeable layer must occur above the reservoir bed, and (4) a favorable structure must exist. Prerequisites three and four function to trap and concentrate the oil (Fig. 33-13).

The task of the geologist is the location of promising structures in regions where rocks are favorable for the occurrence of the other prerequisites. Drilling a hole is then the only known method of determining whether or not oil is present in the structure.

EARTHQUAKES

More than a million natural earthquakes (p. 601) occur each year, and about 700 are strong enough to cause damage in the regions where they

Fig. 33-13. *Above, anticlinal structures in sedimentary rocks are common oil traps. The development of an anticline is shown in three stages in the top inserts: sediments are deposited in the sea, folding occurs, and erosion takes place. The oil occurs near the crest of the anticline, where it is confined in permeable reservoir beds between overlying and underlying impervious shale beds.* [*Courtesy of the American Museum of Natural History.*]

occur. Sensitive instruments called *seismographs* (Fig. 33-14) record the passage of such waves. Even a passing train or heavy truck can cause the ground to tremble locally a little. A major problem for seismologists today is distinguishing waves produced by natural earthquakes from those created by underground nuclear explosions.

Earthquakes have terrified mankind for thousands of years. For many the term evokes a picture of death and devastation: huge waves race across an ocean and overwhelm coastal regions; villages are buried by landslides; fissures open and spout water; rivers and springs go dry; and all of this is imagined to occur amid the rumble of earth sounds. Earthquakes have relatively little effect in shaping landscapes and affect man chiefly in an indirect manner by causing destructive waves, landslides, fires, broken water mains, and diseases produced by contaminated water supplies. The opening and closing of fissures during earthquakes is not so common as is popularly believed, but it grips the imagination: in 1948 in Japan a woman working in a rice paddy sank to her neck in a fissure that opened beneath her and was crushed when it closed. In the center of an earthquake area, people, rocks, and small buildings are said to have been "tossed into the air like peas on an enormous kettle drum." Great loss of life has accompanied some earthquakes; e.g., in the 1923 Japanese catastrophe, approximately 140,000 persons were killed and property damage was estimated at 3 billion dollars.

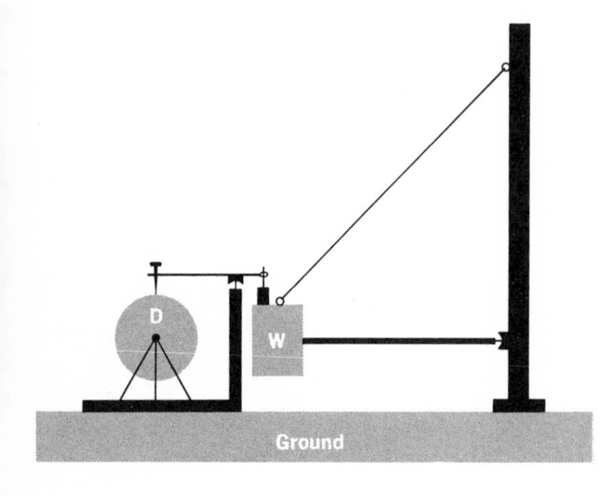

Fig. 33-14. *The seismograph. The main factors in the function of a seismograph are shown, although the model is obsolete. A rotating drum* (*D*) *is attached firmly to the ground and shakes during the passage of earthquake waves. The inertia of a heavy weight* (*W*) *tends to keep it from moving, and thus a pointer attached to it tends to remain steady. A wavy line is recorded on the slowly rotating, vibrating drum. Most seismographs today are kept in darkened rooms; photographic paper is wound around a drum; and a light source is substituted for a pointer.* [*Courtesy of the American Museum of Natural History.*]

Many earthquakes originate beneath the floors of the oceans or near their margins, and some produce a destructive "tidal wave" (a misnomer: tides do not cause them). Such waves are known as *seismic sea waves* (*tsunamis*) and some probably originate when part of the ocean floor moves upward or downward suddenly along a fault surface. The overlying water is thus disturbed violently all the way to the surface. The waves may cross the ocean at speeds up to 500 mi/hr. Wave heights are only a few feet in the open ocean, but wavelengths may exceed 100 miles. Such waves are unnoticed as they pass ships at sea. However, when they strike land, they pile up and move inland as a great destructive flood of water.

Earthquakes are most frequent and violent in elongated belts that coincide approximately with volcanic regions and young mountain systems (p. 713).

Many earthquakes originate within the outer 30 miles of the Earth, but deeper earthquakes are known, and a small percentage are created by disturbances that occur about 400 miles below the surface. Although rocks at such depths are not brittle and can flow plastically, they are also capable of rupture. Perhaps the rocks move so slowly that rupture-producing forces build up and eventually exceed the breaking point. The ultimate forces causing such movements of rocks should probably best be regarded as essentially unknown, although they have been the subject of much speculation.

Measurements have been made of the slow shifting of rocks on opposite sides of a fault surface during the years prior to the sudden rupture which causes an earthquake; e.g., a rate of about two inches per year has been measured for the San Andreas fault in California. However, the amount of bending necessary to cause the next break may be more or less than that which caused the preceding one, and the forces may be applied at an uneven rate. Thus the exact time when a major earthquake will occur in the future is difficult or impossible to predict. Yet people need to realize that earthquakes seem inevitable in certain areas, and precautions should be taken in the construction and location of buildings. At least three factors influence the amount of damage to buildings: (1) the distance from the place of origin of the earthquake; (2) construction—less destruction occurs if a building is constructed so that it vibrates as a unit; and (3) location—less damage occurs if a building is located on bedrock or at least on dry, well consolidated mantle.

THE INTERIOR OF THE EARTH

The study of earthquakes (*seismology*) provides important data concerning the Earth's interior and has shown that it consists of three main units: crust, mantle, and core (Fig. 27-5).

Earthquake waves are of three main types: primary, secondary, and main waves (Figs. 33-15 and 33-16). The *primary* or P wave travels fastest; it passes through the Earth and arrives first at a distant station. It is similar to the familiar sound wave (primary waves are audible at times), and the medium through which it passes is alternately compressed and expanded; an apt term is a push-pull wave. Such waves can travel through solids, liquids, and gases. The *secondary* or S wave is recorded next at a distant station and likewise passes through the Earth. It is a transverse wave that moves because particles in the transporting medium vibrate back and forth in planes that are perpendicular to the direction in which the wave itself moves. Such waves can travel only through solids. The *main* waves are complex and travel around the outside of the Earth confined to the crust.

The distance from a seismograph station to the place of origin of an earthquake can be determined by the time interval between the arrivals of the first P and S waves—the more distant the earthquake, the longer the interval (Fig. 33-15).

The place of origin of an earthquake can be located by the three-circle method if its distance from each of three stations is known. The *focus* is the place within the Earth where the earthquake originated, and the *epicenter* is the place on the Earth's surface directly above the focus. Three

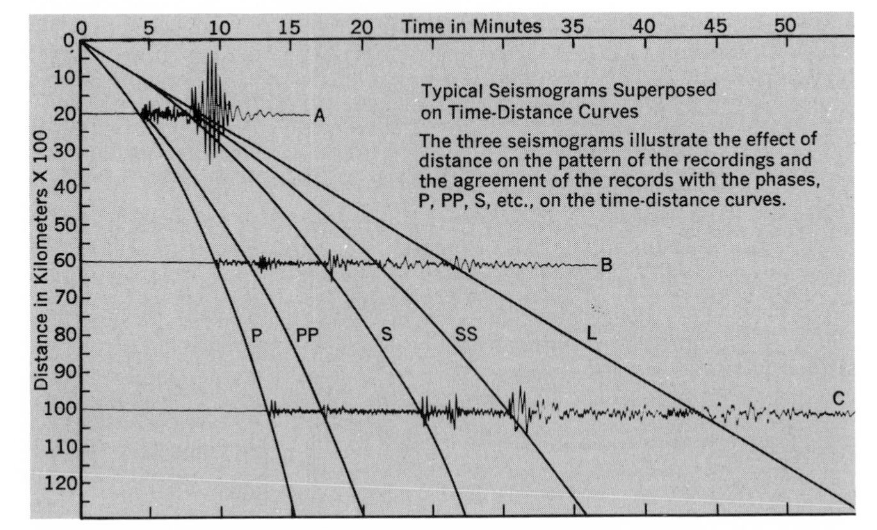

Fig. 33-15. *Travel-time curves are shown with idealized seismograms. PP and SS are reflected P and S waves. Multiply kilometers by 0.62 to convert to miles. The time interval between the first arrival of each of the three principal waves at any one station increases with distance. From the graph then one can determine that if the first S wave arrives 5 minutes after the first P wave, the earthquake originated about 3600 km away; a 10-minute time difference indicates a source about 9300 km away. Since waves travel more rapidly at greater depths, twice the distance does not take twice the time.* [*U.S. Coast and Geodetic Survey.*]

circles are drawn to scale on a map. Each station is the center of a circle, and the radii are equal to the distances to the earthquake. The three circles intersect at the epicenter.

Earthquake waves are important in geology because, in a sense, they X-ray the Earth. The Earth has a core about 4300 miles in diameter; its existence is revealed by a shadow zone that occurs in association with each major earthquake. The surface of the core is located at a depth of about 1800 miles; i.e., the mantle is about 1800 miles thick.

Seismograph stations located in any direction from a major earthquake receive the three types of waves if they are located within a distance of about 7000 miles of the place of origin (we shall ignore the main waves which tell us little about the Earth's interior). However, no direct P or S waves are received by any stations located within a belt nearly 3000 miles wide that is situated approximately 7000 to 10,000 miles from the place of origin of an earthquake. This 3000-mile wide belt extends all of the way around the Earth and is aptly called the P-wave shadow zone (Figs. 33-16 and 33-17). Its location is unique for each earthquake; it is always

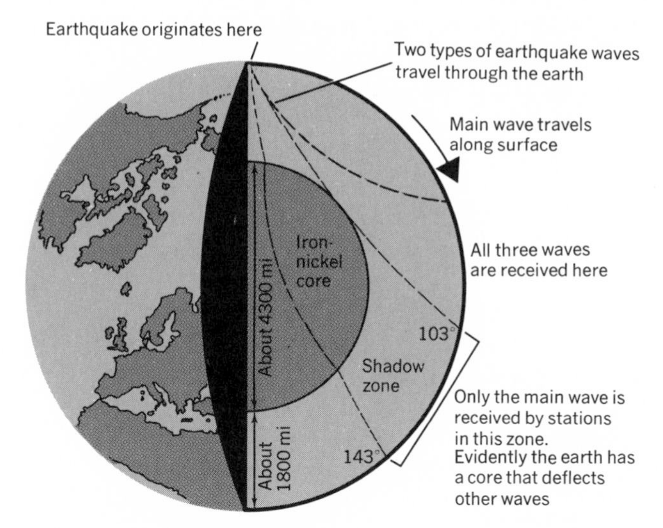

Fig. 33-16. *The Earth's interior. A shadow zone (for primary waves) is a belt nearly 3000 miles wide (about 40 degrees) which extends around the Earth. It begins at about 103 degrees (about 7000 miles) from the place of origin of an earthquake and extends to about 143 degrees from that place. Its location shifts with the location of the earthquake, which indicates the presence of a spherical core within the Earth. The outer part of the core is nonrigid, and the core may consist chiefly of iron.*

the same width and occurs at the same distance. The S-wave shadow zone is more widespread.

What causes shadow zones? P and S waves pass through the Earth along curved paths, and more distant stations receive P and S waves that have traveled more deeply through the Earth. P and S waves tend to travel faster at greater depths until the core is reached (Fig. 33-18). Waves received at a station located about 7000 miles from an epicenter were about 1800 miles below the surface at the deepest part of their curved paths. Thus a change occurs at a depth of about 1800 miles: P waves are deflected at this depth, their speeds are decreased, and the S waves are completely eliminated. Stations beyond 10,000 miles from the place of origin of an earthquake receive both P and main waves, but no S waves. Thus the Earth has a core which is spherical in shape because the same data are obtained at the same distance from earthquakes located anywhere on the Earth.

The change is thought to be both chemical and physical (i.e., from rigid to nonrigid material). The core seems to consist of an inner portion (probably solid) and an outer portion that is nonrigid; i.e., it behaves as if it were a liquid. As evidence, the S waves do not pass through the core, and they cannot penetrate nonrigid material. The existence of an inner core is indicated by the abrupt increase in the speed of the P waves about 800 to 850 miles from the Earth's center.

Both parts of the core may consist chiefly of iron alloyed with some nickel and other heavy elements. This hypothesis is based upn an analogy with meteorites, which are of two main kinds (p. 412): metallic meteorites are chiefly iron and are much less abundant than stony meteorites, which are similar to certain low silica rocks somewhat like gabbro. Perhaps meteorites are representative of the type of debris that makes up the Earth. If

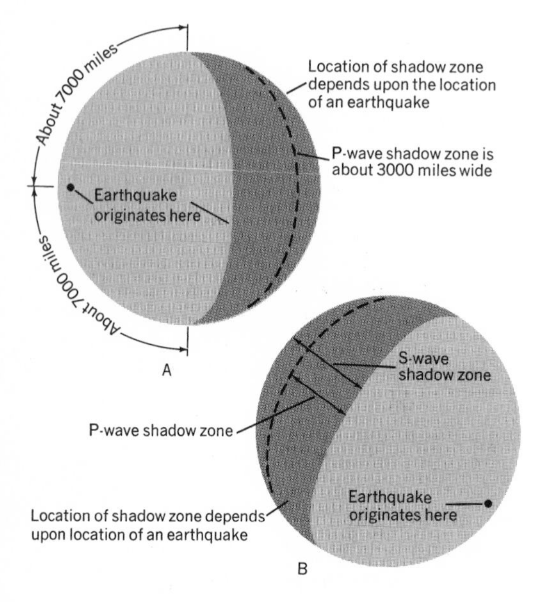

Fig. 33-17. *The locations of the P-wave and S-wave shadow zones depend upon the location of an earthquake.*

so, the core is the logical place for the large volumes of iron that should be present within the Earth, and the composition of the mantle (about 84 percent of the total volume of the Earth) may be similar to that of the stony meteorites.

The specific gravity of the interior is probably high: perhaps it increases gradually with depth to about 5 at the bottom of the mantle; it probably increases abruptly to about 10 in the outer core, and it may reach about 15 at the center of the Earth. We know that the specific gravity of the Earth as a whole (5.5) is about double that of the average rocks found in the Earth's crust (2.7 to 3.0).

Pressures comparable to those at the core-mantle boundary have been produced momentarily in laboratory experiments by shock waves generated by explosives. Materials like those presumed to make up the mantle have

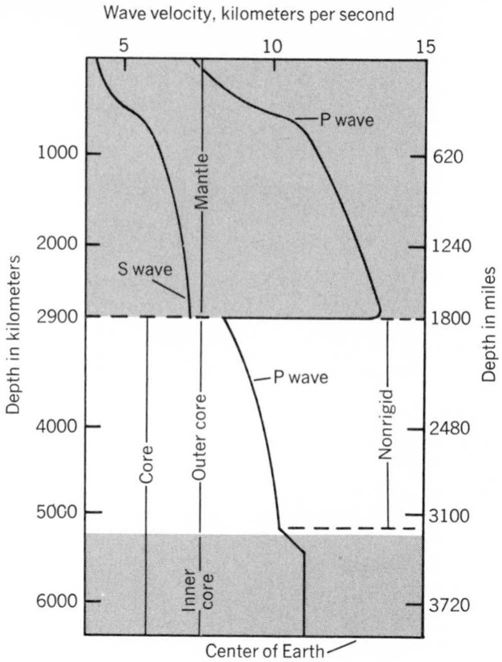

Fig. 33-18. *Estimated travel-time curves for P and S waves at different depths within the Earth. At the boundary zone between the mantle and core, S waves stop and the speed of P waves drops abruptly.* [*After Longwell and Flint.*]

been subjected to these pressures and have failed to develop high enough densities via phase changes to account for the high-density core. This strengthens the hypothesis of an iron core. The magnitude of the Earth's equatorial bulge implies the existence of a heavy core.

MOUNTAINS

Mountains are large isolated land masses which project conspicuously above their surroundings (Fig. 33-19). Diastrophism, igneous activity, and erosion—these are three phases of their origin. The simplest type of mountain is the volcano, the mountain of accumulation. Products of volcanism heap up quietly or explosively about an opening that leads downward to a magma reservoir. As a volcano grows larger, it is attacked ever more vigorously by erosion; its shape at any one time is a compromise between the additions of new material from below and erosion at the surface.

All other types of mountains have originated because part of the Earth's crust was uplifted and subsequently sculptured by erosion into isolated mountainous remnants. The Catskill Mountains in New York furnish a good example. They consist of nearly horizontal sedimentary rocks that were elevated by crustal movements to form a large plateau. Erosion then cut into the plateau, isolated the part in New York, and made it mountainous.

"How old is that mountain?" is a common question, and one often difficult to answer simply. We can tell the age of the rocks making up the mountain, and this yields one answer. However, the mountain did not exist then.

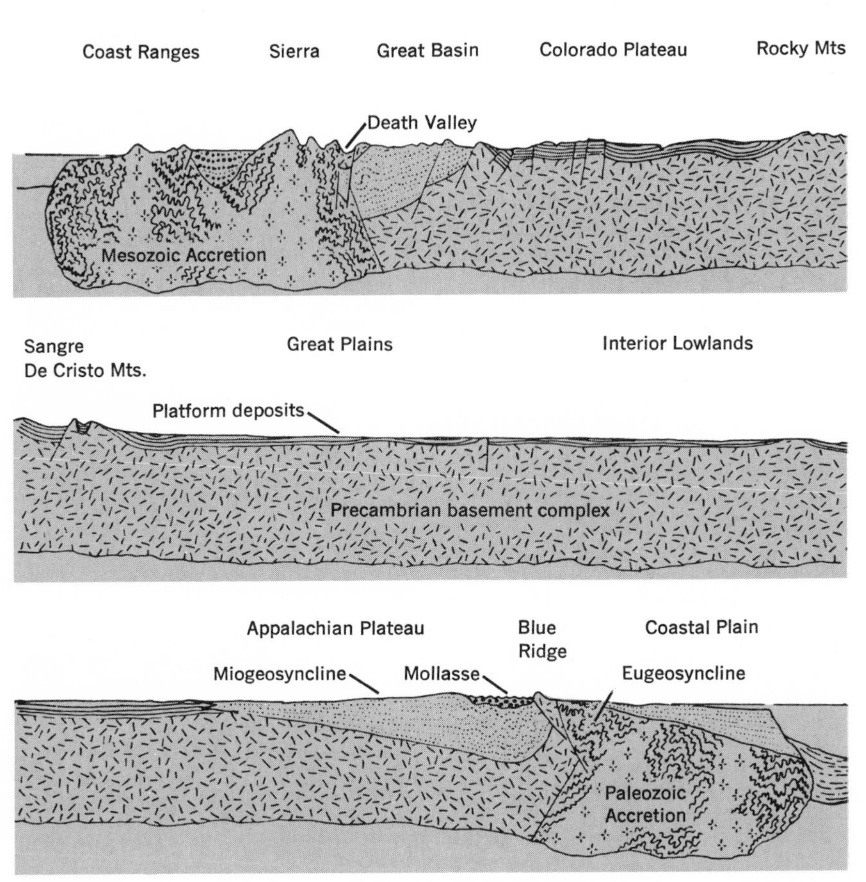

Fig. 33-19. *Diagrammatic structure section from the Pacific Ocean across the United States to the Atlantic Ocean. According to this interpretation, the continent has grown laterally by the addition of marginal mountain belts during orogenies that occurred near the ends of the Paleozoic and Mesozoic eras.* [*After L. H. Nobles in Robert S. Dietz, "Collapsing continental rises: An actualistic concept of geosynclines and mountain building,"* J. Geol., *Vol. 71, May 1963.*]

If "age" refers to a time in the past when diastrophic movements folded, faulted, and uplifted the entire region, a second and different answer is required, but the present mountain had not yet been formed. If "age of the mountain" refers to a time when erosion had proceeded far enough to separate a high land mass into isolated remnants, a third answer is called for. A correct reply manifestly should be an explanation involving all these factors.

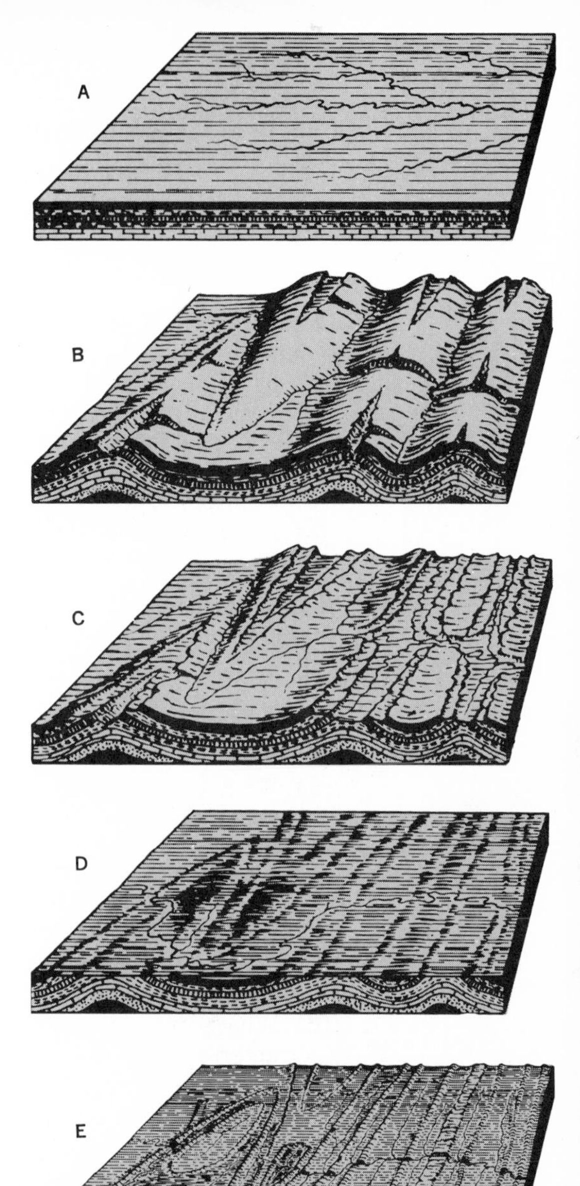

Fig. 33-20. *Five stages in the development of the Appalachians. The diagrams illustrate the erosional development of linear ridges, parallel valleys, enclosed valleys, and water gaps in the Appalachian ridge-and-valley section.* A, *horizontal strata;* B, *anticlinal and synclinal folding, with pitching anticlines;* C, *erosional mountains carved from the folded structures;* D, *the region eroded to old age;* E, *the peneplain slightly elevated and re-eroded.* [*Courtesy of the American Museum of Natural History.*]

Parts of the crust may be elevated along great fractures to form fault block mountains. Compression may wrinkle the crust into huge folds similar to those that form part of the Appalachian Mountains. On the other hand, the Earth's crust may be warped upward in the form of a dome. Still other mountains—great chains like the Appalachians (Fig. 33-20), Rockies, Andes, Alps, and Himalayas—have been called complex because their origins have involved long periods of sedimentation, crustal compression, metamorphism, vulcanism, uplift, and erosion (the geosynclinal cycle).

Exercises and Questions, Chapter 33

1. Draw a structure section which shows that the following events have occurred in the geologic history of a certain area:
   (a) Sandstone formed.
   (b) Basalt dike formed.
   (c) A normal fault occurred.
   (d) Erosion continued for a long time.
   (e) Conglomerate formed.
   (f) A granite dike formed.
   (g) Erosion occurred.

2. Make structure sections to show several different ways in which an unconformity may develop.

3. What kinds of evidence could you look for to determine whether a certain crack in rocks represents a joint or a fault?

4. Folded rocks:
   (a) What is the difference between a topographic basin and a structural basin? Make structure-topographic sections across these features to illustrate differences.
   (b) What is meant by a synclinal mountain?
   (c) Assume that you fly over an area and note nearly parallel ridges and valleys. These trend across the area to form a zig-zag or loop-shaped pattern. What is the probable cause of such features?
   (d) What sort of drainage pattern would you expect to develop in an area where sedimentary rocks have been deformed into a structural dome? Assume that sedimentary beds resistant to erosion alternate with others that are less resistant.

5. What conditions favor rock deformation by folding? by faulting?
   (a) You discover a fault with a total vertical displacement of about 1000 feet. Describe different ways in which this displacement might have taken place.
   (b) Draw a structure section which illustrates the type of drag folds associated with a normal fault; with a reverse fault.

6. In a certain area, a series of sedimentary rocks strike north-south and dip 45 degrees toward the east. The rocks consist of conglomerate (oldest), sandstone, shale, and limestone (youngest).
   (a) Sketch an east-west structure section through these rocks.
   (b) Which type of rock is located farthest toward the east? the west?

(c) If a vertical hole is to be drilled in such a manner that it will penetrate all four sedimentary rocks, where must it be located at the surface?

7. In a certain area you find three types of rocks that trend in a north-south direction and dip about 30 degrees toward the west. Marble occurs in the west; quartzite occurs in the east; and schist occurs between the marble and quartzite.
   (a) Sketch an east-west structure section through these rocks.
   (b) Is the association of marble, schist, and quartzite in this manner one that you might expect to find or is it one that you would regard as quite unusual? Explain.

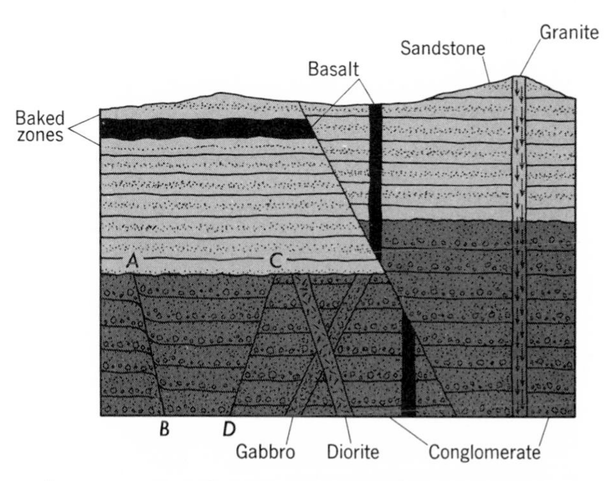

8. Refer to the accompanying structure section and list in chronological order (from earliest to latest) the events that have occurred in the geologic history of this area. Describe the evidence showing that a given geologic event must have occurred before, about the same time as, or after another geologic event. However, not enough information is given in two instances for you to decide which of two or more events came first. Describe these.

9. Answer each of the following concerning seismology and the Earth's interior:
   (a) What is the immediate cause of an earthquake (not the ultimate sources of the forces that deform rocks)?
   (b) What precautions can be taken to minimize damage when buildings are constructed in earthquake areas?
   (c) How can the distance to an earthquake epicenter be calculated from records received at a single seismograph station?
   (d) How can the location of the epicenter of an earthquake be calculated from records received at three stations?

(e) What evidence indicates that the Earth has a core? that this core is spherical? that it may consist chiefly of iron? that its outer part consists of nonrigid material (i.e., that it behaves as a liquid)?

34 The water planet

ABOUT 71 percent of the surface of the Earth is covered by a single, worldwide, interconnected body of sea water, and this abundant supply of water makes the Earth unique among the planets of the solar system. The continents resemble gigantic islands projecting above the surface of this water planet (Fig. 34-1). However, the one ocean has been subdivided by man. Some describe three oceans—the Atlantic, Pacific, and Indian—separated by the meridians that extend southward to Antarctica from the southern tips of Africa, South America, and Australia. The Atlantic and Pacific may each be subdivided at the equator into northern and southern units, and separate Arctic and Antarctic Oceans can be recognized. Thus the number of oceans may range from one to seven.

There are two dominant levels of the Earth's crust: one occurs about 500 yards above sea level and the other about 5000 yards below it (Fig. 34-2). The deepest parts of the oceans (in the marginal trenches, p. 773) approximate 36,000 feet below sea level. By comparison, the top of the highest mountain is about 29,000 feet above sea level. Yet the oceans are proportionally quite shallow; the ratio of mean depth to width is of an order from 1 : 1000 to 1 : 4000. A pond 1 mile wide and 1 to 2 feet deep would be on a similar scale.

The oceans are important to man in many ways that involve science, the military, economics, politics, communication, transportation, natural resources, and recreation. Oceanography is a rapidly growing field but is not a single discipline; rather, it is the coordinated application of mathematics and the sciences to the oceans—astronomy, biology, chemistry, geology, meteorology, and physics. The third dimension of the ocean—its depth, its inner space—is the unknown, mysterious part and is truly a new frontier awaiting adequate exploration.

Fig. 34-1. *Sea smoke 23 miles off Cape Kennedy, Florida, December 22, 1960. Sea smoke is a type of evaporation fog that tends to develop if cold air moves across a warmer water body and rapid evaporation from the water saturates the air. Temperature differences between the air and water probably should be approximately 20 to 25 Fahrenheit degrees or more.* [*Official U.S. Navy photograph.*]

The oceans form a vast storehouse of the foods, chemicals, and minerals needed by man. Ocean water can be desalted and also used as an energy source. While an inadequate protein food supply is a vital world problem, the oceans are ready to furnish far more protein food than is harvested today. Oil occurs beneath the continental shelf, and very large amounts may eventually be discovered and extracted. Materials such as manganese and phosphorite occur on the sea floor. Some valuable substances are present in almost limitless quantities and are accumulating at rates that far exceed total world consumption. Increased international cooperation is required in the study of the global sea, and thus the opportunity is present for improvement in international relations.

SOME OCEANOGRAPHIC EQUIPMENT AND TECHNIQUES

Much ingenuity and imagination have been applied in developing instruments and techniques to sample and probe the oceans. One problem in all ages has been ascertaining the position of a ship at sea. This has been solved by piloting when in sight of land, by dead reckoning, by taking astronomical fixes, and by recording radio and radar signals. Much of the earlier work in oceanography was carried out from ships whose locations

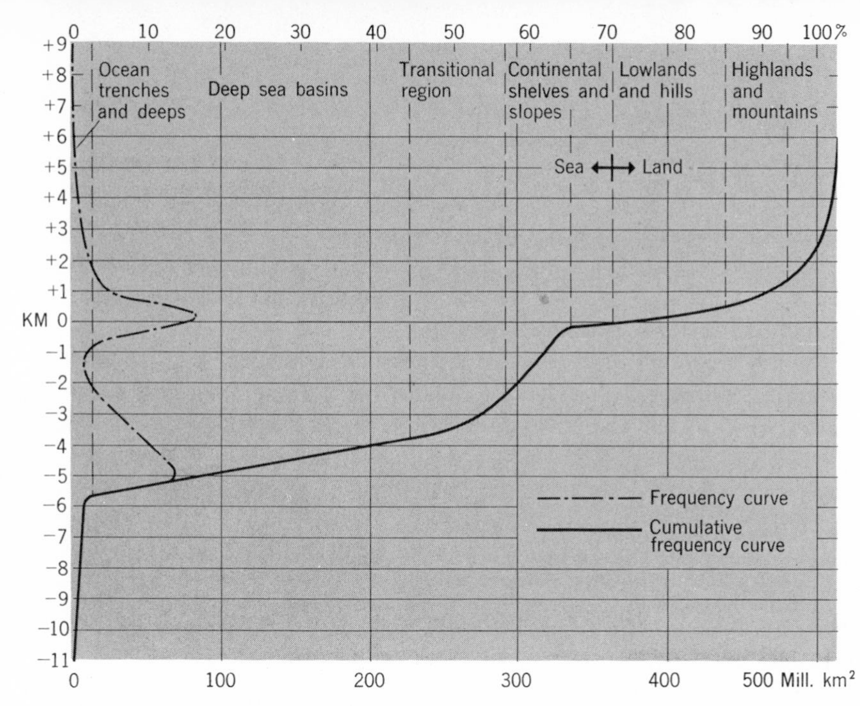

Fig. 34-2. *Estimated percentages of the Earth's surface occurring within certain altitude zones either above or below sea level. Kilometers can be converted into miles approximately by multiplying by 0.62.* [*Courtesy, William Maurice Ewing.*]

when far at sea were not known to the nearest mile or even perhaps within the nearest several miles. But navigation satellites seem to provide a breakthrough here, and accurate positioning—within one-fifth of a mile—is becoming possible.

Water depths were first measured laboriously by lowering a weight attached to a rope or cable, and results were commonly given in fathoms. A

Fig. 34-3. *An echo sounder measures the depth of water beneath a ship.*

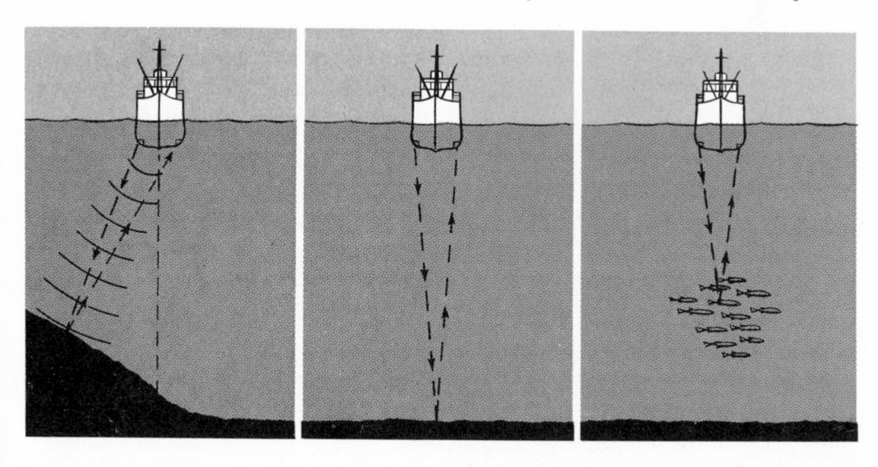

fathom is six feet in length, the approximate distance spanned by the outstretched arms of a man. Thus a sailor formerly measured the length of a rope during a sounding as so many "wingspreads." In deep water $1\frac{1}{2}$ hours might be needed to lower a weight to the bottom. The length of rope involved is so heavy that it continues to run out after the weight touches bottom, but at a measurably slower rate.

The introduction of the echo sounder in the 1920's was a great advance. Ultrasonic signals (sound waves too short for a human ear to detect) are emitted from a ship and are reflected from the ocean floor beneath. Sound waves travel at about 4900 ft/sec through water, although the speed increases with increase in density, salinity, and temperature. Thus the time interval required for an echo to arrive is a measure of the depth, after misleading echoes are discounted that may be received from schools of fish and sloping surfaces (Fig. 34-3). Continuous, generally quite accurate automatic records thus trace out the topography of the sea floor from moving ships and submarines.

During the last decade or two, geophysical techniques utilizing sound, electromagnetic waves, and electric and magnetic fields have contributed greatly to an understanding of the oceans. Reflection and refraction of sound

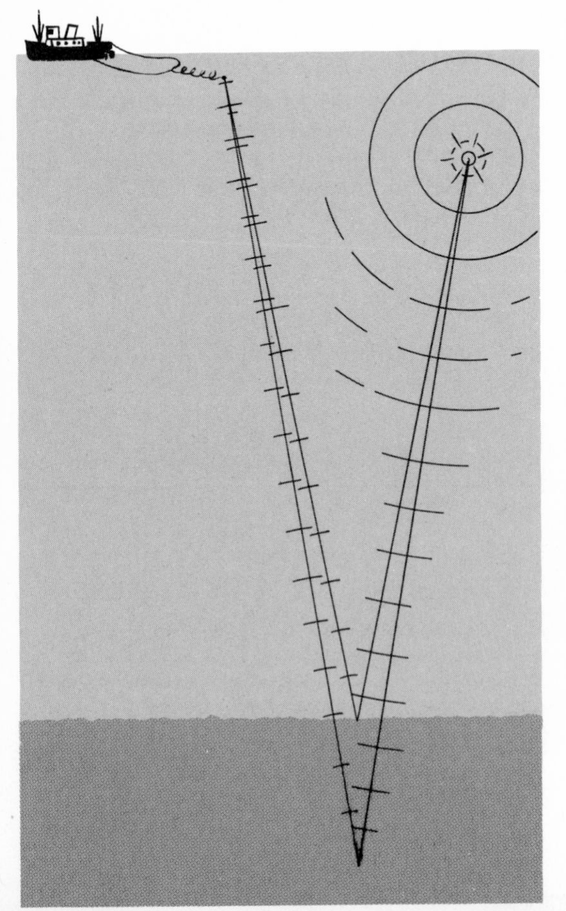

Fig. 34-4. *Reflection shooting at sea. Part of a sound wave signal is reflected from the ocean floor, but some penetrates as much as 3 miles below the sea bottom. Some of these penetrating waves are then reflected from different layers below the sea floor. Recently developed techniques make it possible to obtain a continuous record from a moving ship of the materials beneath the sea floor—the thickness of sediments and other materials forming the upper part of the oceanic crust.* [*After Ewing and Engel.*]

signals by sediments and rocks beneath the sea floor reveal thicknesses and structures of underlying layers (Figs. 34-4 and 34-5).

Undeformed cores of sea-floor sediments, tens of feet in length, have been pulled up from depths of several miles. Older coring methods were liable to compress the sediments as a weighted coring tube punched downward through them. In the modern piston coring method (Fig. 34-6) the piston stops as it strikes bottom, and the coring tube slips downward by it into the sediments. A vacuum thus occurs inside the descending tube and sediment rises upward into it. Friction is thereby reduced. A new type of increment corer may be able to obtain cores 300 feet in length in 6-foot segments, and plans are under way to drill much deeper holes. Sediment samples are also obtained by means of grabs and dredges of various types.

Water samples may be taken at selected depths by means of a Nansen-type bottle. A cable is lowered and bottles are attached to it at designated intervals. During descent the metal bottles are open and water passes freely through them. When all of the bottles have been lowered, a small "messenger" weight is sent down the cable to strike the topmost bottle. This impact releases a catch, the bottle turns upside down, and lids at each end are closed by springs. The impact also releases another messenger weight

Fig. 34-5. *Refraction shooting at sea. Explosives are dropped by a "shooting ship," and signals from sound waves thus produced are picked up by hydrophones at a "listening ship." Waves go out in all directions, and some are bent as they pass from one zone to another. Some of the waves then travel along the boundary zones between layers of different velocity, and some of this signal is refracted upward to the surface. In general, higher velocities occur at greater depths. The hydrophone record must then be unscrambled and interpreted. The differences in arrival times are a measure of the depth below the surface traveled by the waves.* [*After Ewing and Engel.*]

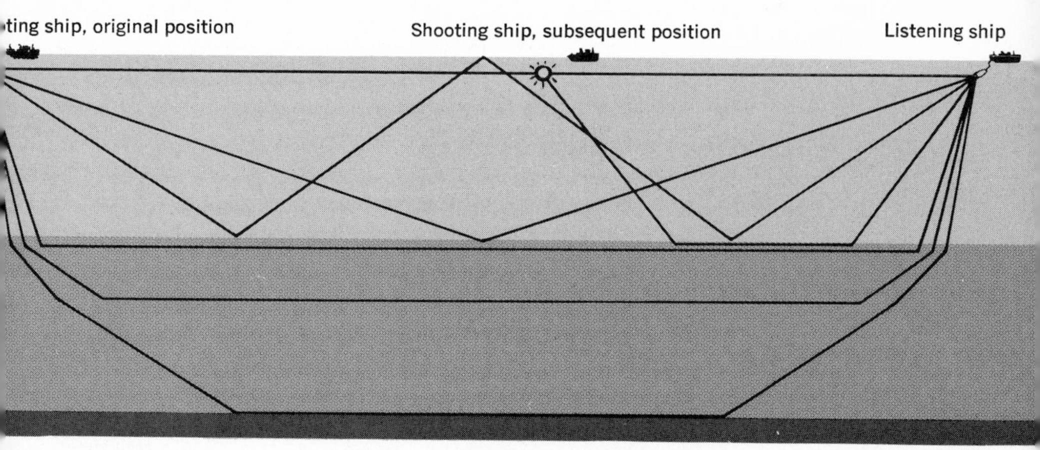

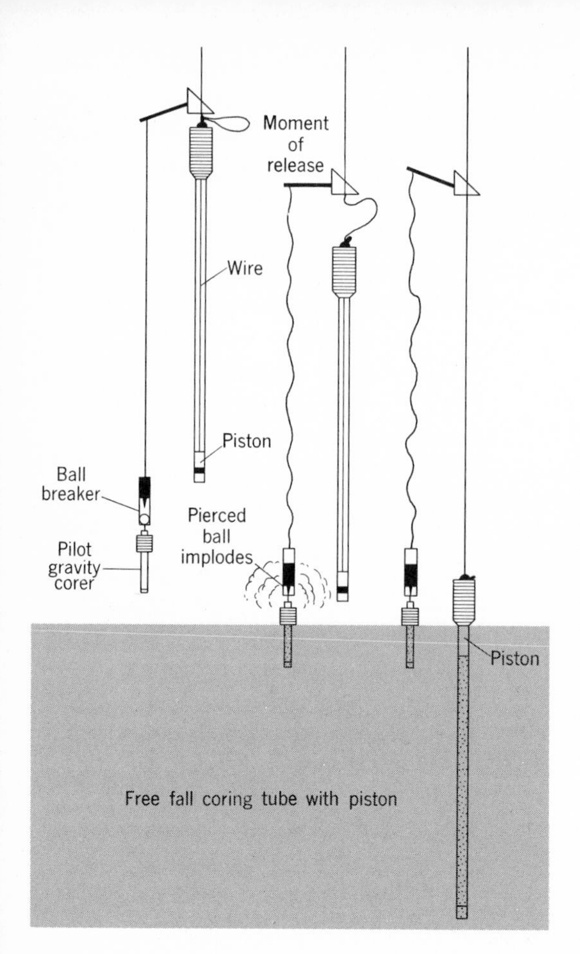

Fig. 34-6. *How the free-fall method of coring operates. The piston makes it possible to secure a long core by reducing friction on the inside of the core tube. A ball breaker is part of the apparatus.* [*Courtesy, Francis P. Shepard,* Submarine Geology, *2d ed., Harper & Row.*]

which had been attached to the bottle. This slides down to the next bottle and the sampling procedure is repeated.

A demand water sampler is a newer version. In this, a number of water bottles are lowered together and instruments record temperature changes continuously during descent. An operator on the ship observes the record and can push a button to collect a sample at any depth he chooses.

A pair of thermometers may also be attached to each bottle to record the temperature at the sample depth. The thermometers are reversed when a bottle is reversed, and the mercury column breaks at a constriction in the tube to record the temperature. One thermometer is protected from water pressure, but the other is not. The unprotected thermometer gives a higher reading because its tube and bulb are compressed. A large difference in the readings between the two thermometers thus means a great depth. In fact, this is probably the most accurate method of measuring water depth—a curved cable does not affect the result.

Ocean currents (Fig. 34-7) may be measured in different ways, but a typical feature of many is an object that drifts with a current and is tracked;

or the flow of a current through a stationary instrument may be measured. The Swallow float is an example: a long, cylindrical metal buoy is weighted to float at some desired depth. A transmitter within it sends out a sound signal periodically, and this is picked up by an instrument on a ship. A number of such buoys floating at different depths provide information about deep-water oceanic circulation.

Various man-in-the-sea projects have involved the establishment of living quarters on the sea floor at depths ranging down to 400 feet and more. Teams of divers have then lived in such quarters and, with the aid of familiar scuba diving equipment, have performed daily work on the sea floor for periods of a few weeks. Pressures inside the living quarters are the same as those outside. Compressed air may be piped down from a surface ship to keep sea water from entering the sea-floor dwelling through an open hatch at the bottom. Men can live, sleep, eat, cook, and relax inside. Thus divers spend more actual working time on the sea floor and at greater depths. At the end of their stay on the sea floor, the divers undergo lengthy decompression, but this lasts no longer after a one-month sea-floor visit than after a visit of a few hours. Most of the continental shelf—an area equal to the entire African continent—can probably be explored using this technique.

Photographs may be taken of the sea bottom at different depths (Fig. 34-8), and underwater vehicles of various types such as the bathyscaph and diving saucer have been developed. The Flip Ship (Fig. 34-9: *fl*oating *i*nstrument *p*latform) is a manned buoy used for observations near the surface.

The great bulk of measurements in oceanographic research have been made, and continue to be made, with instruments that are lowered over the side of a ship at the end of a long wire, suspended for a time, and then pulled back to the surface. The instruments may sample the sediments on the sea floor or the water itself, make collections of organisms at various depths, measure the amount of heat energy escaping through the sea floor, or record the magnetism of the sea floor.

Among the newer developments in oceanographic research are attempts at lowering a number of different instruments at one time, the making of measurements while a ship is under way, and the location of an extensive geophysical observatory on the ocean floor at a 13,000-foot depth off the California coast—data are transmitted to shore via cable. Data are also collected in a form suitable for computer processing, and many of the data are processed at sea during a voyage. Less emphasis is now put upon samples and measurements taken in a gridlike pattern followed by months and years of laboratory study on the results. Although such observations are still important, many are now made specifically to test hypotheses that

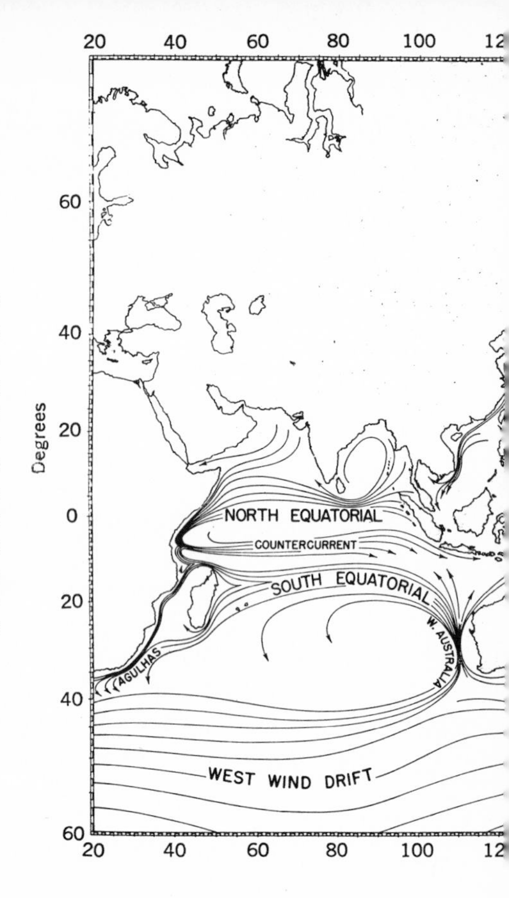

Fig. 34-7. *Major surface currents of the world. Winds are the main cause of surface currents, and the distribution of the major surface currents coincides approximately with the global pattern of atmospheric circulation. Note the existence of the huge, slow-moving, circular whirls (gyres) clockwise in the Northern Hemisphere and counterclockwise in the Southern Hemisphere.* [Oceanography for the Navy Meteorologist, *U.S. Weather Research Facility*.]

develop from the application of the various sciences to oceanography. Imagination and observation are thus closely interwoven, to the enhancement of each.

THE EARTH'S SURFACE BENEATH THE OCEANS

By topography and structure the continents subdivide naturally into major and minor units; the ocean floor does likewise. Its three main units consist of a marginal zone, the ocean basins, and a system of mid-ocean ridges (Fig. 34-10). Lesser units include the continental shelves, continental slopes, continental rises, submarine canyons, deep-sea trenches, abyssal plains, mid-ocean canyons, seamounts, guyots, fracture zones, and rift valleys.

The *continental shelves* are the submerged margins of the continents and constitute about 6 percent of the Earth's surface area. The continental shelf consists of the relatively smooth-floored, shallow-water zone that begins at the coastline and slopes very gradually seaward. Its average slope is about 10 ft/mi; i.e., across a 40-mile width the depth would change 400 feet—a nearly flat surface to the eye. In contrast, the continental slope is much

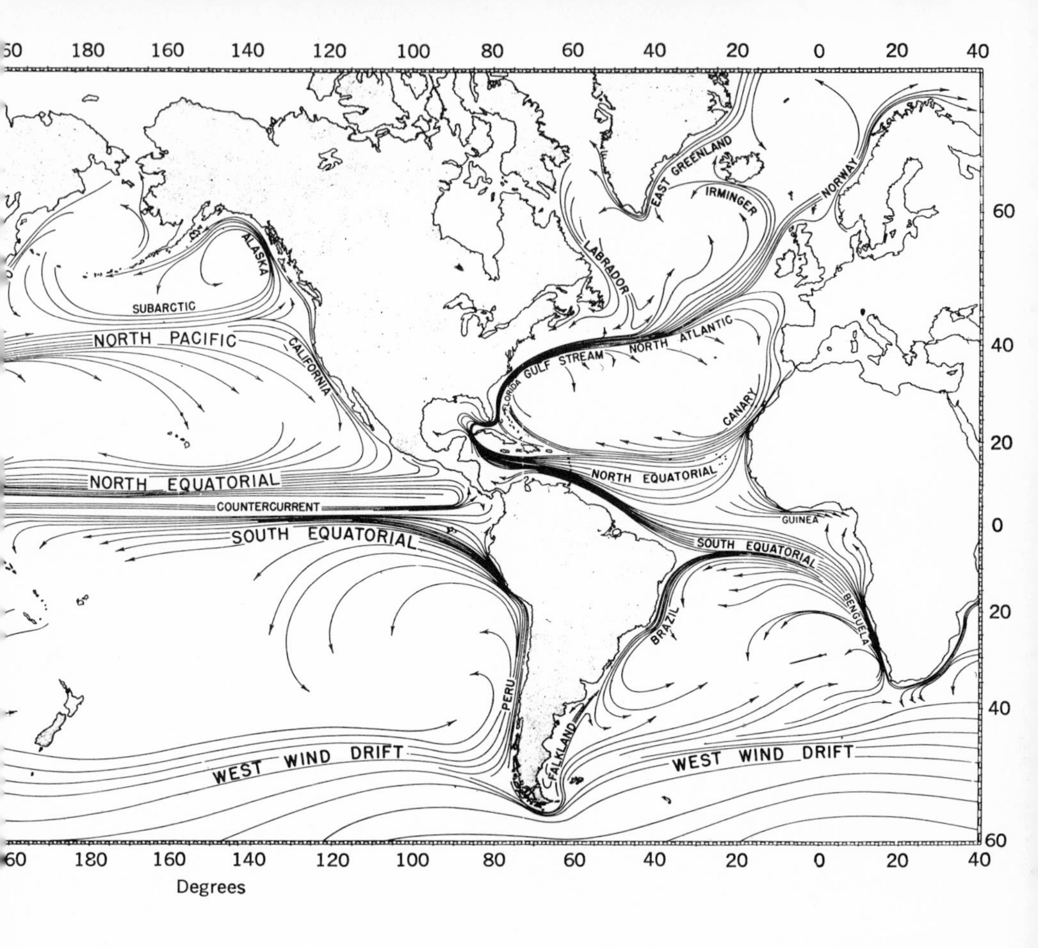

steeper—about 130 to 300 ft/mi or more. By one definition, the seaward termination of the continental shelf occurs at this rather abrupt change in slope. The shelf has an average width that exceeds 40 miles, but it is missing in places and extends outward for several hundred miles in others. Its average depth at its termination is 400 to 450 feet, but it may be more or less than this.

The shelves commonly show a close relationship to the geology and topography of adjacent coastal regions. In some areas, a shelf forms the submerged extension of a coastal plain or a delta. Off the glaciated coast of New England, the sea floor is irregular with basins, troughs, hills, and ridges suggesting glacial erosion and deposition. When sea level was lower during the Pleistocene glacial ages, much of the present continental shelves was probably exposed to the air, and this exposure was a worldwide occurrence.

From the edge of the continental shelf the sea floor commonly descends to the ocean basins along surfaces known as the continental slope (upper steeper part) and continental rise. In other areas, deep-sea trenches, island arcs, and fault surfaces intervene to form a different type of marginal zone. The *continental slope* occurs off all the lands and is one of the most striking

Fig. 34-8. *Sea bottom photograph obtained from U.S. Coast and Geodetic Survey ship* Explorer *in July 1961 about 80 miles south-southwest of Nantucket Island, Massachusetts. Water depth was about 260 fathoms (1560 feet). A skate is located in the upper left (note his protective coloration); the nearby light spot is part of the camera equipment. Some sea anemones are shown.* [*Courtesy, U.S. Coast and Geodetic Survey.*]

relief features on the Earth. It commonly has a straight or gently curved trend, and its average width of 10 to 20 miles is narrower than either the continental shelf or rise. It begins in water some 300 to 600 feet deep and ends at depths of 1 to 2 miles. Submarine canyons (p. 771) cross the continental slopes in places.

A *continental rise* extends from the bottom of a continental slope to the floor of an ocean basin. Although a change in slope is common between a slope and rise, one may merge with the other. The rise has a relatively smooth surface and varies in width from a few miles to a few hundred miles. Its average gradient is about 20 ft/mi, which is midway between the gradients of the shelf and slope. The rise evidently is an accumulation of sediment at the base of the continental slope. Thus it resembles a series of coalesced alluvial fans located along the base of a mountain range.

Abyssal plains form topographic units within the ocean basins and are the flattest areas on the Earth's surface. Apparently they are constructional features like the continental rise and have formed by the accumulation of

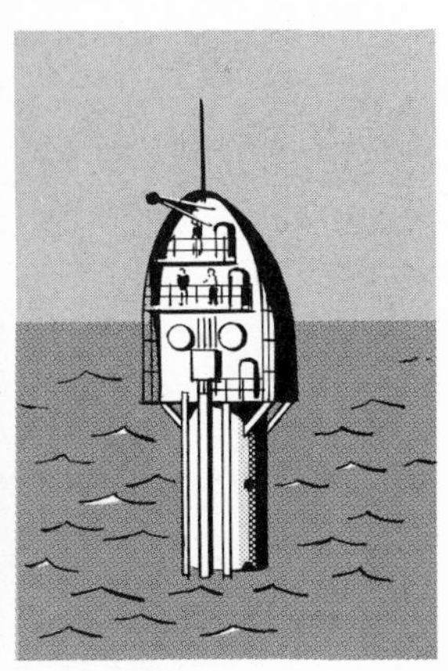

Fig. 34-9. *The Flip Ship is a new type of oceanographic research vessel. It can literally flip from a horizontal to vertical position while at sea if its long aft section is flooded with sea water. Air at high pressure blows out this water to return* Flip *to a horizontal position. Flip is being used for studies of wave motion, marine biology, internal waves, sound waves, and other phenomena. Two watertight cylindrical tubes permit the crew to descend to 150 feet below the ocean surface.*

sediments on a sea floor that was once as rough topographically as the irregular surfaces now surrounding them. Partially buried hills project here and there above the plains. Intrusive and extrusive igneous activity and crustal movements presumably account for much of the irregularity of the ocean floor.

Abyssal plains are absent off continental margins that contain deep-sea trenches not yet filled by sediments. These deep troughs in the sea floor parallel the continental margins and trap the sediments transported by turbidity currents (p. 769) down the adjacent continental slope. Thus abyssal plains are more widespread in the Atlantic, which has fewer of the deep troughs, than in the Pacific.

The *mid-ocean canyons* that occur on the ocean basin floor in places have a misleading name; they are rather shallow, steep-sided, flat-floored troughs, a few miles wide, a few hundred feet deep, with very gentle gradients. However, they may extend for hundreds of miles. Wide, low, leveelike ridges are likely to margin the channels on each side.

A *seamount* has been defined as any isolated elevation which rises 3000 feet or more above the surrounding sea floor, and a *guyot* (pronounced to rhyme with "Leo") is a special type of seamount with a flat top. Most seamounts and guyots seem to be volcanic cones, but they tend to have steeper slopes than comparable land volcanoes because lava cools more quickly under water. A guyot probably formed somewhat as follows. A volcanic

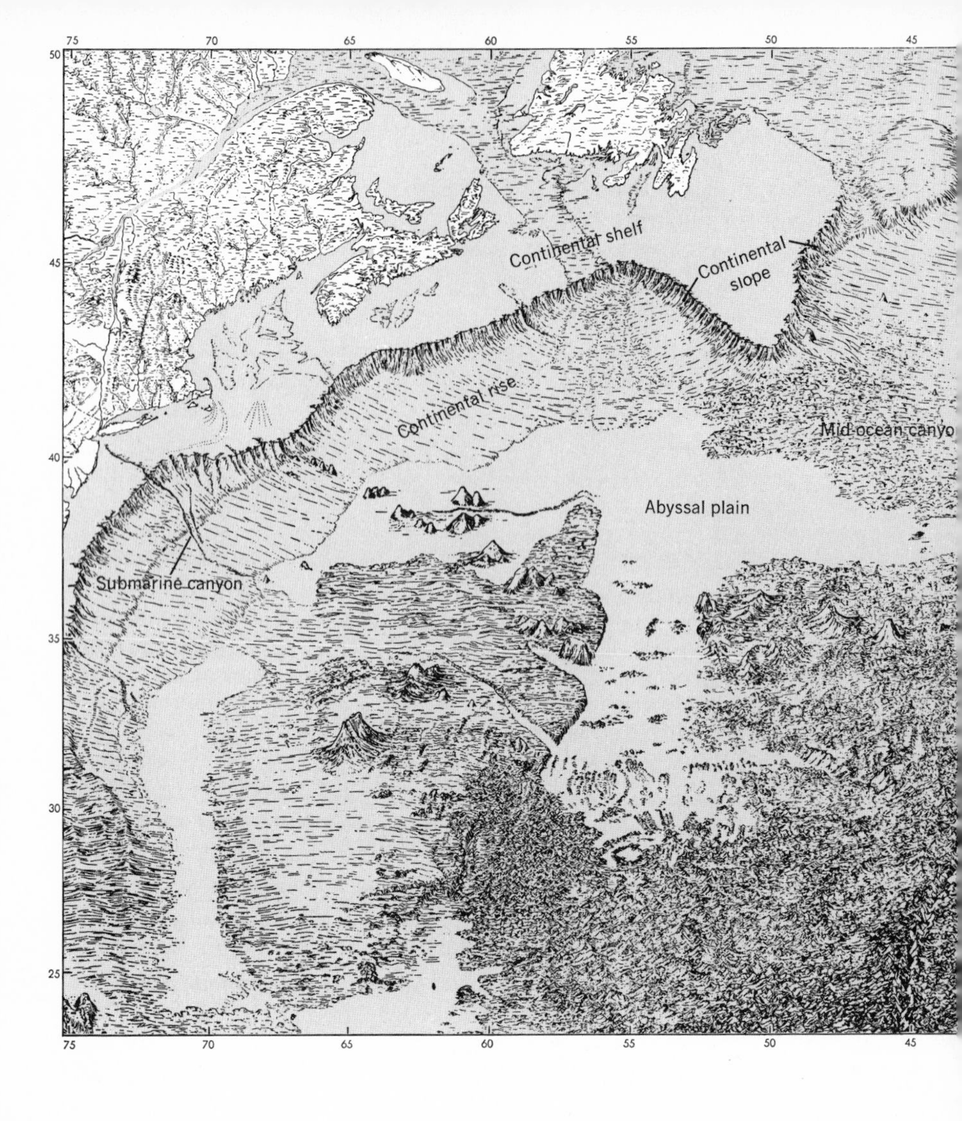

Fig. 34-10. *Physiographic diagram of the bottom of the North Atlantic by Marie Tharp and Bruce C. Heezen. The vertical scale is exaggerated about 20 times.* [*Bruce C. Heezen, "The Rift in the Ocean Floor,"* Scientific American, *October 1960.*]

cone was first built upward from the sea floor to form an island. Then erosion by surf (predominant) and streams produced a nearly flat erosion surface near sea level. Next subsidence occurred, and today we find the flat tops of guyots $\frac{1}{2}$ to 1 mile or more below sea level. Subsidence probably has occurred both locally, thus accounting for a single guyot, and regionally to form whole groups of guyots.

The Mid-Atlantic Ridge (Figs. 34-11 and 34-10) is the best-known

portion of the 40,000-mile-long, world-encircling system of *mid-ocean ridges* (Fig. 34-12) that approximately equals the continents in areal extent. The system may not be continuous, and some sections seem to be older than others. The centrally located Mid-Atlantic Ridge is an enormous broad arch about 1000 miles wide. Its crest rises some two miles above the ocean basin floor on either side. The surfaces that slope away from the crest are jagged and irregular.

A large, deep, relatively narrow valley (10 to 30 miles wide and 1 to $1\frac{1}{2}$ miles deep), or a zone of parallel, steep-sided valleys with intervening ridges, extends along the crest of the ridge. Shallow-focus earthquakes (whose centers are located within about 20 miles of the surface) occur frequently along this central rift zone. In fact, an apparently continuous belt of earthquake epicenters follows the median line of the system of mid-ocean ridges. The central rift zone, like rift valleys on the continents, may result from tensional forces (pulling apart).

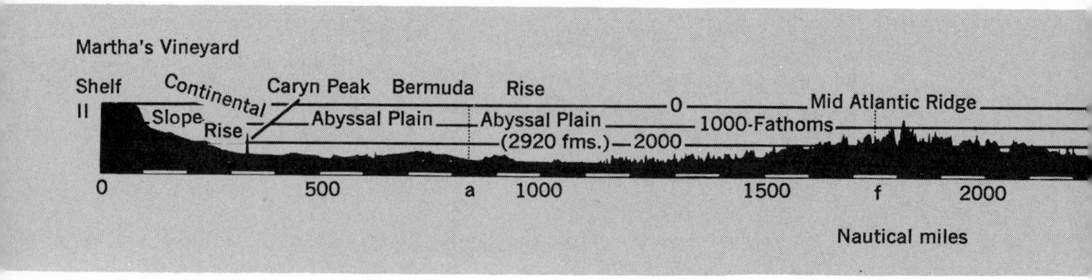

According to one interpretation, a mid-ocean ridge may form a huge upward bulge with a central rift zone because convectional currents within the mantle rise beneath it, diverge horizontally for a distance, and then subside (Fig. 34-13). Additional types of evidence support the convection current concept (p. 774 and 780).

A number of relatively narrow, elongated, east-west *fracture zones* have been mapped on the sea floor. The zones are characterized by irregular topography, fault surfaces, elongated troughs and ridges, and submarine volcanoes. The sea floor tends to be deeper on one side than the other. The highest points within a fracture zone may rise two miles above the lowest portions. Such large linear fracture zones—one in the Pacific has been traced nearly one-fourth of the way around the Earth—are obviously important factors in an understanding of the Earth's structure and history, but their overall interpretation is uncertain. Since they are nearly straight, the fracture zones are presumably vertical. Displacements along them seem to be chiefly horizontal.

Measurements of the magnetism of the sea floor made by towing a magnetometer behind a ship have revealed striking north-south belts of positive and negative magnetic anomalies. Apparently these would extend for thousands of miles if they were not offset along the east-west fracture zones. Attempts to correlate the magnetic anomalies across the fracture zones have suggested to some oceanographers that horizontal displacements ranging up to several hundred miles have occurred along some of the fracture zones. A recently suggested interpretation of the anomalies involves sea-floor spreading (p. 780).

Ocean-Floor Sedimentation Sediments from different sources have followed diverse routes to reach the ocean floor, where deposition predominates over erosion. Ocean sediments are commonly classified as *terrigenous* (from the land) and *pelagic* (belonging to the deep sea), but the two types are sometimes hard to distinguish. The great bulk of the sediment has come from the lands. Rivers carry large quantities and surf erosion in coastal areas produces more. Wind, volcanic eruptions, the melting of debris-laden

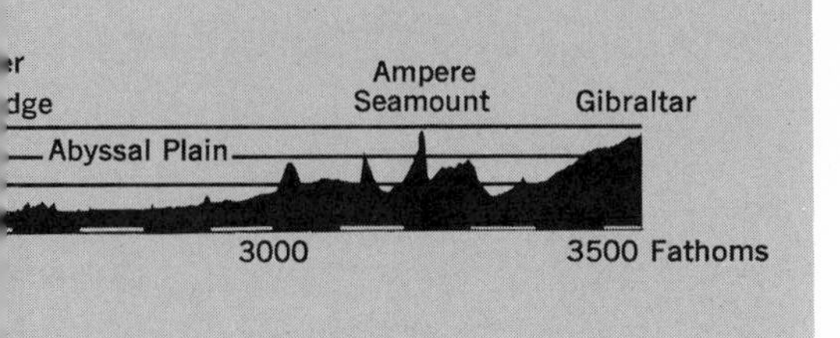

Fig. 34-11. *Trans-Atlantic topographic profile from Martha's Vineyard to Gibraltar. Vertical scale is exaggerated 40 times. Heezen, Tharp, and Ewing, "The Floors of the Oceans,"* Geol. Soc. Am. Special Paper *65*.]

glacier ice, and meteoritic fragments account for lesser amounts. Substances such as calcium carbonate and silica are transported from the lands in solution and extracted by organisms to make shells. These shells—many of them very small—descend one by one to the sea floor and form a large portion of pelagic sediment. These accumulate steadily and can pile up on guyots and ridges as well as in depressions. They are thin or absent where deep-water currents are strong and slopes are steep.

Turbidity current deposits (*turbidites*) form a quite different type of ocean-floor sediment (Fig. 34-14). These are relatively coarse-grained and include silts, sands, and even an occasional gravel. Any one layer tends to show graded bedding. Larger particles at the bottom grade upward into smaller particles at the top. Although the bottom of such a layer may be sharply defined, its top may merge with the tiny pelagic sediments that subsequently cover it. Burrowing organisms may also mix the upper part. Turbidity current sediments may contain fragments of animals and plants that live only in the shallower waters of the continental shelves, and they tend to have a higher lime content than deep-sea clays. Thus geologists have had to restudy many ancient, coarse-grained sedimentary rocks containing shallow-water fossils; some of these originated as turbidites rather than in a near-shore environment.

Sediments may pile up temporarily on the outer part of a continental shelf, on the upper part of a continental slope, or in the head of a submarine canyon. Later, these sediments may be jarred loose by an earthquake, by a storm at sea, or by slumping from a slope that became too steep during sedimentation. Such a mass of sliding turbulent sediment tends to disperse throughout the water immediately adjacent to it, and the suspended material increases the density of the water. This mass of muddy, discolored, sediment-laden water then slides downslope as a gravity-pulled density current. Deposition occurs as a current slows down on lesser gradients at greater depths. Such *turbidity currents* have been very important in the formation of the continental rise and abyssal plains as well as in the origin of sediments elsewhere on the deep-sea floor.

Ocean-floor sediments are surprisingly thin. The Atlantic Ocean has many miles of coastline, large rivers, and few marginal deep-sea trenches. Thus its average sediment thickness (2000 to 3000 feet or less) exceeds that of the Pacific (1000 to 1500 feet or less). However, calculated upon current rates of sedimentation and an estimated age of 4½ billion years for

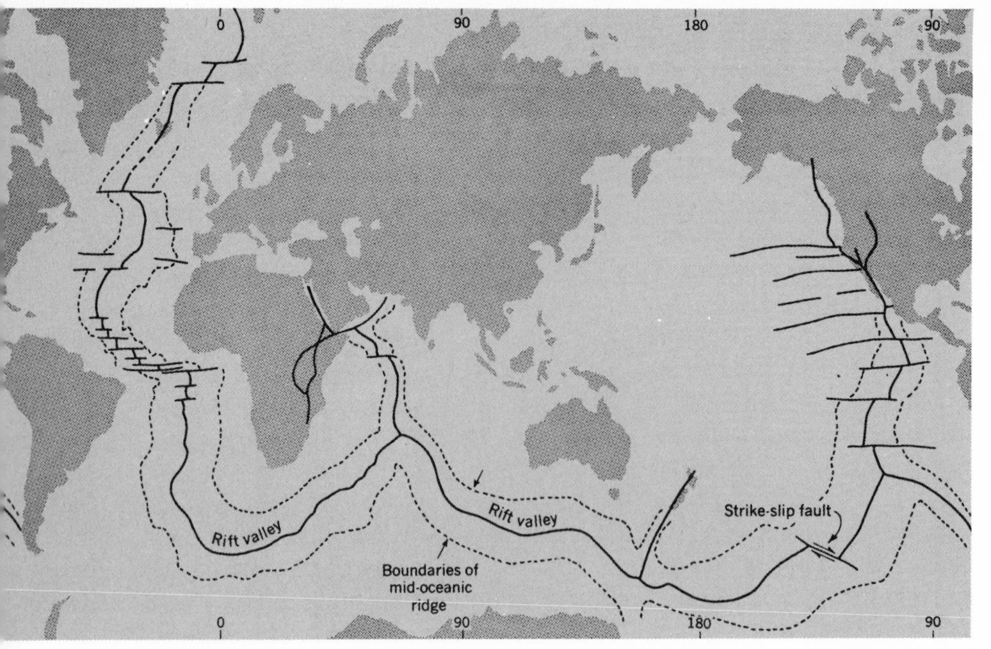

Fig. 34-12. *System of mid-ocean ridges and fracture zones.* [*Courtesy, B. C. Heezen and Lamont Geological Observatory of Columbia University.*]

the Earth, a thickness two or three times this had been expected. Explanations for the apparent discrepancy have involved the nature of the underlying layer 2 (possibly consolidated sediments, at least in part, p. 776) and the concept that the present floor of the ocean may have formed rather recently (p. 780).

The layers in the upper part of a deep-sea core can be dated by the carbon-14 method, and it has been learned that if its texture is uniform, its

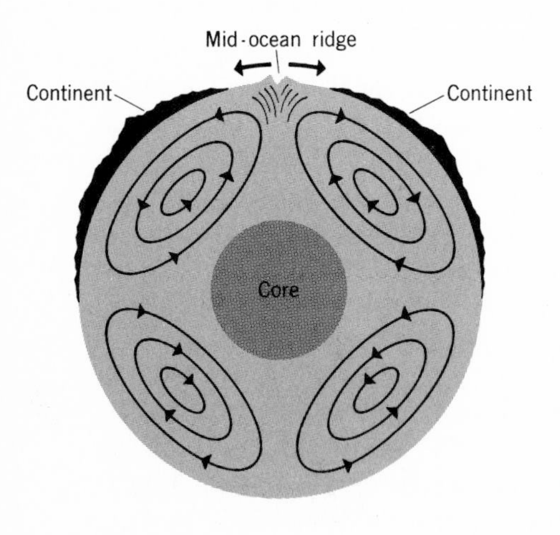

Fig. 34-13. *Convection cells may occur in the Earth's mantle. It is assumed that no major discontinuities occur in the mantle and that very slow plastic flow can occur within it. An upward moving current is initiated if one section of the lower mantle becomes hotter than adjoining sections. This section expands, becomes less dense, and rises. The upward moving current divides and diverges laterally when it reaches the crust; possibly this explains the mid-ocean ridge as an upward bulge and the tension on the central rift zone. Wherever such a current descends, the crust may be pulled downward, possibly in deep-sea trenches and geosynclines.* [*W. M. Ewing.*]

Fig. 34-14. *Part of a deep-sea core 460 cm long, showing a turbidite layer with graded bedding; taken at a depth of 715 meters southwest of Jan Mayen Island (70°33′ N., 11°14′ W.).* [*Lamont Geological Observatory photograph.*]

sediments have piled up at a reasonably constant rate (checked by age determinations at different levels). These rates can be extrapolated downward into the older part of a long core (older than 40,000 years) with confidence if the texture remains uniform.

Submarine Canyons Great canyons, resembling land canyons, cross the continental slopes in places and generally have V-shaped cross sections, concave-up longitudinal profiles, and tributaries, some with a dendritic pattern. Some canyons (Fig. 34-15) can be subdivided into three different parts (on the shelf, slope, and rise) that may have formed in different ways at different times. Rocks, including granite, have been dredged from the walls of some canyons. The canyons may be worldwide in distribution, but some slopes are relatively unexplored. Some occur offshore from major, present-day rivers, but others do not.

The origin of submarine canyons is controversial. One hypothesis, more generally favored at present, suggests that the canyons have been eroded by turbidity currents, and there is little doubt that very large quantities of sediment have been transported through the canyons by turbidity currents. Thick turbidites have been cored in the bottoms of some canyons, and many

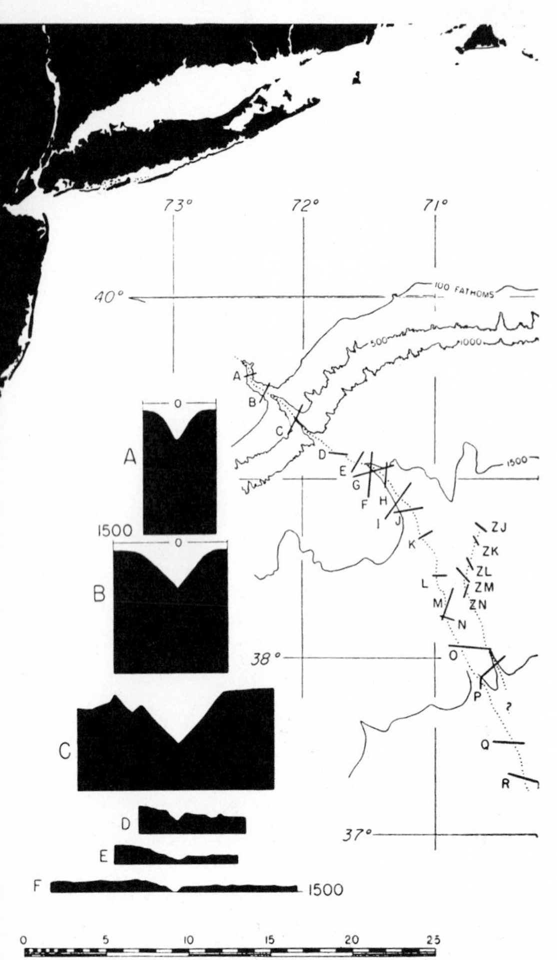

Fig. 34-15. *Cross sections of the Hudson Submarine Canyon. [Heezen, Tharp, and Ewing, "The Floors of the Oceans,"* Geol. Soc. Am. Special Paper *65*.]

canyons terminate at large fan-shaped piles of sediment. In some instances, the volume of the sediments in a fan far exceeds the total volume of the canyon itself. According to one estimate, if every 10,000 grains transported by turbidity currents through a canyon can pick up just one grain from its bedrock floor, then this process can account for the origin of most submarine canyons. In accordance with the logic of this estimate, most of the volume of a land valley is produced by the valley-widening processes of weathering and mass-wasting. Since the valley-widening process should also operate on the sea floor, erosion by turbidity currents must account chiefly for the deepening of a narrow median portion of any one canyon. However, the ability of turbidity currents to carve a canyon out of solid rock is still questioned even though such currents have apparently broken communication cables laid along the ocean floor.

According to the other major hypothesis, the main V-shaped portion of a submarine canyon may have been carved by a river at some long-ago time when the area was land. Subsidence subsequently occurred. Turbidity currents may then have helped to keep the canyon from filling with sediment and may also have eroded the wide, shallow, troughlike outer portion that crosses the continental rise. The inner, shelf portion of such a canyon may have been eroded during the Pleistocene when sea level was lower.

Deep-Sea Trenches and Island Arcs The greatest oceanic depths (about 36,000 feet) occur in long narrow trenches that margin the ocean in places, particularly the Pacific. They are associated with island arcs, volcanoes, earthquakes, young mountain belts, and abnormally low gravitational attraction and are the most unstable regions on the Earth.

Island arcs are elongated, crescent-shaped chains of volcanic mountains which project upward as islands from long narrow submarine ridges (examples: the Aleutians, Japan, the Philippines, the Indonesian States, and the West Indies). The trenches occur on the ocean side of the arcs or parallel coastal mountain belts such as the Andes (which have some features of island arcs). The trenches may extend for 2000 miles, but variations in depth occur along them. Thus the deepest parts form isolated sliver-shaped

Fig. 34-16. *Crustal section for typical island arc (Gutenberg and Richter). In the map* (right) *deep foci are marked by black triangles.* [*Maurice Ewing and Frank Press, "Structure of the Earth's Crust,"* Lamont Geological Observatory Contribution No. *143*.]

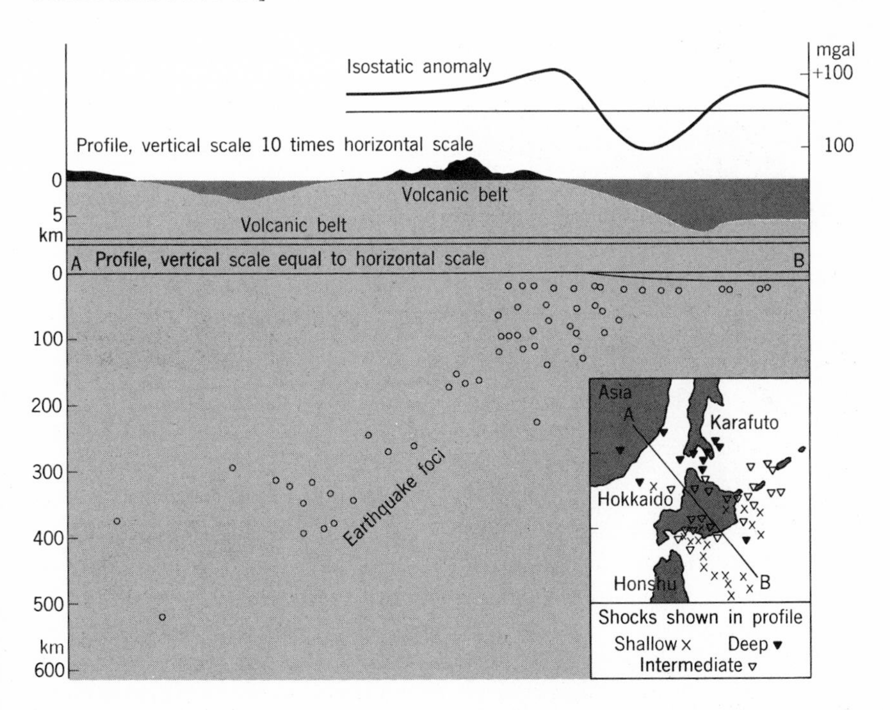

depressions. Some trenches are V-shaped, but others have flat floors apparently formed by the piling up of sediments.

The focal points (places of origin) of the earthquakes that occur frequently in such a region are aligned along a plane that dips about 45 degrees beneath the adjacent continent (Fig. 34-16). Thus shallow-focus earthquakes (within 20 miles or so of the surface) occur primarily beneath a trench; earthquakes of intermediate depth tend to occur beneath an island arc; and deep-focus earthquakes occur farther inland. This plane may represent a major fault zone along which the oceanic crust is being thrust downward beneath a continent. Possibly convection currents account for the downward movement which could also produce a trench and earthquakes. Major fractures might extend upward from such a plane and serve as passageways for the movement of magma to form an island arc. Geometrically, such a plane would form an arc where it intersected the Earth's surface.

The Earth's gravitational pull along a trench is much below normal, whereas island arcs tend to have above-average gravitational attraction. Such regions are not in isostatic balance, and a mass deficiency occurs in the rocks along a trench. This could be explained by an underlying zone of lightweight rocks or by the depression of the surface by a downward moving current.

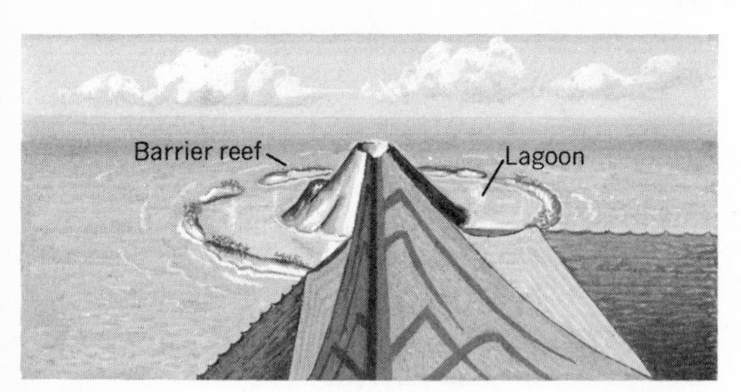

Fig. 34-17. *Darwin's views on the origin of barrier reefs and atolls.*

Coral Reefs and Atolls A *coral reef* is a limestone rock structure that has been built by organisms and fashioned into diverse topographic shapes such as ridges, platforms, and mounds. It is constructed mainly from the skeletal remains of sedentary organisms such as corals and algae, but corals form only a portion of a reef. Reef-building corals are restricted to warm, shallow, relatively clear sea water of normal salinity.

Darwin suggested that the three main types of coral reefs—*fringing, barrier,* and *atoll*—have a genetic relationship (Fig. 34-17). He proposed that a fringing reef developed first around the margin of a volcanic island that had formed in a suitable environment. Next, the island subsides, but slowly enough that the organisms can continue to build the reef upward and outward (they may grow best on the seaward edge of a reef). Thus a barrier reef forms and an atoll may develop from this. Deep drilling at several atolls has shown that subsidence has occurred. One drill hole penetrated some 4000 feet of limestone reef rock before it hit basalt. However, some atolls have probably not passed through the fringing and barrier reef stages, and changes in sea level and climate during the Pleistocene had a modifying effect on some reefs.

SOME GLOBAL GEOLOGY

The Crust and Upper Mantle To interpret the nature of the Earth's interior, seismologists measure the speeds of earthquake waves at various depths. They then make informed guesses concerning the properties that the rocks at such depths must have to explain the behavior of the earthquake waves (at the temperatures and pressures assumed for these depths). Thus interpretations may vary considerably. In some instances, earthquake wave behavior may be explained either by chemical changes or by changes in physical structure (phase changes).

However, it is generally agreed that the oceanic crust differs in important aspects from the continental crust, and these differences seem to extend downward into the upper mantle. The base of the crust is marked by a sharp discontinuity (the *Mohorovičič* or *Moho*) that occurs at an average depth of 20 to 25 miles beneath the continents, but perhaps at nearly twice this depth beneath great mountain belts (Fig. 28-19). The speeds of the P and S waves change abruptly above and below this discontinuity. The Moho is located about 3 to 5 miles below the deep ocean floor, and thus the continental crust on the average is some six times thicker than the oceanic crust.

Beneath the continents the crust seems to consist of two main layers: an upper *sial* (oxides of silicon and aluminum are abundant) and a lower *sima* (oxides of silicon and magnesium are abundant). "Granitic" is a term usually synonymous with sial, but it connotes a wide range of igneous and metamorphic rocks that have chemical compositions somewhat similar to granite. In like manner, "basaltic" is used for sima.

Typical deep-ocean crust seems to consist of three layers. Beneath water that is $2\frac{1}{2}$ to 3 miles deep occurs *layer 1,* which consists of sediments and sedimentary rocks that are commonly less than $\frac{1}{2}$ mile in thickness. The underlying, little known *layer 2* (P-wave speed averages 5 km/sec) is generally about 1 mile in thickness. It may consist chiefly of volcanic rocks or it may be a mixture of igneous and sedimentary rocks. Seismic profiler surveys show the top of layer 2 to be quite irregular. *Layer 3* averages about 3 miles in thickness (P-wave speed is about 6.7 km/sec) and is commonly assumed to have a basaltic-gabbroic compotion. However, serpentine is also a possibility (serpentine can form by alteration of olivine, which may be an abundant mineral in the mantle). Beneath layer 3 is the mantle (P-wave speed is about 8 km/sec).

The continents and ocean basins seem to be in approximate isostatic balance (Fig. 28-19). In other words, the continents project above the surfaces of the oceans because they consist of thick plates of lightweight rocks that are in hydrostatic (actually lithostatic) balance with the heavier oceanic crust. Furthermore, the great mountain belts of the continents are underlain by still thicker zones of lightweight rock. Somewhat analogously, the iceberg that towers highest also extends farthest below the sea surface. This isostatic balance indicates a certain amount of permanence for continents and ocean basins, once formed—i.e., a large part of a continent apparently cannot subside sufficiently to form an ocean basin, nor can a part of the deep ocean floor rise high enough to become a continent. However, this does not rule out the lateral drifting of a continent.

On the average, heat energy is flowing at about the same rate upward through the continents as through the ocean floors (but significant differences occur in different parts of the oceans, p. 781). Heat energy liberated by the radioactive disintegration of elements within the continental crust can account for all of its heat flow; no contributions are needed from the underlying mantle. However, radioactive elements are much less abundant in the thinner oceanic crust. Thus perhaps 90 percent of the heat energy liberated through the oceanic crust must come from the mantle beneath it. Therefore, heat flow measurements indicate that the upper mantle (about 250 miles thick) is not homogeneous: it is different under the continents from what it is beneath the oceans.

The approximate equality of oceanic and continental heat flow may argue against continental drifting, unless the upper mantle drifts also. That is, if oceanic upper mantle (i.e., the type of mantle now beneath the oceans) were to become overlain by continental crust, then heat flow probably would be much higher than average. On the other hand, if continental upper mantle were to become overlain by oceanic crust, then heat flow should become much lower than average.

Although speeds of P and S waves generally increase with depth all of the way to the core, they tend to decrease in the upper mantle within a region known as the *low-velocity zone*. This zone is about 100 miles thick, and its upper boundary is located at a depth of about 35 to 40 miles. The low-velocity zone seems to be more prominent beneath the oceans than beneath the continents and also to be nearer the surface. Perhaps the rocks in the low-velocity zone are closer to their melting points than are rocks located above and below them. If so, the low-velocity zone might be the region where isostatic adjustment occurs (slow plastic flow outward from overloaded areas) and where basaltic magma originates.

Continents and Geosynclines The origin of the continents is basically the problem of the origin of the isolated, sialic portions of the crust. Most hypotheses fall into two groups. In the view of one group, sial may have separated out long ago when the Earth developed a core and mantle. Possibly it was a world wide layer, and subsequently it broke into segments that were then piled up to form the continents.

In another view, the original crust may have been chiefly simatic, perhaps with scattered portions that were somewhat thicker and more sialic. These served as the nuclei of continents that developed later by a process known as *continental accretion*—marginal mountain belts are welded to the edges of the continents and thus enlarge them by the width of each belt.

Complex mountain systems such as the Appalachians and Rockies share a number of sequential steps in their development, although details differ. Such a system initially is an elongated subsiding trough (a *geosyncline*) located along the margin of a continent. Great masses of sediments accumulate in this trough over long periods of time, and lava flows and volcanic debris also pile up in some portions. Just why such a mountain system develops out of the very belt that was once a subsiding trough in the sea floor is a major problem for geologists to solve. According to one suggestion, the sediments beneath the present continental shelf and continental rise may show how geosynclinal sediments accumulated in the past.

Eventually the sediments and igneous rocks in a geosyncline are compressed by powerful lateral crustal forces during a time of mountain building known as an *orogeny*. More than one orogeny may occur, and these may affect different parts of a geosyncline at different times. Granitic

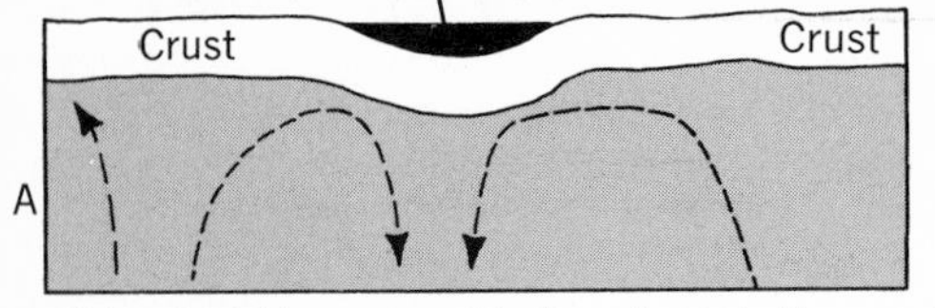

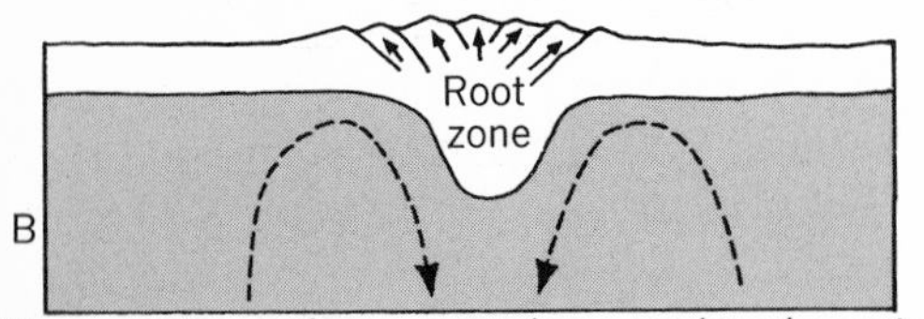

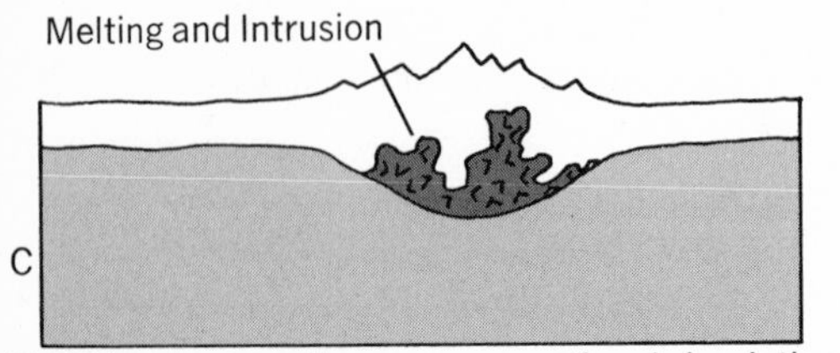

Fig. 34-18. *Mountain building and convection circulation (highly diagrammatic). A. At the beginning of the cycle slowly rising warm currents spread beneath the crust, cool, and descend.—B. The convectional system reaches its maximum speed and the crust is highly compressed. The lightweight rocks of the crust are dragged deep into the mantle. Mountains which result at this time are subsequently eroded to a peneplain as the currents gradually slow down.—C. At the end of the cycle, convection ceases, and the part of the crust which had been pulled downward tends to "float" back up as the subcrustal forces weaken. A mountain belt is formed when this uplifted peneplain is subsequently carved into isolated topographic units by erosion. [After Griggs.]*

batholiths tend to form in parts of a geosyncline during orogenies. Long-continued erosion may then wear away these mountains. Subsequently an entire region may be arched upward many hundreds of feet and then dissected by erosion. Weaker rocks are removed at a relatively rapid rate, and areas underlain by resistant rocks become isolated to form a younger generation of mountains—the ones we see today.

The oldest rocks (igneous and metamorphic types) in North America are somewhat centrally located. In a general way, successively younger rocks are encountered as one goes radially outward from this central zone. Successive belts are arranged in a more or less concentric pattern, with the oldest on the inside and the youngest on the outside. The younger of these concentric belts may be more sialic than the older ones.

Thus the following sequence of events may have been repeated many times as a continent grew. A geosyncline developed along the border of the continental nucleus and was eventually transformed into a marginal mountain belt during one or more orogenies. This transformed crust was thicker and more sialic than the crust that previously existed in the area. Materials of a sialic nature were added (from below by igneous activity) and some materials of a simatic nature were removed by weathering and erosion. Repetition of these events along different margins at different times has resulted in the continent we observe today.

It has been suggested that convectional currents may be responsible for a geosynclinal cycle (Fig. 34-18).

Continental Drifting and Sea-Floor Spreading The possibility that continents may slowly shift their locations has been discussed by geologists for decades. New data have been collected—e.g., data involving paleomagnetism, sea-floor fracture zones, and oceanic heat flow—but the issue has not been settled. Two types of movement seem possible. A continent may shift relative to another. A glance at a map suggests that South America and Africa were once one continent, and the fit is even better along their continental slopes—the true edge of a continent. Some correlations of rocks, fossils, and geologic events in one continent with those in another suggest that the two continents may once have been joined together. However, other comparisons suggest a history of separate continents, and thus uncertainty exists.

The other type of movement might involve the slipping of an outer shell, consisting of the crust and part of the mantle, as a unit on the low-velocity zone or elsewhere within the Earth. The locations of the geographic poles would thus remain unchanged, but different parts of the Earth would pass in turn slowly by these poles.

Paleomagnetism provides some tentative evidence supporting the concept of continental drift. The Earth's magnetic field (p. 490) has been called one of the best described and least understood of planetary phenomena. Apparently the core (chiefly iron?) is not magnetized permanently because it is too hot, and presumably the magnetism is related to slow movements in its outer nonrigid part. When lava cools below a critical temperature (called the Curie temperature; it varies from about 700°C to 200°C) the iron-containing minerals in it become magnetized along the lines of magnetic force acting at this location at this time.

Specimens taken from recent lava flows have recorded the direction of the Earth's magnetic field quite accurately, as have cores taken from deep-sea sediments. Apparently, tiny magnetic particles become aligned along the Earth's magnetic field as they sink through the water and are deposited. Thus the direction of the Earth's magnetism at a particular time and place is frozen into certain rocks and sediments. A similar study of older specimens shows "fossil magnetism," provided the specimens have not been disturbed subsequently by crustal movements or metamorphism.

The magnetic axis is generally assumed to have had about the same orientation through time as the Earth's axis of rotation, although this cannot be checked. So far as is known, the rotation axis remains tilted at the same angle to the ecliptic, although precession does occur (p. 365). The paleomagnetism of ancient rocks of a similar age from different locations on the Earth can thus be consulted to ascertain the location of the Earth's magnetic poles at a particular time in the past. According to some interpretations of these data, polar wandering has occurred, but uncertainties exist.

Specimens from different levels in a pile of superposed lava flows or from different levels in a deep-sea core seem to show that the direction (polarity) of the Earth's magnetic field has been reversed a number of times in the past. In other words, a compass now points north at a particular location, but at one time in the past a compass would have pointed south, and these directions have alternated (nine reversals may have occurred during the last 4 million years or so). As viewed in the perspective of geologic time, the reversals seem to occur rather abruptly, and thus a new method of dating and correlating the past may be at hand.

Reversals of the Earth's magnetic field may explain the north-south magnetic anomalies mapped on the sea floor (p. 768). The magnetism of the sea floor will be higher if magnetic particles beneath it are aligned parallel to the Earth's present field; thus they reinforce it to produce a positive anomaly. However, if an adjoining section of sea floor has a reversed polarity, this will subtract from the Earth's present magnetic field, and a negative magnetic anomaly results.

Recent measurements have shown the sea-floor magnetic anomalies to be symmetrical on either side of a mid-ocean ridge—e.g., a negative anomaly occurs at the same distance east of the Mid-Atlantic Ridge as another negative anomaly does to the west. This has led to a hypothesis of ocean-floor spreading. According to this view, slow convectional currents rise beneath the median portion of a mid-ocean ridge and diverge laterally to produce a rift zone along its crest. New oceanic crust thus forms along this central axis as material from the mantle cools and hardens. The direction of the Earth's magnetic field at this time is frozen into the rocks, and slow lateral creep moves these rocks outward away from the rift zone. This process continues, and reversals of the Earth's magnetic field account for the symmetrical distribution of the magnetic anomalies on either side of a mid-ocean ridge. According to this view, the ocean floor is relatively young even though the ocean basins may be quite old. This might account for the unexpected thinness of the deep-sea sediments.

In some mid-ocean ridges (the East Pacific Ridge is a good example), heat flow is much higher along the crest of the ridge than it is along its flanks. This suggests the presence of a hot upward-moving convection current beneath the crest and a cooler downward-moving one beneath its flanks. Heat flow does not seem to be as high along the crest of the Mid-Atlantic Ridge, a fact which may indicate that mid-ocean ridges have different ages and pass through an evolutionary cycle. The youthful stage may be represented by the East Pacific Ridge. It lacks a central rift zone, although shallow-focus earthquakes do occur along its crest. The Mid-Atlantic Ridge may be mature. Possibly the Mid-Pacific Rise (not now a part of the system of mid-ocean ridges) may represent an old stage—it has subsided and is seismically inactive.

Origin of the Atmosphere and Hydrosphere According to an older view, the Earth began as a large ball of hot gases, cooled to a liquid, and then became solid. The atmosphere and hydrosphere are residual in this model, although some gases may have escaped into space, and the oceans had their present volume at a very early stage in the history of the Earth.

According to a currently favored hypothesis, the Earth probably formed from nebular material that was quite cold (p. 478). Heat energy from radioactive disintegrations eventually produced a molten Earth which differentiated into core, mantle, and crust. The Earth's original atmosphere then escaped, and the present air and water have resulted from degassing of the Earth's interior. Emanations from volcanoes and other sources are apparently being given off at present in suitable proportions and quantities to account for all of the air and water on the Earth, provided that they have continued in this manner since early in the Earth's history.

The following gases may have been present in the early atmosphere: hydrogen, nitrogen, water vapor, carbon dioxide, ammonia, and methane. Free oxygen in the atmosphere may have resulted from two processes: the dissociation of water vapor in the upper air by solar radiation, and as a by-product of photosynthesis by green plants. The ammonia and methane would gradually disappear by reacting with oxygen to produce nitrogen, water vapor, and carbon dioxide.

Exercises and Questions, Chapter 34

1. Describe some of the methods devised by oceanographers to measure or observe each of the following: oceanic depths, sediments on the sea floor, location of an observing ship, and the ocean water at various depths.

2. Go to the library and collect information about one of the following: desalination of sea water, sea-floor manganese deposits, the food resources of the sea, the natural resources of the sea floor, a recent man-in-the-sea project, or the voyages of the *Challenger* or the *Fram*.

3. How does the oceanic crust differ from the continental crust?
4. Why do mountains project higher above sea level than a low plains area?
5. What is the hypothesis of continental accretion and what evidence supports it?
6. What is meant by the geosynclinal cycle?
7. Describe briefly the different types of geologic phenomena that may possibly be explained by convection curents.
8. What is the evidence that a major fault zone may occur beneath a deep-sea trench and island arc?
9. Go to the library and search for evidence for and against the hypothesis of continental drifting.
10. Discuss the origin of submarine canyons.

35 The fossil record

SO FAR we have dealt chiefly with the inanimate side of the Earth's long and eventful past—the advances and retreats of ancient seas, volcanic eruptions, the formation of great mountain ranges, the constant wasting away of higher land masses by the persistent forces of erosion, and the growth and shrinkage of glacier ice. With the entrance of life on this scene, the Earth's history takes on a new focus of interest. Our subject is the countless multitude of animals and plants—huge and tiny, spectacular and commonplace, marine and nonmarine—that lived and died during the immensity of geologic time (Fig. 35-1) that stretches back from the present to the shadowy reach when life itself first arose. The former existence of these organisms is recorded by fossils found in rocks (Fig. 35-2).

Whether the history is physical or biological, its keynote is unending, leisurely change—change that is imperceptible to untrained eyes during any one lifetime, but whose cumulative effect has been very great. These changes, of course, continue at the present time: valleys grow deeper, glaciers shrink, the Earth's crust moves, new types of animals and plants arise. Although the changes of today are not precisely those of the past—in kind, location, magnitude, and rate—their observation and measurement by geologists have resulted in the development of techniques and concepts that have been applied with considerable success in elucidating the events of the past (this is the assumption of uniformitarianism which is basic in geology).

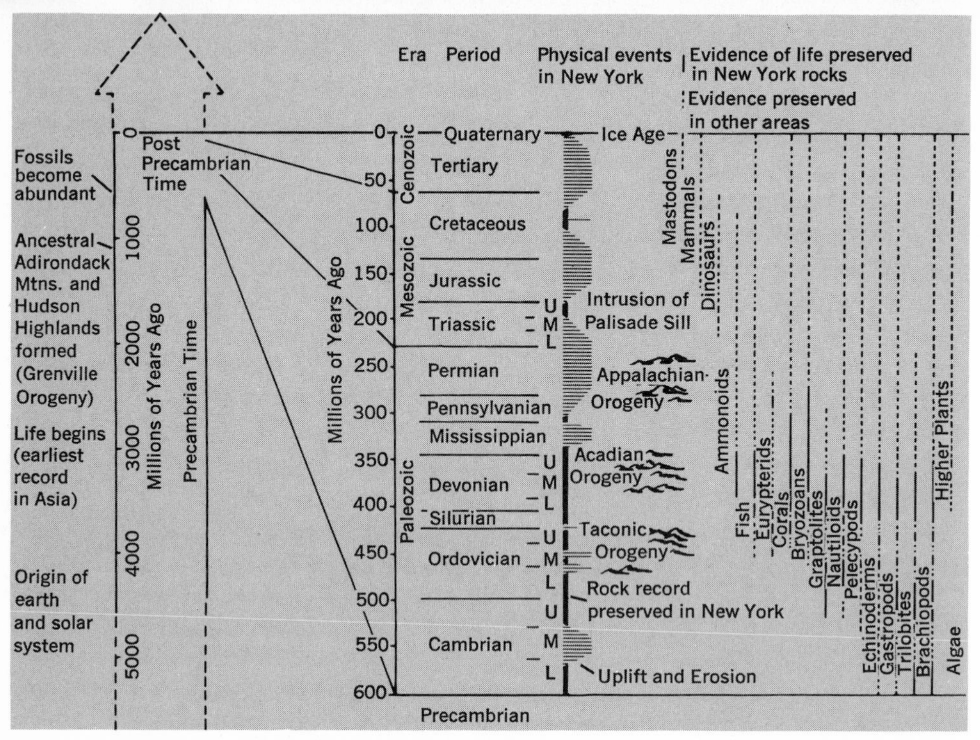

Fig. 35-1. *Geologic chronology, showing major events in the geological history of New York State and elsewhere on the Earth.* [*Broughton, Fisher, Isachsen, and Rickard,* The Geology of New York State, *1962.*]

FOSSILS

Fossils (Latin: "something dug up") are the recognizable remains or traces of animals and plants that were preserved in sediments, rocks, and other materials such as ice, tar, and amber prior to historic times (Fig. 35-3). Fossils provide a record of past life which is at best fragmentary. Consider the multitude of organisms in existence at any one time. A very small fraction of these will subsequently become fossilized, and only a tiny proportion of the fossilized forms will eventually be exposed by erosion at the Earth's surface and be discovered by man (Fig. 27-4).

Two factors are favorable for the preservation of organisms as fossils: one is the possession of hard parts, such as shells and bones; the other is quick burial of the remains to prevent destruction by scavengers and decay. Obviously, animals like worms and jellyfish have much less chance for preservation than animals like clams and oysters. Yet the possession of hard parts does not guarantee fossilization. Until late in the 1800's, huge herds of buffalo roamed the prairies of the western United States. Today the buffalo have all but vanished, and hardly a trace remains of their former presence. Their carcasses were not covered rapidly by

Fig. 35-2. *A large Jurassic fossil cephalopod, about 4½ feet across, discovered near Fernie, British Columbia* (Titanites occidentalis). [*Geological Survey of Canada.*]

sedimentation following death. Scavengers and bacteria quickly destroyed the flesh, and the hard parts soon weathered to dust.

Quick burial may occur in a number of ways, but the sea floor is the environment in which the greatest number of organisms become fossilized. This was especially true in much of the past when ocean water covered larger portions of the Earth's surface than it does at present. Countless animals with hard parts lived on sandy or muddy bottoms. After death, the shells of these organisms accumulated on the sea floor and were covered rapidly by sediments. In addition, many living animals were buried by sudden shifts of sediments during storms at sea. As a result, some marine sedimentary rocks are extremely fossiliferous—they should be observed in the field for a true appreciation of the significance of fossils to geology.

Falls of volcanic ash may also bury organic remains quickly. Much

Fig. 35-3. *Eohippus fossils partially uncovered in the block where they were found. Burlap and plaster are used to protect a fossil-bearing rock slab during transportation and are visible in the lower part of the photograph.* [*Courtesy of the American Museum of Natural History*.]

less frequently, lava flows overwhelm and solidify around organisms before their destruction. A mold in the lava indicates the former presence of the animal or plant and provides an exception to the general rule that igneous rocks do not contain fossils; e.g., the mold of a prehistoric rhinoceros has been discovered in the State of Washington. As indicated by its bloated shape, the animal was dead when it was buried by a lava flow. Its carcass probably floated in a pond of water, and the water chilled the lava quickly enough to preserve a mold. Animals have likewise been buried rapidly when they were trapped in peat bogs and asphalt pits (Fig. 35-4).

Fossilization Fossilization may occur in several ways; preservation of an entire organism, preservation of the hard parts alone, petrifaction, and the formation of impressions, molds, casts, and carbon residues.

Preservation Entire. An entire animal may be preserved without change. This is rare and occurs only for organisms that lived recently (in terms of geologic time). Woolly mammoths have been preserved in the permanently frozen ground of Siberia; e.g., one was discovered in the bank of a stream. Sudden, violent death seemed indicated by the presence of unswallowed food in its mouth, a broken hip, and clotted blood in its chest. Perhaps the scene can be reconstructed: the animal browsed along a high river bank, moved too near the edge, caused a landslide, and was killed and buried during the resulting fall of dirt and debris. Later the loose material froze solid, and the mammoth remained in cold storage until the river eroded laterally and exposed it. When discovered, some of the animal's flesh was still red and was eaten by wild dogs, but the stench was overwhelming. Evidently the flesh had partially rotted before it was frozen because the ground around the specimen also smelled.

Preservation of Hard Parts. More commonly the flesh and soft tissues disappear soon after death, and only the hard parts of animals are preserved unchanged. To date, more than 45,000 pairs of tusks of the woolly mammoth have been sold from Siberia as ivory, whereas the flesh and soft parts of these huge beasts were found preserved in only a few cases. Marine shells have been kept intact in sediments for millions of years.

Petrifaction. Still more commonly the bones and shells of animals are changed into a stony substance and thus preserved. Porous bone structures may be filled by mineral matter, or organic materials may actually be replaced little by little by such substances as silica (Fig. 35-5), calcium carbonate, and iron oxide. Replacement is often complete to the tiniest detail of structure.

Impressions, Molds, Casts, Footprints. Impressions may be formed and preserved in the sand and mud in which animals are buried. Again, a shelled animal, say one of the snail-like gastropods, may die, be buried in sediments, and lose its soft internal parts by decay. An *internal mold*

Fig. 35-4. *The La Brea tar pits in Los Angeles. As oil seeped from the ground the more volatile portions distilled away and left a sticky residue. Here Pleistocene birds and beasts of prey gather around victims mired in the pits, only to find themselves trapped. We see great vultures, saber-toothed cats, the dire wolf, horses. The bones of camels, elephants, giant bison, great ground sloths, bears, and lions have also been found.* [*Chicago Museum of Natural History.*]

(of the inside of the shell) of this animal forms subsequently if mineral matter or sediments fill the interior of the shell and harden before the shell itself is dissolved. An *external mold* may form in the entombing sediments if the entire organism—shell as well as soft parts—disappears. Mineral matter of some kind may later accumulate in the mold and harden to form a cast.

Remarkable lifelike molds of insects turn up in amber. The insects were trapped and buried in the sticky resin of conifer trees, which later hardened and fossilized as amber. The insects decayed and disappeared, but the hollow molds that remain in the amber faithfully reproduce delicate hairlike appendages and other parts of their bodies.

Footprints or trails may form if an animal crosses an area of soft mud or sand that is subsequently baked and hardened by the sun's rays and then buried by additional sediments (Fig. 35-6).

Carbon Residues. Organic materials are composed largely of carbon, hydrogen, and oxygen. The hydrogen and oxygen volatilize and disappear soon after the death of an organism, but a thin film of carbon may be left. It may cover an impression as a "carbon copy."

Fig. 35-5. *A tree in the Petrified Forest in Arizona. The organic material which once made up this log has been entirely replaced by silica. This and other logs occur in stream-deposited sedimentary rocks of early Mesozoic age. Apparently the trees, something like the present pines, grew upstream and were floated down as logs to be buried in sand bars and along the banks of the ancient river. During and after petrifaction, sediments accumulated to a great thickness on top of the layers with abundant logs. Uplift and erosion has subsequently exposed the zone with petrified logs. A few millions of years ago the logs were probably still buried; a few millions of years in the future (or sooner) all traces of the logs will probably be gone.* [*Santa Fe Railroad.*]

Fossils and Ancient Geography Fossils commonly distinguish marine from nonmarine sedimentary rocks. Careful mapping of rock formations and the matching (correlation) of various strata thus enable the geologist to work out past distributions of land and sea. Such evidence shows that North America and other continents were partly submerged by extensive shallow seas a number of times in the past. Perhaps half of North America was flooded by ocean water during one interval in its geologic history.

Fig. 35-6. *Mud cracks and dinosaur footprints in Jurassic rocks in Morocco.* [*Photograph, G. Termier.*]

Fossils show that a land bridge once connected Siberia to Alaska in the Bering Strait area; in fact, a drop in sea level of about 150 feet would produce such a land bridge today. Part of the evidence involves the sudden appearance in North America of elephants and certain other animals after they had developed for a long time in Eurasia. In other words, older rocks in Siberia contain a sequence of fossils of primitive and more advanced types of elephants, whereas these are absent in rocks of a similar age found in Alaska. Therefore, when these rocks formed, the two areas apparently were separated by water as they are today. This conclusion is based upon the general principle that two interconnected areas with similar environments should contain more or less similar types of animals at any one time if they can wander freely back and forth (but not necessarily the same species). Since younger rocks in Alaska and Siberia both contain fossils of the more advanced types of elephants, a land connection must have formed between the two which permitted migration from one continent to the other.

Similar evidence indicates that North and South America were separated several times in the past by a water barrier in the vicinity of the present Isthmus of Panama, and that Great Britain formerly had a land connection with the European coast.

The direction from which the sea flooded an area can sometimes be determined. For example, modern marine shells differ considerably along the Atlantic and Pacific coasts because a land barrier exists. One study in the Panama area had the following results: several hundred species of shell-bearing invertebrates were collected on the Atlantic side, a similar number of species were found on the Pacific side, and only two dozen or so were common to both sides. Thus identical species tend to be restricted to certain faunal realms even though fauna throughout the world have similar characteristics at any one interval in geologic time. Therefore, a comparison of fossil types may indicate the direction from which a sea once advanced upon the interior of a continent.

Interpretation of Former Climates Most animals and plants of today are restricted to certain climates and types of environment, and their ancient relatives presumably existed under similar climatic restrictions. Although animals and plants do adapt themselves to widely different environments as they evolve, yet whole assemblages of animals and plants seem to have strong climatic implications. For example, since large reptiles and amphibia live today only in areas where temperatures do not drop below freezing, the fossil palms and crocodiles found in the Dakotas imply a warm climate in that region at one time in the past.

EVOLUTION

Evolution (Latin: "an unrolling") is the concept that all existing types of animals and plants have developed from previously living types by slow orderly changes which, in general, have produced more complex organisms from more simple. The kinds of changes that produce new types are familiar to us all. Consider the varieties of dogs existing today. They have all originated by selective breeding from common ancestral stock, most of it during the last few centuries. The great variations in other domesticated animals—horse, cow, and chicken—and the diversity of cultivated plants underscore the ability of organisms to change type. A theory of evolution is accepted almost unanimously by all who have examined the evidence bearing on it, even though its exact mechanisms are not fully understood.

According to this concept, if one could go backward into time, he would encounter animals and plants progressively more different from the familiar forms of today. Eventually on this imaginary trip, the traveller would observe remote ancestors very different from their present-day descendants, and he would fail to see the connection between them without the many transitional steps. The in-between forms are needed to show that descent with modification has occurred, and the fossil record furnishes the evidence. On this imaginary journey, one would eventually encounter types of organisms that have no living descendants—they became extinct long ago. On the other hand, certain organisms would show remarkably little change even though the cycle of birth, growth, and death has been repeated a prodigious number of times.

At the conclusion of the journey, a keen observer would realize that certain important trends had occurred.

- Throughout known geologic time, life has tended to increase in total numbers, in the types of organisms in existence at any one time, and in the variety of environments inhabited. There has been a tendency for any one group to spread out, diversify, and occupy different types of environmental niches.
- Older groups of organisms have been replaced by newer and generally more complex groups. Many extinctions have occurred.
- Generalized forms have tended to persist longest on the Earth, whereas other forms, which became more and more specialized for life under certain conditions, could not survive when those conditions subsequently changed.
- Rates of modification have varied from time to time, place to place, and group to group.
- The vast majority of the members of any one phylum and class have remained part of that phylum and class even though new families, genera, and species have been produced within the major group. However, at certain key times in evolutionary development, a few members of a certain group began a series of changes that, after much time, deviation, and random selection, resulted in the development of an entirely different group (p. 799).
- One is led, according to evolutionary theory, backward through time to some obscure, undated time and place where the first one-celled forms of life originated. Prior to this, no organisms existed, only lifeless chemical substances.

The central themes in the current explanation of evolution include the following (modified considerably from Darwin's original proposal).

(1) All organisms tend to produce more offspring than can survive within limitations set by amounts of food and space available.

(2) Because of this overpopulation, a competition for the means of existence occurs, in which some organisms perish without reproducing themselves, or reproduce in inferior numbers.

(3) Although offspring tend to inherit the characteristics of their parents, they vary in many minor ways. No two individuals are exactly alike. From time to time, notable changes show up in offspring, produced apparently in the reproductive cells by a process of *gene mutation*.

(4) Changes caused by gene mutations are transmitted to the next generation of offspring.

(5) Some of the changes produced by gene mutations are advantageous in a given environment, whereas some handicap the offspring. If the variation results in organisms which are better adapted to their environment, they will tend to live longer and produce more offspring than their less fortunate relatives. The interaction of these factors eventually produces changed individuals that are remarkably fitted for survival under the conditions in which they exist.

(6) If the environment changes, or if the organisms move to a different environment, they may not survive under the new conditions unless their mutations produce a new and better-adapted type from the original stock. Environments have changed many times in geologic history and animals and plants have evolved into many diverse forms, as one or another mutated form was favored.

Note that according to current ideas of evolution, the individual does not pass on the qualities he has acquired to his offspring, but only those he has inherited. The modern giraffe has a long neck, but not because its ancestors stretched for foliage high in the air in times of food shortage. A child may develop powerful muscles, but not because his father exercised diligently and passed powerful muscles on to his children. What is thought to have happened is more nearly this: sudden change in germinal material (gene mutation) created a few antelope that varied from their parents by having unusually long necks. These antelope-giraffes were better able to survive periods of food shortage than their relatives with short necks. The long-necked antelope-giraffes produced offspring who inherited the long necks of the parents because gene mutations are transmittable to future generations; and among these, those with the very longest necks thrived best. Thus antelope-giraffes with short necks became scarce, whereas long-necked antelope-giraffes increased in number. The food shortage did not cause the gene mutations which produced long necks, but it did act as a selective factor that eliminated giraffes without long necks (natural selection). Perhaps gene mutations producing long necks had occurred in the antelope-giraffe species a number of times in the past when food was abundant. As there would be no advantage for long necks at such times, a breed of long-necked giraffes could not become established.

An impressive body of data supports this concept of evolution. *Ontogeny recapitulates phylogeny* is an erudite phrase asserting that the life history of the individual (ontogeny) tends to go quickly through the stages (recapitulate) of the evolutionary history of the race (phylogeny). Often this recapitulation is incomplete and imperfect. As an illustration, the tadpole stage of the frog seems to imply descent from a fishlike ancestor. Vestigial structures, such as the vermiform appendix and ear muscles in man, seem to indicate their derivation from an ancestor who found them useful. Dozens of such vestigial structures have been recognized in the human body. Blood tests, taxonomy, and the geographic distribution of animals and plants furnish additional evidence supporting the concept of evolution.

But the most convincing data attesting to evolutionary changes are fossil sequences found in rock layers. The horse furnishes a classic example (Fig. 28-14), and the rhinoceros, elephant, and camel also developed from small unspecialized ancestors.

GEOLOGIC CHRONOLOGY

The Earth's long record of physical and biological changes would be difficult to survey if we could not subdivide it into convenient units. The four major units are called *eras:* Precambrian, Paleozoic, Mesozoic, and Cenozoic (Fig. 35-1). Smaller units—*periods*—subdivide the eras. We can think of the eras as four volumes of a multivolume book and the periods as chapters making up a single volume. Volumes and chapters vary greatly in length, and the number of chapters in a volume is far from uniform.

Pronounced changes in both the physical and fossil records separate the eras of geologic time. Their names show the influence of the prevalent forms of life: Paleozoic means "ancient life," Mesozoic "medieval life," and Cenozoic "recent life." The Precambrian era includes all Earth history which precedes the oldest period (Cambrian) of the Paleozoic era. Precambrian is sometimes divided by geologists into two or more eras. It encompasses an estimated five-sixths or more of all geologic time.

The eras and periods were delineated chiefly by nineteenth-century geologists on the basis of physical and biological changes that they observed in the rocks in the local areas they investigated (largely in western Europe and Great Britain). The most profound changes occurred where major unconformities separated thick units of folded, faulted, deformed rocks from thick sequences of younger and much less deformed rocks. Fossils found in the rocks above the unconformities were much different from those in the older rocks below. The rocks occurring below one unconformity were said to have formed in one era, and those above in another. This constituted a natural basis for subdividing the rock record. Unconformities of lesser magnitude and extent and associated with smaller fossil differences were then used to subdivide the rocks of each era into smaller units, and each such unit was said to have accumulated during a geologic period.

Disagreement and uncertainty subsequently developed when these units were looked for on other lands. Do worldwide physical breaks occur in the rock record and can they be recognized? Some geologists reason somewhat as follows. Times of great mountain building (orogenies) have occurred at irregularly spaced intervals in the past. At such intervals, crustal movements deform rocks, elevate lands, and restrict seas, perhaps in part by deepening the ocean basins. Erosion tends to predominate on the lands; few permanent records are formed which are later accessible to man, and older records are destroyed. Climatic changes caused by such mountain building may speed up evolutionary development somewhat because organisms must adapt themselves successfully to changed environments or perish; e.g., restricted seas may crowd shallow-water marine animals into smaller areas where many perish.

If erosion finally succeeds in wearing down the continents or if subsidence takes place, seas again may encroach upon the lands. Sediments deposited in such shallow seas may be preserved as permanent records. These sediments rest unconformably upon the underlying rocks and contain distinctly different types of fossils from those in the older rocks beneath. Thus great biological changes may tend to correspond to great physical changes on the Earth.

Whether or not such changes have recognizable worldwide effects is disputed. Major and minor changes may well have occurred in different areas at different times, which makes it difficult to match unconformities from one place to another on the Earth. At any rate, the rocks and geologic histories of western Europe and Great Britain have had a great influence on geologic chronology. If rocks had been studied first in other areas (e.g., China or Africa) a quite different chronology might have been proposed.

EVOLUTIONARY DEVELOPMENT OF THE VERTEBRATES

Beginning of the Fossil Record In most areas Cambrian strata are separated by a major unconformity from older, underlying Precambrian rocks which commonly are deformed, crumpled members of the igneous and metamorphic groups. However, unmetamorphosed Precambrian sedimentary strata also occur, and in a few places these appear to merge conformably upward into Cambrian strata. Many Cambrian and younger sedimentary rocks are abundantly fossiliferous, whereas the Precambrian rocks are almost devoid of fossils. However, concentrically layered deposits built by simple, single-celled plants known as calcareous algae have been found in Precambrian rocks, and the oldest of these have an age of approximately $2\frac{1}{2}$ billion years. Similar plants exist today. These algae, a few sponge spicules, some wormlike trails, and a few markings or deposits of a dubious nature formed the only direct evidence of Precambrian fossils prior to some rather recent discoveries.

Indirect evidence of Precambrian life may be furnished by the calcareous materials in ancient limestones and marbles, the carbon in graphite deposits, and huge sedimentary iron ore deposits. Such deposits may owe their origin to the accumulation of organic materials or to the functions of living organisms which caused their precipitation.

Among the more recent discoveries are South Australian fossils that represent soft-bodied creatures: jellyfish, corals, segmented worms, wormlike creatures, and two animals completely different from any known creatures (circular in outline, about an inch or less in diameter, with markings that suggest tentacles, legs, or gills). However, the age of these fossils is somewhat uncertain; possibly they are younger than Precambrian.

Two-billion-year-old Precambrian fossils from Ontario include new species which "range from single-celled isolated spheroidal bodies to filamentous branched and unbranched forms." Their classification is uncertain. Other reported finds include primitive, thin-shelled brachiopods on Victoria Island from shales that may be about 700 million years old, and from other areas objects resembling bacteria, perhaps 3 billion years old.

Paleozoic Forms Many of the Cambrian strata above this major unconformity are abundantly fossiliferous. Furthermore, the creatures who left fossils possessed highly organized systems; they were not simple forms. Nearly every phylum except that of the vertebrates had representatives living near the beginning of the Paleozoic era. In fact, if life has evolved from a one-cell initial stage to present-day complex organisms, more than half of the evolutionary development is estimated to have taken place before Cambrian time.

Animals with hard parts predominate among fossils collected from Cambrian rocks. Trilobites are the most abundant forms, and brachiopods are numerous. Thus, the fossil record suggests that trilobites, brachiopods, and other creatures with hard parts constituted about the only forms of life in existence at that time.

But now an exciting fossil discovery in Cambrian rocks in British Columbia indicates that soft-bodied forms were also abundant. About 130 species of animals have been described from this location: sponges, jellyfish, worms, and shrimplike animals are preserved as thin films of carbon along bedding surfaces.

The advanced stage of development attained by Cambrian forms implies complex ancestors and a long period of evolution in the Precambrian, but no record of such forms has been found. Why? According to one hypothesis such ancestors did exist but had no hard parts and were not preserved as fossils. Perhaps during a relatively short interval that occurred some 600 million years ago, animals developed external shells and hard parts and then could later be preserved as fossils (Fig. 35-7). The first organisms may have lived near the surface of the ocean where hard parts would be a handicap in staying afloat. However, when this zone became too crowded, some organisms shifted to an ocean-floor environment, and certain ones began to prey on others. Under such conditions, protective hard parts would be an advantage, and they might evolve rapidly.

Fossil trees have not been found in rocks older than the Devonian, and the appearance of early Paleozoic lands is uncertain. However, it seems likely that plants more complex than algae were present during the first part of the Paleozoic era and perhaps in the Precambrian, because animals are directly or indirectly dependent upon the plant kingdom for food. Furthermore, Devonian plants were complex forms which imply a long previous period of evolution. Perhaps early Paleozoic plants were restricted to water. At any rate, vegetation flourished during the latter part of the Paleozoic era (Fig. 35-8).

Fig. 35-7. *Restoration of Middle Silurian sea bottom in the Niagara region. The model shows typical forms of crinoids, straight-shelled cephalopods, trilobites, brachiopods, and snails.* [*Prepared by George and Paul Marchand under the direction of Irving G. Reimann. Courtesy, Buffalo Museum of Science.*]

Fig. 35-8. *A late Paleozoic (Pennsylvanian) landscape showing characteristic animals and plants. Fin-backed reptiles and the giant dragonfly move among the scale trees.* [*Courtesy, Peabody Museum, Yale University. From a painting by Rudolph Zallinger.*]

Primitive types of fishes, found as scattered bony plates and scales in Ordovician rocks, constitute the first known vertebrate remains, but their invertebrate ancestors are unknown—a real "missing link" in the evolutionary sequence is awaited here. Succeeding steps in the evolutionary history of the vertebrates are shown by fossils found in Paleozoic and younger rocks.

The order of increasing complexity among the five classes of vertebrates follows: fish (simplest), amphibians, reptiles, and birds and mammals (most complex). This sequence parallels their first appearances in the geologic record: the oldest fossil fish are found in rocks which formed in the Ordovician period; amphibians first became part of the fossil record in the Devonian and reptiles during the Pennsylvanian. Mammals and birds are not found in Paleozoic rocks; they appeared in the Mesozoic era, although they were not prominent in the life of that time. Future fossil discoveries may, of course, produce older representatives of each class, although this appears unlikely in some instances.

These parallel sequences—the order of first appearance and of increasing complexity—suggest convincingly that evolutionary development has taken place. After the first appearance of a vertebrate class, the fossil record shows that the unspecialized ancestral forms spread out widely, were adapted to different environments, and developed numerous new types. Each group of animals has tended to spread, diverge, and specialize until it attained a maximum development. This culmination has generally been followed by a decline and the eventual extinction of many forms.

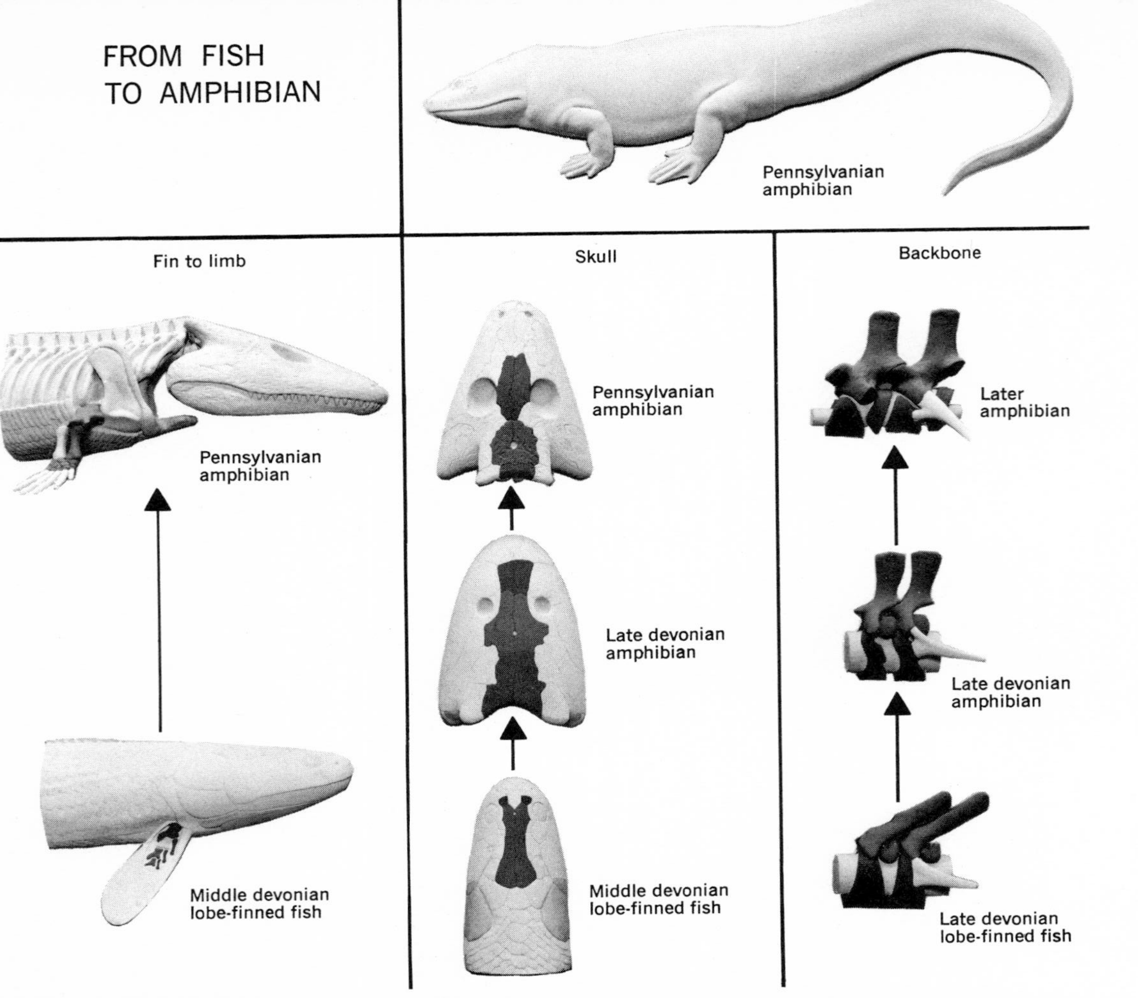

Fig. 35-9 *From fish to amphibians.* [*Courtesy of the American Museum of Natural History.*]

Conservative, unspecialized forms have tended to persist longest on the Earth in each of the phyla, whereas highly specialized forms have tended to become extinct rather rapidly. So far as is known, evolutionary development is not reversible, and evolution has been succinctly termed a one-way road, i.e., more complex forms do not evolve retrogressively into simpler forms, although certain trends, such as a gradual increase in size, may be reversed. But the road winds and detours, and it has many dead-end branches.

Fish Fossil fish are uncommon in rocks formed before the Devonian period, which is sometimes called the Age of Fish because of this initial abundance. Fish become more plentiful as fossils in succeeding periods.

Among the diverse kinds of fish existing in the Devonian were two groups which have special evolutionary significance: the lobe-finned fish (Crossopterygii) and the lungfish (Dipnoi). Although lungfish are rare today, once they were more numerous and varied. A few members live in areas that annually undergo long dry spells (Africa, Australia, and South America), and apparently the type has experienced relatively little change since the Devonian. These lungfish have a swimming bladder opening into the throat so that it can be used as a crude lung. They have been known to live out of water for more than a year.

Amphibians The lobe-finned crossopterygians apparently evolved into a short-legged, sprawling, primitive type of amphibian late in the Devonian period, and the evidence is quite convincing (Fig. 35-9). These lobe-finned fish apparently had three structures that made this evolutionary step possible: an air bladder was connected to the throat and might have functioned as a primitive lung; their lobe fins contained bones that could support primitive limbs; and a nostril-like opening existed from the outside into the mouth.

The evolutionary development of air bladder into lung and of fins into limbs appears to be an adaptation to recurrent conditions of dryness. As the amount of water diminished in an area during its annual dry spell, fresh-water lakes and streams may have been reduced to isolated pools of stagnant water. Many fish would die under these circumstances. According to the modern theory, however, if gene mutations had previously produced a few fish with stronger fins and more efficient air-breathing apparatus, such fish might be able to flop from a waterless depression to a nearby pool of water. The mating of such survivors would produce offspring with the characteristics of their parents. Perhaps the amphibians evolved from the stout-finned lungfish in such a manner, with occasional detours into unproductive forms. As a class, the amphibians culminated late in the Paleozoic era and then declined.

Descent from fishlike ancestors is likewise strongly suggested by the life cycle of living amphibians. Members of this class of vertebrates still return to the water to lay their eggs. These hatch into youthful, appendageless, fishlike forms which breathe by means of gills. Later in life, limbs grow and lungs develop.

The question has been asked: Why do simple forms of life still exist today if the evolutionary trend has been in general from the simple to the complex? For example, why have not all fish evolved into amphibians? This question seems to ask, "Why are there any fish left?" and the answer is not far to seek. So long as the waters of the Earth furnish abundant means of fish life, there will be abundant forms of fish life there. Fish no doubt continue to develop forms even better adapted to life in this and that environment in the water. But why does not a new form of amphibian crawl onto the land and start a new race of land-living forms? There is nothing in the theory of gene mutations to deny this possibility, and yet the opportunity will never again be what it was for the first lung fish that flopped out upon a land where no other vertebrate had its range. Today so ill-adapted a land creature would hardly be in the competition. The first creature to produce the needed mutations for land forms gave its descendants a head start that is not easily overcome. Fish today lack both the opportunity and the necessary organic structures, such as primitive lungs and stout fins, to initiate such an evolutionary change.

Reptiles The stem reptiles (cotylosaurs, Fig. 35-10) evolved from a type of amphibian (labyrinthodonts). Transitional forms found in Pennsylvanian rocks fit so nicely between reptiles and amphibians that their classification is disputed.

From the stem reptiles came the turtles, certain huge marine reptiles, the flying reptiles, mammal-like reptiles (therapsids), the fin-backs, the ancestral stock of the dinosaurs and birds (thecodonts), and other forms. The finbacks (pelycosaurs, Fig. 32-11) were unique in their possession of spines that projected upward from the vertebral column and attained a maximum length of about 3 feet. Membranous material probably connected the spines in life, but their function is uncertain. The group became extinct in the Permian at a time when it had no known enemies.

Fossils taken from Mesozoic rocks, the Age of Reptiles (Figs. 35-11 and 35-12), make spectacular exhibits in museums. Reptiles were present during the latter part of the Paleozoic era and exist today, but they culminated in numbers, varieties, and evolutionary development during the Mesozoic era, when they dominated land, sea, and air. The dinosaurs (Greek: "terrible lizards") were confined to the lands and varied in bulk

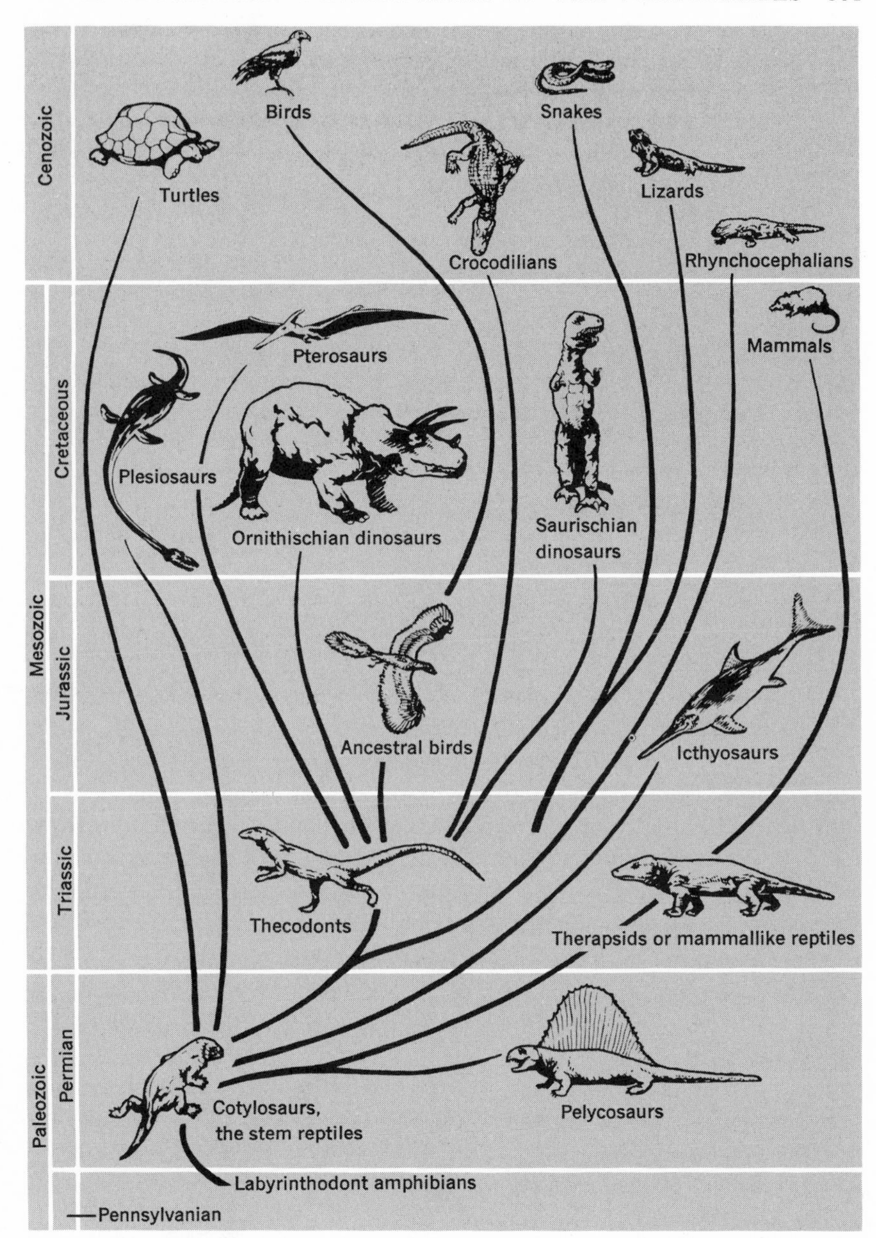

Fig. 35-10. *The family tree of the reptiles: a pictorial diagram summarizing the history of the reptiles. [Courtesy of the American Museum of Natural History.]*

Fig. 35-11. *A Mesozoic (Jurassic) landscape showing characteristic animals and plants. The neck of* Brontosaurus *arches over* Stegosaurus *with his ridge of plates.* Allosaurus *is at the center.* [*Courtesy, Peabody Museum, Yale University. From a painting by Rudolph Zallinger.*]

from the size of a rooster to ponderous giants that approached a length of 85 feet and a weight of 50 tons.

One of the greatest meat-eaters of all times was *Tyrannosaurus rex,* who lived near the end of the era. This creature had two powerful hind legs armed with claws 6 to 8 inches long, a powerful tail, short neck, two tiny forelegs, and a skull which measured 4 feet in length and 3 feet in width. He was armed with jaws that contained recurved teeth as long as 6 inches. This carnivorous monster may have reached a length of 50 feet, walked with his head about 20 feet above the ground, and weighed as much as 10 tons. One set of footprints shows a stride more than 13 feet in length. However, he was quite different from the first carnivorous dinosaurs that appeared early in the Mesozoic. They had fairly long necks in proportion to their small sizes and forefeet which were more nearly equal in size to the

Fig. 35-12. *A Late Mesozoic landscape in Wyoming. We see rhinoceros-like* Triceratops, *the terrible* Tyrannosaurus, *a flying reptile, and spike-armored and duck-billed dinosaurs.* [*Courtesy, Peabody Museum, Yale University. From a painting by Rudolph Zallinger.*]

hindlegs. A dinosaur (*Allosaurus*) that lived during the middle part of the era showed an evolutionary development between the first primitive flesh-eating dinosaurs and the highly specialized *Tyrannosaurus*.

Certain plant-eating dinosaurs, such as *Brontosaurus,* carried a huge body on four elephantlike legs; they had long tails, very long necks, and small heads. If these animals had walked head to tail as in a circus parade, 60 to 70 of them would have spanned a mile. Imagine four of these incredible creatures lined up in this manner on a football field. The tail of the first would project beyond one goal post, and the head of the fourth would extend beyond the other! Probably such heavy creatures lived in swampy areas to take advantage of the buoyant effect of the water in moving about. The combination of such a hulk with the tiny head seems grotesque, indeed ridiculous. Keeping the huge stomach filled was probably difficult even if the animal ate almost continuously and possessed a slow body metabolism. Fortunately, the tiny head with its 2-ounce brain would hardly have the means to worry.

Brachiosaurus was the real giant. Perhaps he attained a length of 80 feet, much of it neck. Probably he could have looked down upon the top of a three-story building! The nostrils of *Brachiosaurus* were located on a sort of

Fig. 35-13. *The first known birds, two small dinosaurs, and a number of flying reptiles (pterosaurs). The trees are cycads. Mesozoic (Jurassic). [From a painting by C. R. Knight. Copyright, Chicago Natural History Museum.]*

bump on the top of the head; evidently the animal could remain submerged and breathe as long as his bump projected above water.

Other types of dinosaurs lived during the Mesozoic era: the duck-billed, plated, armored, and horned varieties (Fig. 35-12). The seas were dominated by several huge reptiles and the air by others (Fig. 35-13). One bat-like flying reptile had wings that measured 26 feet from tip to tip. Its light-weight body was about the size of a goose. Apparently the creature was well adapted for soaring long distances over water because most fossils occur in marine sediments which seem to have been deposited far from land.

At the end of the Mesozoic era, the dinosaurs, flying reptiles, some of the large marine reptiles, the toothed birds, and other forms of life became extinct: "the time of the great dying when the hosts were tried in the balance and found wanting." Many of the forms which became extinct, like *Tyrannosaurus rex,* appeared to be at their zenith point of development: they were not decadent creatures ready for extinction.

Several hypotheses attempt to explain these mysterious disappearances, but none is very satisfactory: some disease may have caused the "great dying"; perhaps a climatic change took place to which these forms could not become adapted; high temperatures may have produced sterility in the huge reptiles; mammals may have developed a fondness for dinosaur eggs left in unguarded nests.

Birds The oldest bird fossils (Jurassic) constitute a convincing link between later birds and their reptilian ancestors. This first bird (*Archaeopteryx,* Fig. 35-14) was about the size of a crow, had feathers, and could fly.

Fig. 35-14. *Cast of a fossil* Archaeopteryx. [*Courtesy of the American Museum of Natural History.*]

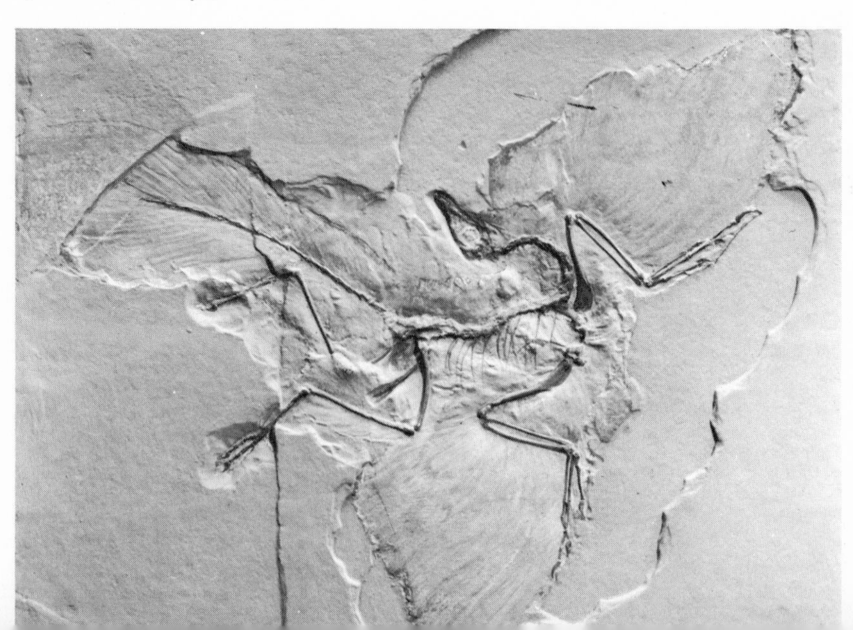

But it had three features retained from its reptilian predecessors: jaws set with true teeth, three claws on each wing, and a long tail on which the feathers were arranged pinnately—not in the fan shape of modern birds.

Apparently *Archaeopteryx* did not evolve directly from the flying reptiles (Fig. 35-10). More modern ancestors of present-day birds had evolved from these toothed birds before their extinction late in the Mesozoic era. Bird fossils are relatively rare.

Mammals Although the first mammals apparently had evolved from reptiles (therapsids) before the middle of the Mesozoic, they remained inconspicuous throughout the era and did not assume their present dominant role in the biological world until the Cenozoic era (Age of Mammals). Four major trends occurred in the mammalian evolution of the Cenozoic—an increase in size and brain power, and a specialization of feet and teeth. The horse series (Fig. 28-14) is a classic illustration.

Some mammals became efficient carnivores, like the lion, tiger, and wolf; others took to the oceans, like the whale and the porpoise; still others developed the ability to run rapidly like the horse and the antelope. Some forms became extremely specialized and then extinct. The ability of a group of animals to adapt and evolve may be greatest early in its history.

Fig. 35-15. *Restoration of* Baluchitherium. *This hornless rhinoceros lived in Central Asia during the middle part of the Cenozoic era.* [*Courtesy of the American Museum of Natural History.*]

During the Cenozoic era, North America was the home of many animals (Fig. 35-15) which are no longer associated with it. The woolly mammoth, the mastodon, and the rhinoceros have vanished from the continent, as have the hippopotamus, camel, giant pig, giant ground sloth, giant bison, and saber-toothed tiger.

MAN

Man is classified as a primate (Latin: "first"), a group of mammals considered to be of foremost importance in the animal kingdom by the biologist

Fig. 35-16. *The heads of some forerunners of modern man as reconstructed from skull fragments. The oldest specimens (early Pleistocene from Africa) are the three heads at the extreme upper right:* Plesianthropus, Australopithecus, *and* Paranthropus. *Just in front of them is a European specimen from Steinheim in Germany, of the middle Pleistocene. Other specimens of the same age are the Asian group of the extreme left:* Pithecanthropus robustus, P. erectus (*Java man*), *and* P. pekinensis. *Of the seven heads in the front tier (late Pleistocene) the two at the left are from a cave in Asia. Then come European types—a skull from Shkul, a Neanderthal man, and a Cromagnon man—the only European type identified as* Homo sapiens. *At the right are two African heads—Rhodesian man and a head from Florisbad.* [*Courtesy of the American Museum of Natural History.*]

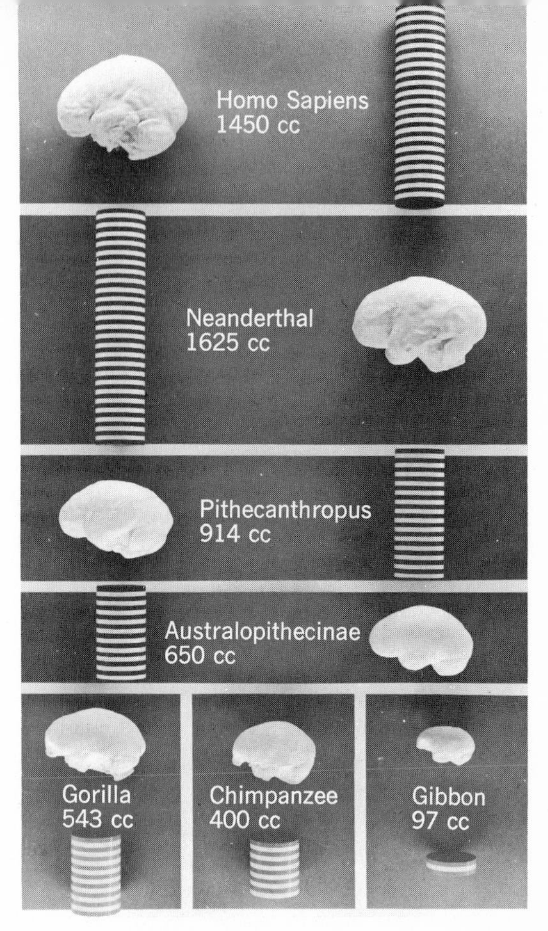

Fig. 35-17. *Increasing brain size: "Intelligence is perhaps the most outstanding trait of the hominids. The best index of it available in the fossil record is brain size as measured by the capacity of the bony brain case. In the earliest hominids or man-like creatures that we know at present, the brain had not yet reached a size much, if at all, superior to that of such large primates as the gorilla. It is, however, highly significant that the brain has been the most dynamic area of growth during human evolution. Recent man now has a brain approximately two and one-half times larger than his earliest known hominid relatives."* [*Courtesy of the American Museum of Natural History.*]

(Linnaeus) who named them. It includes the gorilla, chimpanzee, and monkey. This classification is supported by studies of man's skeletal structures, muscles, organs, embryological development, vestigial organs, blood tests, and the fossil record.

The origin of man and his role upon the Earth are topics of profound significance and have been considered at great length. The fossil record, although fragmentary in nature, indicates that man has evolved in a manner that follows the general pattern of other groups—descent with modification (Fig. 35-16). If one traces man's lineage backward into time (i.e., downward through the thick piles of superposed, fossil-bearing, layered rocks that have been correlated into a composite geologic history), he passes in turn from mammal, to reptile, to amphibian, to fish, to invertebrate. Man and other present-day primates at some remote time in the past had a common ancestor, and each has descended with modification from this ancestor.

The classification of certain fossil discoveries as man or ape depends in some instances upon one's definition of man; the capacity to use tools, the size of the brain, and erect posture are important criteria. An outstanding primate characteristic is a large, well-developed brain (Fig. 35-17). Among the primates, man seems most closely akin to the gorilla and chimpanzee. As compared with the skull of a man, the thicker skull of a gorilla exhibits a protruding mouth with projecting canine teeth, no chin, a prominent eyebrow ridge, low sloping forehead, and a saggital crest.

Primate fossils, particularly those of man, are not common, and the paleontological record of the entire group is fragmentary. Primates tend to inhabit forested areas in which quick burial following death occurs infrequently, and fossilization is rare. Early man's intelligence was his chief advantage over the animals with whom he had to compete. In speed, strength, and ability to withstand cold or heat, he was inferior to many of his contemporaries. When man began to bury his dead and dwell in caves, fossilization occurred more frequently.

Fossil evidence of the primates is confined to the Cenozoic era, and that of man to the last few million years. Man's culture has been subdivided into three main phases: the Stone Age, the Bronze Age, and the Iron Age. At first, stones were used as tools and weapons by man as he found them (eoliths); next they were shaped by chipping (paleoliths); and finally they were ground and polished as well as chipped (neoliths). Next, man used implements made of metal. The three cultures were not attained simultaneously throughout the world. For example, the culture of certain primitive tribes today is still that of the Stone Age. Superposed piles of refuse collected in great amounts around favored camp sites such as caves which were used in succession by different tribes.

Four types of early man are discussed below to illustrate the fossil record; but dozens of additional discoveries are omitted. The peoples represented by such finds have commonly been named for the localities in which they were found. Many of the discoveries are very fragmentary. Part of a jaw and a limb bone, a few teeth, or bits of a shattered skull furnish the only evidence now available for the former existence of many of these types. Reconstructions and interpretations made on such limited data should be flexible enough to allow for future discoveries. Many of the discoveries have been classified as different genera and species and have been given different names. This may have complicated the evolutionary picture unnecessarily and obscured relationships. Perhaps a much wider range of variation should be expected within a genus or species.

Australopithecus ("Southern Ape") The first specimen of this group was discovered in South Africa in 1924, and a sizable number of similar fossils have since been found (Fig. 35-16). Both the dating and classification of *Australopithecus* have been controversial. A typical representative probably was not much more than four feet in height, had a small brain (cranial capacity apparently from 450 to 700 cm^3), large teeth and jaws, eyebrow ridges, protruding mouth, and no chin. On the other hand, he walked in an erect manner and had features of dentition, skull, and pelvis that resemble those of man.

Two groups of australopithecines are recognized: one had exceptionally large jaws and huge grinding teeth and may have lived chiefly on vegetation, whereas the other group had smaller jaws and teeth and presumably had an omnivorous diet.

The time when the australopithecines existed is of importance in assessing their role in human evolution, but their range in time is uncertain. They are generally dated as Early Pleistocene, but perhaps they also existed in the latter part of the Pliocene, and they may have lived more recently than the Early Pleistocene. They may or may not be direct ancestors of man. If the australopithecines are not, they probably at least resemble such an ancestor.

Pithecanthropus Erectus (Java Ape-Man) A widely publicized discovery of a human fossil was made in Java in 1891. It consisted of part of a skull and several bones belonging to a primitive type of man subsequently named *Pithecanthropus erectus* (Greek: "ape-man"). Several additional finds of skull caps and other bones suggest that this creature was a primitive human type with characteristics of both man and ape (upper left in Fig. 35-16). Although the brain cases outlined by these skulls are smaller than those of modern man, they are nearly double the cranial capacity of a great ape. The receding chin and forehead, the heavy brow ridges, and the protruding mouth are apelike. However, the large brain, teeth, and erect posture (shown by the shape of the leg bones) indicate a primitive human type. Associated mammalian fossils show a mid-Pleistocene age for this discovery.

Near Peking, China, fossil fragments have been found of a few dozen individuals of a type similar to the Java ape-man. The remains are chiefly skulls and jaws, and each skull has been fractured in a manner which suggests cannibalism. Apparently fire was used by this type of man; numerous flint tools were found associated with the fossil remains. Later discoveries in other areas suggest that several pithecanthropoid types lived at various times during the Pleistocene, some before *Pithecanthropus erectus,* others later.

Neanderthal Man This prehistoric type of man lived during the latter part of the Pleistocene in Europe, western Asia, and along the northern coast of Africa. According to fairly abundant fossil remains, Neanderthal man had a short stocky body, a receding chin and forehead, and heavy brow ridges. The big toe was offset against the others in an apelike manner. The brain was as large as modern man's, but apparently smaller in the parts devoted to thinking. Neanderthal man used fire, buried his dead, was a good hunter, and made stone implements with considerable skill. According to one view, Neanderthal man was not the direct ancestor of modern man and became extinct. According to an opposing view, modern man in Europe has resulted from intermixing of Neanderthal with Cro-Magnon.

Cro-Magnon Man At some time during the last glacial age, a modern type of man came to southern Europe. Many skeletons show that these individuals were tall and straight, with high foreheads, prominent chins, and modern-size brain cases. The Cro-Magnons appear to be the direct ancestors of the present southern Europeans; therefore, they probably did not become extinct as a number of earlier types seem to have done. The Cro-Magnons used fire, were good hunters, buried their dead, made excellent stone implements, and possessed clothing and ornaments. In combination with the numerous multicolored paintings found on the walls of caves, these show a well-developed art and culture. The bones of about 100,000 horses reportedly were found in piles around one camp site in France.

Early Man in North America The history of early man in the Americas is still quite uncertain but seems to involve only the last part of the Pleistocene epoch. Apparently he came from the Old World (via Siberia?) before the extinction of certain types of elephant, camel, horse, and bison, because his stone implements and some human bones are associated with them. Carbon-14 data indicate that man arrived in North America some 20,000 to 30,000 years ago or earlier.

Tentative Conclusions on the Origin of Man Until recently it was rather generally assumed that man had evolved from a common primate ancestor in an undeviating progressive development from primitive to modern types, and that the most primitive types (i.e., unlike modern types) of fossil men were the oldest and most brutal ones. These ideas led to a search for a series of "missing links" which, when found, would determine the places of all fossil men in this progression. The validity of these assumptions is now questioned; e.g., certain modern types of fossil men are now known to have lived long before other types of a more primitive nature. Since the soft parts have not been fossilized, the appearance of fossil types is largely guesswork.

The common ancestor believed to link the great apes and man has been commonly pictured as much more apelike than manlike. Perhaps this is not correct. Certain primitive fossil men such as the Neanderthals may represent specializations away from a more modern ancestral stock.

Fossils of man illustrate an important principle in evolutionary development. Evolutionary changes apparently do not take place little by little with each part of an organism changing at the same rate. In other words, a transitional stage midway between an ancestral species and a descendant species would not be midway in the development of all of its features. More probably, there would be a number of midway organisms with various combinations of more advanced and more primitive features. For example, in man the limbs apparently reached their present stage of evolutionary development before the brain, skull, and teeth. In fact, the development of an erect posture apparently was a necessary prerequisite, because this freed the hands to fashion and use tools, and this led to a relatively rapid enlargement of the brain.

Dr. and Mrs. Leakey in 1959 found fossils of a primitive, tool-making man or near-man in Tanganyika, East Africa; it has been dated as 1,750,000 years old and called *Zinjanthropus*. It appears closely related to the near-men, the australopithecines of South Africa. The potassium-argon radioactive "clock" was used to date beds of volcanic ash associated with the fossils. Additional fossils of a more primitive type were found in this area in 1960.

Unfortunately, many details concerning man's evolutionary history are still shrouded in the mists of antiquity and the fogs of too few fossil remains. His (and her) present position of dominance in the biologic world on the Earth comes only at the end of a long succession of changes. Apparently man has succeeded to a position formerly held in turn by other mammals, reptiles, amphibians, fish, and invertebrates during an immensity of time that is nearly beyond comprehension.

Exercises and Questions, Chapter 35

1. Describe some of the ways in which fossils form.

2. Fossil collecting:
   (a) Does the appearance of a fossil in a rock tend to be quite different depending upon whether you observe the surface or edge of the enclosing layer?
   (b) How may undamaged silicified fossils be obtained from a limestone? Is an understanding of differential weathering of value in searching for fossils in such areas?
   (c) Make a collection of fossils that illustrates the concept of fossil succession.

3. Describe several different types of fossils which would give useful information about the environment in which they lived.

4. Describe some of the key ideas involved in evolution.

5. What implication does the fossil record of the oldest Paleozoic rocks have for the Precambrian? What unsolved problems occur here?

6. What criteria have geologists used in subdividing the geologic record into eras and periods?

7. Describe briefly some of the important evolutionary steps in the origin of the vertebrates. Which of these steps are well supported by fossil evidence? Which are not?

8. Which came first the chicken or the egg? Does the following statement have some validity: the first bird hatched from an egg laid by a reptile?

9. If evolutionary development has generally proceeded from the simple to the complex, why are any simple forms left?

10. Why are certain organisms today called "living fossils"?

11. Imagine that all of the Earth's history is represented by the length of a football field (each yard represents about 45 million years). Calculate the distances of each of the following from one of the goal posts (assume the goal posts are on the zero yard lines):
    (a) First direct evidence of life.
    (b) Fossils first become abundant.
    (c) Oldest known amphibian.
    (d) Beginning and end of "age of reptiles."
    (e) First man.

Bibliography

Chemistry and Physics

Anthony, H. D., *Science and Its Background,* Macmillan, 1948.

Bassett, L. G., S. C. Bunce, A. E. Carter, H. M Clark, and H. B. Hollinger, *Principles of Chemistry,* Prentice-Hall, 1966.

Bell, A. E., *Christian Huygens and the Development of Science in the Seventeenth Century,* Edward Arnold, 1947.

Bendick, Jeanne, *How Much and How Many,* Whittlesey House, 1947.

Boyle, Robert, *The Sceptical Chymist,* Dutton, 1911.

Bragg, Sir William, *Concerning the Nature of Things,* G. Bell, 1932.

———, *Electricity,* Macmillan, 1936.

Brown, G. Burniston, *Science, Its Method and Its Philosophy,* George Allen & Unwin, 1950.

Buckley, H., *A Short History of Physics,* 2d ed., Van Nostrand, 1929.

Cajori, Florian, *A History of Physics,* Macmillan, 1929.

Caldin, E. F., *The Power and Limits of Science,* Chapman and Hall, 1949.

Cohen, I. Bernard (Ed.), *Benjamin Franklin's Experiments, A New Edition of Franklin's Experiments and Observations on Electricity,* Harvard University Press, 1941.

Cohen, M. R., and I. E. Drabkin, *A Source Book of Greek Science,* McGraw-Hill, 1948.

Conant, James B., *Robert Boyle's Experiments in Pneumatics,* Harvard University Press, 1950.

———, *Science and Common Sense,* Yale University Press, 1951.

———, *The Overthrow of the Phlogiston Theory,* Harvard University Press, 1950.

Cooper, Lane, *Aristotle, Galileo and the Tower of Pisa,* Cornell University Press, 1935.

Crew, Henry, *The Rise of Modern Physics,* Williams & Wilkins, 1935.

Curie, E., *Madame Curie,* Doubleday, 1937.

Dampier, William Cecil, *A History of Science and Its Relations with Philosophy and Religion,* Macmillan, 1936.

Eidinoff, Maxwell L., and H. Ruchlis, *Atomics for the Millions,* McGraw-Hill, 1947.

Farber, Edward, *The Evolution of Chemistry,* Ronald Press, 1952.

Frank, Philip, *Modern Science and Its Philosophy,* Harvard University Press, 1949.

Fraser, Charles G., *Half Hours with Great Scientists* (*The Story of Physics*), Reinhold, 1948.

Freeman, Ira, *Modern Introductory Physics,* McGraw-Hill, 1949.

French, S. J., *Torch and Crucible,* Princeton University Press, 1941.

Garrett, A. B., *The Flash of Genius,* Van Nostrand, 1963.

Glasstone, S., *Sourcebook on Atomic Energy,* 2nd ed., Van Nostrand, 1958.

Hart, Ivor B., *James Watt and the History of Steam Power,* H. Schuman, 1949.

———, *Makers of Science,* Oxford University Press, 1924.

———, *The Mechanical Investigations of Leonardo da Vinci,* Open Court, 1925.

Hecht, Selig, *Explaining the Atom,* Viking Press, 1947.

Heckman, H. H., and P. W. Starring, *Nuclear Physics and the Fundamental Particles,* Holt, Rinehart and Winston, 1963.

Hodgins, Eric, and F. A. Magoun, *Behemoth, The Story of Power,* Doubleday Doran, 1932.

Hoffman, Banesh, *The Strange Story of the Quantum,* Harper, 1947.

Hogben, Lancelot, *Science for the Citizen,* Knopf, 1938.

Holmyard, Eric J., *Makers of Chemistry,* Oxford, 1931.

Holton, G., and D. H. D. Roller, *Foundations of Modern Physical Science,* Addison Wesley, 1958.

Humphreys, R. F., and R. Beringer, *First Principles of Atomic Physics,* Harper, 1950.

Jaffe, Bernard, *Crucibles,* Simon & Schuster, 1930.

Jeans, Sir James, *An Introduction to the Kinetic Theory of Gases,* Cambridge University Press, 1946.

Kaplan, I., *Nuclear Physics,* Addison Wesley, 1963.

Knedler, John W. (Ed.), *Masterworks of Science,* Doubleday, 1949.

Kolin, Alexander, *Physics, Its Laws, Ideas and Methods,* McGraw-Hill, 1950.

Ladenburg, Albert, *Histoire du développement de la chimie depuis Lavoisier jusqu'à nos jours,* Colson, 1911.

Lapp, R. E., and H. L. Andrews, *Nuclear Radiation Physics,* 2d ed., Prentice-Hall, 1954.

Larsen, Egon (Lehrburger), *An American in Europe: the Life of Benjamin Thompson, Count Rumford,* Rider, 1953.

Leicester, Henry M., and Herbert S. Klickstein, *A Source Book in Chemistry,* McGraw-Hill, 1952.

Lemon, H. B., *From Galileo to the Nuclear Age,* University of Chicago Press, 1946.

Luhr, Overton, *Physics Tells Why,* Ronald Press, 1943.
Mach, Ernst, *History and Root of the Principle of Conservation of Energy,* Open Court, 1911.
———, *The Science of Mechanics,* Open Court, 1942.
Magie, W. F., *A Source Book in Physics,* McGraw-Hill, 1935.
McKie, Douglas, *Antoine Lavoisier,* Henry Schuman, 1952.
———, and Niels H. Heathcote, *The Discovery of Specific and Latent Heats,* Edward Arnold, 1935.
Mees, C. E. Kenneth, *The Path of Science,* Wiley, 1946.
Meggers, William F., "Measurement by Mercury," *Scientific American* 179: 49 (Aug. 1948).
Miller, Dayton C., *The Science of Musical Sounds,* Macmillan, 1937.
Moody, Ernest A., and Marshall Claggett (Ed.), *The Medieval Science of Weights,* University of Wisconsin Press, 1952.
More, Louis Trenchard, *Isaac Newton, A Biography,* Scribner, 1934.
———, *The Life and Works of the Honorable Robert Boyle,* Oxford, 1944.
Moreau, Henri, "The Genesis of the Metric System and the Work of the International Bureau of Weights and Measures," *Journal of Chemical Education* 30:3 (Jan. 1953).
National Council of Teachers of Mathematics Twentieth Yearbook, *The Metric System of Weights and Measures,* Bureau of Publications, Teachers College, Columbia University, 1948.
Newton, Sir Isaac, *Opticks,* Dover Publications, 1952.
Ornstein, Martha, *The Role of Scientific Societies in the Seventeenth Century,* University of Chicago Press, 1928.
Overman, R. T., *Basic Concepts of Nuclear Chemistry,* Reinhold, 1963.
Partington, J. R., *A Short History of Chemistry,* Macmillan, 1949.
Phin, John, *The Seven Follies of Science,* Van Nostrand, 1912.
Pledge, H. T., *Science Since 1500,* Ministry of Education, Science Museum, Great Britain.
Quagliano, J. V., *Chemistry,* 2nd ed., Prentice-Hall, 1958.
Read, John, *Humor and Humanism in Chemistry,* G. Bell & Sons, 1947.
———, *Prelude to Chemistry,* Macmillan, 1937.
Roller, Duane, *The Early Development of the Concepts of Temperature and Heat,* Harvard University Press, 1950.
Sebera, D. K., *Electronic Structure and Chemical Bonding,* Blaisdell, 1964.
Sisler, H. H., C. A. Vander Werf, and A. H. Davidson, *College Chemistry: A Systematic Approach,* 2nd ed., Macmillan, 1961.
Skilling, Hugh H., *Exploring Electricity,* Ronald Press, 1948.
Solomon, Arthur K., *Why Smash Atoms,* Harvard University Press, 1946.
Spiers, I. H. B. and A. G. H. (translators), *The Physical Treatises of Pascal,* Columbia University Press, 1937.
Stillman, John M., *The Story of Early Chemistry,* D. Appleton & Co., 1924.
Stimson, Dorothy, *Scientists and Amateurs,* Henry Schuman, 1948.
Sullivan, J. W. N., *Isaac Newton,* Macmillan, 1938.
Taylor, F. Sherwood, *Galileo and the Freedom of Thought,* Watts & Co., 1938.

———, *The Alchemists,* Henry Schuman, 1949.

———, *The March of Mind,* Macmillan, 1939.

Taylor, Lloyd W., *Physics, the Pioneer Science,* Houghton Mifflin, 1941.

Thompson, J. A., *Count Rumford of Massachusetts,* Farrar & Rhinehart, 1935.

Tyndall, John, *Heat a Mode of Motion,* Appleton-Century, 1909.

Weeks, Mary E., *The Discovery of the Elements,* Journal of Chemical Education, 1945.

White, Harvey E., *Classical and Modern Physics,* Van Nostrand, 1940.

———, *Modern College Physics,* Van Nostrand, 1948.

Whittaker, Sir Edmund T., *History of the Theory of Aether and Electricity,* Thomas Nelson & Sons, 1951.

Wightman, William P. D., *The Growth of Scientific Ideas,* Yale University Press, 1951.

Wolf, Abraham, *A History of Science and Technology in the 18th Century,* Macmillan, 1939.

———, *A History of Science and Technology in the 16th and 17th Century,* Allen & Unwin, 1950.

Wood, Alexander, *Joule and the Study of Energy,* G. Bell & Sons, 1925.

General Earth Science

Baird, P. D., *The Polar World,* Wiley, 1964.

Bascom, W., *A Hole in the Bottom of the Sea: The Story of the Mohole Project,* Doubleday, 1961.

Beiser, A., and the editors of LIFE, *The Earth,* Time, Inc., 1962.

Ericson, D., and G. Wollin, *The Deep and the Past,* Knopf, 1964.

Greenhood, David, *Mapping,* University of Chicago Press, 1964.

Investigating the Earth, American Geological Institute's Earth Science Curriculum Project, Houghton Mifflin Co., 1967.

Petrie, W., *Keoeeit—The Story of the Aurora Borealis,* Pergamon Press, 1963.

Scientific American Reprints: see the list published by W. H. Freeman & Co., Publishers.

Shapley, Harlow, *Beyond the Observatory,* Charles Scribner's Sons, 1967.

Spar, J., *Earth, Sea, and Air,* 2d ed., Addison-Wesley, 1965 (paperback).

Spencer, E. W., *Geology: A Survey of Earth Science,* Crowell, 1965.

Strahler, A. N., *The Earth Sciences,* Harper and Row, 1963.

Taylor, E. F., and J. A. Wheeler, *Spacetime Physics,* W. H. Freeman and Co., 1966.

Young, L. (Ed.), *The Mystery of Matter,* American Foundation for Continuing Education, Oxford University Press, 1965.

PERIODICALS

American Journal of Physics, American Institute of Physics, Inc., 335 E. 45th St., New York, N.Y.

American Scientist, Sigma Xi, 155 Whitney Ave., New Haven, Conn.

Bulletin of Atomic Scientists, 935 E. 60th St., Chicago, Ill.

Deep-Sea Research and Oceanographic Abstracts, Pergamon Press, 122 E. 55th St., New York, N.Y.

Geophysical Journal, Royal Astronomical Society, Burlington House, London W.1, England.

Journal of Chemical Education, 20th and Northampton Sts., Easton, Pa.

National Geographic Magazine, 17th and M Sts. N.W., Washington, D.C.

Natural History, American Museum of Natural History, Central Park West at 79th St., New York, N.Y.

Science, 1515 Massachusetts Ave. N.W., Washington, D.C.

The Sciences, The New York Academy of Sciences, 2 East 63 St., New York, N.Y.

Science Digest, 1775 Broadway, New York, N.Y.

Scientific American, 415 Madison Ave., New York, N.Y.

Astronomy

Abell, G., *Exploration of the Universe,* Holt, Rinehart, and Winston, 1964.

Abetti, Giorgio, *The History of Astronomy,* Henry Schuman, 1952.

Ahrendt, Myrl H., *The Mathematics of Space Exploration,* 1965.

Baade, W., *Evolution of Stars and Galaxies,* Harvard University Press, 1963.

Baker, Robert H., *Astronomy,* 8th ed., Van Nostrand, 1964.

Baker, Robert H., and L. W. Fredrick, *An Introduction to Astronomy,* 7th ed., Van Nostrand, 1967.

Baldwin, R. B., *The Measure of the Moon,* University of Chicago Press, 1963.

Baldwin, R. B., *A Fundamental Survey of the Moon,* McGraw-Hill, 1966.

Bergamini and the editors of LIFE, *The Universe,* Time, Inc., 1962.

Bok, Bart J. and Priscilla F., *The Milky Way,* 3rd ed., Harvard University Press, 1957.

Brandt, John C., *The Sun and Stars,* McGraw-Hill, 1966.

Bray, R. J., and R. E. Loughhead, *Sunspots,* Wiley, 1964.

Dietz, David, *All About the Universe,* Random House, 1965.

Gamow, George, *A Planet Called Earth,* Viking Press, 1963.

——— *A Star Called the Sun,* Viking Press, 1964.

Glasstone, S., *Sourcebook on the Space Sciences,* Van Nostrand, 1965.

Goldberg, Leo, ed., *Annual Review of Astronomy and Astrophysics,* Annual Reviews, Inc., 1967.

Goldberg, Lew, and L. H. Aller, *Atoms, Stars, and Nebulae,* Harvard University Press, 1963.

Goodwin, H. L., *Space: Frontier Unlimited,* Van Nostrand, 1962.

Harsanyi, Zsolt, *The Star Gazer,* translated by Paul Tabor, Putnam, 1939.

Hawkins, G. S., *Meteors, Comets, and Meteorites,* McGraw-Hill, 1965.

Hodge, Paul W., *Galaxies and Cosmology,* McGraw-Hill, 1966.

Howard, N. E., *Standard Handbook for Telescope Making,* Crowell, 1959.

Hoyle, F., *Astronomy,* Doubleday, 1962.

Huffer, C. M., F. E. Trinklein, and M. Bunge, *An Introduction to Astronomy,* Holt, Rinehart and Winston, 1967.

Inglis, Stuart J., *Planets, Stars, and Galaxies,* 2nd ed., Wiley, 1967.
Jennison, Roger C., *An Introduction to Radio Astronomy,* Philosophical Library, 1966.
King, Henry C., *The History of the Telescope,* Sky Publishing Corp., 1955.
King-Hele, Desmond, *Observing Earth Satellites,* St. Martin's Press, 1966.
Koestler, A., *The Sleepwalkers,* Macmillan, 1959.
Kopal, Zdenek, *An Introduction to the Study of the Moon,* Gordon and Breach Science Publishers, 1966.
Krinov, E. L., *Giant Meteorites,* Pergamon Press, Inc., 1966.
Kuipfer, G. P. (Ed.), *The Earth as a Planet,* University of Chicago Press, 1955.
———, *The Sun,* University of Chicago Press, 1954.
Larousse Encyclopedia of Astronomy, Putnam, 1959 (English language ed.).
Lovell, B. and J. *Discovering the Universe,* Harper and Row, 1963.
Lundquist, Charles A., *Space Science,* McGraw-Hill, 1966.
Mayall, Wyckoff, and Polgreen, *The Sky Observer's Guide, A Golden Handbook,* 1965.
McLaughlin, D. B., *Introduction to Astronomy,* Houghton Mifflin, 1961.
Mehlin, T. G., *Astronomy,* Wiley, 1959.
Menzel, Donald H., *A Field Guide to the Stars and Planets,* Houghton Mifflin, 1964.
Messel, H., and S. T. Butler, *The Universe and Its Origin,* St. Martin's Press, 1964.
Miczaika, G., and W. Stinton, *Tools of the Astronomer,* Harvard University Press, 1961.
Motz, Lloyd, and Anneta Duveen, *Essentials of Astronomy,* Wadsworth Publishing Co., 1966.
Naugle, John E., *Unmanned Space Flight,* 1965.
Page, Thornton (Ed.), *Stars and Galaxies,* Prentice-Hall, 1962 (paperback).
Page, Thornton, and Lou Williams Page (Eds.), *Neighbors of the Earth,* Macmillan, 1965.
———, *Wanderers in the Sky,* Macmillan, 1965.
Pannekoek, A., *A History of Astronomy,* Interscience Publishers, 1961.
Pfeiffer, J., *The Changing Universe,* Random House, 1956.
Rey, Hans A., *The Stars, a New Way to See Them,* Houghton Mifflin, 1967.
Richardson, R. S., *Mars,* Harcourt, Brace and World, 1964.
Richardson, Robert S., *Getting Acquainted with Comets,* McGraw-Hill, 1967.
Rogers, E. M., *Physics for the Inquiring Mind,* Princeton University Press, 1960.
Ronan, Colin, *The Astronomers,* Hill and Wang, 1964.
Sandage, A., *The Hubble Atlas of Galaxies,* Carnegie Institution of Washington, 1961.
Shapley, H., *Galaxies,* rev. ed., Harvard University Press, 1963.
———, *The View from a Distant Star,* Basic Books, 1963.
———, and H. E. Howarth (Eds.), *Source Book in Astronomy,* McGraw-Hill, 1929.

——— (Ed.), *Source Book in Astronomy, 1900-1950,* Harvard University Press, 1960.

Shaw, R. William, and Samuel L. Boothroyd, *Manual of Astronomy,* Wm. C. Brown Co., 5th ed., 1968.

Smith, Alex G., and Thomas D. Carr, *Radio Exploration of the Planetary System,* D. Van Nostrand Co., Inc., 1964.

Steinberg, J. L., and J. Lequeux, *Radio Astronomy,* McGraw-Hill, 1963.

Stern, Phillip D., *Our Space Environment,* 1965.

Struve, O., *Astronomy of the Twentieth Century,* Macmillan, 1962.

———, B. Lynds, H. Pillans, *Elementary Astronomy,* Oxford University Press, 1959.

Sutton, Richard M., *The Physics of Space,* 1965.

Vehrenberg, Hans, *Atlas of Deep-Sky Splendors,* Sky Publishing Corp., n.d.

von Braun, Wernher, and F. I. Ordway, III, *History of Rocketry and Space Travel,* Thomas Y. Crowell Co., n.d.

Whipple, F. L., *Earth, Moon, and Planets,* Harvard University Press, 1963.

Widger, William K., Jr., *Meteorological Satellites,* 1966.

Woodbury, D. O., *The Glass Giant of Palomar,* Dodd Mead, 1948.

Wyatt, S. P., *Principles of Astronomy,* Allyn and Bacon, 1964.

Young, L. B. (Ed.), *Exploring the Universe,* American Foundation for Continuing Education, McGraw-Hill, 1963.

PERIODICALS

Astronomical Journal, American Institute of Physics, 335 E. 45th Street, New York, N.Y.

Astrophysical Journal, University of Chicago Press, 5750 Ellis Ave., Chicago, Ill.

The Griffith Observer, published monthly by the Griffith Observatory, Los Angeles, Calif.

Icarus: International Journal of the Solar System, Academic Press, Inc., 111 Fifth Ave., New York, N.Y.

Leaflets and *Publications of the Astronomical Society of the Pacific,* California Academy of Sciences, Golden Gate Park, San Francisco, Calif.

The Observer's Handbook for 19—, published annually by the Royal Astronomical Society of Canada, 252 College St., Toronto 23, Canada.

The Review of Popular Astronomy, Sky Map Publications, Inc., St. Louis, Mo.

Sky and Telescope, Sky Publishing Corp., 49-51 Bay State Road, Cambridge, Mass.

Meteorology

Aviation Weather, U.S. Government Printing Office, 1965.

Battan, L. J., *Cloud Physics and Cloud Seeding,* Anchor Books, Doubleday & Co. (paperback), 1962.

The Nature of Violent Storms, 1962.

Radar Observes the Weather, 1962.

The Unclean Sky, 1966.

Bentley, W., and W. Humphreys, *Snow Crystals,* Dover, 1962.

Blair, T. A., and R. C. Fite, *Weather Elements,* Prentice-Hall, Inc., 5th ed., 1965.

Blanchard, Duncan C., *From Raindrops to Volcanoes,* Doubleday & Co., 1967.

Blumenstock, D. I., *The Ocean of Air,* Rutgers University Press, 1959.

Byers, H. R., *Elements of Cloud Physics,* University of Chicago Press, 1965.

———, *General Meteorology,* McGraw-Hill, 1959.

———, *The Thunderstorm,* U.S. Government Printing Office, 1949.

Dobson, G., *Exploring the Atmosphere,* Oxford, Clarendon Press, 1963.

Donn, William L., *Meteorology,* McGraw-Hill Co., 3rd ed., 1965.

Dunn, G. E., and B. I. Miller, *Atlantic Hurricanes,* Louisiana State University Press, 1964.

Flora, S. D., *Tornadoes of the United States,* University of Oklahoma Press, 1953.

———, *Hailstorms of the United States,* University of Oklahoma Press, 1956.

Grant, H. D., *Cloud and Weather Atlas,* Coward-McCann, 1944.

Kimble, G., *Our American Weather,* McGraw-Hill, 1955.

Krick, I., and R. Fleming, *Sun, Sea, and Sky,* Lippincott, 1954.

Lehr, P., R. Burnett, and H. Zim, *Weather,* Simon and Schuster, 1957 (paperback).

Ludlam, F., and R. Scorer, *Cloud Study,* Macmillan, 1958.

Mason, B. J., *Clouds, Rain, and Rainmaking,* Cambridge University Press, 1962.

Neuberger, Hans, and George Nicholas, *Manual of Lecture Demonstrations, Laboratory Experiments, and Observational Equipment for Teaching Elementary Meteorology in Schools and Colleges,* Pennsylvania State University, 1962.

Petterssen, S., *Introduction to Meteorology,* 2d ed., McGraw-Hill, 1958.

Riehl, Herbert, *Introduction to the Atmosphere,* McGraw-Hill, 1965.

Sloane, Eric, *Folklore of American Weather,* Meredith Press, 1963.

Smith, L. P., *Weather Studies,* Macmillan, 1963.

Stewart, G., *Storm,* Modern Library, Random House, 1947.

Sutcliffe, R. C., *Weather and Climate,* W. W. Norton and Co., 1966.

Sutton, O., *The Challenge of the Atmosphere,* Harper and Brothers, 1961.

Tannehill, I., *Hurricanes: Their Nature and History,* 5th ed., Princeton University Press, 1944.

Taylor, G., *Elementary Meteorology,* Prentice-Hall, 1954.

Thompson, P., R. O'Brien, and the editors of LIFE, *Weather,* Time, Inc., 1965.

Trewartha, G., *An Introduction to Climate,* 3d ed., McGraw-Hill, 1954.

Vaeth, Joseph G., *Weather Eyes in the Sky,* Ronald Press, 1965.

Viemeister, P., *The Lightning Book,* Doubleday, 1961.

Weatherwise, October 1965. This issue contains a selected bibliography on meteorology.

World Meteorological Organization, Geneva, Switzerland, International Cloud Atlas, Vols. 1 and 2, 1956.

PERIODICALS

Bulletin of the American Meteorological Society, 45 Beacon St., Boston, Mass.
Daily Weather Map, U.S. Dept. of Commerce, Environmental Science Services Administration, Weather Bureau, Silver Spring, Maryland.
Journal of the Atmospheric Sciences, American Meteorological Society, 45 Beacon St., Boston, Mass.
Monthly Weather Review, U.S. Weather Bureau, Washington, D.C.
Weather, published monthly by the Royal Meteorological Society, London, England.
Weatherwise, American Meteorological Society, 45 Beacon St., Boston, Mass.

Geology

Adams, F. D., *The Birth and Development of the Geological Sciences,* Dover, 1938 (paperback).
Adams, W. M., *Earthquakes,* D. C. Heath and Co., 1964.
Ahrens, L. H., *Distribution of the Elements in Our Planet,* McGraw-Hill, 1965 (paperback).
Albritton, C. C., Jr. *The Fabric of Geology,* Addison-Wesley, 1963.
Ayres, E., and C. A. Scarlott, *Energy Sources—the Wealth of the World,* McGraw-Hill, 1952.
Bateman, A. M., *The Formation of Mineral Deposits,* Wiley, 1951.
Beerbower, J. R., *Search for the Past,* Prentice-Hall, 1960.
Brice, J. C., J. P. Miller, and R. Scholten, *Laboratory Studies in Geology,* Freeman, 1962 and 1964.
Bullard, F. M., *Volcanoes: in History, in Theory, in Eruption,* University of Texas Press, 1962.
Cloos, Hans, *Conversation with the Earth,* Knopf, 1953.
Colbert, E. H., *Dinosaurs: Their Discovery and Their World,* Dutton, 1961.
———, *Evolution of the Vertebrates,* Wiley, 1955.
Davis, W. M., *Geographical Essays,* Dover Publications, 1954.
Dyson, J. L., *The World of Ice,* Knopf, 1962.
Dunbar, C. O., *Historical Geology,* 2nd ed., Wiley, 1960.
Eardley, A. J., *General College Geology,* Harper and Row, 1965.
———, *Structural Geology of North America,* 2d ed., Harper and Row, 1962.
Fagan, J. J., *View of the Earth,* Holt, Rinehart, and Winston, 1965.
Faul, Henry, *Ages of Rocks, Planets, and Stars,* McGraw-Hill Co., (paperback), 1966.
Fenton, C. L., and M. A. Fenton, *Giants of Geology,* Doubleday, 1952.
———, *The Fossil Book,* Doubleday, 1958.
———, *The Rock Book,* Doubleday, 1940.
Flint, R. F., *Glacial and Pleistocene Geology,* Wiley, 1957.
Gilluly, J., A. Waters, and A. Woodford, *Principles of Geology,* 2d ed., Freeman, 1959.
Heller, R. L. (Ed.), *Geology and Earth Sciences Sourcebook,* Holt, Rinehart, and Winston, 1962.

Hodgson, J. H., *Earthquakes and Earth Structure,* Prentice-Hall, 1964 (paperback).
Holmes, A., *Principles of Physical Geology,* 2d ed., Ronald Press, 1965.
Hough, J. L., *Geology of the Great Lakes,* University of Illinois Press, 1958.
Hoyt, W. G., and W. B. Langbein, *Floods,* Princeton University Press, 1955.
Hunt, Charles B., *Physiography of the United States,* W. H. Freeman Co., 1967.
Hurlbut, C. S., and H. E. Werden, *Mineralogy,* D. C. Heath and Co., 1964.
Hurley, P. M., *How Old Is the Earth?,* Anchor Books, 1959 (paperback).
Iacopi, Robert, *Earthquake Country,* Lane Book Company, 1964.
Jacobs, J., R. Russell, and J. Wilson, *Physics and Geology,* McGraw-Hill, 1959.
Kay, M., and E. Colbert, *Stratigraphy and Earth History,* Wiley, 1965.
King, P. B., *The Evolution of North America,* Princeton University Press, 1959.
Kraus, E., W. Hunt, L. Ramsdell, *Mineralogy,* 5th ed., McGraw-Hill, 1959.
Kummell, B., *History of the Earth,* Freeman, 1961.
Lahee, F. H., *Field Geology,* 6th ed., McGraw-Hill, 1961.
Leet, L. Don, *Earthquake—Discoveries in Seismology,* Dell, 1964 (paperback).
———, and Sheldon Judson, *Physical Geology,* Prentice-Hall, 3d ed., 1965.
Leopold, L., M. Wolman, and J. Miller, *Fluvial Processes in Geomorphology,* Freeman, 1964.
Lobeck, A. K., *Geomorphology,* McGraw-Hill, 1939.
———, Things Maps Don't Tell Us, Macmillan, 1956.
Longwell, C., and R. Flint, *Introduction to Physical Geology,* 2d ed., Wiley, 1962.
Mason, Brian, *Principles of Geochemistry,* 2d ed., Wiley, 1960.
Mather, K. F., *The Earth Beneath Us,* Random House, 1964.
———, and S. L. Mason, *A Source Book in Geology,* McGraw-Hill, 1939.
Meinesz, F., *The Earth's Crust and Mantle,* Elsevier, 1964.
Moody, P. A., *Introduction to Evolution,* 2d ed., Harper and Row, 1962.
Moore, G. W., and Brother G. Nicholas, *Speleology,* D. C. Heath and Co., 1964.
National Geographic Society, *America's Wonderlands, the National Parks,* 1959.
Ordway, Richard J., *Earth Science,* D. Van Nostrand Co., Inc., 1966.
Park, C. F., Jr., and R. A. MacDiarmid, *Ore Deposits,* Freeman, 1964.
Palmer, E. L., *Fossils,* D. C. Heath and Co., 1965.
Parson, R. L., *Conserving American Resources,* Prentice-Hall, 1956.
Pettijohn, F. J., *Sedimentary Rocks,* 2d ed., Harper and Row, 1957.
Pirsson, L., and A. Knopf, *Rocks and Rock Minerals,* 3d ed., Wiley, 1947.
Putnam, W. C., *Geology,* Oxford, 1964.
Rittman, A., *Volcanoes and Their Activity,* Wiley, 1962.
Roberts, E., *Our Quaking Earth,* Little, Brown, and Company, 1963.
Rogers, John J. W., and John A. S. Adams, *Fundamentals of Geology,* Harper and Row, 1966.
Schultz, G., *Glaciers and the Ice Age,* Holt, Rinehart, and Winston, 1963.
Schwarzbach, M., *Climates of the Past,* Van Nostrand, 1963.
Shelton, John S., *Geology Illustrated,* W. H. Freeman and Co., 1966.

Shimer, H., and R. Shrock, *Index Fossils of North America,* Wiley, 1944.
Shimer, John, *This Sculptured Earth: The Landscape of America,* Columbia University Press, 1959.
Sinkankas, John, *Mineralogy: A First Course,* D. Van Nostrand Co., 1966.
———, *Mineralogy for Amateurs,* D. Van Nostrand Co., 1964.
Spock, L. E., *Guide to the Study of Rocks,* Harper and Row, 1962.
Stirton, R. A., *Time, Life, and Man: The Fossil Record,* Wiley, 1959.
Stokes, W. L., *Essentials of Earth History,* 2d ed., Prentice-Hall, 1966.
Thornbury, W. D., *Principles of Geomorphology,* Wiley, 1954.
Todd, D. K., *Ground Water Hydrology,* Wiley, 1959.
Trefethen, J., *Geology for Engineers,* 2d ed., Van Nostrand, 1959.
U.S. Dept. of Agriculture, *Water* (*1955 Yearbook*), U.S. Government Printing Office, 1955.
Weitz, Joseph L., *Your Future in Geology,* Richards Rosen Press, Inc., 1966.
White, J. F. (Ed.), *Study of the Earth,* Prentice-Hall, 1962.
Woodford, A. O., *Historical Geology,* Freeman, 1965.
Zumberge, J. H., *Elements of Geology,* 2d ed., Wiley, 1963.

PERIODICALS

American Journal of Science, Box 2161, Yale University, New Haven, Conn.
Bulletin of the Geological Society of America, Colorado Building, P.O. Box 1719, Boulder, Colo.
Earth Science, Earth Science Publishing Corp., Box 1357, Chicago, Ill.
Geological Magazine, Stephen Austin & Sons, Ltd., Caxton Hill, Ware Rd., Hertford, Herts, England.
Geoscience News, bi-monthly, P.O. Box 4428, Pasadena, Calif.
Geotimes, American Geological Institute, 1444 N. St. N.W., Washington, D.C.
Journal of Geological Education, P.O. Box 7909, University Station, Austin, Texas.

Oceanography

Barnes, H., *Oceanography and Marine Biology,* Macmillan, 1960.
Bascom, W., *Waves and Beaches,* Anchor Books, Doubleday, 1964 (paperback).
Carson, Rachel, *The Sea Around Us,* Oxford University Press, 1951.
Coker, R., *This Great and Wide Sea,* Harper Torchbooks, 1962 (paperback).
Cotter, Charles H., *The Physical Geography of the Oceans,* American Elsevier Publishing Co., 1965.
Cowen, R., *Frontiers of the Sea,* Doubleday, 1960.
Cromie, W., *Exploring the Secrets of the Sea,* Prentice-Hall, 1962.
Defant, A., *Physical Oceanography,* Vols. 1 and 2, Pergamon Press, 1961.
Dugan, James, *et al., World Beneath the Sea,* National Geographic Society, 1967.
Engel, L., and the editors of LIFE, *The Sea,* Time, Inc., 1961.
Exploiting the Ocean, Marine Technology Society, 1966.

Groen, Pier, *The Waters of the Sea,* D. Van Nostrand Co., Inc., 1967.

Guilcher, A. *Coastal and Submarine Morphology,* Wiley, 1958.

Heezen, B., M. Tharp, and W. M. Ewing, *The Floors of the Oceans,* Geological Society of America Special Paper 65, 1959.

Hill, M. N. (Ed.), *The Sea,* Vols. 1, 2, and 3, Wiley, 1963.

Hull, Seabrook, *The Bountiful Sea,* Prentice-Hall, 1964.

Interagency Committee on Oceanography, *Bibliography of Oceanographic Publications,* I. C. O. Pamphlet No. 9, April 1963.

King, C., *An Introduction to Oceanography,* McGraw-Hill, 1963.

Long, E. J., Capt. (Ed.). *Ocean Sciences,* U.S. Naval Institute, 1964.

Menard, H., *Marine Geology of the Pacific,* McGraw-Hill, 1964.

Mero, J., *Mineral Resources of the Sea,* American Elsevier Publishing Company, 1965.

Neumann, Gerhard, and Willard J. Pierson, Jr., *Principles of Physical Oceanography,* Prentice-Hall, Inc., 1966.

Pickard, G. L., *Descriptive Physical Oceanography,* Macmillan, 1964.

Science and the Sea, U.S. Naval Oceanographic Office, U.S. Government Printing Office, Washington, D.C., 1967.

Sears, Mary (Ed.), *Oceanography,* International Oceanographic Congress, 1959, A.A.A.S. Publication No. 67, 1961.

———, Progress in Oceanography, Vols. 1, 2, and 3, Pergamon Press, 1963, 1964, and 1965.

Shepard, F. P., The Earth Beneath the Sea, Johns Hopkins Press, 1967.

———, *Submarine Geology,* 2d ed., Harper and Row, 1963.

Stewart, H. B., Jr., *Deep Challenge,* Van Nostrand, 1966 (paperback).

Stommel, H., *The Gulf Stream,* University of California Press, 1958.

von Arx, W. S., *An Introduction to Physical Oceanography,* Addison-Wesley, 1962.

Williams, Jerome, *Oceanography,* Little, Brown, and Company, 1962.

Appendix I : Tables

UNITS CONVERSION TABLE

Length

1 inch = 2.54 cm
1 meter = 3.28 ft = 1.09 yd
1 angstrom unit (Å) = 10^{-8} cm
1 kilometer = 0.621 mi

Area

1 acre = 43,560 sq ft

Volume

1 pint = 16 fluid oz (U.S.)
1 ml = 0.034 fluid oz (U.S.)
1 liter = 0.264 gal (U.S.) = 61.0 cu in
1 gallon (U.S.) = 231.0 cu in
1 gallon (Imperial) = 277.4 cu in

Mass

1 pound = 453.6 g
1 kilogram = 2.21 lb

Speed

60 mi/hr = 88 ft/sec

Pressure

1 atmosphere = 14.7 lb/sq in.
1 atmosphere = 76 cm Hg
1 atmosphere = 1013.2 millibars

Work and Energy

1 calorie = 4.19 joules
1 joule = 10^7 ergs = 1 newton meter
1 erg = 1 dyne cm
1 foot pound = 1.35 joules
1 electron volt = 1.60×10^{-12} erg
1 kilowatt hour = 3.6×10^6 joules

Force

1 gram = 980 dynes
1 pound = 32 poundals
= 16 oz (avoirdupois)
1 newton = 10^5 dynes

Power

1 horse power = 33,000 ft lb/min = 746 watts
1 watt = 1 joule/sec

Time Conversions

Seconds in one year (365 days) 3.15×10^7 sec
Seconds in one day 86,400 sec

TABLE OF CONSTANTS

Avogadro's number	6.02×10^{23}
Charge of the electron (e)	4.80×10^{-10} E.S.U.
Gravitational constant (G)	6.67×10^{-8} (cgs unit)
Planck's constant (h)	6.62×10^{-27} erg sec
Standard temperature and pressure	0°C and 760 mm of mercury
Velocity of light (in vacuo) (c)	186,000 mi/sec or 3×10^{10} cm/sec
Velocity of sound (at 0°C)	1090 ft/sec

Molar ideal gas constant (R)	82.05 ml atm/mole °
	8.31×10^{7} ergs/mole °
Boltzmann constant (k)	1.38×10^{-16} ergs/molecule °
Atomic mass unit (amu) (C^{12} scale)	1.660×10^{-24} g = 931.4 mev
Proton mass (C^{12} scale)	1.00783 amu
Neutron mass (C^{12} scale)	1.00866 amu
Electron mass (rest mass)	9.11×10^{-28} g

TABLE OF DENSITIES

Metals

Aluminum	2.70 g/cc
Copper	8.92 g/cc
Lead	11.4 g/cc
Silver	10.5 g/cc

Liquids

Benzene	0.879 g/ml
Carbon tetrachloride	1.595 g/ml
Mercury	13.6 g/ml
Water at 4°C	1.00 g/ml = 62.4 lb/cu ft

Gases

Air	1.293 g/liter at 0°C and 760 mm Hg

TABLE OF IONIC CHARGES

Name of Ion	*Symbol*	*Ionic Charge*
Ammonium	NH_4^+	+1
Hydrogen	H^+	+1
Potassium	K^+	+1
Silver	Ag^+	+1
Sodium	Na^+	+1
Barium	Ba^{++}	+2
Calcium	Ca^{++}	+2
Copper (cupric) (II)	Cu^{++}	+2
Iron (ferrous) (II)	Fe^{++}	+2
Lead (plumbous) (II)	Pb^{++}	+2
Magnesium	Mg^{++}	+2
Mercury (mercuric) (II)	Hg^{++}	+2
Tin (stannous) (II)	Sn^{++}	+2
Zinc	Zn^{++}	+2
Aluminum	Al^{+++}	+3
Iron (ferric) (III)	Fe^{+++}	+3

Name of Ion	*Symbol*	*Ionic Charge*
Acetate	$C_2H_3O_2^-$	−1
Bicarbonate	HCO_3^-	−1
Bisulfate	HSO_4^-	−1
Bromide	Br^-	−1
Chloride	Cl^-	−1
Fluoride	F^-	−1
Hydroxide	OH^-	−1
Iodide	I^-	−1
Nitrate	NO_3^-	−1
Thiocyanate	SCN^-	−1
Carbonate	$CO_3^=$	−2
Chromate	$CrO_4^=$	−2
Oxalate	$C_2O_4^=$	−2
Silicate	$SiO_3^=$	−2
Sulfate	$SO_4^=$	−2
Sulfide	$S^=$	−2
Sulfite	$SO_3^=$	−2
Arsenate	$AsO_4^{\equiv}$	−3
Phosphate	$PO_4^{\equiv}$	−3

FORMULA WRITING SHEET I

IONIC CHARGES OF NEGATIVE IONS	Silver Ag^{+}	Sodium Na^{+}	Calcium Ca^{++}	Aluminum Al^{+++}	Magnesium Mg^{++}
Nitrate NO_3^{-}					
Carbonate $CO_3^{=}$					
Phosphate $PO_4^{\equiv}$					
Sulfide $S^{=}$					
Chloride Cl^{-}					
Sulfate $SO_4^{=}$					
Hydroxide OH^{-}					
Sulfite $SO_3^{=}$					
Chromate $CrO_4^{=}$					
Bromide Br^{-}					

(Column headings: IONIC CHARGES OF POSITIVE IONS)

FORMULA WRITING SHEET II

OXIDATION NUMBERS OF NEGATIVE IONS	Potassium K^{+}	Cupric Cu^{++}	Ferric Fe^{+++}	Zinc Zn^{++}	Ammonium NH_4^{+}
Nitrate NO_3^{-}					
Carbonate $CO_3^{=}$					
Phosphate $PO_4^{\equiv}$					
Sulfide $S^{=}$					
Chloride Cl^{-}					
Sulfate $SO_4^{=}$					
Hydroxide OH^{-}					
Sulfite $SO_3^{=}$					
Chromate $CrO_4^{=}$					
Bromide Br^{-}					

(Column headings: IONIC CHARGES OF POSITIVE IONS)

URANIUM DISINTEGRATION SERIES

Name of Parent Nuclide	*Atomic Number Z*	*Mass Number A*	*Radiation Emitted*	*Daughter Nuclide Formed*
Uranium I	92	238	α	Th^{234}
Uranium X_1	90	234	β	—
Uranium X_2	—	—	β	—
Uranium II	—	—	α	—
Ionium	—	—	α	—
Radium	—	—	α	—
Radon	—	—	α	—
Radium A	—	—	α	—
Radium B	—	—	β	—
Radium C	—	—	β	—
Radium C′	—	—	α	—
Radium D	—	—	β	—
Radium E	—	—	β	—
Radium F	84	210	α	Pb^{206}
Radium G (stable lead isotope)	82	206	—	—

The student is expected to write in correct values of the blank atomic numbers and mass numbers of the parent elements, the symbol for the daughter nuclide formed and its appropriate mass number. Loss of an alpha particle decreases the atomic number by 2 units (loss of 2 protons) and the mass number by 4 units (loss of 2 protons and 2 neutrons). The nuclide formed by alpha decay of U^{238} is Th^{234} (also called Uranium X_1). When Uranium X_1 emits a beta particle, a neutron is converted into a proton. Hence the atomic number of the daughter product is one unit larger. The mass of the beta particle is negligible; so the mass of the daughter atom is the same as that of the parent.

Appendix II : Atomic

Element	*Symbol*	*Atomic Number*	*Atomic Weight*
Actinium	Ac	89	
Aluminum	Al	13	26.9815
Americium	Am	95	(243)
Antimony	Sb	51	121.75
Argon	Ar	18	39.948
Arsenic	As	33	74.9216
Astatine	At	85	(210)
Barium	Ba	56	137.34
Berkelium	Bk	97	(247)
Beryllium	Be	4	9.0122
Bismuth	Bi	83	208.980
Boron	B	5	10.811 (±0.003, nat.)
Bromine	Br	35	79.909 (±0.002, exp.)
Cadmium	Cd	48	112.40
Calcium	Ca	20	40.08
Californium	Cf	98	(249)
Carbon	C	6	12.01115 (±0.00005, nat.)
Cerium	Ce	58	140.12
Cesium	Cs	55	132.905
Chlorine	Cl	17	35.453 (±0.001, exp.)
Chromium	Cr	24	51.996 (±0.001, exp.)
Cobalt	Co	27	58.9332
Copper	Cu	29	63.54
Curium	Cm	96	(247)
Dysprosium	Dy	66	162.50
Einsteinium	Es	99	(254)
Erbium	Er	68	167.26
Europium	Eu	63	151.96
Fermium	Fm	100	(253)
Fluorine	F	9	18.9984
Francium	Fr	87	(223)
Gadolinium	Gd	64	157.25
Gallium	Ga	31	69.72
Germanium	Ge	32	72.59
Gold	Au	79	196.967
Hafnium	Hf	72	178.49
Helium	He	2	4.0026
Holmium	Ho	67	164.930
Hydrogen	H	1	1.00797 (±0.00001, nat.)
Indium	In	49	114.82
Iodine	I	53	126.9044
Iridium	Ir	77	192.2
Iron	Fe	26	55.847 (±0.003, exp.)
Krypton	Kr	36	83.80
Lanthanum	La	57	138.91
Lawrencium	Lw	103	(257)
Lead	Pb	82	207.19
Lithium	Li	3	6.939
Lutetium	Lu	71	174.97
Magnesium	Mg	12	24.312
Manganese	Mn	25	54.9380
Mendelevium	Md	101	(256)
Mercury	Hg	80	200.59
Molybdenum	Mo	42	95.94

weights and numbers

Element	*Symbol*	*Atomic Number*	*Atomic Weight*
Neodymium	Nd	60	144.24
Neon	Ne	10	20.183
Neptunium	Np	93	(237)
Nickel	Ni	28	58.71
Niobium	Nb	41	92.906
Nitrogen	N	7	14.0067
Nobelium	No	102	(256)
Osmium	Os	76	190.2
Oxygen	O	8	15.9994 (±0.0001, nat.)
Palladium	Pd	46	106.4
Phosphorus	P	15	30.9738
Platinum	Pt	78	195.09
Plutonium	Pu	94	(242)
Polonium	Po	84	(210)
Potassium	K	19	39.102
Praseodymium	Pr	59	140.907
Promethium	Pm	61	(147)
Protactinium	Pa	91	(231)
Radium	Ra	88	(226)
Radon	Rn	86	(222)
Rhenium	Re	75	186.2
Rhodium	Rh	45	102.905
Rubidium	Rb	37	85.47
Ruthenium	Ru	44	101.07
Samarium	Sm	62	150.35
Scandium	Sc	21	44.956
Selenium	Se	34	78.96
Silicon	Si	14	28.086 (±0.001, nat.)
Silver	Ag	47	107.870 (±0.003, exp.)
Sodium	Na	11	22.9898
Strontium	Sr	38	87.62
Sulfur	S	16	32.064 (±0.003, nat.)
Tantalum	Ta	73	180.948
Technetium	Tc	43	(99)
Tellurium	Te	52	127.60
Terbium	Tb	65	158.924
Thallium	Tl	81	204.37
Thorium	Th	90	232.038
Thulium	Tm	69	168.934
Tin	Sn	50	118.69
Titanium	Ti	22	47.90
Tungsten	W	74	183.85
Uranium	U	92	238.03
Vanadium	V	23	50.942
Xenon	Xe	54	131.30
Ytterbium	Yb	70	173.04
Yttrium	Y	39	88.905
Zinc	Zn	30	65.37
Zirconium	Zr	40	91.22

nat. = Variation in atomic weight due to natural variation in the isotopic composition.

exp. = Experimental uncertainty of magnitude given.

Numbers in parentheses are mass numbers of most stable or most common isotope.

Appendix III : Logarithms

When a problem involves much multiplying and dividing, raising to powers, or extracting roots, the calculations can be very time-consuming. To shorten work, it is often convenient to calculate with logarithms. Using Table III-1, we first match each ordinary number with its corresponding logarithm. Then we operate with the logarithms, adding to multiply and subtracting to divide. Finally, we take the resulting logarithm and with the aid of Table III-1 (and often interpolation) we find the ordinary number (the antilogarithm) corresponding to it. This is the answer sought.

Operating with Logarithms A common logarithm is simply an exponent of the number 10. It is easy to see that log 100 is 2 ($10^2 = 100$) and log 10 is 1, and even that log 1 is 0 (since $10^0 = 1$). But what about such numbers as 465 or 1.39 or 0.0743? Is there a logarithm for each of these? Very conveniently, there is a power of 10 corresponding to each of these, too—a fractional power, expressed as a decimal exponent. Table III-1 is a list of such decimal logarithms (to four places) for all three-digit numbers from 1.01 to 9.99. For the moment, let us confine operations to this range.

ILLUSTRATIVE PROBLEM

Multiply 2.06 and 4.35, using logarithms.

Solution

To find the logarithms that correspond to these numbers, look up 2.06 and 4.35 in Table II-1:

$$\log 2.06 = 0.3139$$
$$\log 4.35 = 0.6385$$

To find the product, first add the logarithms:

$$\begin{array}{r} 0.3139 \\ +0.6385 \\ \hline 0.9524 \end{array}$$

Then use Table III-1, and look for 0.9524. If it cannot be found, look in adjacent columns for the four-digit decimal nearest to 0.9524. The nearest decimal listed is 0.9523. Thus the answer, correct to three places, is the antilogarithm of 0.9523, which is 8.96. What is the answer by direct multiplication?

Logarithms of Larger Numbers Every logarithm includes a decimal part, called the *mantissa* (and this is what is listed, to four places, in Table III-1) which is always the same for any given sequence of three digits. The other part of the logarithm, to the left of the decimal point, is called the *characteristic*. This expresses as a power of 10 how many decimal places there are in the number—i.e., it locates the decimal point. Thus the number 896 is 8.96×10^2. Its logarithm is $.9523 + 2$, i.e., 2.9523. The 2 at the left of decimal represents the power of 10. The number 8.96 is 8.96×10^0. Its logarithm is 0.9523. The number 896,312, taken to an accuracy of three places, is 8.96×10^5, and its four-place logarithm is 5.9523. For greater accuracy we would need a larger table of logarithms—say, a 7-place table (which would fill a large volume). Such a table would give accuracy to six successive digits.

ILLUSTRATIVE PROBLEM

Multiply 50 by 3000, using logarithms.

Solution

$$\log 50 = \log (5.00 \times 10^1) = 1.6990$$
$$\log 3000 = \log (3.00 \times 10^3) = 3.4771$$

To multiply, add logarithms;

$$\begin{array}{r} 1.6990 \\ +\ 3.4771 \\ \hline 5.1761 \end{array}$$

The answer will be the antilogarithm of 5.1761. In Table III-1, locate 0.1761; it corresponds to the number 1.50. The answer is 1.50×10^5 or 150,000.

To divide by a number, subtract its logarithm from the logarithm of the dividend:

	LOGARITHMS	
690	2.8388	
÷ 3.00	− 0.4771	
230	2.3617	which corresponds to 2.30×10^2, or 230.

Table III.1—Four-Place Table of Common Logarithms

N	0	1	2	3	4	5	6	7	8	9
55	7404	7412	7419	7427	7435	7443	7451	7459	7466	7474
56	7482	7490	7497	7505	7513	7520	7528	7536	7543	7551
57	7559	7566	7574	7582	7589	7597	7604	7612	7619	7627
58	7634	7642	7649	7657	7664	7672	7679	7686	7694	7701
59	7709	7716	7723	7731	7738	7745	7752	7760	7767	7774
60	7782	7789	7796	7803	7810	7818	7825	7832	7839	7846
61	7853	7860	7868	7875	7882	7889	7896	7903	7910	7917
62	7924	7931	7938	7945	7952	7959	7966	7973	7980	7987
63	7993	8000	8007	8014	8021	8028	8035	8041	8048	8055
64	8062	8069	8075	8082	8089	8096	8102	8109	8116	8122
65	8129	8136	8142	8149	8156	8162	8169	8176	8182	8189
66	8195	8202	8209	8215	8222	8228	8235	8241	8248	8254
67	8261	8267	8274	8280	8287	8293	8299	8306	8312	8319
68	8325	8331	8338	8344	8351	8357	8363	8370	8376	8382
69	8388	8395	8401	8407	8414	8420	8426	8432	8439	8445
70	8451	8457	8463	8470	8476	8482	8488	8494	8500	8506
71	8513	8519	8525	8531	8537	8543	8549	8555	8561	8567
72	8573	8579	8585	8591	8597	8603	8609	8615	8621	8627
73	8633	8639	8645	8651	8657	8663	8669	8675	8681	8686
74	8692	8698	8704	8710	8716	8722	8727	8733	8739	8745
75	8751	8756	8762	8768	8774	8779	8785	8791	8797	8802
76	8808	8814	8820	8825	8831	8837	8842	8848	8854	8859
77	8865	8871	8876	8882	8887	8893	8899	8904	8910	8915
78	8921	8927	8932	8938	8943	8949	8954	8960	8965	8971
79	8976	8982	8987	8993	8998	9004	9009	9015	9020	9025
80	9031	9036	9042	9047	9053	9058	9063	9069	9074	9079
81	9085	9090	9096	9101	9106	9112	9117	9122	9128	9133
82	9138	9143	9149	9154	9159	9165	9170	9175	9180	9186
83	9191	9196	9201	9206	9212	9217	9222	9227	9232	9238
84	9243	9248	9253	9258	9263	9269	9274	9279	9284	9289
85	9294	9299	9304	9309	9315	9320	9325	9330	9335	9340
86	9345	9350	9355	9360	9365	9370	9375	9380	9385	9390
87	9395	9400	9405	9410	9415	9420	9425	9430	9435	9440
88	9445	9450	9455	9460	9465	9469	9474	9479	9484	9489
89	9494	9499	9504	9509	9513	9518	9523	9528	9533	9538
90	9542	9547	9552	9557	9562	9566	9571	9576	9581	9586
91	9590	9595	9600	9605	9609	9614	9619	9624	9628	9633
92	9638	9643	9647	9652	9657	9661	9666	9671	9675	9680
93	9685	9689	9694	9699	9703	9708	9713	9717	9722	9727
94	9731	9736	9741	9745	9750	9754	9759	9763	9768	9773
95	9777	9782	9786	9791	9795	9800	9805	9809	9814	9818
96	9823	9827	9832	9836	9841	9845	9850	9854	9859	9863
97	9868	9872	9877	9881	9886	9890	9894	9899	9903	9908
98	9912	9917	9921	9926	9930	9934	9939	9943	9948	9952
99	9956	9961	9965	9969	9974	9978	9983	9987	9991	9996
N	0	1	2	3	4	5	6	7	8	9

FOUR-PLACE TABLE OF COMMON LOGARITHMS (*cont.*)

N	0	1	2	3	4	5	6	7	8	9
10	0000	0043	0086	0128	0170	0212	0253	0294	0334	0374
11	0414	0453	0492	0531	0569	0607	0645	0682	0719	0755
12	0792	0828	0864	0899	0934	0969	1004	1038	1072	1106
13	1139	1173	1206	1239	1271	1303	1335	1367	1399	1430
14	1461	1492	1523	1553	1584	1614	1644	1673	1703	1732
15	1761	1790	1818	1847	1875	1903	1931	1959	1987	2014
16	2041	2068	2095	2122	2148	2175	2201	2227	2253	2279
17	2304	2330	2355	2380	2405	2430	2455	2480	2504	2529
18	2553	2577	2601	2625	2648	2672	2695	2718	2742	2765
19	2788	2810	2833	2856	2878	2900	2923	2945	2967	2989
20	3010	3032	3054	3075	3096	3118	3139	3160	3181	3201
21	3222	3243	3263	3284	3304	3324	3345	3365	3385	3404
22	3424	3444	3463	3483	3502	3522	3541	3560	3579	3598
23	3617	3636	3655	3674	3692	3711	3729	3747	3766	3784
24	3802	3820	3838	3856	3874	3892	3909	3927	3945	3962
25	3979	3997	4014	4031	4048	4065	4082	4099	4116	4133
26	4150	4166	4183	4200	4216	4232	4249	4265	4281	4298
27	4314	4330	4346	4362	4378	4393	4409	4425	4440	4456
28	4472	4487	4502	4518	4533	4548	4564	4579	4594	4609
29	4624	4639	4654	4669	4683	4698	4713	4728	4742	4757
30	4771	4786	4800	4814	4829	4843	4857	4871	4886	4900
31	4914	4928	4942	4955	4969	4983	4997	5011	5024	5038
32	5051	5065	5079	5092	5105	5119	5132	5145	5159	5172
33	5185	5198	5211	5224	5237	5250	5263	5276	5289	5302
34	5315	5328	5340	5353	5366	5378	5391	5403	5416	5428
35	5441	5453	5465	5478	5490	5502	5514	5527	5539	5551
36	5563	5575	5587	5599	5611	5623	5635	5647	5658	5670
37	5682	5694	5705	5717	5729	5740	5752	5763	5775	5786
38	5798	5809	5821	5832	5843	5855	5866	5877	5888	5899
39	5911	5922	5933	5944	5955	5966	5977	5988	5999	6010
40	6021	6031	6042	6053	6064	6075	6085	6096	6107	6117
41	6128	6138	6149	6160	6170	6180	6191	6201	6212	6222
42	6232	6243	6253	6263	6274	6284	6294	6304	6314	6325
43	6335	6345	6355	6365	6375	6385	6395	6405	6415	6425
44	6435	6444	6454	6464	6474	6484	6493	6503	6513	6522
45	6532	6542	6551	6561	6571	6580	6590	6599	6609	6618
46	6628	6637	6646	6656	6665	6675	6684	6693	6702	6712
47	6721	6730	6739	6749	6758	6767	6776	6785	6794	6803
48	6812	6821	6830	6839	6848	6857	6866	6875	6884	6893
49	6902	6911	6920	6928	6937	6946	6955	6964	6972	6981
50	6990	6998	7007	7016	7024	7033	7042	7050	7059	7067
51	7076	7084	7093	7101	7110	7118	7126	7135	7143	7152
52	7160	7168	7177	7185	7193	7202	7210	7218	7226	7235
53	7243	7251	7259	7267	7275	7284	7292	7300	7308	7316
54	7324	7332	7340	7348	7356	7364	7372	7380	7388	7396
N	0	1	2	3	4	5	6	7	8	9

Logarithms of Decimal Fractions The logarithms of numbers less than 1 (e.g., .00481) are figures less than zero. They are *negative logarithms.* The number 0.230, for example is $2.30/10 = 2.30 \times 10^{-1}$. Its logarithm is $.3617^{-1}$ or -0.6383. A negative mantissa is quite inconvenient in calculations; so negative logarithms are expressed in a special way that preserves a positive mantissa, thus:

$$\log. 0.230 = \log(2.30 \times 10^{-1}) = 0.3617 - 1, \text{ or } 9.3617 - 10$$

The last form is convenient for most calculations.

ILLUSTRATIVE PROBLEM

Multiply 425 by 0.0326, using logarithms.

Solution

$\log 425 = \log(4.25 \times 10^{2}) = 2.6284$
$\log 0.0326 = \log(3.26 \times 10^{-2}) = 8.5132 - 10$

Adding:

$$\begin{array}{r} 2.6284 \\ 8.5132 - 10 \\ \hline 11.1416 - 10 = 1.1416 \end{array}$$

In the body of Table III-1, the mantissas nearest to .1416 are .1430 (for log 1.39) and .1399 (for log 1.38). Thus the answer will be somewhere between 13.9 and 13.8.

Interpolation For greater accuracy in the answer to the problem above, a fourth digit can be estimated by interpolation.

ILLUSTRATIVE PROBLEM

What four-digit number between 13.9 and 13.8 corresponds to the logarithm, 1.1416?

Solution

To the difference of 0.10 in antilogs 13.9 and 13.8 there corresponds a difference of .0031 in mantissas (.1430 – .1399). To what difference in antilogs does the difference between mantissas .1430 and .1416 correspond?

$$\frac{.0014}{.0031} = .5, \text{ nearly, and } .5 \text{ of } 0.10 = .05$$

Thus the four-digit antilogarithm corresponding most closely to 1.1416 is 13.90 – 0.05, or 13.85. The fourth (interpolated) digit though it does not have the assured accuracy of the other three, may be useful in further calculations.

Complex Problems Calculation with logarithms is especially useful when there are several multiplications and divisions.

ILLUSTRATIVE PROBLEM

Find the value of

$$\frac{6.02 \times 10^{23} \times 0.693}{233 \times 24 \times 3600}$$

Solution

Add the logarithms of numbers in the numerator, and add separately the logarithms of numbers in the denominator. Subtract the second sum from the first, and find the antilogarithm of the result. For the numerator

$$\begin{aligned} \log 6.02 &= 0.7796 \\ \log 10^{23} &= 23.0000 \\ \log 0.693 &= \underline{9.8407} - 10 \\ & 33.6203 - 10 = 23.6203 \end{aligned}$$

For the denominator:

$$\begin{aligned} \log 233 &= 2.3674 \\ \log 24 &= 1.3802 \\ \log 3600 &= \underline{3.5563} \\ & 7.6039 \end{aligned}$$

$$\begin{array}{r} 23.6203 \\ -\ \underline{7.6039} \\ 16.0164 \end{array}$$

To find the value sought, which is the antilogarithm of 16.0164, interpolate. The mantissas above and below .0164 are

.0170	for log 1.040
.0128	for log 1.030

Thus to a difference of .010 in the antilogarithms, there corresponds a difference of .0042 in the mantissas. The difference between .0170 and .0164 is .006, which corresponds to a difference in the antilogarithm of .006/.0042 or $\frac{1}{7}$ of .010, i.e., .0014. The antilog of .0164 is thus somewhat less than 1.040 − .001 and the value sought, to four places, is 1.039×10^{16}, which is conveniently left in this form.

Appendix IV: Minerals and rocks

MINERAL PROPERTIES

Each mineral has a distinctive set of physical properties (Tables IV-1, IV-2, and IV-3). Specimens may be large or small, and colors may range from white to black, but there is something diagnostic about each. The principal useful properties for field identification are color, streak, hardness, luster, crystal form, cleavage or fracture, specific gravity, and miscellaneous characteristics involving reaction to acid, taste, feel, odor, and magnetism.

Color and Streak A mineral's color may be useful in mineral identification, or it may be deceptive. An inherent color depends upon the kinds and arrangement of the atoms in the mineral and is diagnostic for some minerals, chiefly for those with a metallic appearance (e.g., pyrite and azurite). Colors of other minerals vary widely because they depend upon the chance presence of impurities or of fractures. A surface alteration may obscure the true color of a mineral.

The color of a mineral in powdered form is called its streak. To find the streak, rub the mineral specimen on a hard rough surface as you would rub chalk on a blackboard. The equivalent of a blackboard is commonly a piece of white unglazed porcelain called a streak plate. The color of the streak tends to be more constant than the color of a mineral.

Luster This is the appearance of the unweathered surface of a mineral in ordinary reflected light. A metallic luster is like that of a shiny polished metal surface. Nonmetallic lusters vary: glassy, dull, greasy, and pearly are examples.

Hardness The hardness of a mineral is its resistance to abrasion (its "scratchability"). The Mohs hardness scale (Table IV-1) is commonly used and has

Table IV–1. The Mohs Scale of Hardness

NUMBER AND MINERAL ON MOHS SCALE	COMPARABLE HARDNESSES
1. Talc	(Softest)
2. Gypsum	A finger nail is 2–2½
3. Calcite	A copper coin is 3
4. Fluorite	
5. Apatite	
	Common window glass is 5½
6. Feldspar	A knife blade is 5½–6
	A steel file is 6½–7
7. Quartz	
8. Topaz	
9. Coroundum	
10. Diamond	(Hardest)

mineral representatives from 1 to 10. However, hardness does not increase by uniform jumps from 1 to 10; e.g., the jump from 9 to 10 may be proportionally as large as the jump from 1 to 9.

A soft mineral is scratched by the fingernail, whereas hard minerals scratch glass and cannot be scratched by a knife. Minerals of medium hardness are scratched by a steel nail or knife blade, but not by the fingernail; they do not scratch glass. Skill in determining the ease with which glass is scratched can be acquired. A soft mineral may leave a mark on a hard mineral, but this mark can be rubbed off. A true scratch can generally be detected with the fingernail as a tiny groove, although observation through a hand lens is sometimes necessary.

Crystal Form A crystal is a solid bounded by smooth plane surfaces that always meet at precisely the same angles for all specimens of any one mineral (similar angles on the different specimens must be compared). However, the plane surfaces or faces may have different areas in different specimens and thus may not have a uniform appearance. If space is available during crystallization, minerals are precipitated from solution with characteristic shapes (Fig. 27-6), but if many tiny crystals form here and there in a solution and grow larger, they evenutally interfere with one another's growth, and a granular aggregate results. This is a more common occurrence than the development of perfectly shaped crystals.

Table IV–2. Simplified Mineral Identification Key

The minerals are subdivided into three main groups on the basis of luster (metallic or nonmetallic) and color (light or dark). Further subdivision is based upon the presence or absence of cleavage. In the last column on the right, the minerals are arranged in order of decreasing hardness. See Table 1–3 for a list of some of the physical properties of a number of the more common minerals. A few minerals are listed in more than one place: e.g., some specimens of a certain mineral may be light, whereas others are dark colored; some specimens of another mineral may have a metallic luster, whereas others do not. You should not expect to detect relatively small differences in hardness: e.g., the difference between 3.5 and 4; but you should easily distinguish between 5 and 7 or 2 and 4.

Nonmetallic, light-colored minerals	Scratch glass	Show cleavage	Sodium plagioclase feldspar (6–6.5 Potassium feldspar (6)
		Show fracture only	Beryl (8) Quartz (7) Olivine (6.5–7) Opal (5–6.5)
	Do not scratch glass	Show cleavage	Fluorite (4) Dolomite (3.5–4) Calcite (3) Biotite mica (2.5–3) Muscovite mica (2–2.5) Halite (2–2.5) Gypsum (2) Talc (1+)
		Show fracture only	Kaolinite (2–2.5) Sulfur (1.5–2.5) Talc (1+)
Nonmetallic dark-colored minerals	Scratch glass	Show cleavage	Corundum (9) has parting which loo like cleavage Calcium plagioclase feldspar (6–6. Amphibole (5–6) Pyroxene (5–6)
		Show fracture only	Corundum (9) Tourmaline (7–7.5) Garnet (6.5–7.5) Quartz (7)
	Do not scratch glass	Show cleavage	Fluorite (4) Sphalerite (3.5–4) Biotite mica (2.5–3) Chlorite (2–2.5) Graphite (1–2)
		Do not show cleavage	Hematite (5–6, but may appear soft Apatite (5) Limonite (1–5.5) Serpentine (2.5–5)
	Minerals with metallic luster	Streak black	Pyrite (6–6.5) Magnetite (5.5–6.5) Chalcopyrite (3.5–4) Galena (2+) Graphite (1–2)
		Streak red or red brown	Hematite (5–6, but may appear soft Copper (2.5–3)
		Yellow, yellowish-brown, or white streak	Limonite (1–5.5) Sphalerite (3.5–4)

Manner of Breaking: Cleavage and/or Fracture The manner in which a mineral breaks is determined by its space lattice structure and may be very useful in identification. If the break produces one or more smooth plane surfaces, the specimen is said to have cleavage (Fig. 27-7). Fracture results in irregular or curved surfaces. Different types of cleavage and fracture occur, and variations in cleavage fall into three categories: minerals may cleave in different directions (from one to six); the angles at which different cleavage planes intersect may be different for different minerals; and the cleavage surfaces range from very smooth planes to slightly irregular surfaces.

The presence or absence of cleavage can be recognized most readily by observing the manner in which light is reflected as a specimen is rotated. If cleavage surfaces are present, light will flash as if a number of tiny mirrors were oriented at definite angles to each other.

IGNEOUS ROCKS

Igneous rocks differ in texture and mineral composition and are classified by these criteria. However, subdivision is somewhat arbitrary because continuous gradations occur in both texture and mineral content. We shall use a simplified classification that is applicable in the field or laboratory without the use of chemical analyses or a microscope. A small hand lens will be useful at times.

Texture The shapes, sizes, and arrangement of individual mineral grains in a rock determine its texture. An interlocking texture is common (Fig. 27-12). As magma cools, the least soluble substances are precipitated first as small crystals, and the most soluble substances crystallize last. Accordingly, the first mineral grains develop with their characteristic shapes, the next grains have modified shapes, and the last mineral grains are formless masses that fill all remaining spaces.

Grain Sizes Slow cooling under a thick cover of rocks produces large mineral grains, whereas the sudden chilling of lava as it reaches the surface results in the formation of many small particles. For large crystals or mineral grains to form, ions must migrate through a magma and encounter and become attached to an enlarging grain. Thus any factor that facilitates the migration of ions also promotes the growth of large mineral grains. Maintaining the magma in a fluid state is most important. The fluid state is prolonged if a magma cools slowly, if it is inherently fluid (low-silica magmas), or if dissolved gases are abundant. In general, intrusive rocks are coarse grained and extrusive rocks are fine grained.

Natural glass (obsidian) results if chilling is so rapid that the ions do not have time to unite into mineral grains; this occurs most commonly when extrusion takes place. Pumice is obsidian froth which is so full of gas bubble holes (vesicles) that it is light enough to float in water.

Table IV–3. Properties of Some Common Minerals

The minerals are arranged alphabetically, and the most useful properties in identification are printed in italic type. Most minerals can be identified by means of two or three of the properties listed below. In some minerals, color is important; in others, cleavage is characteristic; and in others, the crystal shape identifies the mineral.

NAME AND CHEMICAL COMPOSITION	HARDNESS	COLOR	STREAK	TYPE OF CLEAVAGE	REMARKS
Amphibole (complex ferromagnesian silicate)	5–6	*Dark green to black*	Greenish black	Two directions at angles of 56° and 124°	Vitreous luster. Horn-blende is the common variety. Long, slender, six-sided crystals. *Black with shiny cleavage surfaces at 56° and 124°.*
Apatite (calcium fluophosphate)	5	Green, brown, red, variegated	White	Indistinct	Crystals are common as are granular masses; vitreous luster
Beryl (beryllium silicate)	8	*Greenish*	Colorless	None	*Hardness, greenish color, six-sided crystals.* Aquamarine and emerald are gem varieties. Nonmetallic luster.
Biotite mica (complex silicate)	2.5–3	Black, brown, dark green	Colorless	*Excellent in one direction*	*Thin elastic films peel off easily.* Nonmetallic luster
Calcite ($CaCO_3$)	3	Varies	Colorless	*Excellent, three directions, not at 90° angles*	*Fizzes in dilute hydrochloric acid. Hardness.* Nonmetallic luster
Chalcopyrite ($CuFeS_2$)	3.5–4	*Golden yellow*	Greenish black	None	*Hardness and color distinguish from pyrite.* Metallic luster
Copper (Cu)	2.5–3	*Copper red*	Red	None	*Metallic luster on fresh surface. Ductile and malleable. Sp. gr. 8.5 to 9.*

Corundum (Al_2O_3)	8	Dark grays or browns common	Colorless	Parting resembles cleavage	*Barrel-shaped, six-sided crystals with flat ends*
Diamond (C)	10	Colorless to black	Colorless	Excellent, four directions	Hardest of all minerals
Chlorite (complex silicate)	1–2.5	*Greenish*	Colorless	Excellent, one direction	*Nonelastic flakes, scaly, micaceous*
Dolomite ($CaMg(CO_3)_2$)	3.5–4	Varies	Colorless	*Good, three directions, not at 90°*	*Scratched surface fizzes in dilute hydrochloric acid. Cleavage surfaces curved*
Feldspar (Potassium variety) (silicate)	6	*Flesh, pink, and red are diagnostic;* may be white and light gray	Colorless	*Good, two directions, 90° intersection*	*Hardness, color, and cleavage*
Feldspar (sodium plagioclase variety) (silicate)	6	*White to light gray*	Colorless	*Good, two directions, about 90°*	*If striations are visible, they are diagnostic.* Nonmetallic luster
Feldspar (calcium plagioclase variety) (silicate)	6	*Gray to dark gray*	Colorless	*Good, two directions, about 90°*	*Striations commonly visible;* may show iridescence. Associated with augite, whereas other feldspars are associated with hornblende. Nonmetallic luster
Fluorite (CaF_2)	4	Varies	Colorless	*Excellent, four directions*	Nonmetallic luster. In cubes or octahedrons as crystals and in cleavable masses
Galena (PbS)	2+	*Bluish lead gray*	Lead gray	*Excellent, three directions, intersect 90°*	*Metallic luster.* Occurs as crystals and cleavable masses. *Very heavy*
Gold (Au)	2.5–3	*Gold*	Gold	None	Malleable, ductile, *heavy.* Metallic luster

Table IV–3. Properties of Some Common Minerals (cont.)

NAME AND CHEMICAL COMPOSITION	HARDNESS	COLOR	STREAK	TYPE OF CLEAVAGE	REMARKS
Graphite (C)	1–2	*Silver gray to black*	Grayish black	Good, one direction	Metallic or earthy luster. *Foliated, scaly masses common. Greasy feel, marks paper.* This is the "lead" in a lead pencil (mixed with clay)
Gypsum (hydrous calcium sulfate)	2	White, yellowish, reddish	Colorless	*Very good in one direction*	Vitreous luster. *Can be scratched easily by fingernail*
Halite (NaCP)	2–2.5	Colorless and various various colors	Colorless	*Excellent, three directions intersect at 90°*	*Taste, cleavage, hardness*
Hematite (Fe_2O_3)	5–6 (may appear softer)	*Reddish*	*Reddish*	None	Sp. gr. 5.3. Metallic luster (also earthy)
Kaolinite (hydrous aluminum silicate)	2–2.5	White	Colorless	None (without a microscope)	Dull, earthy luster. Claylike masses
Limonite (group of hydrous iron oxides)	1–5.5	*Yellowish brown*	*Yellowish brown*	None	Earthy, granular. Rust stains
Magnetite (Fe_3O_4)	5.5–6.5	*Black*	Black	None	Metallic luster. Occurs in eight-sided crystals and granular masses. *Magnetic. Sp. gr. 5.2*
Muscovite mica (complex silicate)	2–2.5	Colorless in thin films; yellow, red, green, and brown in thicker pieces	Colorless	*Excellent, one direction*	*Thin elastic films peel off readily.* Nonmetallic luster

Table IV-3 (*cont.*)

Olivine (iron-magnesium silicate)	6.5–7	*Yellowish and greenish*	White to light green	None	*Green, glassy, granular*
Opal (hydrous silica)	5–6.5	Varies	Colorless	None	*Glassy and pearly lusters, conchoidal fracture*
Pyrite (FeS_2)	6–6.5	*Brass yellow*	Greenish black	None	*Cubic crystals* and granular masses. Metallic *luster*. Crystals may be striated. *Hardness important*
Pyroxene (complex silicate)	5–6	Greenish black	Greenish gray	*Two, nearly at 90°*	*Stubby eight-sided crystals and cleavable masses. Augite* is common variety. Nonmetallic
Quartz (SiO_2)	7	Varies from white to black	Colorless	None	Vitreous luster. *Conchoidal fracture. Six-sided crystals common*. Many varieties. Very common mineral. *Hardness*
Serpentine (hydrous magnesium silicate)	2.5–4	*Greenish (variegated)*	Colorless	Indistinct	*Luster resinuous to greasy. Conchoidal fracture*. The most common kind of asbestos is a variety of serpentine
Sphalerite (ZnS)	3.5–4	Yellowish brown to black	White to yellow	*Good, six directions*	*Color, hardness, cleavage, and resinous luster*
Sulfur (S)	1.5–2.5	*Yellow*	White to yellow	Indistinct	Granular, earthy
Talc (hydrous magnesium silicate)	1+	White, green, gray	Colorless	Good, one direction	Nonelastic flakes with *greasy feel. Soft*. Nonmetallic luster
Topaz (complex silicate)	8	Varies	Colorless	*One distinct (basal)*	Vitreous. *Crystals commonly striated lengthwise*
Tourmaline (complex silicate)	7–7.5	Varies; *black* is common	Colorless	Indistinct	*Elongated, striated crystals with triangular-shaped cross sections are common*

Porphyritic Texture Some igneous rocks are composed of mineral grains of two sharply different sizes, one much larger than the other. Such a texture is called porphyritic (Fig. 27-10). One manner in which it develops involves two stages of cooling. A rising magma may stop far below the surface and cool slowly. The first-formed minerals grow into large grains which may remain suspended in the fluid magma. Later the magma may rise and cool rapidly

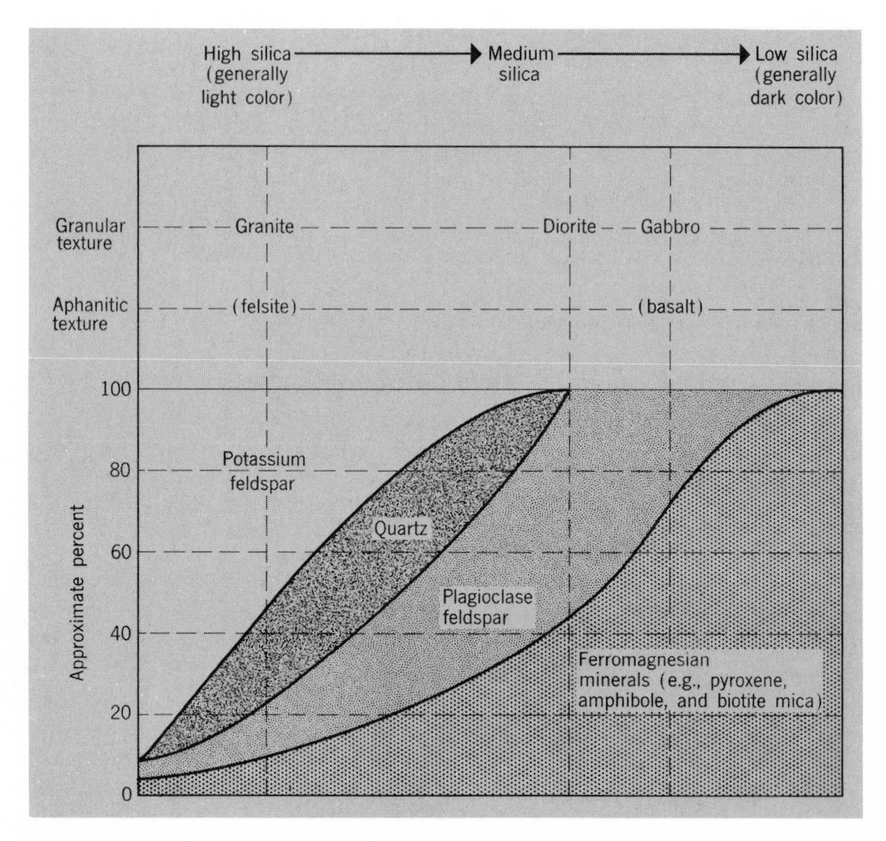

Fig. IV-1. *The gradational nature of the classification of igneous rocks is illustrated schematically. The graph does not show all of the possible combinations of minerals that have been found in igneous rocks. A vertical line through the graph at any one place indicates the approximate mineral content of a certain type of igneous rock. Only a few types are shown. Granite, diorite, and gabbro in the top row are all coarse grained, and felsite and basalt are fine grained. Felsite thus differs from granite only in grain size. From the graph one can determine that granite is a coarse-grained igneous rock composed chiefly of potassium feldspar with some quartz and minor amounts of ferromagnesian minerals and plagioclase feldspar (potassium feldspar and quartz are the diagnostic minerals). Gabbro is made of ferromagnesian minerals with the plagioclase type of feldspar.* [*After Pirsson.*]

Table IV–4. Classification of Igneous Rocks

(AFTER FORBES ROBERTSON IN PART)

1. The feldspars must be identified to use this classification, and the following criteria will be useful.
   (1) POTASSIUM FELDSPARS: diagnostic colors are flesh, pink, and red; may also be white or light gray, but without striations; hornblende and quartz tend to be associated minerals.
   (2) SODIUM PLAGIOCLASE FELDSPARS: colors are commonly white to light gray; striations, if visible, are diagnostic; hornblende is a common associate.
   (3) CALCIUM PLAGIOCLASE FELDSPARS: diagnostic colors are gray to dark gray, some with bluish iridescence; augite is a common associate; striations may be visible.
2. Several additional igneous rocks are described briefly below. They illustrate the gradational nature and diversity of the igneous rocks. All are granular in texture. The first four are included in the term granitic.
   (1) SYENITE: potassium feldspar is dominant; some sodium plagioclase may be present, also hornblende; quartz is missing; however, quartz syenite is a transition to granite; a relatively rare rock.
   (2) MONZONITE: potassium and sodium plagioclase about equal; if quartz is present, the rock is called a quartz monzonite.
   (3) GRANODIORITE: sodium plagioclase is more abundant than potassium feldspar; some quartz is also present.
   (4) QUARTZ DIORITE: feldspar is all plagioclase, and some quartz is present.
   (5) Rock consists almost entirely of ferromagnesian minerals:
      a. PERIDOTITE: made of olivine and pyroxene.
      b. DUNITE: chiefly olivine.
      c. PYROXENITE: chiefly pyroxene.

a content	High (70–80 percent)		Low (40–50 percent)
ture (each in a hori- al row has e texture)	Potassium feldspar is dominant. Quartz is abundant. Some sodium plagioclase, hornblende, and mica may be present	Sodium plagioclase is dominant. Hornblende is abundant	Calcium plagioclase and augite are dominant
ıular (all s in this row be porphyri-		Diorite	Gabbro
anitic hyritic	Rhyolite (phenocrystals of potassium feldspar and/or quartz)	Andesite (phenocrysts of sodium plagioclase and/or hornblende)	Basalt (dark colored)
ssy	Obsidian (although dark colored commonly, most obsidians form from high-silica lavas; they may be porphyritic) Pumice (obsidian froth)		
oclastic	These rocks may be divided into two groups on the basis of particle size—silica and mineral content are not factors: Volcanic tuff: average fragment is less than 4 mm (⅙ inch) Volcanic breccia average fragment is more than 4 mm (⅙ inch)		

upon the Earth's surface, where many small particles crystallize and surround the larger ones (phenocrysts) which had formed previously below the surface.

Classification of Igneous Rocks Figure IV-1 and Table IV-4 have the same basic organization. Rocks with the same texture are aligned horizontally in the same row. (1) Granular (phaneric) texture: mineral grains are large enough to be visible and identified; these rocks are generally intrusive. (2) Aphanitic texture: mineral grains range from specks too small to be identified without a microscope to still smaller invisible grains. However, porphyritic

Table IV–5. The Clastic Sedimentary Rocks

SEDIMENTS	CONSOLIDATED ROCK	AVERAGE DIAMETERS IN MILLIMETERS (25.4 mm = 1 inch)
Gravel	Conglomerate	Greater than 2 mm
Sand	Sandstone	2 mm to $1/16$ mm
Silt	Santstone	$1/16$ mm to $1/256$ mm
Clay	Shale	Less than $1/256$ mm

aphanitic rocks have phenocrysts that are large enough to be identified. (3) Glassy texture: no mineral grains occur unless phenocrysts are present. (4) Pyroclastic texture: fragments were erupted from a volcano as solids or as liquids that cooled in flight. Some pieces of nonigneous rocks torn from the walls of volcanic vents may be included. Pyroclastic rocks accumulated more or less as sediments, but are classified here with the igneous rocks.

Rocks of the same mineral and chemical content (except the glasses and pyroclastic rocks) are aligned in vertical rows. High-silica rocks at the left are commonly light colored and grade gradually into low-silica rocks at the right that commonly are dark colored.

SEDIMENTARY ROCKS

The clastic sedimentary rocks are subdivided into rock types chiefly on the basis of particle size (Table IV-5). Mixtures of different sizes are common and may be described by terms such as sandy shale (chiefly shale) and shaly sandstone (chiefly sandstone).

The chemical and organic sedimentary rocks have formed from sediments that were precipitated from saturated solutions or that accumulated as a result of the actions of living organisms. They have undergone little if any transportation, in contrast to the clastic group. Some, such as coal, represent accumulations of organic material.

Limestones are made almost entirely of calcite and may be clastic, nonclastic, or a combination of the two. Limestones may be so fine grained that individual particles cannot be seen with the unaided eye. Limestones are scratched easily by a knife, do not scratch a glass plate if pure, and effervesce readily in cold dilute hydrochloric acid. They may be almost any color, although grays are common. With increase in clastic material such as clay, they grade into other rocks.

Dolomite is a sedimentary rock similar to limestone, but is composed of the mineral dolomite. It does not effervesce in cold dilute hydrocloric acid, but if scratched by a knife, its powder does fizz.

METAMORPHIC ROCKS

Rocks become adapted to their environments in a slow and sometimes incomplete process. Thus older rocks are changed into new, notably different kinds of rocks. The transformations involve the development of new textures, new structures, new minerals, or any combination of these. The original characteristics of a rock may be destroyed completely by metamorphism while the bulk of the rock materials remains solid at all times. Some minerals constitute valuable "geologic thermometers" because they measure the degree of metamorphism that has occurred: some occur only in intensely metamorphosed rocks, whereas others occur in mildly metamorphosed rocks.

Heat has a major role in metamorphism because a rise in temperature makes most chemical changes proceed at a faster rate, and certain chemical reactions can only occur at high temperatures. Pressure is another factor, and solutions aid in the formation of new minerals by adding and removing various substances or by facilitating the recrystallization and rearrangement of mineral materials already present. Many chemical reactions are slow or impossible between dry substances.

Foliation Foliation or rock cleavage is a new structure produced during metamorphism and refers to the capacity of some metamorphic rocks to split into thin slabs along closely spaced, nearly parallel, relatively smooth surfaces (Figs. 27-20 and IV-2). Metamorphic rocks may be subdivided into two groups by the presence or absence of foliation. Slate, schist, and gneiss are foliated, whereas marble and quartzite are not.

When shale is squeezed by great compressive forces during one kind of metamorphism, the layers are bent and folded. During this folding, existing mineral grains are granulated, recrystallized, and reoriented, and new platy or elongated minerals may develop. Most of the resulting flaky mineral particles are aligned with their flat surfaces parallel to each other and perpendicular to the axis of greatest compression.

Slate is a fine-grained, well-foliated rock which can be split into thin smooth slabs for blackboards, roofing, and other uses. Its excellent foliation is the direct result of the combined cleavages of its tiny flaky constituents. All traces of the original stratification have been destroyed in some slates but not in others. Bedding and foliation may be parallel or intersect at any angle. Gradations between shale and slate occur.

Schist is a coarse-grained, foliated metamorphic rock which splits easily into thin slabs. Schists may be described by the minerals that are abundant in them and cause the foliation (e.g., mica schist).

Gneiss is a coarse-grained metamorphic rock which has poorly developed foliation and a banded, streaked appearance. Gneisses and schists are high-rank metamorphic rocks which grade into each other.

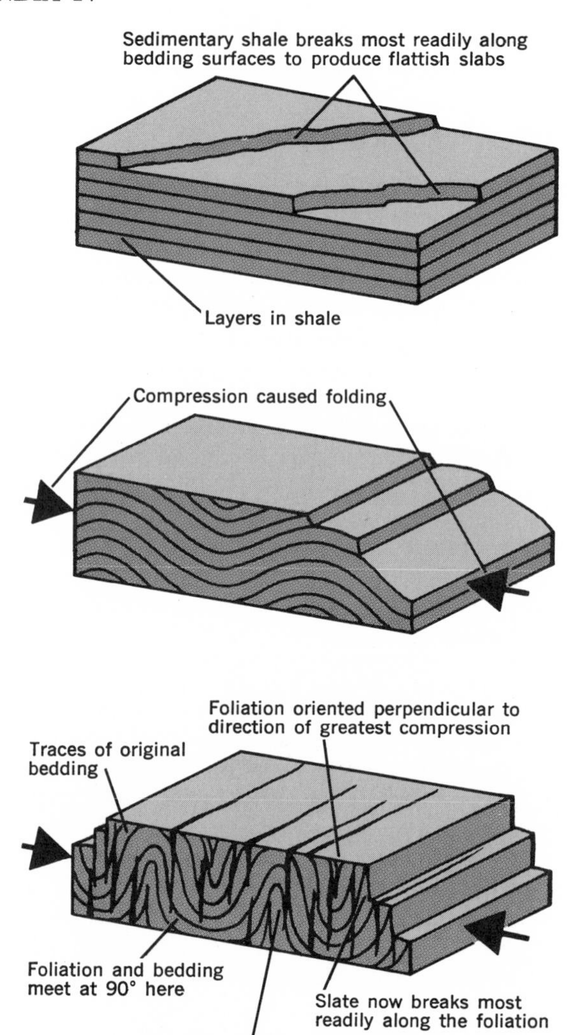

Fig. IV-2. *A shale is metamorphosed into slate.—(A) The sedimentary rock shale breaks most readily along bedding surfaces to produce flattish slabs.—(B) Folding and erosion have occurred. Foliation has not yet developed; the rock is still shale.—(C) After more intense folding, metamorphism, and erosion, the rock is slate and breaks most readily along the newly developed foliation. Imagine the microscopic flaky minerals that cause this foliation to be playing-cards. The flat sides of the cards would be aligned parallel with the foliation and would be seen edgewise in the sketch. Traces of the original bedding are still preserved; these cut across the foliation at the crest or trough of a fold and parallel it along the sides of a fold.*

Quartzite is a nonfoliated rock which is made of quartz grains that have been cemented firmly together. Therefore, the rock breaks through, not around, the individual grains. Some quartzites are actually sedimentary rocks (orthoquartzites) formed by the deposition of a silica cement around the quartz grains in a sandstone. Associated rocks in the field should provide clues in distinguishing them from metamorphic quartzites (metaquartzites).

Marble is commonly a nonfoliated metamorphic rock composed almost entirely of calcite (also of dolomite). It is produced by the metamorphism of limestone and similar rocks and is visibly crystalline. Metamorphism of a clayey limestone may produce a foliated marble; atoms in the clay minerals may recombine into minerals such as those in schist.

Answers to problems

Chapter 1. **1.** 29.9 in. **2.** 242 km. **3.** 28.2 kg. **4.** 28.4 g/oz. **5. a.** 28.1 liters/ft^3. **b.** 3.79 liters. **6.** 59.3 g. **7.** 7.35 cm^3. **8. a.** 177.6 g. **b.** Weighs more in vacuo. Buoyant effect is the weight of the displaced air. **9. a.** 125 cm^3. **b.** 125 g. **c.** 125 cm^3. **d.** 110 g. **e.** 74.1 cm^3. **f.** 74.1 g. **g.** 19.0. cm^3. **h.** 19.0 g. **i.** 19.0 cm^3. **j.** 16.7 g. **k.** 74.1 cm^3. **l.** 65.1 g. **10.** 280 g. **11.** 7580 g. **12.** 745 g. **13.** 2.67. **14.** 168 lb. **15.** 1.200 gal (U.S.)/gal (imperial). **16.** 7.480 gal/ft^3.

Chapter 2. **1.** 100 lb. **2.** 60 lb. **3.** $16\frac{2}{3}$ lb. **4.** 68.6 lb. **5.** 66.6 lb. **6.** 62 lb. **7.** 22.4 lb. **8.** 480 lb. **9.** 12 laborers. **10.** 111 lb, mech. adv. 6.75. **11.** 502.

Chapter 3. **1. a.** r = ft/sec. **b.** d = mi. **c.** t = sec. **2. a.** 248 mi. **b.** 66 ft/sec. **3.** 2400 lb force. **4.** 2000 poundals. **5.** 122.5 cm/sec^2. **6.** 61.2 cm/sec^2. **7.** 167:1 (auto:bullet). **8.** 40 newtons. **9.** 88 ft/sec. **10.** 2.5 m/sec. **11.** 6.74×10^{14} cm/sec^2.

Chapter 4. **1. a.** v = ft/sec. **b.** g = m/sec^2, d = meters. **c.** d = meters. **2. a.** 160 ft/sec. **b.** 320 ft/sec. **3. a.** 2.5×10^6 (or 25×10^5). **b.** 1.37×10^{-8} (or 13.7×10^{-9}). **4.** 4.9×10^{19}. **5.** 1.5×10^8. **6.** 1.35×10^{26} molecules. **7.** 5.03×10^{-9} cm^2. **8.** 6.24×10^{18} electrons/coulomb. **9.** 26.7×10^{-4} dyne. **10.** 225 ft. **11. a.** 1094 ft. **b.** 265 ft/sec. **12.** 384 in/sec^2. **13.** 6.39 sec. **14.** 576 ft.

Chapter 5. **1.** 252 calories. **2.** 144×10^6 ft lb. **3.** 387,200 ft poundals. **4.** 725,000 ft lb. **5. a.** 96.0 kg m. **b.** 0.427 hp. **c.** 0.311 kw. **6.** 2550 g cm. **7.** 27.5°C. **8.** 37°C. **9.** 11.7 calories. **10.** 5 kg m. **11.** 1.64×10^{-6} ergs. **12.** 169°F, 340° Kelvin or absolute. **13.** 4.58×10^{-14} ergs. **14.** −40°.

Chapter 6. **1.** 1026 millibars. **2.** 9.56 liters. **3.** 152°C. **4.** 111 cm Hg. **5. a.** See derivation, page 103. **b.** 1040 mm Hg. **6. a.** $PV = \frac{1}{3}\, mnu^2$; mn = mass of gas; $T = ku^2$; $PV = \frac{1}{3}\, mnT/k$; $PV/T = mn/3k$ (right hand member constant for a given mass of gas); $PV/T = \mathbf{K}$ (a new constant); $P_1V_1/T_1 = P_2V_2/T_2$ (for a given mass of gas). **b.** 23.9 liters. **7.** 0.745. **8.** 0.172 cal/g/C°. **9.** 0.250 mi/sec. **10.** 33.9 ft. **11.** 0.117%. **12.** 176°F, 353°K. **13.** 17.3 liters. **14.** 15.7°C. **15.** 37,200 cal. **16.** 44.3 g.

Chapter 7. **1.** $NaCl = 58.5$; $NaHCO_3 = 84.0$; $NaOH = 40.0$; $CaCO_3 = 100$; $Al_2(SO_4)_3 = 342$; $Ca_3(PO_4)_2 = 310$; $C_2H_5OH = 46.0$; $HNO_3 = 63.0$. **2. a.** $2H_2 + O_2 \rightarrow 2H_2O$. **b.** $N_2 + 3H_2 \rightarrow 2NH_3$. **c.** $CaCO_3 + 2HCl \rightarrow CaCl_2 + H_2O + CO_2$. **d.** $2Al + 3H_2SO_4 \rightarrow Al_2(SO_4)_3 + 3H_2$. **e.** $MgCl_2 + 2AgNO_3 \rightarrow 2AgCl\downarrow + Mg(NO_3)_2$. **f.** $BaCl_2 + K_2CrO_4 \rightarrow BaCrO_4\downarrow + 2KCl$. **g.** $2Na_3AsO_4 + 3Ca(NO_3)_2 \rightarrow Ca_3(AsO_4)_2\downarrow + 6NaNO_4$. **h.** $2NaBr + 2H_2SO_4 \rightarrow Na_2SO_4 + Br_2 + SO_2\uparrow + 2H_2O$. **3. a.** $2AgNO_3 + CaCl_2 \rightarrow 2AgCl\downarrow + Ca(NO_3)_2$. **b.** $3BaCl_2 + Al_2(SO_4)_3 \rightarrow 3BaSO_4\downarrow + 2AlCl_3$. **c.** $2NH_4I + Ca(OH)_2 \rightarrow CaI_2 + 2NH_3\uparrow + 2H_2O$. **d.** $H_2 + Cl_2 \rightarrow 2HCl$. **e.** $CaCO_3 \rightarrow CaO + CO_2\uparrow$. **f.** $Zn + AgNO_3 \rightarrow Zn(NO_3)_2 + Ag$. **g.** $3Cu + 8HNO_3 \rightarrow$

$3Cu(NO_3)_2 + 2NO\uparrow + 4H_2O$. **4. a.** 11 lb. **b.** 12 g. **c.** 5.0 liters. **5. a.** Na = 27.4%; H = 1.19%; C = 14.3%; O = 57.1%. **b.** H = 1.59%; N = 22.2%; O = 76.2%. **c.** Na = 57.5%; O = 40.0%; H = 2.50%. **d.** Cl = 92.2%; C = 7.99%. **6.** 107 lb Zn. **7. a.** 0.915 kg. **b.** 10.5 kg. **8.** 83.2 g. **9.** 43.8; gas is heavier than air; 1.51 times as heavy. **10.** 357 lb CaO. **11.** 63.3. **12. a.** $Cu + 4HNO_3 \rightarrow Cu(NO_3)_2 + 2NO_2 + 2H_2O$. **b.** 149 g. **13. a.** 50.7 g. **b.** 12.5 g.

Chapter 9. **1.** 5.10×10^{14} waves/sec. **2.** 300 Å. **3.** 362 meters; station WGY, 810 kc. **4. a.** 10^{-4} cm. **b.** 1 mμ = 10^{-7} cm. **c.** 60 mμ. **d.** 667 waves/cm. **5.** 27 μsec. **6.** Radio waves 0.0266 sec; sound waves 6.7 hr. **7.** 7.72×10^{14} waves/sec. **8. a.** 4.13 ft. **b.** 1130 ft/sec. **9.** 0.52 ft.

Chapter 10. **1.** 0.33 ampere. **2.** 15 ohms. **3.** 0.101 g. **4.** 3.3×10^{-3} ohm, 455 amperes. **5.** 0.25 dyne. **6.** 0.22 dyne. **7. a.** 6.0 ohms. **b.** 4.8 amperes. **8.** 8280 cal. **9.** 14.4 min. **10.** 0.0014 g. **11.** 0.323 g.

Chapter 11. **1.** 6.72×10^{14} molecules. **2. a.** 0.179 g/liter. **b.** Hydrogen diffuses 1.42 times as fast as deuterium.

Chapter 12. **1.** 3.1×10^6 atoms. **2.** 5.61×10^{-7} sec^{-1}. **3. a.** 1.33×10^{-5} erg. **b.** 8.31×10^6 ev. **c.** 8.31 mev. **4.** 3.71×10^{-13} g. **5.** A (dis/sec) = (sec^{-1})$\cdot$ n (atoms); if the activity (A) is a curie, $A = 3.7 \times 10^{10}$ dis/sec;

$$3.7 \times 10^{10} \text{ dis/sec} = \frac{0.693}{T_{\frac{1}{2}}} \cdot \frac{\text{wt}}{\text{atomic wt}} \cdot 6.02 \times 10^{23};$$

hence

$$\text{wt} = \frac{3.7 \times 10^{10} \times \text{atomic wt} \times T_{\frac{1}{2}}}{0.693 \times 6.02 \times 10^{23}}.$$

6. 0.23 microcurie. **7.** 12.3 yr.

Chapter 13. **1.** $_ZM^A + {}_2He^4 \rightarrow {}_1H^1 + {}_{Z+1}X^{A+3}$. **2. a.** $_{20}Ca^{44} + {}_1H^1 \rightarrow {}_0n^1 + {}_{21}Sc^{44}$. **b.** $_{15}P^{31} + {}_1H^2 \rightarrow {}_{15}P^{32} + {}_1H^1$. **c.** $_5B^{11} + {}_2He^4 \rightarrow {}_7N^{14} + {}_0n^1$. **d.** $_{13}Al^{27} + {}_0n^1 \rightarrow {}_{12}Mg^{27} + {}_1H^1$. **3. a.** Negatron. **b.** Positron. **c.** Negatron. **d.** Negatron. **e.** Positron. **4.** $_6O^{13}, {}_7N^{13}$; $_{18}Ar^{40}, {}_{19}K^{40}$; $_{24}Cr^{51}, {}_{25}Mn^{51}$. **5.** $_{18}Ar^{38}, {}_{19}K^{39}$; $_{29}Cu^{63}, {}_{30}Zn^{64}$; $_{33}As^{75}, {}_{34}Se^{76}$. **6.** $_2He^4, {}_8O^{16}, {}_{20}Ca^{40}, {}_{28}Ni^{58}, {}_{50}Sn^{118}, {}_{72}Pb^{208}$. **7. a.** $_{88}Ra^{226} \rightarrow {}_2He^4 + {}_{86}Rn^{222}$. **b.** $_{90}Th^{234} \rightarrow {}_{-1}e^0 + {}_{91}Pa^{234}$. **c.** $_{86}Rn^{222} \rightarrow {}_2He^4 + {}_{84}Po^{218}$. **d.** $_{82}Pb^{210} \rightarrow {}_{-1}e^0 + {}_{83}Bi^{210}$.

Chapter 14. **1.** 0.025 ev. **2.** 14 lb. **3.** 4×10^{-4} g. **4. a.** 1.13×10^{11} atoms. **b.** 4.4×10^{-11} g.

Chapter 15. **1.** Volume of atom $10^{12} \times$ volume of nucleus. **2.** $\frac{27{,}452}{100{,}000{,}000}$. **3.** 10.7°C.

Index